MW00427564

SUSTAINABLE SOLID WASTE MANAGEMENT

SUSTAINABLE SOLID WASTE MANAGEMENT
A Systems Engineering Approach

NI-BIN CHANG
ANA PIRES

IEEE PRESS

Published by John Wiley & Sons, Inc., Hoboken, New Jersey.
Published simultaneously in Canada.

For general information on our other products and services or for technical support, please contact our Customer Care Department within the United States at (800) 762-2974, outside the United States at (317) 572-3993 or fax (317) 572-4002.

Wiley also publishes its books in a variety of electronic formats. Some content that appears in print may not be available in electronic formats. For more information about Wiley products, visit our web site at www.wiley.com.

Library of Congress Cataloging-in-Publication Data:

Chang, Ni-Bin.
  Sustainable solid waste management : a systems engineering approach / Ni-Bin Chang, Ana Pires.
    pages cm
  Includes bibliographical references and index.
  ISBN 978-1-118-45691-0 (hardback)
  1. Integrated solid waste management.   I. Pires, Ana.   II. Title.
  TD794.2.C45 2015
  628.4′4–dc23   2014034307

Printed in the United States of America

10 9 8 7 6 5 4 3 2 1

CONTENTS

III INDUSTRIAL ECOLOGY AND INTEGRATED SOLID WASTE MANAGEMENT STRATEGIES 301

9 INDUSTRIAL ECOLOGY AND MUNICIPAL UTILITY PARKS 303

10 LIFE CYCLE ASSESSMENT AND SOLID WASTE MANAGEMENT 323

IV INTEGRATED SYSTEMS PLANNING, DESIGN, AND MANAGEMENT   441

13 MULTIOBJECTIVE DECISION-MAKING FOR SOLID WASTE MANAGEMENT IN A CARBON-REGULATED ENVIRONMENT   443

PREFACE

Solid waste management is a significant issue in sustainable development encompassing technical, socioeconomic, legal, ecological, financial, political, and even cultural components. Sustainable solid waste management refers to a mode of waste management sciences in concert with urban development, in which resource use aims to meet human needs of daily consumption while ensuring the sustainability of natural systems and the environment through appropriate waste collection, treatment, resources conservation, and recycling. However, the interactions between human activities and the environment are complicated and often difficult to quantify. In many occasions, judging where the optimal balance should lie among environmental protection, social well-being, economic growth, and technological progress is complex. The use of a systems engineering approach will fill the gap contributing to how we understand the intricacy by a holistic way and how we generate better sustainable solid waste management practices. The book also aims to advance interdisciplinary understanding of intertwined facets between policy and technology relevant to solid waste management issues interrelated to climate change, land use, economic growth, environmental pollution, industrial ecology, and population dynamics. The chapters in the book are grouped into five thematic parts, including

- Part I: Fundamental Background—This part discusses the basic concepts of sustainability science in which more highlighted information is provided on technology matrix and other resources of legal and institutional concerns where social and economic relevance may be interconnected.
- Part II: Principles of Systems Engineering—This part introduces the use of formal systems engineering principles including top-down and bottom-up

approaches that is encouraged to evaluate solid waste management alternatives with respect to the criteria of cost–benefit–risk aspects.

- Part III: Industrial Ecology and Integrated Solid Waste Management Strategies—This part recognizes that sustainable solid waste management is intimately tied to industrial ecology, in which life cycle impact assessments of a product and appraisals of solid waste management processes over or beyond life cycle can be carried out in a more sustainable way.

- Part IV: Integrated Systems Planning, Design, and Management—This part considers connections across resource areas and fosters linkages across agencies, which require a holistic means of integrated sustainability assessment.

- Part V: Uncertainty Analyses and Future Perspectives—This part emphasizes quantitative uncertainty analyses that might be useful in systematically evaluating the possible or plausible changes in decision analysis outcomes due to changes in measurement accuracy, sources of data, communication, and social behavior. Future perspectives of sustainable solid waste management highlight possible movement in the field.

It is our great honor to work with the IEEE Press Series on Systems Science and Engineering to publish this book. We gratefully acknowledge the encouragement of the book series editor, Dr. Mengchu Zhou. Much of the work in preparing this book was supported as part of the educational mission of the University of Central Florida, Orlando, FL, USA; Universidade Nova de Lisboa, Lisbon, Portugal; and UNINOVA (Institute for the Development of New Technologies), Caparica, Portugal. We also gratefully acknowledge the many contributions made by the present and past colleagues, students, and friends around the world. Their sharp insight and recommendations on improvements to this book have been invaluable in framing this publication. Particular thanks are due to Ms. Janice Faaborg and Mr. Steven Mcclure for their insightful editorial proofreading of these manuscripts. Finally, we are exceedingly grateful to our families for their encouragement, patience, and unfailing support, even when they were continually asked to sacrifice, and the end never seemed to be within reach.

NI-BIN CHANG
ANA PIRES

July 30, 2014

PART I

FUNDAMENTAL BACKGROUND

The basic concepts of sustainability science are highlighted and more detailed information is provided on technology matrix and other resources of legal and institutional concerns, where social and economic relevance may be interconnected. The following chapters lead to the holistic discussion of environmental risk assessment and management of risk:

- Introduction (Chapter 1)
- Technology matrix for SWM (Chapter 2)
- The social and economic aspects of SWM (Chapter 3)
- The legal and institutional aspects of SWM (Chapter 4)
- A framework for environmental risk assessment and management (Chapter 5)

Sustainable Solid Waste Management: A Systems Engineering Approach, First Edition. Ni-Bin Chang and Ana Pires.
© 2015 The Institute of Electrical and Electronics Engineers, Inc. Published 2015 by John Wiley & Sons, Inc.

CHAPTER 1

INTRODUCTION

Society is increasingly conscious of the importance of solid waste management (SWM) in the context of sustainable development. The need to operate our waste management activities in a way that minimizes environmental and health risks and ensures economic growth and social progress has been well received by the community. The purpose of this chapter is to emphasize the essence of sustainable development as part of the package of tools for making decisions about SWM. This chapter provides a common framework for sustainable development and relevant basic principles that support such ideas. The guidelines describe possible actions to establish a framework for a wide range of SWM activities across diverse spatial and temporal scales. Case studies that demonstrate how to apply sustainable SWM processes across a variety of activities are introduced sequentially in subsequent chapters.

1.1 THE CONCEPT OF SUSTAINABLE DEVELOPMENT

1.1.1 The Concept Formation

The book "Silent Spring" written by Rachel Carson was published in 1962 (Carson, 1962). The seemingly related connection between the insecticide applications and bird populations was considered a turning point in our basic understanding of the interconnections among the environment, the economy, and social well-being. In 1972, the United Nations Conference on the Human Environment held in Stockholm

Sustainable Solid Waste Management: A Systems Engineering Approach, First Edition. Ni-Bin Chang and Ana Pires.
© 2015 The Institute of Electrical and Electronics Engineers, Inc. Published 2015 by John Wiley & Sons, Inc.

brought the industrialized and developing nations together to delineate the "rights" of the human family to a healthy and productive environment (United Nations, 2013). In the 1980s, human society was increasingly conscious of possible detrimental effects that its economic activities can have on ecosystems and the environment. Note that ecosystems in this context are systems of plants, animals, and microorganisms together with the nonliving components of their environment (UNEP/WWF/IUCCNF, 1980). This book adopts the definition used in the United Kingdom Environmental Protection Act 1990, that the environment "… consists of all, or any, of the following media, namely the air, water and land." Over generations, the loss of quality of life in human society can result from environmental degradation due to past economic activities, as seen in the numerous hazardous waste remediation sites across the United States (US). The "World Conservation Strategy," jointly published by United Nations Environment Programme (UNEP), World Wide Fund for Nature (WWF), and International Union for Conservation of Nature and Natural Resources (IUCNNR), noted that (UNEP/WWF/IUCCNF, 1980):

> The combined destructive impacts of a poor majority struggling to stay alive and an affluent minority consuming most of the world's resources are undermining the very means by which all people can survive and flourish. Humanity's relationship with the biosphere (the thin covering of the planet that contains and sustains life) will continue to deteriorate until a new international economic order is achieved, a new environmental ethic adopted, human populations stabilize, and sustainable modes of development become the rule rather than the exception. Among the prerequisites for sustainable development is the conservation of living resources.

The World Conservation Strategy, which provided a precursor to the concept of sustainable development, aims to (UNEP/WWF/IUCCNF, 1980):

- maintain essential ecological processes and life-support systems (such as soil regeneration and protection, the recycling of nutrients and the cleansing of waters), on which human survival and development depend;
- preserve genetic diversity (the range of genetic material found in the world's organisms), on which depend the breeding programs necessary for the protection and improvement of cultivated plants and domesticated animals, as well as much scientific advance, technical innovation, and the security of the many industries that use living resources;
- ensure the sustainability utilization of species and ecosystems (notably fish and other wildlife, forests, and grazing lands), which supports millions of rural communities as well as major industries.

The United Nations General Assembly convened in 1983 to discuss "The World Commission on Environment and Development" to address concerns about the accelerating degradation of the human environment and natural resources and the consequences of such degradation for economic and social development. Later, the concept of "sustainable development" was formalized by the Brundtland Report published in 1987. Although sustainable development has been defined in many ways, the most frequently quoted definition is from "Our Common Future" in the Brundtland Report (WCED, 1987):

> Sustainable development is development that meets the needs of the present without compromising the ability of future generations to meet their own needs.

Two key concepts are emphasized in the Brundtland Report (WCED, 1987) as excerpted below:

- "needs," in particular the essential needs of the world's poor, to which overriding priority should be given; and
- "limitations" imposed by the state of technology and social organization on the environment's ability to meet present and future needs.

In comparison, sustainable development was defined by the President's Council on Sustainable Development in the United States as (USEPA, 2013):

> ... an evolving process that improves the economy, the environment, and society for the benefit of current and future generations.

In June 1992, the first UN Conference on Environment and Development was held in Rio de Janeiro and adopted an agenda entitled "Agenda 21: A Programme of Action for Sustainable Development" (United Nations, 1992). Agenda 21 states the Rio Declaration on Environment and Development, which agrees to some 27 supporting principles that are abbreviated as the "Rio Principles." Agenda 21 reaffirmed that sustainable development was delimited by the integration of the economic, social, and environmental pillars. This understanding triggers the possible change in consumption and production patterns. Within these 27 supporting principles, principles 3, 4, 6, 8, 10, 11, 13, 14, 15, 16, and 17 are most relevant to waste management, as excerpted below (United Nations, 1992):

Principle 3: The right to development must be fulfilled so as to equitably meet developmental and environmental needs of present and future generations.

Principle 4: In order to achieve sustainable development, environmental protection shall constitute an integral part of the development process and cannot be considered in isolation from it.

Principle 6: The special situation and needs of developing countries, particularly the least developed and those most environmentally vulnerable, shall be given special priority. International actions in the field of environment and development should also address the interests and needs of all countries.

Principle 8: To achieve sustainable development and a higher quality of life for all people, States should reduce and eliminate unsustainable patterns of production and consumption and promote appropriate demographic policies.

Principle 10: Environmental issues are best handled with participation of all concerned citizens, at the relevant level. At the national level, each individual shall have appropriate access to information concerning the environment that is held by public authorities, including information on hazardous materials and activities.

Principle 11: States shall enact effective environmental legislation. Environmental standards, management objectives, and priorities should reflect the environmental and development context to which they apply. Standards applied by some countries may be inappropriate and of unwarranted economic and social cost to other countries, in particular developing countries.

Principle 13: States shall develop national law regarding liability and compensation for the victims of pollution and other environmental damage. States shall also cooperate in an expeditious and more determined manner to develop further international law regarding liability and compensation for adverse effects of environmental damage caused by activities within their jurisdiction or control to areas beyond their jurisdiction.

Principle 14: States should effectively cooperate to discourage or prevent the relocation and transfer to other States of any activities and substances that cause severe environmental degradation or are found to be harmful to human health.

Principle 15: (Precautionary principle)—In order to protect the environment, the "precautionary approach" shall be widely applied by States according to their capabilities. Where there are threats of serious or irreversible damage, lack of full scientific certainty shall not be used as a reason for postponing cost-effective measures to prevent environmental degradation.

Principle 16: (Polluter pay principle)—National authorities should endeavor to promote the internalization of environmental costs and the use of economic instruments, taking into account the approach that the polluter should, in principle, bear the cost of pollution, with due regard to the public interest and without distorting international trade and investment.

Principle 17: Environmental impact assessment, as a national instrument, shall be undertaken for proposed activities that are likely to have a significant adverse impact on the environment and are subject to a decision of a competent national authority.

1.1.2 The Three Pillars in Sustainable Development

In 2002, the World Summit on Sustainable Development was convened in Johannesburg to renew the global commitment to sustainable development. The conference agreed to the Johannesburg Plan of Implementation to follow up on the implementation of sustainable development. It signifies the three pillars approach to illustrate sustainability (Figure 1.1). Sustainable development seeks to achieve economic development, social welfare, and environmental protection, in a balanced manner, from which we start seeing the world as a collection of interconnected systems.

Given that the concept of sustainable development is rooted in systems thinking, definitions of sustainable development in this illustration require that the whole world

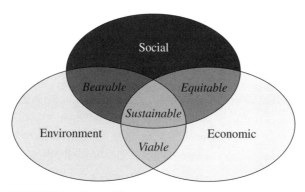

FIGURE 1.1   Three pillars approach to illustrate sustainability

be considered a system over space and time. Hence, sustainable development relies on a systems-based approach that seeks to understand the interactions that exist among the three pillars (environment, social, and economic) in an effort to better realize the unintended consequences of our actions (USEPA, 2013). The United States Environmental Protection Agency (USEPA) has an insightful list that embodies the principles of sustainability via six aspects of each pillar as excerpted below (USEPA, 2013).

1. **Environmental pillar**
   - **Ecosystem services:** Protect, sustain, and restore the health of critical natural habitats and ecosystems (e.g., potential impacts of hydraulic fracturing).
   - **Green engineering and chemistry:** Develop chemical products and processes to reduce/prevent chemical hazards, reuse or recycle chemicals, treat chemicals to render them less hazardous, and dispose of chemicals properly (e.g., life cycle environmental impacts).
   - **Air quality:** Attain and maintain air quality standards and reduce the risk from toxic air pollutants (e.g., investigate potential greenhouse gas emissions reduction strategies).
   - **Water quality:** Reduce exposure to contaminants in drinking water (including protecting source waters) in fish and shellfish and in recreational waters (e.g., pathogen removal in riverbank filtration).
   - **Stressors:** Reduce effects by stressors (e.g., pollutants, greenhouse gas emissions, genetically modified organisms) to the ecosystem (e.g., fate of modified nanoparticles in aqueous media).
   - **Resource integrity:** Reduce adverse effects by reducing waste generation, increasing recycling, and ensuring proper waste management; and restore resources by mitigating and cleaning up accidental or intentional releases (e.g., improving recycling technology to prevent environmental impact of mining).

2. **Economic pillar**
   - **Jobs:** Create or maintain current and future jobs (e.g., create green jobs).
   - **Incentives:** Generate incentives that work with human nature to encourage sustainable practices (e.g., conservation reserve program, encouraging sustainable logging practices).
   - **Supply and demand:** Promote price or quantity changes that alter economic growth, environmental health, and social prosperity (e.g., increasing supply of green energy sources to reduce the need for fossil fuels).
   - **Natural resource accounting:** Incorporate natural capital depreciation in accounting indices and ecosystem services in cost–benefit analysis (CBA) (e.g., green net national product).
   - **Costs:** Positively impact costs of processes, services, and products (e.g., strive to develop a waste-free process for eliminating the need for regulation costs).
   - **Prices:** Promote a cost structure that accounts for externalities to production (e.g., bottle bill—beverage container deposit laws) throughout the United States and around the world).

3. **Social pillar**
   - **Environmental justice:** Protect health of communities over-burdened by pollution by empowering them to take action to improve their health and environment (e.g., establish partnerships with local, state, tribal, and federal organizations to achieve healthy and sustainable communities).
   - **Human health:** Protect, sustain, and improve human health (e.g., parameterize the model to predict developmental toxicology).
   - **Participation:** Use open and transparent processes that engage relevant stakeholders (e.g., develop database of reduced-risk pesticides for commonly used products, create greater public access and understanding about sustainability).
   - **Education:** Enhance education on sustainability to the general public, stakeholders, and potentially affected groups (e.g., provide opportunities for students to learn about sustainability).
   - **Resource security:** Protect, maintain, and restore access to basic resources (e.g., food, land, and energy, and study impacts of dispersants/oil combination on natural water ways).
   - **Sustainable communities:** Promote the development, planning, building, or modification of communities to promote sustainable living (e.g., landscape with native plant species, construct "green" buildings).

1.1.3 Temporal and Spatial Characteristics of Sustainability Goal

Sustainable development concurrently addresses both spatial and temporal characteristics that must be clearly defined from local, regional, and global viewpoints for current and future generations. Sustainability concerned with intergeneration equity may link with a larger time scale than the life cycle of a product, technology, or treatment plant. The time frame employed for project evaluation should be extended to the same time horizon to meet sustainability implications. Long-term projects involve higher complexity and wider ranges of scenarios with uncertainty. This might also be true for time-dependent technology innovation, development, and improvement. Residents who live in the proximity of these developments bear more pollution impact than those farther away, however, and environmental justice may be a sustainability concern from the societal point of view. Sustainability in this regard may be linked with varying spatial scales depending on the types of pollutants of concern. Integrating both spatial and temporal characteristics may generate higher uncertainty anyhow. Assessment of uncertainties and their consequences require a deeper level of risk assessment, which becomes an essential component of sustainability analysis.

1.1.4 The Possible Actions to Achieve the Sustainability Goal

In 2012, United Nations published "Review of Implementation of Agenda 21 and the Rio Principles," which outlined areas that would need to be addressed to enable more rapid progress toward the objectives set during the Rio Earth Summit 20 years earlier. The detailed reviews of Agenda 21 and the Rio Principles and the submission from Stakeholder Forum to the Rio conference (United Nations, 2012) offer some perspectives for action in these areas.

1. **Progressing and protecting human development**
   - **A rights-based approach:** Human development requires having a true rights-based approach to coping with various welfare, well-being, and environmental issues that are essential to sustainable development.
   - **Increasing participation:** All people have the basic right to receive environmental information, participate in transparent decision-making processes, and access judicial and administrative proceedings.
   - **Giving a voice to future generations:** The future needs of next generations are a crucial element of sustainable development; but they are not represented in the relevant decision-making processes.
2. **Sustainable management of the Earth**
   - **Acknowledge environmental limits:** There is an acute need to formally realize key environmental thresholds within which we must count on for our livelihood and to maintain the ecosystem sustainability of our planet.
   - **Sustainable management of natural resources and capitals:** All levels of government should ensure that their accounting efforts may address not only the GDP but also the state of natural assets and ecosystems and their role in sustaining human and economic activity.
3. **The green economy**
   - **Beyond gross domestic product (GDP):** GDP is an indicator of success that is the current reliance on economic growth in most of the developing

countries. This tendency has led to perverse outcomes due to the ignorance of environmental sustainability. A new economic indicator that has correction of environmental costs may better justify the true outcomes.

- **Fiscal reform:** Taxes or other policy instruments should be used to motivate positive behavior and discourage undesirable behavior.
- **Restart a meaningful conversation about the role of corporations in the achievement of sustainable development:** Conversations could take the form of a "Convention on Corporate Social Responsibility" to improve the producer's responsibility.

4. **Sustainable institutions and governance**
   - **Sustainable development goals:** The inclusion of sustainable development goals is a possible foundation for building international consensus, aiming to the provision of quantifiable "tangible goals" for sustainable development.
   - **Improving international cooperation and development aid:** As outlined in the review of Chapter 33 of Agenda 21, future agreements concerning the financing effort for sustainable development should be centered on measurable and time-bound targets.
   - **Reform of international financial institutions:** As discussed in Chapters 33 and 38 of Agenda 21, sustainable development parameters must be better incorporated into the existing international financial institutions.
   - **National, local, and regional governance:** These sustainable development strategies with different scales should be revived and refreshed with full engagement and support from business and all parts of civil society.
   - **International court for the environment:** Environmental problems extend across international boundaries and should be governed globally.

1.2 SUSTAINABILITY IN THE CONTEXT OF SWM

1.2.1 The Possible Conflicts in Achieving the Sustainability Objectives

Achieving sustainability goals involves balancing social, economic, and environmental perspectives constrained by environmental limits over an inter- and intragenerational timeframe, and possible conflicts of objectives related to the three pillars of sustainability would be inevitable. It is necessary to acknowledge and deal with these conflicting objectives across domain boundaries in the diverse spectrum of projects with system thinking. The current waste management industry, which sometimes allows pure commercial opportunism to capitalize promptly on a perceived waste management market, has not completely transformed to embrace or even address sustainability objectives. Actions such as tipping fees, waste stream availability, waste management markets, cost–benefit analyses, competing technologies, longer-term projections, and cross linkages with other industries in relation to supply chain management may be required to aid new systems engineering techniques.

1.2.2 The Possible Sustainability Indicators

In the context of SWM, the concept of sustainability applies to the whole SWM industry sectors, process technologies, and individual process plants. In assessing sustainability performance from storage and collection, to routing and shipping, to separation and treatment, and to final disposal, a system boundary should be well defined. Besides, suitable sustainability indicators to quantify the performance and monitor the progress related to economic, environmental, and social perspectives may be selected for a holistic assessment up front. The perspectives discussed in section 1.1.2 could provide a rational basis to develop appropriate scenarios in SWM. Several key indicators may be considered as options to support a sustainability assessment (Brennan, 2013).

1. **Environmental indicators**
   - **Global-warming potential:** Global-warming potential (GWP) is related to climate change impact and is a relative measure of heat trapped in the atmosphere by greenhouse gases. The GWP value compares the amount of heat trapped by a greenhouse gas to that of carbon dioxide, which has a GWP standard of 1. For example, the GWP of methane is 72 within a 20-year time frame, which means that if the same mass of methane and carbon dioxide were introduced into the atmosphere, that amount of methane will trap 72 times more heat than the carbon dioxide over the next 20 years. The combustion of solid waste may lead to the emission of carbon dioxide and other greenhouse gases.
   - **Ozone layer depletion:** The stratospheric ozone layer forms a thin shield that acts as a sunscreen in the upper atmosphere, protecting life on the surface of Earth from the sun's ultraviolet (UV) rays. Depletion of the ozone layer due to the presence of compounds that contain chlorine and bromine molecules, such as methyl chloroform, halons, and chlorofluorocarbons (CFCs), results in increased UV radiation reaching the Earth's surface, which leads to detrimental health effects such as skin cancer, cataracts, and immune suppression. The final disposal of refrigerant (CFC) at landfills may lead to the impact of stratospheric ozone layer depletion.
   - **Photochemical smog:** Both nitrogen oxides and volatile organic compounds are precursors of photochemical smog in urban regions. High concentrations of nitrogen oxides and volatile organic compounds are associated with industrialization and transportation through fossil fuel combustion. Waste shipping may result in emissions and lead to the generation of photochemical smog.
   - **Human and ecotoxicity:** Human and ecotoxicity indicators are related to public health and risk assessment, exemplified by the toxicity impact on human health from the heavy metal content of organic waste. Air emissions from waste incineration facilities could result in such impacts.
   - **Resources conservation potential:** Separate collection of recyclables from municipal solid waste streams may have greater resources conservation potential.

2. **Economic indicators**
   - **Value-added by-product:** The opportunities of value-added utilization of by-products may be a legitimate sustainability indicator. In waste management, value can be derived at every stage of the chain during collection, shipping, recycling, treatment, and disposal processes. Recyclables, waste heat recovered from waste combustion, compost, as well as the reuse of other residuals may be deemed as value-added by-products.
   - **Contribution to green GDP:** The green GDP is an index of economic growth with the essential correction of environmental consequences of the GDP. Green GDP monetizes the loss of biodiversity and environmental quality and accounts for costs caused by climate change. Environmental costs and benefits of waste management factored into conventional GDP of a country may contribute to the correction of environmental consequences of economic growth.
   - **Environmental costs and benefits:** In CBA of SWM projects, environmental costs and benefits related to waste management may become a set of standalone indicators. CBA is a technique that compares the monetary value of benefits against the monetary value of costs in a series of alternatives to evaluate and prioritize management options. For example, environmental groups in the United States often assert that recycling was doubling energy consumption and pollution while costing taxpayers more money than the potential benefits from value-added by-products.
   - **Environmental liability:** The environmental liability coverage for possible failure of waste management operation tailored to different waste management projects may be deemed as an indicator of sustainability of a waste management project.

3. **Social indicators**
   - **Stakeholder identification and participation:** Stakeholder identification with some analysis techniques is particularly relevant when choosing stakeholders to help waste management agencies organize a participation list. Appropriate forms or channels of participation such as minority group identification in a region would certainly improve the social sustainability.
   - **Income distribution or redistribution through policy instruments:** Income distribution or redistribution measures driven by some policy instruments in SWM projects may be used as an indicator of societal well-being. The distribution or redistribution of compensation or fair fund due to pollution impact caused by waste treatment facilities is a salient example.

1.3   THE FRAMEWORK FOR SUSTAINABILITY ASSESSMENT

The National Research Council in the United States laid out a framework for sustainability assessment structured from the formulation of a problem through achievement of outcomes that warrant a multiagency approach (CSLFG/STSP/PGA/NRC, 2013):

Phase I: Preparation and Planning
- **Frame the problem:** A thorough understanding of the problem is required in all aspects, including environmental resources connections, societal connections, and economic connections. The focus is to determine baseline information, key drivers, metrics, and goals.
- **Identify and enlist stakeholders:** Relevant agency linkages and nonagency stakeholders to serve on the project team must be identified and contacted.
- **Develop a project management plan:** Roles, responsibilities, and accountability of each member must be delineated to create a business plan for project design, implementation, and operation.

Phase II: Design and Implementation
- **Set project goals:** The project team members should formalize the goals together with essential inputs from all stakeholders and relevant members. Evaluation metrics in terms of short-term and long-term outcomes must be outlined in this step.
- **Design an action plan:** The team members should develop a comprehensive plan to elucidate the approaches, strategies, and actions to meet the prescribed goals of the project.
- **Implement the action plan:** At this stage, selecting a boundary organization that bridges scientific and technical experts with policy makers and stakeholders is deemed critical.

Phase III: Evaluation and Adaptation
- **Realize short-term outcomes:** Short-term outcomes that occur on the scale of a year to a few years need to be assessed relative to the baseline information collected in the first phase.
- **Assess and evaluate outcomes:** The knowledge and experience gained is applied to modify problem formulation and adjust approaches, methods, and strategies.

Phase IV: Long-term Outcomes
- **Achieve long-term outcomes:** Short-term outcomes that occur on the scale of a few years or more may be close to the project goals to be achieved. The evaluation plan generated in the second phase may be instrumental to judge if short-term and long-term goals are met.

1.4 THE STRUCTURE OF THIS BOOK

The interactions between human activity and the environment are complicated and often difficult to quantify. In many situations, judging where the optimal balance should lie among environmental protection, social well-being, economic growth, and technological progress is difficult. Decision frameworks refer to principles, processes, and practices to proceed from information and desires to choices that inform actions and outcomes (Lockie and Rockloff, 2005). Decision frameworks may facilitate and enhance decision making by providing conceptual structures and principles for integrating all sustainability dimensions of decisions (CSLFG/STSP/PGA/NRC, 2013). Development of a decision framework to strengthen sustainability linkages is a challenging task. While decision frameworks vary in purpose, common elements include (CSLFG/STSP/PGA/NRC, 2013) the following:

- problem identification and formulation;
- identification of clear goals;
- illumination of key questions that help the decision maker scope problems and management options;
- processes for knowledge-building and application of appropriate analytical tools to assess actions, options, trade-offs, risks, and uncertainties;
- connection of authorities tasked with making decisions to outcomes associated with those decisions.

Because the system thinking of sustainable development has broad international consensus, this book aims to promote a systems engineering approach for SWM and provide useful sources of advice and information in support of sustainable SWM. The book is thus intended to be used in conjunction with existing literature and other relevant guidance, primarily by academic researchers, policy makers, and waste managers in public and private sectors. It also aims to advance interdisciplinary research of policy and technology relevant to SWM issues interrelated to climate change, land use, economic growth, environmental pollution, industrial ecology, population dynamics, and the interactions among these issues.

This book proposes a systematic decision framework consisting of parallel, interlinked, and complementary processes through science-based analyses with various peripheral subtopics, which is organized within the general perspectives of sustainability for SWM. A comprehensive bibliography is provided at the end of each chapter, and case studies are used to illustrate and demonstrate the processes of sustainability assessment and environmental management. This system-based approach is reflected in the structure of the five parts as follows:

Part I: Fundamental Background: The basic concepts of sustainability science are highlighted and more detailed information is provided on technology matrix and other resources of legal and institutional concerns where social and economic relevance may be interconnected. The following chapters lead to the holistic discussion of environmental risk assessment and management of risk.
  - Introduction (Chapter 1)
  - Technology matrix for SWM (Chapter 2)
  - The social and economic aspects of SWM (Chapter 3)
  - The legal and institutional aspects of SWM (Chapter 4)
  - A framework for environmental risk assessment and management (Chapter 5)

Part II: Principles of Systems Engineering: The use of formal systems engineering principles including top-down and bottom-up approaches is encouraged to evaluate SWM alternatives. The following chapters are organized to illuminate the internal linkages among global changes, sustainability, and adaptive management strategies and to introduce systems engineering principles. While such a system-based approach related to the integrated SWM should be the norm, risk assessments may sometimes be applied usefully to aid in the decision-making if uncertainties come to bother the choice of adaptive management strategies.

- Linkages among global change, sustainability, and adaptive management strategies (Chapter 6)
- Systems engineering principles and decision-making (Chapter 7)
- Systems engineering tools for evaluating the significance of alternatives (Chapter 8)

Part III: Industrial Ecology and Integrated Solid Waste Management Strategies: Industrial symbiosis with a particular focus on material and energy exchange in natural ecosystem is the foundation of industrial ecology, which includes the study of material and energy flows through ecoindustrial parks in human society. Sustainable SWM is intimately tied to industrial ecology in which life cycle impact assessments of a product and appraisals of SWM processes over or beyond life cycle can be carried out in a more sustainable way. The processes covered in the following chapters command more specific requirements with respect to life cycle concept combined with risk assessment not covered by the general guidelines of Parts I and II.

- Principles of industrial symbiosis and industrial ecology in support of municipal utility parks (Chapter 9)
- Evaluating the significance of life cycle assessment for SWM (Chapter 10)
- Options appraisal and decision-making based on streamlined life cycle assessment (Chapter 11)
- SWM under a carbon-regulated environment (Chapter 12)

Part IV: Integrated Systems Planning, Design, and Management: Considering connections across resource areas and fostering linkages across agencies requires a unique means of sustainability assessment. When coping with complex sustainability issues such as SWM, which is complicated by the separated and dispersed authorities resulting from the basic legal framework, advances in environmental informatics and system analysis may provide a framework for valuable sustainability assessment.

- Multiobjective decision-making framework for SWM in a carbon-regulated environment (Chapter 13)
- Integrated forecasting and optimization modeling for planning regional material recovery facilities in an SWM system (Chapter 14)
- Optimal waste collection and vehicle routing strategies (Chapter 15)

- Multiattribute decision-making framework (Chapter 16)
- Multiobjective decision-making framework for balancing waste incineration and recycling (Chapter 17)
- Environmental informatics in support of SWM (Chapter 18)

Part V: Uncertainty Analyses and Future Perspectives: Risk analysis that fails to account for measurement uncertainties may produce misleading and sometimes dangerous results. Quantitative uncertainty analyses might be useful in systematically evaluating the possible or plausible changes in decision analysis outcomes due to changes in measurement accuracy, sources of data, communication, and social behavior.

- Evaluating the significance of uncertainty with random phenomenon and game theory for SWM in decision-making (Chapter 19)
- Considering linguistic uncertainty related to institutional settings and social behavior by fuzzy multiattribute analysis for SWM in decision-making (Chapter 20)
- Considering linguistic uncertainty related to institutional settings and technological implications by fuzzy multiattribute analysis for SWM in decision-making (Chapter 21)
- Assessing linguistic uncertainty by fuzzy multiobjective programming for SWM in decision-making (Chapter 22)
- Formalizing grey uncertainty by interval programming for SWM in decision-making (Chapter 23)
- Future perspectives (Chapter 24)

REFERENCES

Brennan, D. 2013. *Sustainable Process Engineering*, Pan Stanford Publishing Pte. Ltd., Singapore.

Carson, R. 1962. *Silent Spring*, Houghton Mifflin, Boston, MA.

Committee on Sustainability Linkages in the Federal Government, Science and Technology for Sustainability Program, Policy and Global Affairs, and National Research Council (CSLFG/STSP/PGA/NRC). 2013. *Sustainability for the Nation: Resources Connection and Governance Linkages,* National Academies Press, Washington, DC.

Lockie, S. and Rockloff, S. 2005. *Decision Frameworks: Assessment of the Social Aspects of Decision Frameworks and Development of a Conceptual Model*, Coastal CRC Discussion Paper, Central Queensland University, Norman Gardens, Australia.

United Nations (UN). 1992. Report of the United Nations Conference on Environment and Development – Annex I Rio Declaration on Environment and Development. Available at: http://www.un.org/documents/ga/conf151/aconf15126-4.htm (accessed August 2013).

United Nations (UN). 2012. Review of Implementation of Agenda 21 and the Rio Principles, Sustainable Development in the 21st Century (SD21), A Study Prepared by the Stakeholder Forum for a Sustainable Future. Available at: http://sustainabledevelopment.un.org/content/documents/641Synthesis_report_Web.pdf (accessed March 2012).

United Nations (UN). 2013. The History of Sustainable Development in the United Nations. Rio +20—United Nations Conference on Sustainable Development. Available at: http://www.uncsd2012.org/history.html (accessed August 2013).

United Nations Environmental Programme, World Wild Fund for Nature, and International Union for Conservation of Nature and Natural Resources (UNEP/WWF/IUCCNF). 1980. *World Conservation Strategy—Living Resource Conservation for Sustainable Development*, IUCN/UNEP/WWF, Gland, Switzerland.

United States Environmental Protection Agency (USEPA). 2013. Sustainability Primer. Funding Opportunities. Available at: http://www.epa.gov/ncer/rfa/forms/sustainability_primer _v7.pdf (accessed August 2013).

World Commission on Environment and Development (WCED). 1987. *Our Common Future* (Ed. Brundtland, G. H.), Oxford University Press, Oxford.

CHAPTER 2

TECHNOLOGY MATRIX FOR SOLID WASTE MANAGEMENT

Technological options available to treat waste streams vary from local-scale facilities, such as waste recycling bins and shipping vehicles, to regional-scale material recovery facilities, municipal solid waste (MSW) incinerators, and landfills, all of which have unique features for every possible solid waste management (SWM) system. In this chapter, technology matrix is systematically introduced in association with types of waste streams, presented separately with featured operational units as well as collectively with synergistic processes in parallel or in sequence. With this knowledge base, waste streams can be regarded as flows connecting various types of system components in urban, man-made networks that contribute to urban sustainability through material and energy recovery and reduce pollution impacts directly and indirectly on a long-term basis. Such a phenomenon can be regarded as an integral part of "urban metabolism."

2.1 WASTE CLASSIFICATION AND TYPES OF WASTE

In practical terms, waste results from any consumption and production process, accompanied by the need to manage it, to reduce its amount, and to avoid pollution problems leading to public health issues and/or environmental degradation. Waste can be characterized according to different features: source, nature, physical and mechanical properties, chemical and elemental properties, biological/biodegradable properties, and combustible properties. Concerning their source, waste can be classified as:

Sustainable Solid Waste Management: A Systems Engineering Approach, First Edition. Ni-Bin Chang and Ana Pires.
© 2015 The Institute of Electrical and Electronics Engineers, Inc. Published 2015 by John Wiley & Sons, Inc.

- MSW, which includes commercial and services waste;
- industrial waste, which can include several activities, like light and heavy manufacturing, construction and demolition activities, refineries, chemical, automotive, energy, mining, and agricultural;
- medical waste;
- other waste not considered municipal, industrial, or medical waste.

Identifying a proper definition of the "source" of various MSW streams is useful to understanding the owner and producer of the waste and also helps to characterize waste in terms of different components. For example, MSW is rich in paper, plastic, and glass, whereas industrial waste streams vary depending on the industrial process. Concerning their nature, solid waste can be differentiated as:

- **Hazardous waste**. Waste that presents at least one hazard to the human health or to the environment.
- **Inert waste.** Waste that does not experience any physical, chemical, or biological transformations.
- **Nonhazardous waste**. Waste that has no hazardous features due to prior physical, chemical, or biological transformations.

Physical properties of waste include waste density, moisture content, and calorific value. Chemical and elemental features of waste are related to chemical composition, carbon/nitrogen ratio, pH, presence of heavy metals, and other hazardous and nonhazardous features. Combustion properties are tied to the latent heat of waste and its calorific value. Biological and biodegradable characteristics are linked with microorganism cultures existing in waste and how they utilize waste materials to survive. Waste characterization is therefore fundamental to SWM (Table 2.1).

Knowing the features of waste streams is vital for planning, design, and operation of an SWM system. The choice of a specific technology matrix for an SWM system and its operation is intimately dependent on the properties of waste materials and the amounts of waste generated, which influence key economic factors of the SWM system and impact the environment and society within that system. This systematic understanding is needed to produce a suite of forward-looking, risk-informed, cost-effective, and environmentally benign SWM solutions.

2.1.1 Municipal Solid Waste and Waste Streams

MSW comprises household waste, commercial waste, and institutional waste. It includes separately collected fractions from public service areas and private sectors, such as:

- garden and park waste (including cemetery waste),
- waste from markets,

TABLE 2.1 Waste properties

Physical and mechanical properties	Description
Waste generation per capita ($\text{kg} \cdot \text{capita}^{-1}$)	$$\text{Waste generation per capita} = \frac{\text{Waste generated}}{\text{Inhabitants}}$$ Fundamental for the waste planning (collection and treatment operations).
Physical composition (%)	$$\text{Composition} = \frac{\text{Component 1}}{\text{Total waste amount}} \cdot 100$$ $$+ \frac{\text{Component 2}}{\text{Total waste amount}} \cdot 100 + \cdots + \frac{\text{Component } n}{\text{Total waste amount}} \cdot 100$$ Can induce the potential recovery of waste materials.
Waste density ($\text{kg} \cdot \text{m}^{-3}$)	$$\text{Density} = \frac{\text{Mass}}{\text{Volume}}$$ Relevant for waste operation equipment, namely vehicles and waste containers, as well conveyors, sorting equipment.
Moisture content (%)	$$W = \frac{A - B}{A} \cdot 100$$ W (%)—percentage of moisture (wet sample) A—weight of wet sample B—weight of dry sample Can influence waste degradation and stabilization, as well energy recovery from waste. The production of leachate during collection but also in waste treatment units is also a factor to be considered when designing installations. Concerning recyclables, moisture can influence paper and cardboard density, which has to be considered in recycling process and acceptance criteria by recyclers.
Particle size and size distribution	To assess and characterize waste concerning materials concentration in a specific size, with implication on equipment's selection.

Chemical properties	Description
pH	It reflects the waste corrosion in waste equipment, and how to prevent it.
Chemical composition	Determination of elements: carbon (C), nitrogen (N), oxygen (O), sulfur (S), calcium (Ca), potassium (K), chlorine (Cl), sodium (Na), aluminum (Al), iron (Fe), magnesium (Mg), silicon (Si), titanium (Ti), and other elements such as heavy metals. Helps to define best waste treatment to be applied. It also indicates the presence of harmful substances, which also influence the adequate waste treatment.
C/N ratio	$$C/N = \frac{\text{Carbon content}}{\text{Nitrogen content}}$$ Useful for the biological waste treatment options as well as the quality of compost produced.

(continued)

TABLE 2.1   *(Continued)*

Combustion properties	Description
Low or net calorific value (or low heating value)	$$NCV = GCV \cdot \left[1 - \left(\frac{W}{100}\right)\right] - 2.447 \cdot \left(\frac{W}{100}\right) - 2.447 \cdot \left(\frac{H}{100}\right)$$ $$\cdot \, 9.01 \cdot \left[1 - \left(\frac{W}{100}\right)\right]$$

NCV—net calorific value ($MJ \cdot kg^{-1}$ wet basis)
GCV—gross calorific value ($MJ \cdot kg^{-1}$ dry basis)
W—moisture content of the fuel in wt% (weight percent) (wet basis)
H—concentration of hydrogen in wt% (dry basis)

(from Rosillo-Calle et al., 2007)

Assess the potential of waste to be subjected to energy recovery.

Biological features	$$DM^t_{degradation} = DM_0 - DM_t = \left[1 - \frac{\left(1 - VS_t\right)}{\left(1 - VS_0\right)}\right] \cdot DM_0$$

$DM^t_{degradation}$—degradation of the dry matter at time t (kg)
DM_0—dry mass at initial time (kg)
DM_t—dry matter at time t (kg)
VS_0—volatile solids content at initial time (wt%)
VS_t—volatile solids content at time t (wt%)

(from von Felde and Doedens, 1997)

Helpful to define biological treatment processes and prevent odor problems.

- street-cleaning residues,
- sludge from sewage-disposal tanks,
- waste from sewage cleaning,
- bulky waste, and
- household hazardous waste (HHW).

The components or fractions of waste materials found in MSW streams are plastics, paper and cardboard, organic waste, textiles, aluminum, ferrous materials, glass, and wood. This composition is influenced by the economic development level (i.e., the income level), educational level (i.e., related to the degree of recycling), and other managerial factors. At a local level, however, the quantity and composition of MSW is also influenced by climate factors such as weather conditions (Gómez et al., 2009). In addition, special seasons like Christmas, Carnival, Easter, and other holidays also influence the generation of waste. The waste generation per capita and waste composition are relevant indicators to be obtained during waste characterization campaigns.

Granulometry can also be used for waste characterization. By comparing the Egyptian and French raw waste (Skhiri et al., 2005), the granulometry can be summarized to show how variational analysis can characterize the waste stream on a comparative basis. According to Skhiri et al. (2005), the plastics category decreases with the decrease of element size, whereas organic waste, glass, stone, and limestone categories increase with decrease of element size. These factors impact technological choices when importing facilities from overseas. In addition to waste recycling and material recovery, heat and/or energy recovery has always been a focus in SWM system planning. Of course, different waste fractions may have quite different features in terms of chemical and calorific characteristics (Table 2.2), which could strongly influence the technological choice of waste treatment and operations in each specific situation.

MSW can also be characterized by its nature, including hazardous or nonhazardous properties, or even inert features. MSW is not inert because its chemical, physical, and biological features can change over time, which is important to consider when planning an SWM system. Hazardous waste materials are sometimes present in MSW streams which are known as HHW. Although various classification systems exist that categorize the relevant household products in a hazardous subcategory of MSW, separate collection of HHW is rare in most countries, these products being discarded alongside nonhazardous household waste (Slack et al., 2004). The improper management of HHW has contributed to specific environmental problems, often because definitions and classifications vary across different countries. For example, in the European community, attempting to include waste of electrical and electronic equipment (WEEE) as HHW would be a challenging issue. Additional environmental problems associated with HHW are related to its disposal in landfills, leading to the complication of leachate composition and subsequent treatment. Slack et al. (2005) demonstrated that a wide diversity of xenobiotic compounds occurring in leachate can be associated to HHW, although the need to evaluate whether such compounds offer a risk to the environment and human health as a result of leakage/seepage or through treatment and discharge has not yet been fully identified. The European list of waste (LoW) (Commission, 2000) classifies solid waste in a code with six digits, including for HHW (Table 2.3).

2.1.2 Industrial Waste

Industrial waste comprises several different waste streams originating from a broad range of industrial processes (EIONET, 2009) (Figure 2.1). In general, mining industry, manufacturing industry, and the construction sector are responsible for the majority of industrial waste generation. Each subsector generates a specific type of waste that may be categorized by, for example, LoW codes (Table 2.4).

The types of waste produced from any industrial process must be characterized and analyzed via in-depth analyses with process knowledge in order to identify the most suitable treatment technology in an SWM system. Process knowledge, as presented by the United States Environmental Protection Agency (USEPA, 2012a), refers to detailed information on processes that generate waste and can be used to partly, or in many cases completely, characterize waste to ensure proper management. Process knowledge includes (USEPA, 2012a):

- "existing published or documented waste analysis data or studies conducted on waste generated by processes similar to those which generated the waste;

TABLE 2.2 Physical, chemical, and calorific features of wet and dried waste

Category	OW	Paper	Plastics	LWTR	Glass	Metals	NC	Miscellaneous	Average
Specific volume (L · kg^{-1}) (wet basis)	4.17	15.14	29.0	9.5	4.3	12.5	2.0	8.4	10.9
Moisture (wt%) (wet basis)	72.9	27.1	21.3	18.6	1.0	1.0	1.0	17.1	40.6
Combustible fraction (wt%) (wet basis)	22.0	63.5	74.8	74.9	0.0	0.0	0.0	37.4	41.6
Ash (wt%) (wet basis)	5.1	9.4	3.9	6.5	99.0	99.0	99.0	45.5	17.8
LHV (kJ · kg^{-1}) (wet basis)	1,486	10,123	29,003	13,810	–	–	–	10,985	8,426
HHV (kJ · kg^{-1}) (wet basis)	4,892	12,037	32,617	15,617	–	–	–	12,820	11,003
C (wt%) (dry basis)	49.2	39.1	81.6	47.3					
H (wt%) (dry basis)	7.0	5.5	13.6	5.9					
N (wt%) (dry basis)	1.94	0.17	1.14	4.88					

Source: From Koufodimos and Samaras (2002).

LWTR, leather-wood-textiles-rubber; LHV, low heating value; OW, organic waste; NC, noncombustible.

TABLE 2.3   Municipal hazardous substances classification according to the LoW

LoW description	LoW code
Solvents	20 01 13
Acids	20 01 14
Alkalines	20 01 15
Photochemicals	20 01 17
Pesticides	20 01 19
Fluorescent tubes and other mercury-containing waste	20 01 21
Chlorofluorocarbons in discarded equipment	20 01 23
Nonedible oil and fat	20 01 26
Paint, inks, resins, and adhesives containing dangerous substances	20 01 27
Detergents containing dangerous substances	20 01 29
Cytotoxic and cytostatic medicines	20 01 31
Batteries and accumulators containing lead, nickel/cadmium, or mercury	20 01 33
Discarded electronic and electrical equipment other than 20 01 21 and 20 01 23 containing dangerous components	20 01 35
Wood containing dangerous substances	20 01 37

Source: From Commission (2000), Slack et al. (2004), and Slack and Letcher (2011).

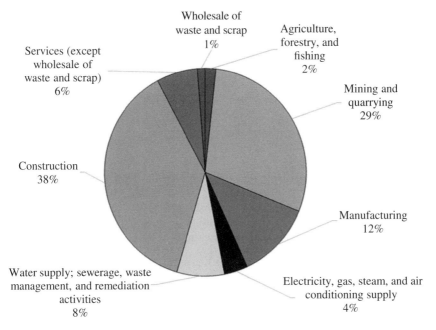

FIGURE 2.1   Industrial waste composition from European Union 27 countries (EU-27) by classification of economic activities in the European Community. *Source*: From Eurostat (2010) (online data code: env_wasgen)

TABLE 2.4   Examples of industrial waste with respect to the type of waste according to LoW code

Industry	Type of waste	Examples
Agriculture and forestry	Code 02	Waste from forestry, sludge from washing and leaning, animal-tissue waste, waste from solvent extraction, waste from spirit distillations.
Mining industry	Code 01	Drilling muds and other drilling waste, waste from physical and chemical processing of metalliferous and nonmetalliferous minerals, waste from mineral extraction.
Manufacturing industry	Codes 03–16	Waste from wood processing, from leather and fur industry, from textile, waste from inorganic chemical processes, waste from photographic industry, waste from manufacture of glass and glass products, sludge and solids from tempering processes, waste hydraulic oils.
Energy sector	Codes 05, 10	Waste from power stations and other combustion plants, waste from petroleum refining.
Waste and wastewater management	Code 19	Waste from incineration or pyrolysis of waste, stabilized/solidified waste, waste from aerobic treatment of solid waste.
Construction sector	Code 17	Concrete, bricks, tile and ceramic insulation materials, gypsum-based construction materials.

Source: From Commission (2000).

- waste analysis data obtained from other facilities in the same industry;
- facility's records of previously performed analysis."

Although industrial waste can be identified through analysis using specific methodologies (e.g., material flow analysis), some waste properties such as chemical and calorific features need to be assessed directly. The determinant used to evaluate these wastes is highly related to the destination of those wastes and even their waste management companies which are committed to processing them. Further, environmental sampling and analysis for waste streams and leachate testing are also essential when the final destination is a landfill or any other facility that poses a risk of water contamination.

2.1.3   Medical Waste

Medical waste (or hospital waste) have a unique classification. In Europe, for example, this type of waste is identified by code 18 on the LoW from Commission Decision 2000/532/EC (Commission, 2000), which is defined as waste from human or animal health care and/or related research. It is further divided into waste from natal care, diagnosis, treatment or prevention of disease in humans, and waste from research, diagnosis, treatment, or prevention of disease involving animals (Commission, 2000). Although all medical waste might be perceived as dangerous, some types are exempt, such as sharps without infection risk, waste whose collection and disposal is not subject to special requirements concerning infection risk, and nonhazardous chemicals. The World Health Organization (WHO) presents another classification of medical or health-care waste (WHO, 2012):

- "**infectious waste**: waste contaminated with blood and its by-products, cultures and stocks of infectious agents, waste from patients in isolation wards, discarded diagnostic samples containing blood and body fluids, infected animals from laboratories, and contaminated materials (swabs, bandages), and equipment (such as disposable medical devices);
- **pathological waste**: recognizable body parts and contaminated animal carcasses;
- **sharps**: syringes, needles, disposable scalpels, blades, and other sharp instruments;
- **chemicals**: mercury, solvents and disinfectants;
- **pharmaceuticals**: expired, unused, and contaminated drugs; vaccines and sera;
- **genotoxic waste**: highly hazardous, mutagenic, teratogenic or carcinogenic materials, such as cytotoxic drugs and their metabolites resulting from their use;
- **radioactive waste**: glassware contaminated with radioactive diagnostic material or radiotherapeutic materials; and
- **heavy metals waste**: broken mercury thermometers," for example.

According to the WHO (2012), the major sources of health-care waste are hospitals and other health-care establishments, laboratories and research centers, mortuary and autopsy centers, animal research and testing laboratories, blood banks and collection services, and nursing homes for the elderly. WHO (2012) further revealed that the infectious and anatomical wastes together represent the majority of the hazardous waste, up to 15% of the total waste from health-care activities; sharps represent about 1%; chemicals and pharmaceuticals account for about 3%; and radioactive matter and heavy metal content account for around 1% of total health-care waste; the remainder fraction, ~80%, is more typical MSW.

2.1.4 Other Wastes

Other wastes not related to any specific source, according to LoW (Commission, 2000), include end-of-life vehicles (ELV) and those from dismantling of ELV and vehicle maintenance; WEEE; off-specification batches and unused products; waste explosives; gases in pressure containers and discarded chemicals; batteries and accumulators; wastes from transport tank, storage tank, and barrel cleaning; spent catalysts; oxidizing substances; aqueous liquid waste destined for off-site treatment; and waste linings and refractories. Like other types of waste, these are also classified as hazardous, nonhazardous, and inert. One type of waste which deserves attention is WEEE, a waste known by its unique physical and chemical composition. Due to its complexity, physical composition can be related to specific materials (Figure 2.2) and specific toxic compounds, such as those observed in Oguchi et al. (2013), who categorized WEEE by concentration and total amount of toxic metals.

2.2 WASTE MANAGEMENT THROUGH WASTE HIERARCHY: REDUCE, REUSE, RECYCLE, RECOVER, AND DISPOSAL

The Western world and parts of Asia have mainly used the waste hierarchy principle to approach waste management since the early 1980s (Christensen, 2011), although the wording used and names may vary. For example, in Japan, the approach is called the 3Rs principle, representing reduction, reuse, and recovery. Different application frameworks with similar philosophies can be summarized with the common emphasis on material conservation (Figure 2.3).

Although the waste hierarchy, or 3Rs perspective, seems to present a common sense approach, the best option from an environmental perspective does not have universal consensus. Finnveden et al. (2005) noted that the positions of recycling and incineration in a hierarchical framework in Figure 2.3 remain contentious, including where to place biological treatments such as the anaerobic digestion and composting in the hierarchy. One way to help waste managers resolve this problem is to use life cycle thinking (LCT), an approach that examines all stages of products and waste to find, in particular, places where waste is generated and how to best implement waste hierarchy. LCT has the ability to show all life cycle stages of waste to determine where inputs and outputs occur, and where waste can be reduced, reused, recycled, and recovered to divert waste from landfills. With this concept, waste is deemed only that fraction of materials that cannot be reintroduced into the human consumption system, such as those destined for landfills. The association of both the strategy and concept of LCT can drive the life cycle of resources to reach zero waste management.

Along this line of system thinking, SWM issues should not be regarded as a public health problem; instead, SWM may be seen as the providers/miners/manufacturers of secondary materials and secondary products. Waste can become a second resource in several phases of products (Figure 2.4), changing an SWM paradigm to a circular economy that would promote the production of secondary raw materials from by-products, waste fuels, and end-of-waste products obtained from new and advanced waste treatment technologies. With recent technological advancements, the quality of secondary materials/resources have improved, triggering an appealing recycling industry in the United States, Europe, and some Asian countries such as Japan and Singapore.

2.2.1 Reduction, Prevention, and Reuse

Goals to reduce, prevent, and minimize waste generation are all faces of the same intention that is to avoid waste generation. In this respect, all possible sources of waste must be considered to develop and promote more efficient processes. First,

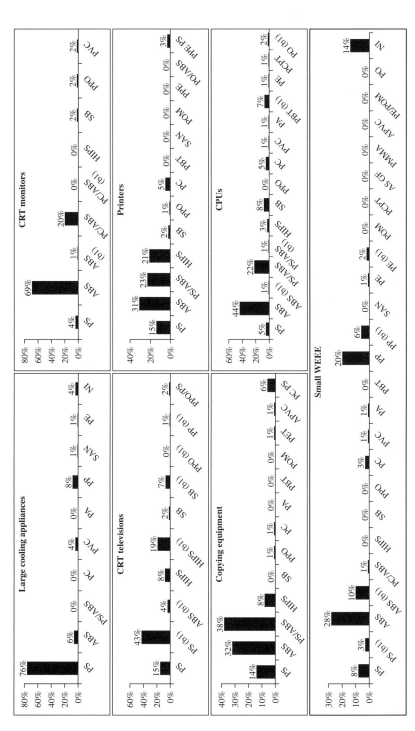

FIGURE 2.2 Plastic compositions for WEEE appliances. *Source:* From Martinho et al. (2012). PE, Polyethylene; ABS, acrylonitrile–butadiene–styrene; HIPS, high-impact polystyrene; PP, polypropylene; PS, polystyrene; SAN, styrene-acrylonitrile; PU, polyurethane; PA, polyamide; PC/ABS, blends of polycarbonate; PPO, poly(p-phenylene oxide; PVC, polyvinyl chloride; SB, styrene/butadiene; PBT, polybutylene terephthalate; POM, polyoxymethylene; PPE, poly(phenyl ether); PMMA, poly(methyl methacrylate); ASGF, acrylonitrile styrene glass fiber; APVC, anti-corrosion PVC; PT, polythiophenes; NI, Not identified; bl, black. Combined names are polymer blends.

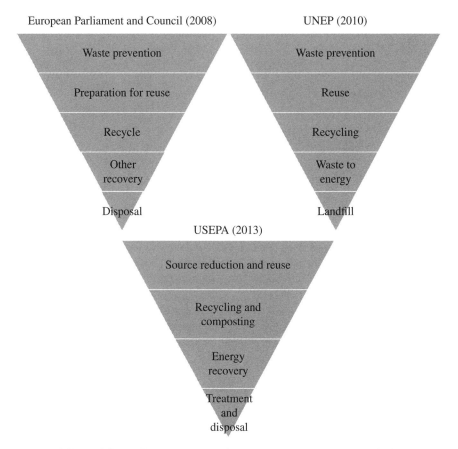

FIGURE 2.3   Different application framework of waste hierarchy principle

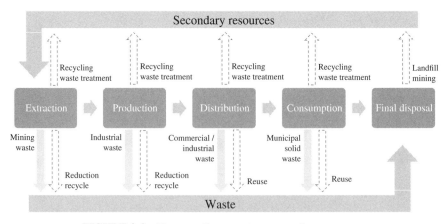

FIGURE 2.4   New paradigm: waste as secondary resources

what is waste prevention? Several legal documents provide examples, such as the EU Waste Framework Directive (WFD) 2008/98/EC (European Parliament and Council, 2008) which states that waste prevention must:

- reduce the quantity of waste, including reusing products or extending the life span of products;
- reduce the adverse impacts of the generated waste on the environment and human health; and
- reduce the content of harmful substances in material and products.

In the United States, the concept of waste minimization was introduced by the USEPA in 1988 with the publication of the *Waste Minimization Opportunity Assessment Manual* (USEPA, 1988). In this concept, the approach to waste prevention and its techniques are defined by conducting waste minimization assessment, on-site operating practices to reduce hazardous waste generation, incorporating waste minimization into the company profile and defining waste minimization programs (USEPA, 1992; UNEP, 2003). Off-site recycling by direct reuse after reclamation is also considered to be a waste minimization technique, but with a distinctly lower priority compared to on-site prevention or minimization of waste.

The concept of waste prevention is, of course, linked with the definition of source reduction, or processes that minimize waste where it is generated. According to Hoornweg and Bhada-Tata (2012), a waste reduction (or source reduction) initiative should reduce the quantity of waste at generation points by redesigning products or changing patterns of production and consumption.

How can waste prevention be applied? Any process that produces waste, at all stages of a product's life cycle, including waste collection, treatment, and disposal, should be taken into account as a whole (Figure 2.4). From a technological point of view, waste prevention/reduction/minimization can be achieved by implementing cleaner production and green manufacturing, designing for the environment (definitions in Box 2.1), and establishing technical features to identify where the waste is generated, including raw waste and by-products.

BOX 2.1   WASTE PREVENTION PARADIGMS AND MEASURES: CLEANER PRODUCTION, GREEN MANUFACTURING, AND DESIGN FOR ENVIRONMENT

- Cleaner production was first introduced by the United Nations Environment Programme (UNEP) in 1989, where a precautionary principle was applied that shifts the focus toward earlier stages in the industrial process, the source of pollution. The concept embraces the notion of efficient use of resources, where waste avoidance is promoted whenever practically possible (OECD, 2009). A recent definition for cleaner production has been proposed by Glavič and Lukman (2007), stating that cleaner production "is a systematically organized approach to production activities, which has positive effects on the environment.

These activities encompass resource use minimization, improved eco-efficiency and source reduction, in order to improve the environmental protection and to reduce risks to living organisms."

- Green manufacturing has been deemed a new paradigm for industry, where different green strategies and techniques are applied to make manufacturing more eco-efficient (Deif, 2011). Strategies can be devoted to the development of new products/systems to consume less energy, to replace input materials, and to reduce waste by reusing it as secondary raw material (Deif, 2011).
- Design for environment, also called green design, considers the product's impact on the environment during the entire life cycle of the product (Glavič and Lukman, 2007). In this respect, every product conceived to reduce waste generated during its life span will contribute to waste prevention.

According to WFD (European Parliament and Council, 2008), by-products are a substance or object resulting from a production process, which is deemed residual materials or objects instead of waste, if the following conditions are met:

- further use of the substance or object is certain;
- the substance of the object can be used directly without any further processing other than normal industrial practice;
- the substance or object is produced as an integral part of a production process; and
- further use is lawful, that is, the substance or object fulfills all relevant products, environmental and health protection requirements for the specific use, and will not lead to overall adverse environmental or human health impacts.

Reuse, which is another measure that can promote waste prevention and minimization, occurs after the product is produced and subsequently used. Depending on the context, the waste hierarchy principle can highlight reuse as an isolated approach or be combined with waste prevention.

According to WFD (European Parliament and Council, 2008), in Europe reuse is identified as any operation by which products or components that are not waste are used again for the same purpose for which they were conceived. A new concept to reinforce reuse is "preparation for reuse," which involves checking, cleaning, or repairing recovery operations to prepare products, or components of products that have become waste, for reuse without further preprocessing.

2.2.2 Recycle

Definition of recycling is more uniform across the globe. According to WFD (European Parliament and Council, 2008), recycling means any recovery operation by which waste materials are reprocessed into products, materials, or substances, whether for the original or other purposes. Recycling includes the reprocessing of organic material, but is not recycling if it is going to be used as fuels or for backfilling operations (European Parliament and Council, 2008).

In recycling economies or circular economies, such as those in Europe and Japan, wastes are reintroduced into the human consumption system, activating a new economic flow. Material recycling therefore becomes more relevant than energy recovery from waste in the context of macroeconomics. The importance of reintroducing recycled materials into the economy has led the European Union to define end-of-waste criteria to channel recycled products out of waste streams.

2.2.3 Biological Recovery: Compost and Methane Gas

Biological recovery involves composting and anaerobic digestion. Composting is a biological aerobic process converting the easily degradable organic waste into carbon dioxide and stabilized organic matter. Anaerobic digestion is a process occurring in the absence of oxygen, where the rapid organic waste is decomposed to produce methane gas. Both processes produce a compost/digestate that can be used as fertilizer (if quality is sufficient) or simply landfilled as daily cover material. The exhaust gases from both processes must be controlled, as well as wastewater and compost/digestate.

These biological recovery procedures are also considered recycling; however, the methods used to reintroduce waste into the circular economy are dependent on the recycled waste properties. From an environmental point of view, anaerobic digestion is more environmentally beneficial than composting, mainly because the biogas production can replace the use of energy from fossil sources to produce vehicle fuel, heat, electricity, and combined heat and power. Evidence also indicates that, for garden waste and mixtures of food with garden waste, dry anaerobic digestion followed by composting is better from an environmental point of view, than just composting those wastes/mixture of wastes (DEFRA, 2012).

2.2.4 Waste-to-energy

Following biological recovery and recycling, the final waste recovery technology available is energy recovery, or waste-to-energy (WTE), a waste treatment process capable of delivering cleaner energy relative to other energy production processes such as coal-fired power plants. Waste can therefore be regarded as a domestic renewable energy source.

WTE is different from waste thermal treatment and elimination because it includes the recovery of the energy produced; however, the "efficiency of energy production from waste is much lower than efficiency of energy generation in conventional plants using fossil fuels" (Pavlas et al., 2010). The specific properties of waste used as fuel create constraints such as reduction of maximum output steam pressure due to corrosion risk and higher flue gas temperature leaving the boiler (Pavlas et al., 2010). With different waste heat utilization strategies, WTE plants can be classified as (Reimann, 2006):

- power plants producing heat only (with 63% production efficiency);

- power plants without heat delivery (18% efficiency); and
- cogeneration systems where heat and electricity are produced simultaneously (43% efficiency).

2.2.5 Disposal

According to WFD (European Parliament and Council, 2008), disposal consists of "any operation that does not involve recovery, even where the operation has a secondary consequence due to the reclamation of substances or energy." This disposal solution is viewed as the last resort according to waste hierarchy principle. In a circular economy where waste is the resource and zero waste philosophy is the goal, disposal is not the last destination. Disposal can be viewed as a storage device for the future, such as storing the refuse-derived fuel (RDF) or solid-recovered fuel (SRF) in landfills. Disposal can be used to save waste streams as a source of fuel, such as the case when landfill is viewed as a source of methane gas in bioreactor landfills. Disposal can even be chosen as a source of secondary materials, claimed through landfill mining with the purpose of recovering waste materials, after the end of landfill operation. The intention of zero waste generation is to avoid landfilling of any resources as long as these resources can be used for human consumption. For instance, food waste recovery would benefit the environment by avoiding landfilling organic waste, reducing greenhouse gases released from landfills, and limiting environmental impacts.

Several disposal operations can be considered according to WFD (European Parliament and Council, 2008) (Table 2.5). Incineration without energy recovery and landfilling are both possible disposal operations; however, in this chapter, incineration is considered a thermal treatment, being described in section 2.4.4.

2.3 WASTE OPERATIONAL UNITS: REAL-WORLD CASES

Waste operational units are the skeleton of an SWM system, consisting of waste collection and transport, mechanical treatment, biological treatment, thermal treatment, and/or disposal. To understand management, features, inputs and outputs of each process, and future developments of these operational units, a short overview of five case studies is presented. The waste management systems in Berlin, Lisbon, Seattle, Copenhagen, and Singapore represent some of the complex SWM systems and the following delineation demonstrates how they can be well operated to meet the goals.

2.3.1 Berlin, Germany

In Berlin, the Act for Promoting Closed Substance Cycle and Waste Management and Ensuring Environmentally Compatible Waste Disposal came into the law in 1999 (Schulze, 2009). In addition, because recyclables (e.g., organic waste, paper, lightpackaging waste) account for approximately 80% of the total amount of commercial waste (Schwilling et al., 2004), great efforts have been made to separate and recycle

TABLE 2.5   Disposal operations from WFD

Identifier	Disposal operation
D1	Deposit into or on to land (e.g., landfill).
D2	Land treatment (e.g., biodegradation of liquid or sludgy discards in soils).
D3	Deep injection (e.g., injection of pumpable discards into wells, salt domes, or naturally occurring repositories).
D4	Surface impoundment (e.g., placement of liquid or sludgy discards into pits, ponds, or lagoons).
D5	Specially engineered landfill (e.g., placement into lined discrete cells, which are capped and isolated from one another and the environment).
D6	Release into a water body except seas and oceans.
D7	Release to seas and oceans, including sea-bed insertion.
D8	Biological treatment not specified elsewhere in this Annex that results in final compounds or mixtures discarded by means of any of the operations numbered D1–D12.
D9	Physicochemical treatment not specified elsewhere in this Annex that results in final compounds or mixtures discarded by means of any of the operations numbered D1–D12 (e.g., evaporation, drying, calcination).
D10	Incineration on land.
D11	Incineration at sea.
D12	Permanent storage (e.g., emplacement of containers in a mine).
D13	Blending or mixing prior to submission to any of the operations numbered D1–D12.
D14	Repackaging prior to submission to any of the operations numbered D1–D13.
D15	Storage pending any of the operations numbered D1–D14 (excluding temporary storage, pending collection, or storage on the site where the waste is produced).

Source: From European Parliament and Council (2008).
© European Union, http://eur-lex.europa.eu/

waste, resulting in a 10-fold reduction in commercial waste from 1992 to 2007 (Zhang et al., 2010) (Table 2.6).

To manage waste, the entity responsible is Berlin's municipality, providing the disposal of the waste generated within the city. *Berliner Stadtreinigungsbetriebe* (BSR), the statutory body of Berlin municipality, has the duty to collect and dispose waste from households, commercial and services (Schulze, 2013).

Collection   At the source location, domestic/household waste is collected by truck fleet regularly, providing service near tenement building areas. In the case of commercial waste, the owner has to ship waste either by itself or by contractual service to the waste treatment facilities operated by the Berlin municipality. Bulky waste is collected separately, incurring a specific charge for such service.

Depending on the type of waste, the collection is made by different standardized bins. Residual waste is collected in grey bins, and organic waste collected in separate brown wheelie bins offered by BSR within the city for recycling. In the suburbs,

TABLE 2.6   Summarized features of SWM system in Berlin, Germany

Summary—Berlin	Year 2012
Population served (inhabitants)	3.5 million
Waste managed (tonnes)	1.42 million
Collection method	Curbside collection of source-separated biodegradable municipal waste (BMW), paper, metals, plastics, and residual waste. Bring system for glass waste. Curbside collection is also applied in suburban dwelling areas.
Total recycled (%)	42.1
Total treated and recovered (includes incineration) (%)	57.4
Total landfilled (%)	0.5

Source: From Zhang et al. (2010) and Schulze (2013).
Note: Tonnes are metric tons.

most organic waste is home composted, while the rest of the organic waste collected separately by the BIOGUT bin is sent to centralized treatment facilities.

The *Duales System Deutschland* (DSD) created in 1990 maintains the collection, sorting and recycling of packaging waste (see Box 2.2). Until December 2012, light packaging waste was disposed in yellow bin (tenement buildings) or in yellow bag (dwellings), to be collected by a company commissioned by the operators of the DSD. In January 2013, a new uniform recycling bin has been created based on the material in general rather than exclusively for packaging waste (Schulze, 2013). Such changes are justified by the new EU recycling targets established by the WFD (European Parliament and Council, 2008) for material types. In the case of glass, the collection is made separately by colors brown and green, in bins (in tenement buildings) and bottle banks (known as bring system). Paper is collected in blue wheelie bins alongside grey residual waste bins, being shipped by waste management companies or by BSR (Schulze, 2013).

BOX 2.2   DUAL WASTE MANAGEMENT SYSTEMS FOR PACKAGING WASTE (OECD, 2002)

In 1990, 95 German enterprises founded the DSD to establish a common take-back system for packaging waste. It was created to be in accordance with government offers to exempt participating producers and retailers from individual take-back responsibilities due to producer-pays principle mentioned in Packaging Avoidance Rule (*VerPackV*). DSD assumes the responsibility for the collection of packaging waste, being financed by the Green Dot seal, which can only be used by those who pay a material-specific fee. These fees are used to pay companies to run the collection and sorting system, for subsidizing plastic recycling, and for recycling campaigns, mostly.

Treatment and recovery   Residual waste is delivered at different treatment plants: one WTE and two RDF/SRF production plants (i.e., one co-owned by a private company). Such technological approach reduces the waste going to landfill. The slags from WTE are used in the waste disposal site to form the substrate for the surface sealing system when upgrading waste depots, and ashes are disposed in underground depots (Schulze, 2013). All collected organic waste is delivered to an anaerobic digestion plant, where the cleaned biogas produced is used in natural gas vehicles of BSR for waste collection (Schulze, 2013). Recyclables are delivered to sorting plants for various sorting purposes, which are operated mostly by private companies. BSR also conduct disassembly of refrigerators and recycling of WEEE.

Disposal   In recent years, due to the disposal prohibition of untreated domestic and commercial waste, the need for landfills has been reduced significantly (Zhang et al., 2010; Schulze, 2013). This prohibition have triggered the recycling and recovery of waste materials, which otherwise would be landfilled. Only slags and ashes from WTE, low calorific fraction waste (which cannot be used for RDF/SRF production), and construction and demolition wastes (CDW) are landfilled nowadays.

2.3.2 Lisbon, Portugal

An association of five municipalities for SWM from the Lisbon district began in 1995. Today, 19 municipalities are integrated together for promoting the power of regionalization for SWM, allowing a better economy of scale (Table 2.7). When landfill is still inexpensive in the Lisbon area, as in the rest of the country, there is no real economic incentive to drive bold recycling programs. Before extended producer responsibility for packaging waste was implemented in Portugal, some separate waste collection existed for packaging waste recycling through the informal sector at a smaller scale.

Collection   While a country-wide, door-to-door (or curbside) or bring collection system exists for mixed MSW, a parallel door-to-door source separation service for

TABLE 2.7   Waste management features in Lisbon area, Portugal

Summary—Lisbon	Year 2009
Population served (inhabitants)	1.6 million
Waste managed (tonnes)	923,000
Collection method	Bring system requiring source separation of paper, plastic, glass, metals and residual waste. Curbside collection for BMW in selected areas.
Total recycled (%)	14.6 (includes the recovery of biodegradable waste)
Total incinerated (%)	65.6
Total landfilled (%)	19.8

Source: From Valorsul (2009).
Note: Tonnes are metric tons.

BMW (mainly food waste) is operated only in Lisbon, Amadora, and Loures cities. Packaging materials like glass, paper and cardboard, plastic, and metals are mostly collected separately using bring systems; one specific neighborhood has pneumatic waste collection.

Treatment and recovery   The waste treatment system includes an automatic material recovery facility (MRF, pronounced as "merf") for packaging materials separation in the Lisbon area, plus a manual sorting plant in the northern municipalities. These MRFs are used to sort all packaging materials except glass, which is stored and sent to reprocessors. Selectively collected BMW is sent to an anaerobic digestion plant, and residual waste is sent to a mechanical biological treatment (MBT) unit (since 2012) and to an incineration plant. The incineration plant produces electric energy from waste combustion; the MBT and the anaerobic digestion plant produce electric energy from biogas burning. MBT and anaerobic digestion plants produce compost from digestate composting, which are sold for agricultural use with restriction. The bottom ashes from incineration are then treated to recover metallic fraction, and the residual inert materials are used for road construction. The flying ashes are stabilized and landfilled.

Disposal   Valorsul possesses two sanitary landfills, and both receive residual waste. One of the landfills also presents a specific cell for stabilized flying ashes. It also receives nonrecyclable CDW.

2.3.3 Seattle, USA

The waste crisis in 1987 in Seattle dictates how waste should be managed, considering citizens' opinion and less expensive solutions. The solution pointed out by the Seattle Public Utilities (SPU), responsible for MSW management, included waste reduction and recycling programs at a large scale over incineration (SPU, 2013). More recently, in 2007, the Mayor and City Council adopted Resolution 30990, the Zero Waste Resolution, where the goal was 60% recycling rate in 2012 and 75% recycling rate in 2025 (SPU, 2013). The features of Seattle waste management are summarized in Table 2.8. SPU possesses two transfer stations, and private sector owns organic composting unit, and two landfills. Other recycling units are operating in Seattle, privately owned, for CDW, and two transfer units for recyclables (SPU, 2013). SPU also has solutions for hazardous waste management.

Collection   Waste collection is divided in four sectors: single family, multi-family, self-haul, and commercial. Single family sector collects all waste fractions (residual waste, recyclables, and food and yard waste). SPU's collection contractors collect residual waste and organics every week, and recyclables every other week. They also provide other collection services for appliances, large volumes, and used motor oil (SPU, 2013). Home composting also occurs. All recyclable materials, including glass bottles and cans are all collected commingled (City of Seattle, 2010). Multi-family

TABLE 2.8 Waste management features in Seattle, US

Summary—Seattle	Year 2010
Population served (inhabitants)	0.6 million
Waste managed (tonnes)	724,468
Collection method	Curbside collection of residual waste, organic waste (food and yard), and recyclables.
Total recycled (%)	53.7 (includes the recovery of biodegradable waste)
Total incinerated (%)	0
Total landfilled (%)	46.3

Source: From City of Seattle (2010) and SPU (2013).
Note: Tonnes are metric tons.

sector that includes apartment and condominium buildings is similar to the single family sector. They use dumpsters instead of tote carts.

Waste collection in commercial sector is similar to that in residential areas. City collection contractors pick up from dumpsters at least weekly and transfer the waste to the two Seattle transfer stations (City of Seattle, 2010). A small part uses the same cart-based, city-contracted, biweekly collection service. Paper and cardboard, being recyclables collected by a wide range of collectors, are not allowed in residual waste. Commercial possessing organic waste can choose either public or private collection service.

Self-haul sector is devoted to material that is produced by residents, business, and governmental agencies which is delivered to the two-city owned recycling and disposal (transfer) stations (City of Seattle, 2010). For this sector, organic waste (food and yard waste, clean wood), appliances and metals, and other recyclables are selectively delivered.

Treatment and recovery   SPU contracts one processor for recyclables – Rabanco Recycling Center – and one processor for organic waste – Cedar Grove. Commercial recyclables are processed by other private facilities. The sorted materials are hauled to a variety of facilities. To ensure that recyclables are neither incinerated nor landfilled, SPU bears 100% of the risk (and benefit) of market price changes for recyclables. The contract between SPU and the private owner of Rabanco MRF sets a base price for the various commodities; if market prices are higher, then SPU receives a "credit" (savings) on the processing bill; if market prices are lower, the processing bill goes up (an extra cost) (SPU, 2013). Composting process intends to produce a marketable product as soil amendment for those waste streams not being possible to end up at a landfill or an incinerator (SPU, 2013). Both public and private entities process 50% of recyclables plus organic waste of Seattle.

Disposal   For disposal, the City of Seattle contracts with a single provider, Waste Management Inc., for the rail haul to and disposal of nonrecyclable waste (the residual waste) at their landfill in Arlington, Oregon (SPU, 2013). To be hauled, waste is compacted into shipping containers at transfer facilities. Double stacked trains leave Seattle six times a week (SPU, 2013).

2.3.4 Copenhagen, Denmark

The Copenhagen metropolitan area has a population of around 1.7 million, including around 600,000 residents in 280,000 apartments and 22,000 dwellings at the city center (Table 2.9). R98 is the nonprofit utility company responsible for the collection of all categories of residential, commercial, and services waste in apartment blocks.

Collection In Copenhagen waste collection is different with varying sources. According to Nilsson and Christensen (2011), household waste is collected in bins, and waste collection frequency is dependent on the source: it is cleaned weekly in dwellings, and two or three times per week in multi-story buildings. In the case of bulky waste, the bulky items are normally placed on the sidewalk by dwellings. Yet the collection is made on-demand in multi-family housing complex. Garden waste as well as paper, glass, and hazardous waste is also collected separately. Glass, being sorted in terms of unbroken bottles and cullet, can be deposited in containers of 2.3 m^3 placed on the sidewalks, and parking lots, and paper is collected via 204 or 660 L bins (Nilsson and Christensen, 2011).

For packaging waste, a deposit-refund system exists for bottles containing carbonated drinks, beer and mineral water, which help return more than 95% of the bottles. Other plastic beverage containers are collected in bring banks together with glass and metal cans in multi-story buildings; but such systems are recently being replaced by a curbside collection scheme (Larsen, 2012). Bulky items are also collected in a curbside system. Copenhagen also has five recycling stations, where citizens can deliver source separated waste.

Treatment and recovery Garden waste collected is composted in centralized composting facilities using the windrow method, which can treat 60,000 tonnes of organic waste (Williams, 2005). Copenhagen municipality is the co-owner of two incineration plants generating both electricity and hot water for district heating. Only

TABLE 2.9 Waste management features in Copenhagen City, Denmark

Summary—Copenhagen	Year 2010
Population served (inhabitants)	0.6 million
Waste managed (tonnes)	820,600
Collection method	Curbside and bring schemes, selective collection of paper, cardboard, glass, gardening waste, and appliances.
Total recycled (%)	58 (includes the recovery of BMW)
Total incinerated (%)	39
Total landfilled (%)	2

Source: From Svendesen (2010) and Nilsson and Christensen (2011).
Note: Tonnes are metric tons. One percent of waste has special treatment.

a very small portion of waste is landfilled. All the rest of the waste streams are sent to WTE. There is also a CDW recycling unit.

Disposal   The city of Copenhagen disposes waste which can be neither recycled nor incinerated. Also, organic waste cannot be landfilled since 1997.

2.3.5 Singapore, Republic of Singapore

The SWM system in Singapore intends to implement the 3Rs by design (Table 2.10). Its MSW streams are classified in terms of domestic, industrial, and institutional waste, and waste management services are fully privatized to improve quality and reduce cost. Two collection methods have been adopted in Singapore (Foo, 1997; Bai and Sutanto, 2002):

- Direct collection: waste is directly collected from single households, particularly from residential and commercial areas. This method requires a considerable deal of time and is labor intensive;
- Indirect collection: includes two methods: (1) waste is stored in old high-rise apartment blocks with the use of bulk containers at the corner of the building block; (2) a centralized refuse chute (CRC) system discharges refuse directly through common hoppers located on individual floors of a building to a central waste container, which is then transferred mechanically from the CRC of each apartment block to the waste collection truck. The introduction of the CRC system to recent communities in 1989 has notably increased the waste collection efficiency and improved the control of smell and leakage during collection and transportation.

Collection   According to Foo (1997) and Bai and Sutanto (2002), waste is collected in two separate systems: a direct/individual collection, where waste from household

TABLE 2.10   **Waste management features in Singapore**

Summary—Singapore	Year 2008
Population served (inhabitants)	4.8 million
Waste managed (tonnes)	5.97 million
Collection method	Dedicated recycling stations (also called centralized recycling depositories) and curbside collections of recyclables: metals (ferrous and nonferrous) paper and cardboard, glass and plastics. Curbside collection with green bags.
Total recycled (%)	56
Total incinerated (%)	41
Total landfilled (%)	3

Source: From Foo (1997), Bai and Sutanto (2002), and Zhang et al., (2010).
Note: Tonnes are metric tons.

shophouses is collected; an indirect collection, or collective collection system, where bulk containers used in high-rise apartment blocks or CRC systems are also in place. In CRC, waste is dropped directly into hoppers located in individual flats to the central refuse container, then transported by truck to treatment facilities. In regard to recyclables, Singapore has centralized selective collection systems in public areas, and a door-to-door collection of recyclables too.

Treatment and recovery Singapore has two recycling parks: one is designed to be a dedicated park of recycling plants and the other is devoted to sort MSW to recover plastics, glass, and metals before incineration (Zhang et al., 2010). CDW is also recycled in Singapore via several existing recycling facilities where secondary aggregates are produced and used in new buildings or as materials for temporary road access at construction sites (Zhang et al., 2010). As of 2012, Singapore has four WTE units (NEA, 2013). The bottom ashes and slags are landfilled in the sanitary landfill.

Disposal Singapore has only one landfill located about 8 km south of Singapore as of 2012, which is located offshore to meet disposal needs (NEA, 2013). The sanitary landfill in Singapore only receives ashes and slags from incineration and other inorganic waste as sanitary landfill is the last means of waste management due to land scarcity and increasing waste generation (Zhang et al., 2010).

2.4 WASTE OPERATIONAL UNITS: EQUIPMENT AND FACILITIES

2.4.1 Collection and Transportation

A waste collection system delivers the inputs to the SWM system. The function of waste collection is to remove waste streams from its origin (e.g., household, commercial shop, industrial park, construction site) and ship the collected waste to the intermediate treatment unit toward the final disposal sites. However, environmental impacts associated with shipping waste streams include exhaust gas emissions, leachate from wastes, smell, noise, and traffic congestion.

Choosing the transport vehicle requires a cross-check with street conditions. The container system can be characterized by the collection method (how the waste container is emptied), the type of container (paper and plastic bags, various types and sizes od metal or plastic cans (Uriarte, 2008), the location of the container (surface/street, underground/buried or semi-underground), and container capacity. All these variables influence the type of vehicles used to collect waste.

Collection in the SWM system plays a critical role. Containers for waste collection can be emptied through simple emptying, one-way, exchange, or pneumatic transfer. Containers can be placed at the street surface, underground, or semi-underground (Figures 2.5 and 2.6). The simple emptying method is a common method, by which the waste containers are mechanically emptied into the collection vehicle and then is returned to the initial location. This method is commonly applied to commingled MSW, residual MSW, biodegradable waste, and small-scale commercial waste. One-way collection method refers to the collection of waste bags, where the bag is carried

FIGURE 2.5   Underground waste containers. *Courtesy*: Valorlis, S.A.

to the vehicle by personnel and removed from the collection site; in this method, the waste container (the bag) is not returned. Although the one-way collection method is easier and faster, it demands more personnel to pick up the waste bags (Bilitewski et al., 1994). Containers' exchange during operation is viewed as a possible alternative to remove the full container and replace it with an empty one. Such an exchange method is applied in cases of hazardous waste including waste oils, CDW, WEEE,

FIGURE 2.6   Semi-underground waste containers – Molok^{TM}. *Courtesy*: Valorlis, S.A.

and other inert waste like glass near restaurants, for example. The full container in either one-way or exchange networks is transported to a temporary storage place or a final destination for processing. Concerning pneumatic transfer, where waste is collected and transported through air, there is the need for vacuum and/or positive pressure conveying equipment (Leverenz et al., 2002).

Container systems can also be characterized by shape, material, and capacity. Small containers made of plastic or galvanized steel reach a capacity up to 100 L being more common between 50–70 L (Diaz et al., 2005). Small-medium containers have a capacity between 100 and 1,100 L, usually plastic or metal with a round or squared base, with or without wheels. These containers are emptied by turning the container into the vehicle. Larger containers such as igloo, prismatic, and Cyclea are specially designed to discharge waste through the bottom and present capacities above 1,100 L. Other larger containers have capacities between 2,000 and 12,000 L (Nilsson, 2011).

A waste collection vehicle system is composed by the truck, the container loader system, and the vehicle body where the garbage is stored, delivered, and dropped. They are further distinguished by the container loader system, which can be manual, semi-automated, or automated (Rogoff, 2014); being automated when the driver does not leave the truck to put the container in position, semi-automated when the driver needs to get out the truck to put the container in position, and manual, when the operator must roll out and return the container by using its force (without mechanical support). Vehicles can be designed with front loaders, side loaders, rear loaders, or cranes.

Front loaders are used in vehicles that can discharge the waste container by using a joystick or levers to grip the container with automated forks, lift the container over the truck, and drop the waste into the vehicle's hopper. A side loader, normally with an automatic system, is similar to a front loader, but the lift equipment is installed on the side of the truck. The advantage of side loader system is that it can be operated with only one worker (Rhyner et al., 1995), when a rear loader needs two or three workers (one driver and two workers) being considered an automated system. The disadvantage of side loaders is that they require specifically designed containers (Figure 2.7). Rear loaders can lift containers over the back of the truck (Figure 2.7).

(a)   (b)

FIGURE 2.7 Vehicles with a side loader (a) and a rear loader (b). *Courtesy*: SUMA – Serviços Urbanos e Meio Ambiente

FIGURE 2.8   Cyclea container discharge to a crane vehicle. *Courtesy*: SUMA – Serviços Urbanos e Meio Ambiente

Modern waste collection vehicles use a compaction system to reduce waste volume, increasing collection capacity.

A crane vehicle system has the opening of the body on the top of the vehicle. The crane on the vehicle lifts everything from packaging waste to multibenne containers, including specifically designed containers such as igloo, prismatic, and Cyclea that open from the bottom, which are elevated and emptied inside the vehicle (Figure 2.8).

Other special collection methods may include pneumatic and hydraulic systems, which require neither a container nor a vehicle. Pneumatic systems consist of pipelines that suck garbage bags into a specific area under the surface, followed by collection with a vehicle. Hydraulic systems are designed to deal with food waste that is ground and flushed with tap water (flushing–hydraulic method) (Bilitewski et al., 1994). These methods are normally applied in the kitchen sink where food waste is discharged together with household wastewater after grinding.

Another important component in solid waste collection systems is the transfer station, an infrastructure needed to improve economic efficiency of waste collection when the intermediate treatment facilities or final destination of waste streams in the network is far away from generation sites, and the amount of waste to be transported is immense. A transfer station can be located between the generation sites and the end points of shipping to minimize shipping costs for MSW collection. At a transfer station, the waste is transferred from collection trucks to larger transportation units that can transport waste via rail, ship, or road. The three types of transfer stations (Tchobanoglous et al., 1993) are direct load, storage load, and combined direct and discharge load. In direct load, waste is discharged directly into an open-top trailer, into compaction facilities, or onto a moving conveyor for transport to processing facilities or compaction facilities; in storage-load transfer stations, waste from a storage pit is pushed into open-top transport trailers, compaction facilities, or a moving conveyor for transport to processing facilities or compaction facilities; in

FIGURE 2.9   Hooklift hoist collection system. *Courtesy*: Valorlis, S.A.

combined transfer station (direct and discharge) is applied when other purposes exist for waste processing, such as for sorting recyclables (Tchobanoglous et al., 1993).

Another possible transport system to take waste from transfer stations to the final destination is the hydraulic hooklift hoist (or only hooklift), where hydraulic arms are used to hook, lift, and hoist the container into the vehicle. The sequence of the lifting process including the container is presented in Figure 2.9. Several suppliers manufacture hooklifts using different brands: Ampliroll is applied by Bennes Marrel company, Multilift Hooklif is used by Cargotec Finland Oy company, and Zetterbergs Industri AB company applies the name LIVAB Load Exchanger.

To clearly characterize the collection schemes, waste collection practices can be divided into two categories: those that collect and ship commingled waste, and those that perform source separation to recycling partial waste fractions. In both cases, waste containers can be located near houses such as curbside collection, or they can be deployed to serve a particular community or neighborhood.

The managerial patterns of MSW collection vary throughout the world, from no obvious managerial control (Mbande, 2003) to full managerial control for the collection of 10 different recyclables at the doorstep using multi-compartment vehicles (Dahlén et al., 2007). Waste materials sorted at the source can be collected separately or commingled; commingled collection can be designed either for manual or mechanical sorting at the MRF (Dahlén and Lagerkvist, 2010). The need to implement source separation at homes is linked to subsequent waste treatment methods. Evolving from a commingled to a source-separation practice for various waste fractions depends on the recycling behavior of residents, but establishing a regular practice that requires extra work is difficult when no policy-driven incentives are offered. The main factors affecting the participation are demographic, such as age, education, income, and household size (Sidique et al., 2010). Other factors highly relevant to the public's involvement and, consequently, the participation rate, are related to opportunities, facilities, knowledge on waste separation at source, as well as strong values and situational factors such as storage convenience and collection times (Ghani et al., 2013).

Several collection schemes are characterized by the location of containers. Containers can be located near households, with one container for each household, which is known as curbside or door-to-door containers. Neighborhood containers (González-Torre et al., 2003) place collection containers in a communal collection point where residents drop off their mixed/residual waste, being recyclables deposited in drop-off points or bring systems, also here considered neighborhood containers. In zone containers (González-Torre et al., 2003) all waste fractions (different recyclables and residual waste) can be deposited in separate containers. Both collection schemes (neighborhood and zone containers) are usually placed in public squares or other spacious areas, with easy access for collection trucks; however, some residents still consider them inconvenient to be used, no matter where the containers are located (González-Torre and Adenso-Díaz, 2005). Clean points or recycling centers/stations are specific sites, usually on the periphery of the city, where residents can drop off waste fractions that are not collected by any of the previously described systems (e.g., HHW, WEEE, CDW) at an affordable cost or for free. Containers of different sizes and shapes are used at drop-off points and at curbsides for building areas, being located in a common area of dwellers. Concerning dwelling curbside, it can be used a combinations of bins, racks, and bags placed either outdoors or indoors (Dahlén and Lagerkvist, 2010).

2.4.2 Mechanical Treatment

Solid waste separation or sorting by different components is an essential step for a strategic MSW management plan, which may be further justified in relation to any stage within the life cycle of waste management. The first stage of waste separation addresses source separation, which is conducted before waste collection. Then, waste components from source separation and also residual waste can be mechanically processed at sorting plants, mechanical (and biological) treatment, RDF (Box 2.3) production plants, energy recovery plants, or even at landfills.

BOX 2.3   REFUSE-DERIVED FUEL

RDF is a fuel obtained from segregated, high calorific fractions of different types of solid waste, including MSW, commercial, or industrial waste (Rotter, 2011). In contrast to commingled MSW, RDF is a uniform fuel, where particle size, calorific value, water, and ash content are taken into account during its production. A mechanical–biological or mechanical–thermal treatment is applied to remove noncombustible components and other contaminants. The RDF produced can be used in cement kilns as a substitute for fossil fuels.

To increase RDF quality (i.e., to reduce heterogeneity and contaminants), standards norms have been established by the American Society for Testing and Materials and European Committee for Standardization (CEN). RDF, when complied with the group of norms CEN/TC 343, is also known as SRF.

Mechanical processing was developed during the 1970s in several countries in Europe, Germany in particular. At that time, waste management in Germany was based on three main goals: produce compost from MSW, recover the high calorific waste fraction to produce RDF, and recycle the waste fraction with economic value (Pretz, 2000). Major mechanical equipment, including sieves, air classifiers, trommels, magnetic separators, eddy current separators, and other tools, allowed the preparation and/or recovery of intended fractions in the fully or partially commingled waste streams. Quality assurance and quality control remain challenging issues, however.

Mechanical sorting or processing of waste has been popularized worldwide, especially in Western Europe and some Asian countries, and is deemed a requisite of an integrated SWM system. The intention is to sort and prepare wastes (separate, comminute, and densify waste) to be used in posterior processing, such as recycling, biological treatment, energy recovery, and even landfilling. The unit operation processes and equipment addressed in this chapter are:

- comminution,
- classification/separation/segregation,
- compaction/densification, and
- internal transportation.

Comminution   Comminution is the process that reduces the size of solid materials by crushing, grinding, and other methods to provide homogeneous waste streams with higher density. In fact, comminution can promote a higher ratio of surface to volume, improving the efficiency of both biological treatment and incineration (Rhyner et al., 1995). There are three primary purposes for size reduction (Kang and Schoenung, 2005):

- production of smaller particles that can be more easily manipulated than bulky parts;
- production of regular sized and well-shaped particles which can be sorted effectively in downstream processes; and
- release of divergent materials from one another.

"Comminution processes are typically classified by methods as follows: compression (jaw crushers, roll mills, gyratory crushers), impact (hammer mills, impactors, pin mills, turbo mills), attrition (ball mills), or cutting (shredders, knife mills, guillotine mills)" (Table 2.11) (Turner et al., 2011). The suitability of a particular process is related to the nature of the material: soft materials like plastic, paper, and rubber can be reduced in size by cutting; hard materials like glass and inert can be comminuted by compression and impaction; and fine waste fractions and other specific waste fractions or components (like printed board circuits) are being treated by attrition.

The most used machines to conduct coarse comminution are mills and shredders. Mills can be hammer mills, cryogenic mills, and ball mills, whereas shredders are mostly shear shredders. Other less common machines include crushers, which are usually applied for CDW.

Hammer mills consist of a device with a rotor (horizontal or vertical) and radial hammers (which can be fixed or swinging) inside a cylindrical housing (Tchobanoglous et al., 1993; Rhyner et al., 2005; Diaz et al., 2005). In horizontal hammer mills, the waste falls by gravity into the chamber where the hammers are installed; the broken waste is discharged through the grate at the bottom of the device; the grid-size determines the size fraction of discharged waste (Tchobanoglous et al., 1993; Rhyner et al., 2005; Diaz et al., 2005). In the vertical hammer device, the material moves by gravity down the sides of the housing (Tchobanoglous et al., 1993; Rhyner et al., 2005; Diaz et al., 2005). According to Worrell and Vesilind (2012), these mills usually have a larger clearance between the housing at the top of the mill and progressively smaller clearances toward the bottom, thus reducing the size of the material in several steps as it moves through the machine. In such a device, there is no discharge grate, and the particle size is mainly controlled by the clearance between the lower hammers and the housing.

According to Anastassakis (2007), flail mills are devices similar to hammer mills in appearance but with differences in construction and operation, because they have no grate at the bottom, making them operate as single-pass machines. Such property makes flail mills application specifically for tearing bags of refuse and breaking up bundles of material in addition to providing some mixing of the waste materials (Anastassakis, 2007). To continue the waste treatment another comminution device such as a shredder is needed, because the product obtained from the flail mill is coarse. Flail mill features the possibility as a front end tool for opening garbage bags.

Shear shredders have been increasingly used to process commingled solid waste because of their lower energy requirements, lower rates of wear, and most important, reduced chance of explosions (Worrell and Vesilind, 2012). Shear shredders are composed of one fixed shaft (Figure 2.10) or two opposite-moving shafts with cutters to shred the material. In this equipment, the cut and deformation are the first mechanisms that promote size reduction (Tchobanoglous et al., 1993; Rhyner et al., 2005; Diaz et al., 2005). The waste material is transported by a conveyor or delivered by a claw and fed into the top of the shredder, where the cutters rotate at low velocity for cutting waste; then, the commingled waste passes through the spaces between the cutters; the distance between the blades determines the particle size (Tchobanoglous et al., 1993; Rhyner et al., 2005; Diaz et al., 2005). Shredders commonly function poorly when fed long, malleable, fibrous materials, which tend to enfold around the cutter shafts (Diaz et al., 2005). Shredders can also be used as bag breakers. The cutters heads are in the shafts that rotate in the opposite direction to the flow of waste, which can be loaded from the top, laterally, or on a conveyor, ripping the plastic bags and releasing the objects (Figure 2.11).

TABLE 2.11 Size reduction equipment

Equipment	Rotation speed	Power	Throughput	Material processed	Output size
Hammer mill	700–3,000 rpm	500–700 kW	20–30 t·h^{-1}	Versatile, clay to leather or steel, can process very-low-density materials.	Pulverized
Shear shredder or rotary shear	60–190 rpm	100–800 kW	–	Tires, refuse bags, bulky waste.	20–250 mm
Flail mill	–	–	–	Card and paper, refuse.	203–305 mm (coarse)
Cascade/ball mill	10 rpm	–	–	Mixed and residual MSW.	35–80 mm[a] (coarse)
Bag opener[b]	13 rpm	17 kW	5–60 t·h^{-1}	Mixed and residual MSW, packaging waste.	Not applicable
Impact crusher	–	250–1,200 kW	130–1,780 t·h^{-1}	CDW.	20–200
Jaw crusher[c]	–	110	205 t·h^{-1}	CDW.	0–40 mm

Source: From Velis et al. (2010) and Metso (2011).
[a]Ball mill coupled with trommel.
[b]From: Matthiessen Lagertechnik GmbH (2013).
[c]Specific case from Coelho and de Brito (2013).

FIGURE 2.10   Shear shredder with one shaft. *Courtesy*: Valorlis, S.A.

A cascade mill (or ball mill) consists of a rotary drum with heavy grinding balls to break up or pulverize the waste (Enviros Consulting Limited, 2007). Ball mills are a type of tumbling pulverizer widely used for mechanochemical processing operations, like minerals processing (Mccormick and Froes, 1998). In ball mills, centrifugal forces lift the metallic balls, in contact with the shell walls and each other, until they lose contact within the shell and drop (Velis et al., 2010). Velis et al. (2010) described the device as "falling balls and other hard objects in the input waste impact the waste feedstock, mainly by striking the bottom of the milling chamber. Pressure and

FIGURE 2.11   Bag opener without and with waste. *Courtesy*: Valorlis, S.A.

shear stresses imposed on the waste constituents result in differentiated comminution, according to their physical–mechanical properties."

Impact crushers are constructed of sectional sheet metal or profiled concrete-welded steel housing lined with breaker parts (Bilitewski et al., 1994). According to Bilitewski et al. (1994), the rapidly rotating roller is equipped with exchangeable hammers constructed of wear-resistant steel in which the roller is carried by roller bearings mounted onto the housing wall and the adjustable breaker plates hung above spindles. When crush-resistant material enters the crusher, the breaker plates withdraw upward and the materials are ejected downward (Bilitewski et al., 1994). Jaw crushers, a variation of an impact crusher, are composed by a fixed series of swinging jaws. Waste to be crushed is fed from above between the two jars, and the swinging jaw crushes waste. The size of the crushed material is defined by the jaw opening.

Classification/separation/segregation   Classification has led to the segregation of several waste fractions based on size, magnetism, density, electric conductivity, shape, color, high-calorific fraction, organic waste, and removal of undesirable particles (Velis et al., 2010). The several types of classification are sieving/screening, densimetric separation, inertial separation, magnetic separation, detection, and removal/automated separation.

Sieving consists on separating waste based on its size, known as size classification, into two or more fractions, using one or more sieving surfaces or screens. Sieving can be dry or wet, and waste fractions resulting from sieving are either undersized or oversized. Possible devices are vibrating screens, trommel screens, and disc screens, among others (Table 2.12).

Vibrating screens consist of a grate or perforated plate capable of separating two different size fractions, however, more than one type of grate can be used to collect fragmented waste with several size fractions (Tchobanoglous et al., 1993). According to Tchobanoglous et al. (1993), an oscillating movement is induced by a motor, which induces particle movement to improve screening, wastes are fed into the upper side onto the angled plates; the vibrating movement oscillates the materials to keep them moving through the screen to retain oversized materials and sieve undersized materials.

Drum sieve, rotary screen, or trommel, one of the most popular devices used in SWM (Figure 2.12), consists of a perforated, cylindrical grate rotating on a horizontal axis inclined. These devices can separate waste materials into several size fractions by employing various mesh-size screening surfaces in different sections of the cylinder (Anastassakis, 2007). They can present ripper devices which can be used as bag openers.

According to Tchobanoglous et al. (1993), Bilitewski et al., 1994, and Diaz et al. (2005), as the rotary screen is fed from the upper extremity, the incline and rotating movement moves waste through the drum, hitting the walls, and smaller-sized materials pass through the grate; the oversized particles go through the drum, exiting the downward side. The efficiency of a rotary screen is dependent not only on geometrical factors (i.e., length, diameter, or shape; and rather than of cylindrical,

TABLE 2.12 Screening equipment

Equipment	Power	Throughput	Material processed	Output
Vibrating screeners	5.5 kW	Up to 120 $m^3 \cdot h^{-1}$	Dry materials like glass or metals, wood chips from compost to be reused.	Fines and coarse fraction; 0–300 mm
Trommel or drum screens	4 kW; 7.5–18.5 kW	0.41–1 $t \cdot m^{-2}$	The equipment is used for RDF production, being placed before comminution step to remove bigger materials. This type of device can process commingled MSW before fragmentation, as well as processed waste. Compost, biomass, waste wood, waste shredder, soil, ravel, sand.	Fines and coarse fraction; 0–10/10–40 mm
Disc (or roller) screens	3–5.5 kW; 4–22 kW; 12–40 kW	Up to 40 $t \cdot h^{-1}$	CDW, commercial waste, organic waste (disc screen). Compost biomass, bark, wood chips, waste wood, shredder, waste shredder (star disc screen).	Fines and coarse fraction; 0–120 mm; 0–50 mm; 0–150 mm; 0–10 mm; 0–20 mm; 0–15 mm. From 0–30 until 0–350 mm

Source: From Komptech (2012) and Hammel (2013).

FIGURE 2.12   Drum sieve (or rotary drum or drum screen) with start ripper devices. *Courtesy*: Valorlis, S.A.

the trommel can by polygonal), but also operational factors such as (Bilitewski et al., 1994; Anastassakis, 2007):

- angle of scope: around 5° inclination is recommended. A higher slope angle would decrease efficiency, but the use of deflectors in the walls of the drum to transport the material through the trommel regardless of the degree of inclination;
- rotational speed: must be maintained at an optimal level to promote a cataract effect; and
- feed rate: should not exceed $1\ t\cdot m^{-2}$ (see Table 2.12).

According to Diaz et al. (2005), disc screens consist of a number of evenly spaced shafts in a horizontal plane fitted with discs that create interference patterns to form openings through which the undersized material can flow. All of the shafts rotate in the same direction, thus carrying feed material from one end of the screen to the other (Diaz et al., 2005). They can have different geometrics (e.g., circular, oblong) to promote tumbling of the particles.

Densimetric separation is based on different weights/volume ratios of waste fraction, which allows them to be separated in two fractions: light and heavy. As in sieving, densimetric separation can be wet or dry, both of which are widely applied in SWM systems. Densimetric separation devices are innumerable. The most popular are dry separators, such as air classifiers, densimetric tables (or stoners), and aqueous separators, such as flotation with air bubbles and sink/float separation.

Air classifiers separate waste using air as separation support, resulting in light and heavy fractions. Light materials are caught in an upward air flow where they are then captured in a cyclone, while the heavy fraction drops (McDougall et al.,

2001). An air knife is similar to an air classifier, except the air is blown horizontally instead of vertically, where the light fraction is drawn through the air stream, and the heavy fraction drops directly to the bottom (McDougall et al., 2001). An intermediate fraction (medium fraction) can also be obtained.

A densimetric table consists of a perforated vibrating plate at a 4° incline, where fluidized air passes through from the bottom (Tchobanoglous et al., 1993). Waste is introduced from the top, closer to one of the sides of the table passing through the fluidized air, where heavy particles that cannot be blown up are left to the higher side of the plate, and light particles leave the table through the lower side (Tchobanoglous et al., 1993).

Flotation is a process that results in selected fine-size particles floating to the surface of the slurry by attaching to bubbles; a common application is the removal of glass from ceramics and other waste contents (Worrell and Vesilind, 2012). The hydrophobic part of the waste to be separated exits, and air bubbles (which are injected into the flotation tank) incorporate into the material, floating it to the top; the remaining waste, which is hydrophilic, will saturate and deposit at the bottom of the tank (Bilitewski et al., 1994; Michaeli e Bittner, 1996). Sink/float separation is based on the liquid (or fluid) density to separate light and heavy fractions. Because the liquid must have an intermediate density to separating both fractions, other substances are added to the water to change its density, like calcium nitrate, calcium chloride, sodium chloride, potassium carbonate, just to name a few used in plastic separation (Fisher, 2003).

Inertial separators create projectile movement to separate fractions with different trajectory paths based on waste density. A typical example of equipment based on inertia is a ballistic separator (Figure 2.13), which is composed of a range of inclined perforated plates (with 15–20% open area), each with individual movement, and the plates are moved by a motor to create projectile movement of the waste material (Bilitewski et al., 1994). According to Bilitewsi et al. (1994), waste is fed into the

FIGURE 2.13   Ballistic separator. *Courtesy*: Valorlis, S.A.

downside of perforated plates and transported in an ascending direction; the heavy and roller materials are transported to the downside due to their shape and gravity, while the light and flat materials (e.g., paper, flattened packages, and plastic bags) are projected in an ascending direction due to the rotating movement of the plates. Fines fraction, which corresponds to the undersized fraction of perforated plates, is also obtained. The equipment is ideal for mixed MSW or light fraction waste from source-segregated collection.

Magnetic separation is based on the magnetic forces of attraction or repulsion, which can be used to separate materials with conductive properties (Figure 2.14). Magnets used for this separation can be permanent or electromagnetic. Magnetic separation devices that use a magnetic belt, most commonly used for SWM separation, can have different configurations. In this separator, the magnet is located between pulleys suspended over a continuously traveling belt (Rao, 2006). The magnet can also be located inside the pulley, being called the separator as magnetic head pulley (Leverenz et al., 2002). In addition to the belt, a magnetic drum can be also applied, but the result is less consistent.

Electric conductivity (nonmagnetic) separation is based on the application of an electrical field, which separates conductive from nonconductive materials. This equipment can separate nonferrous metals from waste and automobile shredder residue by passing a magnetic current through the feed stream and using repulsive forces interacting between the magnetic field and the eddy currents in the metals (Rao, 2006).

FIGURE 2.14 Suspended belt magnetic separator. *Courtesy*: Valorlis, S.A.

TABLE 2.13   Sensors used to detect waste material properties

Sensors	Description
Optical	Used to detect color on glass, plastic, and metal.
Image recognition	Visualize the object and the microprocessor compares the object with the database.
X-ray fluorescence	Scans material surface to detect the presence of chlorine atoms.
X-ray transmission	Chlorine atoms are detected by sending X-ray through the waste material.
Infrared	Allows to distinguish transparent, translucent, and opaque.
Near-infrared	Measures plastics absorbance in near-infrared, distinguishing them.
Electrostatic	Electric conductivity is used to separate nonconducting waste materials, such as plastic and paper.
Eddy currents	Electric conductivity materials are detectable in nonferrous metals like aluminum.

Source: From Leverenz et al. (2002).

Detection and routing systems, or automated sorting, is the key technology to demonstrate how the latest sensing and sensor networks come to play to sort waste. This operation depends on an array of sensors (Table 2.13), such as optical sensors (Figure 2.15) acting upon an individual object. After the material is detected, it is separated by air classifier processes downstream.

Compaction/densification   Waste compaction has been indispensable to promote waste transport optimization, storage optimization (temporary storage and landfilling), resulting in a density increase and volume reduction. Also, compacting

FIGURE 2.15   Optical separator. *Courtesy*: Valorlis, S.A.

has been useful to increase energy density of materials with the purpose of energy recovery (Bilitewski et al., 1994).

According to Bilitewski et al. (1994), the common compaction methods are solid resistance and extrusion molding: the first intends to use chamber walls as solid resistance, although the second method uses the mold to bring resistance. Solid resistance devices are bale presses and compactors, when extrusion molding devices can be presses/pelletizers.

As specified by Leverenz et al. (2002), all bales are composed by a feeding area, hydraulic arm or mechanic (it can be more than one), compression chamber and discharge area; bales can be vertical or horizontal (depending on hydraulic arm position). Waste is fed through the feeding area, then compressed in the compression chamber and, after the needed time to compression have occurred, the bale is tied with steel cables or plastic rips to keep the bale shape (Leverenz et al., 2002). Bale presses are used to bale recyclables, like plastic and paper and cardboard, before being sent for recyclers (see Figure 2.16). The same equipment in a smaller dimension is used to press ferrous and nonferrous metals, resulting in a densified biscuit (Leverenz et al., 2002).

In compactors case, they are similar to bale press with a few differences such as the compactors are composed by the feeding area using mechanical arm with or without a compaction chamber (Bilitweski et al., 1994). The discharger chamber is usually a container, where the shape will be induced. In this device, after waste being introduced through the feeding area, the waste is compressed in the chamber with the aid of the mechanic arm; then, the compacted waste is pushed to the container as this process is repeated until the container reaches the maximum capacity (Bilitewski et al., 1994). When there is no compaction chamber, the container will retain such

FIGURE 2.16   Bale press for plastic materials. *Courtesy*: Valorlis, S.A.

function at the discharge area (Bilitewski et al., 1994). Compactors are used to densify waste to be landfilled (Bilitweski et al., 1994).

Pelletizers are mostly used for the production of RDF or granulated compost. The device comprises two rolls inside of a cylinder perforated all over, when having a blade at the outside of the cylinder (Glorius et al., 2005). When waste is fed by the chute, at the top, the material goes inside the cylinder; the rolls start moving and pressing the material against the cylinder wall; due to the high pressure promoted by the rolls the temperature starts to rise, making the material get melted, just enough to make it pass the holes at the cylinder, reaching the outside; the blade finishes the pellet, cutting it in the intended size (Glorius et al., 2005).

Internal transportation Waste processing is conducted in a series of sequential operations, making it necessary to develop devices that can transport waste between operational units. For that reason, transportation is the unit operation which bonds all mechanical operations already mentioned in this sub-section. The main devices used to promote transportation are belt conveyors. Belt conveyors can be characterized by the position, either horizontal or inclined; or by the style, either drag (Figure 2.17) or auger; or by the drive system which can be friction, chain, or vibratory motion (Leverenz et al., 2002). In addition to belt conveyors, transportation can also be made by pneumatic systems by air pressure or by vacuum.

2.4.3 Biological Treatment

Biological treatment methods are used to treat biodegradable waste fractions, which can be fermentable, green waste, and waste paper. Biological treatments can be divided into two main processes, aerobic and anaerobic, with two popular practical

FIGURE 2.17 An inclined drag belt conveyor. *Courtesy*: Valorlis, S.A.

implementations including composting and anaerobic digestion, respectively. Either can be used independently or both can be used collectively to achieve an integrated goal. The objectives of conducting biological treatment are:

- waste prevention: through the implementation of backyard composting or home composting;
- compost production: in decentralized and centralized composting system and in combined anaerobic digestion-composting units;
- waste treatment before landfilling; through stabilization and volume reduction occurring in mechanical–biological treatment units;
- waste pretreatment: to produce a solid fuel, specifically SRF or RDF in mechanical–biological treatment units with such purposes; and
- fuel production: treat waste to produce biogas, also a fuel, in an anaerobic digestion process.

Biological treatment uses naturally occurring microorganisms to decompose the biodegradable components of waste (McDougall et al., 2001). The most common biological treatment methods are composting via aerobic processes and anaerobic digestion (Figure 2.18). In the case of aerobic processes, oxygen is needed to degrade organic matter, as opposed to anaerobic digestion, where microorganisms can degrade waste without oxygen. In real world applications, composting process can be developed in open-air or enclosed processes. Open-air systems can be categorized as windrow, static pile, and bin composting. Composting converts organic residues of plant and animal origin into compost through a largely microbiological process based on the activities of several bacteria, actinomycetes, and fungi, being released carbon dioxide, water and heat, during the process (Bharadwaj, 1995; Abbasi and Ramasamy, 1999). The main product, the compost, is rich in humus and plant nutrients (Abbasi and Ramasamy, 1999). In the case of anaerobic digestion, however, a post-composting is often needed to turn the digestate into compost with organic matter content (Figure 2.19).

In windrow composting, the organic waste is laid out in parallel rows, 2–3 m high and 3–4 m wide across the base (Gajalakshmi and Abbasi, 2008). Windrows acquire a trapezoidal shape, with angles of repose depending on the nature of materials being composted (Gajalakshmi and Abbasi, 2008). This process involves turning the material to ensure its aeration (i.e., supply oxygen to the process). In a static pile, the supply of air is provided by forced aeration. In the bin composting system, organic waste is fed into a composting box, where the revolving is manual. In all cases, water has to be added to ensure humidity conditions to allow microorganisms to grow and degrade organic matter.

Most composting processes are enclosed processes (also called in-vessel processes) classified as horizontal and batch flow. The advantage of in-vessel processes, as compared to open air, is the control of conditions to ensure temperature and moisture, and to control odor emissions, making it more adequate for large-scale composting units.

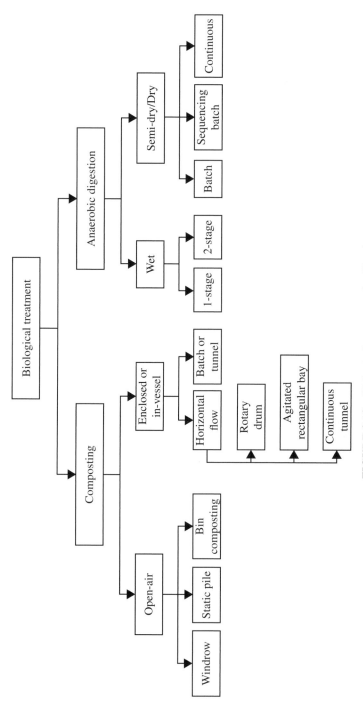

FIGURE 2.18 Biological treatment classification

FIGURE 2.19 Compost produced from digestate of mixed MSW anaerobic digestion. *Courtesy*: Valorlis, S.A.

Horizontal flow composting can be achieved in a rotary drum or an agitated rectangular bay. Rotary drums present a large-diameter (3 m or larger), slowly rotating drum, that are fed at one end with untreated biodegradable waste and water (Gajalakshmi and Abbasi, 2008), resulting compost at the other end. While the waste mixture is tumbled in its passage down the length of the drum, the material is gradually broken down to enhance the mixing with oxygen and water (Gajalakshmi and Abbasi, 2008). Agitated rectangular bays consist of rectangular-shaped, open-topped, agitation bays that operate on a continuous or intermittent feed basis (Bio-Wise, 2001). Material enters the front of the bay on a conveyor, where a spiked rotating drum (the mixer) mounted on wheels at the top of the bay walls moves along the length of the bay; this spiked drum mixes the material and gradually moves it along the length of the bay (Bio-Wise, 2001).

Batch flow systems are contained systems in which rectangular tunnel vessels (reactors) with perforated floors are fed with organic waste. The material does not move, so forced air is used to provide oxygen for the aerobic process.

In anaerobic digestion, also known as biomethanation, the organic waste is converted into energy (methane) and digestate by an association of microbial action in the absence of air, through the processes hydrolysis, acidogenesis, acetogenesis, and methanogenesis. According to Krishania et al. (2013), hydrolysis is an extra-cellular process in which hydrolytic and acidogenic bacteria excrete enzymes to catalyze hydrolysis of complex organic materials into smaller units, resulting hydrolyzed substrates; acidogenesis occurs when such substrates are consumed by acidogenic bacteria, resulting in short chain volatile acids, ketones, alcohols, hydrogen, and carbon dioxide; in acetogenesis, specific bacteria will produce the precursor of methane: acetic acid. This substance, together with carbon dioxide and hydrogen will be used by methanogens to produce methane. The whole process is carried out with the help of microorganisms whose growth depends on various parameters such as pH,

FIGURE 2.20 Single-stage continuous anaerobic digesters (at the left is the digester, at the right is the suspension buffer tank). *Courtesy*: Valorlis, S.A.

temperature, C/N ratio, organic loading rate, reactor designing, inoculums, and hydraulic retention time (Yadav et al., 2004).

The separation of the acidogenic from methanogenic substances results in a two-stage system, in opposition to the single-stage system, where all anaerobic digestion occurs (Figure 2.20). The retention time needed to anaerobic digestion to occur can range from a few days to several weeks depending on the chemical characteristics of the organic material, amount of preprocessing occurred previously, and design of anaerobic digestion system (single-stage, two-stage, multi-stage, wet or dry, temperature, and pH control) (McDougall et al., 2001). In regard to the solid contents of the material digested, anaerobic digestion can be characterized as dry (total solids concentration of more than 20), wet (between 6 and 10% total solids), and semi-dry system (total solids content between 10 and 20%) (Krishania et al., 2013).

Anaerobic digestion reactors can have a batch or continuous processing. In batch processing, organic matter is fed all at once into the reactor and retained for a specific time period, until the process reaches the end of degradation. Although the biogas production rate is not steady, at the end of the process all biogas can be extracted and the effluent discharged. In a continuous system, the digester is fed continuously and biogas production is more consistent. Concerning emissions, both processes release pollutants. Anaerobic digestion will release less air emissions, since the principal gaseous emission (methane) is a desired product (EIPPCB, 2006a), being emissions related to the conversion of biogas into electric energy. In aerobic treatment, specifically plants treating MSW like aerobic MBT, the main pollutants are ammonia, bioaerosols, odors, methane and volatile organic compounds (EIPPCB, 2006a). In such treatment, acid scrubber and a biofilter are used to treat gaseous emissions.

2.4.4 Thermal Treatment

Thermal treatment has been considered an effective way to treat solid waste since the Industrial Revolution era in the United Kingdom. Its significant benefits include (Brunner, 1994; Chimenos et al., 1999; Karak et al., 2012):

- reduction of MSW volume and mass to a fraction of its original size (reduced 85–90% by volume and about 70% by mass);
- possible energy recovery;
- immediate waste reduction not dependent on long biological breakdown reaction times;
- construction of incineration facilities closer to MSW sources or collection points, reducing transportation costs;
- offset of operational costs by energy sales using heat recovery technology; and
- control of air discharges can be controlled to meet environmentally legislated limit values.

All these benefits are achievable, even in cases where moisture content is high, heat content of waste is low, and auxiliary fuel (such as fossil fuel) is required to maintain the combustion temperature. Incineration is applied as a treatment most often for (EIPPCB, 2006a):

- MSW (residual waste—not pretreated),
- pretreated MSW (e.g., selected fraction or RDF),
- nonhazardous industrial waste and packaging,
- hazardous waste,
- sewage sludge, and
- clinical waste.

The most common thermal treatment technology for SWM is incineration with or without energy recovery, pyrolysis, and gasification. Incineration is a full oxidation combustion process, pyrolysis consists of a thermal degradation of organic material in the absence of oxygen, and gasification only occurs at partial oxidation (Table 2.14).

Incineration   Today's incineration facilities are designed for more efficient combustion processes. When designing and operating an incinerator, the limiting aspects to be considered are temperature, turbulence of the mixture being combusted, and residence time at the incineration temperature, which are referred to as the 3Ts of combustion criteria (i.e., time, temperature, and turbulence) (Senkan, 2006).

The spectrum of incineration technologies includes grate incinerators, fluidized beds, and rotary kilns. Grate incinerators, common for treating MSW, are distinguished by rocking, reciprocating, travelling, roller, or cooled incineration grates (EIPPCB, 2006a) that function to transport materials to be incinerated through the bottom of the furnace, allowing waste agitation and mixing with air to enhance combustion. To improve mixing of waste and air, a primary air blower is blown through the grate in an ascending direction, being added extra air to complete combustion

TABLE 2.14 Main features of chemical processes for thermal treatment of solid waste

Features	Combustion or incineration	Gasification	Pyrolysis
Aim of the process	To maximize waste conversion to high-temperature flue gases, mainly CO_2 and H_2O	To maximize waste conversion to high heating value fuel gases, mainly CO, H_2, and CH_4	To maximize thermal decomposition of solid waste to gases and condensed phases
Operating conditions			
Reaction environment	Oxidizing (oxidant amount larger than that required by stoichiometric combustion)	Reducing (oxidant amount lower than that required by stoichiometric combustion)	Total absence of any oxidant
Reactant gas	Air	Air, pure oxygen, oxygen-enriched air, steam	None
Temperature	Between 800 and 1,450°C	500–900°C (in air gasification) and 1,000–1,600°C	Between 250–800°C
Pressure	Generally atmospheric (1 bar)	Generally atmospheric (1–45 bar)	Slight over-pressure (1 bar)
Process output			
Produced gases	CO_2, H_2O, O_2, N_2	CO, H_2, CO_2, H_2O, CH_4, N_2	CO, H_2, CH_4, and other hydrocarbons, H_2O, N_2
Pollutants	SO_2, NO_x, HCl, PCDD/F, particulate	H_2S, HCl, CO_2, NH_3, HCN, tar, alkali, particulate	H_2S, HCl, NH_3, HCN, tar, particulate
Ash	Bottom ash can be treated to recover ferrous (iron, steel) and nonferrous metals (such as aluminum, copper and zinc) and inert materials (to be utilized as a sustainable building material). Air Pollution Control residues (fly ash) are generally treated and disposed as industrial waste	Same as combustion process. Bottom ashes are often produced as vitreous slag that can be utilized as backfilling material for road construction	Often having substantial carbon content. Treated and disposed as industrial special waste. Coke is also produced
Liquid pollutants	–	–	Pyrolysis oil and water
Gas cleaning	Treated in air pollution control units to meet the emission limits and then sent to the stack	Syngas can be cleaned to meet the standards of chemical production processes or those of high-efficiency energy conversion devices	

Source: From Arena (2012).

CO_2, carbon dioxide; H_2O, water; CO, carbon monoxide; H_2, hydrogen; CH_4, methane; N_2, nitrogen; SO_2, sulfur dioxide; NO_x, nitrogen oxides; HCl, hydrochloric acid; PCDD/F, polychlorinated dibenzo-*p*-dioxin and polychlorinated dibenzofuran; H_2S, hydrogen sulfide; NH_3, ammonia; HCN, hydrogen cyanide.

on the top of the grate (EIPPCB, 2006a). Resulting ashes, known as bottom ashes, are then discharged. To burn out the combustion gases, a minimum gas phase combustion temperature of 850°C (1,100°C for some hazardous wastes) and a minimum residence time of the flue gases of 2 seconds above this temperature after the last incineration is required by the European legislation Directive 2000/76/EC (European Parliament and Council, 2000a; EIPPCB, 2006a, 2006b) and USEPA's Good Combustion Practice in the United States (Kilgroe et al., 1992).

Fluidized bed incinerators are simple devices consisting of a vessel lined with heat-resistant material, containing inert granular particles (i.e., sand) (McDougall et al., 2001). Preheated air is introduced into the combustion chamber, a vertical cylinder, via openings in the bedplate, forming a fluidized bed with the sand (EIPPCB, 2006a). The waste, introduced via a pump, a star feeder, or a screw-tube conveyor (EIPPCB, 2006a), is burned in the immediate area of the bed at a temperature between 850 and 950°C. Excess air for normal incineration is usually limited to nearly 40% above the stoichiometric air requirements due to the close contact between combustion gases and waste being burned (McDougall et al., 2001). Yet, according to McDougall et al. (2001), fluidized beds are subject to problems caused by low ash fusion temperatures and materials with low melting points; therefore, MSW to be incinerated in these units must be pretreated, promoting the removal of fractions such as aluminum and glass. Other mechanical waste treatments, such as shredding, are often needed to ensure size specifications of waste going onto the fluidized bed. Due to the homogeneous requirement, waste fractions that are usually burned out in fluidized beds include commercial waste, pretreated construction waste, sorted and pretreated household waste, sewage sludge, and RDF (EIPPCB, 2006a). The following fluidized bed furnace technologies can be differentiated according to the gas speeds and design of the nozzle plate (EIPPCB, 2006a):

- stationary or bubbling fluidized bed under atmospheric and pressurized condition may keep the inert material mixed, but the resulting upward movement of solids is not significant;
- rotating fluidized bed is rotated in the incineration chamber, resulting in longer residence time in the incineration chamber, which have been used for mixed municipal waste for many years; and
- circulating fluidized bed has the higher gas speeds in the combustion chamber which are responsible for partial removal of the fuel and bed material with having a recirculation duct to externally feed the bed material back into the incineration chamber.

Pyrolysis   Pyrolysis is theoretically a zero-air, indirect-heat process, with a resulting formation of a combustible gas used as fuel and a solid coke. In a broader sense, "pyrolysis" is a generic term including a number of different technology combinations that constitute, in general, the following technological steps (EIPPCB, 2006a):

- smoldering process: production of gas from volatile waste particles at temperatures between 400 and 600°C;

- pyrolysis: thermal decomposition of the organic molecules between 500 and 800°C resulting in a gas and a solid fraction;
- gasification: conversion of the carbon share remaining in the pyrolysis coke at 800–1,000°C using a gasification substance (e.g., air or steam) in a process gas (CO, H_2);
- incineration: depending on the technology integration, the gas and pyrolysis coke are combusted in an incineration chamber.

Gasification Gasification, or "indirect combustion" in particular, is the conversion of solid waste to fuel or the synthesis of gas in the presence of low oxygen concentration. Gasification has been defined and described by Arena (2012): "basically, part of the fuel is combusted to provide the heat needed to gasify the rest (auto- thermal gasification), as in the case of air gasification, or heat energy is provided by an external supply (allo-thermal gasification), as in the case of plasma torch utilization." The special features of the gasification process are (EIPPCB, 2006a):

- smaller gas volume compared to the flue gas volume in incineration (by up a factor of 10 by using pure oxygen);
- prevalent production of CO rather than CO_2;
- high operating pressures (in some processes);
- cumulating solid residues as slag (in high-temperature slagging gasifiers);
- small and compact aggregates (especially in pressure gasification);
- synthesis gas utilization energetically; and
- smaller waste water flows from synthesis gas cleaning.

Energy recovery—Energy from waste or waste-to-energy The heat from flue gases can be recovered for heating water or steam generation. Flue gases must be treated and cooled before release, and the resulting heated water can be used for district heating services, mostly in northern countries, and for some industrial applications. Steam generated can be used to produce electricity.

According to McDougall et al. (2001), heat recovery can be achieved by two means: water wall combustion chambers and waste-heat boilers. The walls of the combustion chamber, known as water walls, are made of water-filled heat exchange pipes, usually with a protective coating of some type. Water walls are widely used to cool the combustion gases through heat-exchange bundles located at boiler passes (EIPPCB, 2006a, 2006b). The first pass is usually through an empty chamber because the hot gases are too corrosive and the particulate matter is too sticky for the heat exchange tubes to be effective (EIPPCB, 2006a, 2006b). Water circulating through the boiler tubes is turned into steam, which can be heated further using a superheater to increase its temperature and pressure to increase the efficiency of electricity generation (EIPPCB, 2006a, 2006b). The thermal efficiency of modern boilers is around 80% if steam is to be used directly in heating; however, if the steam is to be used to produce electricity, the overall energy recovery efficiency is around 20% (RCEP, 1993; EIPPCB, 2006a) (Table 2.15).

TABLE 2.15   Energy recovery combination efficiencies

Plant type	Reported potential thermal efficiency % ((heat + electricity)/ energy output from the boiler)
Electricity generation only	17–30
Combined heat and power (CHP) plants	70–85
Heating stations with sales of steam and/or hot water	80–90
Steam sales to large chemical plants	90–100
CHP and heating plants with flue gas condensation	85–95
CHP and heating plants with condensation and heat pumps	90–100

Source: From EIPPCB (2006a), RVF (2002), and Sims (2010).

If wet feedstock is used (e.g., high fractions of food and garden waste) much of the gross calorific content of the waste would be consumed to evaporate the moisture (McDougall et al., 2001). The latent heat contained in the waste is lost and cannot be recovered. Existing measures to improve energy recovery efficiency are (EIPPCB, 2006a):

- waste feed pretreatment (such as homogenization, extraction/separation;)
- improvement of boiler and heat transfer using economizer and superheating;
- combustion air preheating;
- water cooled grates;
- flue gas condensation;
- heat pumps and flue gas recirculation;
- reheating of flue gases to the operation temperature of flue gas treatment devices;
- plume visibility reduction; and
- steam–water cycle improvements.

Emission control   Although thermal treatment will inevitably induce the release of pollutants, mainly atmospheric emissions, flue gas treatments existing today are capable of significantly increasing the removal efficiency of those pollutants from flue gases. An overview of several possible combinations to treat flue gases can be found in EIPPCB (2006a).

The selection of well-known flue gas treatment equipment depends on the pollutant to be removed. For particles, equipment for reducing their emissions are electrostatic precipitators, wet electrostatic precipitators, condensation electrostatic precipitators, ionization wet scrubbers, fabric filters, cyclones, and multi-cyclones (EIPPCB, 2006a). For the reduction of acid gases (e.g., HCl, HF, and SO_X emissions), flue gas cleaning processes can be dry, semi-wet, or wet. Dry processes adopt a dry sorption agent like lime or sodium bicarbonate; semi-wet uses an aqueous solution, like lime milk or a suspension (such as a slurry); wet processes treat the flue gas by spraying water, hydrogen peroxide, and/or a washing solution containing part of the reagent (e.g., sodium hydroxide solution) (EIPPCB, 2006a). For NO_x control, the primary reduction techniques include furnace control measures such as air supply, gas mixing, temperature control, flue gas recirculation, oxygen injection, staged combustion, natural gas injection (for reburning NO_x), and injection of water into the furnace/flame, when secondary techniques to reduce NO_x are selective catalytic and noncatalytic reduction processes (EIPPCB, 2006a).

2.4.5 Disposal

Landfilling is the final operation of an SWM system and can be a solution to any type of waste, including MSW, combustion ashes and slags from incineration. According to Diaz et al. (2005), sanitary landfilling, which is the controlled disposal of waste on ground, is appropriate to many countries as a wherewithal to manage the disposal of waste due to the flexibility and low technology.

Depending on the state of the waste, sanitary landfill can be a waste treatment process with the inputs wastes and water from rain to promote decomposition (McDougall et al., 2001). The process outputs are the final stabilized solid waste and the gaseous and aqueous decomposition products, which emerge as landfill gas and leachate (McDougall et al., 2001) (Figure 2.21).

According to Diaz et al. (2005), all sanitary landfill definitions focus on the landfilled waste isolation from the environment until innocuousness is reached through natural biological, chemical, and physical processes. The objectives of sanitary landfill also include solving public health problems and minimizing environmental pollution risk due to waste exposure by confining the waste at a specific place (Diaz et al., 2005).

A typical gas recovery system at a sanitary landfill produces biogas, which is collected and burned to produce electricity (Figure 2.22). Biogas, which constitutes one of the main sources of greenhouse gases, must also be collected and burned to prevent its contribution to global warming as well as for safety and environmental reasons (McDougall et al., 2001). When sanitary landfills are constructed in places like quarries, they are also regarded as land reclamation projects.

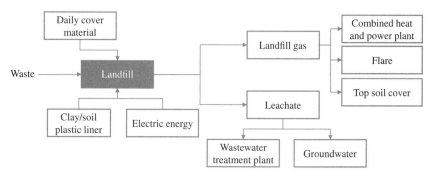

FIGURE 2.21   General structure of the landfilling technologies and boundary of the assessment. *Source*: From Manfredi and Christensen (2009)

FIGURE 2.22 Sanitary landfill with biogas collection system and soil protection system. *Courtesy*: Valorlis, S.A.

Sanitary landfills can be classified according to their size (small, medium, large), by the waste received (hazardous waste, nonhazardous waste, slag, inert waste), and by the construction design (area or trench). A sanitary landfill is composed of a soil protection system, a leachate collection system, a landfill gas collection system, and a capping system.

The soil protection system is composed of liners that protect the landfill bottom and sides. Liners can be one or several layers of materials, with soil comprising one of the layers; above the soil a flexible membrane liner is used as a protective layer (Diaz et al., 2005). For soil to be used as a liner, it must have a low permeability (preferably less than 1 E^{-6} cm $\cdot$ s^{-1}) (Diaz et al., 2005). The landfill liner system includes the leachate collection system, with the leachate collection pipe placed above a geotextile liner, protected by clay, soil or other inert material from mechanical damage. Waste is then partitioned, deposited, and compacted within inert material layers (soil, coarse composted material), added to interpolate the waste (McDougall et al., 2001). The actual working face of the landfill is kept small, and the raw waste is covered by landfill cover material at the end of daily operation to reduce the nuisance from wind-blow material, and to deter rodents, birds, and other potential pathogen-carrying vermin (McDougall et al., 2001).

Leachate production due to rainfall and water content of waste begins shortly after landfilling commences and may continue for hundreds of years. The leachate collection system is composed of a network of perforated pipes that receive leachate from waste mass by gravity or by pump (McDougall et al., 2001). After collection, leachate must be treated; existing options are evaporation (natural or forced), recirculation and recycling, discharge to an offsite wastewater treatment facility, and onsite treatment (Diaz et al., 2005).

Evaporation occurs through the use of evaporation ponds, mostly popular in hot and dry countries or regions. Recirculation of the leachate through the landfilled waste has been applied in several facilities throughout the world as a method of leachate management. Relatively high concentrations of biochemical oxygen demand (BOD), chemical oxygen demand (COD), and, in some cases, heavy metals are found in the leachate soon after the waste is placed in the landfill (O'Leary and Tchobanoglous, 2002). Under certain conditions, the potentially polluting characteristics of organic compounds in the recirculated leachate can be attenuated by the chemical and biological processes occurring in the landfill and, thus, substantial savings can be achieved in terms of the capital and operational expenses of treatment (Hekimian et al., 1976). The temporal variations in biogas and leachate quality during the landfill lifetime can be observed in the work of Kjeldsen et al. (2002).

The landfill gas collection system is a network of vertical or horizontal perforated pipes plus a gravel layer between the cover and the waste, or gravel-filled trenches. Pipes are located in wells, and areas of high permeability channel gases to collection points (Diaz et al., 2005). Pumped extraction of gas is required for efficient collection and results in fewer odor and emission problems (McDougall et al., 2001). The quality of landfill gas is not sufficient to be used directly in motors or to produce electric energy, and additional treatment to remove contaminants is usually required. The rate of gas production depends on many factors such as the weather conditions, temperature variations, and landfill management policy (McDougall et al., 2001).

When the operational period is complete, landfill cells must be capped by inert materials with vegetated cover to prevent or control infiltration of precipitation, which avoids or controls leachate production, and to regulate landfill gas production and release into the atmosphere (Simon and Müller, 2004) (Table 2.16).

Bioreactor landfill   Bioreactor landfills consist of controlled systems where moisture (often leachate recirculation) and/or air injected is used to promote conditions to make waste capable of actively degrading the promptly biodegradable organic fraction of the waste (Berge et al., 2005). Decomposition and biological stabilization of the waste in a bioreactor landfill can occur in a much shorter time frame than occurs in a traditional "dry tomb" landfill, potentially decreasing long-term environmental risks and landfill operating and post-closure costs (USEPA, 2012a, 2012b). Potential advantages of bioreactors include (USEPA, 2012a, 2012b):

- decomposition and biological stabilization in years versus decades in "dry tombs";

TABLE 2.16 Standard capping design for CDW and MSW landfills

Position	System component	CDW landfill	MSW (treated)
Top	Vegetation	Necessary	Necessary
	Restoration layer	>1 m	>1 m
	Drainage layer	$d \geq 0.3$ m; $K \geq 1 \times 10^{-3}$ m·s⁻¹	$d \geq 0.3$ m; $K \geq 1 \times 10^{-3}$ m·s⁻¹
	Protective layer	Not necessary	Necessary
	Geomembrane	Not necessary	$d \geq 2.5$ mm
	Compacted clay liner	$d \geq 0.5$ m, $K \leq 5 \times 10^{-10}$ m·s⁻¹	$d \geq 0.5$ m, $K \leq 5 \times 10^{-10}$ m·s⁻¹
	Gas venting layer	Not necessary	Necessary
Bottom	Regulating layer (foundation)	$d \geq 0.5$ m	$d \geq 0.5$ m

Source: From Simon and Müller (2004).

- lower waste toxicity and mobility due to both aerobic and anaerobic conditions;
- reduced leachate disposal costs;
- a 15–30% percent gain in landfill space due to an increase in density of waste mass;
- significantly increased landfill gas generation that, when captured, can be used for energy use onsite or sold; and
- reduced post-closure care.

A bioreactor can be aerobic, anaerobic, or hybrid. According to USEPA (2012a, 2012b), in an aerobic bioreactor landfill, leachate is removed from the bottom layer, piped to a liquid storage tank, and recirculated into the landfill in a controlled mode; air is injected into the waste mass, using vertical or horizontal wells, to promote aerobic activity and speed up waste stabilization. In an anaerobic bioreactor landfill, moisture is added to the waste mass in the form of recirculated leachate and other sources to obtain optimal moisture levels (USEPA, 2012a, 2012b). Biodegradation occurs in anaerobic conditions and produces landfill gas, being collected to minimize greenhouse gas emissions and to produce energy (USEPA, 2012a, 2012b). In a hybrid system, a sequential aerobic–anaerobic treatment is applied to rapidly degrade organics in the upper sections of the landfill and collect gas from lower sections (USEPA, 2012a, 2012b).

2.5 TECHNOLOGY MATRIX FOR MULTIPLE SOLID WASTE STREAMS

Considering the treatment technologies for handling solid waste streams introduced in previous sections, the key design question is how all waste treatment options can be interwoven smoothly to provide a total solution based on risk-informed, cost-effective, environmentally benign, and forward-looking criteria. For demonstrating various types of SWM systems, several case studies presented in this section emphasize different technology matrixes that may achieve a balanced choice over the abovementioned criteria. The basic assessment tools for technology matrix are related to unit processes, process flow diagrams, mass balance diagrams, layout, and configuration. In principle, the technical planning and unit layouts encompass three main steps (Tchobanoglous et al., 1993; Leverenz et al., 2002):

- Feasibility analysis: analyzes the possibilities for the SWM unit to be built. It should provide decision-makers with clear recommendations on technical and economic aspects of the planned unit;
- Configuration preview: studies the equipment to be selected, flow charts of treatment procedures, rates of recovered materials, the mass balance, the environmental and safety aspects, and the staff; and
- Setting end: provides the final preparation of the plans and specifications used in construction with respect to estimated costs and legal documentation.

Before starting the technical planning it is necessary to define the purpose and function of the unit. According to Leverenz et al. (2002), functions depend on the role of the unit in the SWM system, the type of materials to be processed, how the waste will be received in the shipping network, and the type of containers required for processed materials to be delivered to the end user. With system thinking, a process flow diagram consisting of the aggregation of different operational units is needed to reach the holistic processing goal. The main goals to be included in the development of a process flow diagram are to (Leverenz et al., 2002):

- identify the characteristics of the waste to be processed;
- specify current and future materials to recover; and
- identify available equipment and facilities.

For example, certain types of waste cannot be efficiently separated from commingled waste unless bulky waste is first removed or shredded. Furthermore, bags must be opened to expose waste for separation and recycling; therefore, a bag opening unit must be installed at the beginning of the processing line. One of the most critical elements during capacity design and equipment selection is the mass balance that determines the amount of material to be recovered, treated, and disposed sequentially, given the feed rates to each operational unit and the whole process; therefore, if the design engineers fully understand the characteristics of the waste streams, the functionality of each unit operation and the project goals, proper selection of various unit operations and their associated equipment may reach balanced conditions. The phases involved in the mass balance preparation given the feed rates are (Leverenz et al., 2002):

- Phase 1 defines the system boundary. This border can be set around the entire unit or unit operation individually;
- Phase 2 identifies residues that enter and leave the system boundary and the amount of material within the system;

- Phase 3 promotes the application of material balance obtained from the process that will occur within the boundaries of the system. The mass balance can be defined according to:

$$\text{Material accumulation} = \text{Input material} - \text{Output material}$$
$$+ \text{Production of material within the system}$$

If there is material accumulation in the processing unit, then the mass balance is:

$$\text{Material accumulation} = \text{Material input} - \text{Output material}$$

In a single unit operation where the material is not accumulated, the above equation can be simplified to:

$$\text{Input material} = \text{Output material}$$

Thus, the mass balance of a unit operation is essentially a quantification process to ensure that the materials are all controlled within the unit.

- Phase 4 determines the capacity (loading rate) of unit operations and processing steps of the unit through the mass balance data. Generally, wastes that enter these units are expressed in tonnes per day. Thus, the transport unit operations or separation must be specified in tonnes per hour. Therefore, the capacity of tonnes per day must be converted to tonnes per hour, based on the actual hours of operation per day. The load capacity is given by:

$$\text{Capacity} = \frac{\left(\dfrac{\text{tonne}}{\text{day}}\right)}{\left(\dfrac{\text{hour}}{\text{day}}\right)}$$

The layout and configuration of waste treatment units depends on the type of waste and amounts to be processed. The factors to be considered in the configuration and layout include (Leverenz et al., 2002):

- methods and means by which the waste must be delivered to the unit;
- estimated rates of delivery of materials;
- definition of capacity;
- development of performance criteria for the selection of equipment; and
- space requirements for maintenance and repair.

Due to the vast number of devices available in the market and their possible combinations, several possible technological configurations are possible. The following sections present some layout examples for residues from the main processing units when handling MSW.

2.5.1 Mixed Municipal Solid Waste and Process Residues

Mechanical biological and thermal treatment solutions are often combined to process mixed MSW and process residues. MBT plants serve to reduce the amount and volume of waste, to process organic waste fractions, and to recover some fractions for industrial reuse such as metals, plastics, and RDF as substitute fuel in industrial facilities (like cement plants). Due to the critical functions of MBT plants, they are considered an equivalent solution to incineration in the context of a technology matrix.

Depending on the actual goal, MBT plants can be operated by two ways: a mechanical process followed by a biological process, or a biological process followed by a mechanical process. Various types of equipment with diverse functions are integrated to ensure optimum recovery of materials and to allow adaptation in response to the market trends of recyclables. For example, in the case of MBT (Figure 2.23), the outputs are recyclables (like metals, cardboard, high-density polyethylene (HDPE),

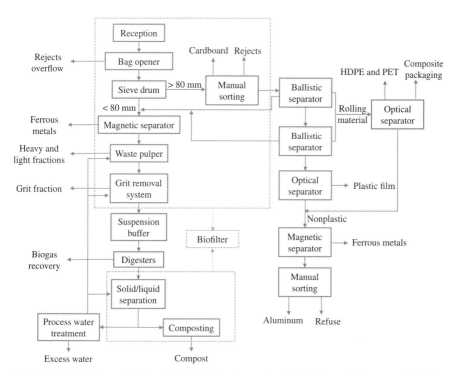

FIGURE 2.23   Anaerobic digestion MBT plant located at Leiria, Portugal. *Courtesy*: Valorlis, S.A.

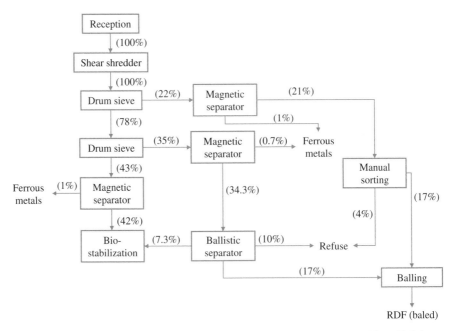

FIGURE 2.24 Aerobic MBT plant with RDF production. *Source*: Adapted from Belgiorno et al. (2007)

low-density polyethylene (LDPE) (mostly plastic film), polyethylene terephthalate (PET)), electric energy (e.g., biogas from anaerobic digestion), and compost. If RDF is favored by the market, the same MBT philosophy can be used to produce RDF from waste components with high heating value, like paper, plastic, and wood. The fraction with low heating value can be treated by aerobic treatment followed by a stabilized biological process in a landfill. An example is found in a unit in Italy (Figure 2.24).

MBT in which biological treatment occurs first (Figure 2.25) is a unique process used in units where the intention is to dry the material by using microorganisms rather than allow it to fully degrade the organic matter. In this unit, the bio-drying process removes moisture, which increases the calorific value of waste, making it more suitable for energy recovery (i.e., to produce RDF). The mechanical process is designed as a follow-up unit to remove some waste contents and recover RDF.

Refuse-derived fuel production In addition to aerobic MBT units, RDF can be produced independently in dedicated units from waste components with high calorific contents from different sources, such as urban, industrial, and commercial sources, or even other recycling processes. A typical RDF processing line (Figure 2.26) has many mechanical devices to remove various waste contents such as metals and low calorific fractions and to comply with requirements of final RDF specifications. RDF produced in these units with high calorific contents can be originated from MSW, CDW, and industrial waste. For future sustainable development of RDF production,

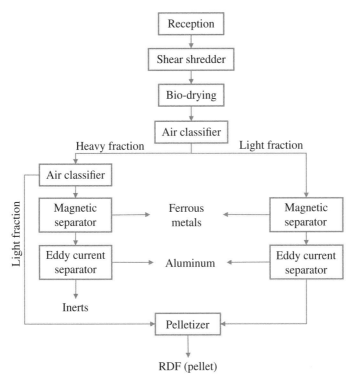

FIGURE 2.25   Aerobic MBT for RDF production

knowing waste composition and predicting how to tune the equipment dimensions and functions to suit the changing demand will be critical (Caputo and Pelagagge, 2002).

Incineration plant   An incineration plant is composed of different stages to ensure that complete combustion occurs in technically sound, environmentally benign, and professionally safe environments. An incineration unit starts with the discharge platform where waste is collected by a crane to feed the furnace (Figure 2.27). The waste heat recovered during combustion may be used to heat the water in the boiler, which can be used further for electricity or steam generation. The gaseous emissions are washed in the gas scrubber, then the bag house filter, and finally discharged via the stack.

2.5.2 Biodegradable Waste

Existing options to process biodegradable waste are basically composting and anaerobic digestion. These options differ from MBT because the material entering the biodegradation process. The organic waste, should have been source separated,

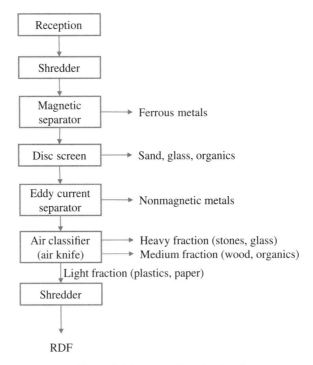

FIGURE 2.26 An RDF production line

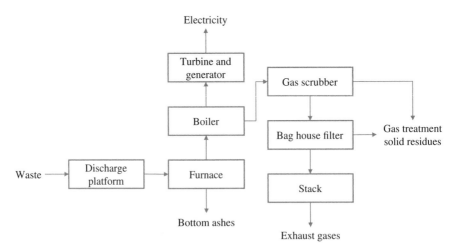

FIGURE 2.27 Incineration plant from Lisbon metropolitan area. *Source*: Adapted from Valorsul (2009)

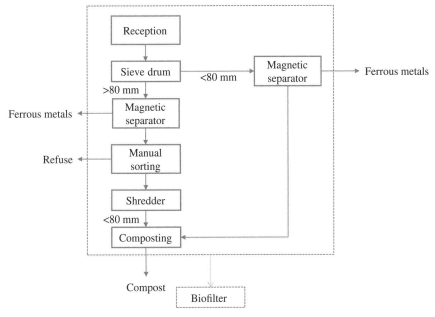

FIGURE 2.28 Composting plant from source-separated biodegradable municipal waste collected from the Madeira Autonomous Region, Portugal

thereby reducing the need to install numerous mechanical sorting units. A typical composting plant manages waste from source-separated collection (Figure 2.28). These concatenated units usually have a refining step as a start-up, mechanized with a drum sieve, vibrating sieves, and a densimetric table to remove stones and sand from compost, improving it visually for the final composting unit. These prior treatment units ensure that the final biological composting process produces high-quality compost.

A real-world anaerobic digestion plant (Figure 2.29) can be found in a metropolitan area in Lisbon, Portugal. Source-separated BMW are collected from two main sources: vegetable markets and restaurants plus hotels. The treatment diagram is similar to the one at plant in the Madeira Autonomous Region in Portugal (Figure 2.28). Note that a manual sorting unit situated somewhere in the middle of the treatment diagram is always required to guarantee the production of a quality compost.

2.5.3 Packaging Waste

As presented in section 2.4.1, packaging waste requires special handling and recycling methods. The preparation unit is usually called a sorting plant or a MRF. The MRF can have different capacities varying from 25,000 to 200,000 tonnes per year, although the average capacity normally lies in between 50,000 and 100,000 tonnes per year (Waite, 2009). One MRF plant can comprise a sequence of operational units, depending on

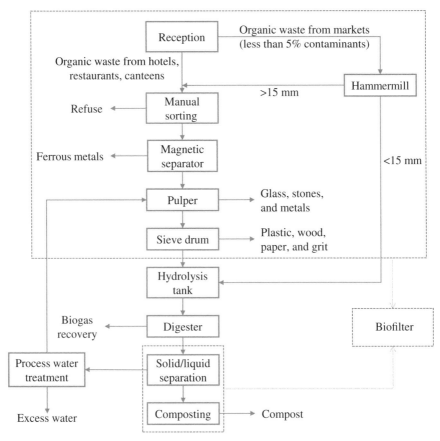

FIGURE 2.29 Anaerobic digestion plant to treated source-separated organic waste from Lisbon metropolitan area

its capacity, the type of equipment to be selected, types of waste to be processed, heterogeneity levels, recycling market demands, and financial factors.

The packaging waste process starts with homogenization of the MSW streams. Separation is then promoted through either manual or mechanized devices, with the choice of units subject to the types of packaging materials, including paper and cardboard, liquid packaging board, ferrous metals, nonferrous metals, plastics as expandable polystyrene, HDPE, LDPE, PET, and polypropylene (PP), and laminated packaging. After segregation, each recycled material is pressed and balled to optimize its storage and transportation to the next recycling organizations. The arrangement of separation units defines the type of MRF. In this respect, MRFs can be viewed as manual, semi-automated, or automated/automatic facilities based on the level of equipment automation (Figures 2.30 and 2.31).

A manual MRF plant uses labor-intensive processes to sort packages. This type of MRF plant is composed of a wheel loader to move the waste from the manual

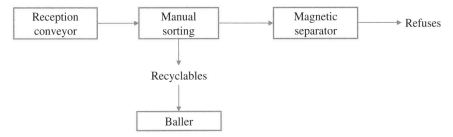

FIGURE 2.30   A manual sorting plant

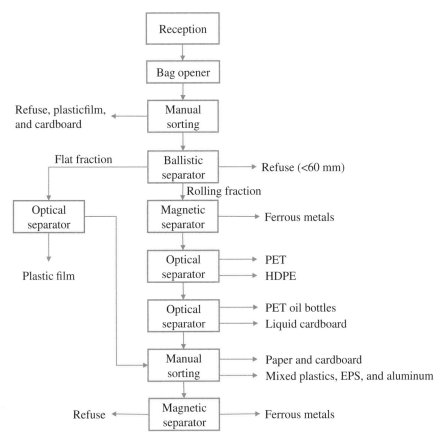

FIGURE 2.31   The schematic of a semi-automatic MRF plant located at Leiria, Portugal. *Courtesy*: Valorlis, S.A.

sorting area onto a conveyor. The workers sort the packaging onto the conveyor belt by positively segregating the intended material or negatively segregating materials undesired for recycling. The selected materials fall from the conveyor into appropriate storage silos along the operational line. Most manual MRF plants work on a continuous model along the sorting conveyor, although batch mode can also be designed for different purposes. A magnetic separator placed at the end of the sorting conveyor is usually present as part of this unit to separate ferrous packaging and improve the separation efficiency. Any waste not sorted at the end of the conveyor is discharged as refuse, usually destined for final disposal at a landfill or incineration plant. Sorted materials are then fed onto the balling line to be sent to recyclers.

A manual sorting plant presents several advantages in terms of separation efficiency when compared to equipment for plastic sorting in particular, including less contamination and reduced investment costs; however, operational costs can be high due to worker salaries and health insurance. In countries where human health legislation imposes limits to human contact with waste, automatic equipment can be a good solution to sort packaging waste.

An automated/automatic MRF works exclusively by mechanical devices; the manual work included in this type of plant is only supplemental for the purpose of quality assurance and process control. Semi-automated units, where manual sorting and mechanical equipment are used together to sort packaging, are a recent trend in some countries. A semi-automatic MRF applies manual sorting to homogenize packaging waste before applying mechanical treatment to sort out a specific fraction of several types of plastic packaging waste (Figure 2.31).

Paper   Waste paper and cardboard from packaging or nonpackaging waste has become vital for the paper manufacturing industry due to the advantageous price of recovered fibers in comparison with the corresponding grades of virgin pulp, and because of the promotion of recovered paper recycling (EIPPCB, 2001). Paper recycling industry quotes for typical energy savings from producing recycled paper ranging from 28 to 70% (Pré Consultants, 1996), depending on paper grade, processing level, mill operation, and proximity to a waste paper source and markets (Zabaniotou and Kassidi, 2003). The benefit of paper recycling is generally assumed to be desirable and necessary, and preference was given to reuse and recycling than landfilling or energy recovery (Zabaniotou and Kassidi, 2003); however, fiber cannot be recycled infinite times, and virgin fiber will be needed at some time points in the recycling process and market. Generally, recycling processes for fiber can be divided into two main categories, including (EIPPCB, 2001):

- processes with exclusively mechanical cleaning without deinking for those products such as testliner, corrugating medium board, and carton board; and
- processes with mechanical and chemical unit processes with deinking for those products such as newsprint, tissue paper, printing and copy paper, magazine papers, some grades of carton board or market deink pulp.

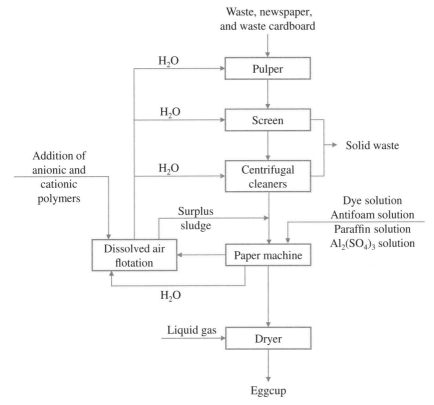

FIGURE 2.32 Process flow-chart for the production of recycled paper egg cups. *Source*: Adapted from Zabaniotou and Kassidi (2003)

The basic steps of wastepaper recycling are soaking, pulping, and screening to remove contaminants; ink flotation to remove ink if there is such intention; cleaning and screening; and thickening where water is removed and recycled paper is dried. The process of recycling wastepaper to produce eggcups (Figure 2.32) is an example that does not involve deinking (Zabaniotou and Kassidi, 2003).

Plastic   Several types of thermoplastic polymers (i.e., those that can be recycled as opposed to thermosetting), mostly known as PET, HDPE, LDPE, PVC, PP, and PS, can be used independently or combined with other plastics or even with other materials like paper and aluminum (i.e., also called laminated film). Before thermoplastics can be recycled, they must be sorted by polymers in the MRF, excepting laminates, which cannot be sorted. Mechanical or chemical process can be used to carry out the recycling of isolated polymers.

According to McDougall et al. (2001), the plastic is shredded or crumbed in mechanical recycling to a flake form, and contaminants such as paper labels are

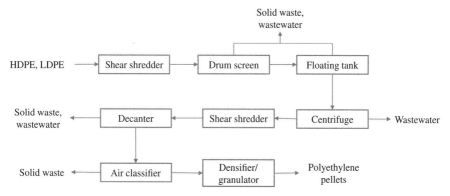

FIGURE 2.33 Mechanical recycling steps for HDPE and LDPE packaging waste

removed using cyclone separators before the flake is washed. This stage may also be used to separate different resins on the basis of density, and then dried and extruded as pellets for sale to the plastic market (McDougall et al., 2001). For instance, this polyethylene recycling process is applied in a facility in Portugal (Figure 2.33).

Chemical recycling involves the conversion of plastic materials into monomers, either liquid or gaseous, which can be used as feedstock for the petrochemical industry or as fuels (Pani, 2007). Depolymerization is the technology responsible for the chemical recycling success (Al-Salem et al., 2009). Within chemical recycling, some advanced processes are employed similar to those in the petrochemical industry, such as pyrolysis, gasification, liquid–gas hydrogenation, viscosity breaking, steam or catalytic cracking, and the use of plastic solid waste as a reducing agent in blast furnaces (Al-Salem et al., 2009). Energy recovery of plastics is also conducted using incineration with energy recovery, such as grate technology, fluidized bed and two-stage incineration, rotary kiln, and cement kiln combustion (Al-Salem et al., 2009).

Glass Glass must be collected separately for recycling because its extraction from mixed or residual waste is difficult. Because recycled glass may be contaminated with metals, plastic, paper as the bottle labels, and ceramics, it cannot be used as a substitute for raw material if it contains these contaminants, which deteriorate glass products. Glass bottles can be source separated by color (e.g., green, brown, and white). The prepared waste glass, called cullet or broken glass, can significantly reduce the energy consumption in the glass furnace, and its reuse is generally applicable to all types of furnaces, such as fossil, fuel-fired, oxy-fuel-fired, and electrically heated furnaces (Scalet et al., 2013). Although using cullet is advantageous, the quality of packaging glass waste makes it adequate only to the production of new glass packaging.

The cullet production process (Figure 2.34) begins with a magnetic separator to remove ferrous contaminants. The manual sorting removes only gross contaminants like plastic bottles, for example. Subsequent processes remove contaminants, break the glass to be homogenized, and improve the separation process, which is essential to ensure the quality of the physical appearance of waste glass.

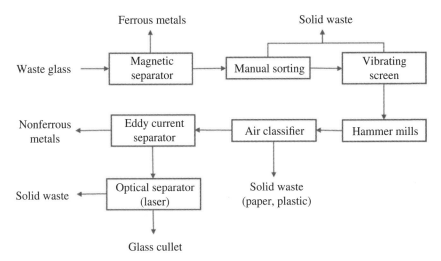

FIGURE 2.34   Glass cullet production process previous to glass recycling

Metals   Metals can be classified based on different features and chemical composition. For the purpose of waste processing, the most relevant classification is related to ferrous and nonferrous properties. Both are meaningful during steel and aluminum production because these industries often result in significant air pollution impacts, water use, wastewater or tailing water pollution, wastes from extracting minerals, and energy consumption; as well to conserve iron and bauxite, coal, and limestone during recycling (McDougall et al., 2001).

According to McDougall et al. (2001), ferrous metals from household and commercial waste are mostly iron and steel scrap; being the majority in the form of tinplate in food and beverage cans. Aluminum in the form of beverage cans is sometimes present in MSW. The processing of scrap metal can be provided from source separation and from the MRF or MBT units that process residual or mixed wastes, all of which are involved with the final delivery of recycled aluminum for recovery. In thermal treatment, such as incineration, metals are recovered from bottom ashes and slags destined for recyclers. Contaminant removal before recycling is always necessary.

2.5.4 End-of-life Vehicles and Scrap Tires

Vehicles and tires become scrap material requiring disposal when they reach end-of-life stages. The sources of both waste types are household, commerce, services, and industry. The recycling process of a scrap vehicle starts with the depollution step, followed by a recycling process consisting of a series shredding and separation units (Figure 2.35).

Depollution, which is often manual, removes all hazardous components of vehicles (e.g., batteries and liquefied gas tanks), removes or neutralizes potential explosives,

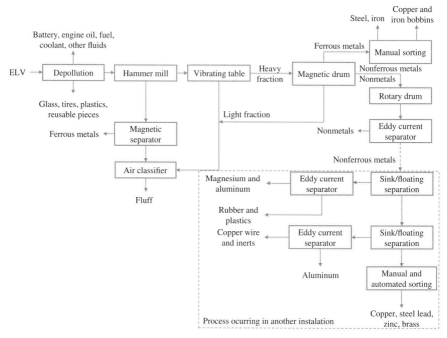

FIGURE 2.35   The schematic of an ELV recycling process

and removes vehicle liquids like fuel, motor oil, gearbox oil, hydraulic oil, cooling oil, antifreeze, brake fluids, air conditioning system fluids, and other fluids with hazardous substances or those that could adversely affect the recycling process. Examples of vehicle components are glass, tires, and large plastic components. A special reference for ELV removal of all components identified as containing mercury can be found in Directive 2000/53/EC (European Parliament and Council, 2000b).

ELV recycling begins with comminution using a hammermill. Some particles may be released and handled by a gas cleaning device during this phase, which often requires a gaseous treatment unit such as a scrubber. Broken materials fall through the 250 mm mesh grate onto the vibrating table. The magnetic drum separates ferrous metals, resulting in a waste stream with nonferrous plus nonmetal fractions. An air classifier system, composed of two cyclones and two venturi scrubbers, acts at the top of the hammermill at the vibrating table to remove the light fraction of plastics, foams, rubbers, and dust. The nonferrous metals plus nonmetals fraction are then separated by a drum sieve (or rotary drum), where three different mesh sizes are used. The resultant outputs of four types of by-products can all pass through an eddy current separator for aluminum and copper recovery. Nonferrous metals are then sent to recycling units where sink/floating separation is used for further separation.

Scrap tire recycling is conducted in three main phases: comminution, sieving, and magnetic separation. Comminution can occur at ambient temperature or negative temperature/cryogenic grinding, and the particle sizes vary depending on the technology applied and the costumers requirements. Each process has an impact on

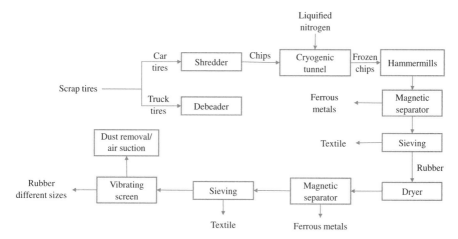

FIGURE 2.36   The schematic of a scrap tire processing plant by cryogenic process. *Source*: Pampulim (personal communication, 2012)

the rubber quality, which can impact further use (Pehlken and Müller, 2009). The comminution allows tires to be reduced to smaller pieces, exposing the different materials such as rubber, metals, and textiles to be sorted. Magnetic separators are used to remove ferrous metals, and sieving is used to separate rubber from textiles (also named fluff). Rubber is smaller in size compared to the fluff. Depending on recycling requirements, a sieving system with different mesh sizes is used to classify recycled rubber. One method, the cryogenic process (Figure 2.36), uses liquid nitrogen to reduce the temperature of tires before comminution, promoting a higher quality of recycled rubber.

2.5.5 Waste Oil

Waste lubricant oils are generated from the use of lubricants in engines and other equipment. Engine oils are required for combustion engines (e.g., vehicle engines, multipurpose diesels, or other engines), gear oils are required as lubricants in automotive gears and shock absorbers (Monier and Labouze, 2001), and grease originates from automotive devices. These three types of oils are managed as "black engine oils," which have homogeneous features and can be adequately re-refined (EIPPCB, 2006b). Black industrial oils can be compressor oils, general machine lubricants, and other oils for nonlubricating uses, and also industrial grease. Due to their content in additives and other substances they are not very attractive to the re-refining process (EIPPCB, 2006b). Turbine oils, electrical oils, and processing (white) oils are called light industrial oils, which can also be re-refined, if no synthetic oils are present (EIPPCB, 2006b). Metal working oils do not generate waste oils because they are consumed or lost during use.

Depending on their features, waste oils can be subject to different treatments. The two generic treatment purposes are production of a fuel or production of base oils to be processed to reproduce lubricant oil, also named re-refining. Fuel is produced from waste oils that are mixed or highly polluted. The processes applied can be thermal cracking, which produces distillated gasoil products; gasification, which produces synthetic gas; severe processing, which can produce demetallized heavy fuel oils; and mild processing, which produces a fuel used in cement kilns, large marine engines, and pulverized coal power stations. A description of the processes and technologies can be found in EIPPCB (2006b).

2.5.6 Waste of Electrical and Electronic Equipment

The increasing diversity and complexity of materials being used in electrical and electronic devices compounds their recycling and reuse processes, creating one of the most important waste streams for the future. WEEE processing activities can be classified into five groups based on the types of WEEE to be processed with convenience, including large household appliances, cooling appliances, information communication equipment with or without display boards, small household appliances, and gas discharge lamps. The first treatment step depollutes WEEE through manual removal of hazardous components. Next step in the treatment is manual dismantling, where recyclable components are removed to prevent contamination during future phases. For example, cathode ray tube (CRT) glass is separated before shredding (Figure 2.37)

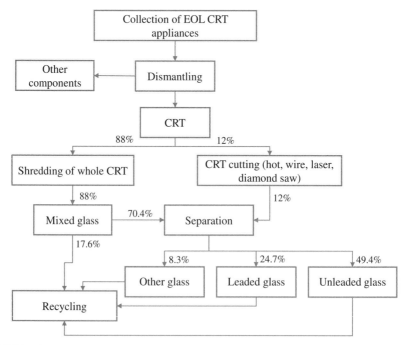

FIGURE 2.37 CRT glass recycling in WEEE recycling plant. *Source*: Adapted from Ecologynet (2010). *Notes*: EOL, end-of-life.

so that the remaining product can be shredded by specific equipment for each classification of WEEE. After shredding, metallic materials can be recovered from magnetic separators and eddy current separators. Combustible fractions like plastics are sent for energy recovery.

2.5.7 Construction and Demolition Wastes

CDW are produced from activities such as the construction of buildings and civil infrastructure, total or partial demolition of buildings, and civil infrastructure, road planning, and maintenance (European Commission, 2012). The treatment and recovery of these wastes depend on the market demand for specific composition of inert materials. According to Mercante et al. (2011), two main types of CDW recycling units can be considered as below (Figure 2.38):

- type I: These plants have a treatment capacity of 500–650 tonnes per day and an installed power of 150–160 kW. Sorted materials are transferred to recycling units. In these plants, sorted inert fraction is transformed into recycled aggregates.

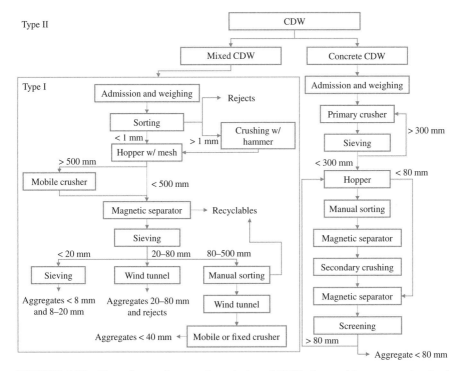

FIGURE 2.38 Flow chart and system boundaries of CDW plants with processes involved for types I and II waste simultaneously. *Source*: Adapted from Mercante et al. (2011)

- type II: These are larger facilities with two lines. The first one processes mixed CDW with a production capacity of 3,000–4,500 tonnes per day; the second one handles source-separated concrete waste, to produce an improved secondary aggregate, with 2,500 tonnes per day capacity. The recovered aggregate materials can be used as a substitute of virgin gravel.

2.6 FINAL REMARKS

This chapter explores the spectrum of technological integration required to process various types of MSW based on different types of unit operation. Schematic flow charts provide a systematic framework for defining and evaluating technology integration that may be applied as alternatives for a diversity of SWM problems. To provide valuable technology solutions at the highest quality and lowest cost through a single effort, there is a need to read subsequent chapters that will offer in-depth discussion with respect to economic, environmental, and societal factors for decision making leading to the provision of a holistic vision of integrated SWM in urban regions.

REFERENCES

Abbasi, S. A. and Ramasamy, E. V. 1999. *Biotechnological Methods of Pollution Control*, Orient Longman (Universities Press India Ltd.), Hyderabad, India.

Al-Salem, S. M., Lettieri, P., and Baeyens, J. 2009. Recycling and recovery routes of plastic solid waste (PSW): a review. *Waste Management*, 29(10), 2625–2643.

Anastassakis, G. N. 2007. Solid waste disposal and recycling: an overview of unit operations and equipment in solid waste separation. In: *Environmentally Conscious Materials and Chemicals Processing* (Ed. Kutz, M.), John Wiley & Sons, Hoboken, NJ, pp. 307–353.

Arena, U. 2012. Process and technological aspects of municipal solid waste gasification. A review. *Waste Management*, 32(4), 625–639.

Bai, R. and Sutanto, M. 2002. The practice and challenges of solid waste management in Singapore. *Waste Management*, 22(5), 557–567.

Belgiorno, V., Panza, D., Amodio, V., and Russo, L. 2007. MSW stabilized organic fraction landfilling. In: Proceedings Sardinia 2007, Eleventh International Waste Management and Landfill Symposium, Cagliari, Italy.

Berge, N. D., Reinhart, D. R., and Townsend, T. G. 2005. The Fate of nitrogen in bioreactor landfills. *Critical Reviews in Environmental Science and Technology*, 35(4), 365–399.

Bharadwaj, K. K. R. 1995. Improvements in microbial compost technology: a special reference to microbiology of composting. In: *Wealth from Waste* (Eds. Khanna, S. and Mohan, K.), Tata Energy Research Institute, New Delhi, India, pp. 115–135.

Bilitewski, B., Härdtle, G., and Marek, K. 1994. *Waste Management*, Springer, Berlin.

Bio-Wise. 2001. *Industrial Solid Waste Treatment: A Review of Composting Technology*, Department of Trade and Industry, London.

Brunner, C. R. 1994. *Hazardous Waste Incineration*, 2nd edition, McGraw-Hill, New York.

Caputo, A. C. and Pelagagge, P. M. 2002. RDF production plants: I Design and costs. *Applied Thermal Engineering*, 22(4), 423–437.

Chimenos, J. M., Segarra, M., Fernandez, M. A., and Espiell, F. 1999. Characterization of the bottom ash in municipal solid waste incinerator. *Journal of Hazardous Materials*, 64(3), 211–222.

Christensen, T. H. 2011. Introduction to waste management. In: *Solid Waste Technology and Management* (Ed. Christensen, T. H.), John Wiley & Sons, Chichester, UK, pp. 3–16.

City of Seattle. 2010. City of Seattle 2010 recycling rate report. City of Seattle. Available at: http://www.seattle.gov/util/groups/public/@spu/@garbage/ documents/webcontent/01 _013797.pdf (accessed February 2014).

Coelho, A. and de Brito, J. 2013. Economic viability analysis of a construction and demolition waste recycling plant in Portugal – part I: location, materials, technology and economic analysis. *Journal of Cleaner Production*, 39, 338–352.

Commission. 2000. Commission Decision of 3 May 2000 replacing Decision 94/3/EC establishing a list of wastes pursuant to Article 1(a) of Council Directive 75/442/EEC on waste and Council Decision 94/904/EC establishing a list of hazardous waste pursuant to Article 1(4) of Council Directive 91/689/EEC on hazardous waste. *Official Journal of the European Union*, L226, 3–24.

Dahlén, L. and Lagerkvist, A. 2010. Pay as you throw Strengths and weaknesses of weight-based billing in household waste collection systems in Sweden. *Waste Management*, 30(1), 23–31.

Dahlén, L., Vukicevic, S., Meijer, J. E., and Lagerkvist, A. 2007. Comparison of different collection systems for sorted household waste in Sweden. *Waste Management*, 27(10), 1298–1305.

Deif, A. M. 2011. A system model for green manufacturing. *Advances in Production Engineering & Management*, 6(1), 27–36.

Department of Environment, Food and Rural Affaris (DEFRA). 2012. Waste hierarchy guidance review 2012. Available at: https://www.gov.uk/government/uploads/system/uploads/ attachment_data/file/69403/pb13530-waste-hierarchy-guidance.pdf (accessed November 2013).

Diaz, L. F., Savage, G. M., and Eggerth, L. L. 2005. *Solid Waste Management*, United Nations Environment Programme, Paris, France.

Ecologynet. 2010. *Closing Material Loops @ WEEE Recycling*. Ecologynet. Available at: http://www.ecologynet-europe.com/site/panasonic-ene/get/params_Dattachment/6574 002/Grieger%20Care%20Innovation%202010.pdf (accessed March 2013).

EIONET. 2009. What is waste? EIONET. Available at: http://scp.eionet.europa.eu/themes/ waste/#treatment (accessed March 2013).

Enviros Consulting Limited. 2007. Mechanical biological treatment of municipal solid waste. Prepared on behalf of DEFRA as part of the New Technologies Supporter Programme.

European Commission. 2012. *Construction and Demolition Waste (CDW)*, European Commission – Waste. Available at: http://ec.europa.eu/environment/waste/construction_dem olition.htm (accessed March 2013).

European Integrated Pollution Prevention and Control Bureau (EIPPCB). 2001. Reference document on best available techniques in the pulp and paper industry. European Commission.

European Integrated Pollution Prevention and Control Bureau (EIPPCB). 2006a. Reference document on the best available techniques for waste incineration. European Commission.

European Integrated Pollution Prevention and Control Bureau (EIPPCB). 2006b. Reference document on the best available techniques for waste treatment industries. European Commission.

European Parliament and Council. 2000a. Directive 2000/76/EC of the European Parliament and of the Council of 4 December 2000 on the incineration of waste. *Official Journal of the European Union*, L332, 91–111.

European Parliament and Council. 2000b. Directive 2000/53/EC of the European Parliament and of the Council of 18 September 2000 on end-of-life vehicles. *Official Journal of the European Union*, L269, 34–42.

European Parliament and Council. 2008. Directive 2008/98/EC of the European Parliament and of the Council of 19 November 2008 on waste and repealing certain Directives. *Official Journal of the European Union*, L312, 3–30.

Eurostat. 2010. Generation of waste. Eurostat. Available at: http://appsso.eurostat.ec.europa.eu/nui/show.do?dataset=env_wasgen&lang=en (accessed January 2013).

Finnveden, G., Johansson, J., Lind, P., and Moberg, Å. 2005. Life cycle assessment of energy from solid waste – part 1: general methodology and results. *Journal of Cleaner Production*, 13(3), 213–229.

Fisher, M. M. 2003. Plastics recycling. In: *Plastics and the Environment* (Ed. Andrady, A. L.), John Wiley & Sons, Hoboken, NJ, pp. 563–628.

Foo, T. K. 1997. Recycling of domestic waste: early experience in Singapore. *Habitat International*, 21(3), 277–289.

Gajalakshmi, S. and Abbasi, S. A. 2008. Solid waste management by composting: state of the art. *Critical Reviews in Environmental Science and Technology*, 38(5), 311–400.

Ghani, W. A. W. A. K., Rusli, I. F., Biak, D. R. A., and Idris, A. 2013. An application of the theory of planned behaviour to study the influencing factors of participation in source separation of food waste. *Waste Management*, 33(5), 1276–1281.

Glavič, P. and Lukman, R. 2007. Review of sustainability terms and their definitions. *Journal of Cleaner Production*, 15(18), 1875–1885.

Glorius, T., van Tubergen, J., Pretz, T., Khoury, A., and Uepping, R. 2005. Solid recovered fuels: contribution to BREF "waste treatment." Report elaborated by European Recovered Fuel Organisation (ERFO) and by Institute and Chair of Processing and Recycling of Solid Waste, RWTH Aachen (IAR).

Gómez, G., Meneses, M., Ballinas, L., and Castells, F. 2009. Seasonal characterization of municipal solid waste (MSW) in the city of Chihuahua, Mexico. *Waste Management*, 29(7), 2018–2024.

González-Torre, P. L. and Adenso-Díaz, B. 2005. Influence of distance on the motivation and frequency of household recycling. *Waste Management*, 25(1), 15–23.

González-Torre, P. L., Adenso-Díaz, B., and Ruiz-Torres, A. 2003. Some comparative factors regarding recycling collection systems in regions of the USA and Europe. *Journal of Environmental Management*, 69(2), 129–138.

Hammel. 2013. HAMMEL-Plant engineering and construction – The complete solution. Hammel. Available at: http://www.hammel.de/downloads/2368.pdf (accessed June 2013).

Hekimian, K. K., Lockman, W. J., and Hest, J. H. 1976. Methane gas recovery from sanitary landfills. *Waste Age*, 7(12), 2.

Hoornweg, D. and Bhada-Tata, P. 2012. *What a Waste: A Global Review of Solid Waste Management*, The World Bank, Washington, DC.

Kang, H. and Schoenung, J. M. 2005. Electronic waste recycling: a review of US infrastructure and technology options. *Resources, Conservation and Recycling*, 45(4), 368–400.

Karak, T., Bhagat, R. M., and Bhattacharyya, P. 2012. Municipal solid waste generation, composition, and management: the world scenario. *Critical Reviews in Environmental Science and Technology*, 42(15), 1509–1630.

Kilgroe, J., Lanier, W., and Alten, T. 1992. *Development of Good Combustion Practice for Municipal Waste Combustor*, US Environmental Protection Agency, Washington, DC, EPA/600/A-92/267.

Kjeldsen, P., Barlaz, M. A., Rooker, A. P., Baun, A., Ledin, A., and Christensen, T. H. 2002. Present and long-term composition of MSW landfill leachate: a review. *Critical Reviews in Environmental Science and Technology*, 32(4), 297–336.

Komptech. 2012. Stationary machines. Komptech. Available at: http://www.komptech .com/uploads/tx_brochuredownload/stationary_machines_eng_2012.pdf (accessed August 2013).

Koufodimos, G. and Samaras, Z. 2002. Waste management options in southern Europe using field and experimental data. *Waste Management*, 22(1), 47–59.

Krishania, M., Kumar, V., Vijay, V. K., and Malik, A. 2013. Analysis of different techniques used for improvement of biomethanation process: a review. *Fuel*, 106, 1–9.

Larsen, A. W. 2012. Action 3.1: Survey on existing technologies and methods for plastic waste sorting and collection. Report elaborated for Plastic ZERO project.

Leverenz, H., Tchobanoglous, G., and Spencer, D. B. 2002. Chapter 8 – Recycling. In: *Handbook of Solid Waste Management,* 2nd edition (Eds. Tchobanoglous, G. and Kreith, F.), McGraw-Hill, New York, pp. 8.1–8.77.

Manfredi, S. and Christensen, T. H. 2009. Environmental assessment of solid waste landfilling technologies by means of LCA-modeling. *Waste Management*, 29(1), 32–43.

Martinho, G., Pires, A., Saraiva, L., and Ribeiro, R. (2012). Composition of plastics from waste electrical and electronic equipment (WEEE) by direct sampling. *Waste Management*, 32(6), 1213–1217.

Matthiessen Lagertechnik GmbH. 2013. Matthiessen bag opener. Products. Available at: http:///www.bagsplitter.com/pdf/bagsplitter-gb.pdf (accessed August 2013).

Mbande, C. 2003. Appropriate approach in measuring waste generation, composition and density in developing areas. *Journal of the South African Institution of Civil Engineering*, 45(3), 2–10.

McCormick, P. G. and Froes, F. H. 1998. The Fundamentals of mechanochemical processing. *Journal of the Minerals Metals & Materials Society*, 50(11), 61–65.

McDougall, F. R., White, P. R., Franke, M., and Hindle, P. 2001. *Integrated Solid Waste Management: A Life Cycle Inventory*, Blackwell Science Ltd, Cornwall, UK.

Mercante, I. T., Bovea, M. D., Ibáñez-Forés, V., and Arena, A. P. 2011. Life cycle assessment of construction and demolition waste management systems: a Spanish case study. *The International Journal of Life Cycle Assessment*, 17(2), 232–241.

Metso. 2011. Impact crushers: NP Series impact crushers. Metso. Available at: http://www .metso.com/miningandconstruction/MaTobox7.nsf/DocsByID/F8EAC8D77753A1114125 6B410031202D/$File/NP_English.pdf (accessed August 2013).

Michaeli, W. and Bittner, M. 1996. Classification. In: *Recycling and Recovery of Plastics* (Eds. Brandrup, J., Bittner, M., Menges, G., Michaeli, W.), Carl Hanser Verlag, Vienna Austria, pp. 231–236.

Monier, V. and Labouze, E. 2001. Critical review of existing studies and life cycle analysis on the regeneration and incineration of waste oils. Final Report Produced by Taylor Nelson Sofres and Bio Intelligence Service for EC-DG Env.

National Environment Agency (NEA). 2013. Waste management—Overview. Energy & Waste. Available at: http://app2.nea.gov.sg/energy-waste/waste-management/overview (accessed February 2014).

Nilsson, P. 2011. Waste collection: equipment and vehicles. In: *Solid Waste Technology and Management* (Ed. Christensen, T. H.), John Wiley & Sons, Chichester, UK, pp. 253–276.

Nilsson, P. and Christensen, T. H. 2011. Waste collection: systems and organization. In: *Solid Waste Technology and Management* (Ed. Christensen, T. H.), John Wiley & Sons, Chichester, UK, pp. 277–295.

Oguchi, M., Sakanakura, H., and Terazono, A. 2013. Toxic metals in WEEE: characterization and substance flow analysis in waste treatment processes. *The Science of the Total Environment*, 463–464, 1124–1132.

O'Leary, P. R. and Tchobanoglous, G. 2002. Chapter 14 – Landfilling. In: *Handbook of Solid Waste Management*, 2nd edition (Eds. Tchobanoglous, G. and Kreith, F.), McGraw-Hill, New York, pp. 14.1–14.93.

Organisation for Economic Co-operation and Development (OECD). 2002. *Towards Sustainable Household Consumption? Trends and Policies in OECD Countries*, OECD, Paris.

Organisation for Economic Co-operation and Development (OECD). 2009. *Eco-Innovation in Industry – Enabling Green Growth*, OECD, Paris.

Pani, B. 2007. *Textbook of Environmental Chemistry*, I.K. International Publishing House, New Dehli.

Pavlas, M., Touš, M., Bébar, L., and Stehlík, P. 2010. Waste to energy - An evaluation of the environmental impact. *Applied Thermal Engineering*, 30(16), 2326–2332.

Pehlken, A. and Müller, D. H. 2009. Using information of the separation process of recycling scrap tires for process modelling. *Resources, Conservation and Recycling*, 54(2), 140–148.

Pré Consultants. 1996. Eco-indicator 95 methodology report (in English) and the manual for designers. Pré Consultants. Available at: http://www.pre.nl/eco-indicator95 (accessed March 2002).

Pretz, T. 2000. Processing of municipal household waste material. In: Proceedings of the of the 8th International Mineral Processing Symposium, Antalya, Turkey, pp. 627–636.

Rao, S. 2006. *Resource Recovery and Recycling from Metallurgical Waste*, Elsevier, Amsterdam, the Netherlands.

Reimann, D. O. 2006. Result of specific data for energy, efficiency rates and coefficients, plant efficiency factors and NCV of 97 European W-t-E plants and determination of the main energy results. CEWEP Energy Report (Status 2001–2004) Bamberg, Germany.

Renhållnings-verksföreningen (RVF). 2002. Energy recovery by condensation and heat pumps at WTE plants in Sweden. RVF – The Swedish Waste Management Association.

Rhyner, C., Schwartz, L. J., Wenger, R. B., and Kohrell, M. G. 1995. *Waste Management and Resource Recovery*, CRC Press, New York.

Rogoff, M. J. 2014. *Solid Waste Recycling and Processing: Planning of Solid Waste Recycling Facilities and Programs*, second edition. Elsevier, Waltham.

Rosillo-Calle, F., de Groot, P., and Hemstock, S. L. 2007. General introduction to the basis of biomass assessment methodology. In: *The Biomass Assessment Handbook: Bioenergy for a Sustainable Environment* (Eds. Rosillo-Calle, F., de Groot, P., Hemstock, S. L., and Woods, J.), Earthscan, Suffolk, UK, pp. 27–68.

Rotter, S. 2011. Incineration: RDF and SRF – Solid fuels from waste. In: *Solid Waste Technology and Management* (Ed. Christensen, T. H.), John Wiley & Sons, Chichester, UK, pp. 486–501.

Royal Commission on Environmental Pollution (RCEP). 1993. *Incineration of Waste – RCEP 17th Report*, HMSO, London.

Scalet, B. M., Muñoz, M. G., Sissa, A. Q., Roudier, S., and Sancho, L. D. 2013. Best Available Techniques (BAT) reference document for the manufacture of glass (Integrated pollution prevention and control). JRC Reference Report. European Union, Spain.

Schulze, C. 2009. *Municipal Waste Management in Berlin*. Senatsverwaltung für Gesundheit, Umwelt und Verbraucherschutz. Berlin Senate Department for Health, the Environment and Consumer Protection, Environmental Policy, Section III B—Waste Management. Available at: http://www.berlin.de/sen/umwelt/abfallwirtschaft/ (accessed October 2009).

Schulze, C. 2013. *Municipal Waste Management in Berlin*, Berlin Senate Department for Urban Development and the Environment Communication, Berlin, Germany.

Schwilling, T., Mehner, H., Faysal, E., and Edel, H. 2004. Draft of waste concept for Berlin City Parliament (in German: *Vorlage des Abfallwirtschaftskonzepts im Abgeordnetenhaus Berlin*). Drucksache 15/3598. Available at: http://www.berlin.de/sen/umwelt/abfall/konzept_berlin/ (accessed October 2009).

Seattle Public Utilities (SPU). 2013. Solid waste management plan. SPU, Available at: http://www.seattle.gov/util/MyServices/Garbage/AboutGarbage/SolidWastePlans/SolidWasteManagementPlan/index.htm (accessed February 2014).

Senkan, S. M. 2006. Incineration and combustion. In: *Encyclopedia of Chemical Processing and Design* (Ed. Lee, S.), Taylor & Francis Group, New York, pp. 1381–1400.

Sidique, S. F., Lupi, F., and Joshi, S. V. 2010. The effects of behavior and attitudes on drop-off recycling activities. *Resources, Conservation and Recycling*, 54(3), 163–170.

Simon, F.-G. and Müller, W. W. 2004. Standard and alternative landfill capping design in Germany. *Environmental Science and Policy*, 7(4), 277–290.

Sims, R. E. H. 2010. Heat and power generation by gasification and combustion. In: *Industrial Crops and Uses* (Ed. Singh, B. P.), CAB International, Chippenham, UK, pp. 35–59.

Skhiri, N., Denois, D., Shaheen, M., and Lagier, T. 2005. Waste characterization in Egypt. In: Proceedings Sardinia 2005, Tenth International Waste Management and Landfill Symposium, Cagliari, Italy.

Slack, R. and Letcher, T. M. 2011. Chemicals in waste: household hazardous waste. In: *Waste: A Handbook for Management* (Eds Letcher, T. M. and Vallero, D. A.), Elsevier, Butterworth-Heinemann, Amsterdam, Netherlands and Boston, MA, pp. 181–195.

Slack, R., Gronow, J., and Voulvoulis, N. 2004. Hazardous components of household waste. *Critical Reviews in Environmental Science and Technology*, 34(5), 419–445.

Slack, R. J., Gronow, J. R., and Voulvoulis, N. 2005. Household hazardous waste in municipal landfills: contaminants in leachate. *The Science of Total Environment*, 337(1–3), 119–137.

Svendesen, J. B. 2010. Waste management system of city of Copenhagen. In: Proceedings of the International Conference on Municipal Waste Management in European Metropolitan Areas: Comparing Models, ATIA ISWA, Florence, Italy.

Tchobanoglous, G., Theisen, H., and Vigil, S. A. 1993. *Integrated Solid Waste Management*, McGraw-Hill, Inc., Singapore.

Turner, T. A., Pickering, S. J., and Warrior, N. A. 2011. Development of recycled carbon fibre moulding compounds – Preparation of waste composites. *Composites Part B: Engineering*, 42(3), 517–525.

United Nations Environment Programme (UNEP). 2003. Related concepts. UNEP. http://www.unep.org/resourceefficiency/Business/CleanerSaferProduction/ResourceEfficientCleanerProduction/UnderstandingRECP/RelatedConcepts/tabid/78838/Default.aspx (accessed May 2014).

United Nations Environment Programme (UNEP). 2010. *Global Trends and Strategy Framework*, International Environmental Technology Centre – UNEP, Osaka, Japan.

United States Environmental Protection Agency (USEPA). 1988. *Waste Minimization Opportunity Assessment Manual*, USEPA, Hazardous Waste Engineering Research Laboratory, Cincinnati, OH.

United States Environmental Protection Agency (USEPA). 1992. *Guides to Pollution Prevention: The Mechanical Equipment Repair Industry*, USEPA, Risk Reduction Engineering Laboratory, Cincinnati, OH.

United States Environmental Protection Agency (USEPA). 2012a. Guide for industrial waste management. USEPA. Available at: http://www.epa.gov/osw/nonhaz/industrial/guide/ (accessed March 2013).

United States Environmental Protection Agency (USEPA). 2012b. Bioreactors. USEPA. Available at: http://www.epa.gov/osw/nonhaz/municipal/landfill/bioreactors.htm (accessed April 2013).

United States Environmental Protection Agency (USEPA). 2013. Non-hazardous waste management hierarchy. USEPA. Available at: http://www.epa.gov/osw/nonhaz/municipal/hierarchy.htm (accessed November 2013).

Uriarte, F. A. 2008. *Solid Waste Management: Principles and Practices*. The University of the Philippines Press, Quezon City.

Valorsul. 2009. Sustainability report (in Portuguese: *Relatório de sustentabilidade*). Valorsul. Available at: http://www.valorsul.pt/media/85979/rs_valorsul_2009.pdf (accessed March 2013).

Velis, C. A., Longhurst, P. J., Drew, G. H., Smith, R., and Pollard, S. J. T. 2010. Production and quality assurance of solid recovered fuels using mechanical—biological treatment (MBT) of waste: a comprehensive assessment. *Critical Reviews in Environmental Science and Technology*, 40(12), 979–1105.

von Felde, D. and Doedens, H. 1997. Mechanical-biological pretreatment: results of full scale plant. In: Proceedings Sardinia, Sixth International Landfill Symposium, Cagliari, Italy.

Waite, R. 2009. *Household Waste Recycling*. Earthscan, Oxon.

Williams, P. T. 2005. *Waste Treatment and Disposal*, 2nd edition, John Wiley & Sons, Chippenham, UK.

World Health Organization (WHO). 2012. Waste from health-care activities. WHO. Available at: http://www.who.int/mediacentre/factsheets/fs253/en/ (accessed March 2013).

Worrell, W. A. and Vesilind, P. A. 2012. *Solid Waste Engineering*, 2nd edition, Cengage Learning, Stamford, CT.

Yadav, Y., Santose, S., SreeKrishnan, T. R., Kohli, S., and Rana, V. 2004. Enhancement of biogas production from solid substrate using different techniques – a review. *Bioresource Technology*, 95(1), 1–10.

Zabaniotou, A. and Kassidi, E. 2003. Life cycle assessment applied to egg packaging made from polystyrene and recycled paper. *Journal of Cleaner Production*, 11(5), 549–559.

Zhang, D., Soon, T., and Gersberg, R. M. 2010. A comparison of municipal solid waste management in Berlin and Singapore. *Waste Management*, 30(5), 921–933.

CHAPTER 3

SOCIAL AND ECONOMIC CONCERNS

In a sustainable solid waste management (SWM) system, not only technical factors but also economic incentives and social constraints need to be considered simultaneously. For each planning scenario of urban SWM, for instance, the calculation of amortized costs and benefits over the planning horizon must be taken into account. At the same time, there is the need to know the willingness to pay in the consumer community to determine an affordable charge system in terms of household income and consumption patterns. With such implementation schemes, the managerial team of an SWM project must assess the financial sustainability of the utility, evaluate public and political acceptability, and provide adequate changes to improve the accountability of the project if necessary. This chapter addresses financial, economic, and social factors in relation to the planning, design, and management of an SWM system. These considerations may be oriented for different projects, programs, and schemes that can be flexibly implemented for any type of SWM system. Whereas financial planning is required in the project planning phase, economic assessment has to be linked to multiple aspects of environmental, resources, and welfare economics of an SWM project. However, social concerns are tied to the legal aspects in decision-making and the attitudes and feedbacks from the community when implementing these SWM projects.

Sustainable Solid Waste Management: A Systems Engineering Approach, First Edition. Ni-Bin Chang and Ana Pires.
© 2015 The Institute of Electrical and Electronics Engineers, Inc. Published 2015 by John Wiley & Sons, Inc.

3.1 FINANCIAL CONCERNS

3.1.1 Financial Concepts

The key issue for any type of SWM is to ensure whether the charge system is affordable, viable, or sustainable in a local community receiving the service. To do so, costs and benefits of the SWM system should be balanced to comply with regulations and policies through proper operation. The core elements of a cost–benefit analysis that concern waste managers are investment costs, operation, and maintenance costs, administrative costs, and possible revenues.

A financial analysis required to smooth out the cash flows and sustain the operation addresses four concepts: net present value (NPV), discount rate, internal rate of return (IRR), and amortized costs/benefits. NPV, IRR, and amortized costs are three alternative ways to calculate whether or not a project is financially viable. The discount rate is a central parameter needed for the calculation of NPV, IRR, and amortized costs/benefits (Jacobsen, 2005).

NPV of a project calculates today's value of all expenditures resulting from the project. By comparing the NPV of different projects, the most valuable project can be determined, expressed as "value today" in terms of necessary revenues and expenditures (Jacobsen, 2005). NPV is calculated by:

$$\text{NPV} = \sum_{i=1}^{n} \frac{\text{Values}}{(1+d)^i},$$

where d is the discount rate; i is counter for the periods; and n is the total number of periods of the payment series.

The discount rate is a kind of interest rate being used to compare revenue and expenditure that occur at different points of time (Jacobsen, 2005). Useful benchmarks of the social discount rate in inflation-adjusted terms may be 10–12% in developing countries and 6–8% in developed countries per annum (Jacobsen, 2005). Theoretically, if there is no inflation concern, the discount rate expresses the rate of compound interest (Jacobsen, 2005). This discount rate can be defined for each person, each company, and each society with differing time frames and financial patterns. Generically, a project can be recommended if NPV is positive; however, the decision maker is also interested in the size of the initial investment and the length of time before the project is fully operating (Ogilvie, 2008). In that case, it becomes necessary to rank project in terms of the "earning power"—placing the project which generates the maximum NPV per dollar invested at the top (Ogilvie, 2008). The way to calculate it is through profitability index, which is the ratio between the present cash value of cash inflows and initial investment, where the higher is the better (Lasher, 2008).

IRR is an economic tool for project evaluation which calculates the discount rate that will make the NPV of a project equal to zero when that discount rate is applied to the NPV calculation (Jacobsen, 2005). The IRR is typically calculated in an iterative procedure that inserts different discount rates into the NPV formula, and the IRR is the one that results in an NPV that converges with zero (Ogilvie, 2008). The IRR of a project can then be directly compared to the cost of capital. A project is financially viable when the IRR is greater than the cost of capital (Jacobsen, 2005; Ogilvie, 2008).

Total amortized costs involve combining the amortized capital investment per annum and annual operating and maintenance (O&M) costs (Gavaskar and Cumming, 2001). Because O&M costs are measured on an annual basis, the capital investment must be amortized to combine these costs on the same basis (Gavaskar and Cumming, 2001). Amortized costs (i.e., annual costs) are calculated using the standard amortization formula:

Total amortized costs = Amortized capital investment + Annual O&M cost.

The amortized capital investment term can be interpreted as a fixed annual payment that a user would have to make every year over the life of the technology (Gavaskar and Cumming, 2001), estimated by:

$$\text{Amortized capital investment} = \frac{\text{Capital investment}}{\sum_{t=1}^{n} \frac{1}{(1+r)^t}},$$

where r is the interest rate, which accounts for the return (or interest) that the money assigned for capital investment would earn if the capital items were paid for over several years, at the end of each year (time, $t = 1, 2, 3, \ldots, n$) (Gavaskar and Cumming, 2001). However, if the inflation rate is taken into account, the adjusted interest rate would become the social discount rate, which would be smaller than the value of interest rate (i.e., if the inflation rate is larger than the interest rate, the economic system is not sustainable by itself). Amortized costs are often used as a shortcut to assess financial viability of an investment project (Jacobsen, 2005) and require less information than the calculation of NPV or IRR, providing a first estimate figure for initial financial assessment of a project.

3.1.2 Waste Management Costs

When planning an SWM system, throughputs, design capacity, generation rates of solid waste streams, recycling potential, and facility locations are all drivers that can define the costs and benefits. Knowing some additional indirect factors influencing the costs and benefits is often necessary. Cost control for SWM has been discussed in various forms, and highlighting the economies of scale would be worthwhile in comprehensive economic modeling. Economies of scale are the output unit cost reduction obtained by companies due to a larger business or process, like a larger facility size, where fixed costs are extended to more output units. According to Wilkinson (2005), economies of scale can be defined as aspects of increasing scale that lead to falling long-run unit costs. But it is not a never-ending advantage. In fact, cost advantage will disappear if a good or service is provided on a larger scale beyond the upper managerial limit. It is known as diseconomy of scale. The concept of a cost advantage that arises with increased output of a product is critical for all unit operations in an SWM system.

Hence, all unit operations for SWM can be analyzed from the perspective of size or the economy of scale. One implementation of an economy of scale effect is the involvement of neighboring municipalities to form a regionalization plan, where their responsibility is tied together for SWM. Economies of scale have been the justification to promote centralized waste management systems that integrate increasing numbers of local municipalities to form a consortium for SWM. Centralized SWM systems can, in reality, be promoted in developed countries where the managerial capacity is mature and forming a consortium will not be difficult via a democratic process. In developing countries with low solid waste generation rates and stringent financial conditions, economies of scale can still be promoted to lower average cost on a long-term basis. From a feasibility point of view, typical low-tech solutions, like neighborhood composting, through a decentralized SWM system that does not earn any economies of scale, are more applicable and feasible in the initial stage in these developing countries facing managerial and technical barriers. However, other socioeconomic reasons can trigger more amenable decentralized SWM systems (Box 3.1).

BOX 3.1　DECENTRALIZED WASTE MANAGEMENT SYSTEMS—AN INDIAN CASE (Zurbrügg et al., 2004)

"Composting, which has a long tradition in India" (Howard, 1943), is quite widespread in rural areas. In the 1970s, managerial and technical obstacles proved that centralized, large-scale composting plants in urban areas were uneconomical (Dulac, 2001), and only a few installations are currently operational (UNEP/WB/RWSGSA, 1991). High operating and transportation costs, poorly developed markets for compost, and low quality of mixed-waste compost (which has a negative effect on compost acceptance) have reduced expected profits (Dulac, 2001). "Since the 1990s, there is a trend towards smaller, manually operated composting plants at the community level, initiated primarily by citizens' initiatives or nongovernmental organizations (NGO) and also supported by international funds" (Furedy, 1992). The following major advantages are generally anticipated from the decentralized approach:

- "In combination with primary waste collection, composting improves the precarious waste situation in the communities, and residents become less dependent on the poor municipal waste collection service;
- Decentralized composting can be operated by an appropriate technology and implemented at reduced investment and operating costs;
- Manual composting in small, decentralized plants is more easily integrated at the community level with common socioeconomic background because it requires labor-intensive processes. It also offers new employment opportunities and a source of income to the underprivileged communities in the Indian society;

- Decentralized composting allows organic waste to be reused where it is generated, thereby reducing waste quantities to be transported as well as transport costs", which positively affects overall costs for SWM.

Several types of composting plants at the community level are available in the Indian community (Table 3.1), where the number of participating households is less than 1000 units. The interviews conducted with the initiators and key persons responsible for community plans reveal that upscaling to a decentralized composting concept organized by citizens and communities is restricted. Numerous examples of community initiatives provide advantages of decentralized composting, such as improved environmental conditions in residential areas, with the aid of a functional waste collection system.

TABLE 3.1   Overview of citizen's initiatives for waste collection and composting in Indian communities

Composting system	Waste quantities $(kg{\cdot}d^{-1})$	Production costs (2004 US\$$\cdot$tonne^{-1} compost)
Box system	50–300	91–1,380
Windrows	200	193
Worm composting in boxes	100	112

Source: Adapted from Zurbrügg et al. (2004).
Tonnes are metric tons.

The consideration of economies of scale may arise from waste collection. Costs for waste collection are influenced by the structure of the collection system and the collection frequency, which includes the crew size, which has a great influence on the cost (O'Leary et al., 1995; Merrild and Christensen, 2011). Collection costs items are mostly containers, collection vehicles, salaries, and fuel for trucks. But these factors can be interrelated; for example, lowering the shipping costs by having a lower collection frequency could result in higher capital costs for containers (O'Leary et al., 1995; Merrild and Christensen, 2011). Savas (1977) noted that the economies of scale in waste collection for a community with 50,000 residents gained the lowest collection costs in their case study. A suite of collection costs by waste stream collection can be observed at an island in Box 3.2.

BOX 3.2   WASTE COLLECTION COSTS AT AN ISLAND (de Gioannis et al., 2006)

In Sardinia Island, the goal of achieving a 50% average efficiency for waste separation and recycling during the collection phase on a regional scale requires the implementation of an integrated SWM system wherever possible. Considering a collection frequency of once per week for the packaging waste, two times per week for the residual municipal solid waste (MSW), and three times per week for the biodegradable fractions, the collection costs for the single waste component may range from 2006 US$69 per tonne for the residual MSW to 2006 US$603 per tonne for the plastic packaging waste.

The overall average collection cost, including the operation of a waste collection center, is roughly US$148 per tonne of the generated MSW, which can be lowered to 2006 US$133 per tonne of the generated MSW if the financial support of the national recyclers' network is considered. Currently the collection cost in Sardinia is about 2006 US$75 per capita$\cdot$y^{-1}.

Because the collection phase is interlinked with the processing and disposal phases in an SWM plan, there is always a trade-off from a system analysis perspective. This implies lower collection costs due to a lower service level which can lead to higher pretreatment costs (Merrild and Christensen, 2011). For example, in a case where the separate collection system is divided into few waste streams (e.g., dry recyclables and residual waste), costs at the material recovery facility to sort or pretreat waste fractions to be sent for recycling will rise. The chosen collection method (i.e., containers and vehicles), also affects the collection cost.

The factors affecting transportation costs are the type of vehicles, shipping distance, personnel, and administrative costs, and type of waste transportation (road, train, or cargo ship).

Transfer stations can optimize costs. Waste collected with smaller capacity vehicles can be transported to an intermediate transfer station and reloaded into larger capacity vehicles for later shipping to the final destination. This intermediate waste station can reduce the average transportation cost because the overall shipping plan would result in a lower cost per unit of mass and shipping distance (Box 3.3).

BOX 3.3   TRANSPORT COSTS AT AN ISLAND (de Gioannis et al., 2006)

Transportation costs may range from 2006 US$0.6 per t$\cdot$km^{-1} for distances of 50 km (heavy compactors) to 2006 US$0.2 per t$\cdot$km^{-1} for distances of 200 km (large self-compacting trailers). Considering that most of the small villages are spread over Sardinia Island, long distance transport is predictable for residual MSW destined for one of the two large-scale waste-to-energy (WTE) plants; therefore a main role could be played by the transfer stations (Table 3.2).

TABLE 3.2   Investment costs of transfer stations

Capacity (t$\cdot$y^{-1})	Investment costs (2006 10^3 US$)
10,000–20,000	213–440
40,000–60,000	565–816
80,000–100,000	879–1,130

Source: From de Gioannis et al. (2006).

TABLE 3.3   Investment and operating costs from temporary storage centers

Capacity (tonnes·y^{-1})	Investment costs (2006 10^3 US$)	Operating costs (2006 US$·tonne^{-1})
5,000–10,000	0.8–0.9	12.6–18.8
10,000–15,000	0.9–1.3	10–12.6
15,000–20,000	1.3–1.6	8.8–10

Source: From de Gioannis et al. (2006).

From an economy of scale perspective, investment and operational costs of temporary storage centers for recyclables (packaging waste) can be estimated (Table 3.3).

Treatment cost factors are related to selection of the best available control technologies throughout the technology matrix, which influences not only the capital cost

TABLE 3.4   Cost data for open-air composting facilities

Country/region	Capacity (10^3 tonnes·y^{-1})	Initial capital investment (10^6 US$)	Operating cost (US$·tonne^{-1})	Annual total cost (US$·tonne^{-1})
Europe (2006)	2.00	0.38	81.62	
Sweden (2006)	3.00	1.26	37.67	
Europe (2006)	5.00	0.75	60.28	
France (2006)	6.00	1.73–2.34	35.16–47.09	72.83–97.95
France (2006)	6.00	2.16–3.11	42.70–70.32	79.11–119.30
Greece (2006)	6.00	1.38		
Finland (2006)	10.00	4.21	46.93	96.69
Europe (2006)	10.00	1.13	50.23	
France (2006)	12.00	2.07–2.94	27.00–50.23	42.70–61.53
Italy (2006)	12.00	3.70	31.53	
Greece (2006)	13.00	1.63		
Sweden (2006)	15.00	3.39	20.93	
UK (2006)	18.00	1.27	21.88	
UK (2006)	18.00	1.27	21.88	
Austria (2006)	20.00	0.00		60.28
Europe (2006)	20.00	1.63	45.83	
Greece (2006)	20.00	2.64		
Sweden (2006)	24.00	4.71	18.84	
USA (2005)	25.00	8.60	26.84	
Greece (2006)	30.00	12.12		
USA (2004)	40.00	3.49	11.82	20.82
Greece (2006)	45.00	6.28		
Europe (2006)	50.00	2.76	37.67	
Greece (2006)	70.00	7.79		
Europe (2006)	100.00	5.65	32.65	
Asia (2008)	182.00	10.00–15.00	15.00–25.00	

Source: From Beck (2005), Tsilemou and Panagiotakopoulos (2006), van Haaren (2009), and UNCRD (2011).

TABLE 3.5   Cost data for anaerobic digestion facilities

Country/region	Capacity (10^3 tonnes·y^{-1})	Initial capital investment (10^6 US\$)	Operating cost (US\$·tonne^{-1})	Annual total cost (US\$·tonne^{-1})
Germany (2006)	2.50	0.64	75.34	79.50
Europe (2006)	5.00	3.64–3.89	30.14	
Europe (2006)	10.00	6.66–7.03	27.63	
Germany (2006)	15.00	9.05	60.28	136.62
Europe (2006)	20.00	11.93–12.56	25.11	
Europe (2006)	20.00	4.65–5.65	8.16	
UK (2006)	20.00	10.61	31.39	
Sweden (2006)	30.00	9.02	0.28	
UK (2006)	30.50	13.30	34.42	
Finland (2006)	44.70	6.92	44.30	
Australia (2006)	50.00	24.44	70.00	
Canada (2000)	50.00	68.32–106.93	34.00	
Europe (2006)	50.00	5.78–6.91	3.77	
UK (2006)	50.50	20.90	25.04	
UK (2006)	61.00	23.28	33.55	
France (2006)	72.00	16.29	66.55	71.58
Canada (2000)	83.00	114.35–130.69	47.00	
Sweden (2006)	100.00	11.62	0.25–0.50	
Australia (2006)	100.00	46.67	65.00	
Europe (2006)	100.00	13.19–15.70	4.40	
Asia (2008)	100.00	29.70–118.81	60.00–100.00	
Australia (2006)	150.00	70.00		
Canada (2000)	150.00	54.95–60.89	28.00	
Canada (2000)	250.00	68.31–99.50	41.00	

Source: From AKA and Enviros RIS (2001), Tsilemou and Panagiotakopoulos (2006), URS (2010), and UNCRD (2011).

but also operational costs. More sophisticated installation and operation is anticipated in the treatment phase, requiring more skilled personnel and higher investment cost (i.e., 2006–2008 US\$0.38–US\$15 million for open-air composting; 2001–2006 US\$0.64–US\$131 million for anaerobic digestion). In this regard, economies of scale are more advantageous when choosing sophisticated technologies relative to small-scale composting units (i.e., decrease from US\$82 per tonne to US\$15 per tonne (years 2006–2008) in open-air composting as compared to the case from US\$75 per tonne to US\$0.25 per tonne (year 2006) in anaerobic digestion plant) (Tables 3.4 and 3.5).

Other factors influencing biological treatments costs (including mechanical biological treatment (MBT)) are pollution control equipment and odor-reducing methods, which can vary in a large scale because no mandatory regulation exists in some countries for these units, and the treatment technologies vary from low-tech bio-filters to regenerative thermal oxidation (Box 3.4). Composting costs are greatly affected by the chosen technology (in vessel or open air), pollution control equipment, front-end separation technologies, and techniques to reduce odors (Hogg et al., 2000).

BOX 3.4   TREATMENT COSTS AT AN ISLAND (MBT AND
COMPOSTING) (de Gioannis et al., 2006)

The need for MBT plants is strictly linked to the efficiency of the separate collection of the biodegradable fractions. Investment and operation costs (Table 3.6) are reported as a function of this potential. Operation costs do not include disposal of the stabilized fraction.

TABLE 3.6   Investment costs and operation costs for MBT plants

Capacity (tonne·y^{-1})	Investment costs (2006 10^6 US$)	Operating costs (2006 US$·tonne^{-1})
10,000–20,000	5.2–9.0	63–75
20,000–40,000	9.0–11.9	50–63
40,000–60,000	11.9–17.3	48–50
60,000–80,000	17.3–20.7	44–48
80,000–100,000	20.7–27.0	40–44
100,000–150,000	27.0–45.8	38–40

Source: From de Gioannis et al. (2006).

Although processes selected for anaerobic digestion can affect the investment and operation costs, many other variables exist, including but not limited to the production of revenues such as electricity from biogas, price of electricity, and digestate market price. These complexities result in lowered economies of scale (Table 3.5).

The investment and operation costs of biological treatments such as composting (Table 3.7) do not take into account the income derived from the sale of compost.

WTE facilities are highly capital-demanding when compared with composting or anaerobic digestion plants (Table 3.8). WTE is the treatment option chosen where economies of scale are not phenomenal in terms of operational costs; however, significant diseconomies exist when choosing small-scale WTE facilities. To be cost-effective, WTE facilities need a guaranteed inflow of waste throughout their useful life and life cycle cost would reflect such a context (Tsilemou and Panagiotakopoulos, 2006). Reduction in either the calorific value or the amount of waste input

TABLE 3.7   Composting operating and investment costs

Capacity (tonne·y^{-1})	Investment costs (2006 10^3 US$)	Operating costs (2006 US$·tonne^{-1})
3,000–5,000	3.1–3.8	88–100
5,000–10,000	3.8–6.8	75–88
10,000–20,000	6.8–10.2	63–75
20,000–40,000	10.2–18.1	50–63

Source: From de Gioannis et al. (2006).

TABLE 3.8 Cost data for WTE facilities

Country/region	Capacity (10^3 tonnes·y^{-1})	Initial capital investment (10^6 US$)	Operating cost (US$·tonne^{-1})	Annual total cost (US$·tonne^{-1})
France (2006)	18.70[a]	14.82–16.70	92.93–99.20	161.99–177.06
Greece (2006)	36.50[a]	29.60–33.94	228.58–236.18	
France (2006)	37.50[a]	22.69–26.45	55.25–60.28	128.09–141.90
France (2006)	37.50[b]	21.37–27.01	74.09–82.88	131.85–150.69
Sweden (2006)	40.00[b]	16.75	43.70	89.66
Sweden (2006)	40.00[c]	30.45	58.58	142.15
Denmark (2006)	40.00	32.65	61.28	
Australia (2010)	50.00	52.78	80.00	
Europe (2006)	50.00	31.39	23.86	
Germany (2006)	50.00[a]			288.82
France (2006)	75.00[a]	42.13–49.69	65.30–70.32	114.27–128.09
Australia (2010)	100.00	100.00	70.00	
Greece (2006)	100.00	43.95		
Germany (2006)	100.00			175.80
UK (2006)	100.00	71.10	45.52	116.16
Europe (2006)	100.00	56.51	21.98	
UK, Ireland (2006)	120.00	84.39	52.74	
Australia (2010)	150.00	150.00	68.00	
France (2006)	150.00[c]	90.74–112.35	60.28–65.30	113.02–130.60
Belgium (2006)	150.00[a]	74.70	48.71	103.69
Belgium (2006)	150.00[a]	80.12	50.23	109.10
Germany (2006)	200.00[a]	153.11	72.41	131.85
Australia (2010)	200.00	200.00	65.00	
Ireland (2006)	200.00	105.78	32.56	
UK (2006)	200.00	102.49	36.43	87.46
Europe (2006)	200.00	113.02	25.11	
Denmark (2006)	230.00	161.74	45.83	
Australia (2010)	250.00	236.11	60.00	
Italy (2006)	300.00	184.36	76.26	160.14
Sweden (2006)	300.00[b]	65.91	24.22	48.35
Sweden (2006)	300.00[a]	119.91	30.35	74.21
Germany (2006)	300.00			106.74
Netherlands (2006)	450.00	582.04	84.13	
UK, Ireland (2006)	420.00	226.79		
Asia (2008)	450.00	30.00–180.00	80.00–120.00	
Europe (2006)	500.00	200.92	17.08	
Italy (2006)	584.00	251.15		
Germany (2006)	600.00			81.62

Source: From Tsilemou and Panagiotakopoulos (2006), URS (2010), and UNCRD (2011).
[a]Electricity; [b]Heat; [c]Cogeneration.

(throughput) would reduce the amount of energy recovered along with the revenues thereof, which could be relatively significant (Hogg et al., 2000; European Commission, 2001).

Landfilling costs are mainly related to capital costs in the form of land and machinery, personnel, and materials and resources (e.g., liners and electricity) (Merrild and

TABLE 3.9   Cost data for landfilling facilities in European countries

Country	Capacity (10^3 tonnes·y^{-1})	Initial capital investment (10^6 US$)	Operating cost (US$·tonne^{-1})	Annual total cost (US$·tonne^{-1})
Greece (2006)	18.25	2.51–2.89	6.53–7.28	
France (2006)	20.00	0.13–0.23	45.21–56.51	90.41–118.04
Luxembourg (2006)	32.20	40.25	61.34	184.59
Greece (2006)	40.15	4.41–4.52	5.15–5.78	
Luxembourg (2006)	40.30	34.91	41.54	154.46
Germany (2006)	50.00			64.04
France (2006)	50.00	0.19–0.38	31.39–38.93	60.28–67.81
Greece (2006)	60.00	9.17–10.42	7.28–8.16	
France (2006)	100.00	0.5–0.75	27.63–31.39	50.23–64.04
Germany (2006)	100.00			42.70
Greece (2006)	120.00	15.07–16.95	6.03–6.78	
Italy (2006)	125.00	43.41	16.45	65.90
Germany (2006)	150.00			35.16
UK (2006)	175.00	27.49	13.78	36.14
Asia (2008)	180.00	5.00–10.00	10.00–20.00	
Greece (2006)	240.00	24.61–27.75	4.90–5.53	
Italy (2006)	300.00	9.04	43.95	
Germany (2006)	300.00	27.27	14.18	25.11
Greece (2006)	480.00	40.18–45.46	4.02–4.52	
Germany (2006)	500.00			21.35
Italy (2006)	1,500.00	18.84–32.65	31.39–37.67	

Source: From de Gioannis et al. (2006), Tsilemou and Panagiotakopoulos (2006), and UNCRD (2011).

Christensen, 2011) (Table 3.9). Landfill costs are dependent on the type of waste landfilled because the density of the waste and the requirements for liners, top covers, and other consumables vary from site to site, all of which affect costs (Merrild and Christensen, 2011). According to Tsilemou and Panagiotakopoulos (2006), whether or not to include a biogas collection system is also a relevant factor in terms of cost-effectiveness, especially for small landfills (up to 60,000 tonnes per year) because the biogas collection system is not a requirement.

3.1.3 Waste Management Revenues

Revenues from an SWM system are obtained from selling by-products such as recyclables, compost, electric energy, or heat. In many countries with an existing polluter-pays principle (PPP), recyclables obtained from the SWM system can be financed by the producer, as in the case of packaging waste managed in response to extended producer responsibility, where collection and sorting are financed by the tax that is paid by packaging producers. However, revenues from recyclables can be difficult to estimate at the planning phase. The recyclables market is quite uncertain due to issues related to the variations of supply and demand in the market, which affects the business framework of an SWM plan (European Commission, 1998; Nicolli et al., 2012):

- On the supply side, recycling is subject to pressure caused by the growing cost of collecting and processing waste, facing direct competition with corresponding virgin materials. In these conditions, structural complexity and technical weaknesses constitute a significant brake on the competitiveness of the recycling sector. Ultimately, the efficiency of recycling could be improved by ensuring that product design takes into account the requirements of post-consumption collection, sorting, and recycling. In addition, the quality of recyclables can affect the market value.

- On the demand side, the competitiveness of recycling is limited by a lack of preference for recyclables (i.e., secondary materials) in processing industries, due to their technical properties, limited applicability, and/or negative image. Furthermore, recycling is likely hampered by the lack of pertinent industrial standards, or even by the tendency for some standards or specifications to ignore or discriminate against recycled materials or products. There is also a fundamental terminology difference between "waste" and "standards" based on inadequate quality criteria.

- Lack of transparency in recycling markets is a major impediment to the investment required to achieve improvements in the industrial structure, improve procedures for treating recyclable waste, and develop new applications for recycled products.

- Usability restrictions are related to technological restriction/incapacities that restrict the market value of recyclable materials; for example, the use of multilayer plastics for food packaging is not mechanically recyclable (Wilson, 2002); the several colors of glass bottles results in costly separation to be recycled (Ecotec, 2000; Porter, 2002); the use of inking technologies in paper printing needs better technical solutions to recycle paper (Apotheker, 1993); metal applications like pigments in paints and constituents in alloys make materials uneconomical to be recycled (ICME, 1996); the use of a great diversity of resins (Sterner et al., 1997) and blow-molding versus injection molding of plastics (Palmer and Walls, 1999); and the use of polymers in cable manufacturing reduces the recovery potential of polyvinyl chloride (Enviros, 2003).

All of these issues (called market failures, explained in section 3.2) trigger the volatility of secondary recyclable trade prices (Figures 3.1–3.3). The most notable change over the last decade is a sharp reduction in secondary material prices during the financial crisis of 2008–2009 (Eurostat, 2011). During this period, anecdotal evidence suggests that, in the short term, some waste management authorities had difficulties selling the materials they had collected for recycling (Eurostat, 2011); however, average annual figures for intra-European Union 27 countries (EU-27) trade suggest that the markets for most secondary materials were not substantially affected.

The data also show that the price for materials often exported out of the EU for recycling recovers well after the sharp reduction seen throughout 2008. EU-27 trade volumes in plastics reduced considerably but have recovered quickly to levels higher

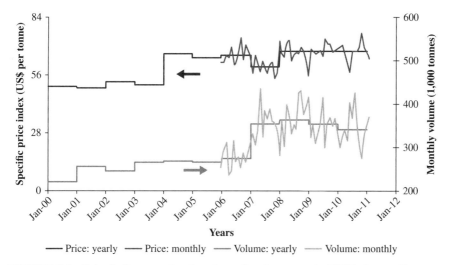

FIGURE 3.1   Price indicator and trade volume for glass waste in EU27. *Source*: Eurostat, http://epp.eurostat.ec.europa.eu, data from 2011. © European Union, 1995–2013 (converted from € to US$)

than before 2008, notwithstanding the price reduction (Figure 3.2). This observation suggests that the waste sector is robust enough to support short-term crises in the secondary material markets. However, prolonged reductions in price will affect the economics of recyclables collection, and, in the longer term, may lead to greater costs for the implementation of the waste and recycling strategy in Europe.

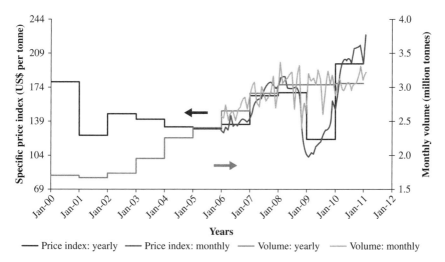

FIGURE 3.2   Price indicator and trade volume for paper and board waste in EU27. *Source*: Eurostat, http://epp.eurostat.ec.europa.eu, data from 2011. © European Union, 1995–2013 (converted from € to US$)

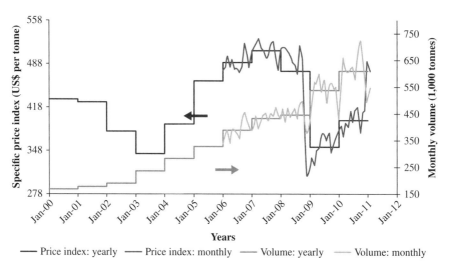

FIGURE 3.3 Price indicator and trade volume for plastic waste in EU27. *Source*: Eurostat, http://epp.eurostat.ec.europa.eu, data from 2011. © European Union, 1995–2013 (converted from € to US$)

In addition to the collection of recyclables, an SWM system can also produce other outputs or by-products, such as compost (including digestate) if biological treatments exist. The factors affecting the compost market are mostly local. The compost quality, presence of a significant agricultural market for compost, subsidies, and existence of competitive products like manure can dictate the role of compost as an SWM revenue. The situation in Europe is quite diverse. In central Europe, bulk compost for agriculture use is rarely higher than US$6.3 per tonne; in Belgium and Netherlands, the compost sales to agriculture have become difficult due to fierce competition with manure; however, high prices from US$113 to US$377 per tonne can be reached when the compost is sold in small amounts, such as for blends, to hobby gardeners, or to wholesalers (IPTS, 2011). Compost is also exported, although there is no uniform regulation to control it. Compost is one of the most common and beneficial garden amendments in the United States and the costs of compost, whether bagged or in bulk, vary widely over different states.

Electric energy produced from SWM is considered a renewable source of energy, including biogas collected from landfill and anaerobic digestion plants and energy recovered from incinerators and refuse-derived fuel (RDF) burning. In some countries, a subsidy exists to promote this source of energy; when applied to the waste sector, this subsidy can considerably increase revenues for SWM.

3.1.4 Public Financial Scheme and Private Sector Financing

In most developing countries, SWM services are under responsibility of local governments (Diaz et al., 2005). There are two critical points where financing is needed:

(1) in the capital investment phase, when infrastructures must be constructed; and (2) in the O&M phase, when the SWM system is functioning.

To financing capital investment costs, current revenues (renewal funds), bonds, loans or grants, and donations can be used. In the case of O&M costs, which are also known as recurrent costs, financing can be provided through different options such as taxes, fees, or user charges reflecting the actual costs for the service, and grants or subsidies from the central government.

The financing of O&M costs through the application of user charges can be a difficult task in countries where the population has low income. In many countries, as indicated by Diaz et al. (2005), residents receive one bill for water, wastewater, solid waste, and other services like television and security. The problem with this simplification is that the community could consider waste collection and treatment a right and not accept a separate bill for waste disposal, which is the case in Portugal and Spain, for example, where specific billing system for waste services is not yet applied.

Another recent measure to receive financing has been the opening of waste services to the private sector, which has been formally conducted in some developed countries through a well-developed waste industry and informally carried out in some developing countries through the informal waste management sector. This transfer and control of a good or a service currently provided by the public sector, either in whole or in part, to the private sector is called a public–private partnership (Massoud and El-Fadel, 2002). The increased interest in public–private partnerships can be attributed to: "(1) improved performance of the public sector by employing innovative operation and maintenance methods; (2) reduced and stabilized costs of providing services by ensuring that work activities are performed by the most productive and cost-effective means; (3) improved environmental protection by dedicating highly skilled personnel to ensure efficient operation and compliance with environmental requirements; and (4) access to private capital for infrastructure investment by broadening and deepening the supply of domestic and international capital" (van de Walle, 1989; Walters, 1989; Ramanadham, 1991; Jefrey, 1996; USEPA, 1998 in Massoud and El-Fadel, 2002).

Public–private partnership contracts for SWM are mainly related to contracting services (e.g., more common waste collection services) and enabling services (e.g., infrastructure construction). Both benefits and drawbacks coexist from those contracts. In the case of contracting services, situations such as the one studied by Antonioli and Filippini (2002) could only achieve optimal collection levels by contracting the service out to firms operating under a franchised monopoly. Empirical studies from the United States and European countries show that privatization does not necessarily provide the lowest cost service delivery, because the expected competition created at the beginning is not maintained over time, not resulting in cost saving (Bel and Warner, 2008). Lombrano (2009) pointed out that no correlation was found between privatization and cost-effectiveness; in fact, evidence shows that both public–private partnerships and publicly owned undertakings are able to sustain development relatively well.

The entrance of the private sector into the SWM sector can be through formal or informal (such as the informal private sector case in developing countries) pathways. The term "informal" sector is applied to refer to the economic activities which has: nonpermanence and casualness profile, occurs outside the scope of existing company law or government regulations, and are operated at a small-scale by less capitalized companies mostly relying on household labor (Salahuddin and Shamim, 1992). Overall, private sector operators may be grouped as waste pickers, itinerant/stationary waste buyers, small-scale recycling industry, large-scale recycling industry, community-based organizations, NGO, and micro-enterprises (Ahmed and Ali, 2004).

The informal sector (or informal recycling systems) can bring significant economic benefits to developing countries (Wilson et al., 2006). From a macroeconomic perspective, the informal sector can provide inexpensive workers, but has no capital for investment. The informal sector is also able to provide a trust and constant supply of secondary raw material for the local manufacturing industry, which stimulates the manufacture of low-cost, affordable products made from recycled materials (Wilson et al., 2006). Other positive economic impacts of an informal recycling sector are the cost reduction of formal SWM systems by reducing the quantity of waste for collection, resulting in less money and time spent on collection and transport; and preservation of void space at disposal sites for wastes with no potential value because all recyclable materials are diverted for reuse (Wilson et al., 2006).

The main challenge for government/municipalities is how to develop waste regulations that integrate an informal waste/recycling sector to constitute a formal public–private partnership. This challenge is significant because the economic benefits cannot be formalized if not appropriately conducted. An example of successful integration of an informal waste sector is the Brazilian model of waste pickers' cooperatives. Waste picker cooperatives were created in the metropolitan areas of Brazil to organize populations and facilitate negotiations with middlemen and companies in the recycling waste market (Pimenteira et al., 2005). In cooperatives, all members are owners of the enterprise (labor, not capital), organizing the negotiations with the recycling industry to optimize process for the recyclables (Tirado-Soto and Zamberlan, 2013). Waste picker cooperatives were promoted by the local government as a way to reduce collection costs and to organize waste pickers, making them get more revenues from their activity (Pimenteira et al., 2005).

3.2 ECONOMIC INCENTIVES AND SOCIOECONOMIC CONCERNS

The previous section addressed SWM from a financial perspective to illustrate how an SWM system can be affordable. Now SWM will be approached from a market perspective. A market exists when goods or services have a price defined by the supply of the good or service and its demand. In a free competitive market, this price reaches a point when supply and demand curves cross; in other words, where the amount demanded by consumers will equal the amount provided by producers, resulting in an economic equilibrium between price and quantity.

Waste management involves the trade of the service (waste collection and treatment) and materials (recyclables, compost, electric energy) to create a market; however, waste management markets are neither free nor competitive. When the price of services and goods cannot reflect the damage of natural resources or human health due to lacking of a pricing mechanism, it is referred to as market failure. Different market failures exist, including public goods, monopoly and oligopoly, externalities, and free riders.

3.2.1 Public Goods

Public goods are neither excludable nor rival to the general public when they consume those goods, whereas common resources are not excludable but rival. In accordance with environmental economics, SWM can be regarded as a common good when there is a competitor for the same resource, such that one person's use of the resource reduces the other's opportunity to use the resource (e.g., as in the case of landfill space). To some extent, however, SWM can also be deemed a public good that everyone can access, and this access does not affect the use by others, as in the case of a waste collection system, which is viewed as a nonexclusive and nonrivalrous service. According to Massoud and El-Fadel (2002), once waste collection is provided to some portions of the community, it benefits the overall public welfare, in such a way that any citizen can enjoy the benefit of the service without diminishing the benefit to anyone else. A main problem is that some users become free riders, a term that describes all the municipalities who will enjoy the waste management service without paying for the cost, thereby deteriorating the budgetary control in the government. One way to avoid the free-rider budgetary crisis is to apply a billing system that integrates waste collection with other public services such as water and television.

3.2.2 Monopoly and Oligopoly

Monopoly in the waste management market is when a single company controls the waste collection service within the city, and citizens must pay the corresponding fees to this company. The overlapped boundary between a monopoly with only one seller and a competitive market with several major sellers is an oligopoly. In the waste management market, oligopolies can be a small group of businesses who limit the number of service providers and dominate the market, which is especially common in the hazardous waste treatment industries and in recycling industry for handling small quantities of waste-derived secondary materials.

3.2.3 Externalities

In economics, an externality is defined as the cost or benefit that affects a group of people which have not considered taking upon itself the cost or benefit. Externalities, which can be described as economic side effect, may generate a market failure if the price mechanism is not considering the full social costs and benefits related to the production and consumption of goods and/or services. Externalities may be classified into positive (an external benefit is imposed) and negative (an external cost is imposed), at consumption and production sides (Parkin et al., 2008). A positive externality occurs when an individual or firm does not receive the full benefit of the decision, having others gaining from that decision or when the marginal social benefit of production and consumption is smaller than the marginal social cost (MSC). In opposition, a negative externality occurs when the externality has the opposite effect. For example, the disposal fees that consumers pay for waste landfilling is not sufficient to clean up the secondary pollution impact caused by the landfill operation on the loss of human health or ecosystem integrity. The charge of such disposal fees is deemed as a market failure resulting in negative externalities. Because the loss is external to the market, it is aptly termed as an externality.

All market failures make SWM a difficult task, raising a number of socioeconomic concerns from a sustainable point of view. Three forms of market failures are of interest, which are:

- **Negative externalities**: It is necessary to reduce negative externalities in such a way that SWM could reach zero waste (e.g., zero environmental burdens).
- **Fairness**: It is necessary to provide the waste management as a public good that everyone can access and avoid the neighboring communities around the SWM facilities having disproportionate environmental impacts due to such operations.
- **Intergenerational equity:** It is necessary to keep today's equity, efficiency, and externality reductions maintained for future generations.

Various instruments can be applied to mitigate these concerns, including command and control, economic instruments, and information instruments. Command and control consist of rules and legislations that impose how much pollution is allowed. Economic instruments apply monetary flows as an incentive to change waste generation and recycling behavior. Finally, information instruments also intend to do the same thing to provide an avenue for reducing the externality effect, via delivering messages to waste management producers. In the next section socioeconomic concerns will be characterized and relevant instruments will be discussed that may be applied to solve the problems of externalities, fairness, and even intergenerational equity to some extent.

Negative Externalities   In many cases, waste collection and treatment are grouped as a public good subsidized by the government, similar to other public goods like public schools, libraries, and safety forces. User charges, which is a kind of price charged by governments according to the market mechanism, are often applied to help finance waste collection (Mikesell, 2007); however, the charges can be applied as a flat rate per month, which is unrelated to the amount of waste generated from the users. Whether the flat rate or a special time-based trash charge is used might not affect the household's trash-generation incentives; no matter which is applied, the end result of total waste generation would be the same, which implies that the marginal private cost (MPC) of putting out an additional unit of waste is zero (Porter, 2004).

In consequence, the application of a flat rate charge does not consider the effect of externalities resulting from the amount of waste generated and disposed of, which is related to the social cost. The difference between MSC and MPC defines the externality in a market (Box 3.5).

BOX 3.5   MARGINAL PRIVATE COSTS AND MARGINAL SOCIAL COSTS

Pollution created by SWM is a negative externality. The actual cost of providing the service is greater than the cost perceived by the SWM service utility. In other words, due to a negative externality, the marginal cost as perceived by the SWM service providers, which is called the MPC, is less than the true marginal cost that is incurred by the society, which is called the MSC. The MSC is the sum of the consumer's MPC and marginal external cost (MEC), being MEC the external cost for society (including waste managers and citizens paying for the SWM service) for the additional unit of SWM service provided:

$$MSC = MPC + MEC.$$

A tax equal to the MEC, called a Pigovian tax, is referred to as the effect of "internalizing the externality" (Pigou, 1920). The tax makes the perpetrator of the external cost think more carefully about continuing the practice (Porter, 2002).

The right price of anything would be the MSC, so people will buy something only if their willingness to pay is greater than the price, which is equal to the MSC.

Note 1: "Social costs represent the expenditures needed to compensate society for resources used to maintain its utility level" (Callan and Thomas, 2007).
Note 2: Marginal cost is the additional cost of producing one extra unit of output, which, in the case of SWM, is the cost of providing the collection of one more ton of waste, for example.

"Marginal private cost is the marginal cost of production as viewed by the private firm or individual" (Taylor and Weerapana, 2012).

"Marginal social cost is the marginal cost of production as viewed by society as a whole" (Taylor and Weerapana, 2012).

All alternative strategies for SWM result in varying degrees of externalities generated at the collection, transportation, and disposal stages (Eshet et al., 2006). These externalities embedded in each SWM operation may be addressed from a demand–supply perspective. Waste collection externalities are mostly related to negative aspects of transportation, like traffic congestion, traffic safety, air pollution, noise, and climate change. Waste incineration presents negative externalities through air pollution and residue generation (slag and ashes), and positive externalities concerning energy recovery and ferrous and nonferrous metals recovery. Waste disposal has negative externalities such as space use, nuisance, air pollution and odors, climate change due to methane emissions, leachate leakage to aquifers, and market value of neighbouring areas; and positive externalities such as energy recovery from landfill gas use.

Waste managers have the challenge to include a Pigovian tax into the price of waste service provided and, at the same time, to know if the society is willing to pay it. Quantification of a Pigovian tax is often complex. Evaluation techniques include stated preference or willingness-to-pay (WTP), revealed preference or willingness-to-accept (WTA), dose–response approach, replacement cost methods, and opportunity cost approach. Most of the direct and indirect techniques of monetary valuation rely on economic welfare theory, that is "based on the values that individuals ascribe to their preferences regarding an environmental good with no observable price in the market" (Eshet et al., 2006).

WTP tries to establish what individuals are willing to pay for a particular environmental good or service, such as preventing polluted air and odors nuisances (O'Riordan, 2001). To quantify WTP, methods such as contingent value and choice modeling can be used. Contingent value is a type of stated-preference approach that employs a hypothetical market system to extract WTP or WTA for environmental goods (Hadker et al., 1997; Carson, 2000). In a contingent valuation process, a hypothetical but plausible scenario proposing environmental improvements (or degradation) is shown to the respondents and then they are asked to declare the amount they are willing to pay (or willing to accept) for this change (Jones et al., 2010). WTA can be defined as the amount of money considered as compensation for foregoing a benefit or for incurring a loss, reflecting the value of such a benefit or loss (Begum et al., 2007). Besides contingent valuation, WTA can be determined by travel costs method and hedonic pricing method. Travel costs method considers the time and the expenses resulting from travelling to an environmental good or service like in the case of a national park or ecosystem; the hedonic pricing method is used to estimate economic values for the ecosystem or environmental services that directly affect market prices (Ahmed and Gotoh, 2006). Choice modeling or conjoint analysis estimates the WTP for the respondents at the individual level based on each respondent's data provided by the whole sample of respondents, which can take the form of contingent rating, contingent ranking, choice experiments, or paired comparison (OECD, 2002; Breidert et al., 2006).

The dose–response approach or function (also named as production function and impact pathways method) is considered the core of the evaluation, once it provides the first information for various techniques associated with the economic damage–cost approach (Eshet et al., 2006). Dose–response function measures the correlation between a unit concentration of a pollutant (dose) and the impact on the receptor based on scientific data analysis (Tellus Institute, 1992; Markandya et al., 2001). Due to the occasional absence of reliable dose–response function, other methods were developed in practice, such as sustainability indicators and linked environmental values or indices (Eshet et al., 2006).

Replacement cost methods use the cost of replacing or repairing an injury asset to its original status as the measure of the benefit, and thus as the cost of the injury, based on the knowledge, experience and judgment of professionals (Eshet et al., 2006). The technique implies that complete replacement is, in fact, feasible (Pearce and Howarth, 2000). Besides, opportunity cost approach is applied to solve a short-term, nonroutine decision-making problem (Siegel and Shim, 2006). It represents the net benefit lost by rejecting some alternative course of action since it considers the cost of the best available alternative not taken (Siegel and Shim, 2006).

Overall, the economic values of externalities obtained by these methods described above allow them to be internalized through economic instruments such as taxes, subsidies, compensations, and tradable emission permits (Eshet et al., 2006). The internalization of externalities will help to prevent damages, leading the SWM options to be capable of promoting a sustainable development of our society.

Pay-As-You-Throw   Pay-as-you-throw (PAYT) is a charge system for SWM based on the amount of waste being generated by households, which create a direct economic incentive to generate less waste and to participate in recycling schemes. PAYT is a variable rate that may be applied based on the amount of the waste being generated. It can be composed of a two-tiered pricing structure in which a flat rate is used to cover part or all fixed costs, and a unit-based component of a variable rate fee is applied in accordance with the amount of waste being generated. PAYT variable rate fees more known are unit pricing, pay-per-bag, pay-per-pound, volume-based (container size), and incentive rate (Taylor, 2000). In accordance with Dijkgraaf and Gradus (2004), weight- and bag-based pricing systems are best succeeded than the frequency- and volume-based pricing systems; the bag-based system seems to be the one with the best performance, due to lower maintenance and administrative costs and to the effects obtained, similar to those of the weight-based system.

Deposit–Refund   In this economic instrument, a payment (the deposit) is made when the product is bought and is fully or partially refunded when the product is sent back to a dealer or specialized treatment facility (OECD, 2001). Deposit–refund schemes are mostly applied to beverages, but more recently have been applied to end-of-life vehicles. Principally, the deposit should include the commercial costs of the container (or specific product), plus the environmental costs associated with the disposal or with littering. Refunds should equal the avoided environmental costs plus the scrap value of the container. To promote the recycling incentives, higher return rates can be achieved when the fee is set at a higher percentage of the price (OECD, 1993). Deposit–refund schemes are often introduced as a means to encourage reuse and the reduction of material inputs (e.g., beverage containers), and/or to maintain a reliable flow of materials for recycling and recovery operations (OECD, 2001).

Taxes, Fees, and Levies   Disposal fees are payments by waste haulers for the ultimate disposal of MSW like incineration plants or sanitary landfills/dumpsites (commonly called "tipping fees"), normally based on weight, which are differentiated according to potential environmental damage by a specific waste, or recovery features (Taylor, 2000; Porter, 2004). For example, recyclable waste will have a higher tipping fee than nonrecyclable waste, because it is not being recycled.

Advanced disposal fees (also called product charges or front-end disposal fees) are designed to internalize the SWM cost of manufactured products and their packaging. These fees are paid at the point of sales and could be levied through the government or by an industry-based private sector organization (OECD, 2001). Product levies are also applied to promote waste reduction, as in the case of the plastic bag levy introduced in several countries, such as Ireland, and some states in the United States. The taxes and levies mentioned so far might or might not be equal to Pigovian taxes (defined in Box 3.5). Pigovian tax is the most efficient and effective way to correct negative externalities whereas general taxes and levies are not.

Subsidies   Public subsidies are used to encourage specific waste prevention, waste diversion, and other needed SWM services that managers are incapable to provide on their own (Taylor, 2000). Such subsidies include provision of outright grants and soft loans to several kinds of waste-related business enterprises to acquire land, buildings, equipment, and other facilities and to meet operating expenses, or to change habits near citizens by subsidizing home composting and the use of reusable nappies (Taylor, 2000; Watkins et al., 2012).

Upstream combination tax/subsidy   An upstream combination tax/subsidy combines a tax on produced intermediate goods with a subsidy to collectors of recyclables, like used beverage cans and old newspapers sold for reprocessing (OECD, 2001). The subsidy finances SWM to some extent for meeting specific goals required by the government. Producers can also be given physical responsibility (full or partial) for treatment of the post-consumer products, such as the Green Dot system in Germany (OECD, 2001).

Tradable Licenses, Permits, or Allowances   Tradable permits can operate in various ways but usually involve "setting statutory pollution targets and the issue or sale of "pollution" permits to affected industries" (Bailey et al., 2004). Polluters may only increase activities provided the total pollution load does not enlarge, so that any increase from one source must be compensated by an equivalent or greater decrease within the prescribed boundary (Barde, 1997). To facilitate this mechanism, trading allows those agents (like companies or waste management authorities) who present lower abatement costs to sell surplus permits or allowances to other agents who have bigger abatement costs. The driving force in such a market for permit trading is to pursue less expensive options. The end result, in theory, is cost-effective compliance with environmental standards through innovation and the cheapest overall abatement efforts region wide (Bailey et al., 2004).

Fairness   The notion of fairness is related to the management of public goods, and in this case, they can be the waste service including collection, treatment, and disposal. A fair waste service means that those who produce more waste should pay more for the pollution they create, which is in line with PPP. Within this context, fairness concerns may be applied to those communities who will be allowed to pay less for SWM service due to bearing more pollution impact from neighboring SWM equipment and facilities. Hence, pricing mechanism is linked with different fairness or equity issues. Some economic instruments used to manage externalities were described in section 3.2.1; however, the implementation of those economic instruments must also be capable of bringing equity to waste generators to compensate for their environmental damage, if any. For example, residents who live closer to an SWM facility might have a lower flat rate than those who live far away. This policy entails fairness in the context of environmental justice. For instance, the PAYT system is fair "when it is in accordance with ideas of what is appropriate, acceptable or expected with regard to the way the system is designed and organized and the manner in which participants are treated" (Batllevell and Hanf, 2008). Three underlying ideas ensure a fair PAYT: equality of cost per unit, equality of opportunity, and equity (Batllevell and Hanf, 2008).

Equality of cost refers to allocation of public expenditure to ensure that all relevant individuals face the same private cost per unit of the service used (Le Grand, 1982). However, implementing the equity of cost (i.e., ensuring that the same unit pricing has the same value at any place in the municipality, at any condition) may not promote equity. If the intention is to implement the PPP, it would be preferable to penalize big producers rather than smaller producers due to factors such as urbanization and distance to disposal place. Active waste charges can be a possible solution where a progressive rate system with marginal increasing charges is established.

Equality that implies fairness to all people receiving SWM service involves avoidance of providing different service to everyone in the same system. For example, if a PAYT is applied in a specific municipality to handle residual waste, then a separate collection system at the source must exist for everyone under the same conditions. This requirement can be difficult to accomplish due to factors such as legal/political, urban, technical, socioeconomic, and environmental factors, which may not abide by the equality of opportunity.

Equity is the concept that warranted or justifiable distinctions or differences cannot be made between persons (Batllevell and Hanf, 2008). In the case of designing a PAYT system, adjustments must be made to ensure that all households have equal opportunities to participate in the system and the benefit of only paying the amount of residual waste actually generated. Differences that must be addressed are related to income, family size, and special groups (such as elderly and handicapped people). According to Batllevell and Hanf (2008), low income households may not be able to decrease the amount of charges paid in some types of PAYT system, due not only to waste generation but also to waste composition. For example, a volume-based PAYT system might be beneficial for low income families because they normally would have less packaging waste and materials like paper, plastic, glass, and metal. Large families would normally produce more waste, paying more PAYT fees than small families. Because special groups might find it difficult to participate in PAYT, municipalities may consider reduced rate or no extra charge is applied (Batllevell and Hanf, 2008).

Another issue related to fairness is regional compliance with a national waste management target, such as the national targets of packaging waste recycling and biodegradable municipal waste diversion from landfills. National authorities often define the same recycling target for every type of SWM system, which may not be fair because the effort to collect packaging waste would be higher (more costly) for low density rural areas than urban communities with high density households.

Intergenerational Justice   The concept of sustainable development in the Brundtland report in 1987 (WCED, 1987) introduced intergenerational concerns in environmental policy, which states that the key to sustainable development is an equitable sharing of benefits and burdens between generations. These concerns can and should be intimately linked with intergenerational justice, a concept first introduced by John Rawls in 1971 as intergenerational distributive justice, which stands for an equal allocation of social benefits and burdens (Rawls, 1971/1999; Taebi, 2010). From an environmental point of view, intergenerational justice in SWM implies:

- Equal access to goods and resources that a future generation has no fewer means to meet their needs than the present generation. Along this line, efforts like reducing waste generation, in quantity and quality, will increase the lifetime of resources, conserving virgin natural resources for the future. In this respect, the overuse of all landfill space within a short period of time is unjust because landfill space is deemed a nonrenewable resource in urban regions.
- Avoiding future unintended consequences in terms of environmental and health risks and/or impacts. This means actions from today will not negatively affect the next generations.
- Avoiding the transfer of actual environmental problems to future generations requires solving current environmental, specifically waste problems, in an efficient way. The life cycle thinking is an appropriate approach to ensure it.

To guarantee the sound implementation of intergenerational justice, the concept of intergenerational externality must first be promoted, which extends the current concept of externalities over generations. Intergenerational externality can be geared toward unique environmental economics issues to prevent the next generation from facing and managing pollution impacts and/or resource overexploitation. With this extended concept, intergenerational justice can be included in the planning phase by using environmental impact assessment (EIA) or strategic environmental assessment (SEA) (Figure 3.4), which include the following premises (Padilla, 2002):

- To halt inappropriate projects: If the project causes irreversible, harmful effects to future generations and these cannot be avoided or compensated, the project should be stopped;
- To undertake precautionary and control measures: If the modification of the structure that the original project would imply is avoidable, the option of "do nothing" is more appropriate;
- Compensation through an associated project: Future generations can be compensated for harmful effects of one project through an associated project; and

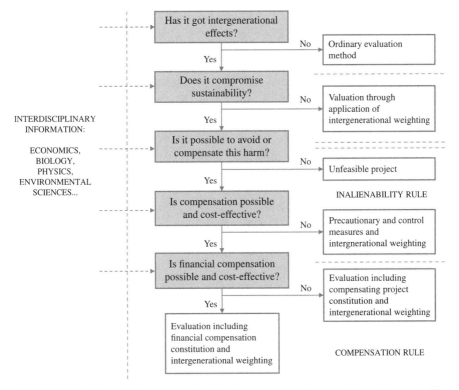

FIGURE 3.4 EIA with intergenerational justice concerns. *Source*: Adapted from Padilla (2002)

- Financial compensation: If not avoidable, the damage caused to future generations by a new project must be quantified in monetary units or, at least, must demonstrate that compensation will, with all probability, be satisfactory to future generations.

3.3 SOCIAL CONCERNS

Citizens' participation in the creation of SWM policy (strategies, planning, and management) is crucial for its success. The public demands environmental information that could affect not only its health but also its local environment. These concerns must be considered in the decision-making process because waste management infrastructures might change the environmental quality of local communities who will pay for the waste management services. The inclusion of residents in decision-making can be more than consultative; it can be determinant to substantiate decisions to be made by waste management officials, not only in the planning phase but also during the system operation phase.

A key social problem is that site selection for landfills and incineration facilities have been receiving negative responses worldwide. To avoid such public objection, planning strategies with appropriate risk communication skills toward social accceptance of SWM solutions is critical during the public hearing process of an EIA or SEA. Specific risk communication toward conflict resolution can be used to seek cooperation with SWM (Ishizaka and Tanaka, 2003).

When an SWM system is already in operation, social concerns are more devoted to the public opinion of the quality of the SWM system in terms of the services provided, odor nuisances, and hazardous emissions from these facilities. In this regard, communication between all stakeholders is vital for success. In addition to technology-driven concerns, public behavior associated with source separation, waste reduction, and recycling in the waste collection systems would also be important.

Environmental impacts have both short- and long-term effects. Changing behaviors focusing exclusively on certain dimensions of change due to social influences would be intimately linked with some long-term SWM projects. For that reason, inter-generation factors are also an issue to be addressed by waste management policy. For example, a social value system influencing waste and resources management based on resource durability, economic viability, technological applicability, environmental friendliness, and public safety and security, is also changeable over time, thereby making such evolutionary pathways of social value systems a valuable long-term social issue to be considered by waste management policy. In the future, the social value system would be highly likely to change in accordance with stakeholders' needs, which deserve our attention. The main social concerns of SWM systems are focused on public acceptance and public behavior. A better participatory process among stakeholders' communities could certainly contribute to an improved societal performance.

3.3.1 Public Acceptance

Many waste treatment technologies such as incineration, landfills, recycling, hazardous waste, and radioactive waste treatment have been facing public opposition from local communities worldwide. Conflicts always exist between municipalities, who prefer to adopt fast technical solutions, and representatives from citizen's groups or ecologically oriented NGO, who prefer to choose waste prevention and recycling as solutions (Salhofer et al., 2007). Objections from local communities where SWM facilities are to be sited are tied to potential risks and environmental disadvantages of these SWM facilities to be constructed near their homes that would relieve waste management problems somewhere else (Petts, 1994). Such physiological reactions have been documented as syndromes:

- Not in my backyard (NIMBY),
- Locally unwanted land uses (LULU),
- Not in anyone's backyard (NIABY),
- Not in my term of office (NIMTOO), and
- Building-anything-at-all-near-anyone (BANANA).

What causes these syndromes? Several complex reasons (or factors) can justify those syndromes/attitudes (Petts, 1994; Johnson and Scicchitano, 2012):

- perceptions of risks to health and the environment;
- a lack of trust in regulatory agencies to monitor and control facilities and in the private sector to manage operations effectively;
- local communities are more likely to be aware of these local environmental issues facing communities;
- lack of trust in government, which makes these SWM project be carried out by local communities directly;
- a paucity of information, especially risk information, by experts; and
- the exclusion of the public from fundamental policy decision-making in the SWM, or allowing public involvement only after initial decisions have been made by waste management officials and regulators.

Efforts to reduce those syndromes have been guided into two phases: (1) in the planning phase, the implementation of public participation during decision-making processes and the risk communication process is essential; (2) in the operation phase, rigorous environmental monitoring is required, and when planning and selecting the waste treatment technologies, public participation must be promoted by providing major EIA procedures for those new SWM facilities such as landfills and incinerators, and the key outputs of SEA for new waste management plans and programs.

According to Canter (1996), a comprehensive definition of public participation in the context of environmental assessment is, "public participation can be defined as a continuous, two-way communication process which involves promoting full public understanding of the processes and mechanisms through which environmental problems and needs are investigated and solved by the responsible agency; keeping the public fully informed about the status and progress of studies and implications of project, plan, program and policy formulation and evaluation activities; and actively soliciting from all concerned citizens their opinions and perceptions of objectives and needs and their preferences regarding resource use and alternatives development or management strategies and any other information and assistance relative to the decision."

Public participation in SWM planning occurs through a democratic planning process, where all stakeholders can contribute, where the objective is to promote a constructive decision-making process, and where more information held by stakeholders can be brought into the process (Figure 3.5). In addition, EIA legal requirements for SWM infrastructures and plans may be open to the public. Public participation can serve three purposes (O'Faircheallaigh, 2010): obtaining public input into decision-making for SWM, sharing decision-making with public, and altering distribution of power and structures of decision-making. Public input can materialize in the field via collecting real-world information to fill information gaps, promote information contestability, or even with the purpose of problem solving and social learning (O'Fairchellaigh, 2010). Sharing decision-making with the public can help

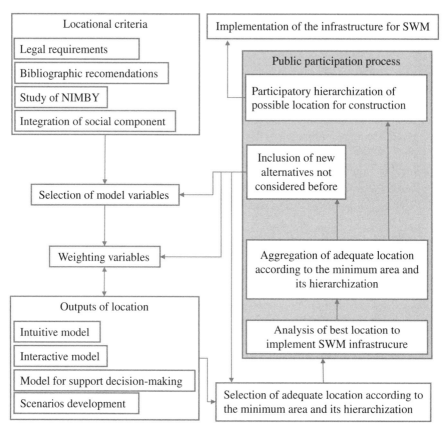

FIGURE 3.5   Illustrative flowchart for public participation for siting an SWM infrastructure. *Source*: Adapted from Vasconcelos et al. (2005)

implement democracy in practice and emphasize pluralist representation. Public participation tends to involve minority groups to shift the locus of decision-making (O'Fairchellaigh, 2010).

In the context of EIA of any SWM project (Figure 3.5), public participation consists of a commitment process, in which the public is invited to contribute to the decision-making process by exchanging information, projections, ideas, interests, and values (Fischer, 2007). In the case of SEA, participation is a key aspect of the process because it intends to be integrative, sustainability-driven, participative, and iterative. Participation techniques applied to SEA can be advisory groups, workshops, visioning exercises, and citizen juries, all of which may reduce NIMBY attitudes if applied at the beginning of the planning stage, helping to define acceptable solutions for all parties involved (Salhofer et al., 2007).

Public participation applied to SWM planning has long been a difficult task, and there are many forms of public participation (Table 3.10). In common with all forms of public participation are the desire to reach a mutual and consensual solution between

TABLE 3.10 Public participation methods

Participation method	Nature of participants	Time scale/duration	Characteristics/mechanism
Referenda	Potentially all members of national or local population: realistically, a significant proportion of these populations.	Vote cast at a single point of time.	Vote is usually the choice of one of two options. All participants have equal influence. Final outcome is binding.
Public hearings/inquiries	Interested citizens, limited in number by the size of the venue. True participants are experts and politicians making presentations.	May last many weeks or months, even years. Usually held during weekdays or working hours.	Entails presentations by agencies regarding plans in open forum; the public may voice options but has no direct impact on recommendations.
Public opinion surveys	Large sample (e.g., hundreds or thousands) usually representative of the population segments of interest.	Single event, usually lasting no more than several minutes.	Often enacted through written questionnaire or telephone survey. May involve variety of question. Used for information gathering.
Negotiated rule making	Small number of representatives of stakeholder groups (may include public representatives).	Uncertain: strict deadline usually set: days/weeks/months.	Working committee of stakeholder representatives (and from sponsor); consensus required on specific question (usually, a regulation).

(continued)

TABLE 3.10 (*Continued*)

Participation method	Nature of participants	Time scale/duration	Characteristics/mechanism
Consensus conference	Generally, 10–16 members of the public (with no knowledge on topic) selected by steering committee as "representative" of the general public.	Preparatory demonstrations and lectures (etc.) to inform panelists about topic, then 3-day conference.	Lay panel with independent facilitator questions expert witnesses chosen by stakeholder panel. Meetings open to wider public. Conclusions on key questions made via report or press conference.
Citizens' jury/panel	Generally 12–20 members of the public selected by stakeholders panel to be roughly representative of the local population.	Not precise but generally involve meetings over a few days (e.g., 4–10).	Lay panel with independent facilitator questions expert witnesses chosen by stakeholder panel. Meetings not generally open. Conclusions on key questions made via report or press conference.
Citizen/ public advisory committee	Small group selected by sponsor to represent views of various groups or communities (may not comprise members of the true public).	Takes place over an extended period of time.	Group convened by sponsor to examine some significant issue. Interaction with industry representatives.
Focus groups	Small group of 5–12 selected to be representative of public; several groups may be used for one project (comprising members of subgroups).	Single meeting, usually up to 2 hours.	Free discussion on general topic with video/tape recording and little input/ direction from the facilitator. Used to assess opinions/ attitudes.

Source: From Rowe and Frewer (2000).

128

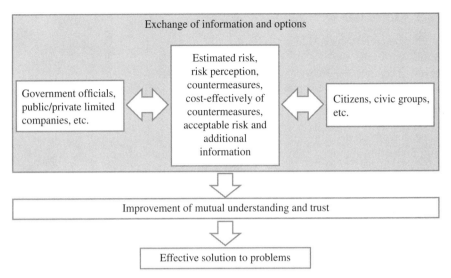

FIGURE 3.6 Objectives of risk communication. *Source*: Adapted from Ishizaka and Tanaka (2003)

stakeholders, and the information shared between stakeholders openly and readily (Webler and Tuler, 2006).

Risk communication, an exchange of information procedure, has been applied to site waste management infrastructures with the primary goal to assist participants to make them cooperate with the decision-making process (O'Leary et al., 1995). The purpose of risk communication is to: (1) include suitable elements in such a way that everyone involved in plan fulfillment knows what is expected to be done, and when; (2) publicize enough elements to allow budget and staff elaborations and to schedule estimates; (3) permit management or policy boards in an agency to evaluate the activities planned capacity in relation to the expected public interest; and (4) clearly divulgate how and when the public will participate (USEPA, 1990; O'Leary et al., 1995). A specific procedure for risk communication was developed by the National Research Council in the United States consisting of an interactive process of exchange of information and opinion among individuals, groups, and institutions (USNRC, 1989) (Figure 3.6).

To aid in risk communication, monitoring committees composed of staff from the SWM system, municipal representatives, experts, and NGO representatives can be formed to analyze monitoring data from problematic waste management facilities. The procedure consists of analyzing monitoring reports delivered to waste management facilities as inputs on physical and chemical properties, air emissions, ashes and slags, and public health control conducted for the population living near the facilities. This type of public health control can be deemed a countermeasure offered during the decision-making process, which could be similar to other possible countermeasures, including monetary ones.

3.3.2 Public Behavior and Participation

Public behavior mentioned in this section is tied to the operational stage of the SWM. In this phase, the SWM plan is often geared toward simultaneously implementing recycling schemes, source separation schemes, and waste prevention programs. The behavior of citizens, such as how they cooperate with these schemes and programs, is a social concern to be considered. To understand public behavior associated with SWM activities (e.g., recycling, waste prevention, reuse), several theories of behavioral studies have been developed over the last 50 years that explain consumers' behavior and provide insights for SWM planning. These well-known theories include theory of reasoned action (TRA), theory of planned behavior (TPB), value-belief-norm (VBN), information–motivation–behavioral skills (IMB) and infrastructure–service–behavior (ISB).

TRA suggests that behavioral intentions depend on the person's attitude with respect to a particular attitude toward performing the behavior and to subjective norms (Fishbein and Ajzen, 1975; Ajzen and Fishbein, 1980; Bortoleto et al., 2012). Thus, if a person evaluates a suggested behavior positively, and if they think their peers wish them to display that behavior, this will result in a higher degree of intention, and the behavior will be displayed (Bortoleto et al., 2012). TPB, proposed by Ajzen (1988, 1991), is an extension of TRA, being a more comprehensive version, which includes measures perceived behavioral control (Armitage and Conner, 2001).

The VBN model proposes that pro-environmental behavior is based on a causal chain of representation variables, where personal norm acts directly on behavior (Stern, 2000; Aguilar-Luzón et al., 2012). VBN describes the impact of personal values in defining a personal norm, or felt obligation, to act in an environment-friendly way (Clayton and Myers, 2009). Personal norms are activated through two other variables: the awareness of the consequences of an action; and the ascription of responsibility, understood as the degree of responsibility that a person assumes over his or her acts concerning the environment, in this case (Stern, 2000; Aguilar-Luzón et al., 2012). Schwartz's altruistic model (Schwartz, 1977) suggests that altruistic behavior occurs when individuals are aware of negative consequences of social conditions for others and attribute responsibility to launch preventive or amendment action to themselves (Guagnano, 2001). In Schwartz's model, personal and social norms affect behavior only when awareness of consequences and denial of responsibilities are activated (Bortoleto et al., 2012).

The IMB model (Fisher and Fisher, 1992) has been extensively utilized to recognize, foresee, and finally modify a diversity of individual social and health behaviors (Seacat and Northrup, 2010). Likewise, the core components of information, motivations, behavior skills, and their inter-relationships are explored to predict waste management behaviors. In this model, information is a prerequisite to conducting correct behavior, and is expected that interventions based on such models are likely to be more effective for producing the desired effect than knowledge-based campaigns alone (Khan et al., 2011). Fisher et al. (2006) noted that individuals may have accurate information that facilitates completion of an intended behavior, and inaccurate information that blocks or serves to deny the impact of an intended behavior (Seacat and Northrup, 2010). The three components can be described as follows (Berlant and Pruitt, 2003; Seacat and Northrup, 2010): information consists of the basic knowledge about a condition and effective strategies for managing it; motivation encompasses personal attitudes toward the behavior, perceived social support for such behavior, and the opinions or perceptions of how others within the same situation might behave; behavior skills include ensuring that the person has the specific behavior tools or strategies necessary to perform the adherence behavior. Motivation can be individual, which is based upon the belief that one can successfully engage in a desired health behavior and that the outcome of engaging in this behavior will be beneficial to the self, or social, which is based upon individuals' perceptions of social norms as well as social support for engaging in a desired behavior (Seacat and Northrup, 2010).

Finally, the ISB model attempts to recognize the range of, and interaction between, individual elements of the system. This behavior model has been developed specifically for assessing recycling behavior in SWM. In this model, motivators and barriers are divided into situational and psychological factors. In the ISB model, infrastructure, service, and behavior are defined as follows (Timlett and Williams, 2011):

- **Infrastructure:** It includes the built environment, products, and objects (e.g., buildings, bins, collection vehicles, civic amenity sites, waste composition, packaging, material recovery facilities, incinerators, landfill, recycling reprocessing facilities and technologies).
- **Service:** It includes the systems, providers, and enablers that allow people to participate in a particular environmental practice (e.g., collections (frequency, method), role of crews, communication materials, perception of customer service and service provider, economic incentives, penalties, markets for reusable and recyclable materials).
- **Behavior:** It relates to the people and disposition toward the environmental practice (e.g., values, attitudes, knowledge, awareness, personalities, lifestyles, communities, social status and norms)".

The applications of public behavior models and theories for SWM have been used to identify the most relevant aspects and factors for changing behavior. Several examples of their applications for SWM exist, but recycling schemes and waste prevention programs (Table 3.11) are the ones for which public behavior theories have been applied.

Recycling schemes are determinants to ensure a bigger landfill life span, increase revenues from SWM, increase the lifetime of materials in the use phase, and contribute to an eco-friendly image of the SWM system. For sound SWM practices, such as a successful recycling program, it is crucial to justify the investment in source separation at collection points and in a dedicated SWM facility. Several studies indicate that recycling behavior is affected by recycling cost, recycling infrastructure and programs availability, environment-related awareness and knowledge, attitudes

TABLE 3.11 **Examples of public behavior theories applied to assess various recycling schemes**

Public behavior theory	Application	References
TPB	To analyze public's view and influencing factors toward participation in source separation of food waste.	Karim Ghani et al. (2013)
VNB	To assess Spanish housewives' recycling behavior.	Aguilar-Luzón et al. (2012)
IMB	To explain curbside recycling behavior in the United States.	Seacat and Northrup (2010)
ISB	To recognize the range of, and interaction between, individual elements of the recycling system, being the case study the city of Portsmouth (UK).	Timlett and Williams (2011)

promoting recycling, social norms and external pressures, and household socioeconomic status (Vining and Ebreo, 1990; Oskamp et al., 1991; Ebreo and Vining, 2001). TPB is recognized as an applicable behavioral model that exhibits a greater degree of fitness and a greater capacity of predicting recycling behavior relative to other methods such as VNB (Aguilar-Luzón et al., 2012).

Many agencies are keen to promote waste prevention due to legal obligations and a strong need to extend landfill life-time, concerns that constitute a great advantage, especially in countries where landfill space can be quite expensive or significantly taxed, like Singapore, Japan, Denmark, and the Netherlands. Factors affecting waste prevention behaviors are people's purchasing behaviors (which is very difficult to change), income per capita, cost of waste collection, collection frequency, and separate curbside collection of organic waste (Gellynck et al., 2011; Kurisu and Bortoleto, 2011). In addition, some countries have initialized zero waste policies that trigger more efficient waste collection, waste prevention, recycling, reuse, and recovery have been promoted by many national and international institutions. Whatever the motivation, including legal obligations, ensuring a waste prevention behavior is a long-term behavior, and obtaining fast results can be difficult.

Tools and techniques are always available to promote public participation. Communication must be developed between the SWM sector and the community to ensure that residents' habits, behaviors, and traditions can be factored into the SWM programs, which in turn enable responsible authorities to achieve local, regional, and national goals of waste prevention and recycling. The diversity of communication strategies ensure that information can be provided in several forms, including mass media, in writing, or verbally (by telephone or face-to-face) (Bernstad et al., 2013). Possible communication vehicles are campaigns, mailings, print advertisements, seminars, education, television/radio, mediation, opinion leaders, teachers, and artists. However, a frequent source of misunderstanding is the different level of technical knowledge and cultural background between the sender and the recipient of the information (Stiglitz, 2005). Defining and tailoring each target group and selecting the best communication vehicle for each target group become critical. Target groups can be industry, unions, citizen groups, NGO, consumers, local government, and political parties, among others. Oral information delivered by informants in "doorstepping" (face-to-face) campaigns has, according to previous studies, led to increased material recycling ranges in areas with curbside recycling (Read, 1999; Timlett and Williams, 2008). Other types of media include newspaper, radio, and television campaigns, signs on transportation vehicles, council magazines and newsletters, bin stickers, fridge magnets, roadshows, displays, posters, talks to schools and other community groups, and websites, with different impact degrees, mostly chosen by the budget, previous experiences and expert opinions (Timlett and Williams, 2008).

Local authorities depend on a diversity of media to communicate the recycling message. Face-to-face approaches have been found to be effective for modifying public behavior; leaflets distributed door-to-door and/or at community points (such as libraries and medical offices) are the medium most commonly well received by the audience (Mee et al., 2004), although the information can be dismissed as junk mail (Read, 1999). In the United States, doorstepping, which has become common has information provider to promote recycling, was found to boost voting turnout by 10–15%, compared to 2.5% for leaflets, and no change with telephone canvassing (Gerber and Green, 2000). In a London borough, doorstepping was attributed with a 2% increase in the recycling rate, even though only 12% of households were investigated (Read, 1999). A campaign in Devon mentioned a 10% increase in household recycling frequency and a 20% increase in the quantity of recyclables collected (Read et al., 2005). Doorstepping campaigns need to have clear and specific aims, with target areas provided with sufficient recycling infrastructure and supported with appropriate communications materials (Read et al., 2005). Nevertheless, research in the literature that measures the effectiveness of communication methods holistically is limited (Timlett and Williams, 2008).

3.4 FINAL REMARKS

This chapter provides an integrative discussion on service provisions for urban SWM, including waste collection, transfer, recycling, and resource recovery. This is a fundamental step to gain awareness for the development of more sustainable plans and illustrate the public concern and economic sensitivity to SWM. The discussion in this chapter includes key elements of institutional and financial arrangements, economic instruments, public acceptance, privatization, public–private partnerships, and possible linkages with regulatory frameworks, and technology choices of SWM. SWM with perspectives of economic incentives, policy context, social implications and socioeconomic assessment processes associated with EIA and SEA were articulated within a highly multidisciplinary framework. Chapters 4–7 will explore efforts that drive a series of expanded viewpoints of sustainable SWM that are vital to understanding additional socioeconomic issues surrounding MSW management.

REFERENCES

Aguilar-Luzón, M. C., García-Martiínez, J. M. A., Calvo-Salguero, A., and Salinas, J. M. 2012. Comparative study between the theory of planned behavior and the value–belief–norm model regarding the environment, on Spanish housewives' recycling behavior. *Journal of Applied Social Psychology*, 42(11), 2797–2833.

Ahmed, S. A. and Ali, M. 2004. Partnerships for solid waste management in developing countries: linking theories to realities. *Habitat International*, 28(3), 467–479.

Ahmed, S. U. and Gotoh, K. 2006. *Cost-Benefit Analysis of Environmental Goods by Applying the Contingent Valuation*, Springer-Verlag, Tokyo, Japan.

Ajzen, I. 1988. *Atitudes, Personality and Behavior*, Open University Press, Milton Keynes.

Ajzen, I. 1991. The theory of planned behaviour. *Organizational Behavior and Human Decision Processes*, 50(2), 179–211.

Ajzen, I. and Fishbein, M. 1980. *Understanding Attitude and Predicting Social Behavior*, Prentice-Hall, Inc., Englewood Cliffs, NJ.

Allen Kani Associates (AKA) and Enviros RIS. 2001. WDO study: implications of different waste feed streams (source-separated organics and mixed waste) on collection options and anaerobic digestion processing facility design, equipment and costs. Report for the City of Toronto. Available at: http://www.nerc.org/documents/toronto_report.pdf (accessed December 2013).

Antonioli, B. and Filippini, M. 2002. Optimal size in the waste collection sector. *Review of Industrial Organization*, 20(3), 239–252.

Apotheker, S. 1993. It's black and white, and recycled all over. *Resources, Conservation and Recycling*, 12(7), 36–44.

Armitage, C. J. and Conner, M. 2001. Efficacy of the theory of planned behaviour: a meta-analytic review. *British Journal of Social Psychology*, 40(4), 471–499.

Bailey, I., Haug, B., and O' Doherty, R. 2004. Tradable permits without legislative targets: a review of the potential for a permit scheme for sterilized clinical waste in the UK. *Waste Management & Research*, 22(3), 202–211.

Barde, J.-P. 1997. Environmental taxation: experience in OECD countries. In: *Ecotaxation* (Ed. O'Riordan, T.), Earthscan, London, pp. 223–245.

Batllevell, M. and Hanf, K. 2008. The fairness of PAYT systems: some guidelines for decision-makers. *Waste Management*, 28(12), 2793–2800.

Beck, R. W. 2005. Carroll County, Maryland—Solid waste management options. Report for the Northeast Maryland Waste Disposal Authority.

Begum, R. A., Siwar, C., Pereira, J. J., and Jaafar, A. H. 2007. Factors and values of willingness to pay for improved construction waste management—a perspective of Malaysian contractors. *Waste Management*, 27(12), 1902–1909.

Bel, G. and Warner, M. 2008. Does privatization of solid waste and water services reduce costs? A review of empirical studies. *Resources, Conservation and Recycling*, 52(12), 1337–1348.

Berlant, N. E. and Pruitt, S. D. 2003. Adherence to medical recommendations. In: *The Health Psychology Handbook: Practical Issues for the Behavioral Medicine Specialist* (Eds Cohen, L. M., McChargue, D. E., and Collins, F. L. Jr.), Sage Publications Inc., pp. 208–224.

Bernstad, A., Cour, J., and Aspegren, A. 2013. Door-stepping as a strategy for improved food waste recycling behaviour—evaluation of a full-scale experiment. *Resources, Conservation and Recycling*, 73, 94–103.

Bortoleto, A. P., Kurisu, K. H., and Hanaki, K. 2012. Model development for household waste prevention behaviour. *Waste Management*, 32(12), 2195–2207.

Breidert, C., Hahsler, M., and Reutterer, T. 2006. A review of methods for measuring willingness to pay. *Innovative Marketing*, 2(4), 8–32.

Callan, S. J. and Thomas, J. M. 2007. *Environmental Economics & Management—Theory, Policy, and Applications*, 4th edition, Cengage Learning, Louiseville, Canada.

Canter, L. W. 1996. *Environmental Impact Assessment*, 2nd edition, McGraw Hill, New York.

Carson, R. T. 2000. Contingent valuation: a user's guide. *Environmental Science & Technology*, 34(8), 1413–1418.

Clayton, S. and Myers, G. 2009. *Conservation Psychology: Understanding and Promoting Human Care for Nature*, Wiley-Blackwell, West Sussex, UK.

de Gioannis, G., Muntoni, A., and Raga, R. 2006. Effects of separate collection efficiency on the costs of the MSW management system—a case study. In: Proceedings Venice 2006, Biomass and Waste to Energy Symposium, Venice, Italy.

Diaz, L. F., Savage, G. M., and Eggerth, L. L. 2005. *Solid Waste Management*, UNEP, Paris.

Dijkgraaf, E. and Gradus, R. H. J. M. 2004. Cost savings in unit-based pricing of household waste: the case of the Netherlands. *Resource and Energy Economics*, 26(4), 353–371.

Dulac, N. 2001. The organic waste flow in integrated sustainable waste management. In: *Integrated Sustainable Waste Management, A Set of Five Tools for Decision-makers, Experiences from the Urban Waste Expertise Programme (1995– 2001)* (Ed. Scheinberg, A.), WASTE, Advisers on Urban Environment and Development, Gouda, the Netherlands.

Ebreo, A. and Vining, J. 2001. How similar are recycling and waste reduction? Future orientation and reasons for reducing waste as predictors of self-reported behavior. *Environment and Behavior*, 33(3), 424–448.

Ecotec. 2000. Policy instruments to correct market failure in the demand for secondary materials. Report prepared by Ecotec Research and Consulting Ltd., Birmingham, UK.

Enviros. 2003. Survey of applications, markets and growth opportunities for recycled plastics in the United Kingdom. WRAP Research Report, August.

Eshet, T., Ayalon, O., and Shechter, M. 2006. Valuation of externalities of selected waste management alternatives: a comparative review and analysis. *Resources, Conservation and Recycling*, 46(4), 335–364.

European Commission. 1998. The competitiveness of the recycling industries. Report prepared for the European Commission, Directorate General for Industry, Brussels, Belgium.

European Commission. 2001. Costs of municipal waste management in the EU. Final report to Directorate General Environment, European Commission, Brussels, Belgium.

Eurostat. 2011. Waste statistics. Eurostat. Available at: http://epp.eurostat.ec.europa.eu (accessed June 2013).

Fishbein, M. A. and Ajzen, I. 1975. *Belief, Attitude, Intention and Behavior: An Introduction to Theory and Research*. Addison-Wesley, Reading, MA.

Fischer, T. B. 2007. *The Theory and Practice of Strategic Environmental Assessment: Towards a More Systematic Approach*. Earthscan, Padstow, UK.

Fisher, J. D. and Fisher, W. A. 1992. Changing AIDS risk behavior. *Psychological Bulletin*, 111, 455–474.

Fisher, J. D., Fisher, W. A., Amico, K. R., and Harman, J. J. 2006. An information–motivation–behavioral skills model of adherence to antiretroviral therapy. *Health Psychology*, 25(4), 462–473.

Furedy, C. 1992. Garbage: exploring non-conventional options in Asian cities. *Environment and Urbanization*, 4(2), 43–61.

Gavaskar, A. and Cumming, L. 2001. *Cost Evaluation Strategies for Technologies Tested under the Environmental Technology Verification Program*, USEPA, Cincinnati, OH.

Gellynck, X., Jacobsen, R., and Verhelst, P. 2011. Identifying the key factors in increasing recycling and reducing residual household waste: a case study of the Flemish region of Belgium. *Journal of Environmental Management*, 92(10), 2683–2690.

Gerber, A. S. and Green, D. P. 2000. The effects of canvassing, telephone calls and direct mail on voter turnout: a field experiment. *American Political Science Review*, 94(3), 653–663.

Guagnano, G. A. 2001. Altruism and market-like behavior: an analysis of willingness to pay for recycled paper products. *Population & Environment: A Journal of Interdisciplinary Studies*, 22(4), 425–438.

Hadker, N., Sharma, S., David, A., and Muraleedharan, T. R. 1997. Willingness to pay for Borivli National Park: evidence from a contingent valuation. *Ecological Economics*, 21(2), 105–122.

Han, H., Hsu, L.-T., and Sheu, C. 2010. Application of the Theory of Planned Behavior to green hotel choice: testing the effect of environmental friendly activities. *Tourism Management*, 31(3), 325–334.

Hogg, D., Favoino, E., Nielsen, N., Thompson, J., Wood, K., Penschke, A., Economides, D., and Papageorgiou, S. 2000. Economic analysis of options for managing biodegradable municipal waste. Final report to the European Commission. Bristol, UK.

Howard, A. 1943. *Agricultural Testament*, Oxford University Press Inc., London, UK.

Institute for Prospective Technological Studies (IPTS). 2011. *Technical Report for End-of-Waste Criteria on Biodegradable Waste Subject to Biological Treatment—First Working Document*. Sevilla, Spain.

International Council on Metals and the Environment (ICME). 1996. *Non-Ferrous Metals Recycling: A Complement to Primary Metals Production*, ICME, Ottawa, ON.

Ishizaka, K. and Tanaka, M. 2003. Resolving public conflict in site selection process—a risk communication approach. *Waste Management*, 23(5), 385–396.

Jacobsen, M. 2005. Project costing and financing. In: *Water and Wastewater Management in the Tropics* (Ed. Lønholdt, J.), IWA Publishing, Cornwall, UK, pp. 51–120.

Jefrey, G. 1996. How much privatization: a research note examining the use of privatization by cities in 1982 and 1992. *Policy Studies Journal*, 24, 632–640.

Johnson, R. J. and Scicchitano, M. J. 2012. Don't call me NIMBY: public attitudes towards solid waste facilities. *Environment and Behavior*, 44(3), 410–426.

Jones, N., Evangelinos, K., Halvadakis, C. P., Iosifides, T., and Sophoulis, C. M. 2010. Social factors influencing perceptions and willingness to pay for a market-based policy aiming on solid waste management. *Resources, Conservation and Recycling*, 54(9), 533–540.

Karim Ghani, W. A., Rusli, I. F., Biak, D. R., and Idris, A. 2013. An application of the theory of planned behaviour to study the influencing factors of participation in source separation of food waste. *Waste Management*, 33(5), 1276–1281.

Khan, B., Alghathbar, K. S., and Khan, M. K. 2011. Information security awareness campaign: an alternative approach. In: *Proceedings of the International Conference of Information Security and Assurance* (Eds Kim, T.-H., Adeli, H., Robles, R. J., and Balitanas, M.), Springer, Berlin, Heidelberg, pp. 1–10.

Kurisu, K. H. and Bortoleto, A. P. 2011. Comparison of waste prevention behaviors among three Japanese megacity regions in the context of local measures and socio-demographics. *Waste Management*, 31(7), 1441–1449.

Lasher, W. R. 2008. *Practical Financial Management*, 5th edition, Thomson South-Western, Willard, OH.

Le Grand, J. 1982. *The Strategy of Equality: Redistribution and the Social Services*, Allen & Unwin, London, UK.

Lombrano, A. 2009. Cost efficiency in the management of solid urban waste. *Resources, Conservation and Recycling*, 53(11), 601–611.

Markandya, A., Perelet, R., Mason, P., Taylor, T. 2001. *Dictionary of Environmental Economics*, Earthscan Publications, London.

Massoud, M. and El-Fadel, M. 2002. Public–private partnerships for solid waste management services. *Environmental Management*, 30(5), 621–630.

Mee, N., Clewes, D., Phillips, P., and Read, A. 2004. Effective implementation of a marketing communications strategy for kerbside recycling: a case study from Rushcliffe, UK. *Resources, Conservation and Recycling*, 42(1), 1–26.

Merrild, H. and Christensen 2011. Introduction to waste economics. In: *Solid Waste Technology and Management* (Ed. Christensen, T. H.), Wiley, Chichester, UK, pp. 29–51.

Mikesell, J. L. 2007. *Fiscal Administration: Analysis and Applications in the Public Sector*, 7th edition, Thomson/Wadsworth, Belmont, CA.

Nicolli, F., Johnstone, N., and Söderholm, P. 2012. Resolving failures in recycling markets: the role of technological innovation. *Environmental Economics and Policy Studies*, 14(3), 261–288.

O'Faircheallaigh, C. 2010. Public participation and environmental impact assessment: purposes, implications, and lessons for public policy making. *Environmental Impact Assessment Review*, 30(1), 19–27.

Ogilvie, J. 2008. *CIMA Official Learning System Management Accounting Financial Strategy Paper P9*, Elsevier, Noida, Italy.

O'Leary, P. R., Walsh, P. W., Ham, R. K., Gruder, S. G., Kohrell, M. G., Johnson, H. J., Pferdehirt, W., and Razvi, A. S. 1995. *Decision-Maker's Guide to Solid Waste Management*, 2nd edition, Environmental Protection Agency, Washington, DC.

Organisation for Economic Co-operation and Development (OECD) 1993. *Applying Economic Instruments to Packaging Waste: Practical Issues for Product Charges and Deposit-Refund Systems*. Environment Monograph n.° 82, Paris.

Organisation for Economic Co-operation and Development (OECD) 2001. *Extended Producer Responsibility: A Guidance Manual for Governments*, OECD Publications, Paris, France.

Organisation for Economic Co-operation and Development (OECD) 2002. *Handbook of Biodiversity Valuation—A Guide for Policy Makers*, OECD Publications, Paris, France.

O'Riordan, T. 2001. *Environmental and Ecological Economics in Environmental Science for Environmental Management*, Prentice Hall Publishers, Upper Saddle River, NJ.

Oskamp, S., Harrington, M. J., Edwards, T. C., Sherwood, D. L., Okuda, S. M., and Swanson, D. C. 1991. Factors influencing household recycling behavior. *Environment and Behavior*, 23(4), 494–519.

Padilla, E. 2002. Intergenerational equity and sustainability. *Ecological Economics*, 41(1), 69–83.

Palmer, K. and Walls, M. 1999. *Extended Product Responsibility: An Economic Assessment of Alternative policies*, Discussion Paper 99–12, Resources for the Future, Washington, DC.

Parkin, M., Powell, M., and Matthews, K. 2008. *Economics*, 7th edition, Pearson Education, Spain.

Pearce, D. W. and Howarth, A. 2000. Technical Report on Methodology: Cost–Benefit Analysis and Policy Responses. Report 481505020 for RIVM, EFTEC, NTUA and IIASA. Available at: http://ec.europa.eu/environment/enveco/priority_study/pdf/methodology.pdf (accessed December 2013).

Petts, J. 1994. Incineration as a waste management option. In: *Waste Incineration and the Environment* (Eds Hester, R. R. and Harrison, R. M.), Royal Society of Chemistry, Cambridge, UK.

Pigou, A. C. 1920. *The Economics of Welfare*, Macmillan, New York.

Pimenteira, C. A., Carpio, L. G., Rosa, L. P., and Tolmansquim, M. T. 2005. Solid wastes integrated management in Rio de Janeiro: input-output analysis. *Waste Management*, 25(5), 539–553.

Porter, R. 2002. *The Economics of Waste*, Resources for the Future, Washington, DC.

Porter, R. C. 2004. Efficient targetting of waste policies in the product chain. In: *Addressing the Economics of Waste* (Ed. Braathen, N. A.), Organisation for Economic Co-operation and Development (OECD), Paris, pp. 117–160.

Ramanadham, V. V. 1991. *The Economics of Public Enterprise*, Routledge, London, UK.

Rawls, J. 1971/1999. *A Theory of Justice*, revised edition, The Belknap Press of Harvard University Press, Cambridge, MA.

Read, A. 1999. A weekly doorstep recycling collection, "I had no idea we could!" Overcoming the barriers to participation. *Resources, Conservation and Recycling*, 26(3–4), 217–249.

Read, A., Harder, M., and Coates, A. 2005. UK best practice in doorstep recycling promotions campaigns. In: Proceedings Sardinia 2005, Tenth International Waste Management and Landfill Symposium, S. Margherita di Pula, Cagliari, Italy.

Rowe, G. and Frewer, L. J. 2000. Public participation methods: a framework for evaluation. *Science, Technology and Human Values*, 25(1), 3–29.

Salahuddin, K. and Shamim, I. 1992. *Women in Urban Informal Sector: Employment Pattern, Activity Types and Problems*, Women for Women: A Research and Study Group, Dhaka, Bangladesh.

Salhofer, S., Wassermann, G., and Binner, E. 2007. Strategic environmental assessment as an approach to assess waste management systems. Experiences from an Austrian case study. *Environmental Modelling & Software*, 22(5), 610–618.

Savas, E. S. 1977. An empirical study of competition in municipal service delivery. *Public Administration Review*, 37(6), 717–724.

Schwartz, S. H. 1977. Normative influences on altruism. In: *Advances in Experimental Social Psychology* (Ed. Berkowitz, L.), Academic Press, New York, pp. 221–279.

Seacat, J. D. and Northrup, D. 2010. An information–motivation–behavioral skills assessment of curbside recycling behavior. *Journal of Environmental Psychology*, 30(4), 393–401.

Siegel, J. G. and Shim, J. K. 2006. *Accounting Handbook*, 4th edition, Barron's Educational Series, New York.

Stern, P. C. 2000. Towards a coherent theory of environmentally significant behavior. *Journal of Social Issues*, 56(3), 407–424.

Sterner, T., Wahlberg, H., Bartelings, H., Belhaj, M., and Fahlberg, A. K. 1997. Waste management and recycling, AFR report 171. Swedish Environmental Protection Agency, Stockholm.

Stiglitz, C. 2005. The achievement of consensus - communication in waste management. In: Proceedings Sardinia 2005, Tenth International Waste Management and Landfill Symposium, S. Margherita di Pula, Cagliari, Italy.

Taebi, B. 2010. Nuclear power and justice between generations: A moral analysis of fuel cycles. Thesis to obtain the degree of Doctor at the Technical University of Delft, Centre for Ethics and Technology.

Taylor, D. C. 2000. Policy incentives to minimize generation of municipal solid waste. *Waste Management & Research*, 18(5), 406–419.

Taylor, J. B. and Weerapana, A. 2012. *Principles of Economics*, 7th edition. South-Western Cengage Learning, Mason, OH.

Tellus Institute 1992. *Tellus Packaging Study – Assessing the Impacts of Production and Disposal of Packaging and Public Policy Measures*, vol. 1, Tellus Institute.

Timlett, R. E. and Williams, I. D. 2008. Public participation and recycling performance in England: a comparison of tools for behaviour change. *Resources, Conservation and Recycling*, 52(4), 622–634.

Timlett, R. and Williams, I. D. 2011. The ISB model (infrastructure, service, behaviour): a tool for waste practitioners. *Waste Management*, 31(6), 1381–1392.

Tirado-Soto, M. M. and Zamberlan, F. L. 2013. Networks, of recyclable material waste-picker's cooperatives: an alternative for the solid waste management in the city of Rio de Janeiro. *Waste Management*, 33(4), 1004–1012.

Tsilemou, K. and Panagiotakopoulos, D. 2006. Approximate cost functions for solid waste treatment facilities. *Waste Management & Research*, 24(4), 310–322.

United Nations Centre for Regional Development (UNCRD). 2011. Market demand as driving force for 3R technology transfer and the role of private sector (Background paper for plenary session 3 of the provisional programme). Third Meeting of the Regional 3R Forum in Asia, 5–7 October, Singapore. Available at: http://www.uncrd.or.jp/content/documents/03_FINAL_Background_Paper_THREE_27Sept2011.pdf (accessed December 2013).

United Nations Development Programme, World Bank, Regional Water and Sanitation Group for South Asia (UNEP/WB/RWSGSA). 1991. Indian experience on composting as means of resource recovery. In: Proceedings of the UNDP/WB Water Supply and Sanitation Program Workshop on Waste Management Policies, Singapore.

United States Environmental Protection Agency (USEPA). 1990. *Sites for Our Solid Waste: A Guidebook for Effective Public Involvement*, Washington, DC.

United States Environmental Protection Agency (USEPA). 1998. *Cost-Effective Environmental Management Case Study: Contract Operations of the Belmont and Southport Advanced Wastewater Treatment Facilities*, EPA Environmental Financial Advisory Board, Indianapolis, IN.

United States National Research Council (USNRC). 1989. *Improving Risk Communication*, National Academy Press, Washington, DC.

URS. 2010. Final report—Supplementary report—Economic modelling of options for waste infrastructure in the ACT. Report prepared for DECCEW. Available at: http://www.environment.act.gov.au/_data/assets/pdf_file/0003/576921/URS_Supplementary-Report-economic-modelling.pdf (accessed December 2013).

van de Walle, N. 1989. Privatization in developing countries: a review of the issues. *World Development*, 17(5), 601–615.

van Haaren, R. 2009. Large scale aerobic composting of source separated organic wastes: a comparative study of environmental impacts costs, and contextual effects. Thesis for the partial fulfillment of requirements for MSc. Degree in Earth Resources Engineering, Columbia University.

Vasconcelos, L., Marques, M. J., and Martinho, G. 2005. Public participation in waste management – overcoming ingrained myths. In: Proceedings Sardinia 2005, Tenth International Waste Management and Landfill Symposium, S. Margherita di Pula, Cagliari, Italy.

Vining, J. and Ebreo, A. 1990. What makes a recycler? A comparison of recyclers and nonrecyclers. *Environment and Behavior*, 22(1), 55–73.

Walters, A. 1989. Liberalization and privatization: an overview. In: *Privatization and Structural Adjustment in the Arab Countries* (Ed. El-Naggar, S.), Abu Dhabi, United Arab Emirates, pp. 18–49.

Watkins, E., Hogg, D., Mitsios, A., Mudgal, S., Neubauer, A., Reisinger, H., Troeltzsch, J., and van Acoleyen, M. 2012. Use of economic instruments and waste management performances. Final report for the European Commission DG-Env. Available at: http://ec.europa.eu/environment/waste/pdf/final_report_10042012.pdf (accessed December 2013).

Webler, T. and Tuler, S. 2006. Four perspectives on public participation process in environmental assessment and decision making: combined results from 10 case studies. *Policy Studies Journal*, 34(4), 699–722.

Wilkinson, N. 2005. *Managerial Economics: A Problem-Solving Approach*, Cambridge University Press, New York.

Scott Wilson, S. 2002. *Plastic Bottle Recycling in the UK*, Waste and Resources Action Programme (WRAP), UK.

Wilson, D. C., Velis, C., and Cheeseman, C. 2006. Role of informal sector recycling in waste management in developing countries. *Habitat International*, 30(4), 797–808.

World Commission on Environment and Development (WCED). 1987. *Our Common Future* (Ed. Brundtland, G. H.), Oxford University Press, Oxford, UK.

Zurbrügg, C., Drescher, S., Patel, A., and Sharatchandra, H. C. 2004. Decentralised composting of urban waste—an overview of community and private initiatives in Indian cities. *Waste Management*, 24(7), 655–662.

CHAPTER 4

LEGAL AND INSTITUTIONAL CONCERNS

Regulatory and policy concerns are an integral part of sustainable solid waste management (SWM) programs. Legislation for SWM is devoted to protecting the environment and public health from detrimental impacts due to waste management practices, which dictates the way how solid waste should be managed and disposed. The first challenge in waste legislation is to propose appropriate definitions of waste including raw waste, recycled waste, reused waste, and recovered waste. The consequences of setting up such definitions point the responsibility for relevant owners or operators who might have liability when carrying out their work. Following this direction, waste legislation is also devoted to defining all stakeholders involved in all operational units for SWM. To ensure the required level of environmental protection, waste legislation can also define the best available control technologies and monitoring methods to implement for improving societal benefit. From economic perspectives, waste legislation can address the fees to be paid by the residents seeking waste management services and how to charge them. In this chapter, we will approach both international and national policies and laws that intend to minimize the negative impacts and promote sustainable SWM, explain the principles applied for SWM, and delineate the real-world economic instruments that have been implemented.

4.1 SWM LEGISLATION

The challenge of developing and implementing waste legislation depends on the legal status in a country. In developing countries, waste legislation is rare; when developing

Sustainable Solid Waste Management: A Systems Engineering Approach, First Edition. Ni-Bin Chang and Ana Pires.
© 2015 The Institute of Electrical and Electronics Engineers, Inc. Published 2015 by John Wiley & Sons, Inc.

specific policy instruments, proper integration with existing legislation is vital for the success and acceptance of the program. In developed countries, where SWM legislation is abundant, the integration has been implemented, and the mechanisms to identify conflicts and incongruence have already been applied.

Legislation has inherently been regarded as the notion of justice and fairness. Legal measures normally have specific arrangements, including geographic scope (i.e., local, regional, national, international, and global), a time window (i.e., legal measures often have an expiration date and must be revised or even replaced by other legal measures after expiration), technical coverage (i.e., they can be only applicable for a specific waste treatment technology), and a relevant duty and right (i.e., they sometimes are only applied to a specific stakeholder). These features help to coordinate legislation at different scales, which can significantly improve SWM.

4.1.1 International Solid Waste Management

Although waste management issues often become transboundary issues, there is no legal framework at the global scale for SWM. Some regions have defined SWM strategies to apply, such as in Antarctica (through Antarctic Treaties), the European Union (EU) Member States (MS) (through Directives; see Box 4.1), and the Organisation for Economic Co-operation and Development (OECD) member countries (through some Council Acts). According to Sands et al. (2012), international measures are related to disposal of wastes at sea, atmospheric emissions, and disposal of wastes in rivers and other freshwaters, although waste generation, which is the source of the problem, is not addressed.

BOX 4.1 EU WASTE AND WASTE STREAMS DIRECTIVES (Chang et al., 2013)

With the onset of the EU, all MS had to comply with common legislation aimed to promote homogeneous conditions across the region. Article 249 European Commission (EC) (ex Art.189) lists the various policy instruments that the EU may use to implement policies included in the European Community Treaty (Hedemann-Robinson, 2007). These policy instruments include but are not limited to directives, regulation, and decisions. Directives are binding on MS; however, each MS may choose how it implements the directive. Regulation is directly applicable to all MS, and a decision is defined as being "binding in its entirety upon those to whom it is addressed" (Hedemann-Robinson, 2007). Strategies defined as the main guidelines are only deemed as EU opinion for future waste management practices. Overall, the EC has mostly promoted directives to enforce the law in MS for SWM legislation (Figure 4.1).

In addition to Regulation on Shipment of Waste, there are considerable regulations and decisions affecting waste management at EU countries. More information can be found at the EU web site (http://europa.eu/legislation_summaries/environment/waste_management/index_en.htm).

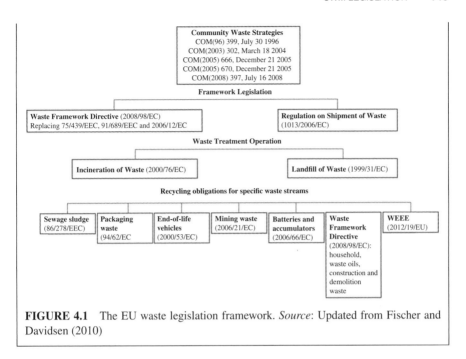

FIGURE 4.1 The EU waste legislation framework. *Source*: Updated from Fischer and Davidsen (2010)

One of the first serious attempts to establish the basis for a more comprehensive international waste management approach was the 1976 OECD Council Recommendation on a Comprehensive Waste Management Policy C(76)155/Final (Sands et al., 2012). This policy recommended that member countries implement waste policies to protect the environment and ensure rational use of energy and resources while taking into account economic constraints (OECD, 1976). Recommended principles included the need to (OECD, 1976; Sands et al., 2012): (1) account for environmental protection, (2) encourage waste prevention, (3) promote recycling, (4) use policy instruments, and (5) ensure access to information. Policy C(76)155/Final also endorsed administrative arrangements, including (OECD, 1976; Sands et al., 2012): (1) inventorying wastes to be disposed, (2) organizing waste collection, (3) establishing disposal centers, (4) promoting research and development on disposal methods and low-waste technology, and (5) encouraging markets for recycled products.

One of the most important international legislations is the Basel Convention, devoted to the movement of hazardous waste. Many international laws require States to explore licensing potentially harmful activities, such as:

- 1972 Oslo Convention for the Prevention of Marine Pollution by dumping waste from Air Craft and Ships;
- 1972 London Dumping Convention (reviewed in 1996 Prevention of Marine Pollution by Dumping Wastes and Other Matter–Thereto Protocol);
- 1973 Marpol Protocol to the International Convention for the Prevention of Pollution from Ships;

- 1974 Paris Convention for the Prevention of Marine Pollution from Land-Based Sources;
- 1974 United Nation Environmental Programme (UNEP) Regional Seas Programs (Conventions and Action Plans);
- 1989 Basel Convention on Hazardous Waste;
- 1991 Bamako Convention on Hazardous Waste in Africa;
- 1991 Antarctic Treaties (The Protocol on Environmental Protection to the Antarctic Treaty);
- 1992 Oslo/Paris (OSPAR) Convention for the Protection of the Marine Environment of the North-East Atlantic;
- 1995 UNEP Global Program of Action (GPA) for the Protection of the Marine Environment from Land-Based Activities; and
- 2009 Hong Kong International Convention for the Safe and Environmentally Sound Recycling of Ship.

In the OSPAR Convention (which updates the earlier Oslo and Paris Conventions), 15 governments (Belgium, Denmark, Finland, France, Germany, Iceland, Ireland, the Netherlands, Norway, Portugal, Spain, Sweden, the United Kingdom, Luxembourg, and Switzerland) work together with the European Community to protect the marine environment of the North-East Atlantic. According to the OSPAR Commission (2013), the convention started in 1972 with Oslo Convention against dumping and was soon joined by the 1974 Paris Convention to cover land-based sources and offshore industry. The most relevant section for waste management is Annex II, which is related to the prevention and elimination of pollution by dumping or incineration. The main aspects are to ensure that:

- incineration is prohibited; and
- dumping of all wastes or other matter is prohibited (including radioactive waste) with exceptions related to dredged material, inert materials of natural origin, fish waste from fishing industry, and carbon dioxide streams from carbon dioxide capture processes for storage.

The last update of OSPAR Convention occurred in 2010, where other challenges were defined in Annex 25. Waste treatment technologies have been particularly addressed regarding the need for developing and applying the life cycle concept of products. Emphasis is placed on waste management technologies which reduce, or better avoid, the use and discharge, emissions, and losses to the environment of hazardous substances. In addition, waste management hierarchy should be considered when developing measures and programs to implement the OSPAR Convention. Important principles stressed by the OSPAR Convention are the polluter-pays principle and the precautionary principle (to be discussed later on).

The London Dumping Convention has been in effect since 1975. According to the International Maritime Organization (IMO, 2013a), the convention intended to promote the control of all sources of marine pollution and to prevent sea pollution by dumping of wastes and other matter into the sea. The convention is currently signed by 87 States. Similar to the OSPAR Convention, exceptions are included to the prohibition of dumping waste, including organic material of natural origin and bulky items primarily comprising iron, steel, and concrete (Keyuan, 2009). The London Convention published guidelines for the assessment of wastes and other matter considered for dumping, and the *de minimis* levels of radionuclides that could be disposed into the sea.

The Marpol Convention is the principal international convention which is dedicated to the pollution prevention of the marine environment resulting from ships (accidental and/or operational triggered) (IMO, 2013b). Annex V is related to the prevention of pollution by garbage from ships and specifies the distances from land and the materials allowed for disposal. The most important feature of the annex, according to the revised version effective on January 1, 2013, is the complete ban of the disposal of all forms of plastics, municipal solid waste (MSW), vegetable oil, incinerator ashes, operational wastes, and fishing gear into the sea. Other specific waste may or may not be discharged depending on ships' locations, which are classified as special areas, outside specific areas, and offshore platforms.

Regional Seas Programs promoted by the United Nations Environment Programme (UNEP) intend to provide a framework at several aspects like legal, administrative, and financial, for Agenda 21 implementation, and its chapter 17 on oceans in particular (Regional Seas Coordinating Office, 2005). The Regional Seas Programs focus on the protection of oceans and coastal areas through their sustainable use, by including countries connected to the same coastal and oceans areas (UNEP, 2012). Action Plans are elaborated to define the program strategy, based on the challenges of each particular region. To date there are 18 seas with Regional Seas Programs. The most significant contribution to the waste management has been the limiting of marine litter; ship-generated marine pollution (oil, chemicals, litter, invasive species) is a priority in the Regional Seas Programs, and activities promoted by these programs are devoted to studies on the global problem of marine litter, development of regional actions on marine litter, and plans for development.

The purpose of the Basel Convention on the Control of Transboundary Movements of Hazardous Wastes and Their Disposal, which has 180 State parties, is to protect human health and the environment from the generation, management, transboundary movements, and elimination of hazardous and other wastes. Principal aims of the convention are (Basel Convention, 2011):

- the reduction of hazardous waste generation and the promotion of environmentally sound management of hazardous wastes, wherever the place of disposal, to promote waste prevention;
- the restriction of transboundary movements of hazardous wastes except where they are perceived to be in accordance with the principles of environmentally sound management, including the idea of treating waste near the place where it is generated; and

- the implementation of a regulatory system applying to cases where transboundary movements are permissible; control mechanisms must be developed by the parties, namely notification system and movement documentation.

The Bamako Convention on Hazardous Waste in Africa intends to secure human health and the environment from threats resulting from hazardous waste management, by reducing their amount and hazardousness to minimum (IMO, 2005). According to IMO (2005), the Bamako Convention defines that all parties are obliged to forbidden the import of all hazardous wastes into Africa from non-contracting parties, without exclusions. Only Organization of African Unity State Members can join this convention.

The Antarctic Treaty and related agreements, generally named the Antarctic Treaty System, regulates how international relations should be conducted concerning Antarctic territory. This treaty, effective since 1961, includes Article 5, which defines the prohibition of nuclear explosions or disposal of radioactive waste (NERC-BAS, 2012). One agreement with implications to waste management is The Protocol on Environmental Protection to the Antarctic Treaty, Annex III, signed in 1991. This annex enforces the need to reduce waste production and disposal and specifies waste storage, disposal, and removal from the Antarctic Treaty Area. All waste operations are defined in the context in this treaty.

The Global Programme of Action for the Protection of the Marine Environment from Land-Based Activities (GPA) aims to prevent marine environment degradation from land-based activities by facilitating the realization of duty of States to preserve and protect the marine environment (UNEP, 2013). GPA is the only global initiative that directly addresses the connectivity between terrestrial, freshwater, coastal, and marine ecosystem (UNEP, 2013). The enforcement of waste management in the GPA targets marine litter. In 2003, Regional Seas Program and GPA entrained on the Global Initiative on Marine Litter through constituting partnerships and cooperative arrangement and coordinating joint initiatives. The initiative intends to build knowledge and understanding about marine litter problem, to develop a common approach to monitor it, to reduce marine litter through the use of economic instruments, to enhance livelihood of fishermen contributing to healthy marine ecosystems, and to promote public awareness and outreach (UNEP, 2009).

The Hong Kong International Convention for the Safe and Environmentally Sound Recycling of Ships (the Hong Kong Convention) is devoted to the end-of-life ships impact. It aims to ensure that such waste does not pose any unnecessary risk to human health and safety or to the environment (IMO, 2013c). Regulations in the new convention cover all steps in ships' life cycle: design, construction, operation, recycling (secure and environmentally sound in accordance with an enforcement mechanism), which incorporate certification and reporting requirements (IMO, 2013c). The convention presents a list of the hazardous substances as well as the installation or use of which is prohibited or restricted in shipyards, ship repair yards, and ships of parties to the convention (IMO, 2013c). An initial survey on ships will be required to verify the hazardous materials inventory, restoration surveys during the life of the ship, and a final survey before recycling (IMO, 2013c; Starling, 2013).

4.1.2 National Solid Waste Management

Depending on the economic development, different countries have different waste management concerns reflected through national legislation. In general, waste legislation at the national level can take two different approaches, one specific for developing countries and the other for developed countries. For the former case, the main purpose is to develop waste legislation where responsibilities, rights, and stakeholders are defined. For example, in Brazil, legislation on waste management specifies that informal waste collection sector (also known as waste pickers) can be included in the integrated solid waste management (ISWM) system planning. In specific, the closing of dumpsites is one of the defined measures of waste legislation in developing countries. For the latter case, in developed countries, the focus on waste management legislation is placed on providing an environmentally sound waste treatment approach to avoid local or regional environmental pollution and promoting a waste hierarchy principle (WHP). To improve our understanding of waste legislation organization, it would be beneficial to gain a comparative viewpoint by exploring several well-known national waste management frameworks, from developed countries such as United States, Japan, and Korea, to developing countries such as China, Brazil, and Angola.

In the United States, enactment of the Resource Conservation and Recovery Act (RCRA, significantly amended in 1984), Pollution Prevention Act (amended in 2002), and Resource Conservation Challenge (RCC) (USEPA, 2004) has led to a waste management policy devoted to resource conservation and pollution prevention (Sakai et al., 2011). In the United States, the RCRA states the fundamental principles for the treatment of solid waste and for the reduction and management of hazardous wastes, being MSW managed by each state regulations based on the delegation from the Federal Government (Sakai et al., 2011). RCC is a program that intends to promote waste reduction and energy recovery activities. Recently, RCC has been changed to the Sustainable Materials Management Program, which intends to reduce the environmental footprint, support state solid waste planning, help local governments aim for zero waste production, reduce food waste, develop measurement, increase certified electronics recycling, and follow stakeholder process on packaging (USEPA, 2011) (Figure 4.2).

In Japan, the basic law for establishing a Material Cycles Society (or Sound Material-Cycle Society) in 2000 defines recyclable resources and states the principles for their utilization (Ministry of the Environment Government of Japan, 2009; Sakai et al., 2011). The law has the goal of a society that restrains the consumption of natural resources and reduces the environmental load as much as possible through promotion of recycling, reuse, and recovery (3Rs) as well as environmentally sound waste management (Sakai et al., 2011). Japanese basic waste management legislation consists of the Waste Disposal and Public Cleansing Law (the basic law for waste management) and the Law for the Promotion of Effective Utilities of Resources (law for recycling of used resources) (Sakai et al., 2011). In addition, the legal framework includes five laws specifically aimed at sectors and products (Ogushi and Kandlikar, 2007) (Figure 4.3).

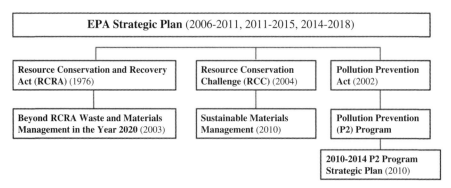

FIGURE 4.2   The waste management framework in the United States. *Source*: From USEPA (2004, 2006, 2010, 2014) and Sakai et al. (2011)

According to Park (2009), policies for solid waste in Korea have been developed through three phases. Before the 1990s, waste policies were devoted to facilities for the post-treatment of the waste generated pursuing safe and clean environment; in the 1990s, Korean governments considered the importance of reducing waste volume before it is generated; in the beginning of the new century, the goal of waste policies was upgraded from small-scale waste minimization to construction of a resource (re)circulation society extending its scope from waste management to enhancing productivity of resource uses (Park, 2009). The waste management framework in Korea (Figure 4.4) is developed based on the Waste Management Act and by the .Act on Promotion of Resources Saving and Recycling, where waste management programs associated with the framework are also included. Programs focus on waste

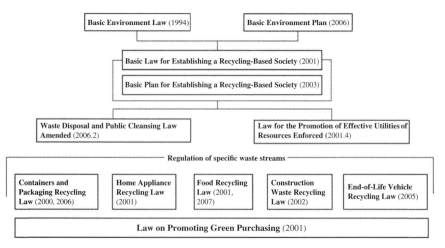

FIGURE 4.3   Japan waste management framework. *Source*: Yoshida (2009) and Sakai et al. (2011)

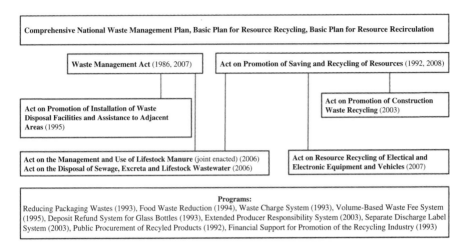

FIGURE 4.4   Korea waste management framework. *Source*: From Ministry of Environment (2007, 2008), Oh (2009), and Sakai et al. (2011)

reduction, promoting reuse (such as restricting the use of disposable products and collection fee deposit programs for empty beverage containers), and reinforcing waste recycling systems, where extended producer responsibility (EPR), eco-assurance, voluntary agreements (VA), separate discharge label, public procurement of recycled products, and financial support for the promotion of recycling industry programs are specified.

The basic environmental legislation for the waste management framework in China (Figure 4.5) is the Environmental Protection Law of People's Republic of China that

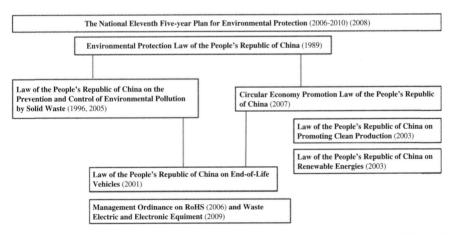

FIGURE 4.5   The waste management framework in China. *Source*: From Li (2009) and Sakai et al. (2011)

establishes relevant regulations and policies (Sakai et al., 2011). The 3Rs policy was established by the Environmental Pollution Prevention and Control Law. The main step has been the adoption of Circular Economy Promotion Law as a priority in the national policy advanced diverse regulations to support and build its implementation (Geng et al., 2012). To support this law, the Cleaner Production Promotion Law was published. Problems arise from several aspects, including a lack of support resources, reduced use of recycled materials, and a missing of national strategy to address the problem of resource depletion, followed by the desire for sustainable economic growth (Sakai et al., 2011). The latest version of the Circular Economy Promotion Law can be effectively categorized to influence three levels of operation (Geng et al., 2012):

- micro- or individual firm level: eco-design and cleaner production strategies and actions;
- meso- or eco-industrial park level: eco-industrial parks and networks which will promote regional economy and environment (Yuan et al., 2006; Geng et al., 2008); and
- macro- or eco-city/eco-province level: sustainable production and consumption activities which intend to create a recycling oriented society (Geng et al., 2008; Li et al., 2009).

Specific regulations similar to end-of-life vehicles (ELV) and waste electrical and electronic equipment (WEEE) in Europe have been promoted to deal with these specific waste streams caused by population expansion and economic development in urban areas (Sakai et al., 2011).

In other developing countries, such as Brazil, waste management legislation has evolved in response to different environmental problems. The first legislative framework is related to waste lubricant oils, in which several laws and decrees are published to regulate regeneration, procedure for its collection, and final destination. The waste oils management framework was promoted in combination with the National Petroleum Agency. MSW was regulated at the country level in 2010. Principles defined to manage solid waste are polluter-pays principle, WHP, eco-efficiency, and EPR through product life cycles (Figure 4.6).

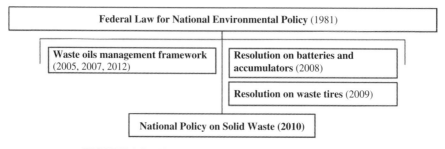

FIGURE 4.6   The waste management framework in Brazil

FIGURE 4.7   The waste management framework in Angola

In many developing countries, however, waste legislation development is slow, as evidenced by the case in Angola (Figure 4.7). Recent changes were made in solid waste legislation, with the advantage of including several policies not only for MSW but for specific waste streams: packaging waste, scrap tires, batteries and accumulators, ELV, waste oils, and construction and demolition waste (CDW). The waste management policies in Angola were developed based on the Portuguese waste legislation and EU strategies, which cover EPR as well as the polluter-pays principle. In such a relatively new waste legislation, emphasis was placed on the creation of regulations for municipalities, goals for waste collection service, and its required efficiency for the implementation of waste management policies.

4.2 SUSTAINABLE WASTE MANAGEMENT PRINCIPLES AND POLICIES

When defining waste legislation, principles and policies dictate the rules to be established. Perceiving the spirit of the law is fundamental to understanding its goals and how to follow them (i.e., how to reach those goals). Several policies and principles exist that may be intimately linked with sustainable SWM.

4.2.1 Waste Hierarchy Principle

WHP refers to the 3Rs of reduce, reuse, and recycle, classifying waste management options in terms of waste minimization goal (Singh, 2010). More components may be added to constitute the hierarchy with varying levels. For example, a more demanding hierarchy is seen with "7Rs": rethink, redesign, reduce, reuse, renew, refurbish, and recycle, and extended WHP that can even reach waste treatment and disposal. The aim of the WHP is to extract the maximum benefits from products and, simultaneously, to generate the smallest amount of waste (Singh, 2010; McElhatton and Pizzuto, 2012).

Although the WHP has been adopted worldwide, its application as the best environmental option has been contested. The first few options, such as prevention, reduction, and reuse, are quite acceptable; yet argument often arises from waste treatment options. The holistic order of recycling and incineration has received attention, as has the holistic order of incineration and landfilling, depending on assumptions and system boundaries being considered (Moberg et al., 2005). To help decision making, waste treatment options must be further evaluated by life cycle assessment (LCA). The actual EU situation can be assessed with regard to the use of LCA to help define waste management policies (Box 4.2).

BOX 4.2   WHP IN EU WASTE FRAMEWORK DIRECTIVE (Herva and Roca, 2013)

EU directives and national policies elaborated for waste management since the 1990s set recycling and recovery targets and restrictions on waste landfilling. WHP defined in the Directive 2008/98/EC (European Parliament and Council, 2008) establishes the following priority order: (1) prevention; (2) preparing for re-use; (3) recycling; (4) other recovery (e.g., energy recovery); and (5) disposal. However, the Directive considers the waste hierarchy order changeable if justified by an LCA (European Parliament and Council, 2008). Thus, LCA can be used to test the waste hierarchy, identifying alternatives where it can be modified (Finnveden et al., 2005; Moberg et al., 2005), depending on the waste itself, the location where the waste arises and its timing, as well as priorities in cases of contrary results. Alternatives should be evaluated consistently so that waste is managed at its most benefic solution considering resource use and environmental impacts, rather than accepting a simple hierarchy, thus pursuing integrative strategies (Clift et al., 2000; Cherubini et al., 2009; Pires et al., 2011).

4.2.2 Polluter-Pays Principle

The polluter-pays principle was defined by OECD in 1972 as the "principle to be used for allocating costs of pollution prevention and control measures to encourage rational use of scare environmental resources and to avoid distortions in international trade and investment" (OECD, 1972). In practice, the polluter-pays principle specifies that the polluter (either the entity that generates waste or emissions from waste management) should support the costs of preventing and controlling pollution. Concerning SWM, the polluter-pays principle entails that all waste generators, including citizens, companies, industries, are responsible for paying the expenses related to the waste they generate (Nahman and Godfrey, 2010). Emissions from waste treatment options are also implied for the consideration of the polluter-pays principle.

4.2.3 Extended Producer Responsibility

OECD (2001) defines EPR "as an environmental policy in which a producer's responsibility, physically and/or financially, for a product is extended to the post-consumer stage of a product's life cycle." EPR is characterized by shifting the responsibility to the producer instead of users, and by the environmental impacts related to the end-of-life, in such a way that could influence the design of the products (i.e., promoting eco-design) (OECD, 2001, 2006a).

Although the EPR has been considered a success from environmental management perspectives, which have promoted several new managerial schemes for different waste streams like packaging waste, batteries and accumulators, WEEE, tires, waste lubricant oils, to name a few, there are still missing links between design for environment and eco-design. Several studies, including Yu et al. (2008), Subramanian et al. (2009) and Brouillat and Oltra (2012), had attempted to identify these impacts. Yu et al. (2008) noted that little evidence exists that EPR has stimulated eco-design in WEEE. Subramanian et al. (2009) noted that in supply chains, where a relation exists between manufacturer and remanufacturer (before going to the final user), the use of charges during consumption and post-consumption phases can be applied as leverage to encourage environmentally favorable product design. Brouillat and Oltra (2012) demonstrated through a simulation that EPR induces eco-design when tax-subsidy systems and stringent norms can lead to such design innovations.

Because EPR is devoted to bringing producers into the chain of responsibility, it can also be used to integrate other stakeholders involved in the end-of-life stage, including the informal recycling waste sector. In the case of Brazil, one purpose of the National Policy on Solid Waste is to integrate waste collectors for reusable and/or recyclable waste into EPR actions and programs. In the same law, the incentive for the creation and development of associations for waste collectors should be promoted in terms of their inclusion from economic to social perspectives related to waste market and societal consensus. The selective waste collection schemes and reverse logistics of those waste collectors are also considered by the law.

4.2.4 Precautionary Principle: Protection of Human Health and Environment

The precautionary principle has its origin in the environmental movement occurred in 1970s and is generally assigned to environmental and health policies' concerns; however, a specific meaning for the precautionary principle has been difficult to reach due to its application to a wide range of contexts (Sandin, 1999; Feintuck, 2005; Kuhkau et al., 2011). One quoted formulation of the precautionary principle was introduced at the Wingspread Conference in 1998 stating that: "Where an activity raises threats of harm to the environment or human health, precautionary measures should be taken even if some cause and effect relationships are not fully established scientifically" (SEHN, 1998). Another formulation was expressed in the 1992 Rio Declaration (UN, 1992), where there are menaces of serious or irreversible damage, the reason to delay cost-effective measures to prevent environmental deterioration should not be based on the lack of entire scientific certainty.

In the context of SWM, the precautionary principle is regarded as its policy when there is insufficient information on waste management operations and their casual effects to human and environment. Applying the precautionary principle can alleviate the burden of evidence and supports those who wish to regulate an action, although normal scientific standards for establishing causal connections between the activity and the harm are not met (Kuhkau et al., 2011).

4.2.5 Principles of Self-sufficiency and Proximity

Self-sufficiency and the proximity principle intend to ensure that a delimitated geographic region could be supported by an integrated solid waste treatment, recovery, and disposal installations network in such a way to minimize or eliminate export waste (hazardous and/or nonhazardous) and ensure its correct end-of-life. This principle has been considered in EU, where the purpose is to ensure that EU countries can find solutions to their solid waste within the community. This policy will favor centralized waste management solutions where economies-of-scale can be reached; it does not require that each EU MS possess the full range of final recovery and disposal facilities.

4.2.6 Zero Waste Principle

According to Glavič and Lukman (2007), the definition of zero waste principle was not found among the UNEP, United States Environmental Protection Agency (USEPA), or European Environment Agency glossaries, but zero waste can be defined as an ideal in which there is no generation of waste since all materials no matter its physical state (solid, liquid or gaseous) will be deviated before they reach the waste stage (May and Flannery, 1995). The principle includes recycling but also takes a holistic approach reaching the flow of resources and waste from antropogenic environment. Zero waste maximizes recycling, minimizes waste to approach zero, reduces consumption, and ensures that products are planned to be reused, regenerated, repaired, and recycled internally or back into nature or the marketplace (Glavič and Lukman, 2007). Furthermore, zero waste does not consider waste a material that must be disposed of or incinerated, but treats it as a resource that can be reused to take full advantage of the waste potential (Glavič and Lukman, 2007). The applications of zero waste principle can be found in some SWM case studies that may be linked to the rationales in industrial symbiosis in the context of industrial ecology and zero waste cities in the context of a circular economy.

4.2.7 Integrated Product Policy

The concept of integrated product policy (IPP) is an approach to reduce the environmental impacts over the life cycle of products from the mining of raw materials to production, distribution, use, and waste management (Commission of the European Communities, 2001). All environmental impacts from product life cycles must be integrated into the product itself, with consequences at the decision level for all stakeholders (Commission of the European Communities, 2001). Five IPP "building blocks" include (European Commission, 1998):

- measures designed for the waste reduction and management resulting from the products consumption;

- measures planned for the innovation of more environmentally friendly products;
- measures to create demand and offer of environmentally sound products;
- measures for disseminate information up and down the product chain;
- measures that designate responsibility for managing the environmental burden of product systems.

4.3 POLICY INSTRUMENTS

Public policy instruments are "the set of techniques by which governmental authorities wield power in attempting to ensure support and effect or prevent social change" (Vedung, 1998). This definition emphasizes the nature of policy instruments to induce changes in a particular way, which is believed to stimulate innovation (i.e., influence the direct innovation policy objectives), and ensure the effectiveness and popular support of policy instruments for innovation (Borrás and Edquist, 2013). The need to implement policy instruments is justified by the necessity to internalize externalities from SWM (and waste generation by itself) and implement possible principles, as defined in Section 4.2.

Although there is no universal classification of policy instruments, they are normally defined by three categories: regulatory instruments, economic and financial instruments, and soft instruments (Borrás and Edquist, 2013), which were derived from Vedung characterization in "sticks," "carrots," and "sermons," respectively; other classifications also exist (Box 4.3).

BOX 4.3   POLICY INSTRUMENTS CLASSIFICATION

Although classification is quite similar, small differences justify the diverse classification of policy instruments. The classification adopted by the Finnish Environment Agency is through four policy instruments categories: legislative controls, economic instruments, informative measures, and voluntary measures (Finnish Ministry of the Environment, 2011).

From the perspective of enforcement, policy instruments can be categorized as mandatory or voluntary (Alberini and Segerson, 2002; Tojo et al., 2006). The key distinction between them is the ability to impose unwanted costs on polluters (Alberini and Segerson, 2002). The stakeholder addressed by the mandatory instruments is forced to fulfill the tasks laid down in regulation, although the private actors can set up the goals themselves and strive to achieve them via voluntary initiatives (Tojo et al., 2006). For instance, negotiated agreements between the government and private actors through a contract, in which the government typically drops the enforcing legislation on the condition that the private actors achieve the accorded goal (Tojo et al., 2006).

Regulatory instruments are also named command and control or administrative control, which constitute all regulatory intervention through legal measures like licensing, laws, products standards, and other regulatory actions. They can be simply defined as laws and regulations for which citizens and waste management services providers must legally comply with. Types of regulatory instruments are bans (e.g., ban on picking up recyclable wastes at landfills, ban on trade of products hard to repair, and ban on disposable packages), standards, requirements (e.g., mandatory source separation), and restrictions (Table 4.1).

The application of these types of instruments is normally conducted with other instruments, so evaluating the effectiveness of an isolated instrument is challenging (Tojo et al., 2006). Regulatory instruments can be applied in all stages of the product life cycle, including production, collection, recycling, and disposal, making them one of the most versatile instruments; however, the application of regulatory instruments needs law enforcement, such as penalties, to make them work. Although regulatory instruments can improve the effectiveness of SWM, in cases where pollution issues and WHP have been conducted, economists have noted their high costs of implementation, inflexibility, and diminishing rates of return (Gunningham, 2007). Other shortcomings from command and control are economic inefficiency, environmental ineffectiveness, and democratic illegitimacy (see Tietenberg, 1988; Eckersley, 1995). "By its very nature, command and control tends toward economic inefficiency by imposing uniform reduction targets and technologies which ignore the variable pollution abatement costs facing individual firms" (Golub, 1998). According to the Institute for Prospective Technological Studies (IPTS, 2007), command and control is still a common and effective policy instrument, specifically when targeting hazardous substance reduction in waste streams; as a result, impact on innovate products and alternative chemicals is also considerably significant, being notorious the impact at the production stage. In fact, the major advantage of the instruments via command and control is that the regulator has a reasonable degree of predictability regarding the level of pollution reduction (UNEP, 2005).

Economic instruments are also called market-based instruments that present a mechanism to change polluters' behavior based on economic motivation (i.e., economic incentive). "The incentive motivates the potential polluters to select the particular means of control that is economically favorable to the potential polluters" (Diaz et al., 2005). The rationale for the innovation associated with market-based instruments is based on the assumption that if properly implemented, these economic incentives may help (Diaz et al., 2005; Tojo et al., 2006) to:

- advance on the use of cost-effective instruments for achieving acceptable levels of pollution control;
- inspire the private sector for the development of pollution control expertise and technologies;
- arrange a source of revenue to pollution control programs, afforded by the government;
- lighten the burden that otherwise would be placed upon the government regarding the data collection and analysis involved in determining the feasible and appropriate level of control for each and every facility or product;

TABLE 4.1   Regulatory instruments for waste management

Regulatory instruments	Description	Examples
Landfill bans or restriction or diversion	Consists of prohibitions on landfill waste disposal. The purpose is to promote a circular/recycling economy, where waste can be reintroduced. It also intends to reduce negative environmental and health impacts.	Existing in several countries for specific waste streams such as biodegradable municipal waste, waste tires, combustible waste, packaging waste, recyclable fraction of CDW.
Substances bans or restriction	To avoid specific substances that could impede recycling or reuse. Promotes waste prevention because it reduces waste hazards.	Bans of certain materials used in cars construction or spare parts (for EU countries), or in electrical and electronic equipment.
Source separation requirement	Consists of a mandate that specific waste cannot be discharged into common waste bin, resulting in specific collection systems. Used to promote WHP because such collection will result in materials better prepared to be recycled and reused, avoiding environmental and public health issues.	In various countries for waste streams such as packaging waste, batteries, WEEE, biodegradable municipal waste, and waste tires. A particular case of this source separation is the mandatory home composting in countries such as Austria, Belgium, and Estonia (Chang et al., 2013) to promote waste prevention.
Producer's take-back	To allow specific waste streams to be collected separately, avoiding their inadequate discharge or damage during specific waste collection.	In EU countries for WEEE and batteries, at least.
Collection, reuse or refill, and recycling targets	Targets are related to WHP options. Targets are normally defined as a function of the market input on an annual basis.	All EU waste directives have collection and recycling targets; some have both reuse and recycling targets.
Use of secondary raw materials, including minimum percentage to be used	To promote a circular economy. Consists of establishing a percentage of inclusion of recycled material into new products.	Common in several countries for recovered/recycled CDW.
Environmentally sound treatment standards	To minimize environmental impacts from waste management operation, normally those with higher environmental and public health issues such as incineration and landfilling. Can be emission standards, operational condition (e.g., temperature and residence time of waste in the incinerator chamber), and construction conditions.	Applicable at all EU MS due to Incineration Directive and Landfill Directive.
By-products and end-of-waste criteria	Parameters and other criteria applied to secondary materials and other outputs from processes to be used as secondary materials.	End-of-waste criteria defined for ferrous scrap, aluminum scrap, and glass cullet.

- decrease externalities so that those who pollute should bear the cost; and
- address problems where current traditional command-and-control instruments often fail.

In SWM systems, economic instruments are applied as fees (charges) (including pay-as-you-throw, known as PAYT), taxes, subsidies, deposit–refund system, and recycling credit schemes (Table 4.2). Their application to SWM has been devoted to promote polluter-pays principle and WHP, as well as supporting implementation of EPR (see Box 4.4). Economic instruments are known to be effective in EU. According to Watkins et al. (2012), the EU evaluated several economic instruments applied to SWM and observed a relationship between higher landfill taxes (and higher total landfill charges, where tax plus charge = total charge) and lower percentages of MSW being sent to landfill; it also found a fairly clear and linear correlation between total landfill charge and the percentage of MSW recycled and composted. Nevertheless, it was difficult to eliminate landfilling through a tax alone due to rates applied (Watkins et al., 2012).

BOX 4.4   USING ECONOMIC INSTRUMENTS TO PROMOTE EPR: THE GREEN DOT SYSTEM CASE (OECD, 2006b)

Like has been mentioned in Chapter 2, Green Dot System (*Duales System Deutschland*, in German) was created in Germany in the 1990s to reduce elimination of packaging either by recycling or by reducing its use, being based on producer responsibility principle. Green Dot System would be capable to make industry to comply with regulatory instruments like packaging waste collection targets and recycling targets.

Companies wanting to participate in the Green Dot System (and comply with German regulation) must apply for permission to use the Green Dot symbol on their packaging, which results in paying a fee. This way, consumers and retailers may dispose their used/sales packaging bearing Green Dot seal into Green Dot System collection bins. The fee paid is for financing the packaging waste collection, sorting, for subsidizing recycling, for disposal costs and for information instruments, where the system is explained to consumers and retailers (the ones who depends collection and recycling targets).

Nowadays, the Green Dot System program, where regulatory instruments are achieved using EPR principle supported by economic instrument, is the most common strategy to implement EPR principle to any waste stream at European countries.

The instruments mentioned so far are mandatory, in which unwanted costs are imposed on polluters. Under a voluntary approach, a polluter will only participate if the payoff is at least as high as it would be without participation (i.e., the firm must perceive some gain, or at least no net loss, from participation) (Alberini and

TABLE 4.2   Overview of the economic instruments

Economic instruments	Description	Examples
Fees or charges (includes variable charges like PAYT)	Apply directly the polluter-pays principle because they are referent to the collection and disposal costs of waste. This fee is charged by the operator/waste service provider.	User charges: applied for waste collection and treatment (does not include marginal cost). Disposal charges (tipping fee or gate fee): applied to waste being disposed at landfill. Product charges: to finance end-of-life product management, applied in EPR context. Products where product charge has been applied are batteries and accumulators, one-way packaging, lubricant oils, plastic bags and tires.
Taxes	To increase the cost of the depositing at the waste destination, such as taxes applied to landfill or incineration processes. Considered as the inclusion of externalities. Levied by public authority.	Landfill tax, incineration tax for all type of waste or for specific waste streams. In countries where the tax is calculated for different final destinations, the tax is called waste disposal tax, such as in Denmark since 1987. Waste disposal in landfills is subjected to the highest tax level, followed by incineration without energy recovery, followed by incineration recovering energy for electric energy production, and incineration recovering electricity and thermal energy; recycling is not taxed (Tojo et al., 2006).
Subsidies	Attributed to industry, business, or private consumers who must decide between buying and using primary products composed of primary (virgin) material or recycled (secondary) products.	Applied to quarry products concerning CDW recycled products.
Deposit–refund	A payment is made when the product is purchased and refunded when the purchaser returns the used product.	Common to beverage containers.
Recycling credits	Attributed to companies that use recycled material in the production of products, which can be trade between companies: those needing credits can buy them to offset their obligation (Finnveden et al., 2013).	Applied in UK to specific packaging waste.

Segerson, 2002). In this context, soft instruments are an umbrella category representing these non-mandatory instruments. They involve different instruments without a direct mandatory or monetary term to promote stakeholders behavior change.

With these instruments, those who are "governed" are not subjected to obligatory measures, sanctions, or direct incentives or disincentives by the government or its public agencies; instead, "the soft instruments provide recommendations, make normative appeals, or offer voluntary or contractual agreements" (Borrás and Edquist, 2013). Soft instruments can be divided in voluntary instruments and informative instruments. Voluntary instruments can be grouped into three broad categories including unilateral commitments, negotiated agreements, and public voluntary programs. Informative instruments are based on the presumption that people will change their behavior differently if they have better information and knowledge (Huhtinen, 2009). Information instruments can include eco-labeling schemes, green shopping guides, marking of products and components, information campaigns to residents, and information provision to treatment facilities (Table 4.3).

According to Borrás and Edquist (2013), soft instruments are diverse but are commonly based on persuasion, mutual exchange of information among actors and less hierarchical forms of collaboration between the public and the private actors. As compared to hard instruments (such as command and control and market based), soft instruments can be devoted to all stages of products simultaneously due to some instruments having a life cycle perspective or focused on specific waste stages such as prevention, recycling, and disposal. Specifically, in cases of VA their success rest on a number of factors such as (Krarup, 2001; Seadon, 2006):

- information available to the public;
- scrutinize and exchange of information that was available to the negotiating actors;
- positive and negative incentives made accessible by the regulator to encourage industry to engage in the process; and
- consumers' demand for environmental quality, either through lobbying or a general demand by consumers.

The advantages of other specific soft instruments such as negotiated agreements over legal or market-based instruments, identified by scholars as well as by practitioners, include their flexibility, their ability to trigger learning processes, their potential for collaboration, and the encouragement of first experiment (Jörgens and Busch, 2005).

More recently, policy instruments are used together as mixed or policy packages that allow different instruments to compensate for the drawbacks of the other instruments, making them more successful. Policy mixes are implemented to replace an individual policy instrument because (Lehman, 2012):

- combining policies can correct multiple market failures, such as pollution externalities, technological spillovers, and asymmetric information; and

TABLE 4.3   Soft instruments applied to SWM

Soft instruments	Description	Examples
Information campaigns	Using communication instruments (such as door-to-door information, television, newspapers, radio internet) to promote waste and recycling behaviors or participation in recycling schemes.	Home composting campaigns, recycling campaigns, labeling on waste bins.
Green public procurement	A process in which goods, services, and works to be bought by public authorities should reduce environmental impact.	Buying 100% recycled and totally chlorine-free paper in such a way to promote recycled goods.
Voluntary environmental standardization	Environmental criteria that classify a product as being environmentally friendly by using certification labels.	German Blue Angel, Nordic Swan, Swedish Good Environmental Choice, EU Flower.
VA	Three types: unilateral commitments, negotiated agreements, and public voluntary program.	–
Unilateral commitments	VA are set only by industries (itself or between industries).	ISO 14001 certification by industries.
Negotiated agreements	"Commitments for environmental protection developed through bargaining between a public authority and industry" (Bauer and Fischer-Bogason, 2011).	Producer responsibility organization from EPR established by negotiated agreements in EU countries for several waste streams like packaging waste, batteries and accumulators, ELV, tires. Outside producers or importers can also negotiate with government directly and establish a negotiated agreement.
Public voluntary programs	"Commitments devised by the environmental agency and in which individual firms are invited to participate" (Bauer and Fischer-Bogason, 2011).	WasteWise from USEPA to reduce municipal and selected industrial wastes. 33/50 Program from USEPA for the reduction of toxic waste.
Marking of products and components	Specific marks/labels that inform the user concerning waste environmental and economic management relevant information.	Directives on packaging, WEEE and batteries, voluntary initiatives by manufacturers. Green Dot System symbol. Plastic polymer symbol.
Information to treatment facilities	Identification of components and materials to be removed by recyclers.	EU WEEE and ELV directives, voluntary initiatives by manufacturers.
Waste exchange program	A web-based structure where, anonymously, waste is traded between stakeholders.	Widespread in EU countries and the United States.

TABLE 4.4   Policy mixes applied to food waste reduction policy in the United Kingdom

Policy instruments options			
Regulatory	Economic	Information	VA
Bans (of landfilling biodegradable municipal waste)	Green public procurement (e.g., food catering)	Indicators, disclosure, and reporting	Grocery supply chain actors agreement
Tradable landfill allowances set by legislation	Landfill Allowance Trading Scheme applied to the quotas	Prevention campaigns, like "Love food hate waste," "Love your leftovers," "Great taste, less waste"	Hospitality and food service sector agreement
Standards (e.g., food catering)		Labeling (e.g., date for food waste)	

Source: From Davis et al. (2013).

- they can reduce the inefficient application of a single policy that has induced noncompliance by polluters, heterogeneity of marginal abatement costs, or even heterogeneity of marginal damages.

EPR is a waste management principle that more leans toward the use of mixed policy instruments. According to OECD (2004), the policy of EPR can be mainly implemented through double regulatory instruments (take-back requirements) and economic instruments (deposit/refund schemes, advanced disposal fees, materials taxes, combined schemes). Other instruments complementary to these are product standards (minimum recycled content), information instruments (eco-labeling), and voluntary approaches. Other waste management policies, such as WHP and polluter-pays principle, can also be implemented via policy mixes, like for example, for food waste reduction (Table 4.4).

4.4   ISWM PLANS

Another way to ensure sustainable SWM is through the elaboration of ISWM plans and programs (P&P). P&P are used to define measures that must be taken to implement waste management policies and identify instruments (regulatory, economic, or soft) to be put into practice to ensure that those policies and strategies are reached. ISWM P&P are so relevant to social sustainability that they constitute, in most countries, a legal document to be followed and fulfilled by stakeholders involved. ISWM plans can have different geographic and time scales/periods; they can be devoted to a specific waste stream like packaging waste or be elaborated for all mixed waste flow, like MSW. Because it includes different policy instruments, P&P can also be regarded as a policy mixes with various supporting instruments. The main elements include (BIS/CRI/ETAGIW, 2012; ETAGIW, 2012):

- general considerations and background;
- status to assess the actual situation and identification of problems;
- planning of objectives, visions, measures, indicators targets, scope and limits, and priority areas;
- implementation via scheduling of actions to be developed;
- monitoring of implemented actions; and
- revision.

P&P is a circular process that must be reformulated at the end to begin again. At the EU level, P&P must be assessed in terms of environmental impact, in accordance with Directive 2001/42/EC (European Parliament and Council, 2001). The most adequate methodology to conduct the assessment is the strategic environmental assessment (SEA (pronounced as 'sea')), which is "a systematic process for evaluating the environmental consequences of a proposed policy, plan or program initiative in order to ensure they are fully included and appropriately addressed at the earliest appropriate stage of decision-making in concert with economic and social considerations" (Sadler and Verheem, 1996). A SEA can be applied to existing P&P or in the preparatory phase, although it is quite time demanding due to the high complexity and the participatory requirements.

4.5 FINAL REMARKS

SWM is a matter of great concern in the urban areas in both developed and developing countries. Whereas developed countries have much matured laws and regulations dedicated for SWM, developing countries have been evolving to charge their obligations effectively through legal framework. This chapter which covers lucidly the present status of legal framework for SWM shares some examples of best practices adopted in various countries on legal issues for SWM and highlights the need of ISWM P&P in the end. It is deemed a key step again to gain the systematic insight for possible development of more sustainable SWM plans. Our discussion in this chapter includes key elements of institutional and legal arrangements with linkages of regulatory frameworks in SWM. It covers the institutional, social, and legal aspects of SWM in support of subsequent discussion of systems engineering approach in the next few chapters, which are intimately linked to urban sustainability.

REFERENCES

Alberini, A. and Segerson, K. 2002. Assessing voluntary programs to improve environmental quality. *Environmental and Resource Economics*, 22(1–2), 157–184.

Basel Convention. 2011. Overview. The convention. Available at: http://www.basel.int/ TheConvention/Overview/tabid/1271/Default.aspx (accessed May 2013).

Bauer, B. and Fischer-Bogason, R. 2011. *Voluntary Agreements and Environmental Labelling in the Nordic Countries*, Nordic Council of Ministers, Copenhagen.

BioIntelligence Service, Copenhagen Resource Institute, and Expert Team for Assessment and Guidance for the Implementation of Waste Legislation (BIS/CRI/ETAGIW). 2012. Preparing a Waste Prevention Programme—Guidance Document.

Borrás, S. and Edquist, C. 2013. The choice of innovation policy instruments. *Technological Forecasting and Social Change*, 80(8), 1513–1522.

Brouillat, E. and Oltra, V. 2012. Extended producer responsibility instruments and innovation in eco-design: an exploration through a simulation model. *Ecological Economics*, 83, 236–245.

Chang, N.-B., Pires, A., and Martinho, G. 2013. Environmental legislation for solid waste management in EU Countries via the use of economic and policy instruments. In: *Encyclopedia of Environmental Management—Volume II* (Ed. Jorgensen, S. E.), Taylor and Francis Group, pp. 892–913.

Cherubini, F., Bargigli, S., and Ulgiati, S. 2009. Life cycle assessment of waste management strategies: landfilling, sorting plant and incineration. *Energy*, 34(12), 2116–2123.

Clift, R., Doig, A., and Finnveden, G. 2000. The application of life cycle assessment to integrated solid waste management. Part 1. Methodology. *Process Safety and Environmental Protection*, 78(4), 279–287.

Commission of the European Communities. 2001. Green paper on integrated product policy. COM(2001)68 final.

Davis, M., Wunder, S., and Lopes, A. F. 2013. DYNAMIX policy mix evaluation—preventing food waste in UK. DYNAMIX. Available at: http://dynamix-project.eu/sites/default/files/ Food%20waste_UK.pdf (accessed February 2014).

Diaz, L. F., Savage, G. M., and Eggerth, L. L. 2005. *Solid Waste Management*, UNEP.

Eckersley, R. 1995. *Markets, the State and the Environment: Towards Integration*, Macmillan, London, UK.

European Commission. 1998. Integrated product policy: a study on analysing national and international developments with regard to integrated product policy in the environmental field and providing elements for an EC policy in this area. Final Report, March, Brussels.

European Parliament and Council. 2001. Directive 2001/42/EC of the European Parliament and of the Council of 27 June 2001 on the assessment of the effects of certain plans and programmes on the environment. *Official Journal of the European Communities*, L197, 30–37.

European Parliament and Council. 2008. Directive 2008/98/EC of the European Parliament and of the Council of 19 November 2008 on waste and repealing certain directives. *Official Journal of European Union*, L312, 3–30.

Expert Team for Assessment and Guidance for the Implementation of Waste Legislation (ETAGIW). 2012. Preparing a Waste Management Plan: A Methodological Guidance Note.

Feintuck, M. 2005. Precautionary maybe, but what's the principle? The precautionary principle, the regulation of risk, and the public domain. *Journal of Law and Society*, 32(3), 371–398.

Finnish Ministry of the Environment. 2011. Policy instruments. Environmental protection. Available at: www.ymparisto.fi./ (accessed May 2013).

Finnveden, G., Johansson, J., Lind, P., and Moberg, Å., 2005. Life cycle assessment of energy from solid waste. Part 1. General methodology and results. *Journal of Cleaner Production*, 13(3), 213–229.

Finnveden, G., Ekvall, T., Arushanyan, Y., Bisaillon, M., Henriksson, G., Östling, U. G., Söderman, M. L., Sahlin, J., Stenmarck, A., Sundberg, J., Sundvist, J.-O., Svenfelt, A., Söderholm, P., Björklund, A., Eriksson, O., Forsfält, T., and Guath, M. 2013. Policy instruments towards a sustainable waste management. *Sustainability*, 5(3), 841–881.

Fischer, C. and Davidsen, C. 2010. Europe as a recycling society—the European recycling map. ETC/SCP working paper 5/2010, Copenhagen, Denmark.

Geng, Y., Zhang, P., Côté, R. P., and Qi, Y. 2008. Evaluating the applicability of the Chinese eco-industrial park standard in two industrial zones. *International Journal of Sustainable Development and World Ecology*, 15(6), 1–10.

Geng, Y., Fu, J., Sarkis, J., and Xue, B. 2012. Towards a national circular economy indicator system in China: an evaluation and critical analysis. *Journal of Cleaner Production*, 23(1), 216–224.

Glavič, P. and Lukman, R., 2007. Review of sustainability terms and their definitions. *Journal of Cleaner Production*, 15(18), 1875–1885.

Golub, J. 1998. New instruments for environmental policy in the EU. In: *New Instruments for Environmental Policy in the EU* (Ed. Golub, J.), Routledge, London, UK, pp. 1–32.

Gunningham, N. 2007. Reconfiguring environmental regulation: next-generation policy instruments. In: *Industrial Innovation and Environmental Regulation: Developing Workable Solutions* (Eds. Parto, S. and Herbert-Copley, B.), United Nations University, Hong Kong, pp. 200–232.

Hedemann-Robinson, M. 2007. *Enforcement of European Union Environmental Law: Legal Issues and Challenges*, Routledge-Cavendish, Oxon, UK.

Herva, M. and Roca, E. 2013. Ranking municipal solid waste treatment alternatives based on ecological footprint and multi-criteria analysis. *Ecological Indicators*, 25, 77–84.

Huhtinen, K. 2009. *Instruments for Waste Prevention and Promoting Material Efficiency: A Nordic Review*, Nordic Council of Ministers, Denmark.

Institute for Prospective Technological Studies (IPTS). 2007. *Waste Policy and Innovation—A Methodological Approach*, European Communities, Spain.

International Maritime Organization (IMO). 2005. Bamako Convention. Available at: http://www.imo.org/blast/mainframemenu.asp?topic_id=1514anddoc_id=7607 (accessed May 2013).

International Maritime Organization (IMO). 2013a. London convention and protocol. Our work—special programmes and initiatives. Available at: http://www.imo.org/OurWork/Environment/SpecialProgrammesAndInitiatives/Pages/London-Convention-and-Protocol.aspx (accessed May 2013).

International Maritime Organization (IMO). 2013b. International Convention for the prevention of pollution from ships (MARPOL). Conventions—list of conventions. Available at: http://www.imo.org/about/conventions/listofconventions/pages/international-convention-for-the-prevention-of-pollution-from-ships-(marpol).aspx (accessed May 2013).

International Maritime Organization (IMO). 2013c. The Hong Kong International Convention for the safe and environmentally sound recycling of ships. Conventions—list of conventions. Available at: http://www.imo.org/about/conventions/listofconventions/page s/the-hong-kong-international-convention-for-the-safe-and-environmentally-sound-recycl ing-of-ships.aspx (accessed May 2013).

Jörgens, H. and Busch, P.-O. 2005. Voluntary approaches in waste management: the case of German ELV program. In: *Industrial Transformation: Environmental Policy Innovation in the United States and Europe* (Eds. de Bruijn, T. and Norberg-Bohm, V.), Massachusetts Institute of Technology, Cambridge, pp. 93–118.

Keyuan, Z. 2009. Regulation of waste dumping at sea: the Chinese practice. *Ocean & Coastal Management*, 52(7), 383–389.

Krarup, S. 2001. Can voluntary approaches ever be efficient? *Journal of Cleaner Production*, 9(2), 135–144.

Kuhkau, F., Höglund, A. T., Evers, K., and Eriksson, S. 2011. A precautionary principle for dual use research in the life sciences. *Bioethics*, 25(1), 1–8.

Lehman, P. 2012. Justifying a policy mix for pollution control: a review of economic literature. *Journal of Economic Surveys*, 26(1), 71–97.

Li, J. 2009. Summary policy and legislative tools in promoting waste management under circular economic model in China. In: International Workshop on 3R and Waste Management, Kyoto City.

Li, H. Q., Bao, W. J., Xiu, C. H., Zhang, Y., and Xu, H. B. 2009. Energy conservation and circular economy in China's process industries. *Energy*, 35(11), 4273–4281.

May, D. R. and Flannery, B. L. 1995. Cutting waste with employee involvement teams. *Business Horizons*, 38(5), 28–38.

McElhatton, A. and Pizzuto, A., 2012. Chapter—Waste and its rational management. In: *Novel Technologies in Food Science—Their Impact on Products, Consumer Trends and the Environment* (Eds. Mcelhatton, A. and Sobral, P. J. A.), Springer, New York, pp. 3–20.

Ministry of Environment. 2007. Resource recirculation policy of Korea. Available at: http:// search.korea.net:8080/intro_korea2008/society/02_re.html (En) (accessed February 2010).

Ministry of Environment. 2008. Korea's 3R policy. Available at: http://search. korea.net:8080/ intro_korea2008/society/02_re.html (En) (accessed February 2010).

Ministry of the Environment Government of Japan. 2009. White paper on the environment. Available at: http://www.env.go.jp/en/wpaper./ (accessed February 2010).

Moberg, Å., Finnveden, G., Johansson, J., and Lind, P. 2005. Life cycle assessment of energy from solid waste. Part 2. Landfilling compared to other treatment methods. *Journal of Cleaner Production*, 13(3), 231–240.

Nahman, A. and Godfrey, L. 2010. Economic instruments for solid waste management in South Africa: opportunities and constraints. *Resources, Conservation and Recycling*, 54(8), 521–531.

National Environment Research Council – British Antarctic Survey (NERC-BAS). 2012. The Antarctic treaty—background information. NERC-BAS. Available at: http://www .antarctica.ac.uk/about_antarctica/geopolitical/treaty/index.php (accessed February 2014).

Ogushi, Y. and Kandlikar, M., 2007. Assessing extended producer responsibility laws in Japan. *Environmental Science and Technology*, 41(13), 4502–4508.

Oh, G. J. 2009. 3R and waste management policy and outcome in Korea. In: *International Workshop on 3R and Waste Management*, Kyoto City.

Organisation for Economic Co-Operation and Development (OECD). 1972. Recommendations of the Council on guiding's principles concerning international aspects of environmental policies C(72)128. Decisions, recommendations and other instruments of the Organization for Economic Co-Operation and Development. Available at: http://acts.oecd.org/Instrum ents/ShowInstrumentView.aspx?InstrumentID=4andLang=enandBook=False (accessed May 2013).

Organisation for Economic Co-Operation and Development (OECD). 1976. Recommendation of the Council on a comprehensive waste management policy C(76)155/FINAL. Decisions, recommendations and other instruments of the Organization for Economic Co-operation and Development. Available at: http://acts.oecd.org/Instruments/ShowInstrumentView.as px?InstrumentID=14andInstrumentPID=12andLang=enandBook=False (accessed May 2013).

Organisation for Economic Co-Operation and Development (OECD). 2001. *Extended Producer Responsibility: A Guidance Manual for Governments*, OECD Publications, Paris.

Organisation for Economic Co-Operation and Development (OECD). 2004. *Economic Aspects of Extended Producer Responsibility*, OECD Publications, Paris.

Organisation for Economic Co-Operation and Development (OECD). 2006a. Working group on waste prevention and recycling—EPR policies and product design: economic theory and selected case studies. Environment Directorate, Environment Policy Committee, ENV/EPOC/WGWPR(2005)9/FINAL, OECD Publications, Paris.

Organisation for Economic Co-Operation and Development (OECD). 2006b. *Improving Recycling Markets*, OECD Publications, Paris.

OSPAR Commission. 2013. About OSPAR. OSPAR Commission—protecting and conserving the North-East Atlantic and its resources. Available at: www.ospar.org (accessed May 2013).

Park, J.-W. 2009. 3R policies of Korea. Available at: http://eng.me.go.kr/board.do?method =viewanddocSeq=195andbbsCode=res_mat_policy (accessed May 2013).

Pires, A., Martinho, G., and Chang, N.-B. 2011. Solid waste management in European countries: a review of systems analysis techniques. *Journal of Environmental Management*, 92(4), 1033–1050.

Regional Seas Coordinating Office. 2005. UNEP regional seas programme, marine litter and abandoned fishing gear (pp. 1–30). Available at: http://www.unep.org/regionalseas/ marinelitter/publications/docs/RS_DOALOS.pdf (accessed May 2013).

Sadler, B. and Verheem, R. 1996. *Strategic Environmental Assessment: Status, Challenges, and Future Directions*, Publication 53, Ministry of Housing, Spatial Planning and the Environment, The Hague, Amsterdam.

Sakai, S., Yoshida, H., Hirai, Y., Asari, M., Takigami, H., Takahashi, S., Tomoda, K., Peeler, M. V., Wejchert, J., Schmid-Unterseh, T., Douvan, A. R., Hathaway, R., Hylander, L.D., Fischer, C., Oh, G. J., Jinhui, L., and Chi, N. K. 2011. International comparative study of 3R and waste management policy developments. *Journal of Material Cycles and Waste Management*, 13(2), 86–102.

Sandin, P. 1999. Dimensions of the precautionary principle. *Human and Ecological Risk Assessment*, 5(5), 889–907.

Sands, P., Peel, J., Fabra, A., and MacKenzie, R. 2012. *Principles of International Environmental Law*, 3rd edition, Cambridge University Press, Cambridge.

Science & Environmental Health Network (SEHN). 1998. Wingspread Conference on the precautionary principle. Precautionary principle. Available at: http://www.sehn.org/wing.html (accessed December 2013).

Seadon, J. K. 2006. Integrated waste management—looking beyond the solid waste horizon. *Waste Management*, 26(12), 1327–1336.

Singh, J. 2010. Different methods in solid waste management. In: *Solid Waste Management: Present and Future Challenges. I.K* (Eds. Singh, J. and Ramanathan, A. L.), International Publishing House Pvt. Ltd., New Delhi, pp. 84–89.

Starling, M. 2013. Ship recycling: from the cradle to the grave. Arabian supply chain. Available at: http://www.arabiansupplychain.com/article-9032-ship-recycling-from-the-cradle-to-the-grave/1/print./ (accessed December 2013).

Subramanian, R., Gupta, S., and Talbot, B., 2009. Product design and supply chain coordination under extended producer responsibility. *Production and Operations Management*, 18(3), 259–277.

Tietenberg, T. 1988. *Environmental and Natural Resource Economics*, 2nd edition, Scott Foresman, Glenview, IL.

Tojo, N., Neubauer, A., and Bräuer, I. 2006. *Waste Management Policies and Policy Instruments in Europe: An Overview*, International Institute for Industrial Environmental Economics, Lund.

United National Environment Programme (UNEP). 2005. Selection, design and implementation of economic instruments in the solid waste management sector in Kenya: the case of plastic bags. United Nations Environment Programme.

United National Environment Programme (UNEP). 2009. UNEP's global initiative on marine litter. UNEP's global initiative on marine litter. Available at: http://www.unep.org/regionalseas/marinelitter/publications/docs/Marinelitter_Flyer2009.pdf (accessed May 2013).

United National Environment Programme (UNEP). 2012. About regional seas. regional seas program. Available at: http://www.unep.org/regionalseas/About (accessed May 2013).

United National Environment Programme (UNEP). 2013. Welcome to the GPA. Global programme of action for the protection of the marine environment from land-based activities (GPA). Available at: http://www.gpa.unep.org/ (accessed May 2013).

United Nations (UN). 1992. Report of the United Nations Conference on Environment and Development (Rio de Janeiro, 3–14 June 1992). Available at: http://www.un.org/documents/ga/conf151/aconf15126–1annex1.htm (accessed May 2013).

United States Environmental Protection Agency (USEPA). 2004. Resource conservation challenge strategic plan: what can you save tomorrow? Five year plan. Available at: http://webapp1.dlib.indiana.edu/cgi-bin/virtcdlib/index.cgi/6825758/FID1/pdfs/rcc-strat-plan.pdf (accessed February 2010).

United States Environmental Protection Agency (USEPA). 2006. 2006–2011 EPA strategic plan. Available at: http://nepis.epa.gov/Adobe/PDF/P1001IPK.pdf (accessed February 2010).

United States Environmental Protection Agency (USEPA). 2010. Fiscal year 2011–2015 EPA strategic plan. Available at: http://nepis.epa.gov/Exe/ZyPDF.cgi?Dockey=P1008YOS.PDF (accessed November 2014).

United States Environmental Protection Agency (USEPA). 2011. Creating an EPA sustainable materials management program. Waste2Resources Advisory Committee. Available at:

http://www.ecy.wa.gov/programs/swfa/w2rac/docs/ 2011SeptEPA.pdf (accessed February 2010).

United States Environmental Protection Agency (USEPA). 2014. Fiscal year 2014–2018 EPA strategic plan. Available at: http://www2.epa.gov/sites/production/files/2014-09/documents/epa_strategic_plan_fy14-18.pdf (accessed November 2014).

Vedung, E. 1998. Policy instruments: typologies and theories. In: *Carrots, Sticks, and Sermons: Policy Instruments and Their Evaluation* (Eds. Bemelmans-Videc, M.-L., Rist, R. C., and Vedung, E. O.), Transaction Publishers, London, pp. 21–58.

Watkins, E., Hogg, D., Mitsios, A., Mudgal, S., Neubauer, A., Reisinger, H., Troeltzsch, J., and van Acoleyen, M. 2012. Use of economic instruments and waste management performances. Final report for the European Commission DG-Env. Available at: http://ec.europa.eu/environment/waste/pdf/final_report_10042012.pdf (accessed December 2013).

Yoshida, H. 2009. Japan's experiences of policy developments on 3rs and waste management. In: International Workshop on 3rs and Waste Management, Kyoto City.

Yu, J., Hills, P., and Welford, R. 2008. Extended producer responsibility and eco-design changes: perspectives from China. *Corporate Social Responsibility and Environmental Management*, 15(2), 111–124.

Yuan, Z. W., Bi, J., and Moriguichi, Y. 2006. The circular economy: a new development strategy in China. *Journal of Industrial Ecology*, 10(1–2), 4–8.

CHAPTER 5

RISK ASSESSMENT AND MANAGEMENT OF RISK

Environmental impacts resulting from solid waste management (SWM) are numerous, although many SWM systems have been designed to minimize or eliminate such environmental impacts through technological solutions. Emissions from landfills, incineration plants, mechanical–biological treatment (MBT) plants, recycling centers, and waste collection points can all impact human health, ecosystems, and material properties to some extent on either a short- or long-term basis. For example, greenhouse gas emissions released from some municipal solid waste (MSW) treatment and disposal facilities have long-term effects on climate change. Handling hazardous waste could result in both short- and long-term negative effects in all managerial phases. This chapter introduces the concept of risk assessment as a tool to assess effects related to human health, environment, and infrastructure, and discuss a framework for management of risk associated with social, economic, and political concerns. Risk assessment for sustainable SWM is an umbrella concept to help formulate problems, to assess detrimental effects, to identify and appraise management options through various policy measures or alternatives, and finally to address management strategies.

5.1 FORMULATE THE PROBLEM: INHERENT HAZARDS IN SOLID WASTE MANAGEMENT

The first stage of risk management is to define risk and hazard concepts. Hazard is a property, situation, substance, or event associated with an accident or biological,

Sustainable Solid Waste Management: A Systems Engineering Approach, First Edition. Ni-Bin Chang and Ana Pires.
© 2015 The Institute of Electrical and Electronics Engineers, Inc. Published 2015 by John Wiley & Sons, Inc.

chemical, or physical agent that may lead to harm or cause adverse effects on people, property, and/or the environment (Asante-Duah, 1998; Gormley et al., 2011; European Parliament and Council, 2012; Theodore and Dupont, 2012). Risk consists of the likelihood/probability of the occurrence of the effect derived from the hazard and consequence caused by it (i.e., the severity of the effect), which occurs in specific circumstances or periods to different receptors (Kaplan and Garrick, 1981; Royal Society, 1992; Asante-Duah, 1998; Slaper and Blaauboer, 1998; Gormley et al., 2011; European Parliament and Council, 2012; Chen et al., 2013).

As discussed in Chapter 2, solid waste can be classified as hazardous waste in some circumstances. According to the European Parliament and Council (2008), for a waste to be considered hazardous it must display one or more of the hazardous properties listed in annex III from the Directive 2008/98/EC or be listed as hazardous in the European List of Waste (Commission Decision 2000/532/EC). In the United States, the Resource Conservation and Recovery Act §1004(5) defines hazardous waste as "solid waste, or combination of solid waste, which because of its quantity, concentration, or physical, chemical, or infectious characteristics may (a) cause, or significantly contribute to, an increase in mortality or an increase in serious irreversible, or incapacitating reversible, illness; or (b) pose a substantial present or potential hazard to human health or the environment when improperly treated, stored, transported, or disposed of, or otherwise managed" (USEPA, 2005).

Hazard embedded in waste streams can be derived from the material/product nature and/or from waste processing. By their nature, waste can be hazardous due to product/material hazard properties, as regulated by the Registration, Evaluation, Authorisation and Restriction Chemicals (REACH) legislation in the European Union (EU). Waste can also be hazardous due to contamination occurring during a product use stage. Products resulting from waste operations such as compost/stabilized residue, refuse-derived fuel, recyclable materials such as glass, paper/cardboard and plastics, ashes from incineration, and biogas from landfills and anaerobic digestion can all potentially be hazardous. If hazard properties are present, those waste products will be difficult to be consumed. In the EU, specific criteria for the consequences of hazard properties in the end-of-waste process have been developed (Box 5.1).

Waste treatment plants may become sources of hazards and, consequently, sources of risks. Waste collection might emit several pollutants dangerous to human health, crops, and ecosystems (e.g., biosphere, water, and soil), especially pollutants such as nitrogen oxides (NO, NO_2, NO_x), particulate matter (PM_{10}, $PM_{2.5}$), sulfur oxides (SO_x), ozone (O_3), carbon dioxide (CO_2), carbon monoxide (CO), nitrous oxide (N_2O), volatile organic compounds (VOC), and ammonia (NH_3). Lead (Pb) is a problematic heavy metal released from gasoline vehicles, consequently from collection and transport vehicles. Direct pollutants from composting and aerobic MBT are CO_2, N_2O, NH_3, VOC, hydrogen sulfide (H_2S), particulates, and bioaerosols. Methane (CH_4) emissions from landfills may also occur. In fact, these treatment and disposal units are known by their odor problems related to the emissions from VOC, NH_3, and H_2S. Biogas from anaerobic digestion (including MBT units), which is rich in CH_4, is one of the main products and is normally burned to produce electricity or collected to be used in other engines. In addition, CH_4, CO_2, CO, NO_x, SO_2, particulate matter, H_2S, VOC, odors, halogenated hydrocarbons, hydrogen chloride

(HCl), hydrogen fluoride (HF), cadmium (Cd), chromium (Cr), mercury, Pb, and zinc (Zn) my also occur (EIPPCB, 2006a). Emissions from water used in waste treatment may be reduced through recirculation, but even so, the presence of chemical oxygen demand, biochemical oxygen demand, NH_3, nitrate, total nitrogen, total phosphorus, and sulfate (SO_4^{2-}) may be detected (EIPPCB, 2006a).

BOX 5.1   END-OF-WASTE CRITERIA AND HAZARDOUSNESS

The Waste Framework Directive 2008/98/EC established certain conditions that must be met by the end-of-waste requisites. A waste may only cease to be a waste if (European Parliament and Council, 2008):

- the substance or object is commonly used for specific purposes;
- a market or demand exists for such a substance or object;
- the substance or object fulfills the technical requirements for the specific purposes and meets the existing legislation and standards applicable to products; and
- the use of the substance or object will not lead to overall adverse environmental or human health impacts.

The fourth criteria aim to ensure zero environmental risk. This translates into several measures that must be applied to ensure a reduced risk from a hazardous substance, with particular emphasis on waste input. According to Joint Research Centre (JRC, 2008), waste streams must be identified as hazard substances that can be controlled or avoided through measures such as listing positive or negative waste streams or specific characteristics, and limiting values of potential pollutants on output material (Figure 5.1).

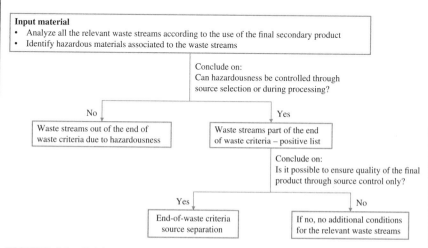

FIGURE 5.1   Guidance to develop end-of-waste input material criteria. *Source*: From JRC (2008)

Incineration with or without energy recovery emits a considerable list of pollutants: CO, CO_2, H_2O, HCl, HF, SO_2, NO_x, N_2O, heavy metals, and other metals, dioxins and furans, VOC, particulate matter, and other pollutants (EIPPCB, 2006b). Solid residues from incineration such as bottom ash or slag, boiler ash, and fly ash must also be considered hazardous. Landfill, which is seen as the most practical and least costly solution, can be a source of pollutants with a long-term release. Gaseous emissions identified from landfill are mainly biogas, which is largely composed of CH_4 and CO_2.

Leachate composition, however, can vary over time in both short and long terms and can include (Kjeldsen et al., 2002):

- dissolved organic matter, quantified as total organic carbon, volatile fatty acids (accumulated during the acid phase of waste stabilization) (Christensen and Kjeldsen, 1989), fulvic-like and humic-like compounds;
- inorganic macro components, including calcium, magnesium, sodium, potassium, ammonium, iron, manganese, chloride, SO_4^{2-}, and hydrogen carbonate;
- heavy metals, including Cd, Cr, copper, Pb, nickel, and Zn; and
- xenobiotic organic compounds from households or industrial chemicals and present in low concentrations (usually <1 mg/L^{-1} of individual compounds), including, among others, a variety of aromatic hydrocarbons, phenols, chlorinated aliphatics, pesticides, and plasticizers.

All these hazards can constitute a risk if released accidentally or even under normal conditions. Without understanding and studying the existing problems of waste management through risk assessment, risk management will be difficult. Relevant regulations can be found in the Organization for Economic Co-operation and Development (OECD) procedure for chemical products, an evaluation tool kit for environmental risk assessment and management of risk (OECD, 2013). Risk assessment and management for SWM are also promoted by the United Kingdom (UK) Environment Agency (DETR et al., 2000) (Figure 5.2).

The key regulatory elements include risk assessment, risk management, and risk communication. This division of elements (which is different from the one presented in Figure 5.2) is also called risk analysis; however, risk analysis and risk assessment are often used interchangeably due to their similar definitions. For example, according to Mullai (2006), risk assessment "combines both risk analysis and risk evaluation, providing practically useful and logically structured inputs and perspectives about risks to the decision-making process, development of policies, strategies and measures for managing risks."

Before conducting a formal risk assessment, the problem must be formulated, which can also be called a preliminary appraisal. Problem formulation consists of framing the problem, developing a conceptual model, planning the risk assessment, and screening and prioritizing the risks to be assessed. In SWM, framing the problem identifies the risk to what or whom, and where and when it will be checked (i.e., discover the case history, identify possible sources of hazards, and screen who or what has been affected). This preliminary appraisal is used to construct the conceptual model for risk assessment, which consists of a representative schematic of the problem

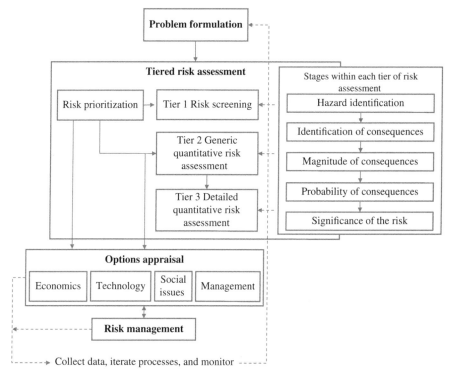

FIGURE 5.2   Framework for environmental risk assessment and management. *Source*: From DETR et al. (2000)

boundaries. The most well-known conceptual model which can be applied to SWM is the source–pathway–receptor model (SPR model); a rapid identification of these three elements is essential during the risk assessment phase. The key factors to construct the conceptual model include the timing, intensity, spatial extent, and duration of the event that control the hazard (Gormley et al., 2011).

More data and other resources are needed to support risk assessment planning. The preliminary appraisal ends with screening and prioritizing the risks to be assessed to help establish a basis for more environmental investigation (Asante-Duah, 1998). Risk screening may rely on the following components, including (Bradford-Hill, 1965):

- the plausibility of relationships between the source of a hazard and a receptor;
- the relative power of a hazard, or vulnerability of a receptor;
- the likelihood of an event, on the basis of historic occurrence or of changed circumstances; and
- the view on the performance of current risk management measures.

In summary, problem formulation includes defining spatial and temporal boundaries, identifying constraints on the assessment, considering uncertainties and assumptions, and developing a conceptual model related to all sources of hazard, all exposed receptors, and all pathways linking them together, which should be organized holistically at this stage. Without proper SPR linkage, impact cannot be quantified. Stakeholders and public participation involvement at this (early) stage can provide useful inputs and views to consider during decision-making.

5.2 RISK ASSESSMENT IN SOLID WASTE MANAGEMENT

Once the problem has been identified, the risk assessment process begins. Risk assessment is a tool or a process used to evaluate the consequence(s) of a hazard and its likelihood/probability in the ecosystem including human society. Both immediate and long-term effects are included to provide a basis for regulatory controls (Asante-Duah, 1998; Gormley et al., 2011; Chen et al., 2013). In this context, risk assessment is more related to environmental risk than other types of risk. Risk assessment can be approached from a simple to a complex level, depending on the nature and complexity of the risk, which will actuate different decision-making needs. If multiple levels of complexity exist, the process becomes a tiered approach.

Uncertainty is always present during risk assessment. According to Pollard et al. (2006), risk assessment has been widely applied in the waste management sector, and cases include but are not limited to risks to groundwater from landfills (Environment Agency, 1996; Hall et al., 2003), potential exposure to human health from continuous stack emissions (Harrop and Pollard, 1998), and potential health impacts from exposures to landfill gas (Attenborough et al., 2002). Risk assessment is conducted using models that relate the hazard to the resulting effects, which are characterized by the probability of an event to release the hazard and the resulting severity (or magnitude) of the event, thereby influencing the ultimate effects. Risk assessment can be divided in two processes: analysis and evaluation. According to Mullai (2006), risk analysis is a scientific process in which risks are identified, estimated, and presented in qualitative and/or quantitative terms by applying a wide range of methods, techniques, and tools. Risk evaluation is the process of comparing estimated risks with defined risk evaluation criteria to determine the level or significance of risks and provide advices for decision makers at various levels (European Commission, 1999).

The most well-known and acceptable risk assessment model was developed by the United States National Academy of Sciences in 1983 (NRC, 1983), who divided risk assessment into four main stages: hazard identification, dose–response assessment (or effects assessment), exposure assessment, and risk characterization. Another similar classification was proposed by Gormley et al. (2011): identify the hazard(s), assess the consequences, assess the consequence probabilities, and characterize risk and uncertainty. Fairman et al. (1998) defined (environmental) risk assessment with additional steps in the context of problem formulation, which include hazard identification, release assessment, exposure assessment, consequence assessment (or dose–response assessment, which includes risk estimation), and risk characterization. The stages describing risk analysis and risk evaluation can be seen in Table 5.1.

TABLE 5.1 Risk assessment steps arranged by different sources

Risk assessment	NRC (1983)	USEPA (1989)	Fairman et al. (1998)	Mullai (2006)	Gormley et al. (2011)
Risk analysis	Hazards identification	Hazards identification	Problem formulation Hazards identification	System definition Hazards identification	Hazards identification
	Dose–response	Dose–response assessment	Hazards identification	Exposure/ consequences analysis	Assessment of consequences
	Exposure assessment	Exposure assessment	Dose–response	Likelihood/ quantification Risk estimation/ presentation	Assessment of consequences probability
Risk evaluation	Risk characterization	Risk characterization	Risk characterization	Risk evaluation	Risk characterization and uncertainty

177

5.2.1 Risk Assessment Steps

Hazard identification establishes agents that could possibly cause harm to the receptor of interest (Fairman et al., 1998). The identification can begin with the existing effects and work backward to discover the source of the effect, or can begin with previous knowledge of the hazards to be studied. This process aims not only to address the hazard itself, but also to observe the situations that cause them (normal and/or accidental situations). Hazard identification is a qualitative assessment of the existing hazards that gathers and evaluates the effects as well as investigates how the exposure occurred. According to Asante-Duah (1998), hazard identification consists of identifying contamination sources, compiling lists of all contaminants present at the local, and selecting chemicals of potential concern based on their hazardous properties.

In SWM, hazard identification can be complex due to the various hazardous substances that can be released from any possible waste operation. If problem formulation is not properly conducted, risk assessment can become difficult and complex, and possibly unsuccessful. The definition of the SWM system and its description can be helpful at this stage. For example, in a composting unit, risk management can be devoted to the compost quality to be applied to agriculture, the odor problems in the neighborhood, or even the public health impact from the composting operation. Although the focus can be linked with, for example, public health impact, risk management can be devoted to one or more substances, such as odors (which can include a long list of VOC compounds) or bioaerosols, which can also include a considerable list of hazards. Sykes et al. (2007) conducted a case study to assess and manage potential public health risks from exposure to bioaerosols from commercial composting activities. They began their hazard identification of bioaerosols process with a literature review to highlight biological hazards such as fungi, bacteria, actinomycetes, endotoxins, and 1–3 β-glucans. In cases where the hazard identification is complex, a screening process can be used to help identify the most important hazards, such as the hazard identification screening process with filters, like the one applied to green waste compost (Box 5.2).

After hazards are identified, the next critical step is to assess the consequences (i.e., also known as exposure and consequences analysis) that may arise from the hazardous release to the receptor. Consequences can be evaluated through dose–response assessment and exposure assessment. A dose–response assessment uses data to estimate the amount of material that produces a known effect in humans or other beings (Vaccari et al., 2005). The results from dose–response assessment are endpoints (i.e., key parameters related to the effect caused), which can be a lethal concentration (LC) that makes 50% of population in the test affected (known as LC_{50}). Exposure assessment determines the extent to which a population is exposed to the hazardous material, the fate and transport of the material in the environment, as well as the media, pathways, and routes of exposure in regard to the extent, character, duration, magnitude, and frequency (Vaccari et al., 2005; DHAHC, 2012).

BOX 5.2   HAZARD IDENTIFICATION AND SCREENING (Hough et al., 2012)

Hough et al. (2012) conducted a quantitative risk assessment for the use of source-separated green waste compost used for livestock, which has been certified by PAS100 standard (UK norm for compost). The first step of the risk assessment is hazard identification. The intent of the research was to ensure an independent and nonbiased hazard identification for a situation in which information was scarce. A successive series of filters were combined to identify the hazards to be studied (Figure 5.3).

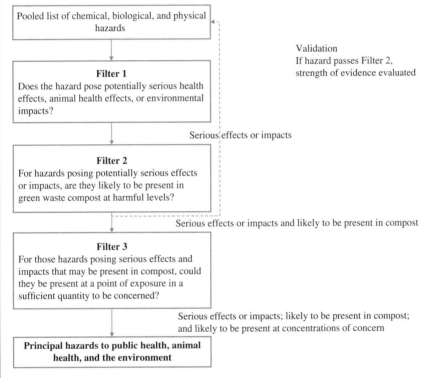

FIGURE 5.3   Flowchart for identifying principal animal health, public health, and environmental hazards from the application of source-segregated green waste compost. *Source*: From Hough et al. (2012)

Following hazard identification, the likelihood that potential impact will occur can be expressed as a probability or frequency. There are three aspects of the likelihood of consequences, including (Gormley et al., 2011):

- the probability of occurrence of an initiating event: people often quantify the risk elements using a probability event tree through probabilistic risk assessment;

- the probability of exposure to the hazard: people often estimates the size/extent of risk that receptors are exposed to using dispersion models such as the Gaussian plume model and geographical information system (GIS) to account for odors from biological treatment units, or groundwater plume dispersion associated with leachate source to obtain factors such as magnitude, duration, and spatial extent of the exposure after the hazard is released; and
- the probability of the receptors being affected by the hazard: people often predict the magnitude of the hazards and the consequences to the receptor through a dose–response relationship associated with predetermined knowledge in health risk assessment.

Probabilistic risk assessment is often used to quantify the frequency of occurrence, the magnitude of consequences, and the response of the system to the hazard. Several different techniques are available, including event tree analysis, fault tree analysis, cause-consequence analysis, failure modes and effects analysis, reliability block diagrams, hazard analysis, and hazard and operability study (also known as HAZOP). The most commonly used analyses are event and fault trees. Event-tree analysis is a diagram that represents the consequence of an event that leads to the release of a hazard. The analysis begins by initiating the event and identifies the subsequent consequences of environmental impact and/or human risk due of the release (Pollard et al., 2006). A fault tree analysis selects an undesired event (such as groundwater contamination; Figure 5.4) and traces it back to the possible causes, which can be component failures, human error, or other factors that could lead to the undesired event (Pollard et al., 2006). The causes are related using Boolean logic relationships (i.e., AND/OR "gates") to identify and model the root causes of the system failure (Pollard et al., 2006).

Risk characterization and uncertainty analysis is the last step of risk assessment, which can be divided into risk estimation, risk presentation, and risk evaluation. According to Gormley et al. (2011), this step will determine the qualitative and, if possible, quantitative likelihood of occurrence of the known event with potentially adverse effects. In this context, an activity or agent is exposed to defined conditions at a receptor location, given the assumptions and uncertainties (OECD, 2011). Frequently, risk characterization summarizes and then integrates outputs of the exposure assessment and likelihoods that define risk levels (Asante-Duah, 2002). Exposures resulting in the greatest risk can be identified in this process, and mitigate measures can then be selected to address the situation (Asante-Duah, 1998). This process has an inherent subjectivity due to the risk perception step that evaluates or categorizes the risk (Power and McCarty, 1998).

Risk characterization or evaluation can be achieved through a reference to some pre-existing measure, like environmental quality standard, or by a reference to a previously defined social, ethical, regulatory, or political standards (Gormley et al., 2011). For example, determining the likelihood of adverse consequences from a lake pollution event by comparing contaminant concentration in lake water with guideline values (Gormley et al., 2011); or evaluating the significance of an environmental toxin

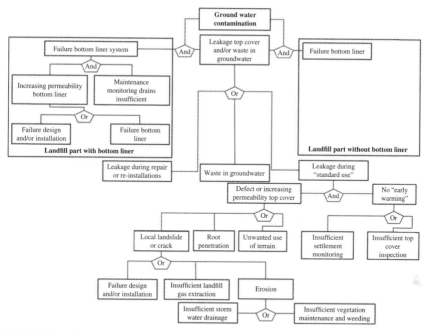

FIGURE 5.4   Fault tree for groundwater contamination from landfill. *Source*: Adapted from Boerboom et al. (2003)

to the ecosystem by calculating the ratio of the measured or estimated environmental exposure concentrations to the predicted no-effect concentration (PNEC). Mullai (2006) proposed several steps to conduct risk evaluation which include (1) choose risk evaluation criteria (such as human risk and environmental risk), (2) compare estimated risks against risk criteria, and (3) prioritize/rank risks according to their significance. The result from risk characterization is used to define risk management strategies, to be addressed in Section 5.3.

Uncertainties are also a concern in risk assessment process, and a reliable risk assessment must recognize and treat the various sources of uncertainty (Hayes et al., 2007) by relying on trusted sources of information, expert judgment, sensitivity analysis, Bayes linear methods, probability analysis techniques (such as the Monte Carlo simulation, described in Box 5.3), and fuzzy sets.

BOX 5.3   UNCERTAINTY QUANTIFICATION IN RISK ASSESSMENT FOR SWM

Sensitivity analysis determines if changing the input variables can change the output results. The analysis is conducted by testing data on the range of values of the model parameters (Asante-Duah, 1998) to determine which variable inputs most strongly influence the results. The Monte Carlo simulation seems to be a simpler procedure, however, and is the most applicable technique to address uncertainty quantification in SWM risk assessment. The use of Monte Carlo simulation is justified by the result presented as frequency or probability distribution graph to support probability risk assessment when modeling event trees and fault trees.

Monte Carlo simulation "is a statistical technique in which a quantity is calculated repeatedly, using randomly selected/generated scenarios for each calculation cycle, and typically presenting the results in simple graphs and tables" (Asante-Duah, 1998). It involves nominating a joint probability distribution to the input variables; the procedure returns a concomitant distribution which is a consequence of the assumed distributions of the model inputs and the considered functional form of the model (Asante-Duah, 1998). An example of a Monte Carlo simulation applied to include uncertainty in risk characterization results was developed by Schuhmacher et al. (2001) to conduct a risk assessment for an MSW incinerator using direct and indirect exposure parameter distributions. Below is an example of direct exposure distribution only (Figure 5.5).

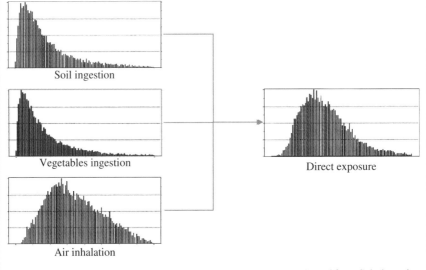

FIGURE 5.5   Illustration of Monte Carlo simulation. *Source*: Adapted from Schuhmacher et al. (2001)

5.2.2 Risk Assessment Models

Assessing and managing risk associated with various SWM technologies is vital for the reduction of impacts on environment and public health. For example, due to potential hazards produced from landfills, many software tools have been developed to provide risk assessment, in full or in part, including LandSim (Environment Agency, 1996), Hydro-geological Evaluation of Landfill Performance (HELP tool) (US Army Corps of Engineers and USEPA, 1998), GasSim (Attenborough et al., 2002), and GasSimLite (Environment Agency, 2002). Some computer-aid software specific for landfills, such as the simulation reactor (Allgaier and Stegmann, 2003), have also been developed and applied.

In a sustainable SWM system, risk assessment and management can play a critical role in the decision-making process during siting waste treatment and disposal units. To improve performance and to answer sustainability needs, integration with other techniques and models is sometimes required. For instance, with the aid of a model/software Environmental and Human Health Risk Assessment Geographical Information System (EHHRA-GIS) applied by Morra et al. (2006), risk assessment has been used with GIS to assess small rural landfills (Victoria et al., 2005) and to prioritize uncontrolled landfills (Masi et al., 2007). In this context, EHHRA-GIS was used to assess human health risk of a delimited region where the pollution sources were a municipal incinerator, a closed landfill, a platform for urban solid waste treatment, and two open pits for the production of bitumen. The integration with GIS results in animation of topical maps at the graphics user interface was deemed extremely useful for an intuitive understanding through visualization (Fedra, 1998) when public participation is required. Risk assessment and management have also been used as decision support tools for site selection of landfills (Nakaishi et al., 2005; Vaccari et al., 2005; Río et al., 2011).

5.3 MANAGEMENT OF RISK

The remaining question is how to manage the identified risk (i.e., determine possible techniques to apply). Stages involved in risk management are to identify the stakeholders and decision makers involved, the important risks to be avoided/eliminated or reduced, and the key strategies/options needed to perform the operation (Mullai, 2006).

The strategies/options assessment is the process where occurs the determination and selection of the most adequate risk management strategy given the constraints from the decision makers (HM Treasury, 2003; Gormley et al., 2011). This stage may involve scoring, weighting, and/or reporting different risk management options. Risk management options can take one of the following forms to (Gormley et al., 2011):

- determine the risk source where possible;
- reduce the effects by developing environmental management techniques or engineered systems;
- transfer the risk through new technology, procedures, or investment;
- accomplish the potential benefits of the risk by embracing new opportunities; and
- tolerate the risk by not interceding with new or existing situations.

To select the favored option, the likely positive and negative impacts correlated with each option are considered based on technical and economic factors, environmental security, social issues, and organizational capabilities (Gormley et al., 2011).

Methods and techniques available to conduct options appraisal can be cost–benefit analysis, trade-off analysis, and multi-criteria analysis. The results can inform decision makers of the risk management measures to be taken at the source, pathway, and/or receptor.

Addressing the risk involves undertaking an action, procedure, or operation to meet the objectives of the risk management strategy (Gormley et al., 2011). The execution of risk management strategy requires a documented rationale, and the actions to be taken must be explicit and lucid. Environmental monitoring plays an important role in controlling the effectiveness of the risk management measures. The results can be used as a comparative basis in future mitigation or avoidance measures.

5.4 RISK COMMUNICATION

If risk assessment is a technical and scientific approach, risk management usually involves subjectivity from stakeholders involved in the process. This multidisciplinary approach, where economic, political, social, and environmental aspects are considered along with subjectivity tasks calls for appropriate and effective facilitator tools/techniques, is known as risk communication (Asante-Duah, 2002).

Risk communication is the exchange of risk information concerning a particular hazard that addresses what can and is being done between risk managers and the public (USEPA, 1990) to manage the hazard and its consequences. Risk communication is defined by Ishizaka and Tanaka (2003) as the process in which each party (government, public/private companies, citizens, civic groups) repeatedly exchanges information and opinions concerning "the risk estimated based on the scientific method," "risk perception," "countermeasures," "risk assessment," and "the extent of acceptable risks" to enhance the level of mutual understanding and trust. Instead of a two-way exchange of information, however, risk communication can also be a one-way transfer of hazard and risk information. In this case, risk communication works as an alert rather than a discussion during the decision-making process for risk assessment and management.

Recently, the role of risk communication has become a vital activity in concert with risk assessment and management that impacts all stages of the process. The International Risk Governance Council Framework (Renn, 2005) requires risk communication from the framing of the risk situation to the implementation and monitoring of measures. During risk assessment, risk communication is used to gain local information concerning potential hazardous effects as well as the concerns of citizens. In risk management, risk communication can provide a forum to discuss the nature, magnitude, significance, or control of risks and related consequences with one another (Asante-Duah, 2002). Risk communication can be a powerful tool to ensure implementation of a risk assessment and management program. Defining all goals of the risk communication is vital to public acceptance of the risk associated with the hazards. Risk communication aims to (USEPA, 1990; Asante-Duah, 2002):

- reach the agreement between the magnitude of a risk and the public's political and behavioral answer to this risk;

- advance public knowledge of environmental and health risks associated with different technologies;
- boost opportunities for public officials to get information from the public regarding their concerns about the potential risk from different technologies;
- enhance public awareness of the safety procedures of the facility;
- assure the public as to how the public can collaborate in the siting process of SWM facilities and what actions they can take to reduce their personal risk; and
- help active participants, and even possible active observers, make informed contributions to the decision-making process and make informed decisions about how to reduce their own risk.

According to USEPA (1990), performing risk communication to increase the likelihood that information will be available to the public is a five-step procedure in order to: (1) identify risk communication, (2) determine the information exchange needed, (3) identify groups or interests with whom information must be exchanged, (4) develop appropriate risk messages for each targeted audience, and (5) search for appropriate channels for communicating risks to different segments of the public. Other risk communication procedures, such as those from OECD (2002) and the Chemical Society of Japan (2001), were developed as a means for conflict resolutions regarding the siting of waste incinerators and landfills due to their environmental and human health impacts. Finally, a typical risk communication procedure (Figure 5.6) provides support for the decision-making process.

An important aspect to consider during risk communication is risk perception. According to Fairman et al. (1998), risk perception "involves people's beliefs,

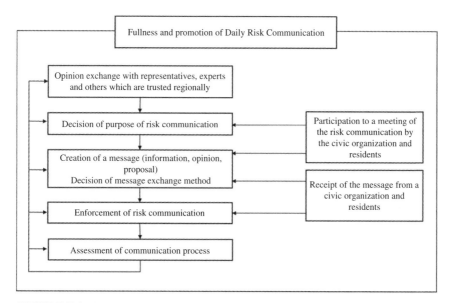

FIGURE 5.6   Process of risk communication. *Source*: From Ishizaka and Tanaka (2003)

attitudes, judgments, and feelings, as well as the wider social or cultural values that people adopt towards hazards and their benefits." Risk perception will be a main element in whether a risk is deemed to be acceptable and measures imposed are seen to resolve the problem (Fairman et al., 1998). Risk communication aims to inform the population related to the risk. Yet, few studies have assessed how this influence is truly affecting people's emotions (Siegrist, 2012). Risk communication is not limited to any group of the public, but it should be devoted to all stakeholders involved in the management of risk.

In a risk communication process, risk perception influences how risk messages are constructed because some risks are more acceptable than others (Sprent, 1988; USEPA, 1990; FACN, 1995; Mullai, 2006):

- voluntary risks can be accepted more readily than those which are imposed;
- risks under individual control are more easily accepted than those under government control;
- fair risks are more acceptable than those that seem unfair;
- risks that are "dreaded" are less acceptable than those that carry less anxiety;
- undetectable risks create more fear than detectable risks;
- physical distance from a site affects the risk acceptability;
- rumor, disinformation, dispute, and the sheer volume of information all may interact to give a misperception of risk;
- familiar risks are regarded as more acceptable;
- risks where uncertainty exists for the effects, severity, or prevalence of the hazard tends to escalate unease;
- for the public, large-scale disasters weigh more seriously than small-scale;
- risks enclosed in relevant events have greater impact than risks that arise in less prominent circumstances; and
- risks with immediate consequences are less accepted than those with delayed effects.

Risk communication is one more aspect that highlights the importance of public participation in decision-making, especially in SWM. The societal component of a decision-making process is becoming vital to ensure a sustainable management. In this particular case of SWM, risk communication is addressed to help assess and manage risk related to solid waste, which can be connected to public participation, addressed in Chapter 3.

5.5 HOW TO PROMOTE A SUSTAINABLE SOLID WASTE MANAGEMENT WITH RISK ANALYSIS?

Environmental impacts resulting from SWM can be addressed by not only risk assessment and management of risk but also life-cycle assessment (LCA) and environmental impact assessment (EIA), and various impact analysis methods that can identify, quantify, and manage risks. Both routes described above present advantages that contribute to a sustainable SWM system; yet they also have drawbacks that can influence their role in the proposed sustainability analysis.

A life-cycle impact assessment (LCIA) is deemed the core of LCA and aims to establish a linkage between a system and potential impacts. A proposed LCIA is comparable to risk analysis (Margni and Curran, 2012). A potential disadvantage is that the "models used within LCIA are often derived and simplified versions of more sophisticated models within each of the various impact categories. These simplified models are suitable for relative comparisons of the potential to cause human or environmental damage, but are not indicators of absolute risk or actual damage to human health or the environment" (Margni and Curran, 2012).

An EIA applies the implications of risk analysis to determine likely environmental consequences associated with a proposed human activity (Asante-Duah, 1998). The principal objective of an EIA is to ensure that environmental considerations are incorporated into the planning, decisions, and implementation of development activities, and to ensure that adverse effects are prevented or at least minimized (Asante-Duah, 1998). EIA deals with future events in which uncertainty is inherent to the decision-making process. The expected impacts of the project are qualitative, but even so, results could also be obtained without a systematic methodology. EIA cannot quantify the potential risks *per se*. In this regard, risk assessment and management of risk can aid in the design of cost-effective sampling, data collection, and data evaluation programs of EIA (Asante-Duah, 1998) as well as assess the same project to specifically address uncertainty and to estimate the probability of event occurrence (Turnbull, 1992). Therefore, the systems engineering features of risk assessment and management of risk can be instrumental in concert with an EIA endeavor because they would incorporate information on impact probabilities, making the EIA quantifiable and less theoretical.

The initial stages of promoting sustainable SWM focus on a systematic and holistic approach with intergenerational concerns, which can be applicable on a significantly large scale of solid waste problems. SWM is, however, a local problem that can eventually have regional and even global impacts and therefore must be addressed by a suite of systems engineering tools and methods. When environmental impacts are estimated and quantified, the importance of the problem will be easier to explain to the public and stakeholders if the analysis includes information such as when the problem may occur, where will occur, and how it can be mitigated. This local-scale impact quantification (even with uncertainty) approach can be observed by risk assessment and management. LCA with synergistic connection to risk assessment and management of risk can be developed to reduce information gaps, aiding the risk communication process with stakeholders. In addition, integration of LCA and EIA under the risk assessment umbrella can be used to identify hotspots and moments for those impacts in a product life cycle, and site-specific tools can be used to understand the severity of the impact and whether it significantly affects the local environment (Hunter et al., 2012). In any circumstances, risk assessment and management of risk for SWM can be the site-specific tool to focus on (Pollard et al., 2006):

- the assessment of an expanded group of hazardous factors (slope, stability, landfill gas, hazardous trace components, issues of odor, risks from unit operations, combined events, such as flooding and inventory loss) from individual facilities;
- aggregated risks at integrated, multi-process facilities; and
- the risks integration that have been evaluated uniquely by reference to individual legislative demands.

5.6   FINAL REMARKS

Environmental and health risk estimates are calculated to be used in setting standards, in cleanup levels for hazardous waste, or in exposure level that is believed to be safe or associated with some risk (Felter and Dourson, 1998). This chapter discusses the general concept of risk assessment and management of risk in support of coupled risk and legal aspects. Given that risk analysis is an inexact science, most risk estimates are calculated in a precaution perspective, rather than predictive of actual toxicity. The most salient example is cancer potency factors that are presented as the 95% upper confidence limit on the dose–response curve, rather than the maximum likelihood estimate in risk assessment (Felter and Dourson, 1998). The values obtained by such risk assessment are imprecise although it can be a critical decision-making factor. Due to such reason, the risk manager has to be able to disseminate information for the general public that wants to know with precision which the exact risks are. To meet this need, scientific judgments that are difficult to gain require developing some professional information for risk measurement, analysis, and communication. Thus, how the environmental informatics may come to play in support of this need would be an essential knowledge based in sustainable SWM.

REFERENCES

Allgaier, G. and Stegmann, R. 2003. Development of a new risk assessment model for small old landfills. In: Proceedings Sardinia 2003, Ninth International Waste Management and Landfill Symposium, Cagliari, Italy.

Asante-Duah, D. K. 1998. *Risk Assessment in Environmental Management*, John Wiley & Sons, Chichester, UK.

Asante-Duah, D. K. 2002. *Public Health Risk Assessment for Human Exposure to Chemicals*, Kluwer Academic Publishers, Dordrecht, the Netherlands.

Attenborough, G., Hall, D. H., Gregory, R. G., and McGoochan, L. 2002. GasSim: Landfill gas risk assessment model. In: *Waste 2002* (Eds. Lencioni, E. and Dhamda, R.), The Waste Conference Limited, Warwick, UK.

Boerboom, A. A. M., Foppen, E., and Van Leeuwen, O. 2003. Risk assessment methodology for after-care of landfills based on the probabilistic approach. In: Proceedings Sardinia 2003, Ninth International Waste Management and Landfill Symposium, Cagliari, Italy.

Bradford-Hill, A. 1965. The environment and disease: Association or causation? *Proceedings of the Royal Society of Medicine*, 58, 295–300.

Chemical Society of Japan. 2001. *Risk Communication Methodology Guide on Chemical Substances*, Gyosei, Tokyo.

Chen, S., Chen, B., and Fath, B. D. 2013. Ecological risk assessment on the system scale: A review of state-of-the-art models and future perspectives. *Ecological Modelling*, 250(19), 25–33.

Christensen, T. H. and Kjeldsen, P. 1989. Basic biochemical processes in landfills. In: *Sanitary Landfilling: Process, Technology and Environmental Impact* (Eds. Christensen, T. H., Cossu, R., and Stegmann, R.), Academic Press, London, pp. 29–49.

Department of Environment, Transport and Regions (DETR), Environment Agency and Institute for Environment and Health. 2000. *Guidelines for Environmental Risk Assessment and Management: Revised Departmental Guidance*, HMSO, London.

Department of Health and Ageing and Health Council (DHAHC). 2012. *Environmental Health Risk Assessment: Guidelines for Assessing Human Health Risks from Environmental Hazards*, Commonwealth of Australia.

Environment Agency. 1996. LandSim: Landfill performance simulation by Monte Carlo method. *LandSim Manual*, ref CWM 094/96.

Environment Agency. 2002. *GasSimLite User Manual*, Golder Associates.

European Commission. 1999. The concerted action on formal safety and environmental assessment of ship operations. Report by Germanischer Lloyd and Det Norske Veritas, project funded by the European Commission under the Transport RTD Programme of the 4th Framework Programme.

European Integrated Pollution Prevention and Control Bureau (EIPPCB). 2006a. Reference document on the best available techniques for the waste treatment industries. European Commission.

European Integrated Pollution Prevention and Control Bureau (EIPPCB). 2006b. Reference document on the best available techniques for the waste incineration. European Commission.

European Parliament and Council. 2008. Directive 2008/98/EC of the European Parliament and of the Council of 19 November 2008 on waste and repealing certain Directives. *Official Journal of European Union*, L312, 3–30.

European Parliament and Council. 2012. Directive 2012/18/EU of the European Parliament and of the Council of 4 July 2012 on the control of major-accident hazards involving dangerous substances, amending and subsequently repealing Council Directive 96/82/EC. *Official Journal of the European Communities*, L197, 1–37.

Fairman, R., Mead, C. D., and Williams, W. P. 1998. *Environmental Risk Assessment - Approaches, Experiences and Information Sources*, European Environment Agency, Copenhagen.

Fedra, K. 1998. Integrated risk assessment and management: Overview and state of the art. *Journal of Hazardous Materials*, 61(1–3), 5–22.

Felter, S. and Dourson, M. 1998. The inexact science of risk assessment (and implications for risk management). *Human and Ecological Risk Assessment*, 4(2), 245–251.

Foundation for American Communications and National Sea Grant College Program (FACN). 1995. *Reporting on Risk: A Handbook for Journalists and Citizens*, Annapolis Center.

Gormley, A., Pollard, S., Rocks, S., and Black, E. 2011. Guidelines for Environmental Risk Assessment and Management - Green Leaves III. Cranfield University and DEFRA.

Hall, D. H., Drury, D., Smith, J., Potter, H., and Gronow, J. 2003. Predicting the groundwater impact of modern landfills: Major developments in the approach to landfill risk assessment in the UK (LandSim 2.5). In: Proceedings Sardinia 2003, Ninth International Waste Management and Landfill Symposium, Cagliari, Italy.

Harrop, D. O. and Pollard, S. J. T. 1998. Quantitative risk assessment for incineration: Is it appropriate for the UK? *Water and Environmental Journal*, 12(1), 48–53.

Hayes, K. R., Regan, H. M., and Burgman, M. A. 2007. Introduction to the concepts and methods of uncertainty analysis. In: *Environmental Risk Assessment of Genetically Modified Organisms* (Eds. Kapuscinski,A. R., Hayes, K. R., Li, S., and Dana, G.), vol. 3, pp. 188–208.

HM Treasury. 2003. *The Green Book Appraisal and Evaluation in Central Government*, The Stationary Office, London.

Hough, R. L., Booth, P., Avery, L. M., Rhind, S., Crews, C., Bacon, J., Campbell, C. D., and Tompkins, D. 2012. Risk assessment of the use of PAS100 green composts in sheep and cattle production in Scotland. *Waste Management*, 32(1), 117–130.

Hunter, S., Helling, R., and Shiang, D. 2012. Integration of LCA and life-cycle thinking within the themes of sustainable chemistry and engineering. In: *Life Cycle Assessment Handbook – A Guide for Environmentally Sustainable Products* (Ed. Curran, A. M.), Wiley – Scrivener, Beverly, MA.

Ishizaka, K. and Tanaka, M. 2003. Resolving public conflict in site selection process – a risk communication approach. *Waste Management*, 23(5), 385–396.

Joint Research Centre (JRC). 2008. End of waste criteria—Final report. European Communities, Seville, Spain.

Kaplan, S. and Garrick, B. J. 1981. On the quantitative definition of risk. *Risk Analysis*, 1(1), 11–27.

Kjeldsen, P., Barlaz, M. A., Rooker, A. P., Baun, A., Ledin, A., and Christensen, T. H. 2002. Present and long-term composition of MSW landfill leachate: A review. *Critical Reviews in Environmental Science and Technology*, 32(4), 297–336.

Margni, M. and Curran, A. M. 2012. Life cycle impact assessment. In: *Life Cycle Assessment Handbook—A Guide for Environmentally Sustainable Products* (Ed. Curran, A. M.), Wiley – Scrivener, Beverly, MA.

Masi, S., Caniani, D., Trulli, E., Perilli, G., Sdao, F., and Zirpoli, P. 2007. Environmental risk assessment and prioritization of uncontrolled landfills by using GIS technology. In: Proceedings Sardinia 2007, Eleventh International Waste Management and Landfill Symposium, Cagliari, Italy.

Morra, P., Lisi, R., Spadoni, G., and Maschio, G. 2006. An integrated quantitative approach for the assessment of human health risk in the Pace Valley of Messina with the EHHRA-GIS Tool, in Venice 2006. In: Biomass and Waste to Energy Symposium, Venice, Italy.

Mullai, A. 2006. *Risk Management System – Risk Assessment Frameworks and Techniques*, Safe and Reliable Transport Chains of Dangerous Goods in the Baltic Sea Region (DaGoB) Project Publication Series 5:2006, Turku School of Economics, Logistics, Turku, Finland.

Nakaishi, K., Igari, F., Kuwamoto, K., and Hanashima, M. 2005. The study on method of appropriate site selection considering environmental risk management. In: Proceedings Sardinia 2005, Tenth International Waste Management and Landfill Symposium, Cagliari, Italy.

National Research Council (NRC). 1983. *Risk Assessment in the Federal Government: Managing the Process*, National Academy Press, Washington, DC.

Organisation for Economic Co-operation and Development) (OECD). 2002. OECD Guidance Document on Risk Communication for Chemicals Risk Management - ENV/JM/MONO (2002)18. Environment Directorate Joint Meeting of the Chemicals Committee and the Working Party on Chemicals, Pesticides and Biotechnology.

Organisation for Economic Co-operation and Development (OECD). 2011. The OECD environmental risk assessment toolkit: Steps in environmental risk assessment and available OECD Products. OECD. Available at: http://www.oecd.org/document/46/0,3746,en_2649 _34373_44915438_1_1_1_1,00.html (accessed April 2011).

Organisation for Economic Co-operation and Development (OECD). 2013. The OECD environmental risk assessment toolkit: Tools for environmental risk assessment and management. OECD. Available at: http://www.oecd.org/chemicalsafety/risk-assessment/theoec denvironmentalriskassessmenttoolkittoolsforenvironmentalriskassessmentandmanagement .htm (accessed May 2013).

Pollard, S. J. T., Smith, R., Longhurst, P. J., Eduljee, G. H., and Hall, D. 2006. Recent developments in the application of risk analysis to waste technologies. *Environment International*, 32(8), 1010–1020.

Power, M. and McCarty, L. S. 1998. A comparative analysis of environmental risk assessment/risk management frameworks. *Environmental Science and Technology*, 32, 224A–231A.

Renn, O. 2005. *White Paper on Risk Governance – Towards an Integrative Approach*, International Risk Governance Council, Geneva, Switzerland.

Río, M., Franco-Uría, A., Abad, E., and Roca, E. 2011. A risk-based decision tool for the management of organic waste in agriculture and farming activities (FARMERS). *Journal of Hazardous Materials*, 185(2–3), 792–800.

Royal Society. 1992. *Risk Analysis, Perception and Management*, The Royal Society, London.

Schuhmacher, M., Meneses, M., Xifró, A., and Domingo, J. L. 2001. The use of Monte-Carlo simulation techniques for risk assessment: Study of a municipal waste incinerator. *Chemosphere*, 43(4–7), 787–799.

Siegrist, M. 2012. Special issue on the conference "Environmental Decisions: Risks and Uncertainties" in Monte Verità, Switzerland. *Journal of Risk Research*, 15(3), 235–236.

Slaper, H. and Blaauboer, R. 1998. A probabilistic risk assessment for accidental releases from nuclear power plants in Europe. *Journal of Hazardous Materials*, 61, 209–215.

Sprent, P. 1988. *Taking Risks: The Science of Uncertainty*, Penguin Books, Canada.

Sykes, P., Jones, K., and Wildsmith, J. D. 2007. Managing the potential public health risks from bioaerosol liberation at commercial composting sites in the UK: An analysis of the evidence base. *Resources, Conservation and Recycling*, 52(2), 410–424.

Theodore, L. and Dupont, R. R. 2012. *Environmental Health and Hazard Risk Assessment: Principles and Calculations*, CRC Press, Boca Raton, FL.

Turnbull, R. G. H. 1992. *Environmental and Health Impact Assessment of Developmental Projects: A Handbook for Practitioners*, Elsevier Applied Science, London (on behalf of the Centre for Environmental Management, Aberdeen and the World Health Organization, Geneva).

United States (US) Army Corps of Engineers and United States Environmental Protection Agency (USEPA). 1998. Hydrologic Evaluation of Landfill Performance (HELP). Landfill models. Available at: http://el.erdc.usace.army.mil/elmodels/helpinfo.html (accessed June 2013).

United States Environment Protection Agency (USEPA). 1989. *Risk Assessment Guidance for Superfund (RAGS): Volume I. Human Health Evaluation Manual (HHEM) (Part A, Baseline Risk Assessment),* Interim Final, Office of Emergency and Remedial Response, Washington, DC (USEPA/540/1–89/002, NTIS PB90—155581).

United States Environment Protection Agency (USEPA). 1990. *Sites for Our Solid Waste: A Guidebook for Effective Public Involvement,* USEPA.

United States Environment Protection Agency (USEPA). 2005. Introduction to United States Environmental Protection Agency Hazardous Waste Identification (40 CFR Parts 261). Information resources. USEPA. Available at: http://www.epa.gov/wastes/inforesources/pubs/training/hwid05.pdf (accessed June 2013).

Vaccari, M., Collivignarelli, C., and Vercesi, P. L. 2005. Risk analysis as a decisional tool for the location of landfills. In: Proceedings Sardinia 2005, Tenth International Waste Management and Landfill Symposium, Cagliari, Italy.

Victoria, I. N., Wenig, D., Strudwick, D., and Schroeder, S. 2005. Environmental risk assessment of landfills exempt from licensing. In: Proceedings Sardinia 2005, Tenth International Waste Management and Landfill Symposium, Cagliari, Italy.

PART II

PRINCIPLES OF SYSTEMS ENGINEERING

The use of formal systems engineering principles including top-down and bottom-up approaches is encouraged to evaluate solid waste management (SWM) alternatives. The following chapters are organized to illuminate the internal linkages among global changes, sustainability, and adaptive management strategies and to introduce systems engineering principles. While such a system-based approach related to the integrated SWM should be the norm, risk assessments may sometimes be applied usefully to aid in the decision-making if uncertainties come to bother the choice of adaptive management strategies.

- Linkages among global change, sustainability, and adaptive management strategies (Chapter 6)
- Systems engineering principles and decision-making (Chapter 7)
- Systems engineering tools for evaluating the significance of alternatives (Chapter 8)

Sustainable Solid Waste Management: A Systems Engineering Approach, First Edition. Ni-Bin Chang and Ana Pires.
© 2015 The Institute of Electrical and Electronics Engineers, Inc. Published 2015 by John Wiley & Sons, Inc.

CHAPTER 6

GLOBAL CHANGE, SUSTAINABILITY, AND ADAPTIVE MANAGEMENT STRATEGIES FOR SOLID WASTE MANAGEMENT

The global changes facing us today include population growth and migration, economic development and globalization, rapid urbanization, resources overexploitation and consumption, and climate change impacts. Integrated solid waste management (ISWM) is deemed a long-standing strategy in dealing with a part of the collective impact of these global changes. The concept of systems engineering can simultaneously address more internal and external factors that influence the decision-making process for ISWM, from planning to design and to operation. The purpose of this chapter is to reinvent the wheel of ISWM and discuss holistic concepts and strategies with respect to recent impacts of global changes from sustainability science and sustainable engineering perspectives. It leads to further magnifying the importance of ISWM in which waste collection, recycling, treatment, recovery, and disposal are handled together at different scales with respect to pollution prevention, material conservation, energy recovery, and ecosystem conservation.

6.1 GLOBAL CHANGE IMPACTS

The world has faced tremendous global challenges over the past two decades, a period of increased volatility and rapid change in all aspects of society, including population growth and migration, economic development and globalization, rapid urbanization, resources overexploitation and consumption, and climate change impacts. To meet these challenges, now and in the future, we must develop and enhance our ability

Sustainable Solid Waste Management: A Systems Engineering Approach, First Edition. Ni-Bin Chang and Ana Pires.
© 2015 The Institute of Electrical and Electronics Engineers, Inc. Published 2015 by John Wiley & Sons, Inc.

to think critically, assess relevant information thoroughly, and develop profound and reasoned arguments in the context of the global challenges associated with solid waste management (SWM). This is largely due to that SWM can be influenced not only by local, regional, and national drivers but also by global change impacts, which in turn affect ISWM systems at different scales, such as global technology advancements, national socioeconomic impacts, regional adaptive management strategies, and local cost–benefit–risk tradeoffs.

The status of SWM worldwide has been highlighted by several international organizations. Current global municipal solid waste (MSW) generation is around 1.3 billion tonnes per year and is expected to rise to almost 2.2 billion tonnes per year by 2025 due to global change, reflecting the increase in per capita MSW generation, from 1.2 to 1.42 kg per person per day over the next 15 years (Hoornweg and Bhada-Tata, 2012) (tonnes are metric tons). Nonetheless, global averages are expansive estimates only because rates vary greatly by region, country, and city, and even within cities (Hoornweg and Bhada-Tata, 2012). According to Hoornweg and Bhada-Tata (2012), costs for SWM globally will increase from today's annual US$205.4 billion to about US$375.5 billion in 2025, being most severe in lower income countries (more than fivefold) and lower middle income countries (more than fourfold). The Environmental Outlook 2030 (OECD, 2008a) presents key messages in regard to waste and material problems facing Europe in 2030 due to:

- illegal shipments and incorrect management of waste materials and products, which represent a significant risk for human health and the environment;
- management of rapidly increasing MSW in countries outside the OECD, which will become an enormous challenge in the coming decades;
- increasing MSW generation in OECD countries since 2000. A relative decoupling of MSW generation in OECD countries from economic growth has been observed, given that waste generation is continuing to increase without regard to economic fluctuations (EEA, 2007);
- extended growth in the global demand for materials and the amounts of waste generated. Ordinary waste management policies alone may not be enough to improve the required production efficiency with different materials and offset the consequent waste-related environmental impacts.

Understanding how global change impacts are responsible for the increasingly complex issues in SWM will aid the search for possible solutions. In the next few sections, the major global change impacts will be characterized with regard to their influence on waste generation and waste management strategies.

6.1.1 Economic Development and Globalization

Economic development can be defined as a broad-based, steady increase in the comprehensive standard of living for individuals within a community (Greenwood and Holt, 2010). The concept goes beyond pure economic growth, which only relates to the increased total output or income at the national or global level. Although economic growth is measured through gross domestic product, economic development is usually measured through gross national income (GNI), formerly gross national product. According to World Bank (2013), GNI is the sum of value added by total citizens producers plus any product taxes excluded in the valuation of output, plus net receipts of primary income from outside. GNI, calculated in national currency, is usually converted into US dollars at official exchange rates for comparisons across economies (World Bank, 2013).

GNI and waste generation are strongly correlated (Table 6.1). Higher income countries produce almost five times more waste on average than lower income countries. Not only is waste generation affected by economic development, but waste composition can be influenced as well. Lower income countries and the lowest among high income countries usually have higher amounts of biodegradable waste (Table 6.2).

Total amount of organic waste tends to increase steadily as affluence increases, but at a slower rate than the nonorganic fraction (Hoornweg and Bhada-Tata, 2012). Waste streams in high income countries are rich in paper, plastics, and other inorganic materials, in opposition to lower income countries, where organic waste can present 64% of total waste. These differences in waste composition increase the importance of knowing specifically the waste being managed, more in a weight-based than in a volume-based composition, due to the implications in waste collection and transport and in treatment projects.

The final impact of economic development in SWM is the destination. Countries with higher income are characterized by landfilling and incineration, while lower and lower middle income countries tend to dispose waste in open dumps. The final destination of waste disposal can be categorized in terms of the income levels of different countries (Table 6.3). Waste collection also differs among countries in terms of income level; lower income countries have a 41% waste collection rate compared to 100% in higher income countries (Figure 6.1).

In addition to globalization, economic development also affects SWM across several economic, social, and political dimensions by influencing how waste is managed and its generation. Globalization is the integration of countries for the exchange of economic goods via bilateral or multilateral treaties, a process that embeds national economies into the international economic system. In the economic dimension, globalization can stimulate economic development through direct foreign investment by creating direct, stable, and long-lasting links between economies (OECD, 2008b). The social dimension reflects cultural flows between countries and information exchange. The political dimension is the intensive interactions of political systems through global or regional directives and conventions working at a global scale.

Globalization influences the flow of materials and products, which will eventually become waste in the receiving country; yet waste itself is also traded all over the world. Looking specifically at SWM, globalization influences:

- global recycling markets, which enforce recycling in developing countries; and
- transboundary shipment of waste.

TABLE 6.1   Waste generation relation with GNI

	Country	GNI (2011) (US$)	Waste generation (2012)
Lower income (US$1,025 or less)	Nepal	540	0.12
	Mozambique	470	0.14
	Bangladesh	780	0.43
	Sierra Leone	340	0.45
	Cote d'Ivoire	1,090	0.48
	Rwanda	570	0.52
	Zimbabwe	660	0.53
	Mali	610	0.65
Average		633	0.42
Lower middle income	India	1,410	0.34
(US$1,026–4,035)	Philippines	2,210	0.50
	Indonesia	2,940	0.52
	Nigeria	1,280	0.56
	Lao PDR	1,130	0.70
	Albania	3,980	0.77
	Pakistan	1,120	0.84
	El Salvador	3,480	1.13
Average		2,194	0.67
Upper middle income	Russia Federation	10,730	0.93
(US$4,036–12,475)	China	4,940	1.02
	Brazil	10,720	1.03
	Latvia	12,350	1.03
	Romania	7,910	1.04
	Venezuela	11,820	1.14
	Turkey	10,410	1.77
	South Africa	6,960	2.00
Average		9,480	1.25
Higher income (US$12,476 or more)	Korea, Rep.	20,870	1.24
	Qatar	80,440	1.33
	Japan	44,900	1.71
	Germany	44,270	2.11
	Portugal	21,210	2.21
	Australia	49,130	2.23
	Denmark	60,120	2.34
	US	48,620	2.58
	Norway	88,890	2.80
Average		49,813	2.07

Source: From Hoornweg and Bhada-Tata (2012) and World Bank (2013).
PDR, People's Democratic Republic.

TABLE 6.2   **Waste composition by income**

Waste composition (%)	Lower income countries	Lower middle income countries	Upper middle income countries	Higher income countries
Organic	64	59	54	28
Paper	5	9	14	31
Plastic	8	12	11	11
Glass	3	3	5	7
Metal	3	2	3	6
Other	17	15	13	17

Source: From Hoornweg and Bhada-Tata (2012).

TABLE 6.3   **MSW disposal categorized by income (million tonnes)**

Waste destination	Lower income	Lower middle income	Upper middle income	Higher income
Dumps	0.47	27[a]	44	0.05
Landfills	2.2	6.1	80	250
Composting	0.05	1.2	1.3	66
Recycling	0.02	2.9	1.9	129
Incineration	0.05	0.12	0.18	122
Other	0.97	18	8.4	21

[a]This value is relatively high due to the inclusion of China.
Source: From Hoornweg and Bhada-Tata (2012).

Due to globalization, Asian countries such as India and China have been growing economically and increasing their demand for raw materials at the same time that global resource prices are rising. Such phenomena have promoted a global recycling market where Asian countries are receiving considerable amounts of waste from all over the world for recycling; for example, the majority of waste plastics from the European Union (EU) is exported to Asia (Figure 6.2).

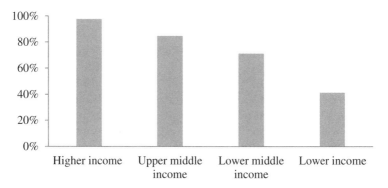

FIGURE 6.1   Waste collection in different income countries. *Source*: From Hoornweg and Bhada-Tata (2012)

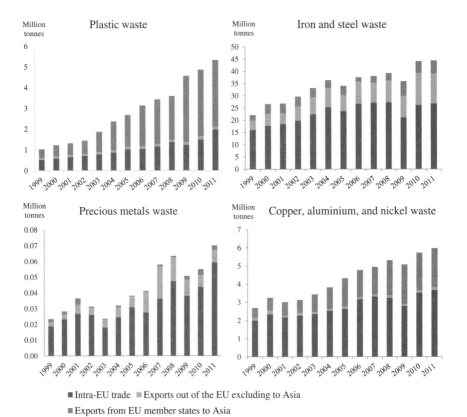

FIGURE 6.2 Exports of waste plastics and selected waste metals from EU Member States, 1999–2011. *Source*: Data from Eurostat (2012)

Transboundary shipment of waste is also a consequence of globalization and economic development. The impacts of the North America Free Trade Agreement (NAFTA) and trade liberalization on the generation, management, and shipment of industrial hazardous wastes in Mexico, Canada, and the United States (US) confirm this observation (Jacott et al., 2001). The loopholes that led to transboundary flows of e-waste (i.e., discarded electrical and electronic equipment, frequently referred to as waste electrical and electronic equipment (WEEE) in Europe) are also a result of the globalization impact on waste management (Salehabadi, 2013). The Basel Convention was established worldwide to control and supervise waste shipments from developed to developing countries to prevent damage to the environment (Basel Convention, 1989). The Convention defines notification requirements for the movement of hazardous waste and obliges the parties to reduce the generation of this particular waste and to guarantee its management is environmentally adequate (European Commission, 2012). In the EU, if an old or used electrical and electronic equipment (EEE) is functioning, it can be traded internationally for reuse because it is not yet being deemed waste. For this reason, e-waste shipments, which include computer components, cell phones, and televisions, exported for reuse in developing countries, have been highlighted by nongovernment organizations, Basel Convention competent authorities, INTERPOL, and other institutions due to improper dismantling (and recycling) conducted in lower income countries. Such recycling is attractive because e-waste contains valuable materials that can be recovered including iron, aluminum, copper, gold, silver, platinum, palladium, indium, gallium, and rare earths metals.

Processing e-waste inevitably generates toxins, due to brominated flame retardants, and heavy metals releases, making them hazardous (Box 6.1); however, "traffickers typically mislabel containers and mix electronic components with legitimate consignments" (Liddick, 2010). In the United Kingdom (UK), "e-waste tourists" visit the country to purchase e-waste to extract precious metals, and then dump the leftovers because proper disposal would eliminate their profits (Liddick, 2010). According to INTERPOL (2009), US electronics recyclers who charge a fee for disposal mention that rivals who offer free disposal or that pay for electronic waste are likely disposing off the e-waste improperly, or else they cannot remain in business.

BOX 6.1 WHERE ARE WEEE IN AFRICA? (Schluep et al., 2011)

The publication *Where are WEEE in Africa?*, funded by the European Commission, the governments of Norway, UK, and Dutch Recyclers Association, assessed and evaluated trends of EEE imports, use, and e-waste generation in West African countries: Benin, Côte d'Ivoire, Ghana, Liberia, and Nigeria (Schluep et al., 2011). The study results showed that used EEE from industrialized countries is entering African countries. Schluep et al. (2011) mentioned that the majority of this imported EEE is destined for reuse after testing and repair, however, significant volumes are unsuitable for reuse and add e-waste to local generation, and also that West Africa serves as the major trading route of used EEE into the African continent, being the important countries Ghana and Nigeria. In Ghana in 2009, for example, a majority of imports was used EEE, 30% of which was determined to be nonfunctioning (i.e., should have been defined as e-waste); half of this amount was repaired locally and sold to consumers, and the other half was irreparable (Schluep et al., 2011).

Such e-waste can reach the informal recycling sector, which collects and provides manual dismantling, open burn sites to recover metals, and open dumps for the residual fractions. This kind of inappropriate waste management releases dangerous gases that affect human health and the environment. In addition, several critical raw materials are not recovered. At the socioeconomic level, e-waste mainly affects the informal sector, where refurbishing generates income for more than 30,000 people in the cities of Accra (Ghana) and Lagos (Nigeria) (Schluep et al., 2011).

The WEEE report highlights the challenges from illegal transboundary shipment of e-waste that African countries must confront, especially how to prevent the import of e-waste without hampering the socioeconomically valuable trade of high quality used EEE (Schluep et al., 2011).

A new trend in the global market is the waste trade for energy recovery of mixed MSW, refuse-derived fuel (RDF) and solid recovered fuel (SRF). The waste-to-energy market has been increasing in Europe mainly promoted by the overcapacity of incineration plants at Northern Europe, like has been brought by Sora (2013). Overcapacity has occurred due to increased selective waste collection and landfill bans, which have promoted the recycling sector and reduced the amount of residual waste. As a consequence, gate fees for incineration and waste-to-energy plants have decreased, making them affordable for countries that have problems reaching landfill deviation targets or simply for economic reasons. In this way, mixed and residual MSW have been traded between European countries, although there is no full inventory of such waste amounts. A similar picture has been highlighted for RDF. UK has been producing RDF/SRF and has exported it to countries like Denmark, Germany, the Netherlands, Norway, Sweden, Estonia and Latvia in a total of 890,000 tonnes as of 2012 (Ovens et al., 2013).

6.1.2 Population Growth and Migration

According to the "2010 Revision of World Population Prospects" (UNDESA, 2011), the global population is predicted to increase from 6.9 billion in mid-2011 to 9.3 billion in 2050 and in 2100 could reach 10.1 billion. In 2011, 52.1% of the world's population was living in urban areas, and by 2050 that number is expected to increase to 67.2% (UNDESA, 2012). This migration of population from rural to urban areas will increase the phenomenon of megacities, high density metropolises of more than 10 million inhabitants.

Today, 21 megacities across the world are home to less than 10% of the global urban population (UN-Habitat, 2008). In 2025, that number is expected to increase to 29%, mostly located in Asian and other developing countries. This accelerated urbanization in developing countries will increase waste generation in places with no waste collection systems, and urban residents produce about twice as much waste as their rural counterparts (Hoornweg and Bhada-Tata, 2012). Intensive urbanization and growing populations also lead to higher incomes for urban dwellers than their rural counterparts, leading to more consumption of goods and services in highly urbanized areas. However, waste collection and recycling may or may not be relatively easier to organize in these populated areas where efficient waste management is most critical (OECD, 2004).

The informal sector in most developing countries is represented by waste pickers, rag pickers, scavengers, junk shops, and street vendors with a decentralized operational pattern. They are single residents or enterprises engaged in recycling and waste management but are not supported, financed, acknowledged, or permitted by the formal SWM authorities, operating in violation of or in competition with formal authorities (Wehenpohl and Kolb, 2007; Scheinberg et al., 2011; Wilson et al., 2012).

The modernization of the waste sector occurred at a time when the informal sector ceased to be robust, partially due to diseconomy of scale that cause such smaller firms to produce goods and services at increased per-unit costs (UN-Habitat, 2010). The same evolutionary process is difficult in developing countries because of the geographic heterogeneity of the disseminated informal sector devoted to recyclable materials collection. In Bamako, Mali, more than 120 self-employed microenterprises collect approximately 300,000 tonnes of waste annually, while in Lusaka, Zambia, informal service providers reach out to 30% of the city (UN-Habitat, 2010). The presence of such a robust recycling informal sector is not only due to the global market for recyclables, but also due to high density areas where the low income population needs economic resources.

Waste management in megacities is a difficult task that has been addressed only recently by the study of UN-Habitat (2010). Yet, megacities nowadays have largely adopted a centralized ISWM, although the informal sector might have dominated in the early stages. The growing amount of waste generated by growing populations in megacities is not being properly handled oftentimes by outdated infrastructure, which in turn creates public health issues. Several megacities have been promoting decentralized SWM systems, which have lower investment and operation costs and are based in waste recycling, including biological recycling of organic waste.

6.1.3   Resources Overexploitation and Limitations

Over the course of the twentieth century, the world increased its fossil fuel consumption by a factor of 12 while extracting 34 times more material resources (European Commission, 2011, Box 6.2). On a per capita basis, resource extraction levels are highest in the OECD area, specifically in North America and the Asia-Pacific region, and are predicted to rise further to reach around 22 tonnes per capita in 2020, mainly due to growing demands for coal, metals, and construction materials (OECD, 2008a). EU-25 countries rank second, presenting 16 tonnes per capita with more or less stable figures over the course of the twentieth century. Countries such as China, India, Indonesia, Pakistan, Thailand, Egypt, Nigeria, South Africa, Brazil, Mexico, and Turkey show the highest growth in per capita resource extraction (by 60%, up to 9 tonnes in 2020) due to rapid economic development and lower population growth comparatively to other developing countries (Giljum et al., 2008).

BOX 6.2   CRITICAL RAW MATERIALS FOR THE EUROPEAN UNION (European Commission, 2010)

The framework of the EU Raw Materials Initiative identifies a list of critical raw materials at the European Union level. To do so, two types of risks are identified related to the supply chain: (a) the supply-side risk due to recycling and substitution, taking the production into account with the potential for substitution and the corresponding recycling rate of the material of interest; and (b) the "risk in relation to environmental resources countries" that assesses the risk that some resources-rich countries with weak environmental performance might take specific measures to protect the environment that will endanger the supply of raw materials to the European Union.

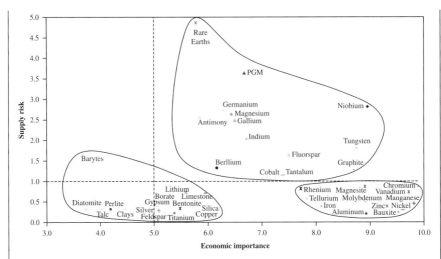

FIGURE 6.3   Economic importance and supply risk of the 41 materials. *Source*: From European Commission (2010)

The report "Critical Raw Materials for the EU" (European Commission, 2010) analyzes 41 minerals and methods (Figure 6.3) with regard to the economic importance and supply chain risk. The critical list identifies 14 raw materials located at the upper right side of the Figure 6.3. In 2013, the list was revised by the Ad Hoc Working Group of the European Commission (European Commission, 2014a), being now considered 20 critical raw materials: antimony, beryllium, borates, chromium, cobalt, coking coal, fluorspar, gallium, germanium, indium, magnesite, magnesium, natural graphite, niobium, platinum group metals, phosphate rock, rare earth elements heavy, rare earth elements light, silicon metal, and tungsten. Such list increase results on a widening scope of raw materials analysed, a bigger availability of additional data and a preserving comparability with the 2010 study (European Commission, 2014b).

In recent decades, like is shown in Table 6.4, OECD countries have decreased their resources extraction intensity. According to OECD (2008b), such decrease results from the decoupling of resources extraction from economic growth, which is due to the deviation of waste management away from the primary and secondary sectors toward the service sector, more material-efficient technologies application, and to the material-intensive imports increase due to outsourcing of material-intensive production stages to other parts of the world.

Resources overexploitation and overconsumption influence SWM in two ways: (1) waste is viewed as a potential alternative and secondary source of materials, and (2) waste is generated due to resources extraction. Industries have begun viewing waste as a secondary resource due to the high volatility of resource prices and heavy pollution of primary production. In this context, the concern is not that overconsumption of resources could promote geological scarcity, but that it could promote technical scarcity (i.e., limit the availability of resources for technical use).

TABLE 6.4   Global resource extraction, by major groups of resources and regions, for the years 2002 and 2020

	2002			2020		
	ROW	OECD	BRICS	ROW	OECD	BRICS
Metal ores	32%	30%	38%	34%	27%	39%
Fossil energy carriers	27%	38%	35%	31%	29%	40%
Biomass	33%	29%	38%	38%	23%	39%
Non-metallic minerals	19%	55%	26%	21%	43%	36%

Source: From OECD (2008b).

ROW, rest of the world countries; OECD, OECD countries; BRICS, Brazil, Russia, India, China, and South Africa.

The competition for primary resources from the anthroposphere exists because of their comparable technical material qualities (Wittmer and Lichtensteiger, 2007). Today, secondary resources from recycling are potential resources for the raw materials supply, mainly in economies with imminent stock saturation (Lichtensteiger, 1998, 2006). Consequently, the competition between anthropogenic and geogenic deposits will be progressively determined by the future availability, planning reliability, and the material classification of the raw materials (Wittmer and Lichtensteiger, 2007). Recycling, as is known, has to change, to be a consistent and possible driver to fill the gap. The stocks and flows of materials, products, and wastes include direct recycling, urban mining, and enhanced landfill mining (or landfill mining) (Figure 6.4). Two additional urban sinks, including urban mining and landfill mining, provide resources near anthropogenic systems.

When resources are extracted or harvested, huge amounts of materials are moved (e.g., mining overburden, by catch from fishing, harvest losses), but not all enter into the economy. Although this waste issue is not visible in production statistics, these movements of unused materials (or resources) may add to the environmental burden of resource extraction, disrupt habitats or ecosystems, and alter landscapes in the supplying region (OECD, 2008b).

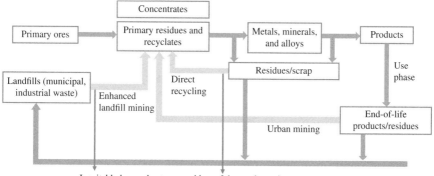

FIGURE 6.4   Closing material flows. *Source:* Adapted from Jones et al. (2011)

6.1.4 Climate Change and Sustainability

Climate change refers to "a statistically significant variation in either the mean state of the climate or in its variability, persisting for an extended period (typically decades or longer). Climate change may be due to natural internal processes or external forcing, or to persistent anthropogenic changes in the composition of the atmosphere or in land use" (IPCC, 2001). Although this definition considers both causes as probable influences on climate, the United Nations Framework Convention on Climate Change, in Article 1 (UNFCCC) defines that "climate change means a change of climate which is attributed directly or indirectly to human activity that alters the composition of the global atmosphere and which is in addition to natural climate variability observed over comparable time periods." For UNFCCC "climate change" is attributable to human activities, and "climate variability" is attributable to natural causes (UN, 1992).

Other sources also help define climate change, but its effects are undeniable. Possible climate change effects, such as temperature variations, precipitation increases or decreases, and sea-level rise and storm surges, will have significant impacts on SWM, depending mostly on selected technologies and managerial practices.

According to Bogner et al. (2007), most waste-management operations and waste-treatment technologies have low vulnerability. The exception is uncontrolled disposal (open dumping and burning), which will become more vulnerable in warmer temperatures that promote pathogen growth and disease vectors. In summary, the climate change will potentially influence SWM in terms of three aspects including (Zimmerman and Faris, 2010):

- **Temperature:** Long-term modifications in average annual temperature and increments in the frequency, intensity, and duration of heat waves;
- **Precipitation:** Long-term modifications in average annual precipitation and more recurrent and intense precipitation events and drought;
- **Sea-level rise and associated storm surge:** These hazards might flush out coastal landfills and inundate waste-treatment facilities.

Higher temperatures may induce the following effects on SWM through (Zimmerman and Faris, 2010; Winne et al., 2012):

- improving biological treatments such as composting and microbial methane oxidation in landfill cover soils;
- increasing rates of waste decomposition and degradation (also dependent on moisture), including temporary deposition in containers, adding to problems of odors;
- increasing health risks (e.g., disease transmission from putrescible waste);
- increasing fire risk from combustibles (also dependent on moisture);
- increasing risk of combustion at open sites and composting; and
- disrupting drainage and surface water flow around landfill sites.

Precipitation (increase or decrease) may have the following effects on SWM through (Zimmerman and Faris, 2010; Winne et al., 2012):

- increasing risk of flooding (fluvial and flash floods), affecting facilities, access, and use of mobile plants;
- increasing risk of flood-related disruption to critical infrastructure and suppliers (mainly transport, energy, and information and communication technology);
- increasing risk that site drainage systems will be overwhelmed during heavy rainfall;
- increasing potential for waterlogging of open containers, with impacts on processing of materials;
- increasing risk due to gas and leachate collection/control during heavy rainfall or floods;
- increasing risk of erosion and instability of bunds and capping layers;
- increasing volume of leachate peaks;
- increasing frequency of low flows in rivers and canals during summer, affecting riverine and canal transport;
- reducing water availability for wet processes and site management (particularly during summer);
- increasing health and safety risk to employees exposed to extreme weather conditions;
- saturating soils and decrease stability of slopes and landfill linings at waste management sites (if clay or soil based);
- enlarging flood areas with untreated, dumped waste, increasing the risk of groundwater contamination;
- disrupting the removal and transportation of solid waste; and
- increasing leachate production and changes in its chemical composition.

Sea-level rise can have the following effects on SWM through (Zimmerman and Faris, 2010; Winne et al., 2012):

- increasing risk of flooding/inundation at low lying coastal sites, waterways, pathways, which may affect facilities access and use of mobile plant;
- increasing risk of erosion in coastal sites (e.g., erosion of bunds);
- increasing risk of seawater intrusion to coastal landfill;
- increasing risk of flood-related disruption to critical infrastructure and suppliers (mainly transport, energy, and information and communication technology);
- disrupting marine transport potentially;
- requiring more flood-proofing facilities and mobilized basement/underground-level equipment;
- increasing incidence of floating wastes that wash ashore with high precipitation or storm surges;

- impacting coastal docking and transfer facilities; and
- creating pools of standing contaminated water that promote water- and vector-borne diseases.

6.2   SUSTAINABILITY CONSIDERATIONS AND CRITERIA

As discussed in Chapter 1, the most frequent and consensual definition of sustainable development is from the Brundtland Report, also known as "Our Common Future," a report published by the World Commission on Environment and Development in 1987: "Sustainable development is development that meets the needs of the present without compromising the ability of future generations to meet their own needs (WCED, 1987)." Chapter 1 also highlights two key concepts including (WCED, 1987):

- the concept of needs, in particular, the crucial demands of the world's poor, to which primordial priority should be given, and
- the concept of limitations appointed by the technology status and social organization on the environment's capability to meet present and future needs.

The definition of sustainable development imposes the idea of planet Earth as a global system, where biotic and abiotic resources are used by innumerous processes, resulting in outputs with impacts over space and time. Over space, impacts will be global, where emissions in one place will affect the environment at another place on the planet. Over time, effects will be intergenerational, where the decisions of today will have impacts on the future.

Although there is a broad acceptance that sustainable development calls for a confluence between economic development, social equity, and environmental protection, the concept continues to be vague (Drexhage and Murphy, 2012). Since the Brundtland report and the Rio Summit, public agencies and organizations have taken up sustainable development as a wanted goal and developed metrics for assessing it, although application has proven to be troublesome (Drexhage and Murphy, 2012). The possible criteria available to assess sustainability of an SWM system (Table 6.5) are not all inclusive and can be subject to future revision and expansion. Matthews and Hammil (2009) noted that, since the Rio Summit, design has moved from theory to practice, but the problems from the tenacious grip of technological, political, and other constraints persist (Drexhage and Murphy, 2012).

6.3   ADAPTIVE MANAGEMENT STRATEGIES FOR SOLID WASTE MANAGEMENT SYSTEMS

SWM requires a multifaceted and holistic approach in response to significant global changes, and the traditional philosophy of ISWM is not sufficient to deal with those challenges. A system thinking philosophy, which includes sustainability criteria as technical, economic, environmental, and social dimensions can help shape modern

TABLE 6.5   Proposed sustainability criteria

Sustainability criteria			
Technical	Economic	Environmental	Social
Waste composition	Capital investment and operational costs	Impacts at global, regional, and local scale	Cultural and behavioral changes
Waste generation	Investment costs	Carbon, water, and ecological footprint	Facility siting and public participation
Appropriate planning	Economic analysis	Environmental impact assessment	Social consensus and public hearing
Technology selection	Life cycle cost and benefit analysis	Secondary pollution	Resources conservation
Construction and operation management	Financial management	Air, water, and land pollution	Technocracy movement[a]
Policy and legal framework	Economic and policy instruments	Life cycle impact assessment	Better risk assessment and communication
Institutional arrangement	Economic or financial risk factors	Environmental risk factors	Public health risk factors

[a]Technocracy movement proposes replacing business people with scientists and engineers who have the technical expertise to manage the SWM facilities and can do a better job at avoiding risk.

sustainable ISWM systems. Adaptive management in this context is based on the concept that future influences/disorders to an ISWM system are predictable. The adaptive management purpose is therefore to manage the ISWM system to maintain the greatest level of functionality when altering management practices based on new experiences and insights gained over time. The aim is to identify uncertainties in ISWM while employing hypothesis testing to further comprehend the system and encourage learning from the conclusions of formerly implemented management actions. Adaptive management strategies are a suite of purposeful activities that maintain and improve the state of environmental resources and ecosystem services, economic driving forces, and social harmony affected by global changes in the context of ISWM plans or programs. To meet the challenges of global changes, however, managers and planners of SWM must use a holistic approach to find adaptive management strategies that help mitigate its impacts.

Methodologies that support essential analyses with a series of interconnected system engineering approaches should be applied at all stages to identify various adaptive management strategies (these methodologies will be introduced in subsequent chapters). In general, life cycle assessment, life cycle cost–benefit analysis, environmental impact assessment, strategic environmental analysis, risk assessment, and multi-criteria decision analysis are all effective options to analyze adaptive management strategies. Financing through private investment, international funding, cost recovery from users, and hybrid financing schemes for SWM can be worthwhile variables to factor into the dedicated systems analysis. Economic instruments like landfill taxes and pay-as-you-throw are sound policy tools that promote better SWM. Policy and regulatory measures, such as regulated targets for waste minimization, reuse and recycling, regulation relevant to the waste management "market," and land-use policies and planning, may be smoothly built up to support waste management markets. Institutional arrangements between formal and informal sectors allow a socially sustainable SWM system to be functional.

6.4   FINAL REMARKS

SWM requires a multifaceted and holistic approach in response to significant global changes including population growth and migration, economic development and globalization, rapid urbanization, resources overexploitation and consumption, and climate change impacts. Both waste generation and composition can profoundly affect technology selection, which could in turn affect water, carbon, and ecological footprints in a region. Adaptive management strategies in response to the impacts of these global changes may vary at different scales, both spatially and temporally. Several core sustainability criteria related to technology as the environment, the economy, and society are intimately linked with adaptive management strategy options. ISWM plans that address sustainability concerns would have operational implications: (1) ISWM should be place based, and the frontiers of the place must be clearly and correctly determined; (2) adaptive management strategies for ISWM reflect a stage in the continuing evolution of social values, environmental goals, economic benefits, and technology priorities as they are neither a beginning nor an end; (3) adaptive management strategies should maintain ISWM systems in an adequate condition to meet desired technical, environmental, economic, and social criteria; (4) adaptive management strategies should end up offering the essential capability of ISWM to ease several stressors, natural and man-made; (5) adaptive management strategies may or may not result in optimal levels of water, carbon, and ecological footprints simultaneously; (6) the term sustainability associated with adaptive management strategies should be delineated over the time frame of concern, the benefits and costs of interest, and the relative priority of the benefits and costs; and (7) adaptive management strategies must include scientific information, but it is only one element in the decision-making process.

REFERENCES

Basel Convention. 1989. Overview. Basel Convention. Available at: http://www.basel.int/theconvention/overview/tabid/1271/default.aspx (accessed February 2014).

Bogner, J., Abdelrafie-Ahmed, M., Diaz, C., Faaij, A., Gao, Q., Hashimoto, S., Mareckova, K., Pipatti, R., and Zhang, T. 2007. Waste management. In: *Climate Change 2007: Mitigation. Contribution of Working Group III to the Fourth Assessment Report of the Intergovernmental Panel on Climate Change* (Eds Metz, B., Davidson, O. R., Bosch, P. R., Dave, R., and Meyer, L. A.), Cambridge University Press, Cambridge, UK, and New York, pp. 585–618.

Drexhage, J. and Murphy, D. 2012. Sustainable development: from Brundtland to Rio 2012. Background paper, prepared for consideration by the High Level Panel on Global Sustainability of United Nations.

European Commission. 2010. Critical raw materials for the EU: report of the ad-hoc working group on defining critical raw materials. European Commission. Available at: http://ec.europa.eu/enterprise/policies/raw-materials/files/docs/report-b_en.pdf (accessed March 2013).

European Commission. 2011. Roadmap to a resource efficient Europe COM(2011)571 final. European Commission. Available at: http://ec.europa.eu/environment/resource_efficiency/pdf/com2011_571.pdf (accessed March 2013).

European Commission. 2012. Waste shipments. European Commission. Available at: http://ec.europa.eu/environment/waste/shipments/background.htm (accessed March 2013).

European Commission. 2014a. Report on critical raw materials for the EU: report of the Ad hoc Working Group on defining critical raw materials. European Commission. Available at: http://ec.europa.eu/enterprise/policies/raw-materials/files/docs/crm-report-on-critical-raw-materials_en.pdf (accessed November 2014).

European Commission. 2014b. Communication from the Commission to the European Parliament, the Council, the European Economic and Social Committee and the Committee of the Regions on the review of the list of critical raw materials for the EU and the implementation of the Raw Materials Initiative – draft version. European Commission. Available at: http://ec.europa.eu/enterprise/policies/raw-materials/files/docs/crm-communication_en.pdf (accessed November 2014).

European Environment Agency (EEA). 2007. Municipal waste generation—outlook from OECD (Outlook 013). EEA. Available at: http://www.eea.europa.eu/data-and-maps/indicators/municipal-waste-generation-outlook-from-oecd/municipal-waste-generation-outlook-from (accessed August 2013).

Eurostat. 2012. International trade – International trade detailed data – EU27 trade since 1988 by CN8 (DS_016890). Eurostat. Available at: http://epp.eurostat.ec.europa.eu/portal/page/portal/international_trade/data/database (accessed September 2012).

Giljum, S., Behrens, A., Hinterberger, F., Lutz, C., and Meyer, B. 2008. Modelling scenarios towards a sustainable use of natural resources in Europe. *Environmental Science & Policy*, 11(3), 204–216.

Greenwood, D. T. and Holt, R. P. F. 2010. *Local Economic Development in the 21st Century: Quality of Life and Sustainability*, M.E. Sharpe, Inc., New York.

Hoornweg, D. and Bhada-Tata, P. 2012. *What a Waste: A Global Review of Solid Waste Management*, The World Bank, Washington, DC.

Intergovernmental Panel on Climate Change (IPCC). 2001. Working group I: the scientific basis. Available at: http://www.ipcc.ch/ipccreports/tar/wg1/518.htm (accessed March 2013).

Interpol. 2009. Electronic waste and organized crime: assessing the links. Phase II Report for the INTERPOL Pollution Crime Working Group. Available at: http://www.interpol.int/Crime-areas/Environmental-crime/Resources (accessed March 2013).

Jacott, M., Neta, L, Reed, C., and Winfield, M. 2001. *The Generation and Management of Hazardous Wastes and Transboundary Hazardous Waste Shipments between Mexico, Canada and the United States, 1990–2000*, Texas Center for Policy Studies, Austin, TX. Available at: http://ban.org/library/haznafta.pdf (accessed March 2013).

Jones, P. T., van Gerven, T., van Acker, K., Geysen, D., Binnemans, K., Fransaer, J., Blanpain, B., Mishra, B., and Apelian, D. 2011. CR3: cornerstone to the sustainable inorganic materials management (SIM2) research program at K.U. Leuven. *Journal of Metals*, 63(12), 14–15.

Lichtensteiger, T., 1998. *Resources in Building: Aspects of Sustainable Resource Management in the Construction Industry* [in German: Ressourcen im Bau: Aspekte einer nachhaltigen Ressourcenbewirtschaftung im Bauwesen], ETH Swiss Federal Institute of Technology Zürich, Zurich, Switzerland.

Lichtensteiger, T. 2006. *Structures as Users of Resources and Donor Resources in the Long-Term Development of Urban Systems* [in German: Bauwerke als Ressourcennutzer und Ressourcenspender in der langfristigen Entwicklung urbaner Systeme]. ETH Swiss Federal Institute of Technology Zürich, Zurich, Switzerland.

Liddick, D. 2010. The traffic in garbage and hazardous wastes: an overview. *Trends in Organized Crime*, 13(2–3), 134–146.

Matthews, R. A., and Hammil, A. 2009. Sustainable development and climate change. *International Affairs*, 85(6), 1117–1128.

Organisation for Economic Co-operation and Development (OECD). 2004. *Addressing the Economics of Waste*, OECD, Paris, France.

Organisation for Economic Co-operation and Development (OECD). 2008a. *OECD Environmental Outlook 2030*, OECD, Paris, France.

Organisation for Economic Co-operation and Development (OECD). 2008b. *OECD Benchmark Definition of Foreign Direct Investment*, 4th edition, OECD, Paris, France.

Ovens, L., Blackburn, S., Green, A., Baldwin, J., Williams, A., and Garsed, R. 2013. Research into SRF and RDF exports to other EU countries. Final techinal report for Chartered Institution of Wastes Management. Availanle at: http://www.ciwm.co.uk/web/FILES/Technical/FINAL_SRF_RDF_REPORT_FOR_PUBLICATION_JULY_2013_(2).pdf (accessed November 2014).

Salehabadi, D. 2013. *Transboundary Movements of Discarded Electrical and Electronic Equipment.* The StEP Green Paper Series, United Nations University Initiatives, Germany. Available at: http://isp.unu.edu/publications/scycle/files/ewaste_flow.pdf (accessed March 2013).

Scheinberg, A., Spies, S., Simpson, M. H., and Mol, A. P. J. 2011. Assessing urban recycling in low- and middle-income countries: building on modernised mixtures. *Habitat International*, 35(2), 188–198.

Schluep, M., Manhart, A., Osibanjo, O., Rochat, D., Isarin, N., and Mueller, E. 2011. Where are WEEE in Africa? Findings from the Basel Convention E-Waste Africa Programme. Secretariat of the Basel Convention. Available at: http://www.basel.int/Portals/4/Basel%20Convention/docs/pub/WhereAreWeeInAfrica_ExecSummary_en.pdf (accessed March 2013).

Sora, M. J. 2013. Incineration overcapacity and waste shipping in Europe: the end of the proximity principle? Global alliance for incinerator alternatives. Available at: http://www.no-burn.org/press-release-more-incineration-than-trash-to-burn-threatens-recycling-in-europe-1 (accessed March 2013).

United Nations (UN). 1992. Article 1. United Nations framework convention on climate change. United Nations. Available at: http://unfccc.int/essential_background/convention/background/items/2536.php (accessed March 2013).

United Nations-Habitat (UN-Habitat). 2008. Urbanization: mega and meta cities, new city states? UN-Habitat state of the world's cities 2006/7. UN-Habitat. Available at: http://ww2.unhabitat.org/mediacentre/documents/sowcr2006/SOWCR%202.pdf (accessed March 2013).

United Nations-Habitat (UN-Habitat). 2010. *Solid Waste Management*. Earthscan, Malta.

United Nations Department of Economic and Social Affairs (UNDESA). 2011. World population prospects: the 2010 revision, volume I: Comprehensive tables. ST/ESA/SER.A/313. United Nations. Available at: http://esa.un.org/wpp/documentation/pdf/WPP2010_Volume-I_Comprehensive-Tables.pdf (accessed March 2013).

United Nations Department of Economic and Social Affairs (UNDESA). 2012. World urbanization prospects: the 2011 revision. United Nations. Available at: http://www.un.org/en/development/desa/publications/world-urbanization-prospects-the-2011-revision.html (accessed March 2013).

Wehenpohl, G. and Kolb, M. 2007. The economical impact of the informal sector oin solid waste management in developing countries. In: *Proceedings Sardinia 2007, Eleventh International Waste Management and Landfill Symposium*, Santa Margherita di Pula, Cagliari, Italy.

Wilson, D. C., Rodic, L., Scheinberg, A., Velis, C. A., and Alabaster, G. 2012. Comparative analysis of solid waste management in 20 cities. *Waste Management and Research*, 30(3), 237–254.

Winne, S., Horrocks, L., Kent, N., Miller, K., Hoy, C., Benzie, M., and Power, R. 2012. Increasing the climate resilience of waste infrastructure. Final report for DEFRA. AEA Technology Plc.

Wittmer, D. and Lichtensteiger, T. 2007. Development of anthropogenic raw material stocks: a retrospective approach for prospective scenarios. *Minerals and Energy – Raw Materials Report*, 22(1–2), 62–71.

World Bank, 2013. GNI per capita, Atlas method (current US$). Available at: http://data.worldbank.org/indicator/NY.GNP.PCAP.CD (accessed March 2013).

World Commission on Environment and Development (WCED). 1987. *Our Common Future* (Ed. Brundtland, G. H.), Oxford University Press, Oxford, UK.

Zimmerman, R. and Faris, C. 2010. Chapter 4: infrastructure impacts and adaptation challenges. *Annals of the New York Academy of Sciences*, 1196, 63–85.

CHAPTER 7

SYSTEMS ENGINEERING PRINCIPLES FOR SOLID WASTE MANAGEMENT

Since the 1980s, the concept of integrated solid waste management (ISWM) system has emerged as the best strategy to manage waste streams through a holistic approach. Because collection, treatment, recovery, and disposal are flexibly arranged, system boundaries may vary from time to time with respect to many internal and external factors. The concept of system of systems (SoS) engineering provides a soft, flexible approach that allows both internal and external factors from global to local scales to influence decision-making in planning, design, and operation stages for waste management. External factors at a global scale can be economic development, globalization, climate changes, population growth and urbanization, and resources over-exploitation and overconsumption, whereas internal factors may include waste generation, waste composition, and the willingness to regionalize. Blending ISWM, SoS, and internal/external factors allows us to elucidate the pros and cons of centralized versus decentralized systems. The purpose of this chapter is to approach ISWM problems from systems engineering perspectives by presenting and discussing methods and strategies to adapt ISWM to various global change impacts in a sustainable way.

7.1 SYSTEMS ENGINEERING PRINCIPLES

7.1.1 The Definition of a System

The term "system" covers both the elements of function and their interactions. Each element may have unique attributes, and the functional relations of these elements

Sustainable Solid Waste Management: A Systems Engineering Approach, First Edition. Ni-Bin Chang and Ana Pires.
© 2015 The Institute of Electrical and Electronics Engineers, Inc. Published 2015 by John Wiley & Sons, Inc.

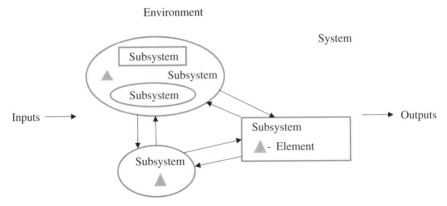

FIGURE 7.1   The fundamental configuration of a system. *Source*: Adapted from Chang (2011)

result in the specific properties and capacities of a subsystem. Different subsystems may have varying role at differing stage when the holistic system is taken into account over time. In systems science, a system is a set of elements interacting with one another purposefully to achieve some common goals, "making the whole functionality greater than the sum of the individual parts" (Haskins et al., 2007).

For a better understanding, the single system elements can be conveniently grouped together in a block diagram (Figure 7.1). In Figure 7.1, a system is separated from the rest of the universe via the boundary, and what is situated outside of the system is known as the surroundings. In complex systems theory, a system may consist of a number of system elements or subsystems. Within this defined boundary of the system, the elements of the system, each of which has a set of attributes or states, are the basic units comprising the system. A subsystem could also own several elements that possess different attributes—a bigger subsystem may even contain smaller subsystems in its hierarchy—which all function as a group to form a unique structural relationship implying the transfer of some materials, energy, and information via a set of network flows. Multilateral causalities exist among elements, subsystems, systems, and the environments in which they function hierarchically; thus, system elements that range at the same level must be distinguished from those of higher or lower levels in the hierarchy. Systems of lower levels are subsystems of a higher system. Between any two or more elements, specific relationships define the aggregation of several elements (subsystem) or the organization of the whole system.

The state of the system, subsystem, or element can be defined when each of its properties has a definite value associated with parameters and/or variables. Given that a system or subsystem could have several system elements to assist in the essential interchanges of material, energy, and information flows, achieving this function requires the presence of some driving force or source of energy across the system boundary. Yet there are situations for not being able to realize the systems dynamics of each subsystem because one set of initial conditions can give rise to different final states due to the interactions among several elements. Thus, the inherent nature of a system may feature system operation linking different states to achieve a delicate equilibrium.

In accordance with these broad observations, all systems have common characteristics, including organization, generalization, and integration as described below (White et al., 1984; Pfirman and AC-ERE, 2003) (Figure 7.1):

- All systems have some structures or organizations.
- All systems function in some way.
- All systems are to some extent creations, abstractions, or idealizations of the real world.
- Functional as well as structural relationships exist between the units.
- Function implies the flow and transfer of materials.
- Function requires the presence of a driving force or source of energy.
- A subsystem or system could own several elements that possess different attributes or aggregation of different attributes.
- System elements or subsystems could range from higher level to the same level to lower levels in a hierarchical way.
- A system or subsystem could have several elements.
- All systems or subsystems show some degree of integration or aggregation.

To further improve our understanding, several distinct types of system or subsystem can be distinguished on the basis of the behavior of the system boundary (Chang, 2011). They include isolated systems, closed systems, and open systems. Isolated systems do not interact with the surroundings across the boundary; these systems are encountered only in the laboratory for the development of thermodynamic concepts. In comparison, closed systems are closed with respect to material, but energy may be transferred between the system and its surroundings. Open systems are those in which both matter and energy can cross the boundary of the system, and information flows between elements, subsystems, or systems.

All environmental systems, including ISWM systems, are open systems characterized by continuous throughputs of information, material, and energy. One of the merits of building systems analysis models is their ability to simplify the complexity of ISWM in the real world. Extended simplifications may even allow complex systems analysis to be formulated in a closed system rather than an open system by defining an appropriate boundary and generating a domain problem of concern. A fundamental configuration of a closed system hierarchy (Figure 7.1) illustrates the system functions at different scales and times. All closed systems must have at least four key features to function independently.

- **Uniqueness:** Each element performs independently in the organization and interrelates with other components to yield an integrated functionality.
- **Connectivity:** Each subsystem or element must, in some way, exhibits external structural relationships with others in the organization.

- **Hierarchy:** All subsystems or elements must be subject to a series of hierarchical interactions to maintain proper transfer of matter, energy, and information flows essential for supporting the basic functionality within the organization.
- **Adaptation:** Each system must be adaptive in response to any challenge of resources relocation in the environment.

7.1.2 Model-Based Systems Engineering Approach

Systems engineering is both an art and a science, and the principles of systems engineering apply at all levels from local to global systems (Ryschkewitsch et al., 2009). The aim of systems engineering is to construct the proper design from a system perspective, analogous to an orchestra performing a symphony, in which each instrument (element) must play its respective part to produce the intended result (Ryschkewitsch et al., 2009). The International Council on Systems Engineering (INCOSE) defines systems engineering as (Haskins et al., 2007):

> "Systems Engineering is an interdisciplinary approach and means to enable the realization of successful systems. It focuses on defining customer needs and required functionality early in the development cycle, documenting requirements, then proceeding with design synthesis and system validation while considering the complete problem. Systems Engineering integrates all the disciplines and specialty groups into a team effort forming a structured development process that proceeds from concept to production to operation. Systems Engineering considers both the business and the technical needs of all customers with the goal of providing a quality product that meets the user needs."

Systems engineering knowledge is composed of three dimensions: (1) domain knowledge of relevant disciplines or subjects, systems theories, and applications/practices, where successful development of synergies between planning, design, and operation in a facility's life cycle will depend on the ability of the systems engineers to master all three dimensions; (2) knowledge of systems theories includes a host of scientific theories that enable systems engineers to analyze specific problems from a holistic point of view; (3) knowledge of relevant disciplines is needed when we attempt to achieve synergy in structuring a multidisciplinary or interdisciplinary approach for systems analysis.

Systems engineering practices should be a set of solution- and technology-independent techniques that have evolved over time into a unique interdisciplinary field in engineering science. Such a set of solution- and technology-independent techniques may be co-located at a particular space within the three-dimensional framework as described in the last paragraph. These practices represent a rational response or methodology to handle increasingly complex situations in a modern society, involving deeper multidisciplinary consideration of not only technical, but also environmental, socioeconomic, and managerial factors. Hence, the basic activities of systems engineers are usually concentrated on the evolution of an appropriate process to enable the design, production, and deployment of an engineered system or on the formulation, analysis, and interpretation of issues associated with one of these phases (Sage and Armstrong, 2000). Systems design may be accomplished through abstraction and formulation at varying levels from management disciplines, to economic and policy aspects, to technical requirements, to environmental standards, and to sociopolitical conditions of the engineering process. As a consequence, the system design process for a specific facility often proceeds by a series of stages and combinations of alternatives, leading to increasingly elaborate refinement up to the point of construction and operation. To empower systems analysis, analytical procedures of these abstraction and formulation frequently involve the proper assembly of different methods, procedures, or techniques flexibly that are integrated, synthesized, or regulated to form an organized system.

Developing a set of solution- and technology-independent techniques in the systems engineering regime should be oriented toward helping decision-making. Any decision-making process inevitably involves metrics composed of various direct and indirect costs and benefits, which may be quantified for trade-offs when balancing priorities. To serve the interests of stakeholders, systems engineers may be asked during the planning process to generate and present alternatives with respect to sustainability criteria to select and rank choices. For a successful decision analysis, technical, economic, environmental, and social considerations must be employed simultaneously in the planning analysis. Every stakeholder has a unique set of constraints and perspectives that might affect the overall goals of decision-making, depending on their relevance to the private or public sectors (or both) of concern. Technical, physical, managerial, legal, institutional, and financial constraints associated with all stakeholders must be included to propose a set of meaningful alternatives in the planning process.

With the potential to use different types of models, including simulation, forecasting, control, and optimization models, model-based systems engineering (MBSE) gains more insight into complex industrial projects, such as ISWM projects, promoting more effective collaborative development of complex systems. To optimize the efficiency, equity, reliability, and risk reduction while balancing cost–benefit–risk trade-offs, systems analysis always requires system engineers to foresee any synergistic possibility of planning, design, and operation in facility's life cycle for solid waste management (SWM) that could result in substantial add-on values or benefits in the long run.

Two alternative system design approaches have been available to systems engineers. They are top-down and bottom-up approaches. In the former approach, the system design process starts with specifying the global system state and assuming that each component has global knowledge of the system subject to a centralized management. The solution is to seek for decentralized communication for replacing global knowledge with local understanding. In the latter approach, the system design starts with specifying requirements and capabilities of individual components leading to pursue the global behavior with a rigorously pre-decided set of rules for the individual behaviors and local interactions among constituent components and between components and the environment, and then proceeds with the inference of the global emergent behavior or patterns. What is in the middle for decision analyses is called hybrid approach with the generic characteristics of either the top-down approach or the bottom-up approach.

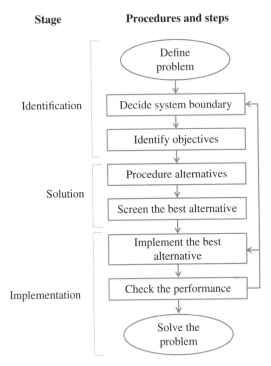

FIGURE 7.2 Top-down analytical framework of competing objectives for systems analysis. *Source*: Adapted from Chang (2011)

Top-Down Approach The top-down approach in systems analysis develops a system's functional and physical requirements from a basic set of mission objectives (NRC, 1998). The approach's goal is to organize information and knowledge for those who manage, direct, and control the planning, development, and operation of the systems necessary to accomplish the mission (Sage, 1992). The result should be a set of traceable requirements used in design and procurement and in system verification and validation (NRC, 1998) (Figure 7.2). The methodology to search for the integrated optimal management strategies over several competing objectives is a system-based top-down approach (i.e., hard-system approach) that usually requires a three-stage analysis from identification to solution and to implementation. The following steps are a prescription of the logic flow of a top-down approach that capture the essence of competing objectives in an organization through optimization or optimal control analyses:

- Define an environmental problem suitable for applying systems analysis.
- Decide the relevant system boundary and identify the objectives of decision makers.
- Produce alternatives based on simulation, optimization, or forecasting techniques.

- Evaluate these alternatives for meeting the objectives.
- Screen the most feasible alternative.
- Implement the most feasible alternative, and if the result is unsatisfactory, go back and modify the previous steps.
- Check the performance of the implementation.

Bottom-Up Approach   The bottom-up approach can be used to identify critical limits for a natural or an engineered system, which to date have been explored using various mathematical tools such as ordinary differential equations (ODEs), partial differential equations (PDEs), agent-based models, system dynamics models, or game theory/conflict resolution to delineate the threshold and criticality of a system. While the ODEs and PDEs may illustrate a relatively simplified system with analytical or numerical solutions, the system dynamics models and agent-based models apply a more sophisticated formulation to delineate the profound interactions among a plethora of components and factors from technical to social, to environmental, and to economic dimensions, such as those complex situations involved in an ISWM system. Examples of relevant modeling analyses in an ISWM system include: (1) the simulation of incompatible incentives between organizations and society to introduce environmentally preferable recycling products to the marketplace under conditions of limited information, (2) simulation of supply chain relationships to minimize carbon footprints, and (3) simulation of the preconditions for radical innovations of waste sorting processes that reduce adverse environmental impacts.

Further distinction is needed to build credible simulation models. First, if an "a priori" structural knowledge exists, which is a deductive reasoning method delineated upfront by the planners, the model relationships for the specific case study can be deduced based on an existing general theory. Otherwise, an "a posteriori" structural knowledge exists, which is an inductive reasoning method delineated by a kind of Monte Carlo simulation. For example, to account for the uncertainty of simulating discrete events in these systems the use of Monte Carlo simulation based on random numbers in digital computers could further contribute to the enhancement of "a posteriori" knowledge. On many occasions, coupled mathematical systems of PDE or ODE may fully or partially delineate the system behavior in the natural environment, whereas the use of multiagent models to simulation interactions of stakeholders' responses without having competing objectives may pin down the social significance in decision-making.

Hybrid Approach   New instrumentation, data handling, model evolution, and integrative methodological capabilities have expanded our understanding of society and the environment allowing more sophisticated practices. The agent-based approach focuses on the interactions between limited individuals in the system of interest, whereas the system dynamics approach focuses on the overall interactions among different components to address temporal changes of target variable. For example, a few states in the United States have mandated up to a 50% reduction in waste going to landfills, such as in California, where a 50% statewide solid waste diversion from landfill disposal was achieved. In addition, some communities are establishing mandatory recycling programs at the local scale. These relatively new regulations and policies will result in profound impacts to our society because the waste management issues are now visible as part of public policy, and concern regarding the appropriateness of various treatment and disposal methods that require systematic assessment. Various modeling analyses will be needed to test possible methods with plausible hypotheses to achieve the overarching goal.

With identification of stakeholders, formulation of agent-based models to create functional assessment may become feasible. With changing local conditions, system response may be characterized by system dynamics modeling. Such advancements allow us to investigate some classic problems in business—environment relations in ISWM, such as investigating behavior changes that favor recycling products as a means to understand consumer's behavior and, at the same time, to reduce environmental impacts. This new, innovative movement relies on experiments, models, and their interactions to understand systems at multiple scales to develop scenarios and projections relevant to sustainable policy and practice (Pfirman and AC-ERE, 2003).

Hence, MBSE may be extended to integrate, couple, or synthesize various types of top-down and bottom-up approaches to analyze an extremely large-scale, complex system. It will lead to the development of decision support systems (DSS). In some DSS, forecasting and control models may be deemed critical to provide foreseen values for the future that may be helpful for both simulation and optimization analyses. Besides, high-end systems engineering practices may be geared toward using high performance computing facilities to perform cloud computing analyses with cloud sourcing support and big data analytics that would have been impossible one or two decades ago. Examples include replicating familiar patterns of environmental–economic interactions, and then testing the efficacy of economic instruments, green policies, alternative behavioral rules, and information pathways on environmental and economic outcomes. This information can be valuable for an agency or organization to optimize its goals and achieve recycling targets within a clearly defined time and space in a systems analysis.

7.2 SYSTEM OF SYSTEMS ENGINEERING APPROACHES

The convolution of high-end systems analysis generates the concept of "systems of systems," or SoS, defined as systems that describe the large-scale integration of many independent self-contained systems to satisfy common goals or global needs in a multi-functional system. SoS are typically characterized by the geographic proximity, distribution, and patterns of the overall system with respect to their operational and managerial independence, evolutionary pathways, and emergent behavior in the holistic system.

SoS are often used to support today's high-end systems analysis. While systems engineering is recognized as a key contributor to successful systems development, SoS engineering (SoSE) may be attributed to solve more complex issues. This observation can be evidenced according to the "Systems Engineering Guide for Systems of Systems" published by the United States Department of Defense

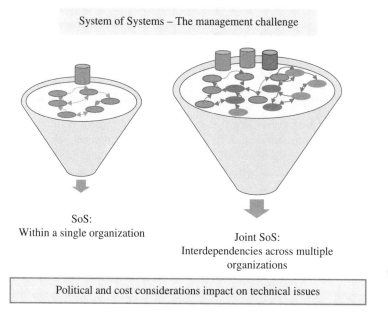

FIGURE 7.3 Political and management consideration that affect SoS. *Source*: Adapted from ODUSD(A&T)SSE (2008)

(ODUSD(A&T)SSE, 2008). Overall, SoS emphasizes a collection of a few dedicated systems or subsystems that pool resources and capabilities to connect a more complex "meta-system," which offers more functionality and performance than simply the sum of the constituent systems (Maier, 1998). SoSE therefore provides a service or solution for complex, distributed problems with four major elements: project planning, SoSE environment, SoSE modeling and simulation, and SoS analysis (Kovacic et al., 2007).

The SoSE discussion in this chapter integrates independent metasystems within the SWM system into a larger, functional system with unique capabilities to solve more complex ISWM issues at the system level, such as global change impacts. It also compares alternative training philosophies including (ODUSD(A&T)SSE, 2008) (1) SoS engineers must be able to function in an environment where the SoS manager does not control all of the systems that impact the SoS capabilities, and stakeholders have interests beyond the SoS objectives (Figure 7.3); (2) SoS engineers must balance SoS needs with individual system needs; (3) SoSE planning and implementation must consider and leverage the development plans of the individual systems; and (4) SoSE must address the end-to-end behavior of the ensemble of systems, addressing the key issues that affect each behavior. A detailed SoSE checklist based on five dimensions, including technology, context, operation, geography, and a conceptual project framework (Allen et al., 2006), can be used by project managers in real-world applications to determine if a problem falls into the SoSE realm.

The goal of an ISWM system can be viewed as an SoS mission before integration. Each of the nearby SWM units in a localized region can be a single unit or several collaborative units, each of which may require lower-level systems to support the intelligent and adaptive activities required to complete their missions. Each SWM has specific actions to take, resources to allocate, and decisions to make. The top-level mission for a localized SWM may not vary from time to time, yet the other missions must adapt to practical situations and complex events that arise from the lower levels. Optimal decisions must be based not only on the commands/missions from the upper level, but also on particular local situations for each SWM. Optimization analysis (i.e., top-down approach) for final decision-making in this network is intimately linked with interdependencies between decision scales and decision sequences. Trade-offs between centralized versus decentralized decision are made possible through conflict resolution strategies (i.e., a bottom-up approach). A changing decision scale and decision sequence through a conflict resolution-based multi-agent simulation (i.e., a bottom-up approach) must be integrated with optimization analysis (i.e., top-down approach) in a complex temporal, dynamic environment. Such a systems analysis is normally managed with heterogeneous regulatory policy tools and supported by an open-source, dedicated database enabling high-end decision analyses in an ISWM system.

An ISWM system can be further analyzed at any level using five aspects of SoSE, including autonomy, belonging, connectivity, diversity, and emergence. Then, the results are characterized by modules, which in turn are converted into objective functions and constraints for various trade-offs in proposed optimal decision-making problems or general simulation frameworks as discrete events for impact assessment. These modules may include but are not limited to (Figure 7.4) the following.

- **Organization module** incorporates knowledge relating to the organizational context in which the system at differing scales is intended to operate. This module will be converted into constrains in the proposed optimization problems if a top-down approach is preferred.
- **Mission module** incorporates knowledge relating to missions to be achieved by the system, tasks, and functions undertaken by its agents, including the human operators. This module changes dynamically to respond to the upper-level commands, information from its collaborative partners, and situations detected by the system. It will be converted into the objective functions in the proposed optimization problems if a top-down approach is preferred.
- **Agent module** incorporates knowledge relating to the participants of a system (i.e., computer and human agents), as well as their roles and responsibilities, which can be adjusted when reacting to the events and particular situations if a bottom-up approach is preferred.
- **User module** incorporates the knowledge of human operators' or decision makers abilities, needs, and preferences to formulate decision alternatives for decision analysis.
- **Data module** enables automated systems to automatically analyze multiple sources of data generate interpretations to define the ISWM within the context of sustainability science.

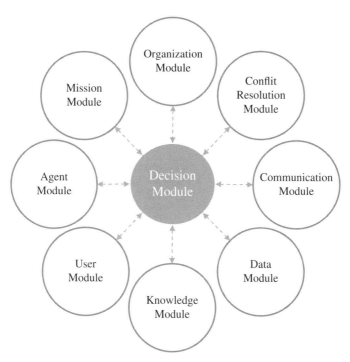

FIGURE 7.4   The proposed SoSE management framework for an ISWM system in a changing environment and the relationship among its constituent modules

- **Decision module** incorporates the knowledge of the systems resource assets, abilities, means, risks, costs, and benefits, which are modeled as lower-level agents. This information can be integrated to help local operators and decision makers minimize direct and indirect costs over the life cycle; maximize the global, local, strategic, and tactic benefits over both long- and short-term perspectives; and minimize the risk associated with global and local loss.

- **Knowledge module** incorporates the knowledge of the external world, such as physical (e.g., principles of waste controls), psychological (e.g., principles of human behavior under policies and regulations), or cultural (e.g., rules associated with policies and regulations) environments, which could affect the system boundaries of SoSE.

- **Communication module** incorporates the knowledge of how communication takes place among the agents and among the systems themselves. The exchanged conveyed information/commands are critical to decision-making. Communication delays and bandwidth constraints must be fully considered to support the bottom-up approach.

- **Conflict resolution module** performs conflict resolution with the aid of the agent module. Conflict resolution can be configured with a multicriteria decision analysis model or a graph model in association with fuzzy preference over the available alternatives, which are linked with a decision scale and decision sequence.

- **Decision module** is designed for ranking alternative trade-offs among cost–benefit–risk criteria in an interactive fashion to achieve goal optimization.

The proposed ISWM networks (Figure 7.5) can be used to further illustrate the concept of SoSE. A source-separated organics composting and incineration treatment train (Figure 7.5a) can be used for source-separated organic streams from a composting plant to produce compost for land application; residuals may be delivered for incineration at a waste-to-energy (WTE) facility. An alternative network (Figure 7.5b) delineates a treatment train for a single stream of recyclables in a material

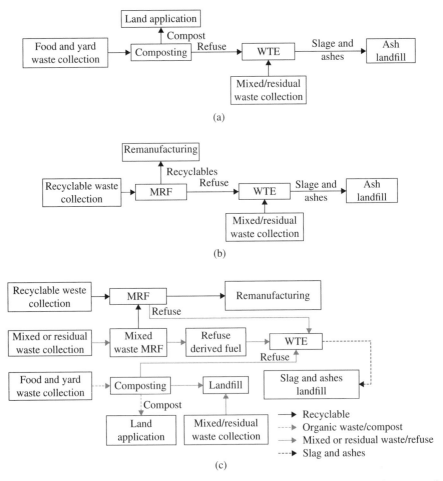

FIGURE 7.5 The proposed ISWM networks in which both (a) and (b) are subsystems of (c). (a) A treatment train for source-separated organics composting and incineration. (b) A treatment train for recycling and incineration. (c) The proposed ISWM networks

recovery facility (MRF); these residuals may also be delivered for incineration at a WTE facility. Because both systems (Figures 7.5a and 7.5b) depend on WTE as a final destination, both can be viewed as subsystems (Figure 7.5c). An ISWM system can then be configured to allow the residuals from both systems (Figures 7.5a and 7.5b) to be delivered for incineration at a common WTE facility (Figure 7.5c), which could be a refuse-derived fuel (RDF) incinerator with a WTE unit. This system expansion can elucidate the concepts of decision scale and decision sequence in the context of SoSE, so that different planning scenarios can be formulated for various purposes. Within these scenarios in Figure 7.5, countless opportunities may be identified to optimize or simulate the proposed ISWM system at different levels with different scales following top-down, bottom-up, or hybrid approaches.

7.3 CENTRALIZED VERSUS DECENTRALIZED APPROACHES

Global changes have been affecting waste generation, composition, and management strategies for the past two decades. In Chapter 2, a thorough review of technology matrix was presented for waste management that supports the planning of collection, sorting, treating, recycling, recovering, reusing, and disposing waste streams. In Chapters 3 and 4, socioeconomic and legal aspects in relation to SWM provided a multifaceted approach to support the planning, design, and operation of SWM. The risk assessment methodology and sustainability criteria in Chapters 5 and 6 allow us to integrate appropriate waste collection, recycling, treatment, and disposal technologies to tackle particular issues within a well-defined system boundary. System boundaries can vary over time due to changing systematic conditions and different involvement of stakeholder groups at differing levels of the decision-making arena. In general, two concepts are used to integrate those SWM technologies into an ISWM system, and they are the concepts of centralized and decentralized systems, although the distinction between them is not always obvious.

How can a choice be made between decentralized and centralized SWM systems (or even both) concerning sustainability, and how can resources be sustainably managed to reflect a well-defined ISWM system relating to technology screening and scaling? The answer is tied to the willingness to regionalize. One organization must have an exemplary willingness to lead and encourage regional organizations to meet the region's manifold SWM challenges, including all aspects of technical, environmental, economic, and social areas of sustainability. Clusters and consortiums in the context of a regionalization plan or program could be a part of this team to perform ISWM. The willingness of regional chiefs to forego positional leadership and individual agency priorities will also likely add to the possibility of success of a regionalization plan or program. Regionalization is different from the concept of regional planning, which addresses the efficient placement of land-use activities, infrastructure systems, and ease of settlement growth across a region; yet, regionalization can undoubtedly be part of the regional planning. For example, the Rio Grande Valley or the Lower Rio Grande Valley is an area located in the southernmost tip of South Texas. Triggered by cultural homogeneity, geographical proximity, and economic development under the North America Free Trade Agreement, the 11 major cities formed a consortium to promote the regionalization for ISWM.

Whatever the criteria applied for a centralized system, the following major economic, environmental, and social criteria should be considered.

- **Economy of scale**: Some sort of economic incentive is essential for the longevity of the regionalization plan or program.
- **Common environmental concern**: Sharing the concerns for environmental quality (e.g., landfill space exhaustion) may produce a cohesive atmosphere to unify efforts.
- **Life cycle perspective**: A holistic view of all phases occurring during the lifetime of SWM.
- **Waste hierarchy principle**: Preference is given to prevention and preparation for reuse, recycling, recovery, and final disposal.
- **Polluter pays principle**: Management by waste producers and processors to guarantee a high level of environmental and human health protection through a user-sustainable mechanism.
- **Precautionary principle**: Consortium members must address any action or policy that has a suspicious risk of causing harm to the human health or environment, even in the absence of scientific consensus that proves such risk.
- **Principle of preventive action (or preventive principle)**: Any environmental damage should be rectified at the source to the greatest possible extent.
- **Principles of proximity and self-sufficiency**: Waste should be treated or disposed of close to the generation point, thus aiming to achieve self-sufficiency in waste treatment and disposal.

Centralized and decentralized systems can be developed that consider all sustainability criteria and global change impacts for today and into the future. Designing a system boundary and making decisions needed for sustainability is a compelling issue for the planning stage. Yet, most decentralized systems do not result from planning; rather, they are usually driven by experience in managing local waste streams, even in the absence of a planning stage. Such experience can even present an opportunity to develop local businesses that promote decentralized systems if they are operational, functional, and lucrative. Overall, centralized systems are those that present a large-scale capacity for waste operations because they have lower marginal operation costs and are organized by a large corporative or hierarchical structure; in contrast, decentralized systems are those that operated and managed locally as small-scale solid waste operations that provide solutions without the regional-scale complexity.

Centralized systems present a large infrastructure, with medium or large processing capacity, organized in a specific geographic area. In these systems, all wastes collected are delivered for combined processing through either an existing administrative boundary in a megacity or in a regional program across several counties. Due to the augmented capacity, these facilities might present higher environmental,

TABLE 7.1 **Large-scale centralized versus decentralized composting system**

Large-scale centralized composting system	Decentralized composting system
Highly mechanized technology, enclosed	Simple technology and labor intense, open piles
Larger investments for advanced, automated machineries	Lower capital cost and locally available materials
Higher operation and maintenance costs and a higher degree of specialized skills to operate and maintain	Comparatively lower operational and maintenance costs and lower level skills required
Lower interaction and involvement of residents	Residents separate their own waste, which reduces the volume of solid waste earmarked for disposal, increases the value of recyclables, and enhances the environmental awareness of the community
Higher transportation cost because all wastes must be transported to disposal facilities, often located far from the city. Compost is trade in regional marketing system	Reduced transportation costs. Compost traded at a local marketing as soil conditioner
Quality of compost is poor due to large quantity of unseparated waste with high risk of contamination	Quality of compost is good because wastes are efficiently separated, and risks of contamination are minimized

Source: From Kolb (1996) and UN, 2005.

ecological, and human health risks. In decentralized systems, infrastructures with neighborhood/small-scale capacity are distributed across cities (or megacities) near population centers where people can deliver specific waste to be processed without the burden of transportation. This local advantage reduces both the risk to the population and environment because the amounts of waste required to be managed and the types of waste to be processed by the facility are also reduced. Due to relatively larger initial investment costs, centralized systems are more common in high income countries and big cities, whereas decentralized systems are more common in low and medium income countries and small communities. Comparison for composting in centralized and decentralized systems is presented in Table 7.1.

The type of technology applied to both systems is also different. In decentralized systems, low cost and lower-tech solutions with reduced resources are preferable because these systems do not require highly skilled management staff. A typical case for a decentralized system is a community-driven composting facility because of its low risk to human health. Anaerobic digestion systems are also frequently present in small- to medium-sized communities that accept feces, agriculture waste, or manure waste, although the facilities are not designed exclusively to receive the organic fraction of MSW. Other common cases for a decentralized option are recycling centers that collect waste through local waste pickers.

According to Liyala (2011), the choice between centralized and decentralized options is a crucial issue in many debates on environmental performance of urban infrastructure and service provision (Tjallingii, 1996). The centralized mode as discussed in Hegger (2007) is seen to have a certain degree of security because the critical tasks are in the hands of few people, in this case, the municipal authorities, who receive constant directives and resources from the central governments. These systems are considered robust due to minimal interference from the larger public, while also allowing economies of scale. This centralized and hierarchical organization is typical of large-scale socio-technical systems characterized by a dominant perspective on the role and implementation of technology (Guy et al., 2001).

In East Africa, however, a considerable mismatch has been reported between the centralized approach and the actual situation because large, centralized technological systems make strong assumptions about the presence of homogeneity in housing stock, density, degree of urbanization, accessibility, related infrastructure (such as street paving and drainage), and other factors (Spaargaren et al., 2006). East Africa's actual systems have decentralized initiatives, such as small-scale neighborhood waste collection driven by community groups, and are considered flexible and able to reach low income and unplanned neighborhoods. Oosterveer and Spaargaren (2010) noted that these decentralized systems are more robust, economical, and better able to deal effectively with the existing environmental challenges. Decentralized technologies offer solutions for individual households, but they do not solve the massive challenge of scaling up low cost technologies to address SWM problems of large cities in developing countries. Large cities in developing regions such as in East Africa are therefore faced with the dilemma of choosing a large-scale centralized system or a small-scale decentralized system because both have serious weaknesses. The modernized hybrid approach argues for optimizing the best mixture of centralized and decentralized systems.

Centralized systems have the advantage of implementing more robust and sophisticated control and monitoring programs, which are vital to increasing the reliability of products made from waste. This advantage is critical when waste products must comply with quality standards for trade, such as in European Union countries and the United States. Three salient examples are RDF from waste streams with high paper and plastic contents, compost derived from the organic fraction of MSW, or even recycled aggregates from construction and demolition waste. Nevertheless, environmental, economic, and social constraints will come into play when making a final decision for waste control.

7.4   SENSITIVITY ANALYSIS AND UNCERTAINTY QUANTIFICATION

7.4.1   Sensitivity Analysis

Sensitivity analysis is used to determine how "sensitive" a model is to changes in the value of a parameter in the model, and even to changes in the structure of the model. The model may be a simulation, forecasting, control, or optimization model.

Parameter sensitivity for a model is usually performed as a series of sequential tests by the modeler who sets different parameter values for a target parameter in sequence to determine how a change in the parameter causes a change in the dynamic behavior of the modeling outputs. In an optimization analysis, for instance, sensitivity analysis tests the parameter sensitivity embedded in either the objective function or a constraint, or both. Because much of the information used in formulating the model is uncertain, the sensitivity of an optimal or simulated solution to changes in those parameter values must be determined to understand how the solution varies when an actual experience deviates from the values used in the original model. If sensitivity is high, reoptimization or resimulation after model reformulation is sometimes required in decision-making.

7.4.2 Uncertainty Quantification

Uncertainty quantification is the science of quantitative characterization and reduction of uncertainties in applications to reduce risk in decision-making. Uncertainties embedded in an assessment of ISWM projects could arise from data inaccuracy, data gaps, unrepresentative data, model uncertainty, uncertainty due to choices, epistemological uncertainty caused by the lack of knowledge on system behavior, plausible mistakes, and estimation of uncertainty. In systematic uncertainty analysis, uncertainty due to choices, epistemological uncertainty caused by lack of knowledge on system behavior, and plausible mistakes are classified as "decision-making uncertainty." In comparison, data inaccuracy, data gaps, unrepresentative data, model uncertainties, and estimation of uncertainty are classified as "data and model uncertainties."

Decision-Making Uncertainties Uncertainty due to choices is related to risk preference, expected utility theory, and the anchorage effect. Risk aversion is a concept in psychology, economics, and finance based on the behavioral patterns of humans exposed to uncertainty while attempting to reduce that uncertainty. An investor with a risk aversion attitude prefers the option with the lower risk when faced with two investments with a similar expected return. The risk-neutral investor is in the middle of the continuum represented by risk-seeking investors on one end, and risk-averse investors on the other. The risk-averse decision maker is reluctant to accept a risk or a bargain with an uncertain payoff, however. The focusing effect (i.e., anchorage effect or focusing illusion) is a cognitive bias that occurs as a common human tendency to rely too heavily on the first piece of information offered (the "anchor") when making decisions. This effect would also affect decision-making under uncertain conditions.

Epistemological uncertainty is caused by the lack of knowledge about system behavior or random variability. In the context of ISWM, epistemic uncertainty is the scientific uncertainty in the model due to limited data and knowledge. In epistemic uncertainty, plausible mistakes or plausible errors imply that the decision maker did not know the correct choice but was applying some known rules to attempt the correct choice.

Data and Model Uncertainties   In parallel with subjective decision-making uncertainties, data and model uncertainties involve computational limitation. Whether the modeling system involves the top-down, bottom-up, or hybrid approach, three sources of uncertainty can be quantified and transferred to the output via modeling:

- **Input data uncertainty.** In an ISWM model with a nonlinear structure, the average of the responses under various conditions is not the same as the response for average conditions. Differences among sources of the data may possibly introduce biases due to model nonlinearity. To reduce potential biases, an ensemble approach (Box 7.1) may be implemented to sample fine-scale heterogeneities in various datasets with multiple Monte Carlo simulations.

- **Model parameter uncertainty.** Model parameter uncertainty is tested with the aid of data assimilation in an ISWM model and can be derived using Monte Carlo simulation from local observations or long-term inventory plots before applying this information to model optimization and/or simulation.

- **Model structure uncertainty.** An ultimate comparison is required among several models with the same site scale and similar structures to compare the target model of interest to other similar models. Once compared, issues and uncertainty related to model structure and mathematical representations of waste management processes can be addressed.

BOX 7.1   ENSEMBLE PREDICTIONS AND UNCERTAINTY ANALYSIS

Ensemble prediction is a numerical prediction method used to generate a representative sample of the possible future states of a dynamical system based on interval values. Ensemble prediction is a form of Monte Carlo analysis in which multiple numerical predictions are conducted to investigate how far the uncertainty could appear in dealing with the same prediction issue. These multiple numerical predictions may include but are not limited to slightly different initial conditions or different forecast models for different members, or different formulations of a forecast model. In short, these multiple simulations with respect to different technical settings are conducted to investigate the two critical sources of uncertainty: (1) the errors introduced by varying initial conditions, amplified by the chaotic nature of the evolution equations of the dynamical system, and (2) errors introduced due to inherent issues in the model structures or equations. With such multiple simulations, ensemble predictions may provide uncertainty interval estimation to provide various tests of hypotheses in decision analysis.

7.5   FINAL REMARKS

Checkland (1981) described a system topology that comprises natural systems, physical systems, and human activity systems. The first two can be characterized as hard systems where the well-established system engineering principles can be applied smoothly. However, the last one that is difficult to define can be classified as soft systems. Thus, systems theory has two main approaches. The original hard systems approach is leaner toward integration of technical and engineered systems, and the more recent soft systems approach is more relevant to the emphasis of human and social systems. ISWM is a major issue in the context of urban sustainability, which requires proper synthesis, analysis, and application of both hard and soft systems.

Systems engineering approaches provide some essential strategies of both hard and soft systems used to develop a better ISWM system by identifying the level or levels at which the highest values of decision analyses can be achieved under given uncertainties. Carrying out MBSE analyses is by no means an easy task, however. The success depends on improved understanding of the multifaceted nature of ISWM as well as the endeavor of appropriate data-model fusion toward final meaningful solutions, which oftentimes requires knowledge from both hard and soft systems. Finally, concerning the implementation, an ISWM project may follow the Systems Engineering Management Plan (SEMP), produced by INCOSE (originally referred to as the Systems Engineering Plan) (Haskins et al., 2007). In the domain of systems engineering, a solid SEMP describing the overall technological effort in a project is instrumental to integrating required planning schemes, design ideas, risks involved, opportunities, and specialty engineering areas. The SEMP should therefore include or reference technical risks embedded in all planning, design, and operation stages along with the process to manage them.

REFERENCES

Allen, P. D., Cury, M., and Orzechowski, R. 2006. A SOSE checklist and mitigation list applied to three case studies. *Systems Research Forum*, 1, 47–54.

Chang, N. B. 2011. *Systems Analysis for Sustainable Engineering*, McGraw Hill, New York.

Checkland, P. B. 1981. *System Thinking, System Practices*, John Wiley & Sons, Chichester, UK.

Guy, S., Marvin, S., and Moss, T. 2001. *Urban Infrastructures in Transition: Networks, Buildings, Plans*, Earthscan Publications Ltd., London, UK.

Haskins, C., Forsberg, K., and Krueger, M. 2007. *INCOSE Systems Engineering Handbook*, Version 3.1, INCOSE. Available at: http://www.gobookee.net/incose-systems-engineering-handbook (accessed March 2013).

Hegger, D. 2007. Greening sanitary systems: an end-user perspective. PhD Thesis, Wageningen University. Available at: http://www2.gtz.de/Dokumente/oe44/ecosan/en-greening-sanitary-systems-2007.pdf (accessed March 2013).

Kolb, A. 1996. Decentralized compost-management: case study of a district of 77.000 inhabitants – the 'Kulmbach Model'. In: *The Science of Composting – Part 1* (Eds. de Bertoldi, M., Sequi, P., Lemmes, B., and Papi, T.), Blackie Academic & Professional, Glasgow, UK, pp. 1188–1192.

Kovacic, S. F., Sousa-Poza, A., and Keating, C. 2007. The National Centers for System of Systems Engineering: a case study on shifting the paradigm for system of systems. *Systems Research Forum*, 2, 52–58.

Liyala, C. M. 2011. *Modernising Solid Waste Management at Municipal Level: Institutional Arrangements in Urban Centres of East Africa*, Wageningen Academic Pub., Wageningen, Netherlands.

Maier, M. W. 1998. Architecting principles for system of systems. *Systems Engineering*, 1(4), 267–284.

National Research Council (NRC). 1998. *Systems Analysis and Systems Engineering in Environmental Remediation Programs at the Department of Energy Hanford Site*, Committee on Remediation of Buried and Tank Wastes, National Research Council, Washington, DC.

Office of the Deputy Under Secretary of Defense for Acquisition and Technology, Systems and Software Engineering (ODUSD(A&T)SSE). 2008. *Systems Engineering Guide for Systems of Systems*, Version 1.0. Washington, DC. Available at http://www.acq.osd.mil/se/docs/SE-Guide-for-SoS.pdf (accessed March, 2013)

Oosterveer, P. and Spaargaren, G. 2010. Meeting social challenges in developing sustainable environmental infrastructures in East African cities. In: *Social Perspective on the Sanitation Challenge* (Eds. Vliet, B., Spaargaren, G., and Oosterveer, P.), Springer, Dordrecht, Netherlands, pp. 11–30.

Pfirman, S. and The Advisory Committee for Environmental Research and Education (ACERE). 2003. Complex environmental systems: synthesis for earth, life, and society in the 21st century. A report summarizing a 10-year outlook in environmental research and education for the National Science Foundation.

Ryschkewitsch, M., Schaible, D., and Larson, W. 2009. The art and science of system engineering. *Systems Research Forum*, 3(2), 81–100.

Sage, A. P. 1992. *Systems Engineering*, John Wiley & Sons, New York.

Sage, A. P. and Armstrong, J. E. Jr. 2000. *Introduction to Systems Engineering*, John Wiley & Sons, New York.

Spaargaren, G., Oosterveer, P., van Buren, J., and Mol, A. P. J. 2006. *Position paper on Mixed Modernities: Towards Viable Urban Environmental Infrastructure Development in East Africa*. Wageningen University, Wageningen, Netherlands.

Tjallingii, S. P. 1996. *Ecological Conditions*, Institute for Forestry and Nature Research (IBN-DLO), Wageningen, Netherlands.

United Nation. 2005. Improving the lives of the urban poor: Case studies on the provision of basic services through partnerships. Available at http://www.unescap.org/sites/default/files/Improving-ubp-lives.pdf (accessed March, 2014).

White, D., Mottershead, D. N., and Harrison, S. J. 1984. *Environmental Systems*, Chapman & Hall, London, UK.

CHAPTER 8

SYSTEMS ENGINEERING TOOLS AND METHODS FOR SOLID WASTE MANAGEMENT

Solid waste management (SWM) is a significant issue in sustainable development encompassing technical, socioeconomic, legal, ecological, financial, political, and even cultural components. Systems analysis has provided unique, interdisciplinary support for SWM policy analysis and decision-making over the last few decades. A variety of recent systems engineering tools and methods promise to provide forward-looking, cost-effective, risk-informed, and environmentally benign decisions rooted in the operation research regime. This chapter introduces three types of top-down modeling approaches and two types of bottom-up modeling. These three types of top-down approaches include (1) single objective programming models for cost–benefit–risk trade-offs, (2) multiobjective decision analysis, and (3) multiattribute decision analysis, all of which may be applied to support-integrated solid waste management (ISWM). In this context, it is necessary to discuss the holistic view of systems analysis for ISWM, which emphasize possible trade-offs among cost, benefit, and risk criteria as a unique example of multicriteria decision-making (MCDM). Next the discussion moves to multiobjective decision-making (MODM) with the inclusion of two popular types of models: compromise and goal programming (GP) models, followed by a complementary discussion of multiattribute decision-making (MADM). Both MODM and MADM are major components in the MCDM regime. The two types of bottom-up modeling approaches include multiagent decision models and system dynamic models. The former can be formulated based on game theory, leading to the evaluation of specific ISWM problems. The latter is introduced to delineate the waste flow control in the last part of the chapter.

Sustainable Solid Waste Management: A Systems Engineering Approach, First Edition. Ni-Bin Chang and Ana Pires.
© 2015 The Institute of Electrical and Electronics Engineers, Inc. Published 2015 by John Wiley & Sons, Inc.

8.1 SYSTEMS ANALYSIS, WASTE MANAGEMENT, AND TECHNOLOGY HUB

Systems analysis techniques with a variety of technical and nontechnical implications have been applied to enhance a range of analyses for SWM over the last few decades (Pires et al., 2011). To simplify the discussion, the first part of this chapter highlights 14 of these formally classified tools and methods to illustrate the SWM challenges, trends, and perspectives (Chang et al., 2011). The spectrum of these methods and tools is classified into two domains, although some may be interconnected with each other (Figure 8.1). They are (1) systems engineering methods, including cost–benefit analysis (CBA) or benefit–cost analysis (BCA), forecasting model (FM), simulation model (SM), optimization model (OM), MCDM, and integrated modeling system (IMS); and (2) system assessment tools, including management information systems (MIS), decision support systems (DSS), expert systems (ES), scenario development (SD), material flow analysis (MFA), life cycle assessment or life cycle inventory (LCA or LCI, respectively), risk assessment (RA), environmental impact assessment (EIA), strategic environmental assessment (SEA), socioeconomic assessment (SoEA), and sustainable assessment (SA). OM may be further divided into linear programming (LP), nonlinear programming model (NLP), dynamic programming (DP), and mixed integer programming (MIP) (Chang et al., 2011). IMS may combine any type of

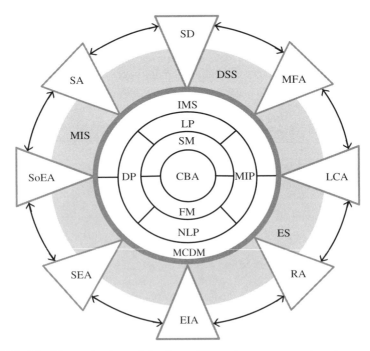

FIGURE 8.1 The technology hub for solid waste management systems analysis. *Source:* Adapted from Chang et al. (2011)

methods and tools in one system to achieve a particular goal (Figure 8.1). All of these 14 systems engineering methods and system assessment tools are delineated below.

- **CBA:** Such systems engineering method consists in economic modeling technique for decision makers to assess the positive and negative economic effects of a project or policy in which all relevant impacts are measured in both physical and monetary values. For example, Dewees and Hare (1998) analyzed different packaging waste reduction programs and the associated direct costs and benefits of several policy options for packaging waste reduction. Although indirect cost and benefit terms are deemed important factors in special cases, they have rarely been considered in previous SWM systems.

- **FM**: Both planning and design of SWM systems require an accurate prediction of solid waste generation, and FMs may support some of the cost–benefit–risk trade-offs in the context of single objective or multiobjective programming models. Single and multiple regression analyses are the most common forecasting methods for estimating solid waste generation, although other models featured by specific modeling structures may be available to support the same practices (Chang et al., 1993; Chen and Chang, 2000). These modes therefore can determine which variables are closely related to solid waste generation (Bach et al., 2004).

- **SM**: These models use digital computers to trace lengthy chains of continuous or discrete events based on the cause-and-effect relations, describing the operations in complex systems and investigating the dynamic behavior of the system. SMs in the SWM field can be logistic simulation, single and multimachine processes, simulation of the environmental fate and transport of harmful constituents in waste management practices, and simulation of costs and schedules for waste management projects or programs (Miller et al., 2003).

- **OM**: Models developed to optimize systems are one of the cores of systems engineering approach. Single-objective programming (SOP) models search for the optimal solution associated with a well-defined SWM problem with a single objective having several technical and managerial constraints. These models are often applied to solve cost minimization issues and are normally formulated by deterministic methods, including LP, NLP, DP, and MIP models (Chang, 2010). For example, OM can optimize economic issues through the minimization of total costs or maximization of the total benefits to optimize vehicle routing (Liebman et al., 1975), or to decide an SWM system design through a series of shipping strategies with the locational screening of landfill, incinerators, material recovery facilities, and transfer stations (Chang and Lin, 1997a, 1997b; Chang and Lu, 1997; Chang et al., 2005).

- **IMS**: By nature, different IMS have different features, scales, and complexity. Models applied to IMS therefore may integrate simulation, forecasting, and/or optimization analyses to address the forcing of human-induced impacts, identify responses or feedbacks from the SWM systems, and assess consequences due to society-wide disturbances resulting from new policies (Chang, 2008; Chang et al., 2009).

- **MCDM**: This method can solve two different types of mathematical problems. Those that select/evaluate a finite number of alternatives or those that find the "best" alternative. The method used to solve the first and second types of problem in above is MADM and MODM, respectively. MCDM can be viewed as a core part of a DSS suitable for addressing complex problems featuring high uncertainty, conflicting objectives, and different forms of data. MADM methods analyzes explicit alternatives' attributes, whereas MODM considers various interactions within the design constraints that best satisfy the decision maker by attaining acceptable levels of a set of quantifiable objectives (Hwang and Yoon, 1981).

- **MIS**: Such management systems of information are an organized combination of people, hardware, software, communications networks, and data resources that collect, transform, and disseminate information in an organization (Kumar and Mittal, 2004; Whitten and Bentley, 2008). Specifically, an MIS is an information system that provides management-oriented reporting based on transaction processing and operation of the organization (Whitten and Bentley, 2008).

- **DSS**: These models consist in computer-based information systems are designed to affect and improve the process of decision-making. They emphasize ideas that collectively use data and models to solve unstructured problems (Sprague and Carlson, 1982).

- **ES**: These are computer programs designed to imitate the advice of a human expert. They draw conclusions from imprecise, ambiguous information (AEA Technology, 1998).

- **SD**: Such method is one of the system analysis tools that can predict future SWM conditions to assess possible prescribed SWM problems. This methodology applies available resources to show how alternative policy decisions may reach specific goals and purposes (Chang, 2008).

- **MFA**: They consist in systematic assessments of the flows and stocks of materials within a system defined in a space and time. MFA diverges somewhat from the traditional SWM boundary to focus on product consumption patterns, waste generation, recycling, recovery, and reuse (Boelens and Olsthoorn, 1998; Brunner and Rechberger, 2003; Chanchampee and Rotter, 2007).

- **LCA or LCI**: Life cycle assessments or inventories address environmental aspects and potential environmental impacts (e.g., use of resources and the environmental consequences of hazardous releases) throughout a product's life cycle from raw material acquisition through production, use, end-of-life treatment, recycling, and final disposal (i.e., cradle-to-grave) (Azapagic and Clift, 1998; Banar et al., 2009).

- **RA**: These assessment methodologies quantitatively relate environmental and human health risk to accidents from a system perspective. The assessment begins at a risk source, either an operational unit or an infrastructure, and follows the possible consequent chain of accidents, ending with an assessment of possible damage to the population and environment (Bote et al., 2003; Cangialosi et al., 2008).

- **EIA**: This assessment ensures that the decision-making process for the proposed activities has a significant emphasis on the environment (Tukker, 2000). EIA must perform a systematic process that examines the environmental consequences of development actions in terms of physical, biological, cultural, economic, and social factors (Lenzen et al., 2003).

- **SEA**: It can be defined as a formalized systematic and comprehensive process of evaluating the environmental effects of a policy, plan, or program (Thérivel and Partidário, 1999).

- **SoEA**: It applies integrated market-based and/or policy/regulation requirements for SWM, such as waste-to-energy taxation. System engineering models and assessment tools, such as CBA, LCA, IMS, MFA, and SD, can perform various types of integrative analyses of SoEA at large scales (Aprilia et al., 2012).

- **SA**: They integrate different methodologies to obtain an analysis, an evaluation, or a plan for several management aspects that emphasizes sustainability (Menikpura et al., 2012).

The interrelationships among the systems engineering methods and system assessment tools can be holistically connected through a technology hub (Figure 8.1). At the hub center, the six systems engineering models (i.e., CBA, OP, SM, FM, IMS, MCDM) serve as core technologies where the DSS is constructed for separate or collective applications (Box 8.1). Yet an ES can still be formed through heuristic approaches with the aid of the eight system assessment tools at the rim of the technology hub (triangles in Figure 8.1). Information flows among the eight triangles improve the formulation of the five types of systems engineering models. The integration of these systems analysis techniques plays an important role in ISWM systems and may lead to search for environmentally benign, cost-effective, risk-informed, and socially acceptable alternatives (Morrissey and Browne, 2004). These alternatives enable us to tackle challenges and observe trends in advancing ISWM strategies for cost–benefit–risk trade-offs through a suite of comparative analyses while also illuminating future perspectives.

BOX 8.1   DECISION SUPPORT SYSTEMS

A DSS can be an interactive software-based computerized information system that gathers and presents data from a wide range of sources to help decision makers compile useful information from raw data, documents, generate personal knowledge, and create business models to identify and solve problems. It can provide users with the essential software ability to raise queries on an ad-hoc basis, analyze information, and predict the impact of possible decisions. A DSS framework or typology can be characterized with respect to purpose, targeted users, and enabling technology. The major types of DSS include (a) communications-driven, (b) data-driven, (c) document-driven, (d) knowledge-driven, and (e) model-driven

DSS. Communications-driven DSS emphasize communicating, collaborating, and shared decision-making support using technology. A data-driven DSS is a type of DSS that emphasizes access to and analysis of internal and/or external databases through data warehousing and data mining. A document-driven DSS focuses on the retrieval and management of unstructured documents to aid in decision-making. A knowledge-driven DSS suggests or recommends actions to targeted users with the aid of specialized problem-solving expertise relevant to a specific narrow task. In a model-driven DSS, a preprogrammed model is applied to a relatively limited dataset for answering the inquiry in regard to what-if scenarios.

8.2 COST–BENEFIT–RISK TRADE-OFFS AND SINGLE-OBJECTIVE OPTIMIZATION

8.2.1 Basic OMs

The most basic technique for solving OMs is a set of "mathematical programming," a representative technique in the field of operational research. Each mathematical programming is based on a different mathematical structure that can only be applied to optimize a specific problem with unique properties. Solving these mathematical programming models must rely on well-defined algorithms designed to find the global optima in the problem domain. Four useful core methods in mathematical programming are LP, integer programming (IP), NLP, and DP methods which have been used to analyze a wide range of environmental management problems. Each of the four techniques has unique characteristics and application potentials.

Linear Programming　LP provides a powerful means to analyze linear system with a well-defined objective function and constraints. Optimal solutions can be rapidly determined by the simplex method using readily available computer programs, even for large-scale models (Box 8.2). More effort of real-world application has often been devoted to LP because it has a wider applicability in many fields than other techniques.

BOX 8.2　SIMPLEX METHOD

George B. Dantzig is best known as the father of LP and the inventor of the simplex method (Cottle et al., 2007). The simplex algorithm is defined in a standard form in dealing with LP problems for seeking the optimal solution. A set of deviational variables, which are called "slack" or "surplus" variables, were defined to transform all constraints in an LP model into a linear simultaneous equation system to simplify the solution procedure so that conventional theorems of linear algebra may be applied. More theoretical details can be found in the book "Linear Programming and Extensions", a classic work published by George B. Dantzig in 1963, some 16 years after his formulation of the LP problem and discovery of the simplex algorithm for its solution. Various case studies in sustainability science and sustainable engineering can be found in Chang (2010).

LP models are OMs that can be solved by LP algorithms, most commonly the simplex method. This procedure is based on the theory of simultaneous linear equations and is available as a preprogrammed package at most computer installations. Once an LP model is formulated, it can be rapidly and easily solved by software packages such as LINDO (Linear INteractive Discrete Optimizer) or LINGO (Linear, INteractive and General Optimizer) in personal computer (LINDO Systems Inc., 2001).

The constraint set of any linear model can be written as a set of simultaneous linear equations. In general, any linear OM can be written to maximize or minimize the objective function subject to (in abbreviation: s.t.) a set of constraints.

$$\text{Max (Min)} \, Z = c_1 X_1 + c_2 X_2 + \cdots + c_n X_n, \tag{8.1}$$

$$\text{s.t.} \quad
\begin{aligned}
a_{11} X_1 + a_{12} X_2 + \ldots a_{1n} X_n &\leq, =, \geq b_1 \\
a_{21} X_1 + a_{22} X_2 + \ldots a_{2n} X_n &\leq, =, \geq b_2
\end{aligned}, \tag{8.2}$$

$$\cdots\cdots\cdots$$

$$
\begin{aligned}
a_{m1} X_1 + a_{m2} X_2 + \ldots a_{mn} X_n &\leq, =, \geq b_m \\
X_1, X_2, \ldots, X_n &\geq 0,
\end{aligned}, \tag{8.3}$$

where $X_1, X_2, \ldots,$ are variables; $b_1, b_2, \ldots$ are non-negative constants; and $c_1, c_2, \ldots,$ and $a_{11}, a_{12}, \ldots,$ are constants that may be positive or negative. Equation (8.1) is an objective function, and Equations (8.2) and (8.3) are called constraints; the number on the right of each constraint is called the constraint's right-hand side; the coefficients of the decision variables in the objective function are called objective function coefficients; and the coefficients of the decision variables in the constraints are called technological coefficients. If a decision variable X_i can only assume non-negative values, the sign restriction $X_i \geq 0$ is added. If a variable X_i can assume both positive and negative (or zero) values, X_i is unrestricted in sign. Before an OM is solved using the simplex method, it is usually transformed to a standard form:

A more compact and equivalent notation is

$$\text{Max} \, Z = \sum_{j=1}^{n} c_j X_j, \tag{8.4}$$

$$\text{s.t.} \, \sum_{j=1}^{n} c_{ij} X_j = b_i \quad i = 1, 2, , \ldots, m \tag{8.5}$$

$$X_j \geq 0 \quad \forall j. \tag{8.6}$$

In constraint (8.6), the symbol "$\forall$" represents "for all"; that is, X_j must be non-negative for all values of j. An LP model can be written using matrix notation as

$$\text{Max} \, Z = \overline{CX} \tag{8.7}$$

$$\text{s.t.} \quad \overline{AX} = \overline{b} \tag{8.8}$$

$$\overline{X} \geq 0, \tag{8.9}$$

where $\overline{C}$ is a $1 \times n$ row vector,

$$\overline{C} = \left[c_1, c_2, \ldots, c_n\right],\tag{8.10}$$

$\overline{X}$ is an $n \times 1$ column vector,

$$\overline{X} = \begin{bmatrix} X_1 \\ X_1 \\ \cdot \\ \cdot \\ \cdot \\ X_n \end{bmatrix},\tag{8.11}$$

$\overline{A}$ is an $m \times n$ matrix,

$$\overline{A} = \begin{bmatrix} a_{11} & a_{12} & \cdots & a_{1n} \\ a_{21} & a_{22} & \cdots & a_{2n} \\ \cdot & \cdot & & \cdot \\ \cdot & \cdot & & \cdot \\ \cdot & \cdot & & \cdot \\ a_{m1} & a_{m2} & \cdots & a_{mn} \end{bmatrix},\tag{8.12}$$

and $\overline{b}$ is an $m \times 1$ column vector,

$$\overline{b} = \begin{bmatrix} b_1 \\ b_2 \\ \cdot \\ \cdot \\ \cdot \\ b_m \end{bmatrix}.\tag{8.13}$$

When the objective function and/or constraints become nonlinear, the mathematical programming model is defined as nonlinear programming model.

Integer Programming Models   An IP is an LP in which some or all of the variables are required to be non-negative integers. With this structure, the divisibility assumption no longer holds in an IP problem. The IP models can be generally classified into four groups in the context of mathematical programming models:

1. A "pure integer programming problem" is a problem in which all variables are required to be integers. For example,

$$\begin{aligned} \text{Max} \quad & z = 4x_1 + 2x_2 \\ \text{s.t.} \quad & 3x_1 + 2x_2 \leq 6 \\ & x_1, x_2 \geq 0 \text{ and } x_1, x_2 \text{ integer.} \end{aligned}$$

2. A "mixed integer programming problem" is a problem in which only some of the variables are required to be integers. For example,

$$\text{Max} \quad z = 4x_1 + 8x_2$$
$$\text{s.t.} \quad x_1 + 2x_2 \leq 6$$
$$x_1, x_2 \geq 0 \text{ and } x_1, x_2 \text{ integer.}$$

3. A "binary integers programming problem" is a problem in which all variables must be equal to 0 or 1, labeled as "0-1 IP." For example,

$$\text{Max} \quad z = 6x_1 + 9x_2$$
$$\text{s.t.} \quad x_1 + 2x_2 \leq 3$$
$$3x_1 - 2x_2 \leq 4$$
$$x_1, x_2 = 0 \text{ or } 1$$

4. A "nonlinear integer programming problem" is an optimization problem in which either the objective function or the left-hand side of some of the constraints is nonlinear functions, and some or all of the variables must be integers. In this case, the model is not limited by the linearity requirement. For example,

$$\text{Max} \quad z = 6x_1^2 + 9x_2^2$$
$$\text{s.t.} \quad x_1 + 2x_2 \leq 3$$
$$3x_1 - 2x_2 \leq 4$$
$$x_1 = 0 \text{ or } 1; x_2 \text{ is an integer.}$$

Dynamic Programming DP was invented by Bellman (1957), which is closely related to the optimal control theory invented by Pontryagin (Pontryagin et al., 1962). Based on the structure of model formulation, the dynamic programming problem can be classified by its continuous and discrete versions. The model with continuous state variables is formulated for solutions to the trajectory problem where a continuous function is needed. The model with discrete state variables is formulated for solving a multistage problem. From the mathematical standpoint, Bellman's dynamic programming is favored when handling the issues with discrete version, and Pontryagin's optimal control model is more suitable to solve the issues with continuous version of the optimum trajectory issues.

In general, discrete dynamic programming converts a large-scale optimization problem into a series of interconnected smaller ones making the mathematical form to be a multistage decision analysis problem. The entire model formulation is based on the principle of optimality (Bellman, 1961):

An optimal policy has the property that, whatever the initial state and optimal first decision maybe, the remaining decisions constitute an optimal policy with regard to the state resulting from the first decision.

With such philosophy, the solution procedure of discrete dynamic programming issues becomes an enumerative technique that can be represented as a multi-stage decision-making process. As a result, the solution procedure involves using a series of partial optimizations requiring a number of reduced efforts stepwise to find the global optimum. Some of the variables may have to be enumerated throughout their range based on the essential domain judgment. As compared to the other linear programming models, dynamic programming has a unique nomenclature. To introduce the nomenclature, the concept of stage and partial optimization at a stage by decision variables has to be illustrated at first. The use of state variables to link the stage and serve as the path for the dynamic programming algorithm is the central idea to complete the optimization steps of the entire process. Hence, in a dynamic programming problem formulation, the dynamic behavior of the system is expressed by using three types of variables.

- **Stage Variables**: The essential feature of the dynamic programming approach is the structuring of optimization problems into multiple stages allowing the problem to be analyzed one stage at a time in sequence. Although each one-stage problem itself is deemed as an ordinary optimization problem independently, its solution aid in identifying the characteristics of the next one-stage problem sequentially no matter the problem is defined as a forward-looking or backward-looking problem. In most cases, time would be the stage variable in the planning horizon. Yet the stages do not always have time implication. A finite number of possible states can be associated with each stage in decision-making. Such structure reflects the order in regard to which events occur in the time horizon and the domain of interest in the system holistically.

- **State Variables**: Associated with each stage of the ordinary optimization problem are the states of the decision process. They define the condition of each stage of the system and reflect the information required to fully assess the consequences that the current decision has upon future decisions. A problem with one state variable per stage is called a one-dimensional problem. Hence, this complexity ends up a multidimensional problem that has more than one state variable per stage to be taken care of.

- **Control Variables or Decision Variables**: They represent the control measures applied at a particular stage to transform the state of the system from one to the other.

The concept of functional diagrams represent the function equations in the multistage process of dynamic programming that may enable engineers to convert a process flow diagram to a dynamic programming functional diagram. It is necessary to begin with the definition of an individual process unit that can be represented as a stage as shown diagrammatically in Figure 8.2. If it is deemed as an economic model, the return function R gives the measure of profit or cost for the stage. State variables, S_i, are inputs to the stage from an adjacent stage that may be viewed as the flow rate of feed from an upstream unit. Decision variables, D_i, are ones that can be manipulated or controlled independently within this stage. Each stage will have outputs, S_{i+1}, which are inputs to adjacent stages. They may be viewed as products

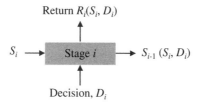

FIGURE 8.2   Diagrammatic representation of a dynamic programming stage

from the stage going to a downstream unit for further processing. There are transition functions, $S_{i-1} = T_i(S_i, D_i)$, at each stage that gives S_{i+1} based on the decision chosen at the current stage. For example, the transition function could represent material and energy balances at the stage.

Recursive equations are tied with fundamental model formulation of dynamic programming. Each recursive equation represents a stage at which a decision is required within the "multistage decision-making procedure." Within this context, a multistage decision process using a dynamic programming model as a means can then be solved by either a backward or a forward induction solution procedure depending on the logic applied in search of the global optimal solution embedded in the recursive equations.

If the model has a total of N stages in the process in which n as the number of stages remaining in the process (see Figure 8.3), we may assume that the state S_n of the system with n stages to go is a full description of the system for decision-making. For decision analysis, the next state of the process depends entirely on the current state and decision of the process. In general, the set of feasible decisions, D_n, available at a given stage depends on the state of the process at that stage, S_n, and could be written as $D_n(S_n)$. To formalize the decision analysis, a transition function, T_n, can be defined such that, given S_n, the state of the process with n stage to go, the subsequent state of the process with $(n - 1)$ stages to go is given by

$$S_{n-1} = T_n(D_n, S_n),$$

where D_n is the decision chosen in the feasible set for the current stage and state. T_n is the transition function that is a function in terms of D_n and S_n. Besides, the return function is given by

$$R_n(D_n, S_n)$$

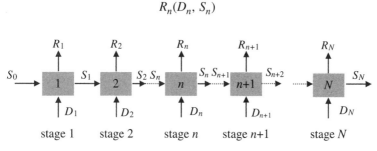

FIGURE 8.3   Multi-stage decision process

For the general optimization problem in this context, the objective is to maximize the sum of the return functions (or minimize the sum of cost functions) over all stages of the decision process. The constraint set is composed of the transition equations and the decisions taken for each stage, which belong to some set D'_n of permissible decisions. Hence, from stage n point of view, the dynamic programming approach to optimizing the decision for the remaining stages can be expressed as

$$V_n(S_n) = \text{Max}[R_n(D_n, S_n) + R_{n-1}(D_{n-1}, S_{n-1}) + \cdots + R_0(D_0, S_0)],$$

$$\text{s.t.:} \quad \begin{aligned} S_{m-1} &= T_m(D_m, S_m) \quad (m = 1, 2, \ldots, n) \\ D_m &\in D'_m \qquad\qquad (m = 1, 2, \ldots, n) \end{aligned},$$

where $V_n(S_n)$ is the optimal value function, since it represents the maximum return possible over the n stages to go. It is the optimal value of all subsequent decisions, given that we are in state S_n with n stages to go. Since $R_n(D_n, S_n)$ involves only the decision variable D_n and not the decision variables $D_{n-1}, \ldots, D_0$, we could first maximize over this latter group for every possible D_n and then choose D_n so as to maximize the entire expression. Therefore, we can rewrite the model above as follows:

$$V_n(S_n) = \text{Max}\{R_n(D_n, S_n) + \text{Max}[R_{n-1}(D_{n-1}, S_{n-1}) + \cdots + R_0(D_0, S_0)]\}$$

$$\text{s.t.:} \quad \begin{aligned} S_{n-1} &= T_n(D_n, S_n) \\ D_n &\in D'_n \end{aligned} \quad \text{s.t.:} \quad \begin{aligned} S_{m-1} &= T_m(D_m, S_m) \quad (m = 1, 2, \ldots, n-1) \\ D_m &\in D'_m \qquad\qquad (m = 1, 2, \ldots, n-1). \end{aligned}$$

To search for the optimal solution for the entire problem, the second part of the model above is simply the optimal value function for the $n - 1$ stage dynamic programming problem. Thus, it can be reiterated by replacing n with $n - 1$ when we move from stage n to stage $n - 1$.

$$V_n(S_n) = \text{Max}[R_n(D_n, S_n) + V_{n-1}(S_{n-1})]$$
$$\text{s.t.:} \quad S_{n-1} = T_n(D_n, S_n)$$
$$D_n \in D'_n.$$

To emphasize that this is an optimization over D_n, the model formulation can be simplified as follows:

$$V_n(S_n) = \text{Max}[R_n(D_n, S_n) + V_{n-1}(T_n(D_n, S_n)]$$
$$\text{s.t.:} \quad D_n \in D'_n.$$

The model defined above is a mathematical statement of the principle of optimality. Regardless of the current decision D_n and current state S_n in a recursive optimization procedure, an optimal sequence of decisions for a multistage problem can be made possible with respect to that all subsequent decisions must be optimal, given the state

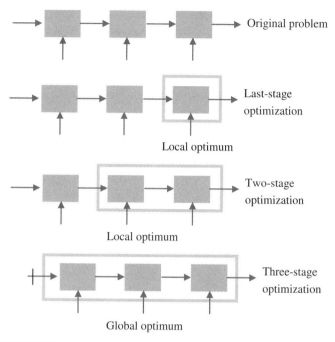

FIGURE 8.4   Backward induction for a three-stage decision process

S_{n-1} resulting from the current decision. Since $V_n(S_n)$ is defined recursively in terms of $V_{n-1}(S_{n-1})$, it is necessary to initiate the computation by solving the "stage-zero" problem. It is no longer defined recursively:

$$V_0(S_0) = \text{Max}[R_0(D_0, S_0)]$$
$$\text{s.t. :} \quad D_0 \in D'_0.$$

Based on such a recursive optimization scheme, an important feature of DP is that nonlinearity can be readily and conveniently accommodated. The DP formulation is the same for linear and nonlinear problems. Thus, no extra effort is required for solving nonlinear problems. Then the sequence to seek out the final global optimal solution in a three-stage decision problem can be shown in Figure 8.4. The recursive relationship is solvable for the optimal solution at each stage, and then a global optimal solution to the overall problem is determined (Bradley et al., 1977).

8.2.2   Trade-offs and Cost–Benefit–Risk Evaluation Matrix

In a traditional CBA matrix, benefits must at least balance or outweigh costs, which might require trade-offs. Cost–benefit tradeoff analysis, which has been the main-stream of optimization analysis for decades, is made possible by formulating the cost and benefit functions as an integral part of the objective function in single-objective mathematical programming models, such as LP, NLP, DP, and IP, as introduced in previous sections. In general, the decision analysis formulated as optimization problems may quickly screen out proposed alternative solutions using well-known algorithms, such as the simplex method, to identify the optimal solution that presents the best strategy to drive benefits to balance out costs (Chang, 2010).

To make more logical and informed decisions, however, a wide range of influential factors must be considered, such as the societal values, priorities, and perceptions. These issues also require consideration of risk factors other than technical and scientific. Questions concerning the level of risks, trade-offs in risk control, costs, and benefits may be factored into the proposed mathematical programming or OMs. Risk control can be in the form of regulation or policy with standards based on pros and cons of the best available control technology and its related costs and benefits. In the optimization context, trade-off designs may also lead the objective function, a CBA-driven framework to be weighed against the risk acceptance criteria described by a risk control or regulatory constraint in an LP, NLP, or IP model. Trade-off of paying minimal costs or receiving maximal benefits emphasizes the risks required to improve existing safety and health, environmental quality, property protection, and ecological conservation in the SWM.

Models used in the context of IMS cover the possible integration of simulation, forecasting, and optimization analyses; therefore, a series of separate or independent cost, benefit, and risk analyses may become background knowledge before the formulation of these OMs can be performed. Information on cost, benefit, and risk may become inputs to the OM, whereas the output from the OM may become inputs to these separate cost, benefit, and risk analyses seeking feedbacks iteratively. In other words, FM and/or SM applied to cost, benefit, and risk analyses in changing situations could address the essential complexity of those independent analyses. These cost, benefit, and risk analyses may also be directly defined and embedded as an integral part of the OM for more cohesive trade-offs. The merit of this numerical scheme is that it avoids the inputs and outputs associated with these separate analyses.

Risk analysis procedures might have to meet regulatory, legal, or stakeholder requirements in some countries. For example, certain stakeholder groups might be predetermined through regulatory requirements. The overall computational effort required to perform these decision analyses is proportional to the number and complexity of the system elements and problems, types, and numbers of events encountered for SWM.

8.3 MULTICRITERIA DECISION-MAKING

8.3.1 Basic Principles

Zeleny (1982) showed that multiple criteria include both multiple attributes and multiple objectives resulting in two theoretical variations, divided into MADM and MODM in decision science. Since then, MCDM has emerged as the common nomenclature for all decision analysis models and approaches; hence, MCDM research problems can be categorized as MODM or MADM. MODM is applied to problems with multiple, usually conflicting, objectives to identify a feasible alternative that yields the most preferred or satisfactory set of values for the objective functions. This setting differs from a single objective problem, as introduced in last section, because subjective methods are required to determine which alternative(s) out of infinite options in the working domain are most preferred or satisfactory within the decision domain.

In parallel to multiobjective optimization problems, MADM issues are based on classical decision analysis, such as utility theory, to carry out trade-offs among a finite number of alternatives with respect to a set of prescribed attributes in a system. These attributes can be related to multiple factors such as costs, benefits, and risks. One extension in this field is the development of multiattribute utility functions consistent with rational decision maker's behavior. The utility function derived in the paradigm of MADM can be used as a caliber for the decision maker to choose among several discrete alternatives or be combined into an MODM problem as objective function(s) for preference evaluation.

Decision analysis often requires distinguishing among goals, criteria, objectives, and attributes, some of which are similar in dictionary definition, such as goals and objectives. Clarifying the differences among the common yet interchangeable terminologies in MCDM will facilitate the subsequent discussion (Tabucanon, 1988; Karpak and Zionts, 1989):

- **Objective** indicates the desired direction, such as maximization or minimization, in decision-making. Multiple objectives are aspirations that also indicate directions of improvement for selected attributes, such as maximized profits and minimized losses. The achievement limits of these objectives are defined by the constraints.

- **Goal** is synonymous with target value, something that is either achieved or not. While "objectives" are aspirations without levels specified by decision makers, "goals" are aspirations with given "a priori" levels of desired attributes. If a goal cannot be achieved, it may be converted to an objective in decision analysis, which implies that an objective with a prescribed target value should be defined as a goal.

- **Attribute** is a measure used to evaluate whether a specific goal can be met, given a particular set of constraints. Attributes can be objective traits such as age, wealth, height, and/or weight, or they can be subjective traits such as prestige, goodwill, and/or beauty.

- **Criterion** is a measure, rule, or standard that guides decision-making and forms the basis for evaluation in MCDM. All attributes, objectives, goals, and even constraints that are judged as relevant in a given decision problem are criteria. For example, a master plan for SWM in a metropolitan region has three criteria— administrative feasibility, cost effectiveness, and lowest environmental impact— to be followed in the planning process. Although the problems of MCDM are diverse, they share the following common characteristics (Hwang and Yoon, 1981):

TABLE 8.1 MADM versus MODM

	MADM	MODM
Criteria	Attributes, small number	Objectives
Objective	Implicit (ill-defined)	Explicit (well-defined)
Attribute	Explicit	Implicit
Constraint	Inactive (incorporated into attributes)	Active
Alternative	Finite number, discrete prescribed	Infinite number, discrete, continuous and binary values, emerging as process goes
Interaction with decision maker	Not often	Sometimes or even mostly
Usage	Selection/evaluation, limited control by the user, outcome-oriented	Design, significant control by the user, process-oriented

Source: From Starr and Zeleny (1977), Hwang and Yoon (1981), and Munier (2011).

- **Multiple objectives and attributes**: A decision maker or a group of decision makers must generate relevant objectives and attributes for each problem setting.
- **Conflict among criteria**: Multiple criteria usually result in conflict.
- **Incommensurable units**: Each objective or attribute has a different unit of measurement, some of which may be expressed in an incommensurate way.
- **Design and selection**: Solutions to these problems are either to design the best alternative or to select the best design among previously specified finite or infinite number of alternatives.

To clarify the conceptual framework, the MCDM framework can be considered to have two categories of decision space. The first corresponds to a decision-making situation with a discrete number of feasible solutions to be ordered according to different attributes of concern, which requires MADM techniques. The multiattribute utility function representing the preferences of the decision maker must be constructed at the beginning of the process and may even be a function of cost, benefit, and risk factors. The second decision space corresponds to a decision-making situation with an infinite number of decision alternatives, which requires MODM techniques. Comparisons can clarify general differences among several features of MADM and MODM (Table 8.1). MODM is not applied to problems with predetermined alternatives; rather, MODM selects the "best" alternative by considering the various interactions between the design constraints and the objectives, which are then used to identify the compromised solution by attaining partially quantifiable objectives. A limited number of predetermined alternatives in the MADM are used to make final selections from the alternatives with the aid of inter- and intra-attribute comparisons via explicit or implicit trade-offs.

In a modern society, planners can usually assume that (1) with a good communication skills in the decision circle, a group of analysts can work together to gather data, identify objectives, build models, generate alternatives, and perform sensitivity analyses; (2) several decision makers working at various levels will choose from a set of possible decisions with respect to the pros and cons of alternatives provided by analysts; and (3) no single person in the decision circle with alternative decision weights and objectives can dominate the selection of a final solution or action.

To model an MCDM problem decision-making process, necessary and sufficient conditions must be identified. A necessary condition of MCDM is the presence of more than one criterion or objective, and a sufficient condition is that the criteria or objectives must be in conflict. In summary (Tabucanon, 1988), (1) a decision-making problem can be considered an MCDM problem if and only if at least two conflicting criteria and at least two alternative solutions exist; and (2) criteria are in conflict if the full satisfaction of one will result in impairing or precluding the full satisfaction of the other(s). In more precise terms, criteria are considered to be "strictly" conflicting if the increase in satisfaction of one may result in a decrease in satisfaction of the other(s). The sufficient condition of MCDM does not require "strictly" conflicting criteria. Conflict may arise in a group decision-making process due to intra- and interpersonal reasons and prevent consensus within the allotted time. To resolve the conflict, an MCDM problem can be further designed as an "interactive" as opposed to "once-through" process to simultaneously maximize several conflicting objectives over the long run.

8.3.2 Multiobjective Decision-Making

The general single-objective, $Z(x)$, mathematical program with n decision variables, x_j, and m constraints, $g_i(x)$, may be defined mathematically as:

$$\begin{aligned}
\text{Optimize} \quad & Z(x_1, x_2, \ldots, x_n) \\
\text{s.t.:} \quad & g_i(x_1, x_2, \ldots, x_n) \leq 0; \ i = 1, 2, \ldots, m \\
& x_j \geq 0; \ j = 1, 2, \ldots, n
\end{aligned}$$

or in vector notation

$$\begin{aligned}
\text{Optimize} \quad & Z(x) \\
\text{s.t.:} \quad & g(x) \leq 0 \\
& x \geq 0,
\end{aligned}$$

where x is the set of decision variables, Z is the objective function, and g is the constraint set.

The use of a single objective optimization approach has been recently recognized to limit the applicability of mathematical programming models, and many multiobjective situations exist in real-world systems. In this section, two frequently used techniques of multiple objective programming models, GP and compromise programming, are introduced in the MODM regime. The first challenge in model formulation in this paradigm is to identify multiple objectives. In SWM systems, MODM models can be designed to provide a strategic planning for recycling at the regional level. The model might include several objectives, to

- maximize stability of long-term recyclable markets;
- minimize recycling costs;
- minimize adverse environmental impacts;
- maximize production efficiency of recycling programs;
- minimize energy requirements; and
- minimize risk or maximize the operational safety.

Another example of the MODM model for raw solid waste collection at a regional scale can be designed to provide better managerial planning. The model may include several objectives, to:

- maximize total amount of waste collection;
- minimize operational cost in vehicle routing;
- minimize operational time (using collection and shipping time);
- maximize operational efficiency (using least labor forces and vehicles); and
- maximize service quality (using derived service index).

Complexity arises because some objectives permit a precise performance measurement, such as economics, but others do not, such as aesthetics and convenience. Although all objectives are quantifiable in a framework, what do we do if some goals or targets can be identified and others cannot? And, if some of the goals are not simultaneously achievable, can we change goals to objectives? To solve these problems, information gathering, identification of achievable objectives, and model building should be addressed as a whole to bridge the gaps between real-world problems and mathematical programming models.

The challenge is to adequately identify the tangible objectives by choosing not only the appropriate criteria but also the acceptable number of objectives. Some guiding principles to identify objectives can be drawn from the literature including (Keeney and Raiffa, 1976):

- Complete criterion: Objectives should cover all aspects of a problem;
- Operational criterion: Objectives can be meaningfully used in the analysis;
- Decomposable criterion: Objectives can be broken into parts to simplify the process;
- Nonredundant criterion: Objectives avoid the problem of double counting;
- Minimal criterion: The number of attributes should be kept small.

All objective functions may be assumed to be maximized without loss of generality because any objective can be minimized by maximizing the value of its negative. Hence, in an MODM model, the general k-objective, $Z(x)$, mathematical program with n decision variables, x_j, and m constraints, $g_i(x)$, can be defined mathematically in a vector notation as

$$\text{Maximize} \quad Z(x) = [Z_1(x), Z_2(x), \dots Z_h(x), \dots, Z_k(x)],$$
$$\text{s.t.:} \quad g(x) \leq 0$$
$$x \geq 0.$$

Once the gathered information, identified objectives, and formulated model are integrated, determining the set of alternatives may require considerable effort. As long as they can be generated, a scenario of management decisions requires decision maker(s) to choose a satisfactory or preferred solution that maximizes a decision maker's utility or satisfaction. Overall, as the decision-making process has continued to evolve over the last decades, MODM problems normally reflect the following needs to identify the best policy for the public.

- **Multiple objectives with or without priority**: In most engineering planning programs, managerial agencies have relied on the use of a benefit–cost ratio to select a recommended plan from a list of alternative plans in a project. Much progress has been made by the agencies to include objectives other than economic considerations.

- **Multiple decision makers in a participatory process**: The majority of the optimization methods presented in earlier times deal with a single decision maker, but real-world decision-making frequently involves multiple decision makers, such as a variety of stakeholders. Behavioral and analytical models need to be developed that reflect various thinking and preferences at different stage of decision-making associated with MODM.

- **Risk and uncertainty analysis in preventive, predictive, and preparatory processes**: Based on the problem statement, many decisions are conducted under the assumption of the "most possible" or "expected" value of costs, risks, and benefits. A better process might be to develop a method that can accommodate ranges and fluctuations of values for all input parameters to aid in the development and evaluation of alternative plans. Even when parameter ranges and fluctuations are considered, addressing how to proceed with the analysis of risk or uncertainty in decision-making is even more complex. In the past, probabilistic approaches provided opportunities to address potential random outcomes that describe the situation of risk; but these approaches could not address the uncertainties in decision-making in terms of the objectively known probability distributions. More recently, fuzzy sets theory and grey systems theory are more advanced tools that exhibit substantial progress to accommodate essential uncertainty analysis in decision-making. Later chapters will entail theories and concomitant applications.

Dominance and Efficiency   In an MODM model, the objective, "maximize $Z(x)$," requires further justification because the set $\{Z(x)\}$ for all feasible x lacks a natural ordering whenever $Z(x)$ is vector-valued. Thus, given any two feasible alternatives x_1 and x_2, it may not be possible to definitely determine whether $Z(x_1)$ is "better than" or "worse than" $Z(x_2)$. Because simultaneously maximizing all objectives is not possible, any meaningful definition of multiobjective evaluation requires the incorporation of some subjectivity with the possibility of trade-offs. This implies that two decision makers could produce different rankings for an identical set of outcomes $Z(x)$. In general, simultaneously maximizing several objectives requires finding all nondominated solutions to an MODM problem between these two decision makers.

From a general perspective, one workable definition of "optimize" is to find all nondominated solutions (alternatives) to an MODM problem. Hence, one property commonly considered as necessary for any candidate solution to the MODM problem is that the solution is not dominated. This property requires the assumption of increasing monotonicity. That is, for every objective function, Z_h, it is assumed that more of Z_h is always preferred to less of Z_h when all other objectives are held at constant levels; however, this may not hold true in special cases.

Again, the general k-objective, $Z(x)$, mathematical program with n decision variables, x_j, and m constraints, $g_i(x)$, may be defined mathematically in a vector notation as

$$\text{Max } Z(x) = [Z_1(x), Z_2(x), \dots Z_h(x), \dots, Z_k(x)],$$
$$\text{s.t.: } g(x) \leq 0$$
$$x \geq 0.$$

The objective functions can always be expressed in their "maximize" form because a minimization problem can always be transformed to a maximization problem by proper sign manipulations. Likewise for constraints, "greater than" and "equal to" constraints are always convertible to their equivalent "less than" or "equal to" constraints. To define dominance mathematically, a feasible point x is dominated by a second feasible point x_2 if and only if

$$Z_h(x_2) \geq Z_h(x_1); \, h = 1, 2, \dots, k$$

and

$$Z_h(x_2) > Z_h(x_1); \text{for at least one h.}$$

When more than one objective function is used in a systems engineering problem, they often conflict. In other words, an increase in one may cause a decrease in the other and vice versa. Therefore, the optimum solution no longer remains unique, and the concept of a unique optimum no longer holds because simultaneously maximizing all the conflicting objectives is impossible. In view of the conflicting nature of the objectives involved in MODM, the increase in any one of the objectives will decrease the others. Optimality is thus replaced by the satisfying (often termed as satisficing) solution or compromise solution. Methodologies need to systematically guide the selection of a "compromise" or "satisficing" solution rather than a "best" solution. In this case, the satisfying logic underlying GP, or the Paretian logic underlying MODM, are the approaches applied to obtain orderings and special subsets from the feasible continuous set of alternatives.

The concept of Pareto optimality was first introduced within the framework of welfare economics by the Italian economist Vilfredo Prato, who illustrated that a collectivity is in an optimum state if no person of that collectivity can improve his or her situation without worsening the situation of any other person of that collectivity. This kind of optimality also receives the name of Paretian efficiency (Box 8.2). An efficient solution, which is also called noninferior solution or Pareto optimal solution, is one in which no increase can be obtained in any of the objectives without causing a simultaneous decrease in at least one of the objectives (Hwang and Yoon, 1981).

To define efficiency, a feasible solution, x_j, is efficient if and only if there is no other feasible solution that dominates it. In other words, the solution x^* is efficient to the problem defined if and only if there does not exist any $x \in S$ such that $Z_h(x) \geq Z_h(x^*)$ for all h and $Z_h(x) > Z_h(x^*)$ for at least one h. This can be determined in the "decision space," but the terms noninferior solution or Pareto optimal solution can only be determined in the "objective space." Pareto-optimality (Box 8.3) involves the simple notion of dominance versus nondominance of the candidate solutions. Thus, the concept of "Pareto-optimality" is a way to identify the set of nondominated solution. If we compare two possible solutions (alternatives) and one is better in terms of Z_1 but inferior in terms of Z_2, while the second is better in terms of Z_2 but inferior in terms of Z_1, then they are considered to be of equal merit and, in other words, nondominating to each other. In a real-world situation, a host of solutions could be nondominating to each other but are not dominated by any other possible candidate solutions. The set of these solutions constitutes the Pareto-optimal solution to the problem, making up the so-called Pareto-frontier (Box 8.3), as it is often described in the literature.

BOX 8.3   PARETO EFFICIENCY AND PARETO-OPTIMAL SOLUTION

Pareto efficiency, or Pareto optimality, named for Vilfredo Pareto (1848–1923), is an important concept in economics with broad applications in game theory, economics, engineering, and the social sciences. Given a set of alternative allocations of, for example, goods or income for a set of individuals, a movement from one resources allocation to another that can make at least one individual better off without making any other individual worse off is called a Pareto improvement. A resources allocation scheme is Pareto efficient or a Pareto optimal solution when no further Pareto improvements are possible. The set that presents the Pareto optimal solution is called Pareto-frontier collectively in multiagent and multiobjective decision-making. More details can be found in Chang (2010).

An illustrative example for the concept of efficient solution and noninferior solution may be presented below to improve the understanding.

Example 8.1   Please find the satisfactory solution for the following multiobjective linear programming problem.

$$\begin{aligned}
\text{Max} \quad & Z = [Z_1, Z_2] = [2x_1 - 6x_2, -x_1 + 0.25x_2] \\
\text{s.t.:} \quad & g_1(x) = -0.5x_1 + 0.5x_2 - 1.8 \leq 0 \\
& g_2(x) = 2x_1 + 2x_2 - 11 \leq 0 \\
& g_3(x) = x_1 + 0.5x_2 - 4.6 \leq 0 \\
& g_4(x) = x_1 - 4 \leq 0 \\
& x_1, x_2 \geq 0.
\end{aligned}$$

Solution: Decision space (Figure 8.5a) is defined in terms of decision variables. Objective space is defined in terms of objective function values in which the Pareto-frontier can be delineated in the objective space, thick line located at the upper right part from A to D. No optimal solution exists for this example due to the nature of conflicting objectives. The line segment A-E-D (Figure 8.5b) is the closest area to the idea solution. Efficient solutions can be decided indirectly in the decision space once the set of non-dominated solutions are found in the objective space.

Because of the conflicting nature of the objectives in multiobjective optimization, achieving the individual optimum of each objective by a single solution is not possible. Each objective function has its own "ideal" solution that is different for all other objectives; in other words, using one of the ideal solutions as the final solution for the multiobjective problem would only achieve an individual optimum within all the conflicting objectives. The matrix, often termed as a "pay-off" table (Table 8.2), helps illustrate the entire set of ideal solutions. The diagonal of the matrix therefore constitutes individual optimal values of the k objective functions.

Methods for various MCDM problems can be described based on whether preference information articulation can be acquired from the decision makers in advance. Methods for MODM are therefore classified with respect to considerations consistent with methods in the classical literature (Charnes and Cooper, 1961; Goicoechea et al., 1982): (1) multiobjective programming with prior articulation of preferences, (2) multiobjective programming with progressive articulation of preferences, and (3) multiobjective programming with posteriori articulation of preferences. The first two methods are relatively popular in real-world applications and thus are discussed in detail.

Prior Articulation of Preferences—Goal Programming   Ordinary linear programming seeks an optimal solution for a single objective, such as maximizing the net benefit or minimizing the net cost or risk. GP attempts to achieve a satisfactory level in relation to multiple, often conflicting goals. The GP model was the first technique capable of handling decision problems with single or multiple goals (Charnes and Cooper, 1961) and can be viewed as an extension of single objective linear (nonlinear) programming. GP found relatively wide acceptance for applications in the context of public decision-making and industrial and business management problems in the 1970s and 1980s.

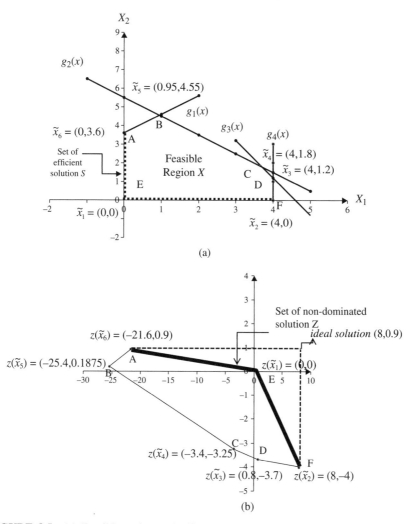

FIGURE 8.5 (a) Feasible region and efficiency solution in multiobjective programming model. (b) Nondominated solution in multiobjective programming model

TABLE 8.2 The pay-off table in a general MODM problem

	Z_1	Z_2	...	...	...	Z_k
$\tilde{x}_1^*$	$f_1\left(\tilde{x}_1^*\right)$	...	...	...	...	$f_1\left(\tilde{x}_k^*\right)$
$\tilde{x}_2^*$	$f_2\left(\tilde{x}_1^*\right)$	$f_2\left(\tilde{x}_2^*\right)$				
$\vdots$	$\vdots$		$\ddots$			$\vdots$
$\vdots$	$\vdots$			$\ddots$		
					$\ddots$	
$\tilde{x}_k^*$	$f_p\left(\tilde{x}_1^*\right)$	...	...	...	...	$f_k\left(\tilde{x}_k^*\right)$

Note: The ideal vector is the one in gray.

Linear GP is essentially a specially designed, multiobjective linear programming problem capable of considering multiple goals simultaneously. Each goal is expressed in the form of a goal constraint and combined under a single-objective function formulated to minimize the absolute value of total deviations from all target values, regardless of whether they deviate positively or negatively from the target values. GP, as in other multiobjective programming techniques, is not a means for optimizing, but for satisfying the decision maker's goals. Thus, GP is a "satisfying" procedure in which decision makers attempt to achieve a "satisfactory" level of multiple objectives rather than the best possible outcome for any single objective considered. In any circumstance, the simplex method is still the applicable solution procedure for solving GP problems.

Formulating a GP model is similar to formulating a linear programming model, but GP modeling incorporates all managerial goals into the system model formulation. The first step is to identify the managerial goals and then define the target values of the specified managerial goals, which are ranked in order of priority (if any). In model formulation, instead of attempting to maximize or minimize the objective function directly, as in linear (nonlinear) programming, the deviations among goals and the achievable limits dictated by the given set of resource or other constraints are minimized. These deviational variables, called "slack" or "surplus" variables in a linear programming model, take on a new meaning in GP. They are partitioned into positive and negative deviations from each subgoal or goal. Finally, all decision variables are summarized before performing the solution procedure by simplex algorithm.

1) Formulation of a Nonpreemptive GP Model: The general nonpreemptive GP model can be expressed mathematically as follows:

$$\text{Min} \quad z = \sum_{k=1}^{K} w_k^- d_k^- + \sum_{k=1}^{K} w_k^+ d_k^+,$$

s.t.

1. Goal constraints

$$\sum_{j=1}^{n} c_{kj} X_j + d_k^- - d_k^+ = g_k \qquad \forall k$$

Attribute ⎯ Deviational variable ⎯ Target value

2. Functional or resource constraints

$$\sum_{j=1}^{n} a_{ij} x_j \leq (=, \geq) b_i \qquad \forall i$$

3. Nonnegative constraint

$$x_j \geq 0 \qquad \forall j$$
$$d_k^-, d_k^+ \geq 0 \qquad \forall k,$$

where x_j represents a decision variable; w_i represents the weights (ordinal and/or cardinal) attached to each goal; a_{ij} is the technological coefficient defined in the resource constraint; and d_k^- and d_k^+ represent the degree of under- or overachievement of a goal, respectively. In other words, d_k^- and d_k^+ can be defined as

$$d_k^+ = \begin{cases} \sum_{j=1}^n c_{kj}X_j - g_k & \text{if } c_{kj}X_j > g_k \\ 0 & \text{otherwise} \end{cases}$$

$$d_k^- = \begin{cases} \sum_{j=1}^n g_k - c_{kj}X_j & \text{if } g_k > c_{kj}X_j \\ 0 & \text{otherwise} \end{cases}$$

Then, it can be further confirmed that

$$\left| \sum_{j=1}^n c_{kj}X_j - g_k \right| = d_k^+ - d_k^-.$$

Because a goal cannot simultaneously under- and overachieve, either one or both of these variables will be equal to zero. That is,

$$\left(d_k^+\right)\left(d_k^-\right) \equiv 0.$$

If the problem of interest can be formulated as a non-preemptive GP model, deviational variables at the same priority level may be given different weights in the objective function so that deviational variables with the same priority have different cardinal weights. The goals given the same priority level must be commensurable, however, to allow a possible trade-off in the solution procedure. To solve the non-preemptive GP model, the solution procedure is almost identical to the simplex method of linear programming. It will move the values of these deviational variables as close to zero as possible within the feasible region as limited by resource constraints and the given goal structure outlined in the model. Once a satisfactory solution is obtained, the decision maker must reexamine each of the goals in the model to see if under- or overachievement of the goal is acceptable.

If the decision maker is particularly concerned about the over- or underachievement for a specific goal, the corresponding goal should be formulated as a one-sided goal. A one-sided goal implies that either underachievement deviation, d_k^- (called a slack

TABLE 8.3   The formulation requirements of the types of goals

Objective	Implication	Constraint formulation	Deviational variable to be included in the objective function	The type of goal
$F(\tilde{x}) = T_i$	Simply close to the target	$F(\tilde{x}) + d_i^- - d_i^+ = T_i$	$d_i^- + d_i^+$	Two-sided
$F(\tilde{x}) \geq T_i$	Close to the target but not to be lower than it	$F(\tilde{x}) + d_i^- - d_i^+ = T_i$	d_i^+	One-sided
$F(\tilde{x}) \leq T_i$	Close to target but not to exceed it	$F(\tilde{x}) + d_i^- - d_i^+ = T_i$	d_i^-	One-sided

variable in linear programming), or overachievement deviation, d_k^+ (called a surplus variable in linear programming), can be eliminated in the model in the beginning of formulation (Table 8.3). If the decision maker reveals a preference to hit the target as close as possible, then a two-sided goal could be the most appropriate one. Conversely, if the decision maker indicates a preference to be close to the target within an upper or lower boundary, then a one-sided goal might be better fitted into the need for model formulation.

2) Formulation of a Preemptive GP Model: In a preemptive GP model, the objective is to minimize deviations within the preemptive priority structure if they are assigned to these deviations. To further prioritize the goals, one more symbolic notation of priority, P_k, is used to modify the general preemptive GP model that can be expressed mathematically as follows:

$$\text{Minimize} \quad z = \sum_{k=1}^{K} P_k \left(w_k^- d_k^- + w_k^+ d_k^+ \right)$$

s.t.

1. Goal constraints

$$\sum_{j=1}^{n} c_{kj} x_j + d_k^- - d_k^- = g_k \qquad \forall k$$

2. Functional or resource constraints

$$\sum_{j=1}^{n} a_{ij} x_j \leq (=, \geq) b_i \qquad \forall i$$

3. Nonnegative constraint

$$x_j \geq 0 \qquad \forall j$$
$$d_k^-, d_k^+ \geq 0 \qquad \forall k.$$

Using a stepwise simplex method, goals that must be ranked in order of priority (importance) by the decision maker are satisfied sequentially by the solution algorithm. Lower-priority goals are considered only after higher-priority goals, so in this sense, GP can be viewed as a lexicographic procedure.

3) Variations of GP Model Formulation: Obviously, all three GP variants can be solved by a conventional simplex method. Relative importance of goals may be expressed in terms of "priority" or "weight." Priority refers to goals that are ordered according to relative importance; unless the higher-level goals are considered, the lower ones do not come into application. Weights are attached to differentiate the relative importance of several goals with the same priority. For convenience in mathematical manipulations, weights can be normalized to 1.0, indicating that the following relationships among each weight must hold true.

$$0 < w_k^-, w_k^+ < 1 \text{ and } \sum_{k=1}^{K} \left(w_k^- + w_k^+ \right) = 1.$$

In a GP problem with a two-sided goal, once the set of goal constraints is complete then the objective function of the model is formulated in one of the following forms.

$$\text{Minimize} \quad z = \sum_{k=1}^{K} \left(d_k^- + d_k^+ \right),$$

$$\text{Minimize} \quad z = \sum_{k=1}^{K} \left(w_k^- d_k^- + w_k^+ d_k^+ \right),$$

$$\text{Minimize} \quad z = \sum_{k=1}^{K} P_k \left(d_k^- + d_k^+ \right),$$

$$\text{Minimize} \quad z = \sum_{k=1}^{K} P_k \left(w_k^- d_k^- + w_k^+ d_k^+ \right).$$

Example 8.2 A municipality proposed a master plan for SWM in which resources conservation and pollution prevention using benign environmental technologies are considered. There are three goals to achieve in one year.

Goal 1: Collect 500 tonnes (metric tons) of organic wastes to produce compost for landscape design, stormwater management and other uses.

TABLE 8.4   Treatment train production capacity based on processing 1000 tonnes per year of waste

	Compost (t.)	Energy production (t.)	Material recovery (t.)	Cost (US$·t^{-1})
Integrated Technology I	100	300	200	220,000
Integrated Technology II	80	200	300	240,000

Goal 2: Collect 4000 tonnes of residential and commercial waste and recover high quality steams through an incineration process for energy conservation.

Goal 3: Collect 2000 tonnes of residential and commercial wastes for material recovery using a material recovery facility (MRF, pronounced as "merf"), curbside recycling programs, and others.

The total budget is US$6 million that can be spent on achieving all of the three goals. Assume there are two novel types of integrated technologies that can achieve these three goals simultaneously. The potential production rate on a per treatment train basis (Table 8.4) can be used in the analysis.

The decision analysis will determine the possible combination of both types of integrated technologies that could improve the city's plan. These three goals are nonpreemptive goals. Please devise a plan by a "once-through" approach assuming that both deviational variables are equally important and all goals are also equally important.

Solution: Definitions of decision variables:

$X1$ is the number of treatment train of technology I required;

$X2$ is the number of treatment train of technology II required;

S_i^+ is the amount by which we numerically exceed the ith goal; and

S_i^- is the amount by which we are numerically under the ith goal.

(a) Model formulation for nonpreemptive GP: Assume that weighting factors associated with both deviational variables are the same.

$$\text{Min} \quad z = \sum_{i=1}^{3} \left(S_i^- + S_i^+ \right)$$

$$
\begin{aligned}
\text{s.t.:} \quad & 100X1 + 80X2 + S_1^- - S_1^+ = 500 \\
& 300X1 + 200X2 + S_2^- - S_2^+ = 4,000 \\
& 200X1 + 300X2 + S_3^- - S_3^+ = 2,000 \\
& 220,000X1 + 240000X2 <= 6,000,000.
\end{aligned}
$$

TABLE 8.5   LINDO inputs and outputs of Example 8.2

LINDO inputs	LINGO outputs		
min $S11 + S12 + S21 + S22 + S31 + S32$ s.t. $100X1 + 80X2 + S11 - S12 = 500$ $300X1 + 200X2 + S21 - S22 = 4,000$ $200X1 + 300X2 + S31 - S32 = 2,000$ $220,000X1 + 240,000X2 \leq 6,000,000$	LP optimum found at STEP 3 Objective function value 1) 1,500.000		
	Variable	Value	Reduced cost
	$S11$	0.000000	2.000000
	$S12$	500.000000	0.000000
	$S21$	1000.000000	0.000000
	$S22$	0.000000	2.000000
	$S31$	0.000000	2.000000
	$X1$	10.000000	0.000000
	$X2$	0.000000	180.000000
	Row	Slack or surplus	Dual prices
	2)	0.000000	1.000000
	3)	0.000000	−1.000000
	4)	0.000000	1.000000
	5)	3800000.000000	0.000000

Web site resources of software packages—LINDO and LINGO—can be found in the Appendix of this chapter. Findings from the software program LINDO (Table 8.5) suggest that investing 10 treatment trains of technology I is the most desirable option that will generate 1,000,000 tonnes of compost, recover high quality steam using 3,000,000 tonnes of waste, and recycle 2,000,000 tonnes of materials per year.

Progressive Articulation of Preferences—Compromise Programming Model Compromise programming is an interactive method widely used for solving multiple objective linear, nonlinear, or integer programming problems. The solution procedure of compromise programming identifies solutions that are closest to the ideal solution as estimated by some measures of distance. The solutions identified as being closet to the ideal solution are called "compromise solutions" and constitute the "compromise set," an exact subset selected from the nondominated solution set. Thus, to provide an index for decision-making, a distanced-based assessment function must be defined to fulfill the selection. The determination of ideal solutions in compromise programming in some ways resembles the selection of target values in GP. Yet the ideal solutions herein will never be achieved in real-world applications.

A plethora of multiobjective decision-making techniques are found in the literature. One thorough assessment included 15 techniques that were systematically evaluated in terms of 24 criteria, forming an evaluation matrix (Tecle, 1992). Using the criteria of minimum distance from the ideal solution, compromise programming became the favored tool to solve various multiobjective water resource problems in the 1980s, and since then has become popular in many applications for environmental systems analysis (Chang and Wang, 1996; Chang et al., 2006). Thus, compromise programming is frequently used to examine noninferior solutions and trade-offs among the objectives for a variety of large-scale complex decision-making issues.

To understand the meaning of a compromise solution, the implications of ideal solution and distance-based function must be examined. The ideal solution is defined as $Z^*(x) = (Z_1^*, Z_2^*, \ldots, Z_k^*)$, in which Z_i^* are the solutions of the following problems:

$$\text{Max} \quad Z_i(x)$$

$$\text{s.t.:} \quad x \in X.$$

The ideal solution can serve as a standard or reference base to evaluate the attainable nondominated solutions. Because the choice of an ideal solution is meant to increase the underlying utility, finding a compromise solution as close as possible to the ideal solution within the nondominated solution set is a reasonable surrogate for the maximization of utility function. The solution procedure to evaluate the subset of nondominated solutions uses a distance-based function to assess how close these points come to the ideal solution. A GP model measures this distance based on the weighted sum of absolute deviations from all given goals. The compromise programming model measures this distance based on Minkowski metric, L_a, which defines the distance between two points, $Z_k^*(x)$ and $Z_k(x)$:

$$L_a = \left\{ \sum_{k=1}^{p} \pi_k^a \left[Z_k^*(x) - (Z_k(x) \right]^a \right\}^{1/a},$$

where $1 \leq a \leq \infty$. As a result, a compromise solution with respect to distance parameter "a" is defined as xa^* such that

$$\text{Min} \quad L_a(x) = L_a(x_a^*)$$

$$\text{s.t.:} \quad x \in X.$$

The use of incommensurable units in the objectives prevents use of the above distance-based assessment function for decision analysis. Rescaling is normally needed before the optimization analysis is performed. Values after rescaling or normalization can be confined to a given range, such as [0, 1], to avoid the possible bias within the trade-off process, which could result in an inherently ambiguous compromise solution. Several scaling functions described in the literature can be applied, but the two types recommended hereafter include

$$Z = \frac{Z_k^*(x) - Z_k(x)}{Z_k^*(x)} \quad \text{or} \quad Z = \frac{Z_k^*(x) - Z_k(x)}{Z_k^*(x) - Z_k^{**}(x)},$$

in which $Z_k^*(x)$ and $Z_k^{**}(x)$ are the maximum and minimum values of each individual objective, respectively, both of which can be obtained from the payoff table. Thus, assessing a compromise programming model in decision analysis is equivalent to solving a dimensionless distance-based function that reflects the relative measure of the decision maker's preference. After normalization, the distance family becomes

$$\text{Minimize} \quad d_a = \text{Min} \left\{ \sum_{k=1}^{p} \pi_k^a \left[\frac{Z_k^*(x) - Z_k(x)}{Z_k^*(x)} \right]^a \right\}^{1/a}$$

$$\text{s.t.:} \quad x \in X,$$
$$\text{where} \quad 1 < a < \infty, \pi_k^a > 0$$
$$\text{and} \quad \sum_{k=1}^{p} \pi_k^a = 1.$$

The parameter p represents the total number of objectives, and π_k^a is the corresponding weight associated with each objective. Operationally, three points of the compromise set, those corresponding to $a = 1$, 2, and ∞, are usually calculated for decision analysis.

1. When $a = 1$, the above OM yields the minimization of the absolute derivations from the ideal solution, and the objective function is defined as the "Manhattan distance."

$$\text{Minimize} \quad d_1 = \text{Max} \sum_{k=1}^{q} \pi_k \left(\left| Z_k^*(x) - Z_k(x) \right| \right)$$

$$\text{s.t.:} \quad g(x)(\geq, =, \leq)0$$
$$x \geq 0.$$

For this particular definition of the ideal solution, however, we can drop the absolute value sign. For the d_1 metric, the best compromise solution is found by solving the following linear programming model.

$$\text{Minimize} \quad d_1 = \text{Max} \sum_{k=1}^{q} \pi_k^a Z_k(x)$$

$$\text{s.t.:} \quad g(x)(\geq, =, \leq)0$$
$$x \geq 0.$$

From the standpoint of decision analysis, these objectives are deemed competitive and compensatory. The LINGO software package can be employed as a solver in this analysis because if all functions are linear, the final model is a linear programming model.

2. When $1 < a < \infty$, the solution of the above OM will be the noninferior feasible solution closest to the ideal solution Z_k^* in terms of a weighted geometric distance. In particular, when $a = 2$, the objective function is specifically called the "Euclidean distance."

$$\text{Minimize} \quad d_2 = \text{Min} \left\{ \sum_{k=1}^{p} \pi_k^2 \left[\frac{Z_k^*(x) - Z_k(x)}{Z_k^*(x)} \right]^2 \right\}^{1/2}$$

$$\text{s.t.:} \qquad g(x)(\geq, =, \leq)0$$
$$x \geq 0.$$

In decision analysis, these objectives are deemed relatively competitive and compensatory; however, the problem becomes a nonlinear programming model. In this situation, LINGO can be employed as a solver.

3. When $a = \infty$, the largest deviation completely dominates distance determination, and the objective function is defined as the "Tchebycheff distance":

$$d_\infty = \max_k \left| Z_k^*(x) - Z_k(x) \right|.$$

For this particular definition of the ideal solution, however, we can drop the absolute value sign. The model formulation for the best compromise solution is

$$\text{Minimize } d_\infty = \text{Min } V$$

$$\text{s.t.:} \qquad g(x)(\geq, =, \leq)0$$
$$\pi_k^a [Z_k^* - Z_k(x)] \leq V \qquad \forall k = 1, \dots, p$$
$$x \geq 0.$$

The problem can be transformed into a linear programming model in which the situation among trade-off mechanics is not only noncompetitive, but also noncompensatory. The min–max criteria would ensure that the maximum deviation from the ideal solution could be minimized in the solution procedure. LINGO can be employed as a solver in this analysis because the final model is a linear programming model if all functions are linear (LINDO, 2001).

The initial effort for solving a compromise-programming model is to determine the decision weights once the mathematical formulation is developed. The weight of each objective is assumed to be equally important in the initial decision-making profile and could be altered in later stages of the decision-making process. A final comparison can be made between different decision scenarios in terms of various a values when dealing with complex environmental management planning issues. Overall, d_1 (the Manhattan distance) and d_2 (the Euclidean distance) are the longest and shortest distances in the geometrical sense, respectively; d_∞ is the shortest distance in the numerical sense.

Example 8.3   Please solve the following problem using the compromise program-ming approach.

$$\text{Max}\quad Z_1(x) = -10x_1 + 8x_2$$
$$\text{Max}\quad Z_2(x) = 4x_1 + 3x_2$$

$$\text{s.t.}\quad x_1 + x_2 \geq 4$$
$$-3x_1 + 6x_2 \leq 18$$
$$3x_1 + 0.5x_2 \leq 21$$
$$x_2 \leq 8$$
$$x_1, x_2 \geq 0.$$

Solution:

Step 1: $\text{Max } Z_1(X) = -10x_1 + 8x_2$

$$\text{s.t.}\quad x_1 + x_2 \geq 4$$
$$-3x_1 + 6x_2 \leq 18$$
$$3x_1 + 0.5x_2 \leq 21$$
$$x_2 \leq 8$$
$$x_1, x_2 \geq 0.$$

LINGO inputs and outputs of step 1 can be used as values in the equations (Table 8.6).

TABLE 8.6   LINGO inputs and outputs of step 1 in Example 8.3

LINGO Inputs	LINGO outputs		
Max = −10 * x1 + 8 * x2;	Objective function value		
!subject to;	1) 20.00		
x1 + x2 > = 4;			
−3 * x1 + 6 * x2 < = 18;	Variable	Value	Reduced cost
3 * x1 + 0.5 * x2 < = 21;			
x2 < = 8;	X1	0.6666667	0.000000
x1 > = 0;	X2	3.333333	0.000000
x2 > = 0;			

Step 2: $\text{Max } Z_2(X) = 4x_1 + 3x_2$

$$\text{s.t.}\quad x_1 + x_2 \geq 4$$
$$-3x_1 + 6x_2 \leq 18$$
$$3x_1 + 0.5x_2 \leq 21$$
$$x_2 \leq 8$$
$$x_1, x_2 \geq 0.$$

LINGO inputs and outputs of step 2 can be used as values in the equations (Table 8.7).

TABLE 8.7   LINGO inputs and outputs of step 2 in Example 8.3

LINGO Inputs	LINGO outputs		

Max = 4 * x1 + 3 * x2;
!subject to;
x1 + x2 > = 4;
−3 * x1 + 6 * x2 < = 18;
3 * x1 + 0.5 * x2 < = 21;
x2 < = 8;
x1 > = 0;
x2 > = 0;
	Objective function value
	  1) 42.00		

	Variable	Value	Reduced cost
	X1	6.000000	0.000000
	X2	6.000000	0.000000

Step 3: The payoff table (Table 8.8) can be summarized.

TABLE 8.8   The pay-off table in Example 8.3

Extreme points	Ideal solution	
	$Z_1^*(\tilde{x})$	$Z_2^*(\tilde{x})$
(0.6666667, 3.333333)	20.000	12.66667
(6.000, 6.000)	−12.00	42.000

Step 4: Finding compromise solution.

Distance-based formula is used to find compromise solution for the multiobjective programming model. Assume two objectives have equal weights.

$$\mathbf{a} = 1: \text{Min } d_1 = \text{Max } \Sigma \, \pi_k \mid Z_k^*(x) - Z_k \, (x) \mid$$

Which reduces to, $\text{Min } d_1 = \text{Max } \Sigma \, \pi_k \cdot Z_k \, (x)$

$$\text{Min } d_1 = \text{Max } (Z_1 \, (x) + Z_2 \, (x))$$
$$\text{Min } d_1 = \text{Max } (-6x_1 + 11x_2)$$

So the model becomes

$$\text{Max} \quad (-6x_1 + 11x_2)$$
$$\text{s.t.} \quad x_1 + x_2 \geq 4$$
$$-3x_1 + 6x_2 \leq 18$$
$$3x_1 + 0.5x_2 \leq 21$$
$$x_2 \leq 8$$
$$x_1, x_2 \geq 0.$$

LINGO inputs and outputs of $a = 1$ can be used in the equations (Table 8.9).

A linear programming model and the optimal solution can found by using LINDO: $(x_1^*, x_2^*) = (0.6666667, 3.333333) \, (Z_1^*, Z_2^*) = (20, 12.66667)$

TABLE 8.9 LINDO inputs and outputs of $a = 1$ in Example 8.3

LINGO inputs	LINGO outputs		
Max = −6 * x1 + 11 * x2;	Objective function value		
!subject to;	1) 32.66667		
x1 + x2 > = 4;			
−3*x1 + 6 * x2 < = 18;	Variable	Value	Reduced cost
3 * x1 + 0.5 * x2 < = 21;			
x2 < = 8;	X1	0.6666667	0.000000
x1 > = 0;	X2	3.333333	0.000000
x2 > = 0;			

$a = 2$:

$$\text{MIN } d_2 = \text{Min} \left\{ \left(\frac{Z_1^*(x) - Z_1(x)}{Z_1^*(x)} \right)^2 + \left(\frac{Z_2^*(x) - Z_2(x)}{Z_2^*(x)} \right)^2 \right\}^{1/2}$$

$$= \text{Min} \left\{ \left(\frac{20 + 10x_1 - 8x_2}{20} \right)^2 + \left(\frac{42 - 4x_1 - 3x}{42} \right)^2 \right\}^{1/2}$$

$$= \text{Min} \left\{ \left(1 + 0.5x_1 - 0.4x_2 \right)^2 + \left(1 - 0.0952382x_1 - 0.071429x_2 \right)^2 \right\}^{1/2}$$

$$\text{s.t.} \quad x_1 + x_2 \geq 4$$
$$-3x_1 + 6x_2 \leq 18$$
$$3x_1 + 0.5x_2 \leq 21$$
$$x_2 \leq 8$$
$$x_1, x_2 \geq 0.$$

LINGO inputs and outputs of $a = 2$ can be used as values in the equation (Table 8.10).

TABLE 8.10 LINGO inputs and outputs of $a = 2$ in Example 8.3

LINGO inputs	LINGO outputs		
Min = ((1 + 0.5 * x1 − 0.4 * x2)^2 +	Objective value: 0.6400875		
(1 − 0.095238 * x1 − 0.071429 *			
x2)^2)^0.5;	Variable	Value	Reduced Cost
!subject to;			
x1 + x2> = 4;	X1	1.520236	0.0000000
−3 * x1 + 6 * x2 < = 18;	X2	3.760118	0.0000000
3 * x1 + 0.5 * x2 < = 21;			
x2 < = 8;			
x1 > = 0;			
x2 > = 0;			

TABLE 8.11   LINGO inputs and outputs of $a = \infty$ in Example 8.3

LINGO inputs	LINGO outputs		
MIN = V;	Objective function value		
!subject to;	1) 15.30435		
x1 + x2 > = 4;			
−3 * x1 + 6 * x2 < = 18;	Variable	Value	Reduced cost
3 * x1 + 0.5 * x2 < = 21;			
x2 < = 8;	V	15.30435	0.000000
x1 > = 0;	X1	3.217391	0.000000
x2 > = 0;	X2	4.608696	0.000000
10 * x1–8 * x2 − V < = − 20;			
− 4 * x1–3 * x2 − V < = − 42			

The case of $a = 2$ ends up a nonlinear programming model and the optimal solution can found by using LINGO: $(x_1, x_2) = (1.520236, 3.760118)$ $(Z_1, Z_2) = (14.87858, 17.3613)$

$a = \infty$:

$$\text{Min} \quad d_\infty = \text{Min } V$$

$$\begin{aligned} \text{s.t.} \quad & 20 + 10x_1 - 8x_2 \le V \\ & 42 - 4x_1 - 3x_2 \le V \\ & x_1 + x_2 \ge 4 \\ & -3x_1 + 6x_2 \le 18 \\ & 3x_1 + 0.5x_2 \le 21 \\ & x_2 \le 8 \\ & x_1, x_2 \ge 0. \end{aligned}$$

LINGO inputs and outputs of $a = \infty$ can be used as values in the equations (Table 8.11).

The optimal solution is $(x_1, x_2) = (3.217391, 4.608696)$ and $(Z_1, Z_2) = (4.696, 26.696)$.

Overall, three scenarios can be summarized (Table 8.12 and Figure 8.6). The compromised set represents the most appealing solutions that the SWM system can choose.

TABLE 8.12   The information of compromised solution summarized in Example 8.3

	x_1	x_2	Z_1	Z_2
$a = 1$	0.6666667	3.333333	20	12.66667
$a = 2$	1.520236	3.760118	14.87858	17.3613
$a = \infty$	3.217391	4.608696	4.696	26.696

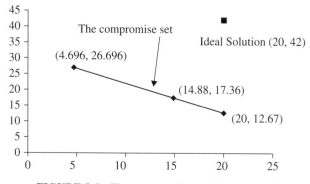

FIGURE 8.6   The compromise set in Example 8.3

8.3.3 Multiattribute Decision-Making

More than 20 different methods have been developed to conduct MADM analyses up to the present. A classification of MADM methods was provided by Hwang and Yoon (1981), distinguishing their attribute in terms of information processing approaches, including compensatory and non-compensatory methods.

Compensatory/additive models allow trade-offs between attributes (Hwang and Yoon, 1981). These models are index or trade-off methods in which individual attributes are transformed into units in a usual notional scale, classifying them concerning their respective importance. The units are then manipulated mathematically to compute indices that allow the relative evaluation of alternatives (Saegrov, 2005). These models are cognitively more difficult; however, it can guide to results near to the optimal result, or at least more logical choices than those identified by non-compensatory models (Yoon and Hwang, 1995). Compensatory models reduce option evaluation and selection so that each of the alternative options is classified using a single score, representing the attractiveness or utility of an option (Saegrov, 2005). Examples of compensatory models are multiattributive utility theory (MAUT) methods (Keeney and Raiffa, 1976), simple additive weighting (SAW) (Barron and Schmidt, 2002), analytical hierarchy processes (AHP) (Saaty, 1980), analytical network processes (ANP) (Saaty, 1996), and techniques of ranking preferences by similarity to the ideal solution (TOPSIS) (Hwang and Yoon, 1981).

Non-compensatory models do not allow trade-offs between attributes (Kalbar et al., 2012). For such models, a drawback (or adverse value) in one attribute cannot be counterbalance by an advantage (or favorable value) in some other attribute (Kalbar et al., 2012); thus, comparisons are made on an attribute-by-attribute basis. The most commonly used noncompensatory methods, known as "outranking methods" (Saegrov, 2005), compare pair-wise alternatives to check which are favored considering each criterion (Løken, 2007). "When aggregating the preference information for all the relevant criteria, the model determines to what extent one of the alternatives can be said to outrank another" (Løken, 2007). An alternative *a* outranks an alternative *b* if there is sufficient confirmation to conclude that *a* is at least as good as *b* when considering full criteria (Belton and Stewart, 2002). The best-known outranking methods are the *Elimination Et Choix Traduisant la Realité* (or Elimination and Choice Expressing Reality (ELECTRE)) methods proposed by Roy (1973). Examples of non-compensatory MADM methods applied in solid waste issues are ELECTRE I, II, III, and IV (Roy, 1973, 1991), and preference ranking organization method for enrichment of evaluations (PROMETHEE) (Brans et al., 1985).

Whether a compensatory or non-compensatory approach is appropriate for SWM decision-making is subject to various factors. The conditions to be considered when choosing an MADM method include (Pérez-Fortes and Bojarski, 2011)

- **Single, overwhelming indicator:** If the importance of one criterion is regarded to be primordial, then any compensation is prohibited; therefore, lexicographic methods should be applied.
- **Uncertainty analysis:** Ranked importance criteria combined with performance uncertainty may be considered, and quantified uncertainty can assist decision makers in establishing threshold values of difference and confidence needed to distinguish between alternatives (eliminate possible ties). After ties are excluded, non-compensatory methods requiring rank order should be applied.
- **Performance thresholds:** The performance thresholds assessment can be applied to identify conditions where compensation does not hold.

Decision analysis involving global sustainability issues might not have many differences between compensatory and non-compensatory approaches. For example, can a higher tipping fee be compensated by a lower climate change impact from a specific waste treatment alternative? Can climate change impact from incineration be compensated with electricity production to be consumed by population or industry? If attributes are substitutable, then one attribute can compensate for another, as in compensatory approaches, also known as the weak sustainability perspective (Munda, 2005). A strong sustainability perspective is the opposite, where, for example, certain sorts of natural capital are deemed not substitutable by man-made capital (Barbier and Markandya, 1990). To execute strong sustainability, a non-compensatory aggregation method has to be utilized (Munda, 2005). In any case, the decision maker determines whether the sustainability criteria/attributes are substitutable or not. The following subsections introduce some major MCDM approaches.

Multiattributive Utility Theory "Utility theory describes the selection of a satisfactory solution as the maximization of satisfaction derived from its selection. The best alternative is the one that maximizes utility for the decision maker's stated preference structure" (Kahraman, 2008). MAUT are established on a full compensatory aggregation of criteria and commensurable judgments, leading to high level trade-offs between criteria (Sadok et al., 2009). According to Kijak and Moy (2004), MAUT helps integrate the qualitative and quantitative information, supplying a well-structured perspective to the information assessment, and in achieving objective, transparent decisions. The most frequently applied perspective is an additive value function (multiattributive value theory (MAVT)) (Løken, 2007).

$$V(a) = \sum_{i=1}^{m} w_i v_i(a),$$

where $v_i(a)$ is a partial value function reflecting alternative a's performance on criterion i, which has to be normalized to some convenient scale (e.g., 0–100) (Løken, 2007). Using the above equation, a full value score $V(a)$ is established for each alternative a, and the alternative with the highest value score is favored (Løken, 2007). MAUT is a relatively simple and user-friendly approach in which the decision maker need only specify value functions and define weights for the criteria in cooperation with the analyst to receive useful feedback for his or her decision (Belton and Stewart, 2002). MAUT was first proposed in detailed steps by Keeney and Raiffa (1976) and Kahraman (2008) for

1. establishment of utility functions for individual attributes;
2. establishment of weighting/scaling factors;
3. establishment of the utility model type;
4. for each alternative conduct the measurement of the utility values with respect to the considered attributes; and
5. choice of the best alternative.

Simple Additive Weighting   SAW generates the result of an alternative equal to the weighted sum of its cardinal evaluation/preference ratings, where the weight reflects the relative importance related with each attribute, being used to rank, screen, or choose an alternative (Kahraman, 2008). This is a widely utilized method for the calculation of final grading values in multiple criteria problems. The mathematical formulation of the method is described by (Yoon and Hwang, 1995).

$$V_i = \sum_{j=1}^{n} w_j v_{ij},$$

where V_i is the suitability index for area i, w_j is the weight of criterion j, v_{ij} is the grading value of area i under criterion j, and n is the total number of criteria.

Analytical Hierarchy Process   AHP, developed by Saaty (1980), is one of the most well-known MADM methods to rank alternatives for achieving an overall goal, consisting in an additive weighting method. It is delineated to cope with both the rational and the instinctive uncertainty sources to select the best alternative evaluated for several criteria (Saaty and Vargas, 2001). The steps involved during an AHP process are (Saaty, 1980) (1) establish the problem and the objective, (2) develop the hierarchy from the top between the intermediate levels to the lowest level,

(3) apply simple pairwise comparison matrices for each of the lower levels, (4) conduct a consistency test, and (5) determine relative weights of the components of each level.

In this process, the problem is "represented using a hierarchy that is an abstraction of the whole problem. Entities are arranged in levels so that entities of the same level are not connected to each other but they are fully connected with entities of adjacent levels" (Gomez-Ruiz et al., 2010). The hierarchy elements are compared in pairs to evaluate the relative preference concerning each elements at the next higher level (Gomez-Ruiz et al., 2010). To elaborate the pairwise comparison matrix, the decision hierarchies are decomposed. The procedure to conduct an AHP is described in three steps as follows (Saaty, 1980):

Step 1: Compose a pairwise comparison decision matrix (A).

$$A = [a_{im}] = \begin{bmatrix} 1 & a_{12} & \cdots & a_{1n} \\ 1/a_{12} & 1 & \cdots & a_{2n} \\ \vdots & \vdots & \ddots & \vdots \\ 1/a_{1n} & 1/a_{2n} & \cdots & 1 \end{bmatrix} \quad i, m = 1, 2, \ldots, n$$

Let $C_1, C_2, \ldots, C_n$ indicate the group of elements, while a_{im} constitutes a quantified judgment on a pair of elements. Saaty (1980) establishes a measurement scale for pairwise comparison; thus, verbal judgments can be expressed by preference degree (Table 8.12). Although Saaty have defined this numerical scale, several authors have also proposed other scales, arguing that Saaty's scale is not transitive (Dong et al., 2008). Other possible scales are presented in Table 8.13.

Step 2: Calculate the normalized decision matrix. Each column values group is added, then each value is divided by its corresponding column total value. In the end, the rows average is calculated, and the weights of the decision-maker's objectives and a group of n numerical weights $w_1, w_2, \ldots, w_i$ are acquired.

Step 3: Conduct consistency analysis.

$$A \times w_i = \lambda_{max} \times w_i, a = 1, 2, \ldots, n$$

The consistency index (CI) is then calculated as:

$$CI = \frac{\lambda_{max} - n}{n - 1}$$

The consistency index of a randomly generated reciprocal matrix is called the random index (RI), with reciprocals forced. An average RI for the matrices of order 1–15 was generated using a sample size of 100.

TABLE 8.13 Different scales to compare two alternatives

Scales	Equal importance	Weak/moderate importance of one over another	Essential or strong importance	Very strong or demonstrated importance	Absolute importance
Linear (Saaty, 1977)	1	3	5	7	9
Power (Harker and Vargas, 1987)	1	9	25	49	81
Geometric (Lootsma, 1989)	1	4	16	64	256
Logarithmic (Ishizaka et al., 2006)	1	2	2.58	3	3.32
Root square (Harker and Vargas, 1987)	1	1.73	2.23	2.65	3
Asymptotical (Dodd and Donegan, 1995)	0	0.24	0.46	0.63	0.76
Inverse linear (Ma and Zheng, 1991)	1	1.29	1.8	3	9
Balanced (Salo and Hämäläinen, 1997)	1	1.5	2.33	4	9

There are also intermediate values between the two adjacent scale values. Also reciprocals exist, if an alternative i has one of the above numbers of the scale assigned to it when compared with alternative j, then j has the reciprocal value when compared with i. The reciprocal scale is given by $y = 1/x$.

The last value to calculate is the consistency ratio (CR). Generally, if CR is <0.1, the judgments are consistent, so the derived weights can be used.

$$CI = \frac{CI}{RI}$$

The AHP method can be difficult to decision makers to determine cardinal importance weights simultaneously for several attributes.

Kahraman (2008) pointed out advantages and disadvantages in the AHP method. Decision makers often find AHP difficult to accurately determine cardinal importance (Kahraman, 2008). As the number of attributes enlarges, improved results are acquired when the problem is translated to one that makes a series of pairwise comparisons. AHP formalizes the translation of the attribute weighting problem into the more tractable problem of making a series of pairwise comparison among competing attributes and summarizes the results in a "matrix of pairwise comparison." For each pairwise comparison of attributes, the decision maker must judge how much more important attribute A is than attribute B, relative to the overall objective.

Analytical Network Process   ANP, proposed by Saaty (1996), is a relatively new MADM method of mathematical theory that can systematically deal with a variety of dependence (Saaty, 2004). A methodological perspective is applied to establish priorities and trade-offs between goals and criteria, and also measure all tangible and intangible criteria in the model (Tseng, 2009). "ANP incorporates the influences and interactions among the elements of the system (criteria and alternatives) as perceived by the decision maker, and groups them into clusters" (Aragonés-Beltrán et al., 2010). Decision-making problems are represented as a network of criteria and alternatives (all called elements), grouped into clusters (Aragonés-Beltrán et al., 2010). All the network elements can be associated in any possible way, making network be capable to incorporate feedback and complex inter-relationships within and between clusters, providing a more accurate modeling of complex settings (Aragonés-Beltrán et al., 2010). A super matrix can be used to represent the influence of the elements in the network on other elements in that network, which consists in a two-dimensional element-to-element matrix that adjusts the relative importance weights in individual pairwise comparison matrices (Aragonés-Beltrán et al., 2010).

Saaty (1996) proposed this MADM model to surpass the problems of interdependence and feedback between criteria and alternatives (Tzeng and Huang, 2011). According to Saaty (2001), the ANP model includes the following steps: (1) identify the components and elements of the network and their relationships, (2) conduct pairwise comparisons on the elements, (3) place the resulting relative importance weights (eigenvectors) in pairwise comparison matrices within the super matrix (unweight super matrix), (4) conduct pairwise comparison on the clusters, (5) weight the blocks of the unweighted super matrix, by the corresponding priorities of the clusters, so that it can be column-stochastic (weighted super matrix), and (6) raise the weighted super matrix to restrict powers until the weights intersect and prevail stable (limit super matrix).

The application of ANP is devoted to the dependencies among groups of criteria and the alternatives under evaluation (Aragonés-Beltrán et al., 2010). The decision makers must carefully reflect on their project priority approach with regard to the detailed analysis of the interrelationships between cluster forces and the decision-making problem itself, allowing the decision maker to better understand the problem and to make a more reliable final decision (Aragonés-Beltrán et al., 2010). The main drawback in the practical application of ANP is a consequence of the complexity of the decision-making problem to be analyzed, which must be incorporated into the ANP model (Bottero and Ferretti, 2011). To this end, ANP prescribes a high number of comparisons that occasionally are too complex for decision makers not familiar with the ANP method to understand (Bottero and Ferretti, 2011). Hence, a facilitator is often needed to elaborate questionnaires and assist with the comparison process (Gómez-Navarro et al., 2009).

The treatments of interdependences in relation to ANP were not deemed complete and perfect (Tseng, 2009), however. Although ANP can be applied to find solutions to waste management issues, interdependences among elements must be considered (Tseng, 2009). ANP was used by Khan and Faisal (2008) for prioritizing and selecting appropriate municipal solid waste (MSW) disposal methods.

Technique for Order Preference by Similarity to an Ideal Solution

TOPSIS, created by Hwang and Yoon (1981), is established on the geometric concept that the chosen alternative should have the shortest distance from the ideal solution and the farthest from the negative-ideal solution. A utility value $D(i)$ for each alternative i is obtained by calculating the relative distance for i to the ideal solution, as described by Hwang and Yoon (1981).

The ideal solution is formed as a composite of the best performance values exhibited (in the decision matrix) by any alternative for each attributive. The negative-ideal solution is the composite of the worst performance values (Kahraman, 2008). Consider an MCDM problem with n alternatives $(A_1, A_2, \dots, A_n)$ and m criteria $(C_1, C_2, \dots, C_m)$. Criteria are used to characterize and evaluate alternatives. A decision matrix $X(x_{ij})_{n \times m}$ is built to rank alternatives and their values. To apply TOPSIS, the procedure is as follows (Hwang and Yoon, 1981; Jahanshahloo et al., 2006):

Step 1: Calculate the normalized decision matrix. The normalized value n_{ij} is calculated as

$$n_{ij} = \frac{x_{ij}}{\sqrt{\sum_{j=1}^{m} x_{ij}^2}}, i = 1, \dots, n, j = 1, \dots, m.$$

Step 2. Calculate the weighted normalized decision matrix. The weighted normalized value v_{ij} is calculated as

$$v_{ij} = w_i n_{ij}, i = 1, \ldots, n, j = 1, \ldots, m,$$

where w_i is the weight of the ith attribute or criterion, and $\sum_{i=1}^{n} w_i = 1$.

Step 3. Determine the positive ideal (A^+) and negative ideal (A^-) solutions.

$$A^+ = \{v_1^+, \ldots, v_n^+\} = \left\{ \left(\max_i v_{ij} \mid j \in J_1 \right), \left(\min_i v_{ij} \mid j \in J_2 \right) \right\}$$
$$A^- = \{v_1^-, \ldots, v_n^-\} = \left\{ \left(\min_i v_{ij} \mid j \in J_1 \right), \left(\max_i v_{ij} \mid j \in J_2 \right) \right\} \quad i = 1, \ldots, m,$$

where J_1 is associated with benefit criteria, and J_2 is associated with cost criteria.

Step 4. Calculate the separation measures using the n-dimensional Euclidean distances. The distance of each alternative for positive ideal solution (d_j^+) and for negative ideal solution (d_j^-) are given as, respectively,

$$d_j^+ = \left\{ \sum_{i=1}^{n} (v_{ij} - v_j^+)^2 \right\}^{1/2}$$
$$d_j^- = \left\{ \sum_{i=1}^{n} (v_{ij} - v_j^-)^2 \right\}^{1/2}, \quad i = 1, \ldots, m.$$

Step 5. Calculate the relative closeness to the ideal solution R_j.

$$R_j = \frac{d_j^-}{d_j^+ + d_j^-}, j = 1, \ldots, m$$

If $d_j^- \geq 0$ and $d_j^+ \geq 0$, then $R_j \in [0, 1]$.

Elimination and Choice Translating Algorithm   The ELECTRE methods were originally introduced by Benayoun et al. (1966) and Roy (1968, 1973, 1991) with several extended versions such as the ELECTRE III and ELECTRE IV. The method constitutes in a pairwise comparison of alternatives established on the degree to which alternatives evaluation and preference weight confirm or contradict the pairwise dominance relationship between the alternatives (Kahraman, 2008). An alternative is considered dominated if another alternative overcomes it in at least one criterion and equals it in the other criteria (Kahraman, 2008). The decision maker may declare its level of preference, such as strong, weak, or indifferent, or may even be unable to express his or her preference between two compared alternatives (Kahraman, 2008).

ELECTRE III is based on binary outranking relations in two major concepts (Achillas et al., 2010): "concordance" (c_j) when alternative A_1 outranks alternative A_2 if a sufficient majority of criteria are in favor of alternative A_1, and "non-Discordance"

(d_j) when the concordance condition holds and none of the minority criteria are too strongly opposed to the outranking of A_2 by A_1. The assertion that A_1 outranks A_2 is characterized by a credibility index that indicates the true degree of this assertion (Roussat et al., 2009). To compare a pair of alternatives (A_1, A_2) for each criterion, the assertion "A_1 outranks A_2" is evaluated with the help of pseudo-criteria (Achillas et al., 2010). The pseudo-criterion is built with two thresholds, indifference (q_j) and preference (p_j), which are applied to determine concordance and discordance indices as follows (Achillas et al., 2010; Vlachokostas et al., 2011):

Step 1. Construct the concordance matrix for each criterion $c_j\left(A_1, A_2\right)$.
- When $g_j(A_1) - g_j(A_2) \le q_j$, then no difference between alternatives A_1 and A_2 for the specific criterion j under study is identified. In this case $c_j(A_1, A_2) = 0$.
- When $g_j(A_1) - g_j(A_2) \le p_j$, then A_1 is strictly preferred to A_2 for criterion j. In this case $c_j(A_1, A_2) = 1$.

Step 2: Gather results in a global concordance metric $C_{A_1 A_2}$.
For a criterion J and a pair of alternatives (A_1, A_2), the concordance index is defined as follows:

$$
\begin{cases}
g_j(A_1) - g_j(A_2) \le q_j \Leftrightarrow c_j(A_1 A_2) = 0 \\
q_j < g_j(A_1) - g_j(A_2) < p_j \Leftrightarrow c_j(A_1 A_2) = \dfrac{g_j(A_1) - g_j(A_2) - q_j}{p_j - q_j} \\
g_j(A_1) - g_j(A_2) \ge p_j \Leftrightarrow c_j(A_1 A_2) = 1
\end{cases}
$$

A global concordance index, $C_{A_1 A_2}$, for each pair of alternatives (A_1, A_2) is computed with the concordance index $c_j\,(A_1, A_2)$ of each criterion j:

$$
C_{A_1 A_2} = \frac{\sum_{j=1}^{n} w_j \times c_j(A_1, A_2)}{\sum_{j=1}^{n} w_j},
$$

where w_j is the weight of criterion j.

Step 3. Construct the discordance matrix for each criterion $d_j\,(A_1, A_2)$.
As stated earlier, the discordance index $d_j\,(A_1, A_2)$ is also considered for all pairs of alternatives and each criterion j. Discordance index (d_j) is evaluated with the help of pseudo-criteria with a veto threshold (v_j), which represents the maximum difference $g_j(A_1) - g_j(A_2)$ acceptable to not reject the assertion "A_1 outranks A_2", as follows:
- When $g_j(A_1) - g_j(A_2) \le p_j$, then there is no discordance and therefore $d_j\,(A_1, A_2) = 0$.

- When $g_j(A_1) - g_j(A_2) \le v_j$, then $d_j(A_1, A_2) = 1$. Discordance index (dj) can be represented as follows:

$$\begin{cases} g_j(A_2) - g_j(A_1) \le p_j \Leftrightarrow d_j(A_1 A_2) = 0 \\ p_j < g_j(A_2) - g_j(A_1) < v_j \Leftrightarrow d_j(A_1, A_2) = \dfrac{g_j(A_2) - g_j(A_1) - p_j}{v_j - p_j} \\ g_j(A_2) - g_j(A_1) \ge v_j \Leftrightarrow d_j(A_1, A_2) = 1 \, . \end{cases}$$

In ELECTRE III, an outranking credibility degree is determined by the combination of the discordance index and the global concordance index. The degree of credibility is equal to concordance index when concordance index is higher or equal to the discordance index; if it is lower, the credibility degree will equal to the concordance degree lowered in direct relation to the importance of those discordances (Giannoulis and Ishizaka, 2010).

Step 4. Calculate the credibility index $\delta_{A_1 A_2}$.
The credibility index of assertion "A_1 outranks A_2" is defined as follows:

$$\delta_{A_1 A_2} = C_{A_1 A_2} \prod_{j \in F} \frac{1 - d_j(A_1, A_2)}{1 - C_{A_1 A_2}}$$

$$\bar{F} = \left\{ j \in F, \quad d_j(A_1, A_2) > C_{A_1 A_2} \text{ with} \right.$$

When a veto threshold is exceeded for at least one of the selected criteria, the credibility $\bar{F} = \{j \in F, \quad d_j(A_1, A_2) > C_{A_1 A_2}$ index is zero. The assertion "A_1 outranks A_2" is rejected.

Step 5. Ranking algorithm (ascending/upward and descending/downward distillations).
To rank all alternatives of A_j, two complete pre-orders are elaborated through a descending and an ascending distillation procedure. Briefly, descending distillation refers to the ranking from the best available alternative to the worst, while ascending distillation refers to the ranking from the worst available alternative to the best.

Step 6. Rank alternatives according to their ranks in each distillation.
ELECTRE is the most applied method in SWM systems, but it sometimes fails to find the preferred alternative and produces a crucial of leading alternatives (Wang et al., 2009). Nevertheless, the ELECTRE III method can be used in the start of the decision process to produce a shortlist of the best alternatives (Løken, 2007), which then be further analyzed by other more detailed methods.

Preference Ranking Organization Method for Enrichment Evaluation

PROMETHEE, developed by Brans and his co-workers (Brans, 1982; Brans et al., 1984, 1986; Brans and Vincke, 1985), is an outranking method based on positive and negative preference flows for each alternative that ranks the SWM alternatives corresponding to the selected preferences (weights) (Herva and Roca, 2013). "The positive outranking flow, $\Phi^+(a)$, expresses how an alternative a is outranking all the others, whereas the negative outranking flow, $\Phi^-(a)$, expresses how an alternative a is outranked by all the others" (Herva and Roca, 2013). The PROMETHEE I partial ranking is acquired from the positive and negative outranking flows; PROMETHEE II results in a full ranking based on the net outranking flow $\Phi(a)$ (Herva and Roca, 2013). Several factors must be defined for every alternative, including (Vego et al., 2008)

- preference ranking sense (maximizing or minimizing);
- preference function $P(a,b)$ that determines how one object is to be chosen concerning another, applied to compute the degree of preference related with the best alternative in pair wise comparisons (Geldermann and Zhang, 2001); and
- selected criteria weighting.

According to Vego et al. (2008), the value of preference functions is among 0 and 1, determined in such a way that the preference of the decision maker grows as the function approaches value 1. In the case of strict preference, the preference function is 1, and in the case of indifference, the preference function is 0 (Vego et al., 2008). Preference functions can be of different types, including usual criterion, quasi-criterion (U-shape), criterion with linear preference (V-shape), level criterion, criterion with linear preference and indifference area, and Gaussian criterion (Brans et al., 1986). PROMETHEE is conducted through the following steps (Vego et al., 2008).

Step 1. Establish an impact matrix/double entry table. The impact matrix for the selected criteria and alternatives can be established by using cardinal (quantitative) and ordinal (qualitative) data.

Step 2. Apply the preference function $P(a,b)$. For each criterion, the selected preference function $P(a,b)$ is applied to decide how much outcome a is preferred to b.

Step 3. Calculate an overall or global preference index $\Pi(a,b)$ that represents the intensity of preference of a over b.

Step 4. Calculate outranking flows. In the PROMETHEE I method, for each $a \in A$, there is a leaving flow (outranking),

$$\phi^+(a) = \frac{1}{n-1} \sum_{b \in A} \Pi(a, b),$$

and entering flow (being outranked),

$$\phi^-(a) = \frac{1}{n-1} \sum_{b \in A} \Pi(b, a).$$

Basically, the higher the leaving flow and the lower the entering flow, the better the alternative is considered to be. PROMETHEE I provides a partial alternatives ranking, but more realistic information about incomparability. PROMETHEE II supplies a total ranking of the alternatives by determining the net flow: $\phi(a) = \phi^+(a) - \phi^-(a)$. Part of the information on mutually incomparable alternatives is lost in PROMETHEE II.

Step 5. Compare outranking flows.

1. a outranks b, if $\phi(a) > \phi(b)$, or
   - (1.1) $\phi^+(a) > \phi^+(b)$ and $\phi^-(a) < \phi^-(b)$;
   - (1.2) $\phi^+(a) > \phi^+(b)$ and $\phi^-(a) = \phi^-(b)$;
   - (1.3) $\phi^+(a) = \phi^+(b)$ and $\phi^-(a) < \phi^-(b)$.
2. a is indifferent to b, if $\phi(a) = \phi(b)$, or $\phi^+(a) = \phi^+(b)$ and $\phi^-(a) = \phi^-(b)$.

Both partial ranking (PROMETHEE I) and complete ranking (PROMETHEE II) of the group of alternatives can be proposed to the stakeholders to solve the problem in the decision-making process (Brans and Vincke, 1985).

The Geometrical Analysis for Interactive Aid (GAIA) is used as a visualization tool to complement the PROMETHEE ranking method. A GAIA matrix is elaborated from $\phi(a)$ net outranking flows decomposition (Keller et al., 1991), which is the matrix data processed by a principal component analysis algorithm and presented on the GAIA biplot (Vego et al., 2008). "This transformation of a multicriteria problem to a two-dimensional space and geometrical representation of relations between alternatives and criteria provides a new perspective to the problem with the inevitable loss of some relation characteristics" (Vego et al., 2008).

Cases in which PROMETHEE has been continuously applied to SWM are few in the literature, some of them mentioned by Behzadian et al. (2010). Briggs et al. (1990) employed PROMETHEE I and II based on a small number of strongly conflicting criteria to obtain a complete ordering of 27 actors that included electric companies, consumers, public bodies, and other entities such as nuclear waste management companies. Vuk and Kozelj (1991) applied PROMETHEE methods and GAIA to select the location for the disposal of MSW in Slovenia. Vaillancourt and Waaub (2002) used PROMETHEE to rank waste management facilities. Kapepula et al. (2007) applied PROMETHEE to rank nine areas of the city in terms of nuisance to improve waste management in the city of Dakar. Queiruga et al. (2008) applied PROMETHEE in combination with a survey of experts to rank Spanish municipalities concerning their suitability for waste recycling plants installation. Rousis et al. (2008) compared 12 alternative systems for waste electrical and electronic equipment management in Cyprus and Vego et al. (2008) used it to focus on ranking SWM alternatives.

In real-world applications, challenges might begin with the interpretation of the ranking results the decision-making practitioner must use to understand the optimal solution, the second and third solutions, and even the worst one, and to determine the reliability of the results. The average ranking is the simplest procedure, even if two alternatives have the same average rank, which would select the alternative with the smallest standard deviation (Hwang and Yoon, 1981).

8.4 GAME THEORY AND CONFLICT RESOLUTION

As more units or people have to share fewer resources, the question of strategic behavior associated with resource competition or benefit protection for competing resources becomes critical. In fact, strategic conflict is a common phenomenon in multiagent and multiobjective decision-making processes (Bashar et al., 2012). Salient human activities exhibiting strategic conflict include bargaining, meetings, military actions, and peace-keeping activities (Kilgour and Hipel, 2005). This phenomenon is often observed when decision makers interact through decision-making processes. For example, two or more individuals or groups may have (1) opposing objectives, as when a suburban community tries to reject a landfill being sited in its backyard, while others aim for siting it in a suburban area, or (2) differing strategies, as when one political party wants to increase tipping fees to reduce waste generation while the other want to enhance household education to promote waste recycling.

A number of mathematical methodologies, including game theory (von Neumann and Morgenstern, 1944), metagame analysis (Howard, 1971), conflict analysis (Fraser and Hipel, 1984), the graph model for conflict resolution (Fang et al., 1993), and confrontation analysis (Howard, 1999), were developed to facilitate conflict resolution and to search for possible solutions. Game theory is a fundamental study of strategic multiagent and multiobjective decision-making. This section introduces game theory and its possible applications for SWM.

Game theory is the study of mathematical modeling of strategic behavior of players (decision makers) in situations where players' decisions may affect with each other. Game theory is essentially the theory for interactive decision-making that entails competition (i.e., noncooperation) and non-competition (i.e., cooperation) between intelligent and rational players (stakeholders). The two main branches of game theory, including cooperative game theory (CGT) and non-cooperative game theory (NCGT), may illustrate how strategic interactions among stakeholders produce ultimate payoff outcomes with the preferences or strategies by a differentiated procedure. The main distinction between NCGT and CGT is that the former models situations where players can only see their own strategic objectives and associated benefit (reward or interest) such that binding agreements among those players are impossible. On the contrary, the latter models situations based on the agreements to allocate cooperative gains (i.e., it could be transformed into a traditional resources allocation issue). In other words, although CGT ignores the strategic competition and focuses on building coalition to maximize cooperative gains with respect to equitable and fair sharing rules, NCGT takes into account the strategic interactions among players to minimize the loss. The cooperative bargaining problem first introduced in Nash (1950) is a typical example of CGT.

Basic terminologies in game theory can be summarized as follows:

- **Strategies:** Strategies are options, moves, or actions available to a set of players in a game. Pure strategy represents the situation in which each play's strategy is not random and should follow the predefined game table, something like the one

		Player 1	
		A	B
Player 2	A	(4,4)	(0,0)
	B	(0,0)	(2,2)

FIGURE 8.7   A two-person game with a normal form

in Figure 8.7. However, mixed strategy represents a game in which choosing each individual strategy might be subject to a probability.

- **Payoffs:** The payoff functions represent each player's utilities or preferences over action profiles consisting of a list of actions, one for each player. The payoffs to players influence the decisions to be made and the type of game a player wishes to play.

- **Rules:** Game theory sets the rules of the game as given, including how many players in a game, strategies, and payoffs to be applied. If the game follows sequential moves, then each play will choose a strategy sequentially and the follower may observe what happened prior to his or her choice, but if the game follows simultaneous moves, then all players must choose their strategies at the same time with no knowledge of previous strategies of others players.

- **Rational behavior:** Game theory assumes that human behavior is guided by instrumental reason; hence, the game theory of rational choice is a basic component of game-theoretical models. In social interactions, rationality must be enriched with further assumptions about individuals' common or mutual knowledge and beliefs, which is the epistemic foundations of game theory (i.e., social dilemma games could show that people do not behave as rationally as expected).

- **Cooperative game:** A game is cooperative if the players are able to form binding commitments to maximize the benefits as a group. In other words, cooperative games focus on the game at large. Communication among players is allowed in cooperative games.

- **Noncooperative game:** A game is noncooperative if the players are not able to form a coalition or binding agreements and are only able to model individual situations to seek the best response without regard to others' benefits. No communication among players is allowed in noncooperative games.

The normal game is usually represented by a decision matrix that shows the players, strategies, and pay-offs. In this context, games are defined mathematical models, consisting of a set of players, a set of strategies (options or actions) available to them, and specifications of players' payoffs for each combination of strategies (possible outcomes of the game). Hence, a game (in strategic or normal form) consists of the following three elements: a set of players, a set of actions (or pure-strategies) available to each player, and a payoff (or utility) function for each player. In a two-person bargaining problem, for example, two players have access to a set of alternatives (i.e., a feasible set) and each player has preference, represented by a utility function, over some of the alternatives. The so-called Nash equilibrium is a solution concept of a noncooperative game involving two or more players, in which each player is assumed to know the rules of the game. The optimal outcome of a game is the one where no player has an incentive or motivation to deviate from his or her chosen strategy after considering an opponent's choice. In other words, a pure-strategy Nash equilibrium is an action profile with the property that no single player can obtain an incremental payoff by deviating unilaterally from this profile or from changing alternatives, assuming other players remain constant in their strategies. To test for a Nash equilibrium, each player may simply reveal his or her own strategy to all other players, and the Nash equilibrium exists if no players change their strategy, in spite of knowing the moves of their opponents in the game.

Seeking the conflict resolution strategies is by no means an easy task due to the increasing complexity in decision analysis. Games in relation to environmental resources management, such as SWM, are often tied to multiobjectives facing uncertainty in a real-world system, given that these multiple objectives are often conflicting and competitive. With uncertainty involved, such as the probability of choosing an option, a transformed multiobjective programming model for conflict resolution can be solved by an approach to maximize the minimum semantic payoff (i.e., a generalization of the minimax theory) over different game scenarios. In this context, the MCDM is proposed as a practical framework for the relevant stakeholders (Banville et al., 1998), and MCDM models may consider stakeholder preferences when assigning values in the criteria weights (Garfi et al., 2009; De Feo and De Gisi, 2010). To reflect a more accurate socioeconomic implication, stakeholders' preferences can be taken into account through surveys of willingness to pay based on public consciousness in the analysis outcomes. These efforts may lead to the generation of the bargaining game framework for SWM (Karmperis, et al., 2013).

To analyze the conflicting objectives with varying payoff conditions in a game, Nash equilibrium is widely used as the delineation of the strategic interaction among several decision makers. The Nash equilibrium is proposed based on the assumption that each player had a well-defined quantitative utility function (Nash, 1951), in which the equilibrium solution is that no player can do better by unilaterally changing his or her strategy. In other words, the key characteristic is that if any player in a game unilaterally deviates from the Nash equilibrium, he or she cannot be better off. For mixed strategies, which have probability distributions over the pure strategies, the payoffs are the expected value of the players, and the analytical problem then becomes polylinear functions in the probabilities with which all kinds of players play their different pure strategies (Nash, 1950).

To elucidate a game play, a two-player normal-form (simultaneous move) game with several strategies for each player can be created. Consider a game in Figure 8.7, each of two players has two available moves or actions, including A and B. The rule is that if the players choose different actions, they each get a payoff of 0. If they both choose A, they each get 4, and if they both choose B, they each get 2. Assume that this is a "cooperative" game represented by player 1 choosing a row, player 2

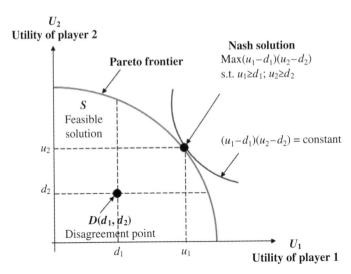

FIGURE 8.8   Illustration of the Nash solution

choosing a column. The resulting payoffs (within parentheses in Figure 8.7) show the first component corresponding to player 1's payoff and the second component corresponding to player 2's payoff. The action profile (B, B) is equilibrium because a unilateral deviation to A by any one player would result in a lower payoff for the deviating player. Similarly, the action profile (A, A) is a Nash equilibrium.

It is noticeable that not all games have a Nash equilibrium. Sometimes, other equilibrium concepts such as perfect Bayesian equilibrium, sequential equilibrium, correlated equilibrium, and sub-game perfect equilibrium can be proposed in various form of game theory (Okada, 2010). Nevertheless, the most widely used solutions in waste management game are the Nash solution (Karmperis et al., 2013). This Nash solution is unique in satisfying five axioms (Karmperis et al., 2013), and the specific solution is the function that maximizes the geometric average of the players' payoffs through the negotiation, instead of settling for the disagreement point D, as illustrated in following Figure 8.8 (Karmperis et al., 2013). In Figure 8.8, the Nash equilibrium for a two-person bargaining game is the solution maximizing the product of the excesses: $(u_1 - d_1) \cdot (u_2 - d_2)$, subject to constraints: $u_1 \geq d_1$ and $u_2 \geq d_2$ (Roth, 1979).

Nash equilibrium assumes that players always make a best response. However, people sometimes cooperate even it is not a best response to do so. A well-known example is the "Prisoner's Dilemma," in which two prisoners can choose to either defect or cooperate with payoffs as shown in the following figure (Figure 8.9). In this "Prisoner's Dilemma," the only best response here is to play "Defect" (alternative B) to reduce the penalty to be on year in prison no matter what the other player does. Yet people often do play "Cooperate" together (alternative A) in which Nash equilibrium does not predict actual behavior well.

	Prisoner 1	
	A	B
Prisoner 2 A	(2,2)	(0,4)
Prisoner 2 B	(4,0)	(1,1)

FIGURE 8.9   A two-person game to illustrate Prisoner's Dilemma

Some strategic game theory can be summarized below:

- **Zero-sum game**: Zero-sum games are those in which one person's gains exactly equal the net losses of the other participant(s) in the game. Zero-sum games are a special case of constant-sum games, when players can neither increase nor decrease the available benefits or payoffs or the total benefit to all players in the game; every combination of strategies always adds to zero. In this case, the game table is the same for all players. A zero-sum game is also called a strictly competitive game that are most often solved with the minimax theorem which is closely related to linear programming duality, or with Nash equilibrium.
- **Generalized game theory**: This extension of game theory incorporates social theory concepts such as norm, value, belief, role, social relationship, and institution. The generalized game theory is designed to address certain perceived limitations of game theory by formulating a set of rules that develop a more robust approach to psychological and sociological phenomena.

8.5 SYSTEM DYNAMICS MODELING

The method of system thinking has been used for more than 30 years (Forrester, 1961). System dynamics, designed based on system thinking, is a well-established methodology for studying and managing complex feedback systems for natural systems as well as the built environments. System thinking provides us with effective tools to better understand large-scale, complex SWM problems by addressing simulation and forecasting issues as well as large-scale optimization issues. System dynamics modeling has been used to address almost all feedback systems, including SWM (Sudhir et al., 1997; Karavezyris et al., 2002; Dyson and Chang, 2005).

To build a credible system dynamics model, a problem must be identified and a dynamic hypothesis developed that explain the cause of the problem, a process that requires constructing unique "causal loop diagrams" or "stock and flow diagrams" to form a system dynamics structure with respect to possible feedback via possible interactions among system components. Many proposed systems can be interconnected with subsystems to account for known interdependencies and hierarchies. Simulation runs in a system dynamics model are governed entirely by the passage of time based on a set of first-order ordinary differential equations in a simultaneous system. The initial conditions should be assigned to variables that denote the state of the system. The model may then begin to produce the related consequences for those system variables based on the prescribed initial conditions and the possible flow of information across the simulation skeleton. Time-step simulation analysis takes a number of computational steps along the timeframe to update the status of system behavior of concern as a result of system impetus stepwise. The entire model formulation is normally designed to test a simulation target with regard to alternative policies in the prescribed problem.

Most computer simulation applications rely on the use of software packages such as Vensim® (Ventana System Inc., 2012) and Stella® (High Performance System, Inc, 2002). The mechanisms of system dynamics can be handled by a user-friendly interface, and the dynamic relationships among these elements, consisting of variables, parameters, and their linkages, can be created onto the interface using user-friendly visual tools. Within the entire structure of a system dynamics model, the feedback loops associated with the employed variables can be visualized at every step throughout the modeling processes. The model development procedures are designed based on a visualization process that allows the modeler to conceptualize, document, modify, simulate, and analyze models of dynamic systems and their sensitivity associated with parameter settings. In any circumstance, these computer simulation tools offer a plethora of tools to build and assess a variety of SMs from customized casual loops or stock and flow. Simulation runs can be carried out entirely along the prescribed timeline. At the end of the modeling process, some designated system variables of interest can be chosen for demonstration and policy evaluation. The aim of this chapter was solely to introduce the fundamental ideas of system dynamics modeling, but relevant theories of how to develop system dynamics models can be found in the literature (Forrester, 1968; Randers, 1980; Richardson and Pugh, 1981; Morecroft, 1981; Mohapatra, 1994; Cellier, 2008).

Stella® software can be used to make basic building blocks of system dynamics models, such as stocks, flows, and converters, assembled to simulate the dynamic processes of the system (Figure 8.8). Stocks represent the accounting of a system component, either spatially or temporally; flows are the rate at which the component flows in or out of the stock; and converters modify rates of change and unit conversions. This flow chart in Figure 8.8 delineates the "principle of accumulation," the fundamental principle in system dynamics modeling that implies all dynamic behavior in the world occurs when flows are accumulated in stocks. A stock can be thought of as a reservoir, and a flow can be thought of as a pipe and faucet that either fills or drains the reservoir in a period of time affected by several external factors, such as the rate of rainfall and evaporation. A connector (thin line between Converter 1 and Waste flow 1 in Figure 8.10) is required to get the appropriate converter connected to the stock. In some cases, a connector is also required between stock and Waste flow 1 and/or Waste flow 2 if a feedback relationship exists.

Example 8.4   The city is building a new MRF to be in concert with the existing household recycling program. The city council wants to know how much residual waste that will be dumped into a neighboring landfill over the next few years, and

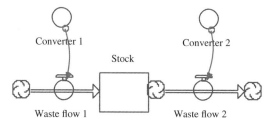

FIGURE 8.10   Stella® diagram showing stocks, flows, and converters

the possible recycling efforts that could prolong the life cycle of the landfill. The recyclables in the waste inflow, such as glass, paper, metal, and plastic, can be partially recycled by manual in household and automatic separation processes at a MRF. The city council predicts that the citizens will generate approximately 1000 tonnes of household waste stream every day over the next few years. Assume that 20% of household waste stream can be recycled and 70% the rest of household waste stream can be fully recycled at a MRF. Calculate how many tonnes of residual waste will be produced over time for the final disposal at a landfill.

Solution: The model can be built according to the following steps using Stella® 9.1.3 version:

Step 1: Click the "stock" icon at the tool bar to create a stock and rename the stock as the reservoir—Waste in a MRF.

Step 2: Place two icons of flow to the left and right of the stock and drag the flow into and out of the stock. Rename them as inflow (Waste flow 1) and outflow (Waste flow 2).

Step 3: Place one converter icon to the right of the outflow, rename them MRF Recycling Rate, and then place one converter icon to the left of the inflow, rename them Household Recycling Rate.

Step 4: Place one connector between the stock and the outflow.

Step 5: Double-click on outflows and the dialog box will appear. Type Waste_in_Material Recovery Facility*(1 − MRF_Recycling_Rate) as the equation for outflows.

Step 6: Double-click on MRF Recycling Rate and the dialog box will appear. Type 0.7 as the equation for recycling.

Step 7: Double-click on inflows and the dialog box will appear. Type 1000*(1 − MRF_Recycling_Rate) as the equation for outflows.

Step 8: Double-click on inflows, and the dialog box will appear on the screen. Type 1000*(1 − Household_Recycling_Rate) as the equation for inflow.

Step 9: Click the button Run Spec and the dialog box will appear to allow you to define the simulation period. Place 0 and 20 as the starting and ending year. Place 1 in DT (Duration of Time). Maintain the Euler's Method as the Integration Method. Click OK to end.

Step 10: Drag one graph from the Graph Pad at the tool bar and double click on Graph icon. Double-click page 1 at the lower left corner to define the simulation variable, which is the Waste flow 2.

Step 11: Click the Run button to trigger the simulation at the Map tab.

The resulting stock and flow model diagram (Figure 8.11a) then appears on the screen. If we switch to the Equation tab at the left margin, a series of equations (Figure 8.11b) appear. If we double click the graph pad, a curve of the residual waste flow destined for landfilling over time will be generated (Figure 8.11c), indicating that the cumulative volume of waste flow destined for landfilling. This result provides the city council with the information it requires to determine how large the landfill capacity should be to handle waste disposal needs for the length of the operational period if the anticipated recycling rate occur as initially predicted.

8.6 FINAL REMARKS

Within the MCDM paradigm, Paretian efficiency is a condition required to guarantee the rationality of any solution provided by any approach. Yet Romero (1991) noted that that solutions provided by GP models can be non-efficient (i.e., the solutions obtained via GP are not necessarily Pareto optimal), and therefore, GP should not be considered for MCDM analysis. GP was invented to obtain satisficing solutions rather than nondominated solutions. A possible method to improve this situation in dealing with the scaling issue over target values was found by Romero (1991). In addition, the inappropriate setting of goals in different scales, the naïve setting of decision weight, and problems associated with the unnecessary formulation of one-sided or two-sided goals can lead to poor modeling practices. The most difficult problem, however, is possibly the simultaneous consideration of several objectives/goals in which some have target values while the others do not. This formulation problem may lead to an assumption of subjective boundaries for objectives without target values.

Sensitivity analysis illustrates the degree to which the results, and especially the final ranking or optimal solution, are influenced by fluctuations of the parameter values, such as the weight coefficients of the criteria in MADM or right-hand side value of a constraint (Rousis et al., 2008). In addition, uncertainty analysis or quantification of uncertainty is usually conducted in association with various schemes of MODM and MADM. Uncertainty is often addressed with the use of fuzzy numbers, interval numbers, or random numbers established in the beginning of the model formulation. Sensitivity analysis is fundamental to providing credible decision analysis, especially when considerable budgets and public resources are a concern, such as in SWM systems, but it also must provide uncertainty analysis for possible inherent variations of decision variables to choose more reliable and justifiable decisions. Later chapters will discuss these relevant topics of sensitivity analysis and uncertainty analysis in greater detail.

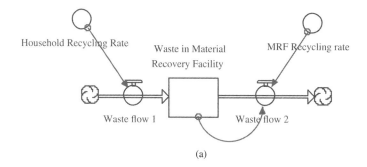

(a)

Waste_in_Material_Recovery_Facility(t) = Waste_in_ Material_Recovery_Facility(t-dt) +
(Waste_flow_1 - Waste_flow_2) * dt
INIT Waste_in_ Material Recovery Facility = 0
INFLOWS:
Waste_flow_1 = 1000*(1-Household_Recycling_Rate)
OUTFLOWS:
Waste_flow_2=Waste_in_Material_Recovery_Facility*(1-MRF_Recycling_Rate)
Household_Recycling_rate=0.2
MRF_Recycling_Rate=0.7

(b)

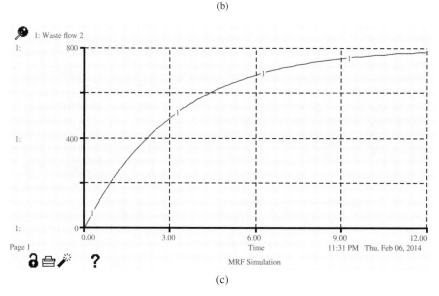

(c)

FIGURE 8.11   (a) The stock and flow diagram on the screen of Stella®. (b) The equations associated with the stock and flow diagram on the screen of Stella®. (c) The output graph pad from the Stella® package

REFERENCES

Achillas, Ch., Vlachokostas, C., Moussiopoulos, N., and Banias, G. 2010. Decision support system for the optimal location of electrical and electronic waste treatment plants: a case study in Greece. *Waste Management*, 30(5), 870–879.

AEA Technology. 1998. Computer-based models in integrated environmental assessment. Technical Report 14. European Environmental Agency.

Aprilia, A., Tezuka, T., and Spaargaren, G. 2012. Household solid waste management in Jakarta, Indonesia: a socio-economic evaluation. In: *Waste Management—An Integrated Vision* (Ed. Rebellon, L. F. M.), InTech.

Aragonés-Beltrán, P., Pastor-Ferrando, J. P., García-García, F., and Pascual-Agulló, A. 2010. An Analytic Network Process approach for siting a municipal solid waste plant in the Metropolitan Area of Valencia (Spain). *Journal of Environmental Management*, 91(5), 1071–1086.

Azapagic, A. and Clift, R. 1998. Linear programming as a tool in life cycle assessment. *International Journal of Life Cycle Assessment*, 3(6), 305–316.

Bach, H., Mild, A., Natter, M., and Weber, A. 2004. Combining socio-demographic and logistic factors to explain the generation and collection of waste paper. *Resources, Conservation and Recycling*, 41(1), 65–73.

Banar, M., Cokaygil, Z., and Ozkan, A. 2009. Life cycle assessment of solid waste management options for Eskisehir, Turkey. *Waste Management*, 29(1), 54–62.

Banville, C., Landry, M., Martel, J.-M., and Boulaire, C. 1998. A stakeholder approach to MCDA. *Systems Research and Behavioral Sciences*, 15(1), 15–32.

Barbier, E. B. and Markandya, A. 1990. The conditions for achieving environmentally sustainable growth. *European Economic Review*, 34(2–3), 659–669.

Barron, H. and Schmidt, C. P. 2002. Sensitivity analysis of additive multi attributes value models. *Operations Research*, 46(1), 122–127.

Bashar, M., Kilgour, M., and Hipel, K. 2012. Fuzzy preferences in the graph model for conflict resolution. *IEEE Transactions on Fuzzy Systems*, 20(4), 760–770.

Behzadian, M., Kazemzadeh, R. B., Albadvi, A., and Aghdasi, M. 2010. PROMETHEE: a comprehensive literature review on methodologies and applications. *European Journal of Operational Research*, 200(1), 198–215.

Bellman, R. E. 1957. *Dynamic Programming*, Princeton University Press, Princeton, NJ.

Bellman, R. 1961. *Adaptive Control Processes: A Guided Tour*, Princeton University Press, Princeton, NJ.

Belton, V. and Stewart, T. J. 2002. *Multiple Criteria Decision Analysis: An Integrated Approach*, Kluwer Academic Publishers, Boston, MA.

Benayoun, R., Roy, B., and Sussman, B. 1966. ELECTRE: une méthode pour guider le choix en présence de points de vue multiples. SEMAMETRA International.

Boelens, J. and Olsthoorn, A. A. 1998. Software for material flow analysis. In: *Sustainable Sustainability* (Eds Vellinga, P., Berkhout, F., and Gupta, J.), Kluwer Publishers, Dordrecht, pp. 115–130.

Bote, T., Nilausen, L., Kjeldsen, P., Andersen, K., and Andersen, L. 2003. Danish guidelines for investigation and risk assessment of gas producing landfills. In: Proceedings Sardinia 2003, Ninth International Waste Management and Landfill Symposium, S. Margherita di Pula, Cagliari, Italy.

Bottero, M. and Ferretti, V. 2011. Assessing urban requalification scenarios by combining environmental indicators with the Analytic Network Process. *Journal of Applied Operational Research*, 3(2), 75–90.

Bradley, S. P., Hax, A. C., and Magnanti, T. L. 1977. *Applied Mathematical Programming*, Addison-Wesley Publishing Company, Reading, MA.

Brans, J. P. 1982. Lingenierie de la decision. Elaboration dinstruments daide a la decision. Methode PROMETHEE. In: *Laide a la Decision: Nature, Instruments et Perspectives Davenir* (Eds. Nadeau, R. and Landry, M.), Presses de Universite Laval, Quebec, Canada, pp. 183–214.

Brans, J. P. and Vincke, Ph. 1985. A preference ranking organization method: the PROMETHEE method for multiple criteria decision-making. *Management Science*, 31(6), 647–656.

Brans, J. P., Mareschal, B., and Vincke, Ph. 1984. *PROMETHEE: A New Family of Outranking Methods in Multicriteria Analysis*. In: Operational Research'84 (Ed. Brans, J. P.), North-Holland, Dordrecht, pp. 447–490.

Brans, J. P., Vincke, Ph., and Mareschal, B. 1986. How to select and how to rank projects: the PROMETHEE method. *European Journal of Operations Research*, 24(2), 228–238.

Briggs, T., Kunsch, P. L., and Mareschal, B. 1990. Nuclear waste management: an application of the multicriteria PROMETHEE methods. *European Journal of Operational Research*, 44(1), 1–10.

Brunner, P. H. and Rechberger, H. 2003. *Practical Handbook of Material Flow Analysis*, CRC Press, Boca Raton, FL.

Cangialosi, F., Intini, G., Libertini, L., Notarnicola, M., and Stellacci, P. 2008. Health risk assessment of air emissions from a municipal solid waste incineration plant – a case study. *Waste Management*, 28(5), 885–895.

Cellier, F. E. 2008. World3 in Modelica: creating system dynamics models in the Modelica framework. In: Proceedings of the Modelica'2008 Conference, Bielefeld, Germany, pp. 393–400.

Chanchampee, P. and Rotter, S. 2007. Material flow analysis as a decision support tool for waste management in growing economies. In: Proceedings Sardinia 2007, Eleventh International Waste Management and Landfill Symposium, S. Marguerita di Pula, Cagliari, Italy.

Chang, N. B. 2008. Economic and policy instrument analyses in support of the scrap tires recycling program in Taiwan. *Journal of Environmental Management*, 86(3), 435–450.

Chang, N. B. 2010. *Systems Analysis for Sustainable Engineering*, McGraw-Hill, New York.

Chang, N. B. and Lin, Y. 1997a. Optimal siting of transfer station locations in a metropolitan solid waste management system. *Journal Environmental Science and Health*, A32(8), 2379–2401.

Chang, N.-B. and Lin, Y. 1997b. Economic evaluation of a regionalization program for solid waste management in a metropolitan region. *Journal of Environmental Management*, 51(3), 241–274.

Chang, N. B. and Lu, H. Y. 1997. A new approach for long term planning of solid waste management systems using fuzzy global criterion. *Journal of Environmental Science and Health*, A32(4), 1025–1047.

Chang, N. B. and Wang, S. F. 1996. Solid waste management system analysis by multi-objective mixed Integer programming model. *Journal of Environmental Management*, 48(1), 17–43.

Chang, N. B., Pan, Y., and Huang, S. 1993. Time series forecasting of solid waste generation. *Journal of Resource Management and Technology*, 21(1), 1–10.

Chang, N. B., Davila, E., Dyson, B., and Brown, R. 2005. Optimal design for sustainable development of a material recovery facility in a fast-growing urban setting. *Waste Management*, 25(8), 833–846.

Chang, N. B., Parvathinathan, G., and Dyson, B. 2006. Multi-objective risk assessment of freshwater inflow on ecosystem in San Antonio Bay, Texas. *Water International*, 31(2), 169–182.

Chang, N. B., Chang, Y. H., and Chen, H. W. 2009. Fair fund distribution for a municipal incinerator using GIS-based fuzzy analytic hierarchy process. *Journal of Environmental Management*, 90(1), 441–454.

Chang, N. B., Pires, A., and Martinho, G. 2011. Empowering systems analysis for solid waste management: challenges, trends and perspectives. *Critical Reviews in Environmental Science and Technology*, 41(16), 1449–1530.

Charnes, A. and Cooper, W. 1961. *Management Models and Industrial Applications of Linear Programming*, John Wiley & Sons, New York.

Chen, H. W. and Chang, N.-B. 2000. Prediction of solid waste generation via grey fuzzy dynamic modeling. *Resources, Conservation and Recycling*, 29(1–2), 1–18.

Cottle, R., Johnson, E., and Wets, R. 2007. George B. Dantzig (1914–2005). *Notice of AMS*, 52(3), 344–362.

De Feo, G. and De Gisi, S. 2010. Using an innovative criteria weighting tool for stakeholders involvement to rank MSW facility sites with the AHP. *Waste Management*, 30(11), 2370–2382.

Dewees, D. and Hare, M. 1998. Economic analysis of packaging waste reduction. *Canadian Public Policy—Analyse de Politiques*, 24(4), 453–470.

Dodd, F. and Donegan, H. M. 1995. Comparison of priotization techniques using interhierarchy mappings. *Journal of the Operational Research Society*, 46(4), 492–498.

Dong, Y., Xu, Y., Li, H., and Dai, M. 2008. A comparative study of the numerical scales and the prioritization methods in AHP. *European Journal of Operational Research*, 128(1), 229–242.

Dyson, B. and Chang, N. B. 2005. Forecasting of solid waste generation in an urban region by system dynamics modeling. *Waste Management*, 25(7), 669–679.

Fang, L., Hipel, K. W., and Kilgour, D. M. 1993. *Interactive Decision Making: The Graph Model for Conflict Resolution*, Wiley, New York.

Forrester, J. W. 1961. *Industrial Dynamics*, The MIT Press, Cambridge, MA.

Forrester, J. W. 1968. *Principles of System*, Productivity Press, Cambridge, MA.

Fraser, N. M. and Hipel, K. W. 1984. *Conflict Analysis: Models and Resolutions*, North Holland, New York.

Garfi, M., Tondelli, S., and Bonoli, A. 2009. Multi-criteria decision analysis for waste management in Saharawi refugee camps. *Waste Management*, 29(10), 2729–2739.

Geldermann, J. and Zhang, K. 2001. Software review: "Decision Lab 2000". *Journal of Multi-Criteria Decision Analysis*, 10(6), 317–323.

Giannoulis, C. and Ishizaka, A. 2010. A web-based decision support system with ELECTRE III for a personalized ranking if British universities. *Decision Support Systems*, 48(3), 488–497.

Goicoechea, A., Hansen, D. R., and Duckstein, L. 1982. *Multiobjective Decision Analysis with Engineering and Business Applications*, John Wiley & Sons, New York.

Gómez-Navarro, T., García-Melón, M., Acuña-Dutra, S., and Díaz-Martín, D. 2009. An environmental pressure index proposal for urban development planning based on the analytic network process. *Environmental Impact Assessment Review*, 29(5), 319–329.

Gomez-Ruiz, J., Karanik, M., and Peláez, J. I. 2010. Estimation of missing judgments in AHP pairwise matrices using a neural network-based model. *Applied Mathematics and Computation*, 216(10), 2959–2975.

Harker, P. T. and Vargas, L. G. 1987. The theory of ratio scale estimation: Saaty's analytic process. *Management Science*, 33(11), 1383–1403.

Herva, M. and Roca, E. 2013. Ranking municipal solid waste treatment alternatives based on ecological footprint and multi-criteria analysis. *Ecological Indicators*, 25, 77–84.

High Performance System, Inc. 2002. Stella®, Available at: http://www.hps-inc.com/stellavpsr.htm

Howard, N. 1971. *Paradoxes of Rationality: Theory of Metagames and Political Behavior*, MIT Press, Cambridge, MA.

Howard, N. 1999. *Confrontation Analysis: How to Win Operations Other Than War*, CCRP Publication Series, Washington, DC.

Hwang, C. L. and Yoon, K. S. 1981. *Multiple Attribute Decision Making: Method and Application*, Springer-Verlag, New York.

Ishizaka, A., Balkenborg, D., and Kaplan, T. 2006. Inuence of aggregation and preference scale on ranking a compromise alternative in AHP. In: Multidisciplinary Workshop on Advances in Preference Handling, Riva del Garda, pp. 51–57.

Jahanshahloo, G. R., Lotfi, F. H., and Izadikhah, M. 2006. Extension of the TOPSIS method for decision-making problems with fuzzy data. *Applied Mathematics and Computation*, 181(2), 1544–1551.

Kahraman, C. 2008. Multi-criteria decision making methods and fuzzy sets. In: *Fuzzy Multi-Criteria Decision Making: Theory and Applications with Recent Developments* (Ed. Kahraman,C.), Springer Science+Business Media.

Kalbar, P. P., Karmakar, S., and Asolekar, S. R. 2012. Selection of an appropriate wastewater treatment technology: a scenario-based multiple-attribute decision-making approach. *Journal of Environmental Management*, 113, 158–169.

Kapepula, K. M., Colson, G., Sabri, K., and Thonart, P. 2007. A multiple criteria analysis for household solid waste management in the urban community of Dakar. *Waste Management*, 27(11), 1690–1705.

Karavezyris, V., Timpe, K.-P., and Marzi, R. 2002. Application of system dynamics and fuzzy logic to forecasting of municipal solid waste. *Mathematics and Computers in Simulation*, 60(3–5), 149–158.

Karmperis, A. C., Aravossis, K., Tatsiopoulos, I. P., and Sotirchos, A. 2013. Decision support models for solid waste management: review and game-theoretic approaches. *Waste Management*, 33(5), 1290–1301.

Karpak, B. and Zionts, S. (Eds.) 1989. *Multiple Criteria Decision Making and Risk Analysis using Microcomputers. Computer and Systems Sciences*, Vol. 56, NATO ASI Series, Springer-Verlag, Berlin, Germany.

Keeney, R. L. and Raiffa, H. 1976. *Decisions with Multiple Objectives Preferences and Value Tradeoffs*, John Wiley & Sons, New York.

Keller, H. R., Massart, D. L., and Brans, J. P. 1991. Multicriteria decision making: a case study. *Chemometrics and Intelligent Laboratory Systems*, 11(2), 175–189.

Khan, S. and Faisal, M. N. 2008. An analytic network process model for municipal solid waste disposal options. *Waste Management*, 28(9), 1500–1508.

Kijak, R. and Moy, D. 2004. A decision support framework for sustainable waste management. *Journal of Industrial Ecology*, 8(3), 33–50.

Kilgour, D. M. and Hipel, K. W. 2005. The graph model for conflict resolution: past, present, and future. *Group Decision and Negotiation*, 14(6), 441–460.

Kumar, R. and Mittal, R. 2004. *Management Information System*, Anmol Publications Pvt. Ltd, Delhi.

Lenzen, M., Murray, S. A., Korte, B., and Dey, C. J. 2003. Environmental impact assessment including indirect effects – a case study using input-output analysis. *Environmental Impact Assessment Review*, 23(3), 263–282.

Liebman, J., Male, J., and Wathne, M. 1975. Minimum cost in residential refuse vehicle routes. *Journal of Environmental Engineering*, 101(3), 399–411.

LINDO Systems Inc. 2001. *LINDO User's Manual: The Optimization Standard*, Chicago, IL.

Løken, E. 2007. Use of multicriteria decision analysis methods for energy planning problems. *Renewable and Sustainable Energy Reviews*, 211(7), 1584–1595.

Lootsma, F. 1989. Conflict resolution via pairwise comparison of concessions. *European Journal of Operational Research*, 40(1), 109–116.

Ma, D. and Zheng, X. 1991. 9/9–9/1 Scale method of AHP. In: Proceedings of the Second International Symposium on the AHP, University of Pittsburgh, Pittsburgh, PA, pp. 197–202.

Menikpura, S. N., Gheewala, S. H., and Bonnet, S. 2012. Framework for life cycle sustainability assessment of municipal solid waste management systems with an application to a case study in Thailand. *Waste Management & Research*, 30(7), 708–719.

Miller, I., Kossik, R., and Voss, C. 2003. General requirements for simulation models in waste management. In: Proceedings of the Waste Management 2003 Symposium, WM SYMPOSIA, INC., Tucson, AZ.

Mohapatra, P. K. 1994. *Introduction to System Dynamics Modeling*, Orient Longman Ltd., Hyderabad, India.

Morecroft, J. D. W. 1981. System dynamics: portraying bounded rationality. In: Proceedings of 1981 Conference in Systems Dynamics Research, Renesselaerville, NY.

Morrissey, A. J. and Browne, J. 2004. Waste management models and their application to sustainable waste management. *Waste Management*, 24(3), 297–308.

Munda, G. 2005. Multiple criteria decision analysis and sustainable development. In: *Multiple Criteria Decision Analysis—State of the Art Surveys* (Eds Figueira, J., Greco, S., Ehrgott, M.), Springer, New York, pp. 953–986.

Munier, N. 2011. *A Strategy for Using Multicriteria Analysis in Decision-Making: A Guide for Simple and Complex Environmental Projects*, Springer, Dordrecht.

Nash, J. F. 1950. Equilibrium points in n-person games. *Proceedings of the National Academy of Sciences*, 36, 48–49.

Nash, J. 1951. Non-cooperative games. *The Annals of Mathematics*, 54(2), 286–295.

Okada, A. 2010. Perfect Bayesian equilibrium and sequential equilibrium. In: *Wiley Encyclopedia of Operations Research and Management Science* (Eds Cochran, J. J., Cox, L. A., Keskinocak, P., Kharoufeh, J. P., and Smith, J. C.), John Wiley & Sons, New York, pp. 1–7.

Pérez-Fortes, M. and Bojarski, A. D. 2011. Global clean gas process synthesis and optimization. In: *Syngas from Waste: Emerging Technologies, Green Energy and Technology Series* (Ed. Puigjaner,L.), Springer-Verlag, pp. 227–252.

Pires, A., Martinho, G., and Chang, N. B. 2011. Solid waste management in European countries: a review of systems analysis techniques. *Journal of Environmental Management*, 92(4), 1033–1050.

Pontryagin, L. S., Boltyanskii, V. G., Gamkrelidze, R. V., and Mischenko, E. F. 1962. *The Mathematical Theory of Optimal Processes*, John Wiley & Sons, New York.

Queiruga, D., Walther, G., Gonzalez-Benito, J., and Spengler, T. 2008. Evaluation of sites for the location of WEEE recycling plants in Spain. *Waste Management*, 28(1), 181–190.

Randers, J. (Ed.) 1980. *Elements of the Systems Dynamics Method*, Productivity Press, Cambridge, MA.

Richardson, G. P. and Pugh, A. L., III 1981. *Introduction to System Dynamics Modeling with DYNAMO*, Productivity Press, Cambridge, MA.

Romero, C. 1991. *Handbook of Critical Issues in Goal Programming*, Pergamon Press, Oxford, UK.

Roth, A. E. 1979. *Axiomatic Models of Bargaining*, Springer-Verlag, Berlin.

Rousis, K., Moustakas, K., Malamis, S., Papadopoulos, A., and Loizidou, M. 2008. Multi-criteria analysis for the determination of the best WEEE management scenario in Cyprus. *Waste Management*, 28(10), 1941–1954.

Roussat, N., Dujet, C., and Méhu, J. 2009. Choosing a sustainable demolition waste management strategy using multicriteria decision analysis. *Waste Management*, 29(1), 12–20.

Roy, B. 1968. Classement et choix en présence de points de vue multiples (la méthode ELECTRE). *RAIRO-Operations Research—Recherche Opérationnelle*, 8, 57–75.

Roy, B. 1973. How outranking relation helps multiple criteria decision making in topics in multiple criteria decision making. In: *Multiple Criteria Decision-Making* (Eds Cochrane, J. and Zeleny, M.), University of South Carolina Press, Columbia, SC, pp. 179–201.

Roy, B. 1991. The outranking approach and the foundations of ELECTRE methods. *Decision Theory*, 31, 49–73.

Saaty, T. L. 1977. A scaling method for priorities in hierarchical structures. *Journal of Mathematical Psychology*, 15(3), 234–281.

Saaty, T. L. 1980. *The Analytical Hierarchy Process*, McGraw-Hill, New York.

Saaty, T. L. 1996. *The Analytical Network Process-Decision Making with Dependence and Feedback*, RWS Publications, Pittsburgh, PA.

Saaty, T. L. 2001. *Decision Making with Independence and Feedback: The Analytic Network Process*, RWS Publications, Pittsburgh, PA.

Saaty, T. L. 2004. The analytic network process: dependence and feedback in decision making (Part 1). In: Proceedings of the 17th International Conference on Multiple Criteria Decision Making, Whistler, DC.

Saaty, T. L. and Vargas, L. G. 2001. *Models, Methods, Concepts & Applications of the Analytic Hierarchy Process*, Kluwer Academic Publishers, Norwell, MA.

Sadok, W., Angevin, F., Bergez, J. E., Bockstaller, C., Colomb, B., Guichard, L., Reau, R., and Doré, T. 2009. Ex ante assessment of the sustainability of alternative cropping systems: implications for using multi-criteria decision-aid methods—A review. In: *Sustainable Agriculture* (Eds Lichtfouse, E., Navarrete, M., Debaeke, P., Souchère, V., and Alberola, C.), Springer, pp. 753–768.

Saegrov, S. 2005. *CARE-W: Computer Aided Rehabilitation for Water Networks*, IWA Publishing, London.

Salo, A. A. and Hämäläinen, R. P. 1997. On the measurement of preferences in the analytic hierarchy process. *Journal of Multi-Criteria Decision Analysis*, 6(6), 309–319.

Sprague, R. and Carlson, E. 1982. *Building Effective Decision Support Systems*, Prentice Hall, Englewood Cliffs, NJ.

Starr, M. K. and Zeleny, M. 1977. MCDM: state and future of the arts. In: *Multiple Criteria Decision Making* (Eds Starr, M. K. and Zeleny, M.), North-Holland, Amsterdam, pp. 5–29.

Sudhir, V., Srinivasan, G., and Muraleedharan, V. R. 1997. Planning for sustainable solid waste in urban India. *System Dynamics Review*, 13(3), 223–246.

Tabucanon, M. T. 1988. *Multiple Criteria Decision Making in Industry*, Elsevier, Amsterdam.

Tecle, A. 1992. Selecting a multicriterion decision making technique for watershed resources management. *Water Resource Bulletin*, 28(1), 129–140.

Thérivel, R. and Partidário, M. 1999. *The Practice of Strategic Environmental Assessment*, Earthscan Publication Ltd., London.

Tseng, M. L. 2009. Application of ANP and DEMATEL to evaluate the decision-making of municipal solid waste management in Metro Manila. *Environmental Monitoring and Assessment*, 156(1–4), 181–197.

Tukker, A. 2000. Life cycle assessment as a tool in environmental impact assessment. *Environmental Impact Assessment Review*, 20(4), 435–456.

Tzeng, G. H. and Huang, J.-J. 2011. *Multiple Attribute Decision Making: Methods and Applications*, CRC Press, New York.

Vaillancourt, K. and Waaub, J. 2002. Environmental site evaluation of waste management facilities embedded into EUGÈNE model: a multicriteria approach. *European Journal of Operational Research*, 139(2), 436–448.

Vego, G., Kucar-Dragicevic, S., and Koprivanac, N. 2008. Application of multi-criteria decision-making on strategic municipal solid waste management in Dalmatia, Croatia. *Waste Management*, 28(11), 2192–2201.

Ventana System Inc. 2012. Vensim®. Ventana Systems, Inc. Available at: http://www.vensim .com (accessed January 2012).

Vlachokostas, Ch., Achillas, Ch., Moussiopoulos, N., and Banias, G. 2011. Multicriteria methodological approach to manage urban air pollution. *Atmospheric Environment*, 45(25), 4160–4169.

von Neumann, J. and Morgenstern, O. 1944. *Theory of Games and Economic Behavior*. Available at: http://press.princeton.edu/titles/7802.html (accessed November 2013).

Vuk, D. and Kozelj, B. 1991. Application of multicriterional analysis on the selection of the location for disposal of communal waste. *European Journal of Operational Research*, 55(22), 211–217.

Wang, J.-J., Jing, Y.-Y., Zhangm,C.-F., and Zhao, J.-H. 2009. Review on multi-criteria decision analysis aid in sustainable energy decision-making. *Renewable and Sustainable Energy Reviews*, 13(9), 2263–2278.

Whitten, J. and Bentley, L. 2008. *Introduction to Systems Analysis & Design*, The McGraw-Hill Companies, Inc., New York.

Yoon, K. P. and Hwang, C.-L. 1995. *Multi Attribute Decision Making: An Introduction*, Sage Publications, Inc., Thousand Oaks, CA.

Zeleny, M. 1982. *Multiple Criteria Decision Making*, McGraw-Hill, New York.

APPENDIX WEB SITE RESOURCES OF SOFTWARE PACKAGES OF LINDO AND LINGO

(A) The following are the instructions for installing LINGO on your computer:

1. Go to http://www.lindo.com/index.php?option = com_content&view = article&id = 35&Itemid = 20

2. Select the version compatible with your operating system.

3. Unzip the file that you downloaded in step 4.

4. Install the exe file.

5. After installation, open LINGO, select Demo and when asked if you want a full capacity license select yes, and then close LINGO.

6. Navigate to the install location you installed LINGO to.

7. There should be a file named userinfo.txt, email that file to sales@lindo.com using your knights.ucf.edu email address.

8. They will send you an email with the full capacity license and the instructions for installing it on your computer.

(B) The following are the instructions for installing LINDO on your computer:

1. Go to http://www.lindo.com/index.php?option = com_content&view = article&id = 34&Itemid = 14

2. Select "Download Classic LINDO^{TM}."

3. Unzip the file that you downloaded in step 2.

4. Install the lnd61.exe.

PART III

INDUSTRIAL ECOLOGY AND INTEGRATED SOLID WASTE MANAGEMENT STRATEGIES

Industrial symbiosis with a particular focus on material and energy exchange in natural ecosystem is the foundation of industrial ecology, which includes the study of material and energy flows through ecoindustrial parks in human society. Sustainable SWM is intimately tied to industrial ecology in which life cycle impact assessments of a product and appraisals of SWM processes over or beyond life cycle can be carried out in a more sustainable way. The processes covered in the following chapters command more specific requirements with respect to life cycle concept combined with risk assessment not covered by the general guidelines of Parts I and II.

- Principles of industrial symbiosis and industrial ecology in support of municipal utility parks (Chapter 9)
- Evaluating the significance of life cycle assessment for SWM (Chapter 10)
- Options appraisal and decision-making based on streamlined life cycle assessment (Chapter 11)
- SWM under a carbon-regulated environment (Chapter 12)

Sustainable Solid Waste Management: A Systems Engineering Approach, First Edition. Ni-Bin Chang and Ana Pires.
© 2015 The Institute of Electrical and Electronics Engineers, Inc. Published 2015 by John Wiley & Sons, Inc.

CHAPTER 9

INDUSTRIAL ECOLOGY AND MUNICIPAL UTILITY PARKS

This chapter introduces the conceptual framework of industrial ecology, which draws biological analogies from natural ecosystems to human industrial systems. Here we describe eco-industrial parks and summarize the design principles of industrial symbiosis, leading up to municipal utility parks (MUPs), in which solid waste management (SWM) facilities are seamlessly integrated with other urban facilities to form the next-generation green infrastructure in response to global change impacts.

9.1 INDUSTRIAL SYMBIOSIS AND INDUSTRIAL ECOLOGY

9.1.1 The Concept of Industrial Symbiosis

Any associative relationships between two species populations that live together through a form of mutualism, commensalism, and/or parasitism is "symbiotic," whether the two species benefit, harm, or have no actual effect on one another (Box 9.1). A positive analogy in the technological world is industrial symbiosis, the simultaneous sharing of services, utilities, and by-product resources among industries to create value-added economic production, reduce production costs, and improve environmental quality. Industrial symbiosis particularly focuses on material and energy exchange among industrial sectors in a geographically proximate region or linked remotely as a cluster. Symbiotic relationships reshape traditional linear production systems into circular industrial production systems (Figure 9.1), deemed

Sustainable Solid Waste Management: A Systems Engineering Approach, First Edition. Ni-Bin Chang and Ana Pires.
© 2015 The Institute of Electrical and Electronics Engineers, Inc. Published 2015 by John Wiley & Sons, Inc.

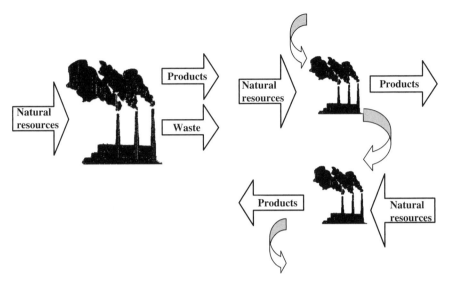

FIGURE 9.1   Fundamental configuration of a system with changing system boundary

as prototype circular economies on a local scale (Box 9.2). Inspired by the concept of symbiosis in biological ecosystems, industrial ecology mirrors the cyclical use of biological food web resources in industrial systems to exhibit similar traits to biological ecology.

BOX 9.1   IMPLICATIONS OF SYMBIOSIS IN BIOLOGY

Mutualism is a type of symbiosis characterized by a mutually beneficial relationship between two species of organisms. An example of mutualism in the Arctic tundra is the lichen, a composite organism consisting of a fungus (the mycobiont) and photosynthetic partner algal cells. Fungal hyphae surround the algal cells to protect and provide them with water and salts; the algal cells in return provide the hyphae with sugars and oxygen.

Commensalism is a symbiotic relationship between two species populations in which one species obtains food, such as nutrients, or other benefits, such as shelter, support, or locomotion, from the host species without either harming or benefiting the host species. An example of commensalism is epiphytic plants, which depend on larger host plants for support but do not exploit them as a source of nutrients.

Parasitism, a non-mutual symbiotic relationship, is a unique relationship between two species of organisms in which one benefits at the expense of the other. Parasitism is different from parasitoidism, a relationship in which the host is always killed by the parasite.

BOX 9.2   CIRCULAR ECONOMY

A circular economy relies on the following five founding principles (Ellen Macarthur Foundation, 2013).

- **Waste is food:** Waste should be eliminated. The biological parts (nutrients) and technical component parts of any product should be designed for disassembly and repurposing.
- **Diversity is strength:** Facing external impacts, diverse systems with many connections and scales are more resilient than those built just for self-efficiency.
- **Energy must come from renewable sources:** Any system should ultimately generate energy through renewable sources.
- **Prices must tell the truth:** The rational use of natural resources must reflect the real cost of the activity, including the environmental cost.
- **Thinking in terms of systems:** Understanding how things influence one another within a whole is key.

In short, circular economy is a generic term for an industrial economy that provides a coherent systems-level design framework to harness innovation and creativity. Ensuring that the whole is greater than the sum of the parts will enable a positive, restorative economy.

9.1.2   The Onset of Industrial Ecology

An industrial park is defined as "a large tract of land, sub-divided and developed for the use of several firms simultaneously, distinguished by its shareable infrastructure and close proximity of firms" (Peddle, 1993). Types and synonyms of industrial parks include industrial estates, industrial districts, export processing zones, industrial clusters, business parks, office parks, science and research parks, and biotechnology parks (Cote and Cohen-Rosenthal, 1998). Eco-industrial parks (EIPs) have now been added to this list (Côté and Cohen-Rosenthal, 1998). The oldest EIP worldwide, designed in the early 1980s in Kalundborg, Denmark, involves five industrial enterprises and the municipality. The history of the Kalundborg symbiosis (i.e., a prototype of industrial symbiosis) began in 1961 with a project to use surface water from Lake Tissø to replace the limited supplies of groundwater for a new oil refinery plant. Beginning with this initial collaboration, several public and private enterprises were subsequently introduced to symbiotic relationships of buying and selling waste products in a closed cycle (i.e., a local-scale circular economy). By the end of the 1980s, these partners realized that they had become an archetype analogy for biological symbiosis, creating a potential model to support the principles of industrial ecology. The model continued growing from 1975 (Figure 9.2a) until the mid-1980s (Figure 9.2b), and into the late 1990s (Figure 9.2c).

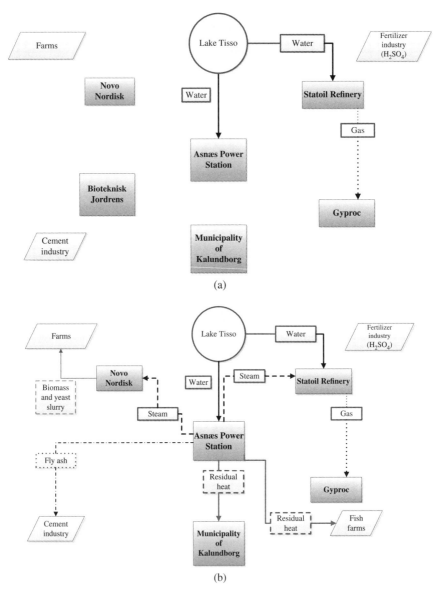

FIGURE 9.2 Material and energy flows of the growing eco-industrial network in Kalundborg, Denmark. (a) The Kalundborg symbiosis in 1975. (b) The Kalundborg symbiosis in 1985. (c) The Kalundborg symbiosis in 1999

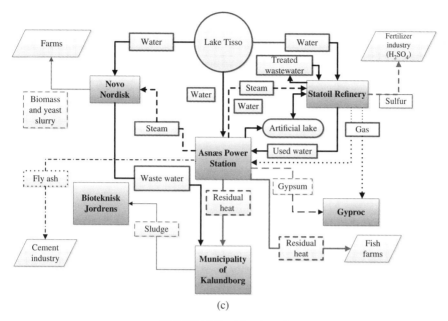

(c)

FIGURE 9.2 (*Continued*)

Today, the scale of the Kalundborg symbiosis has evolved into a formal EIP that trades residual products including steam, dust, gases, heat, slurry, and other commodities. In addition to several companies that participated as recipients of materials or energy, the industrial park consisted of six more main partners in 1999. (Figure 9.2c).

- **Asnæs Power Station:** Part of SK Power Company and the largest coal-fired plant producing electricity in Denmark.
- **Statoil:** An oil refinery belonging to the Norwegian State Oil Company.
- **Novo Nordisk:** A multinational biotechnology company that is the largest producer of insulin and industrial enzymes.
- **Gyproc:** A Swedish company producing plasterboard for the building industry.
- **The town of Kalundborg:** Receives excess heat from the Asnæs Power Station for its residential district heating system.
- **Bioteknisk Jordrens:** A soil remediation company that joined the Kalundborg symbiosis in 1998.

A series of by-product synergy projects triggered the success of the eco-industrial network in Kalundborg, Denmark. First, water is deemed a scarce resource in this part of Denmark, and the reduction in the use of groundwater due to the Kalundborg symbiosis has been estimated at close to two million metric tons (i.e., tonnes) per year (Grann, 1997; Symbiosis Institute, 2013). To reduce overall water consumption, the Statoil refinery delivered its treated wastewater and used cooling water to the Asnæs Power Station for reuse, thereby saving an additional one million metric tons of water per year (Grann, 1997; Symbiosis Institute, 2013). The Asnæs Power Station supplies steam to both Statoil (a refinery plant) and Novo Nordisk (a pharmaceutical plant) for heating their production processes. At the same time, the power station is able to increase its efficiency by functioning in a cogeneration mode. Residual heat in the cogeneration system of the Asnæs Power Station is delivered to support the district heating of the town and a neighboring fish farm.

In addition to water and energy flows, material flows also play a critical role. Flue gas from the operations at the Statoil refinery plant was collected, treated to remove sulfur, and sold as a raw material for the production of sulfuric acid. The clean gas was then piped to the Asnæs Power Station and to Gyproc as an energy source before 1993. In 1993, the Asnæs Power Station installed a desulfurization unit to remove sulfur from flue gas, creating the by-product gypsum (calcium sulfate) by the addition of limestone. Gypsum then became a raw material in the production of plasterboard at Gypro, saving 190,000 metric tons per year of the natural gypsum that was previously imported from Spain. Within the Kalundborg symbiosis, Novo Nordisk created a large amount of used biomass from its synthetic processes, which became a source of fertilizer due to the abundant nitrogen, phosphorus, and potassium content. More than 800,000 cubic meters of liquid and 60,000 metric tons of solid fertilizer were used by the local farming communities to improve production, according to 1998 statistics. In addition, fly ash collected from the flue gas at the Asnæs Power Station became the raw material in a cement production plant.

The Kalundborg operations fulfilled the vision of industrial symbiosis, and many residual products originating from one enterprise became the raw material of another enterprise, benefiting both the economy and the environment significantly reducing the environmental impact (Table 9.1) (Erkman, 1998). Healthy communication and smooth collaboration among the participants were critical elements in the success of the Kalundborg symbiotic system and the resulting social sustainability. From this example, a new subject, "Industrial Ecology," was born in the context of sustainable development.

TABLE 9.1   Environmental aspects of the Kalundborg symbiosis in 1997

Reduction in consumption of resources	
Oil	19,000 tonnes per year
Coal	30,000 metric tons per year
Water	1,200,000 m³ per year
Input chemicals	800 metric tons nitrogen and 400 tonnes phosphorous per year
Value-added by-products	
Sulfur	2,800 tonnes per year
Calcium sulfate (gypsum)	80,000 tonnes per year
Fly ash (for cement production)	200,000 tonnes per year

Source: From Ehrenfeld and Gertler (1997).
Tonne is metric ton.

9.2 CREATION OF ECO-INDUSTRIAL PARKS AND ECO-INDUSTRIAL CLUSTERS

9.2.1 The Conceptual Framework

With the concept of industrial ecology, both EIP and eco-industrial clusters (EICs) are management strategies to innovate industrial systems. In the context of environmental management, an EIP is an industrial community of manufacturing and service businesses seeking to improve the economic performance of the participating companies while minimizing their environmental impact through voluntary collaboration. Industrial communities manage environmental and resource issues in a geographically proximate region (i.e., in an industrial park) involving a network of participating firms and organizations with the synergistic potential to create symbiotic relationships among energy, water, and material flows. By working together, the industrial community realizes a collective benefit greater than the sum of the benefits that companies would obtain through independent optimization of individual performances (Research Triangle Institute and Indigo Development International, 1996). Collectively, EIPs simultaneously reduce waste streams, increase resource efficiency, reduce infrastructure requirements, provide access to better information about partnerships, and reduce costs of regulation (Research Triangle Institute and Indigo Development International, 1996).

In some developing countries, however, numerous types of environmental and economic linkages exist among local industries located in urban–suburban areas that rely on the import of renewable inputs, raw materials, and labor from neighboring areas. In contrast to the idea of centralized industrial zones and export processing zones, EIC may link a group of small and medium industrial clusters established in towns and rural areas to assemble similar production enterprises that stimulate development and rural industries. These EICs create opportunities to form symbiotic relationships via appropriate eco-industrial networking. As a result, both EIP and EIC can be regarded as different forms of eco-industrial networks (Box 9.3).

BOX 9.3   ECO-INDUSTRIAL NETWORKING

Eco-industrial networking develops new symbiotic relationships among private sectors, government agencies, and educational institutions. Eco-industrial networking is deemed an important new approach for communities and businesses to weave various energy, material, water, human, and infrastructure resources together to improve production efficiency, economic viability, and investment competitiveness, while promoting community and ecosystem health.

9.2.2 The Design Principles of an Eco-industrial Park

The Industry and Environment Office of United Nations Environment Programme (UNEP) released a technical report on Environmental Management of Industrial

Estates in 1997 (UNEP, 1997) noting that "industrial estates have become common features of the global landscape." Today, many industrial parks and export processing zones exist worldwide, and colocation of different industrial plants can facilitate management of materials, energy, and wastes (Côté and Cohen-Rosenthal, 1998). Research Triangle Institute and Indigo Development International (1994) suggested that an EIP should be more than

- a single by-product exchange pattern or network of exchanges;
- a recycling business cluster (resource recovery, recycling companies, etc.);
- a collection of environmental technology companies;
- a collection of companies making "green" products;
- an industrial park designed around a single theme; or
- a park with environmentally infrastructure or construction; a mixed use development (industrial, commercial, and residential).

When compared with a typical industrial park, Côté and Cohen-Rosenthal (1998) proposed that an EIP in general should

- define the community of interests and involve that community in the design of the park;
- reduce environmental impact or ecological footprint by providing alternatives to toxic materials, absorbing carbon dioxide, exchanging materials, and integrating waste treatment;
- maximize energy efficiency through facility design and construction, cogeneration, and cascading;
- conserve materials through facility design and construction, reuse, recovery, and recycling;
- link companies with suppliers and customers in the wider community in which the EIP is situated;
- continuously improve the environmental performance of the individual businesses and the community as a whole;
- build a regulatory system that permits some flexibility while encouraging companies to meet performance goals;
- implement economic instruments that discourage waste and pollution;
- adopt an information management system that facilitates the flow of energy and materials within a more or less closed loop;
- innovate a mechanism to train and educate managers and workers in new strategies, tools, and technologies to improve the system; and
- promote marketing to attract companies that fill niches and complement other businesses.

Many EIPs require synthesizing a rich menu of design components, including but not limited to a green design of park infrastructure and plants, cleaner production, waste minimization and pollution prevention, energy and water efficiency, and inter-company partnerships (Indigo Development, 2006) (Box 9.4). Five principles of natural systems, green engineering, energy, material flows, water flows, and park management and support services with emphasis on cycles, webs, and networks are as follows.

- **Natural Systems:** An EIP should establish or maintain the ecological balance of the site via natural landscape settings to minimize environmental impacts. Examples may include the use of native plant species in landscape design, the creation of wetlands to minimize storm water run-off impact, and the construction of flood protection for buildings.
- **Green Engineering:** An EIP should apply a systems engineering approach to the sustainability of manufacturing processes, green buildings, and infrastructures that would affect more than one chemical or manufacturing process to achieve pollution prevention and cleaner production. For example, personal computer remanufacturing could save a substantial amount of energy and raw materials, thereby achieving the sustainability goal (Box 9.4). Additionally, pollution prevention and environmental management of stormwater, recycling, and reuse of drinking water to support sustainable construction projects could also be part of the green engineering initiatives.
- **Energy:** An EIP should have an energy resources management plan for lowering operational costs and reducing environmental burdens, such as recovery of energy by flowing steam or heated water from one plant to another (i.e., energy cascading) or to a district heating system.
- **Material Flows:** An EIP should have a sustainable waste management plan that ideally turns waste streams into potential products for recovery and reuse internally or marketed to someone else. For example, an EIP may include the channels for delivering by-products from one plant to another or warehousing by-products for shipment to external customers.
- **Water Flows:** An EIP should establish a sound water resources management plan. For example, process water, such as cooling water, used by one plant may be reused by another (i.e., water cascading), passing through a water pretreatment plant, if needed.
- **Logistics Services and Park Management:** An EIP should have strong park management support for not only environmental management but also promoting cooperation and interaction among industries. These promotional actions may include the exchange of by-products information among companies and help them adapt to dynamic partnership changes in the EIP through recruitment and/or regrouping responsibilities. Ecological engineering approaches may be favored for green infrastructure design (Box 9.5). For example, in environmental management, the EIP infrastructure system may collect and use stormwater run-off for water conservation, run a shared air quality monitoring network, or anchor the park around resource recovery companies recruited to the EIP for dedicated, long-term resources recovery.

BOX 9.4   CONCEPTS OF GREEN MANUFACTURING IN INDUSTRIAL ECOLOGY

- The concept of cleaner production was introduced by the Industry and Environment Office of United Nations Environment Programme (UNEP) in 1989. Cleaner production is an integrated strategy applied to processes, products, and services to increase eco-efficiency and reduce risks for humans and the environment. Cleaner production is implemented as a preventive, company-specific environmental protection initiative intended to minimize waste and emissions and maximize production and resources consumption efficiency.
- According to United States Environmental Protection Agency (USEPA), Pollution prevention (P2) aims to reduce or eliminate waste at the source by modifying production processes, promoting the use of non-toxic or less-toxic substances, implementing conservation techniques, and reusing materials rather than putting them into the waste stream. P2 minimizes environmental impacts by focusing on the design of a product or operation of a manufacturing process.
- According to USEPA, waste minimization reduces the amount of waste generated and lowers its toxicity and persistence in the manufacturing processes.
- Remanufacturing is a green engineering process by which a previously sold, worn, or non-functional product can be returned to the manufacturing process after disassembly and recovery at the module level and, eventually, at the component level of a product.

BOX 9.5   ECOLOGICAL ENGINEERING AND GREEN INFRASTRUCTURE

- Ecological engineering is the process of restoring ecological function to natural systems, and enhancing natural capital to foster sustainable development via engineering practices.
- Green infrastructure weaves natural processes into the built environment, providing not only stormwater management, flood mitigation, SWM (i.e., composting and bioreactor landfill), and air quality management, but also much more for solving global change impacts (i.e., urbanization effects and climatic challenges).

9.2.3 The Linkages with Solid Waste Management

With global change impacts and sustainability requirements, both EIP and EIC must evolve in response to the impacts of technology advancements, eco-efficiency of the goods and services, population growth and migration, urbanization, climate variability and change, economic development and fluctuations, and dematerialization (Box 9.6). With these impacts, a sustainable SWM system may be stimulated through a strong biological analogy in which everything is managed with the integrated concept of industrial symbiosis, pollution prevention, waste minimization, and design for the environment (DfE) (Box 9.6). Cleaner production starts from consideration of environmental efficiency, which have positive economic benefits while eco-efficiency starts from consideration of economic efficiency which have positive environmental benefits (Box 9.7). Such integration from cleaner production to DfE and to eco-efficiency elucidates how sustainable engineered systems for SWM can support human well-being while also sustaining environmental (natural) systems. The creation of MUPs discussed in the next section is a salient example of this movement.

A sustainable SWM system typically considers long time horizons and may incorporate contributions from the social sciences and environmental justice (Chang et al., 2009), supported by modeling tools in industrial ecology such as life cycle assessment (LCA) (Ning et al., 2013), materials flow analysis (Chang et al., 1997), input/output economic models (Wu and Chang, 2007, 2008; Chang, 2008), and novel metrics for measuring sustainable systems (Chen et al., 2010). To evaluate the potential of various management strategies to meet the sustainability goal, for example, one of the core tenets is LCA, which can screen, identify, and assess relevant environmental impacts of a material, process, product, or system across its life span from cradle to grave (i.e., from creation to disposal, discussed in Chapter 10). The holistic use of these modeling tools may substantially enhance sustainability in SWM.

BOX 9.6   DESIGN FOR THE ENVIRONMENT

- Design for the environment (DfE), a program created by USEPA in 1992, is an engineering perspective focused on energy efficiency, materials innovation, and design for recyclability. DfE is related to a number of product attributes that modern product design engineers must consider, such as assembly, compliance, disassembly, environment, manufacturability, reliability, safety, and serviceability.
- The core philosophy is that engineers should improve the environmentally related attributes of a product while not compromising other design attributes.
- DfE is intimately related to the potential of remanufacturing.

BOX 9.7   ECO-PRODUCT DESIGN IN INDUSTRIAL ECOLOGY

- The concept of eco-efficiency was first coined in 1992 by the Business Council for Sustainable Development in its landmark report Changing Course.
- Eco-efficiency is a business strategy to produce goods with fewer materials and lower energy demands to realize the economic benefits of environmental improvements. Considering the Earth's estimated carrying capacity, eco-efficiency can be attained by the delivery of competitively priced goods and

services that satisfy societal demand, while reducing ecological footprints and resource consumption throughout the life cycle of goods and services.

- The World Business Council for Sustainable Development has identified seven success factors for eco-efficiency.
  - Reduce the material intensity of goods and services.
  - Reduce the energy intensity of goods and services.
  - Reduce toxic dispersion.
  - Enhance material recyclability.
  - Maximize sustainable use of renewable resources.
  - Enhance material durability.
  - Increase the service intensity of goods and services.
- Dematerialization, regarded as a synonym of eco-efficiency, can be achieved by making the products lighter and smaller, or by replacing a product with an immaterial substitute.

9.3 MUNICIPAL UTILITY PARKS IN URBAN REGIONS

Emerging paradigms affecting the evolution of SWM are being driven by a growing awareness of sustainability concept, greenhouse gas emissions, adaption strategies for climate change, the need for renewable energy, and perhaps more importantly, the need for sustainable economic development at local and regional levels (Hauck and Parker, 2012). An MUP is defined as a park that combines several utility components in an urban region, such as SWM, drinking water treatment, wastewater treatment, and stormwater treatment, which work well together based on the concept of industrial symbiosis. In the context of such a water–energy nexus, water and energy are inextricably linked together to provide a clean and affordable municipal water supply. For example, lower quality water supply sources require higher levels of treatment, and higher levels of treatment require greater inputs of energy to pump from greater depths/distances to pressurize membrane treatment processes and to power more sophisticated disinfection treatments (e.g., ultraviolet light, ozone) (Hauck, 2011). Therefore, colocating a municipal incineration facility (waste-to-energy (WTE) facility) that has energy recovery potential with a drinking water treatment plant (WTP) with multiple sources of tap water would have positive salient effects. Such synergistic opportunities have helped mitigate some of the future competition for water and energy in both water and energy sectors, and many existing municipally owned WTE facilities are colocated adjacent to existing wastewater treatment plants (WWTPs), or served by piping systems to allow reclaimed water to be used for cooling water (Hauck and Parker, 2012). If additional energy sources are required, local wind and solar energy could be recovered in concert with the fixed energy supply from WTE facilities to satisfy the overall energy consumption in the MUP, providing an exemplary model of green infrastructure linking urban sustainability with municipal utility design.

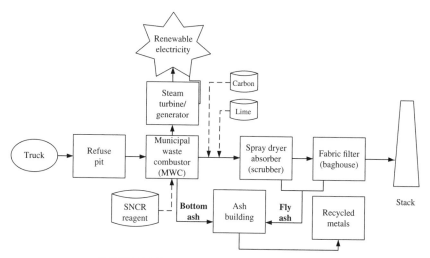

FIGURE 9.3   A typical WTE treatment process

A typical WTE process (Figure 9.3) includes the generation of electricity and steam to support the energy consumption needs for WTPs and/or WWTPs. Most WTE facilities use wet cooling systems, and typical sources of makeup water include (1) groundwater withdrawals, (2) remote surface water from nearby rivers or lakes, (3) on-site surface water storage systems, (4) reclaimed wastewater after appropriate treatment, and even (5) potable water sources if there is no other option. Significant quantities of water consumed in WTE processes such as cooling, irrigation, wash downs, and fire protection may be sustained by pumping the treated wastewater back from WTPs and/or WWTPs located nearby to meet the greatest water consumption demands. The effluents from a WWTP with low diurnal and seasonal fluctuations can be discharged into a nearby groundwater aquifer and recovered later as a source of tap water for water treatment if the groundwater aquifer has a geological structure providing an adequate filtration capacity. In addition, reclaimed wastewater is a viable alternative makeup water source used in multiple WTE uses.

With industrial symbiosis, siting WTE facilities near WTPs and/or WWTPs can be a catalyst for sustainable water resources management, and conversely, siting water and/or wastewater treatment plants near a WTE facility can be a catalyst for sustainable SWM. From a societal point of view, technological synergy embedded in an MUP can prove more acceptable than regular industrial complexes in local communities. A typical MUP symbiosis operation (Figures 9.4 and 9.5) has expansive pathways of synergistic effect between the WTE facility, the WTP, and the WWTP, each with various levels of energy intensity of water treatment technologies (Table 9.2) (Hauck, 2011). Because water and energy issues are inextricably linked, an MUP provides

- emerging SWM and water resource paradigms;
- various proven waste conversion technologies;

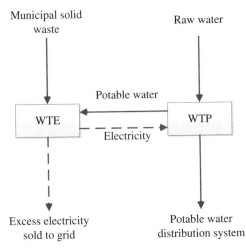

FIGURE 9.4   WTE facility integrated with water treatment plant

- improvement of energy intensity of water treatment technologies; and
- synergistic opportunities for integration of WTE with municipal water utilities.

Proven waste conversion technologies include those that have been successfully implemented on a commercial basis (Hauck, 2011). The following waste conversion technologies have had successful commercial experience with municipal feedstock wastes in the United States (Hauck and Parker, 2012).

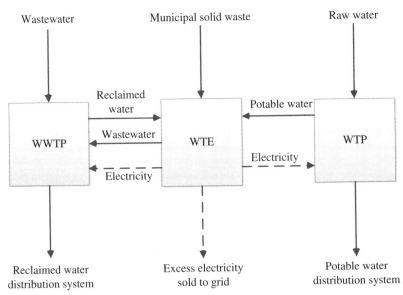

FIGURE 9.5   WTE facility integrated with water treatment and wastewater treatment plants

TABLE 9.2   Various levels of energy intensity of water treatment technologies

Water resource	Treatment technology	Energy intensity (kWh/MG)
Groundwater	Conventional softening, filtration, and disinfection	150–750
Surface water	Conventional softening, filtration, and disinfection	150–750
Brackish water	Reverse osmosis/membrane	4,000–10,000
Seawater	Reverse osmosis/membrane	10,000–20,000
Seawater	Multistage flash evaporation/multiple effect distillation	20,000–100,000
Reclaimed water	Reverse osmosis/membrane	10,000–15,000
Reclaimed water	Multi-stage flash evaporation/multiple effect distillation	15,000–20,000
Wastewater	Biological treatment/disinfection	1,000–5,000

Source: From Hauck (2011).
kWh/MG, kilowatt hour per million gallons.

- Source separated biomass/yard waste composting.
- Advanced mass-burn combustion WTE for production of steam and electricity.
- Refuse-derived fuel combustion WTE for production of steam and electricity.
- Landfill gas to energy for production of electricity.

For communities that cannot find favorable rates for the sale of their renewable energy produced from proven waste conversion technologies to the local grid, there is an option to use their own power internally, "behind the meter," for other vital municipal services (Hauck and Parker, 2012). Examples include using electricity, steam, and/or waste heat for (Hauck and Parker, 2012)

- treatment/pumping of water, wastewater, reclaimed water, and stormwater resources;
- treatment of wastewater biosolids (e.g., drying, pelletizing);
- operation of other recycling processes (e.g., material recovery facility (MRF), electronic waste recycling, construction and demolition waste recycling); and
- operation of municipal buildings (e.g., municipal service, city hall).

Within the larger scope of a formidable MUP with three types of integrated municipal utility units (Figure 9.5), the synergy between WTE and WTP may be environmentally highlighted and analyzed to reduce potable water demands, reduce surface water discharges, and maximize the water conservation effect based on a "system of systems engineering" (SoSE) approach (Figure 9.3). In comparison, MUP can be scaled up by the inclusion of the advanced water treatment processes,

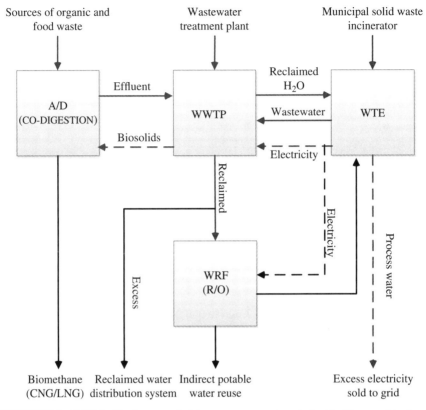

FIGURE 9.6 WTE facility integrated with wastewater treatment plant (with anaerobic digestion) and water reclamation facility (for indirect potable reuse). A/D, anaerobic digester; WRF (R/O), water reclamation facility (reverse osmosis); CNG, compressed natural gas; LNG, liquid natural gas

such as reverse osmosis (RO) if reuse of treated wastewater is required for drinking water purposes (Figure 9.6). A key criterion for an RO layout is the specific electricity consumption (i.e., energy intensity), which manifests the value of the energy supply from a WTE facility. Because MUP symbiosis is flexible, it can be extended to include methane gas recovery produced by the anaerobic digesters in a WWTP to support the operation of an advanced WTP based on the concept of SoSE (Figure 9.6).

Reclaimed water used at Florida-based WTE facilities has been a popular strategy in sustainable SWM (Table 9.3). The MUP located at Pasco County, Florida, is an excellent example (Figure 9.7). Commissioned on May 15, 1991, the current WTE (the Pasco County Solid Waste Resource Recovery Facility (RRF)) has three 350 ton-per-day water-wall furnaces with Martin® reverse-reciprocating grates, an ash handling system with a capacity of 1050 tonnes per day, and a boiler system that runs in efficient superheater outlet conditions. Energy generation at rated capacity

TABLE 9.3   Reclaimed water used at Florida-based WTE facilities

WTE facility	Tampa	Pasco	Hillsborough	Lee County	Broward North	Pinellas
Cooling Tower Makeup	×	×[a]	×	×	×	×[b]
Boiler Water Makeup					×[c]	×[d]
Fire Protection		×	×			×[e]
Irrigation	×	×	×	×	×	Minimal

Source: From Hauck (2011).
gpd, gallons per day ($= 0.0037$ m$^3 \cdot$ per day).
[a]Blended water.
[b]2,459.8 m$^3 \cdot$ per day (650,000 gpd) reclaimed water blended with surface water.
[c]Microfiltration and RO.
[d]529.6 m$^3 \cdot$ per day (140,000 gpd) from microfiltration and RO.
[e]5.48 m$^3 \cdot$ per day (1,450 gpd).

can be up to 31.2 megawatts from one condensing steam turbine generator, which is sold to "Progress Energy," a WTE power company equipped with air pollution control equipment with semi-dry flue gas scrubbers injecting lime, a baghouse filter, nitrogen oxide and mercury control systems, and a continuous emissions monitoring system. The neighboring WWTP provides a synergistic effect with reclaimed water for possible reuse as make up water in the WTE facility. Wastewater treatment is currently provided at the Shady Hills Subregional WWTP, part of the reuse system maintained by Pasco County. Reclaimed water is sent to the RRF (e.g., a WTE facility) (Pasco County, 2013), which has percolation ponds that serve as a backup effluent disposal system (Pasco County, 2013).

The West Pasco Class I and Class III Sanitary Landfills are located adjacent to the RRF and provide support to the facility. The Class I facility includes development of a Solid Waste-1 (SW-1) cell, and Ash Waste-1 (AW-1) and Ash Waste-2 (AW-2) cells. The SW-1 cell provides emergency overflow storage for the RRF if collected municipal solid waste (MSW) exceeds the operating rate for MSW processing (Pasco County, 2013). This facility is permitted for up to six months of storage of MSW if required (Pasco County, 2013). AW-1 and AW-2 cells accommodate ashfill from the RRF and are classified monofill facilities (Pasco County, 2013). The West Pasco Class III Sanitary Landfill is a lined facility that accommodates construction and demolition debris (Pasco County, 2013). A future MRF is planned near the RRF (Figure 9.7).

9.4   FINAL REMARKS

Industrial ecology offers a realm of concepts, methods, and tools to analyze environmental burdens at various spatial and temporal scales from product to processes, facilities, regions, nations, and even the globe. It looks beyond the action of single firms to those of groups of firms or to society as a whole. Beginning with

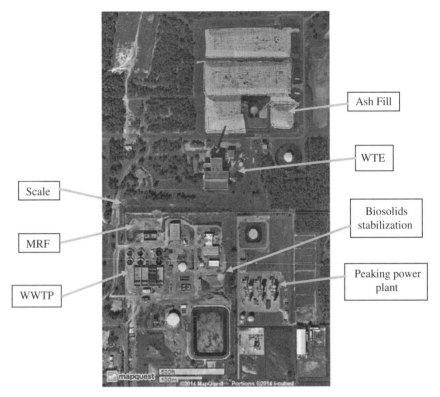

FIGURE 9.7   The MUP located at Pasco County, Florida

the Kalundborg symbiosis in Denmark, a basic awareness of integrated economic and environmental analysis was born and promoted an effective concept of circular economy at a local scale. The use of SoSE approach to analyze such complex systems would be promising. In summary, several core elements characterize the discipline of industrial ecology (Lifset and Graedel, 2002).

- Biological analogy
- Use of systems perspectives
- Role of technological change
- Role of companies
- Eco-efficiency and dematerialization
- Forward-looking research and practices

Continued development of synergistic systems will contribute to future MUP designs in urban regions to promote urban sustainability.

REFERENCES

Chang, N. B. 2008. Economic and policy instrument analyses in support of the scrap tires recycling program in Taiwan. *Journal of Environmental Management*, 86(3), 435–450.

Chang, N. B., Chang, Y. H., and Chen, Y. L. 1997. Cost-effective and workload balancing operation in solid waste management systems. *Journal of Environmental Engineering*, ASCE, 123(2), 178–190.

Chang, N. B., Chang, Y. H., and Chen, H. W. 2009. Fair fund distribution for a municipal incinerator using a GIS-based fuzzy analytic hierarchy process. *Journal of Environmental Management*, 90(1), 441–454.

Chen, H. W., Chen, J. C., and Chang, N. B. 2010. Environmental performance evaluation of large-scale municipal incinerators using Data Envelopment Analysis (DEA). *Waste Management*, 30(7), 1371–1381.

Côté, R. P. and Cohen-Rosenthal, E. 1998. Designing eco-industrial parks: a synthesis of some experiences. *Journal of Cleaner Production*, 6(3–4), 181–188.

Ehrenfeld, J. and Gertler, N. 1997. Industrial ecology in practice: the evolution of interdependence at Kalundborg. *Journal of Industrial Ecology*, 1(1), 67–80.

Ellen Macarthur Foundation. 2013. Chapter II—The circular model's founding principles. Ellen Macarthur Foundation. Available at: http://www.ellenmacarthurfoundation.org /circular-economy /circular-economy/part-ii-the-circular-models-founding-principles.pdf (accessed August 2013).

Erkman, S. 1998. *Vers Une Écologie Industrielle*. Éditions Charles Léopold Mayer, Paris, France.

Grann, H. 1997. The industrial symbiosis at Kalundborg, Denmark. In: *The Industrial Green Game* (Ed. Richards, D. J.), National Academy Press, Washington, DC.

Hauck, P. L. 2011. Integration of water treatment system with energy derived from municipal wastes. In: Proceedings of the 2012 World Environmental and Water Resources Congress, Palm Springs, CA.

Hauck, P. L. and Parker, T. 2012. Closing the loop in pursuit of a sustainable integrated solid waste management system via proven and emerging waste conversion technologies. In: Proceedings of the 2012 World Environmental and Water Resources Congress, Albuquerque, NM, pp. 849–858.

Indigo Development. 2006. Eco-industrial parks. Indigo Development. Available at: http://www.indigodev.com/Ecoparks.html (accessed August 2013).

Lifset, R. L. and Graedel, T. E. 2002. Industrial ecology: goals and definitions. In: *Handbook of Industrial Ecology* (Eds. Ayres, R. U. and Ayres, L. W.), Edward Elgar, Northampton, UK, pp. 3–15.

Ning, S. K., Chang, N. B., and Hong, M. C. 2013. A comparative streamlined life cycle assessment of two types of municipal solid waste incinerator. *Journal of Cleaner Production*, 53(15), 56–66.

Pasco County. 2013. Comprehensive plan – 2025. Pasco County. Available at: http:// fl-pascocounty.civicplus.com/DocumentCenter/View/2216 (accessed August 2013).

Peddle, M. T. 1993. Planned industrial and commercial developments in the United States: a review of the history, literature and empirical evidence regarding industrial parks. *Economic Development Quarterly*, 7(1), 107–124.

Research Triangle Institute and Indigo Development International. 1994. Eco-industrial parks and industrial ecosystems: a technical memorandum. http://www.indigodev.com/Ecoparks .html.

Research Triangle Institute and Indigo Development International. 1996. Eco-industrial parks: a case study and analysis of economic, environmental, technical, and regulatory issues. RTI Project Number 6050 FR. https://www.rti.org/pubs/case-study.pdf.

Symbiosis Institute. 2013. A circular ecosystem of economy. Kalundborg Symbiosis. Available at: www.symbiosis.dk (accessed August 2013).

United Nations Environmental Programme (UNEP). 1997. The environmental management of industrial estates. Industry and Environment Technical Report No. 39, Paris.

Wu, C. C. and Chang, N. B. 2007. Evaluation of environmentally benign production program in the textile dying industry (I): an input-output analysis. *Civil Engineering and Environmental Systems*, 24(4), 275–298.

Wu, C. C. and Chang, N. B. 2008. Evaluation of environmentally benign production program in the textile dying industry (II): a multi-objective programming approach. *Civil Engineering and Environmental Systems*, 25(1), 299–322.

CHAPTER 10

LIFE CYCLE ASSESSMENT AND SOLID WASTE MANAGEMENT

Life cycle assessment (LCA), also known as life cycle analysis, is a method applied to assess potential environmental impacts related with a product, process, or service. The process includes assembling a life cycle inventory (LCI) of pertinent energy and material inputs and emissions, assessing the potential environmental impacts associated with recognized inputs and outputs in LCI, and explaining the results to help decision makers achieve more risk-informed and forward-looking decisions. This chapter provides a knowledge base of LCA applied to solid waste management (SWM), beginning with an introduction describing major terminologies widely used in LCA, followed by an explanation of the steps required to conduct an LCA, and descriptions of the major LCA software packages available to help develop real-world LCA case studies. The final part of this chapter explains how the current methodological approaches provide a wealth of realistic viewpoints with the aid of software packages, and the chapter concludes by highlighting the changing nature of LCA.

10.1 LIFE CYCLE ASSESSMENT FOR SOLID WASTE MANAGEMENT

LCA is an assessment method that holistically compiles all pertinent materials and energy consumptions as well as emissions that occur during the life cycle of a product or a service. The assessment quantifies the environmental impacts, including global warming potential (GWP) generated during the life cycle of a product or a service, for decision-making. The key to understand LCA is the life cycle concept (Figure 10.1)

Sustainable Solid Waste Management: A Systems Engineering Approach, First Edition. Ni-Bin Chang and Ana Pires.
© 2015 The Institute of Electrical and Electronics Engineers, Inc. Published 2015 by John Wiley & Sons, Inc.

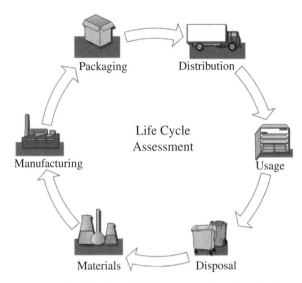

FIGURE 10.1 A symbolic representation of the LCA

over a specified time horizon of a product or a service, applied to a coupled natural system and the built environment, such as an SWM system, to generate risk-informed and forward-looking solutions. In system thinking, the boundaries defined for an LCA begin with natural resources entering a predefined system in dynamic ways. The holistic processes occurring in the life cycle include extraction, manufacturing, use, and disposal. During each process, anthropogenic systems consumptions and emissions that interact with the natural system have various environmental impacts that must be quantified.

Because LCA can be used to examine each process, it can identify where the environmental impacts occur and which stakeholders are responsible. In general, several stakeholders exist in a resource life cycle, including raw material extractors, raw material sellers, manufacturers, users/re-users, waste collectors, waste recyclers, and waste eliminators. With life cycle thinking (LCT), LCA may (1) minimize the magnitude of pollution at hot spots and during hot periods, (2) conserve nonrenewable resources, (3) conserve ecosystem integrity, and (4) develop and utilize cleaner technologies. LCT avoids the "shifting of burdens" between life cycle stages, or regions, or generations or even between environmental impact categories (EC-JRC-IES, 2011). LCA can greatly contribute to a sustainable life cycle management in concert with other tools to quantify socioeconomic burdens, such as life cycle cost analysis and social LCA.

LCA applied to SWM is mainly focused on the end-of-life product phase, which is also the phase responsible for closing the cycle, a phase when prolonging the lifetime of the product is possible. LCA applied to SWM can serve two purposes: (1) to analyze the end-of-life phase of a product and evaluate the environmental impacts and possible benefits via reuse, recovery, recycling, or elimination at the existing

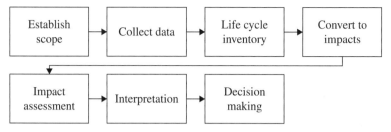

FIGURE 10.2   An LCA process supporting decision-making

destination of a product, and (2) to evaluate the waste management service provided by a private entity or a government agency. LCA aids the decision-making process by introducing environmental factors, like is highlighted in Figure 10.2. LCA has been applied to SWM since the 1990s, and today it is used to create better management strategies, compare alternative technologies, and provide environmental performance evaluation of various types of SWM systems. The ultimate goals of LCA applied for SWM are to aid decision-making and analyze policy options.

10.2   PHASES OF LIFE CYCLE ASSESSMENT

Over the past few decades, several professional organizations, such as the International Organization of Standardization (ISO) and the Society of Environmental Toxicity and Chemistry (SETAC), have contributed to the development of the methodology used to perform LCA, a product-oriented method for sustainability analysis (Cleary, 2009). ISO 14040 family norms are the resultant effort of these two institutions. In addition, other guidelines include Nordic Guidelines 1995:20 (Lindfors et al., 1995a), Life Cycle Engineering Guidelines EPA/600/R-01/101 (Cooper and Vigon, 2001), Guidelines for Assessing the Quality of Life Cycle Inventory Analysis EPA/530/R-95/010 (Bakst et al., 1995), Canadian LCA standard CAN/CSA-Z760 (CSA, 1994), and relevant handbooks published to define and recommend procedures to carry out meaningful LCA practices (Guinée et al., 2002). Recently, several ongoing international initiatives including the SETAC, the Life Cycle Initiative of United Nations Environment Programme (UNEP), the European Platform for LCA of the European Commission, and the emerging International Reference Life Cycle Data System (ILCD) (Finnveden et al., 2009) have been created to help build consensus and provide recommendations.

In this chapter, LCA applied to SWM follows the family of ISO 14040 and the guidance of waste management LCA published by the European Commission, through the Joint Research Centre in the Institute for Environment and Sustainability where the ILCD is funded. Other relevant information can be found in the literature published in the last decades (Table 10.1).

The LCA methodology is divided into four consecutive phases (ISO, 1999): (1) goal and scope definition, (2) LCI, (3) life cycle impact assessment (LCIA), and (4) interpretation. All phases are interconnected (Figure 10.3, left side) to maintain

TABLE 10.1 Publications concerning LCA

LCA publications	1990–1999	2000–2009	2010–present
LCA norms	CSA (1994) and Lindfors et al. (1995a, 1995b)	ISO (2006a, 2006b)	
Models applied to SWM	White et al. (1995), Dalemo et al. (1997), and Björklund et al. (1999)	McDougall et al. (2001), Diaz and Warith (2006), Christensen et al. (2007), den Boer et al. (2007), and Thorneloe et al. (2007)	
Impact factor models for LCIA		Goedkoop and Spriensma (2000), Jolliet et al. (2003), Hauschild and Potting (2005), Potting and Hauschild (2005), and Goedkoop et al. (2009)	
LCIA impact characterization			Hauschild et al. (2012)
Guidelines and manuals		Guinée et al. (2002), Baumann and Tillman (2004), and Frischknecht et al. (2007)	EC-JRC-IES (2010a, 2010b, 2011), Hischier et al. (2010), and Blengini et al. (2012)
Reviews		Cleary (2009) and Finnveden et al. (2009)	Damgaard (2010) and Gentil et al. (2010)
Case studies		Corti and Lombardi (2004), Bovea and Gallardo (2006), and Buttol et al. (2007)	Cleary (2010), Gentil et al. (2011), Pires et al. (2011), Grosso et al. (2012), and Pires and Martinho (2013)

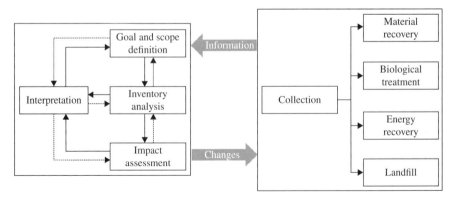

FIGURE 10.3 Connection between LCA and waste management system

an interactive mode for information flows and essential changes over time. Using SWM as an example, if the environmental impacts considered in the LCA are not sufficient to respond to the goal, the breadth of the environmental impact assessment may be increased and/or additional data may be collected to quantify the additional environmental impacts as required. In some cases, even the goals can be revised due to unexpected limitations or restrictions due to the impact assessment. The interactions among the four phases therefore comprise the LCA methodology. Following the impact assessment, the interpretations contained in the report may lead to changes in the technology options, capacity expansion, and other possible options in the SWM system. The following subsections discuss each phase in a greater detail.

10.2.1 Goal and Scope Definition

For an SWM system, the goal of an LCA study can be tied to a wealth of decision-making processes associated with long-, medium-, or short-term consequences. According to the European Commission (EC-JRC-IES, 2011), there are three different decision support situations: micro-level, meso/macro-level, and an accounting/monitoring situation.

A micro-level decision support has only a small-scale consequence on the background system. The modeling principle applied for this type of LCA is attributional modeling, explained by its emphasis on the environmentally important physical flows to and from a life cycle and its subsystems (Ekvall et al., 2005; Finnveden et al., 2009).

A meso/macro-level typically involves decision support with large-scale consequences in a background system. The effects of the decision are sufficient to cause structural changes to the installed capacity of at least one process situated outside the foreground system (EC-JRC-IES, 2011). Rather than conducting a stand-alone LCA, goal comparisons among various scenarios can be performed for meso/macro-level decision support. To achieve adequate LCI goal comparisons, "consequential modeling" is applied to describe how flow into and out of the environment will change as a result of dissimilar decisions (Curran et al., 2005).

The accounting/monitoring case proposed by EC-JRC-IES (2011) is a decision-perspective/retrospective accounting/documentation related to what has happened (or will happen based on extrapolation forecasting) without accounting for any consequences that target system may have on the background system or other systems. The accounting/monitoring case described by EC-JRC-IES (2011) considers two situations, C1 and C2, where C1 expresses an existing system but reports the interactions with other systems, and C2 expresses an existing system in isolation without reporting interactions with other systems. Because EC-JRC-IES (2011) considers C2 as conditions that rarely occur and C1 is identical to a micro-level situation, this chapter focuses on decision support situations for both micro-level and meso/macro-levels (Table 10.2).

The scope of the LCA requires several choices (ISO, 2006b): the system to be analyzed, functions of the product system, functional unit, system boundary, allocation approaches, LCIA methodology and variety of impacts, interpretation to be applied, data demands, presumptions, value choices and optional elements, constraints, data

TABLE 10.2 Goal definition under various LCA purposes associated with SWM

Situation	Intended application	Reasons	Audience
Micro-level decision support	• Technology analysis • Sites/companies • Local/regional studies with no consequences on background system or other systems • Environmental management systems • Environmental inventory for waste management sector	• Waste management system improvement at local, regional, or site-specific level, benchmarking, development of indicators • Studying a specific waste process, such as incineration, landfill, and recycling, being defined as streamlined LCA	• Waste management organization, advisory board, decision makers at waste management system • Waste managers and technicians
Meso/macro-level decision support	• Strategies with large-scale effects on the background or other systems • Decision may generate structural modifications of installed capacity of one process outside the foreground system, at least	• Policy development, policy information, development of datasets, waste management planning at national scale, national and international strategies elaboration	• Government, public, non-governmental organizations, industry (manufacturer and recycling), that is, decision makers at a national and international scale

quality requirements, critical review structure (if any), and report type and format. The relevant aspects to define goal and scope of the LCA study are discussed in the following sections.

Functions of the system, the functional unit, and reference flow   An SWM system is capable of performing different functions simultaneously, including waste collection and transportation, material recovery through recycling, energy recovery from burning biogas and raw solid waste, biological treatment, and landfilling. The function to be studied and its functional unit must be defined in an LCA at first (i.e., the quantified amount selected in a system to assess its performance); for example, 1 metric ton (or tonne) of waste streams destined for incineration can be selected as the functional unit (e.g., reference flow) in an LCA to provide a reference for the normalization of input and output data. In this context, the management of waste streams is the main function, and the duration of the service provided must be taken into consideration. Once the functional unit in SWM systems has been clearly defined, this information can be transposed into the reference flow, which is quantitatively related to all other input and output flows. In the context of a comparative LCA, the selection of a functional unit can be 1 tonne of waste streams managed or the total amount of waste treated in 1 year. Waste streams considered can be municipal solid waste (MSW), industrial waste, hazardous waste, packaging waste, or biodegradable waste, or any other waste stream which is intended to be studied in parallel. Yet this choice may vary within different cases depending on the goal of the study.

Functional unit and reference flow must be characterized and selected. Waste composition (e.g., components in plastic, paper/cardboard, glass, fermentable, metal, diapers) and waste chemical characterization of each waste component (e.g., moisture, organic dry matter, biodegradable organic dry matter, biogenic carbon, hydrogen, oxygen, nitrogen, chlorine, sulfur, heavy metals content such as arsenic, cadmium, chromium, copper, mercury, nickel, zinc) is assessed for mechanical, biological, and thermal waste treatment technologies, including disposal. Physical properties such as particle size distribution and density of waste components are also needed when analyzing mechanical treatment options and collection and transport systems for waste management.

System Boundaries The SWM systems studied in the LCA can be:

- the service provided by a municipality or a company, as described in many real-world SWM systems;
- the end-of-life phase of a product that determines how much will become waste, depending on whether the waste product is destined for incineration or landfilling, material recycling and separation, reuse and prevention of waste, or other fate; and
- the waste treatment technologies from which a product is produced, for example energy and compost.

Possible flowcharts developed at this early stage to define the product system do not need to contain all the details of a system; it may be advantageous if the flowchart is generic enough to include all possible options studied (Figure 10.4).

Although the SWM system is generically defined, its boundaries are not. The boundaries are related to a coupled natural system and the built environment, which must include the processes related to different geographic locations, time scales, and technical components that help define the boundaries. For considering a reasonable system boundary, LCA studies can be classified by their technical components:

- **Cradle-to-grave:** The system begins with resource extraction and progresses to the use and disposal phase, encompassing the entire life cycle;
- **Cradle-to-gate:** This sectional LCA examines only the value-added process in the entire product/service chain, encompassing raw materials extraction through a processing chain to a product leaving the factory gate;
- **Gate-to-gate:** This sectional LCA encompasses a manufacturing process at a particular site;
- **Cradle-to-cradle:** It is a holistic framework of social, industrial, or economic nature that tries to elaborate systems efficient and mainly waste free, suggesting that industry have to preserve and enhance the metabolism of the ecosystems and of the nature.

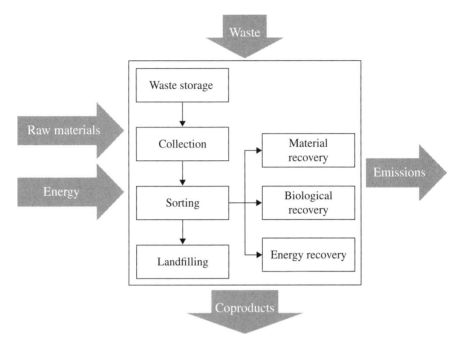

FIGURE 10.4 Primordial flowchart of an MSW system

In general, an SWM system can be subject to a cradle-to-grave LCA if a holistic assessment is required, or it can be a cradle-to-gate process if a specific process is analyzed (i.e., an incineration process). Therefore, a cradle-to-gate LCA is also called a streamlined or partial LCA, whereas a cradle-to-cradle LCA can be a case for pollution prevention and product reuse analysis (Box 10.1).

BOX 10.1 FULL LCA AND STREAMLINED LCA

A full LCA (also called full-scale LCA), a complete LCA of a system, can be time and resource intensive, expensive, take years to conduct, and require qualitative and updates/new data inventory. A full LCA examines each process and becomes even more difficult when processes are external to the entity requiring the study. The intrinsic difficulty of conducting a full LCA in SWM is practical if looking at specific waste streams, such as packaging, batteries, and organic waste fraction. Moreover, a full LCA results can be complex and demanding for decision makers to comprehend, whether in the industry or in the public sector.

The inadequacies of LCA have promoted other approaches that non-experts could implement in a more cost-effective and understandable way. Streamlined LCA is a possible solution to make LCA more popular without losing accuracy and significance. Streamlined LCA (also named screening and matrix) involves an analysis if the range of life cycle stages, environmental parameters, impact categories, or processes (such as transportation or manufacturing of specific materials or components) are limited (Crawford, 2011). Quantitative data are usually required, but readymade databases are also available that avoid new inventory calculations; therefore, streamlined life cycle approaches can be qualitative, quantitative, or semi-quantitative (Pesonen and Horn, 2013).

The primordial flowchart for a typical MSW system (Figure 10.4) is composed by different components. In assessments of the environmental impacts resulting from these components, an LCA helps enlarge the view beyond the SWM system boundary. Yet, a streamlined LCA does not begin with the extraction of raw materials, and product reuse is not included in the assessment. The process can be simplified based on the "zero burden assumption," which focuses solely on the SWM system, assuming that the waste carries no upstream environmental burden into the SWM system (Ekvall et al., 2007). In other words, all product life cycle phases previous to the waste phase can be excluded if they are repeated to all following waste management alternatives (Finnveden, 1999; Buttol et al., 2007). The zero burden approach, as defined by the only inclusion of solid waste treatment and recovery phase in an LCA, has been applied in several previous LCA case studies related to SWM (McDougall et al., 2001; Blengini 2008; Scipioni et al., 2009; Cleary, 2010; Gentil et al., 2010; Rigamonti et al., 2010; Williams et al., 2010; Pires et al., 2011; Grosso et al., 2012; Pires and Martinho, 2013).

SWM systems can be regional, national (between regions), and even international because waste can be sent out of a region or a country through trades. For example, hazardous waste treatment technology does not exist in all European countries; in comparison, recycling can be conducted in one country, and recycled material can be exported to other countries in the European Union for processing. These downstream systems (i.e., operations that occur after leaving the target SWM system), such as recycling and thermal recovery in cement kilns, are difficult to characterize in terms of technical aspects, pollutant emissions, and energy/material consumptions. Relevant upstream processes might not be directly related to the SWM system, yet without them the SWM is not functional. An example is electricity generation facilities, which can be operated in other countries. But relevant information in regard to when pollution prevention and product reuse were applied to electricity generation facilities in those countries affecting the LCA would be difficult to obtain. Geographic discrepancies are also tied to environmental impacts because sensitivity of some environmental pollutants varies among regions. Such complexity partially justifies the use of the zero burden approach.

Functional units associated with environmental impacts are linked to a specified time horizon, which is the period when all environmental aspects (e.g., inputs and outputs) are considered (i.e., the accounting period) according to EC-JRC-IES (2011). A common example of a functional unit is the quantity of waste treated in a year. Choosing the proper LCA time horizon is a compromise between the need to include most (virtually all) of the emissions and the availability of sufficiently accurate data for that period. A challenging waste operation for setting a time horizon is landfilling. Gas and leachate emissions from the landfill can be regarded as system output, but not the waste itself (Finnveden et al., 1995); although waste can be landfilled within a specific year (i.e., in accordance with the functional unit), the emissions will continue long after the closure of the landfill. The most common practice is to include the emissions in the inventory, such as in a 100-year time frame, but this simplification inevitably limits the interpretation of the LCA results by excluding the possibility of long-term emissions. One solution is to cover the long-term emissions after 100 years in a separate assessment (Finnveden et al., 1995). An optional solution is to insert an impact category called "stored toxicity" that monitors toxic loads that remain in waste disposal sites at the end of the chosen time period (Christensen et al., 2007; Hauschild et al., 2008). In either case, the problem of weighing the impacts during LCIA over different time scales remains (Finnveden et al., 2009).

The next issue examines the relevance of processes of concern. Environmental impacts associated with capital goods such as buildings, machinery, vehicles, and personnel are often not considered relevant processes, and to simplify the method, they are sometimes disregarded when the selected functional unit is the main focus of an LCA. In a complete LCA study, however, the guideline is to include all environmental impacts associated with production and maintenance of capital goods (Baumman and Tillman, 2004). A compromise solution is to consider nothing below a cutoff value of 1% of the overall functional flow.

Another aspect to be defined during goal and scope definition is the allocation procedure to conduct during LCA. According to ISO (2006b), allocation represents the portioning of input and output flows of a process or a system concerning the system under analysis and one or more other systems. There are two situations where allocation procedure needs to be solved (Ekvall and Tillman, 1997):

- when the system has multifunctional processes; and
- when occurs open recycling inside of the system.

Procedures Developed for Handling Multifunctionality   When considering different technical systems with multifunctional flows with multiple products, the selection of the system boundary is vital to both the upstream supply chain and downstream networks (Box 10.2). A coproduct is defined as two or more products resulting from the same unit process or system (ISO, 2006b). A process is often shared among several product systems in an LCA study, and allocating environmental impacts to the proper product becomes complicated (Finnveden et al., 2009).

BOX 10.2   MULTIFUNCTIONALITY PROBLEMS

Functional flow is defined as any of the flows of a unit process which compose its goal as product outflows of a production process or waste inflows of a waste treatment process (Guinée et al., 2004). Multifunctional process includes a unit process yielding more than one functional flow, or possess coproduction (multi-output) (Guinée et al., 2004). It also covers the cases when combined waste processing (multi-input) or recycling (input-output) is taken into account. Two examples are shown in Figure 10.5. However, the key challenges are (1) how the environmental impacts of processes should be allocated or participation to the different products engaged? and (2) Which processes belong to the target functional flow studied and which do not?

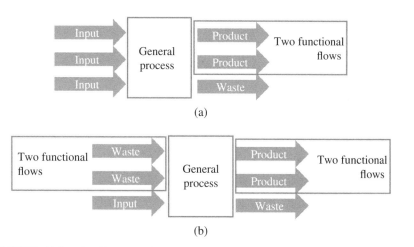

(a)

(b)

FIGURE 10.5 Identification of functional flows in multifunctionality problems: (a) multi-output and (b) multi-input

When solving a multifunctionality problem (Box 10.2) that involves multiple products and multiple environmental impacts, allocation of environmental impacts to different products becomes much more complicated. Analyzing broader issues that encompass the entire LCI model must simultaneously consider: (1) which processes are multifunctional?, (2) what are the functional flows of each process?, and (3) how to resolve the problems using the proper allocation approach or other related methods to account for environmental burden? One possible solution is to choose a more refined data collection approach during the first attempt; however, when no data and no timeframes are available, methods to solve multifunctionality include the system expansion/substitution method and the allocation method (also known as the partition method) (Table 10.3). System expansion and substitution can be presented as the same method because recent developments indicate that both methods are conceptually equivalent (Tillman et al., 1994; Lindfors et al., 1995b; Heijungs and Guinée, 2007), reinforced by the denomination of substitution by system expansion or avoided burden method (Finnveden et al., 2009). However, for higher complex product systems, system expansion use by adding more functions into the system can induce more allocation problems. In any circumstance, system expansion needs to be applied by a more cautious way to avoid such complexity in LCA studies.

TABLE 10.3 Comparisons among methods for solving multifunctionality problems

Method	Description	Functional unit	Flow diagram	Required data	Remained problem
Allocation	Partitioning of the system inputs and outputs associated with the environmental impacts	No change	No change	Changed	Varying set of the allocation factors may yield differing results
System expansion/ substitution/ avoided burden method	The process refers to a "regular" production that is substituted or expanded by adding extra functional unit or by removing the avoided processes to address coproduct impact	No change	Enlarged	No change	Identifying the avoided processes may be debatable and unclear if the coproduct is not common

In an SWM system, waste management services commonly exhibit multifunctionality throughout several technical systems producing recyclables, electricity, compost, and methane gas. This raises the question of whether or not a particular environmental impact is related to the prescribed functional unit in cases such as multi-output and multi-input.

A multi-output system (Figure 10.6a) consists of multiple MSW processes that produce more than one product (i.e., recyclables, compost, and electricity) across the system boundary. To manage this complexity, subdivision of multifunctional processes through a system expansion/substitution and allocation may be an applicable approach. System expansion accepts extra function(s) as composite reference flow to the functional unit. For example, when dealing with ash recovery from an MSW incinerator, a composite functional unit of two reference flows, steam and electricity, can be defined by changing the system boundary to include additional functional unit.

The substitution method subtracts "avoided" burdens (substitution) in an "avoided" process with subsequent "avoided" interventions/impacts. For example, substituting recyclables for virgin materials in a close-loop recycling system avoids the consumption of resources needed to produce a new equivalent. The key question is which process can be avoided. The allocation problem complicates the inputs and outputs of unit processes partition between product systems. The allocation procedure is designed to eliminate extra functions by an extra modeling step (allocation) that effectively splits the multifunctional process into several monofunctional processes. In other words, only a part of the process is allocated to the function with upstream to downstream consequences. Yet, the question of how to split and allocate only part of burden to the relevant function (i.e., partitioning) remains.

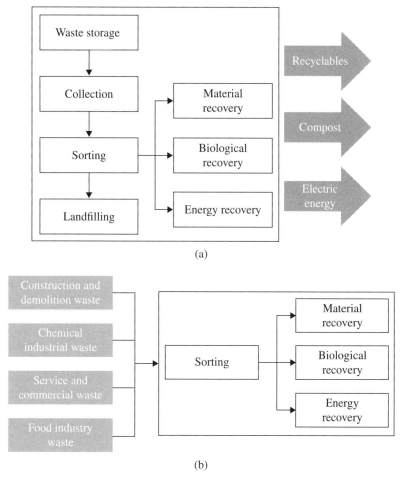

FIGURE 10.6   (a) A multi-output LCA. (b) A multi-input LCA

In SWM systems, emphasis is on the amount of resources and emissions that occur during the treatment of waste associated with each one of the products. For example, in a multi-input situation (Figure 10.6b), waste streams from different sources enter the system boundary of an SWM system, complicating the process of allocating environmental impacts to specific waste streams. Hence, the allocation problem is actually an artifact of isolating a single function, and the artificial methods required to resolve artifacts might not have theoretical support. Solutions should be consistent with LCI modeling principles. The LCI discussion (Section 10.2.2) further describes approaches to system expansion/substitution.

Recycling Allocation   When analyzing a product system which is recycled when reaches its end-of-life, becoming a new product which can avoid the use of virgin material, allocation problem exists. The two types of recycling processes are closed-loop and open-loop. Closed-loop recycling occurs when a secondary good is returned to an earlier process in the same system (e.g., recycled gypsum from waste plasterboard is introduced into a new plasterboard production line) (EC-JRC-IES, 2011). Without losing the inherent properties of the product, this replacement occurs in the same cycle, directly replacing or substituting the input in the primary production line of the same material/product. Recycled glass is also an example of closed-loop recycling (e.g., recycled glass from waste bottles is introduced into a new glass production line). In comparison, open-loop recycling occurs when the secondary material is used for a different product system than the original, where loss of intrinsic recycled materials properties may, or not, occur. An example of open-loop recycling is the transformation of food packaging materials into other types of plastic for different uses.

When considering the use of recycling materials in one product, cutoff method (Box 10.3) (Ekvall and Tillman, 1997; Norgate, 2004) and avoided burden method (Ozihel, 2012) are two commonly used approaches in the LCA with different principles (other methods also exist like 50/50 method (Lindfors et al., 1995a), which is applied when the inclusion of recycled material into the product is meaningful). Note that a cutoff value or criterion described above in the context of system boundaries is not related with the cutoff method defined in Box 10.3. Both cutoff and avoided burden methods capture the range of options for modeling the recycling systems in the ISO recycling allocation hierarchy (ISO, 2006b).

Under the cutoff method, a boundary is drawn between the primary and secondary materials used in the product studied. All environmental burdens associated with virgin material (i.e., primary material) production are associated with the first use of the primary material. Yet the environmental burdens assigned to the use of secondary material begin with the recycling of the postconsumer material involving material recovery, transport, separation and sorting, and reprocessing. The avoided burden incorporates total recycling burden in the first life cycle as recycling of a material avoids extraction and manufacturing of raw material. All avoided expenses and emissions are completely attributed to the product that includes the recycled materials after its service life.

BOX 10.3   CUTOFF METHOD (Ligthart and Ansems, 2012)

Cutoff method considers that the environmental impacts caused by the product's manufacturing employing primary and secondary materials are allocated to the product itself. Besides, a possible waste treatment, other than recycling, is allocated to the product too. There is no need for data from outside the life cycle of the analyzed product. In this method, the recycled content is important as this may reduce the environmental impact of the system because the use of secondary materials has a minor impact than the primary material. An increase in recycling rate is favorable as this recycling effort reduces the impact of the waste disposal.

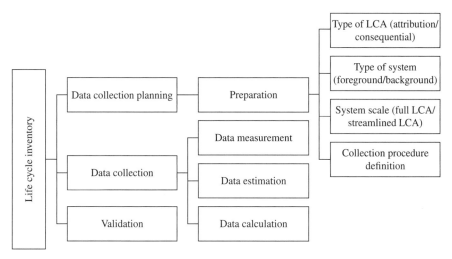

FIGURE 10.7   Life cycle inventory steps

10.2.2 Life Cycle Inventory

LCA has evolved into many extensions, such as attributional LCA and consequential LCA. LCI analysis to support an LCA creates an inventory of flows from and to nature for a product system. LCI associated with different types of LCA have three steps: data collection planning, data collection, and validation (Figure 10.7). Data collection planning prepares the data collection based on the goal and scope definition phase for the background and foreground systems. For example, in attributional modeling, data collection planning is focused on data related to the average status of the SWM system. In this context, the average, generic data that best represent the waste management practices are collected. In consequential modeling, data collection planning uses domain experts' inputs for technology development forecasting (Shen et al., 2011), scenario development (Björklund and Finnveden, 2007), market forecasting (Münster and Meibom, 2010), and general-equilibrium and partial-equilibrium modeling (Ekvall, 2000) to identify and model large-scale consequences.

Consequential modeling determines the effects of a small change in product outputs and/or services from a system on the environmental burdens of the analyzed system (Zamagni et al., 2008). The data collected in consequential modeling are marginal rather than average. There are two possible drawing tools to develop consequential LCA, including scenario development and market forecasting. Scenario development models future systems, but the construction of these scenarios is critical (Börjeson et al., 2006; EC-JRC-IES, 2010a, 2011) to their quantitative and qualitative application in workshops, time series modeling, and optimization modeling (Table 10.4). In comparison, market forecasting generally follows the current generic investigation of the existing and potential markets to identify possible inputs.

Foreground data are also called primary data because they are related to the direct control or decisive influences such as on-site separation activities, logistical collection arrangements, and reprocessing operations. To collect primary data, the focus must

TABLE 10.4   Classification of scenarios for scenario development

Scenarios	Methods	Examples
Predictive: "What will happen?"	Forecast, What-if	To model background processes
Explorative: "What can happen?"	External, Strategic	To model substitution or avoided products (in cases of multifunctionality)
Normative: "How can a specific target be reached?"	Preserving, Transforming	Relevant to reach recycling targets imposed by legislation, for example

Source: Adapted from Börjeson et al. (2006).

therefore be on raw materials, energy, water, chemicals, wastes, wastewater, and gaseous emission from the foreground system. Background data are secondary data because they are related to processes with no obvious direct influence on the core system of interest, such as diesel production, or avoided production of primary materials replaced by recycling.

Choosing a full LCA or a streamlined LCA will also influence how LCI is conducted. A streamlined model (Box 10.1) involves limiting the scope of the LCI if some of the background systems are deemed insignificant. A number of factors can be relevant in the streamlined approach (Crawford, 2011), including regulatory, marketing, and other internal or external demands; some intended applications of the streamlined approach are related to a particular environmental parameter, life cycle stage, impact category, or the importance of particular environmental issues. For example, a previous full LCA may have identified particular areas where significant issues exist that allow future improvement efforts to be targeted and separately arranged as a streamlined LCA in the next stage.

LCI databases provide ready-made inventory datasets of emissions and consumed resources (e.g., electricity, virgin materials, and other intermediate goods and services) generally associated with the background system. Users can refer to several existing databases for individual LCI (Table 10.5). One of the first LCI databases is the Global Emission Model for Integrated Systems (GEMIS), a German energy modeling tool with LCI data on energy systems, was created in 1989. Ecoinvent database, a worldwide diffused LCI/LCA database, started to be developed in the 1990s. Since 2011, the Global Guidance Principles for LCA Databases developed by the SETAC has become a common reference in this field. Well-known LCA software such as SimaPro and Umberto have LCI that integrates data from various industrial sources and other well-known data providers.

Data for each unit process within the systems boundaries can be classified under important headings, including (ISO, 2006b):

- inputs, such as energy, raw materials, ancillary, and others;
- outputs, such as products, coproducts, waste, emissions to air, releases to water and soil, and;
- other environmental aspects.

TABLE 10.5 LCI databases in the literature

Name and creation time	Processes covered	Availability	Geographic coverage
Ecoinvent (since 1994)	More than 2,700 datasets: electricity mixes, power plants, fuels and heat supply, construction materials, basic chemicals, metals, transport, waste management, wood materials, agriculture	On request and with cost for single-user and multi-user license or through LCA tools such as Umberto, SimaPro, GaBi	Switzerland, western Europe, global
National Renewable Energy Laboratory (NREL) (since 2001)	For commonly used materials, products and processes	Cost free	United States (US), Canada
GEMIS (since 1989)	Energy, materials transport, recycling and waste treatment processes. More than 10,000 processes and some 1,000 products	Cost free	Germany and global
ProBas (since 2000)	Over 7,000 unit processes for energy, material	Cost free	Germany, Europe, and European countries
Sustainable Product Information Network for the Environment (SPINE) (since 1995)	More than 500 LCI datasets	Cost free, provided by Center for Environmental Assessment of Product and Material Systems (CPM)	Europe
Option data pack (since 1995/1998)	Over 900 processes on chemical production, iron and steel, waste management	A user license is required, being included in AIST-LCA	Japan
Australian LCI (since 2003)	Materials, agriculture, food, energy, transport and services	Cost free	Australia

(continued)

TABLE 10.5 *(Continued)*

Name and creation time	Processes covered	Availability	Geographic coverage
European Reference Life Cycle Database (ELCD) (since 2005)	End-of-life treatment, energy carriers and technologies, materials production, systems and transport services	Cost free	Europe
Thai National LCI Database (Since 2005)	Processes are focused on energy, utilities and transportation, industrial materials, agriculture, recycled and waste management, chemicals, and building and construction materials	(No information)	Thailand
European Aluminum Association LCI (since 2008)	Industrial database for aluminum	On request	Europe
Deutsches Kupferinstitut LCI (since 2005)	Industrial database for copper	Cost free	Germany
Nickel Institute (since 2000)	Industrial database for nickel	Free	World
International Stainless Steel Forum (ISSF) LCI (since 2003)	Industrial database for steel	On request but is free	World
Plastics Europe LCI (since 1991)	Industrial database for plastic in a cradle to factory gate environmental data	Cost free, available at the site	western Europe
Canadian Raw Materials Database (CRMD) (since 1998)	Industrial processes	Cost free	Canada
LCA food database (since 2003)	Food	Free, provided that proper reference to the source is given	Denmark

Korea National Cleaner Production Center (since 1998)	Chemical, electronics, and automotive industries, energy, polymer, process, transport, infrastructure, raw materials, incineration, recycling	(No information)	Korea
GaBi LCI (since 1990)	More than 4,500 processes for several materials, energy carriers, services, and processing technologies	With LCA software	Global
Data for Environmental Analysis and Management (DEAM™) (since 1994)	Over 3,000 modules, from fuel production to transportation, from chemical production to plastic molding	Data for Environmental Analysis and Management (DEAM). Sold with TEAM™ and WISARD software developed by Ecobilan	United Kingdom
DIM (since 2004)	Background systems: energy, materials, transport packaging. Product chains: lamps, printed circuit boards, kitchen furniture, office desks, textile finishing processes	Through eVerdEE tool only	Italy, global
ECODESIGN X-Pro database (since 2007)	150 industrial processes: cold/hot forming, electronic appliances manufacturing, plastics processing machining, joining, heat treatment, recycling processes, transport, logistics	Through ECODESIGN X-Pro tool	Global
EIME (since 1996)	Database dedicated to description of electrical, mechanical, and electronic products	Annual license, being diffusion restricted to EIME software uses	Canada, France

(continued)

TABLE 10.5 *(Continued)*

Name and creation time	Processes covered	Availability	Geographic coverage
Eurofer (since 2002)	Steel industry	Cost-free	Global, Europe
University of Amsterdam (IVAM) LCA (since 1996)	1,300 processes, focused on building and construction sector, food production and waste management	License for indefinite period, including version updates	Global, Dutch
Keskuslaboratorio-Centrallaboratorium (KCL) EcoData (since 1994)	300 modules: energy production, chemicals manufacturing, wood growth and harvesting operations for spruce, pine and birch, pulp, paper and board mills, deinking process for different printing papers and tissue papers, printing, waste management operations and plastics, transport	Full database or groups of modules or only one data module can be purchased to database developed KCL	Global
LC Data (since 2008)	The German Network on LCI Data provides basic life cycle inventory datasets in the field of materials, energy, transport, and end-of-life treatment	Cost free	Germany
The Boustead Model (since 1992)	More than 33,300 unit operations, which include fuel production and processing operations for almost every country	License for commercial purposes, for education purposes is cost free	Global
Database Waste Technologies Data Centre (since 2006)	Around 40 waste treatment technologies are covered	Cost free	United Kingdom

Source: Adapted from JRC (2013).

TABLE 10.6   Data collection registration

Name of material to be transported	Road transport			
	Distance (km)	Vehicle capacity (tonnes)	Load (tonnes)	Unload return (Yes/No)
.....	.....	.....	.....	.....
.....	.....	.....	.....	.....

How and which data must/should be collected in an LCI are determined on a case-by-case basis. In this context, a unique procedure is needed to identify data and information sources and to assess data quality requirements as defined in goal and scope to aid the interpretation phase (Table 10.6). Units should be from International System of Units (abbreviated SI from French: *Le Système International d'Unités*) to minimize conversion efforts and potential errors (ISO, 2006b).

In summary, data collection planning determines where the sources are registered, how the interim quality control can be implemented, and how to manage missing inventory data. The aggregated inventory should verify if the reference flow(s) is related to the product(s) and/or waste flow(s) in the LCI. Quality control must consider aspects of time, geographic reference, and technology. Any product and waste flow related to non-functional flows must be highlighted in the report and/or dataset and requires modeling when these datasets are used to interpret and draw conclusions from the LCA study (EC-JRC-IES, 2010a). Special procedures might be needed to address gap or missing data to support the quality of the LCA.

Legal limits can be checked during the data collection planning stage. An interim quality control procedure can be developed to check validity, to deal with missing data and data with irregularities, to fill in data gaps or estimates with minimum quality, and to register the remaining unit process data. Strategies to deal with missing information (EC-JRC-IES, 2010a) include:

- calculating from other, known information;
- adopting information from similar processes or regions with similar process operation (and background process) or older data;
- estimating the value based on specific expertise;
- applying methodologies that are sufficiently consistent with acquisition of required data; and
- accepting and documenting information gaps.

Measurement, estimation, and calculation are all valid procedures to collect data. Measured data are preferred to estimated and calculated data because they better represent reality and are less uncertain. Measured data resulting from internal audits are expensive and time-consuming if the waste operations are not yet fully managed; therefore, estimates can be provided from literature or even from sparse information.

Data calculation methods can be retrieved from various published references tailored to meet the local need for varying types of LCI. Complex, system-level calculations must consider the functional unit and problem-solving methods for multifunctionality issues. This is especially true when dealing with micro-level decision support and accounting/monitoring cases, in which subdivision of multifunctional processes might be preferable at the beginning stage, followed by system expansion/substitution of avoided burdens; for consequential modeling subdivision should be used and followed by system expansion/substitution at the macro-level decision support stage, and sometimes must exclude existing interactions with other systems (EC-JRC-IES, 2011).

Two methodologies can be applied to conduct the calculations, including the sequential approach, also known as process flow diagram, and the matrix method. The first approach relies on unit processes scaled in sequence, beginning with the process for supplying the reference flow, then scaling up the unit processes supply to be commensurate with the unit process of interest, then scaling up to unit processes dedicated to supplying downstream units, and continuing up the sequence (Sonnemann and Vigon, 2011). This intuitive perspective has the advantage of making interpretation easier because the contribution of individual supply chains can be easily assessed, but has the disadvantage to dealing with feedback loops because it is a fully or partially terminated system. This implies that the resulting inventory might not be exact (Sonnemann and Vigon, 2011). This problem is less relevant to datasets aggregated for confidentiality reasons on a gate-to-gate or partially aggregated basis (Sonnemann and Vigon, 2011). So does the case for dynamic system modeling, like can be conducted through Umberto software (IFU Hamburg, 2009). The matrix method arranges the inputs and outputs of a unit process in a matrix representing the coefficients of a group of linear equations (Heijungs and Suh, 2002). In real-life product systems, matrix algebra can obtain precise inventories in all cases, even when there are many feedback loops; nevertheless, the matrix inversion approach complicates the investigation of individual branches and supply chains (Sonnemann and Vigon, 2011).

Different approaches exist to solve the issue of multifunctionality depending on the type of LCA to be conducted at either micro-level or meso/macro-level. Choosing a method to calculate a corresponding LCI depends on the required dimensions and data availability of the LCA. The sequential approach is the simplest (Sonnemann and Vigon, 2011) and favors the application of system expansion for micro-level decision support. The LCI of the superseded process(es) or product(s) can be equivalently subtracted (substituted) from that of the analyzed system (i.e., it is "credited"), being a special (subtractive) case when applying the system expansion principle (EC-JRC-IES, 2010a). Secondary materials (e.g., those resulting from waste treatment, including recyclables and energy recovery) can be substituted for primary materials (i.e., virgin materials) to evaluate the substitution effect in closed- or open-loop recycling (Table 10.7).

In a case of multi-output multifunctionality where one of the products can promote a closed-loop recycling, secondary materials substitute the virgin material in the same system; in open-loop recycling, secondary materials enter the same system after losing their inherent characteristics rather than replacing the virgin materials as equivalents. The substitution effect can be quantified by a substitution ratio. According to

TABLE 10.7   Product obtained from the material recovery (secondary product) and its corresponding substituted product (primary product) plus the substitution ratio

Material	Secondary product	Primary product	Substitution ratio
Iron	Liquid iron	Liquid iron	1:1
Aluminum	Aluminum ingot	Aluminum ingot	1:1
Glass	Generic glass container	Generic glass container	1:1
Wood	Particle board	Plywood	1:1
Paper	Pulp from recovered paper	Thermo-mechanical pulp	<1
Plastic	Granulates of PET, HDPE, and mixture of LDPE, LLDPE, PP	Granulates of PET, HDPE, LDPE, LLDPE, PP	<1

Source: From Rigamonti et al. (2009a).
HDPE, high-density polyethylene; LDPE, low-density polyethylene; PP, polypropylene; LLDPE, linear low-density polyethylene; PET, terephthalate.

Rigamonti et al. (2009a, 2009b), the substitution ratio is 1:1 in closed-loop recycling and <1 in open-loop recycling. Cases with a substitution ratio <1 are challenging because the quantity of the replacement product is required, such as for products with a limited recycling capacity. A common example is paper (Rigamonti et al., 2009a), which can only be recycled about five times (Comieco, 2008).

According to Rigamonti et al. (2009a), virgin pulp (one "entity") can only be used to produce five secondary pulps (five entities). All inputs used in the production of virgin pulp must be divided among six entities rather than infinite entities (like happens for metals recycling), so the production of 1 kg of secondary pulp adds one-sixth of the energy and material inputs from the production of 1 kg of virgin pulp to the energy and material inputs for the recycling activity (Rigamonti et al., 2009a). From this presumption, 1 kg of secondary pulp plus 0.167 (1/6) kg of virgin pulp is supposed to substitute for 1 kg of virgin pulp (i.e., 1 kg of secondary pulp replaces 0.833 (1 − 0.167) kg of virgin pulp), which yields a substitution ratio of 1:0.833 (Rigamonti et al., 2009a).

When solving multifunctional systems for meso/macro-level decision support, allocation can be avoided using the same substitution principle. In this case, the substitution ratio is not based on the loss of inherent characteristics, but is associated with the market value of secondary materials produced. The value-corrected substitution ratio is therefore applied if there is a difference between the market value of the primary material and that of the corresponding recycled material (Rigamonti et al., 2009a). According to the European Aluminium Association (2007), this method presumes that the substitution capacity is reflected by the ratio between the market prices of the recycled and primary materials (Rigamonti et al., 2009a). If the recycled material market price is 90% of the primary material market price, 1 kg of recycled material will replace just 0.9 kg of primary material (Rigamonti et al., 2009a).

The final step in LCI is about validation (Figure 10.7). To date, processes must be validated individually or collectively verifying would be difficult, if not impossible if all inventoried emissions and consumptions are close to a real-world scale.

Information collected from various sources can be validated by experts in the subject area. Partial systems can also be measured if the budget and time are sufficient, but all measurements must be documented. To produce the inventory results of the determined system for each unit process and to all systems, the following procedures should be followed: validation of collected data, the relation of data and unit processes, and the relation of data and the reference flow or the functional unit. For the LCA to be robust and trustworthy, calculation procedures must be consistent, especially when aggregating processes within the system boundary to guarantee that reference flow represents simply the product and waste flow.

10.2.3 Life Cycle Impact Assessment

A typical LCI often has many substance flows, resulting in a myriad of impacts on the environment. The LCIA phase classifies and characterizes the LCI results for all predetermined environmental impact categories related to human health, natural environment, natural resources, and the built environment. LCIA does not necessarily quantify specific impacts associated with a product, process, or activity; rather, it establishes a systematic linkage between a studied system and all potential impacts (USEPA, 2006). With this concept, the models used within LCIA are often simplified within each of the various impact categories, although unlike traditional risk assessments, the LCIA models often cannot be used to delineate absolute risk or actual damage to human health or the environment (USEPA, 2006). Key steps in an LCIA include (USEPA, 2006): (1) selection and definition of impact categories to identify relevant environmental impact categories (e.g., GWP, acidification, eutrophication) (Table 10.8), (2) classification to assign LCI results to the impact categories (e.g., classifying carbon dioxide (CO_2) emissions to GWP), and (3) characterization of LCI impacts within impact categories using science-based conversion factors (e.g., modeling the impact of CO_2 and methane (CH_4) on GWP).

In all real-world applications, including well-defined cases of SWM systems, the major LCIA task is transforming hundreds of substances and materials into a few environmental stressor impact categories to simplify the impact assessment communication to technical and non-technical stakeholders. As a first step, the LCIA uses a holistic approach on environmental impacts by modeling any impact from the product or service system predicted to damage one or more areas of protection (i.e., receptor locations) (Finnveden et al., 2009). Modeling these impacts requires two types of indicators, including midpoint and endpoint indicators. Midpoint indicators, based on the problem-oriented approach (Guinée et al., 2002), emphasize the environmental impacts that occur somewhere between the emissions and the areas of protection. They convert into equivalents of a reference substance the several substances responsible for a specific environmental impact (Schulz et al., 2011). The endpoint indicators, based on the damage-oriented approach (Guinée et al., 2002), emphasize impacts on the areas of protection. Endpoint effects, such as health and ecosystem damage, comprise a weighting and aggregation of the several impact categories into a unique score indicator based on a damage model (Schulz et al., 2011) make the LCIA much more understandable, both for technical as well as non-technical stakeholders, facilitating the assessment of environmental impacts.

TABLE 10.8 Commonly used life cycle impact categories

Impact category	Examples of LCI data (i.e., classification)	Common possible characterization factor	Description of characterization factor
Global warming	CO_2 Nitrogen dioxide (NO_2) CH_4 Chlorofluorocarbons (CFC) Hydrochlorofluorocarbons (HCFC) Methyl bromide (CH_3Br)	Global warming potential	Converts LCI data to CO_2 equivalents Note: global warming potentials can be 50, 100 or 500 year potentials
Stratospheric ozone depletion	Chlorofluorocarbons (CFC) Hydrochlorofluorocarbons (HCFC) Methyl bromide (CH_3BR)	Ozone depleting potential	Converts LCI data to trichlorofluoromethane (CFC-11) equivalents
Acidification	Sulfur oxides (SO_x) Nitrogen oxides (NO_x) Hydrochloric acid (HCl) Hydrofluoric acid (HF) Ammonia (NH_3)	Acidification potential	Converts LCI data to hydrogen (H+) ion equivalent
Eutrophication	Phosphate (PO_4^{3-}) Nitrogen oxide (NO) Nitrogen dioxide (NO_2) Nitrates Ammonia (NH_3)	Eutrophication potential	Converts LCI data to phosphate (PO_4^{3-}) equivalents
Photochemical Smog	Non-methane hydrocarbon (NMHC)	Photochemical oxidant creation potential	Converts LCI data to ethylene (C_2H_4) equivalents

(continued)

347

TABLE 10.8 (*Continued*)

Impact category	Examples of LCI data (i.e., classification)	Common possible characterization factor	Description of characterization factor
Terrestrial toxicity	Toxic chemicals with a reported lethal concentration to rodents	LC_{50}	Converts LC_{50} data to equivalents; uses multi-media modeling, exposure pathways
Aquatic toxicity	Toxic chemicals with a reported lethal concentration to fish		
Human health	Toxic releases to air, water and soil		
Resource depletion	Quantity of minerals used Quantity of fossil fuels used	Resource depletion potential	Converts LCI data to a ratio of quantity of resource used versus quantity of resource left in reserve
Land use	Quantity disposed into landfill and other land modification	Land availability	Converts mass of solid waste into volume using an estimated density
Water use	Water used or consumed	Water shortage potential	Converts LCI data to a ratio of quantity of water used versus quantity of resource left in reserve

Source: USEPA (2006).

Note: Global impact: 1 and 2; Regional impact: 3 and 5; Local impact: the rest of impact categories.

LC_{50}: The lethal concentration 50 (LD_{50}) is the dose of a given chemical product which causes the mortality of 50% of a group of test animals in a specified period.

Analyzing hundreds or thousands of substances from several midpoints and end-points begins with the selection of impact categories (already stated in goal and scope definition phase), followed by the selection of category indicators and characterization models, and ending with the attribution of elementary flows to each impact category (classification). By following this procedure from ISO (2006b), the impact associated with each category indicator can be calculated (characterization method). Impact category selection must consider whether midpoint impact categories (e.g., climate change, acidification, and eutrophication) or endpoint categories (e.g., human health, quality of the built environment, and status of natural systems) are needed.

In addition to the compulsory key steps, other optional steps can be added to the metrics; for example, normalization, grouping, and weighting facilitate comparison in the impact assessment and interpretation phases (ISO, 2006b). According to ISO (2006b), normalization calculates the magnitude of indicator category respective to a reference basis; grouping sorts or ranks indicators of category impact; and weighting of relevant category indicators converts or aggregates the impacts using numerical factors based on value-choices. In some cases, endpoints can be reached simply by conducting the optional steps based on category impacts. Different endpoints expressed as a single number can be subject to the choice of varying optional steps, which facilitate comparison among scenarios or even across different LCA studies. Attempting to make LCA results more understandable and accessible does not avoid uncertainty, however. Although uncertainty can be mainly associated with current knowledge limitation about the environmental systems, a common agreement of weighting factors among different objectives or goals could be quite difficult to reach and would introduce additional uncertainty in terms of management perspectives (Hauschild and Barlaz, 2011). In many cases, not all the LCIA steps are necessary, as long as the mandatory steps are completed. These findings may be useful for evaluating and reporting final LCIA results.

Currently Available LCIA Methods and Models   The first LCIA methods to consider are those recommended by ISO 14044 (ISO, 2006b), which relate impact categories to environmental relevance, scientific validation and reproducibility, independence to avoid double-counting, representativeness, as well as internationally accepted agreements or criteria approved by a competent international body. Midpoint impact assessment models address the relative potency of the stressor at a common midpoint within the cause–effect chain (USEPA, 2006). Ozone depletion potentials produced by the World and Global Meteorological Organization and GWP produced by the Intergovernmental Panel on Climate Change are common factors selected for the midpoint impact category of stratospheric ozone depletion and the climate change midpoint impact category, respectively (Forster et al., 2007). Midpoint analyses minimize the amount of forecasting and effect modeling incorporated into the LCIA, thereby simplifying the model and its communication, leading to a more comprehensive endpoint estimate (USEPA, 2006).

Numerous LCIA methods have been developed and applied for various SWM LCA studies (Table 10.9), including Eco-indicator 99 (EI'99) (Goedkoop and Spriensma, 2000), Environmental Priority System (EPS 1999) (Steen, 1999), Institute of

TABLE 10.9 **Features of the main LCIA methods**

Type of environmental impact model	EDIP 2003	CML 2002	EI'99	EPS 2000	IMPACT 2002+	ReCiPe	TRACI
Midpoint/problem-oriented	x	x					x
Endpoint/damage-oriented			x	x			
Both					x	x	

Environmental Sciences 2002 (in Dutch: *Centrum voor Milieuwetenschappen Leiden* (CML 2002)) (Guinée et al., 2002), Environmental Design of Industrial Products 2003 (EDIP 2003) (Hauschild and Potting, 2005; Potting and Hauschild, 2005), Impact Assessment (IMPACT 2002+) (Jolliet et al., 2003), Dutch National Institute for Public Health and the Environment (in Dutch: *Rijksinstituut voor Volksgezondheid en Milieu* (RIVM)), Radboud University, CML and PRé Consultants impact method (ReCiPe) (Goedkoop et al., 2009), and Tool for the Reduction and Assessment of Chemical and Other Environmental Impacts (TRACI) (Bare et al., 2003). Some employ all aspects of both midpoints and endpoints in LCIA.

EI'99 was developed under the Dutch National Reuse of Waste Research (in Dutch: *NOH*) Programme by PRé Consultants as a new version of EI'95. The new version is based on the damage approach and sorts inventory results into three damage categories: human health, ecosystem quality, and resources (Bovea and Gallardo, 2006). The characterization factors are acidification, carcinogenic substances, climate change, eco-toxicity, eutrophication, ozone depletion substances, radiation, summer smog, and winter smog. These factors are consistent with those employed in the CML method, although specific characterizations for heavy metals, winter smog, and pesticides have been added in EI'99 for toxicity assessment. Normalization is based on the total inventory of mass and energy used across western Europe by one person per year (population of 495 million assumed) (Bovea and Gallardo, 2006). Mass may imply any commodity or waste flow. Three weighting perspectives can be applied: individualist (gives higher weight to human health), egalitarian (gives higher weight to ecosystem quality), and hierarchist (equal weight distribution) (Bovea and Gallardo, 2006), each of which alters the LCIA results.

EDIP 2003, an evolution of the EDIP 97 method, includes spatially differentiated characterization modeling (Frischknecht et al., 2007; Hischier et al., 2010). Characterization impact categories included in EDIP 2003 are global warming, ozone depletion, acidification, terrestrial eutrophication, aquatic eutrophication (nitrogen and phosphorous based), ozone formation (for human and vegetation impact), human toxicity (through air, water, and soil exposure), eco-toxicity (water acute, water chronic, and soil chronic), hazardous waste, slags/ashes, bulk waste, radioactive waste, and resources. Normalization is based on equivalent relative to 1990 on a capita basis, whereas weighting is based on the distance-to-target approach relative to the target emissions in 1990 (Bovea and Gallardo, 2006).

CML 2002 is a problem-oriented method that uses category indicators at the midpoint level, including depletion of abiotic resources, global warming (with different time frames), ozone layer depletion (different time frames), stratospheric ozone depletion, human toxicity (different time frames), freshwater aquatic ecotoxicity, marine aquatic ecotoxicity, marine sediment ecotoxicity, terrestrial ecotoxicity, photo-oxidant formation, acidification, eutrophication, and malodorous air, among others. Previous versions of CML existed like CML 2000 and CML 2001. The normalization factors are calculated from total substance emissions and characterization factors per substance, and hence follow the substance-level updates. Normalization factors are available from the Netherlands 1997, Western Europe 1995, the European Union (EU) 1995, and the World 1990. No grouping or weighting procedure is included in this methodology.

EPS method, developed in 1990–1991 and updated in 2002 as a conceptual tool for LCA (Ryding and Steen, 1991; Frischknecht, 2007), describes environmental impacts on specific subjects, including biodiversity, production, human health, resources, and aesthetic values (Hischier et al., 2010). With a damage-oriented model, normalization and weighting are evaluated through valuation. The willingness to pay to avoid changes is the factor applied, and the indicator unit is environmental load unit.

IMPACT 2002+ is an impact assessment methodology originally developed by the Swiss Federal Institute of Technology (Humbert et al., 2012). The present version is a feasible approach combining midpoint and damage methods, sorting all types of LCI results (elementary flows or other interventions) via 14 midpoint categories into four damage categories: human health, ecosystem quality, climate change, and resources (Jolliet et al., 2003). Midpoint categories include human toxicity, respiratory effects, ionizing radiation, ozone layer depletion, photochemical oxidation, aquatic ecotoxicity, terrestrial ecotoxicity, aquatic acidification, aquatic eutrophication, terrestrial acidification/nutrification, land occupation, global warming, nonrenewable energy, and mineral extraction (Jolliet et al., 2003; Humbert et al., 2012). This methodology takes advantage of midpoint-based indicators such as CML 2000 (Guinée et al., 2002), and damage-based methodologies such as EI'99 (Goedkoop and Spriensma, 2000). Normalization is conducted based on a per capita basis in Europe (Humbert et al., 2012). The weighting procedure can be self-determined, or a default weighting factor can be applied, unless other social weighting values are available (Humbert et al., 2012).

ReCiPe 2008 is a fusion of backgrounds models CML 2001 and EI'99 methodologies, combining the midpoint indicators from CML 2001 and the endpoint indicators from EI'99 (Goedkoop et al., 2009). The user can opt for 18 unweighted midpoint indicators. If calculating endpoints, not only the three damage categories (human health, ecosystems, and resources), but also contributions of various midpoint indicators (from respective damage categories) as well as the overall single score can be applied (Hischier et al., 2010). The 18 midpoint indicators have low uncertainty but are difficult to interpret, however, the three endpoint indicators are much easier to interpret but have higher uncertainty.

TRACI is a midpoint-oriented LCIA method, developed specifically by users in the United States, includes impact categories of ozone depletion, global warming, acidification, eutrophication, tropospheric ozone formation, eco-toxicity, human health air pollutants effects, human health cancer effects, human health non-cancer effects, fossil fuel depletion, and land-use effects (Bare et al., 2003). According to Frischknecht et al. (2007) and Hischier et al. (2010), however, the TRACI methodology does not take into account impact categories of resource consumption; also, normalization and weighting are not included, which might lead to misinterpretation and misuse in real-world applications.

10.2.4 Interpretation

According to ISO (2006b), life cycle interpretation is the final phase of the LCA procedure, in which the results of an LCI or LCIA, or both, are summarized for conclusions, recommendations, and decision-making in accordance with the goal and scope definition. As the final step, data from previous LCA phases are holistically analyzed. Because an LCA may be a looping and iterative procedure, the discussion conducted during interpretation phase can determine valuable choices such as allocation rules, system boundaries, and options identified during the goal and scope definition phase. Feedback from the discussion might also affect the data collected during the LCI phase, the environmental impact categories and models chosen during the LCIA phase, and findings based on data quality during the interpretation phase of LCA (ISO, 2006b). Interpretation must also consider the role and responsibilities of the all parties identified in the goal and scope definition phase in concerning the application (ISO, 2006b). Interpretation of the results from a concurrent critical review process, if conducted, are also helpful (ISO, 2006b).

Interpretation emphasizes the identification of significant issues and the evaluation of key results. The issues to be identified are related to the goal and scope defined in the first stage and can focus on life cycle stages (e.g., collection, sorting, treatment, recycling, and landfilling) in an SWM system in a carbon-constrained and resource-limited environment; groups of processes related to energy transport and supply in a municipal utility park (see Chapter 9); levels of different management policies on power grid mix; or even the individual unit process with marginal impacts. The most common procedures used to identify significant issues are contribution analysis, dominance analysis, and anomaly assessment.

Contribution analysis (also called gravity analysis or weak point analysis) determines which issue is most responsible for a specific feature in the analysis (environmental impact, inputs, and outputs) using statistical tools or other techniques, such as quantitative or qualitative ranking and remarkable or significant contributions (USEPA, 2006). Contribution analysis results are typically interpreted through stacked columns and pie charts, such as those in Blengini (2008) and Battisti and Corrado (2005).

Dominance analysis determines which issue prevails related to the others. According to ISO 14044 (ISO, 2006b), the difference between contribution and dominance analyses is the application of statistical tools or other techniques such as quantitative or qualitative ranking (e.g., ranking A-B-C). Before ISO 14044 (ISO, 2006b), Baumann and Tillman (2004) proposed other methods for differentiation devoted to studying life cycle stages and groups of processes, as opposed to contribution analysis, which is used almost exclusively for environmental loads contributing to the environmental impact.

Anomaly assessment is used to observe results outside the usual expectations, such as emissions that are too high or too low. This comparative assessment can serve as a control for LCI/LCIA results, so that a problem could be addressed by reexamining the three LCA phases. Unjustified anomalies must be reported in the assessment.

A streamlined LCA simplifies the analysis by limiting the scope of the LCA. In real-world applications, a streamlined LCA is usually conducted through the zero burden assumption. For example, the conventional LCA for SWM considers waste as a certain (zero burden assumption), and the environmental performance of SWM systems is evaluated by omitting all environmental emissions upstream from entering into SWM system. As a consequence, conventional LCA studies for SWM do not include the waste prevention effects. In reality, waste prevention and product reuse should be included in a total assessment of the waste hierarchy using LCA; therefore, a methodological procedure consistent with ISO 14044 (ISO, 2006b) to support various SWM scenarios incorporates both treatment as well as prevention through reduced consumption, dematerialization, and product reuse (Cleary, 2010). These treatment scenarios for different quantities of waste versus product reuse are not functionally equivalent to one another, however, and they also differ in the amount of product services supplied to the population during the waste prevention stage. The cutoff criterion in the context of zero burden assumption must be justified. Another LCA limitation often present in SWM evaluation is missing data, which should be documented when possible. The structural approaches for interpretation are dependent on the type of LCA study (Table 10.10).

A key element of the interpretation phase is evaluating the robustness of LCA results to establish confidence. Completeness, sensitivity, consistency, and any other data validation that may be required must be crosschecked according to the goal and

TABLE 10.10 Interpretation to be conducted depending on the type of LCA study

Situation	Interpretation	Evaluation options
Micro-level decision support—inside waste management company and decision board	Identify significant issues for improvements that must be made in the SWM system by determining which step is responsible for additional environmental impacts and should be reduced. Both quantitative and qualitative approaches can be used	Uncertainty and/or sensitivity analysis
Macro-level decision support—several stakeholders involved, non-technical stakeholders	Identify significant issues for the improvements	Levels of management policy
Accounting/monitoring— inside waste management company	It is required in the first time of an LCA to achieve some quantifiable results	Robustness of results is mandatory

scope definition of the study (ISO, 2006b). According to ISO 14044 (ISO, 2006b), a completeness check verifies whether information from all LCA phases is sufficient to draw conclusions; a consistency check verifies that the assumptions, methods, and data are regularly applied all over the assessment; a sensitivity check verifies that information acquired from sensitivity analysis is important for reaching conclusions and offering recommendations. A list of categories is proposed by both ISO 14044 (ISO, 2006b) as well as the US Environmental Protection Agency (USEPA, 2006).

A sensitivity analysis can be conducted by comparing results obtained using different assumptions, methods, or data (ISO, 2006b). Sensitivity can be described as percentage of change or as absolute deviation of the results, being capable of identifying significant changes (e.g. >10%), where assumptions or data are verified by defining a range (e.g., ±25%) and determining the influence on the results from the changes (ISO, 2006b). Sensitivity analysis is valuable for allocation rules, cutoff criteria, boundary setting and system definition, data assumptions, impact category selection, assignment of inventory results, calculation of category indicator results, normalized data, weighted data, weighting method, and data quality (ISO, 2006b).

According to Baumann and Tillman (2004), a Monte Carlo simulation can be used to conduct a sensitivity analysis. Random numbers are multiplied by standard errors (geometric standard deviation) for each input data. Several hundred simulations are then made to determine variations in the final result. Other methodologies for checking sensitivity are uncertainty calculations that assess how imprecise data, with associated error, can influence the LCA results. Most data used to conduct LCA are average data, with an associated interval that could influence the final results, especially in intercomparisons. Uncertainty can be defined as the inconsistency between a measured or calculated quantity and the true value of that quantity (Finnveden et al., 2009). Uncertainty can arise from, for example, data variability, misspecified data, data errors, incomplete data, round-off errors, incorrect relations, and inaccuracy. Uncertainty can be assessed using various statistical tools, such as (Finnveden et al., 2009):

- Monte Carlo simulations (previously mentioned in sensitivity analysis), bootstrapping, and other sampling approaches;
- parameter variation and scenario analysis;
- classical statistical theory based on probability distributions and tests of hypothesis;
- analytical methods based on first-order error propagation;
- use of less conventional methods such as non-parametric statistics, Bayesian analysis, and fuzzy set theory; and
- use of qualitative uncertainty methods such as those based on data quality indicators.

Data can be validated by justifying the specific data chosen for the model. At the end of the interpretation phase, conclusions, limitations, and recommendations are obtained through a logical procedure defined in the ISO 14044 (ISO, 2006b) to:

- identify the important issues;
- assess the methodology and results for completeness, sensitivity, and consistency;
- draw introductory conclusions and check that are consistent with the requirements of the goal and scope of the study, in detail data quality requirements, predefined presumptions and values, limitations in methodology and study, and application-oriented requirements; and
- report conclusion if the conclusion is consistent; otherwise return to previous steps as appropriate.

10.3 LCA WASTE MANAGEMENT SOFTWARE

Software applications to conduct LCA were developed to support and facilitate procedures and calculations that avoid time-consuming steps. An effective LCA software package must be user-friendly, stable, reliable, flexible, and fast when calculating, as well as have strong documentation support. The producer should have a support and maintenance service to help users develop models based on their needs.

Researchers have developed specific LCA tools to evaluate environmental impacts resulting from SWM systems, including but not limited to IWM-1 and IWM-2 (White et al., 1995; McDougall et al., 2001), MIMES/Waste (Sundberg, 1993, 1995), WASTED (Diaz and Warith, 2006), WISARD/WRATE (Ecobilan, 2004), and EASE-WASTE (Christensen et al., 2007). Others have combined LCA tools with economic aspects, such as LCA-IWM (den Boer et al., 2007) and MSW-DST (Thorneloe et al., 2007; Weitz et al., 1999), or even integrated with modeling systems, such as ORWARE (Dalemo et al., 1997; Björklund et al., 1999). These LCA tools are designed for various levels of SWM decision makers and staff.

When applied to SWM, however, these specific LCA software packages do not allow practitioners to implement meaningful changes to reflect system specificities, such as treatment technologies, waste composition, and existing waste operations, which alter the LCA results. The general-purpose LCA software packages can better assess the SWM system than those dedicated to SWM only; this is due to that LCA packages applied to SWM have professional databases, thereby reducing lack of information and allowing LCA to carry out the additional assessment that would otherwise not be possible.

Several general-purpose LCA software packages are available on the market (Table 10.11), organized based on information from a European website (http://lca.jrc.ec.europa.eu) and an USEPA website (www.epa.gov/nrmrl/std/lca/resources.html). These software packages, such as GaBi, SimaPro, and Umberto, include other functions such as life cycle sustainability assessment, life cycle engineering, product stewardship, supply chain management, social LCA, substance/material flow analysis, life cycle cost accounting, strategic risk management, and energy efficiency studies. Other general-purpose LCI software packages, such as GEMIS, include possible extensions for LCIA. The software packages selected for further study in this

TABLE 10.11 Generic features of LCA software packages

Name of software	Developer	Type of user	Result level	Demo available
AIST-LCA (before was JEMAI LCA Pro)	National Institute of Advanced Industrial Science and Technology (AIST) (Japan)	Environmental engineer LCA expert	Inventory Characterization and weighting Evaluation of completeness, sensitivity and consistency, data quality, data type, and age of data	No
eVerdEE	Italian National Agency for New Technology, Energy and the Environment (Italy)	Design engineer Environmental engineer LCA experts	Inventory Characterization Data quality estimation	Software free of cost
GaBi (and related applications)	PE International (Germany)	Design engineer Environmental engineer LCA experts	Inventory Characterization and weighting Data analysis Uncertainty through Monte Carlo analysis/simulation	Yes
LCA Evaluator	GreenDeltaTC (Germany)	Environmental engineer LCA expert	Inventory Characterization. Weighting and normalization are done manually Uncertainty analysis through Monte Carlo simulation	Only by contact

Software	Company (Country)	User	Functions	
REGIS	SINUM (Switzerland)	Environmental engineer LCA expert	Inventory Characterization and weighting	Yes
SimaPro	PRé Consultants (Netherlands)	Design engineer Environmental engineer LCA expert	Inventory Characterization and weighting Uncertainty analysis through Monte Carlo simulation, data quality	Yes
Tools for Environmental Analysis and Management (TEAM™)	Ecobilan—PricewaterhouseCoopers (France)	Environmental engineer LCA expert	Inventory Characterization and weighting Sensitivity analysis Uncertainty with Min-Max and Monte Carlo simulation	Yes
Umberto	IFU (Germany)	Design engineer Environmental engineer LCA expert	Inventory Characterization and weighting Data quality Uncertainty analysis through Monte Carlo simulation	Yes
The Boustead model	Boustead Consulting Ltd. (UK)	Design engineer Environmental engineer LCA expert	Inventory Characterization, automatic GWP calculation Sensitivity analysis	Yes

357

chapter were Umberto, GaBi, and SimaPro, because they are well established, available in English, provide demo versions, and are well supported and maintained for different LCI calculation models. For example, the design of GaBi is based on linear equation systems, Umberto on Petri nets, and SimaPro on inverse matrix.

10.3.1 Umberto Software

Umberto, first presented at Hannover, Germany, in 1994 (Brunner and Rechberger, 2004), was developed by the Institute for Energy and Environmental Research (IFEU) in cooperation with the Institute for Environmental Informatics Hamburg Ltd (IFU) and has been in use for more than 15 years (JRC, 2013). Several versions are available, including Umberto NXT LCA, Umberto for Efficiency, Umberto for Eco-Efficiency, and Umberto for Education (for more information or a free web demo, see http://www.umberto.de/en).

An additional version, Umberto 5.5, allows visualization of both material and energy flow systems. The strength of this software package is its versatility. It can adapt to meet users' specific needs, including complex processes with the level of detail needed for each decision (JRC, 2013). The main task of Umberto is to model and optimize production processes based on a hierarchical network model approach (JRC, 2013). The user can identify different levels of detail in the SWM, beginning with the overarching level and progressing stepwise to a specific waste treatment process (Figure 10.8).

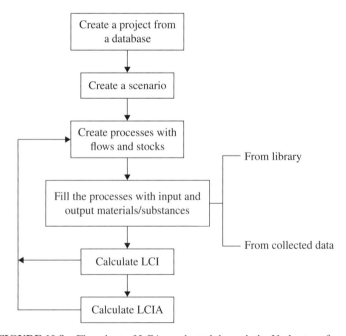

FIGURE 10.8 Flowchart of LCA conducted through the Umberto software

Umberto can accommodate several databases, such as Ecoinvent or those built by the user. The LCI is constructed on material flow analysis principles, where flows, stocks, and processes are drawn, quantifying the amounts obtained and discharged from the built environment into the natural system. Umberto 5.5 can construct SWM systems with the concept of avoided burden and substitution based on comparisons of LCI and present the LCIA results by either aggregated or disaggregated processes and substances flexibly. Other functionalities of Umberto 5.5 are making Sankey diagrams, creating systems for different time periods, enabling updated flow data, and modeling time-related questions. Individual allocation rules can be defined for transitions representing coupled projects. Additionally, Umberto 5.5 is capable of modeling SWM recycling loops and closed-loop dynamic models.

In the complex Umberto structure, the SWM system can be modeled simply as a network with few components; the detail degree can be assorted concerning data quantity and data quality and the goal of the study (JRC, 2013). A process is defined applying the coefficients between input and output flows, which are linearly dependent but can also contain nonlinear functions or parameters, and formulation of dynamic and time-dependent process specifications (JRC, 2013). For each value, the quality of data may supply information about the data source (JRC, 2013). The calculations are sequential and local, independent of the flow direction; in other words, the calculation step does not rely on the material flow direction, but is only related to known information (JRC, 2013).

10.3.2 SimaPro Software

SimaPro, a Dutch LCA software package developed by PRé Consultants, is a flexible application that allows parameterized modeling and interactive results analysis with the aid of a large database (Figure 10.9). SimaPro can rapidly evaluate an SWM system because there is no need to construct the network; the user need only to register and choose the process of each life cycle stage. Evaluating different impact assessments with this application is intuitive and simple. A results screen can be viewed through a user-friendly graphic support to identify emissions and consumptions related to the process. Monte Carlo simulation can be conducted for various parameter values.

The allocation of a process, when presents multiple outputs, can be extended to avoided products via system boundary expansion or can have an allocation percentage. Inputs can be associated with either the natural system or the built environment (JRC, 2013), expressed in terms of physical units and financial terms, allowing us to make hybrid data models that combine input and output with traditional processes. Note that calculation routines use matrix inversion methods, allowing loops to be modeled. In each process record, emissions are specified in terms of the reception ground (air, water, and soil) and waste streams linked to a waste treatment facility can also be specified as gas, liquid, or solids. Emissions are defined using the sub-compartments via Ecoinvent. Parameters are defined directly by the user, or as the result of an expression, can be linear and nonlinear, or conditional expressions can be defined

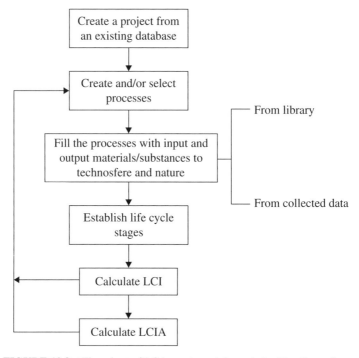

FIGURE 10.9   Flowchart of LCA conducted through the SimaPro software

(JRC, 2013). In some versions, an amount in a study domain can be directly linked to an external data source such as Excel or an SQL dataset, allowing data storage outside SimaPro with versatile documentation fields (JRC, 2013).

Impact assessment methods are defined in SimaPro as a series of tables associated with impact categories and normalization and weighting procedures, existing more than 10 different impact assessment methods, which can be edited, extended, and used (JRC, 2013). Once an impact assessment method is selected, all impact categories can be organized and displayed appropriately as a profile in support of the designated LCA (JRC, 2013).

10.3.3 GaBi Software

The GaBi 5 software system was developed by the Institute for Polymer Testing and Polymer Science at the University of Stuttgart in the cooperation with PE Europe GmbH in Leinfelden-Echterdingen, Germany (Brunner and Rechberger, 2004). GaBi, which stands for *ganzheitliche bilanzierung* or, in English, life cycle engineering, is an LCA tool with additional functionality for material flow analysis, similar to Umberto (Figure 10.10).

GaBi has one of the most globally comprehensive, consistent, high quality database systems (JRC, 2013) that can create models based on physical process chains and

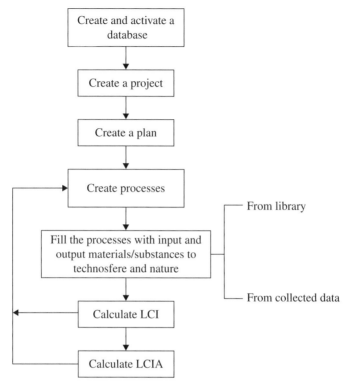

FIGURE 10.10   Flowchart of LCA conducted through the GaBi software

integrated parameter functionality, resulting adaptable systems, incorporating circularity effects. Monte Carlo simulation is also included in the software. GaBi 4.2 easily models realistic process chains that describe a specific production technology or service with the input and output flows, and links them with the product system (JRC, 2013). GaBi also has a user-friendly graphic interface providing a full overview and insight of complex product systems, and hence supports an efficient workflow (JRC, 2013).

10.4   PUTTING LCA INTO PRACTICE

A well-planned LCA study estimates the workload, outlines the operational procedures, and understands the structure of the database and the LCIA to be applied (PRé Consultants, 2004). This section describes a comparative LCA of 18 SWM scenarios in the Setúbal peninsula, Portugal (Figure 10.11a). The analysis compares the sustainability performance of SWM alternatives in the Setúbal peninsula and identifies key elements of each option based on the prescribed metrics of impact categories. Umberto 5.5 software package was used to support this LCA.

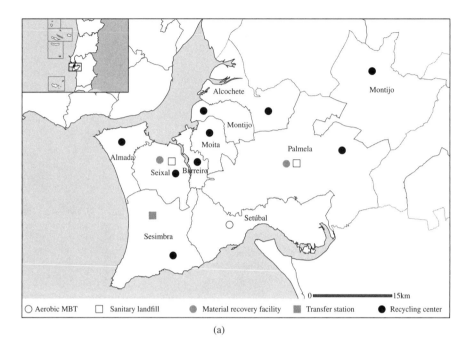

(a)

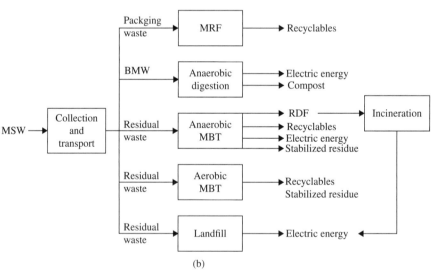

(b)

FIGURE 10.11 (a) The geographical location of the Setúbal Peninsula SWM system. *Source*: From Pires et al. (2011). (b) Schematic representation of the MSW management system in the Setúbal Peninsula study. *Source*: From Pires et al. (2011). (c) The alternatives of MSW management system in the Setúbal Peninsula case. PW, packaging waste; RW, residual waste; BMW, biodegradable municipal waste (*continued*)

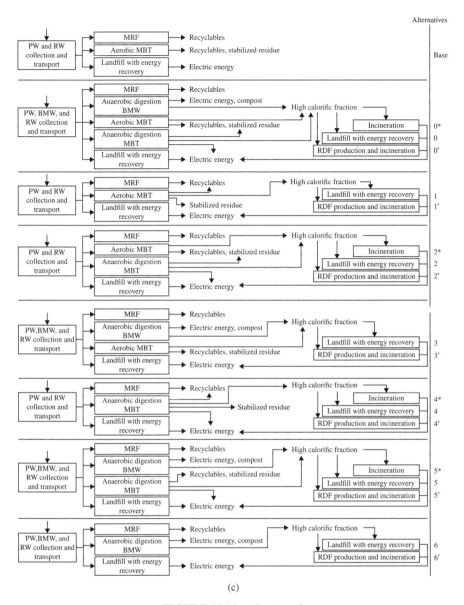

(c)

FIGURE 10.11 *(Continued)*

10.4.1 Goal and Scope Definition

Amarsul, the management company of the SWM system in the Setúbal peninsula (Figure 10.11a), promotes the separation of paper/cardboard, glass, and light packaging (plastics, metals, and composites) waste. Each type of waste is collected separately in three specific containers and then sent directly to the material recovery facility (MRF) for recycling. A small part of the remaining waste fractions, normally destined for final disposal at landfills, was processed by an aerobic mechanical biological treatment (MBT) plant. The area covered by Amarsul has around 700 thousands inhabitants in 1500 km^2, being divided in nine municipalities (Figure 10.11a) (Amarsul, 2009). The SWM system is composed of nine recycling centers, two MRF, two landfills, one transfer station, and one MBT (Pires et al., 2011; Chang et al., 2013).

Some recent changes to the Setúbal SWM system have occurred to comply with the Packaging and Packaging Waste Directive (European Parliament and Council, 2004), Landfill Directive (Council of the European Union, 1999) and National Plan for SWM. To respond to all the goals in these strategic and mandatory documents being studied, the construction of an anaerobic digestion (AD) MBT unit with a mechanical treatment to separate recyclables and high calorific material to produce refuse-derived fuel (RDF) to generate electricity is under planning (Pires et al., 2011). This unit is expected to work with two separate lines, one for biodegradable municipal solid wastes (BMWs) and the other for residual waste streams. Both MRF plants, which are currently fitted with manual sorting, are expected to be later fitted with two automatic sorting units.

Our aim of applying LCA into this waste management system is to analyze and compare different waste management alternatives subject to the targets associated with waste management directives and national plan in such a way that could promote sustainable management. LCA applied to this purpose can contribute with the environmental assessment of those waste management alternatives helping on the decision-making process of the waste management alternative. An LCA with multiple products, as in this Setúbal study, must have a methodological framework. The functional unit considered at the LCA was 421,726 tonnes, the amount of MSW received and treated in 2008. This study assumes "zero burden assumption." According to the ISO 14044 (ISO, 2006b), the system boundary should be geared toward expanding the product system to include the additional functions related to the coproducts. In this LCA, the material recycling, energy recovery, and fertilizers application (i.e., stabilized residue waste from aerobic MBT and anaerobic MBT) of MSW were included in the LCA as coproducts, which collectively resulted in an expansion of the system boundary (Chang et al., 2013). Besides, the emissions from the referred operations were included as baseline comparative information with emissions from the competing products and energy recovery potential from the alternative operations (Pires et al., 2011). In this context, the system can be expanded to include additional burdens of coproduct processing and the avoided burdens of any avoided processes (i.e., substitution or avoided burden method) (Tillman et al., 1994; Guinée et al., 2002; Thomassen et al., 2008; Finnvedden et al., 2009; Pires et al., 2011).

Other boundaries of the system are the geographic boundaries of Portugal and Spain due to the exportation of recyclables. Capital goods are excluded from the study, but upstream systems are included. As an attributional LCA, the electricity considered is average Portuguese mixed consumption which is composed of 28.1% coal, 8.37% fuel oil, 30.5% natural gas, 0.55% biomass, 25% hydro, 7% waste, 0.33% geothermic, and 0.15% wind (Pires et al., 2011).

According to Pires et al. (2011), average MSW composition is 31.69% putrescibles, 14.13% paper and cardboard, 11.35% plastics, 5.83% glass, 4.14% composites,

1.82% metals, 2.07% wood, 11.72% textiles, 15.33% fine particles, and 1.92% other (EGF, personal communication, 2009). Waste composition from different processes, such as BMW, another alternative studied, was obtained from other characterizations of separate collection.

In accordance with the scope of the study, an LCI was prepared for the specified waste management activities (Figure 10.11b). A short description of the data and assumptions considered for prescribed scenarios is provided for each operational unit analyzed in the SWM system (Pires et al., 2011). Some information applied to our systems analysis was provided by the *Empresa Geral do Fomento* (EGF), co-owner of the SWM system responsible for the managing this MSW system, and the Portuguese Environment Agency (in Portuguese: *Agência Portuguesa do Ambiente* (APA)). The remaining information was supplied by the Umberto software library and selected data sources, such as machinery specifications provided by the vendors (Pires et al., 2011). The MSW management system in this case study has seven alternatives, indexed from zero to six with or without an RDF production option (Figure 10.11c), described in the following sections.

10.4.2 Life Cycle Inventory

The SWM processes analyzed for the Setúbal peninsula include collection and transportation of residual waste and recyclables, waste treatment, waste transport from waste treatment facilities to the final destination, energy-from-waste or waste-to-energy, and landfilling. Several final destinations for recyclables are located in Spain rather than Portugal, specifically for cases that handle composite packaging and ferrous and nonferrous metals packaging materials. This system has a current situation (base scenario) plus 18 management alternatives for assessment (Table 10.12), which include waste collection and separate recycling of the three packaging materials through bring system that handle 12.4% of the current MSW in the study area. This

TABLE 10.12 **Distribution of waste streams associated with each alternative in the SWM system**

Fraction option (%)	Alternatives							
	0/0[b]/0[a]	1/1[a]	2/2[b]/2[a]	3/3[a]	4/4[b]/4[a]	5/5[b]/5[a]	6/6[a]	Base
MRF	12.4	12.4	12.4	12.4	12.4	12.4	12.4	4.8
Anaerobic digestion BMW	5.4	0	0	13.3	0	7.5	28.7	0
Anaerobic digestion MBT	28.2	0	33.9	0	49.6	38.9	0	0
Aerobic MBT	13.2	49.7	15.8	32.6	0	0	0	13.8
Landfill with ER	40.8	37.9	37.9	41.7	38.0	41.2	58.9	81.4

Source: Pires et al. (2011).

[a] Alternatives considering RDF production plus incineration of high-calorific fraction.

[b] Alternatives not considering RDF production but considering incineration of high-calorific fraction from MBT.

MRF system is responsible for compliance with the prescribed target in the Packaging Waste Directive.

According to Finnveden (1999), using identical amounts of treated waste in different scenarios simplifies the comparative analysis by neglecting the production and use of the materials. Alternative 0 refers to the predicted change that will take place in the Setúbal Peninsula waste management system. The remaining alternatives were designed to examine special options for complying with the Landfill Directive. For example, alternative 1 includes aerobic MBT; alternative 4 includes AD MBT; alternative 6 includes a BMW AD line. In general, alternatives 0, 3, and 5 are options for differing intermediate processing. Separation of high calorific fractions of waste for energy recovery was considered through the production of RDF and the direct burning of high calorific fractions in a municipal incinerator.

To ensure proper implementation of the avoided burden through successful waste recycling and reuse, the coproducts in the expanded system boundary should have the same function as the raw products. The substitution ratios are then applied considering closed-loop and open-loop procedures (Section 10.1.2) (Table 10.13) (Rigamonti et al., 2009a). Specifically, 15% of the electricity consumed in Portugal for the year 2008 was purchased from Spain, so a ratio of 85/15 can be used to conduct the LCA. HDPE and LDPE (here defined generically as polyethylene (PE)), expandable

TABLE 10.13   Products obtained from the SWM system and the assumptions for LCA

Products obtained	Substitutes assumed	Substitution ratio assumed
Cardboard from recovered paper and cardboard	Cardboard from virgin pulp	1:0.833
Glass produced from recovered glass processed	Glass from virgin materials	1:1
Tubes from PE recycled	Tubes from virgin PE	1:1
Multilayer packaging materials from recycled PET	Multilayer packaging from virgin PET	1:0.625
Recycled EPS lightweight soil	Virgin EPS lightweight soil	1:1
Paper from composite packaging materials recycled	Paper from virgin pulp	1:0.625
Outside furniture blocks from recycled mixed plastics	Outside furniture blocks from wood	1:1
Ferrous metals from recycled ferrous metals	Pig iron	1:1
Aluminum ingot from recycled aluminum metals	Aluminum ingot from virgin aluminum	1:1
Compost	N, P, K, Ca, and Mg fertilizers	1:1 (based on nutrient content)
Electricity	Electricity mix consumed in Portugal	1:1

Source: From Pires et al. (2011).

TABLE 10.14   Data requirement for collection and transport waste life cycle stage

Waste collection and transport	MSW	BMW	Packaging waste	Paper/cardboard waste	Glass waste
Distance (km)	1,699,646	121,355	641,334	446,296	179,672
Diesel fuel consumption ($L \cdot 100 \text{ km}^{-1}$)	49.6	49.6	65.0	94.6	78.3
References	C. Aleixo, personal communication, 2010; N. Canta, personal communication, 2010; J. Didelet, personal communication, 2010; E. Gomes, and P. V. N. Rodrigues, personal communication, 2010; P. Pinto, personal communication, 2010; and N. M. Valério, personal communication, 2010		EGF, personal communication, 2009, and Gomes (2009)		

Source: From Pires et al. (2011).

polystyrene (EPS), and plastic wood are specific cases having a 1:1 substitution ratio because degradation of the material is not considered.

Waste Collection and Transport   Municipal waste collection in the Amarsul area is routinely performed by the municipalities or by hiring private collection companies. Solid waste is temporary discarded into roadside containers (bins) and periodically removed by collection vehicles (Table 10.14). Transportation between operational units inside the SWM facilities was not considered. In the case of BMW collection the approach used in this LCA assigned the same shipping distance and diesel fuel consumption to the municipalities that would treat BMW in a future AD MBT unit in parallel.

According to Pires et al. (2011), "BMW composition in the Setúbal Peninsula system is 70% food waste, 15% green waste, 5% plastics, 1.9% glass, 0.25% ferrous metals, 0.15% nonferrous, 7.05% fine particles, and 0.65% other, adapted from a BMW characterization program in the Lisbon metropolitan area (Vaz, 2009). Packaging waste is composed of 2.45% putrescibles, 10.58% paper and cardboard, 60.8% plastics, 3.98% glass, 12.71% composites, 4.98% ferrous metals, 0.21% nonferrous metals, 0.02% wood, 1.01% textiles, 1% fine particles, and 0.53% other (EGF, personal communication, 2009)." Other default characteristics were collected from literature values reported by Rotter (2004), Dehoust et al. (2002), and Fricke et al. (2002). Emissions resulting from waste collection and shipping were modeled based on Borken et al. (1999), Knörr et al. (1997), Schmidt et al. (1998), and EEA (2009).

Sorting Plants According to Pires et al. (2011), the packaging waste materials to be sorted mechanically are HDPE, LDPE, PET, EPS, mixed plastics, glass, composites, and ferrous and nonferrous materials. Data derived were based on processing 1 tonne of packaging waste in this recycling operation (Table 10.15); however, manual sorting will still be employed when handling paper/cardboard waste streams.

Anaerobic Digestion A combined MBT unit is expected to be adopted, with one line to process MSW and the other line to process BMW. A small mechanical treatment processor in the BMW line will be installed to remove inorganic matter such as metals and plastic waste. Organic waste portions delivered to the BMW unit will be decomposed in a thermophilic, dry AD process, resulting in a digestate that will be sent to a post-composting unit to decompose the residual organic waste into fresh compost, to become a mature compost to be used in agriculture (Pires et al., 2011) (Table 10.15). Emissions during AD mainly result from biogas burning to produce electricity and heat (Soyez et al., 2000; Vogt et al., 2002), wastewater and gas treatment (Loll, 1994, 1998; Martinho et al., 2008; Yamada and Jung, 2007), and biofilter (den Boer et al., 2005).

Anaerobic Digestion MBT The future AD MBT plant will be composed of a mechanical sorting component to remove recyclables and combustible fraction for RDF production, and of an anaerobic digestion component to treat biologically degradable remaining fractions. The mechanical sorting process usually includes flail mills, trommels, magnetic separators, eddy current separators, and ballistic separators (Pires et al., 2011). Each operational unit has material consumptions and requirements needed to simulate the process (Table 10.15). The AD is identical to the one designed to treat BMW, including post-composting to obtain mature compost. All data needed to model the inventory are listed in Table 10.15. With biogas used to generate electricity, the engineering design used to model the emissions from AD MBT was considered the same as those applied to AD of BMW.

Aerobic MBT An aerobic MBT is composed of a mechanical sorting processing unit and a biological treatment processing unit which is an aerobic biological process. The mechanical processing unit, which can also include manual sorting, is designed to remove the waste stream unsuited for the biological treatment unit, mainly ferrous and nonferrous metals as well as some glass and plastics (Pires et al., 2011).

The requirements applied to decompose the organic fraction of waste in an aerobic treatment process include several biological processes (Table 10.15). The main output from an MBT process is "stabilized residue," which must be landfilled or used as the daily cover materials in landfills (Pires et al., 2011). This MBT produces no wastewater, and contaminated air is treated by a biofilter. The engineering design applied to model this biofilter was considered the same as those applied for other similar biological treatment processes.

Landfill Sanitary landfill receives waste from different sources, like mixed MSW and residuals associated with several operational units in the MSW management

TABLE 10.15 Consumptions and requirements of operational units

Operational units	Operational requirements	Auxiliary materials (per tonne waste input in the operation)	References
Packaging MRF	Material recovery rate: 90%	Electricity (kWh): 20.92 Diesel (L): 2.01 Lube oil (L): 0.20 Steel (kg): 1.20	Rodrigo and Castells (2000); EGF, personal communication, 2009; and Rodrigues (2009)
Paper/cardboard MRF	Material recovery rate: 90%	Electricity (kWh): 5.35 Diesel (L): 0.64 Lube oil (L): 0.01 Steel (kg): 1.20	Rodrigo and Castells (2000) and EGF, personal communication, 2009
AD	*Mechanical step* Refuse: 2.8% Ferrous metals recovery rate for recycling: 99% *Biological process* Biogas production: 380 m³ per tonne organic waste *Post-composting* Decomposition rate: 30% *Maturation step* Rejects (%): 5	*Mechanical step* Electricity (kWh): 34.8 Diesel (L): 1.16 Lube oil (L): 0.12 *Biological process* Water (L): 279 *Post-composting* Electricity (kWh): 10 Structural material (%): 5 *Maturation step* Electricity (kWh): 10 Water (%): 20	EGF, personal communication, 2009 Vogt et al. (2002); APA (2009); and EGF, personal communication, 2009

(continued)

TABLE 10.15 *(Continued)*

Operational units	Operational requirements	Auxiliary materials (per tonne waste input in the operation)	References
AD MBT	Material recovery for recycling: mainly metals, 95% Material recovery for RDF (when applied): 98 of high calorific material *Biological process* Biogas production: 380 m³ per tonne organic waste *Post-composting* Decomposition rate: 50% *Maturation step* Rejects (%): 10	Electricity (kWh): 34.8 Diesel (L): 1.16 Lube oil (L): 0.12 *Biological process* Water (L): 279 *Post-composting* Electricity (kWh): 10 Structural material (%): 5 *Maturation step* Electricity (kWh): 10 Water (%): 20	EGF, personal communication, 2009 Vogt et al. (2002) and EGF, personal communication, 2009
Aerobic MBT	Material recovery for recycling glass: 1%, plastic: 7%, ferrous metals: 97%; nonferrous metals: 14% *Biological process* Decomposition rate: 65% *Maturation process* Decomposition rate: 20%	Electricity (kWh): 34.8 Diesel (L): 0.5 Lube oil (L): 0.12 Electricity (kWh): 10 Diesel (L): 0.12 Water: 2% for biological step; 20% for maturation Structural material: 8.2%	Wallmann and Fricke (2002) and EGF, personal communication, 2009 Fricke and Müller (1999), Vogt et al. (2002) and EGF, personal communication, 2009
Landfill	Annual precipitation (JNS): 1550 mm Leachate production during phase A (N24TI): 40% Leachate production during phase B (N25TI): 8% Duration phase A (PHAA): 10 years Duration phase B (PHAB): 20 years	Electricity (kWh): 0.002 Mechanical energy (kJ): 10.99 Heat energy (kJ): 1.6	Weber (1990), Eggels and van der Ven (1995), Rettenberger (1996), Rettenberger and Stegmann (1997) and BUWAL (1998)

Source: From Pires et al. (2011).

system. The emissions from landfills diffuse into air, soil, and water. Typical sanitary landfills have a collection system for leachate and a biogas collection. The existing Amarsul system is for the collection of biogas (i.e., methane gas) to produce electricity, being the inventory provided by the Umberto module, which is based on several sources of information (Rettenberger, 1996; Rettenberger and Stegmann, 1997; Weber, 1990; Eggels and van der Ven, 1995; BUWAL, 1998). The formula used to quantify the methane gas production was derived based on values from the literature and adapted as (Tabasaran and Rettenberger, 1987):

$$Ge = 1,868 \; Co(0.014 \; T + 0.28), \tag{10.1}$$

where Ge is the potential long-term methane gas production (m^3 per tonne waste); 1868 is the gas production rate resulting from decomposition per kilogram of organic waste (m^3 biogas $\cdot kg^{-1}$ C) (note that $(22.4 \; L \; biogas \cdot mol^{-1})/(12 \; g \; C \cdot mol^{-1}) = 1.868 \; L \; biogas \cdot g^{-1}$ C); and Co is the content of the organically degradable carbon in MSW (kg Co per tonne waste) (i.e., typical figures of the production rate of 170–220 kg per tonne). This way, Co calculation is based on the carbon content of biologically degradable organic waste. The temperature-dependent decomposition rate (in °C) is $(0.014 \; T + 0.28)$ (note that for household waste landfill T is between 30°C and 35°C).

Air emissions resulting from landfill can be released into the air through a direct source from burning biogas and diffuse source (from landfill itself). Diffused emissions are linked with the arrangement of the biogas collection system during landfill operation, phase A, and post-closure, phase B. Based on the Umberto module and described by Pires et al. (2011), "25% of the biogas collected was considered direct emissions throughout the operation and post-closure. During phase A, 30% of the biogas was considered released emissions, whereas during phase B, this number is potentially as high as 70%. The entire landfill life cycle is assumed to produce approximately 50% of biogas from phases A and B; hence, in phases A and B, $(75/100)(30/100)(50/100) = 11.25\%$ and $(75/100)(70/100)(50/100) = 26.25\%$ of diffused biogas, respectively. The amount of biogas collected can be estimated as $(75/100)(50/100)(1-70/100)$ for phase A and $(75/100)(50/100)(1-30/100)$ for phase B".

Landfill gas energy recovery is performed using a gas turbine, being emissions calculated by the den Boer et al.'s (2005) data. Landfill leachate production was quantified for the phases A and B over a planning horizon of 100 years. The leachate production level depends on the annual average precipitation as well as the water content inside landfills. According to Schwing (1999), in operation phase A, leachate production can be estimated between 10% and 50% of the total annual precipitation; after the closure phase B, leachate production can be as low as 5–10% of total annual precipitation. In the Umberto module, the default values are 40% and 8% for phases A and B, respectively (Rettenberger and Schneider, 1997), applied for every type of waste (Pires et al., 2011). Based on German landfills, MSW is assumed to have a residual water content of 15% by weight, of which 76% may be collected as leachate (Schwing, 1999), and leachate collection systems at landfills are assumed to collect

TABLE 10.16   Distances between MSW management system and final deposition sites for products

	Distances (km)	
Products transport	Pre-processors	Recyclers/Incineration[1]/Agriculture[2]
Ferrous metals	241.3	521.5
Nonferrous metals	259.3	592.2
PE	0	238.6
PET	0	210.7
EPS	0	293.0
Mixed plastics	0	524.0
Paper/cardboard	339.9	811.2
Composites	210.2	1116.5
Glass	233.0	60.5
RDF[1]	0	45.4
Compost[2]	0	73.7

Source: From Pires et al. (2011).
Note: Superscript 1 associated with RDF stands for incineration. Superscript 2 associated with compost stands for agriculture applications.

90% of the leachate produced. In regard to land use required to landfill MSW, 1 tonne per m^3 is assumed to be the density of MSW, which was used to determine the required land-use area based on the ratio of the volume of waste landfilled to the soil within a 20-meter of height area wide (Pires et al., 2011).

Products Shipping   All products resulting from each alternative must be transported to their final destination, being distances obtained using the Google maps tool (Google, 2010) (Table 10.16). Diesel consumption records collected from transportation companies based on $25 \, L \cdot 100 \, km^{-1}$.

Auxiliary Materials and Recyclables   All the sources of information are presented in Table 10.17. Auxiliary materials such as electricity, diesel production and burning, and lubricating oil consumption in MSW management systems were discussed by Frischknecht et al. (1996), GEMIS database (Oeko-Institut, 2001), IFU (2009), and EEA (2007, 2009).

Life Cycle Inventory Remarks   Many key parameters for waste collection and transport databases are vehicle features, transport distance, fuel type, waste density, driving practices, and vehicle maintenance, datasets that would require an audit of the nine Amarsul municipalities. The software itself performed the selective packaging waste collection; and information to characterize the system in terms of consumption $(L \cdot 100 \, km^{-1})$, distance traveled, and amounts collected for the most recent year was possible to obtain.

The MRF (or sorting plant) mechanically separates different packaging waste, similar to MBT units; however, some waste fractions such as paper/cardboard are

TABLE 10.17 Summary of LCI data sources for expanded systems and avoided products

Type of data	Sources of data
PET recycling, mixed plastics recycling, glass pre-processing, and glass recycling	Mata (1998), ProBas (2004), APA (2009), and Alves, personal communication (2010)
RDF production	Fricke et al. (2003)
RDF incineration	Schäfl (1995), Achernbosch and Richers (1997), UBA (1999), and Valorsul (2008)
Paper and cardboard pre-processing, composites packaging pre-processing	Rodrigo and Castells (2000)
Paper and cardboard recycling	ProBas (2004) and APA (2009)
PE recycling	Arena et al. (2003)
EPS recycling	Silva, personal communication (2010)
Composites recycling	Stora Enso (2008)
Ferrous metal pre-processing	Rodrigo and Castells (2000)
Ferrous recycling	ETH Zürich (2008)
Aluminum metal pre-processing	Rodrigo and Castells (2000)
Aluminum recycling	Boustead (2000)
Auxiliary material production	APME (1995), Patyk and Reinhart (1997), BUWAL (1998), Oeko-Institut (2001), Ecoinvent (2006), APA (2009), and Martinho and Pires (2009)
Avoided products, including fertilizers	IFEU (1994), BUWAL (1998), Mata (1998), ProBas (2004), and APA (2009)

Source: Pires et al. (2011).

still manually separated. Requirements for these processes are related to energy consumption, fuel consumption, emissions from maintenance components, and recovery rates, which may contribute to calculate possible benefits and costs during sorting process. Most of the information used to calculate LCIA is from literature and local SWM system.

MSW collected is destined for landfill and MBT through aerobic and anaerobic digestion treatment. The requirements for biological and thermal treatment or landfilling to support the LCA could affect how goal and scope are defined and how the LCI is conducted. Emissions related to CO_2 and CH_4 and other emissions discharged to air and water must be measured. Energy produced from biological treatment (from AD) or thermal treatment (from incineration) must also be correctly included once it is substituted for other types of energy sources, such as electricity or heat energy. Carbon storage obtained through compost production must also be modeled, in addition to nutrient supply: the application of the nutrient supply via compost production will avoid/substitute the use of chemical fertilizers. Most relevant data on auxiliary materials such as electricity, diesel production and burning, and lubricating oil are from literature.

Recycling, one of the most demanding waste operation processes such as paper, plastic, metal, and glass recycling plants, occurs inside or outside the SWM system, which has shown complexity to obtaining detailed information. Determining the type of recycling to be modeled (e.g., closed loop or open loop) also influences LCA implementation. Recyclability, changes in inherent technical properties, identification of substituted/avoided processes, and, in the case of a change-oriented LCA, the market availability must be obtained. Information must be gathered on transportation of sorted recyclables, including the location of the recycling plants, and on consumptions and emissions that result in recycled material.

10.4.3 Life Cycle Impact Assessment

Based on the ISO 14040–44 standards (ISO 2006a, 2006b), environmental indicators were attained for different impact categories, including abiotic depletion, acidification, eutrophication, global warming, human toxicity, and photochemical oxidation. The characterization factors applied to each impact category are proposed by the CML 2000 method (Figure 10.12).

10.4.4 Interpretation of LCA Results

For each alternative, several waste management operations responsible for each environmental impact can be analyzed. For example, for cases of depletion of abiotic resources, acidification, eutrophication, global warming, and photochemical oxidation, the options that derive electricity from direct burning of high calorific fraction of MSW, RDF, and biogas combustion (A4′, A4*; A5′, A5*; A0′, A0*; and A2′, A2*) are preferred. To assess human toxicity, options with the lowest concentration of heavy metals in compost due to selective collection of BMW (A6 and A6*) are preferred.

Uncertainty and sensitivity analyses can also be conducted during the interpretation phase, where databases can be tested, and Monte Carlo simulation can determine data variability for biogas production, electricity consumption, the Portuguese electricity grid, selected substitution ratios of recyclables, and the amount of waste recovered for RDF production. Finally, AD and MBT followed by energy recovery of the high calorific fraction of waste is an environmentally benign option.

10.5 LIFE CYCLE MANAGEMENT

Although LCA is an iterative approach in which assumptions and measures taken in one iteration can be assessed and changed in another iteration leading to more accurate and robust results, the LCA conceptual procedure has not yet reached maturity. In terms of the goal and scope definition, for instance, the life cycle of an SWM begins the moment that a consumer disposes of a product, complicating the analysis of waste streams production. Once the waste stream enters the SWM system, cost- or benefit-driven processes trigger the optimal method to deliver raw waste streams, RDF, recyclables, incineration ash, and other secondary waste materials to

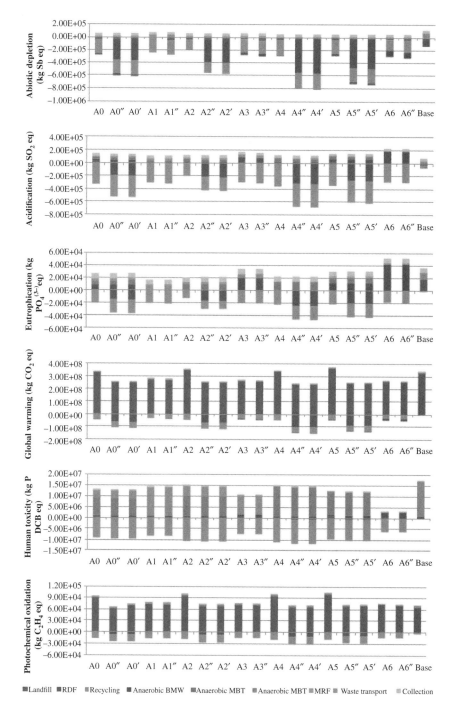

FIGURE 10.12   Contribution made by each stage of the waste management life cycle associated with each impact category. *Source*: From Pires et al. (2011)

various destinations, where interactions with other type of waste management facilities and natural systems may be salient. This simplification results in challenges for implementing LCIA and interpreting LCA results.

The waste hierarchy principle has been applied for SWM in the EU based on the Waste Framework Directive 2008/98/EC (European Parliament and Council, 2008). LCT is an alternative waste management philosophy in connection to the waste hierarchy principle, helping assess the benefits and trade-offs associated with different options. LCT has also been promoted through EU legislation on Waste Framework Directive 2008/98/EC (European Parliament and Council, 2008), stating that all Member States must take measures to motivate the options with the best total environmental outcome when applying the waste hierarchy principle. The LCT philosophy may depart from the traditional viewpoints beginning with source generation and ending with final disposal, and lead to the investigation of consumer and producer responsibility or behavior. This new LCT philosophy results in a well-known managerial approach—Life Cycle Management (LCM). By getting beyond short-term thinking, holistic LCA in the future may be tied to different options at a system level to determine alternative methods to conform with the waste hierarchy principle while LCM can help streamline and clarify priorities. These impacts will have various consequences in system analysis for advanced SWM such as carbon-regulated SWM strategies. A municipal utility park could be one of the most complicated practices, in which the LCM approach may have competing advantages.

10.6 FINAL REMARKS

A major challenge in LCI is to integrate various sources of data from different databases and/or literature to build up a firm foundation for LCA. In addition, many LCIA methods or models have varying assessment structures with similar concept in which damages can be categorized as human health, ecosystem quality, and resources. Performance comparisons related with different methods or models for LCIA are essential in various types of streamlined LCA practices. This complexity which compounds the screening and selection of appropriate LCIA methods or models impacts the application potential of LCA. In the context of LCM, multiple issues related with public health and environmental quality at different LCA scales can be addressed through environmental benefits accumulated from effective waste management practices. Such goals require having high-end system analysis techniques to support.

REFERENCES

Achernbosch, M. and Richers, U. 1997. *Comparison and Assessment of Material Flows of Sewage Production Method and of Wet Flue Gas from Municipal Incinerators* [German: Vergleich und Beurteilung von Stoffströmen der abwasserfreien und abwassererzeugenden Verfahren der "nassen" Rauchgasreinigung von Hausmüllverbrennungsanlagen], Wissenschaftliche Berichte FZKA 5874, Forschungszentrum Karlsruhe, Karlsruhe, Germany.

Agência Portuguesa do Ambiente (APA). 2009. Environmental licenses [Portuguese: Licenças ambientais emitidas]. APA. Available at: http://sniamb.apambiente.pt/portalmetadados/index.php?option=com_content&view=article&id=14&Itemid=10&lang=pt   (accessed June 2009).

Amarsul. 2009. Annual report 2009 [Portuguese: Relatório e contas 2009]. Amarsul.

Association of Plastics Manufacturers in Europe (APME). 1995. Eco-profiles: life cycle assessment. APME. Available at: http://www.plasticseurope.org (accessed July 2009).

Arena, U., Mastellone, M. L., and Perugini, F. 2003. Life cycle assessment of a plastic packaging recycling system. *International Journal of Life Cycle Assessment*, 8(2), 92–98.

Bakst, J. S., Lacke, C. J., Weitz, K. A., and Warren, J. L. 1995. *Guidelines for Assessing the Quality of Life Cycle Inventory Analysis*, United States Environmental Protection Agency.

Bare, J. C., Norris, G. A., and Pennington, D. W. 2003. TRACI, the tool for the reduction and assessment of chemical and other environmental impacts. *Journal of Industrial Ecology*, 6(3–4), 49–78.

Battisti, R. and Corrado, A. 2005. Evaluation of technical improvements of photovoltaic systems through life cycle assessment methodology. *Energy*, 30(7), 952–967.

Baumann, H. and Tillman, A.-M. 2004. *The Hitch Hiker's Guide to LCA—An Orientation in Life Cycle Assessment Methodology and Application*, Studentlitteratur, Lund, Sweden.

Björklund, A. E. and Finnveden, G. 2007. Life cycle assessment of a national policy proposal—the case of a Swedish waste incineration tax. *Waste Management*, 27(8), 1046–1058.

Björklund, A., Dalemo, M., and Sonesson, U. 1999. Evaluating a municipal waste management plan using ORWARE. *Journal of Cleaner Production*, 7(4), 271–280.

Blengini, G. A. 2008. Using LCA to evaluate impacts and resources conservation potential of composting: a case study of the Asti District in Italy. *Resources, Conservation and Recycling*, 52(12), 1373–1381.

Blengini, G. A., Garbarino, E., Šolar, S., Shields, D. J., Hámor, T., Vinai, R., and Agioutantis, Z. 2012. Life Cycle Assessment guidelines for the sustainable production and recycling of aggregates: the Sustainable Aggregates Resource Management project (SARMa). *Journal of Cleaner Production*, 27, 177–181.

Börjeson, L., Höjer, M., Dreborg, K.-H., Ekvall, T., and Finnveden, G. 2006. Scenario types and techniques: towards a user's guide. *Futures*, 38(7), 723–739.

Borken, J., Patyk, A., and Reinhardt, G. A. 1999. *Database for Ecological Balances* [German: Basisdaten für ökologische Bilanzierungen – Einsatz mobiler Maschinen in Transport, Landwirtschaft und Bergbau], Vieweg & Sohn Verlagsgesellschaft mbH, Braunschweig.

Boustead, I. 2000. *Environmental Profile Report for the European Aluminium Industry*, European Aluminium Association, Brussels.

Bovea, M. D. and Gallardo, A. 2006. The influence of impact assessment methods on materials selection for eco-design. *Materials and Design*, 27(3), 209–215.

Brunner, P. H. and Rechberger, H. 2004. *Practical Handbook of Material Flow Analysis*, CRC Press, Boca Raton, FL.

Bundesamt für Umwelt, Wald und Landschaft (BUWAL). 1998. Life cycle inventories for packaging [German: Ökoinventare für Verpackungen, Bd.II Schriftenreihe Umwelt Nr. 250]. Bundesamt für Umwelt, Wald und Landschaft (Swiss Agency for the Environment, Forests and Landscape), Berne, Switzerland.

Buttol, P., Masoni, P., Bonoli, A., Goldoni, S., Belladonna, V., and Cavazzuti, C. 2007. LCA of integrated MSW management systems: case study of the Bologna District. *Waste Management*, 27(8), 1059–1070.

Canadian Standards Association (CSA). 1994. *Life Cycle Assessment, Standard CAN/CSA-Z760*, CSA.

Chang, N.-B., Pires, A., and Martinho, G. 2013. Chapter 17: Impacts of life cycle assessment on solid waste management. In: *Encyclopedia of Environmental Management* (Ed. Jorgensen, S. E.), Vol. IV, Taylor and Francis Group, pp. 2399–2414.

Christensen, T. H., Bhander, G., Lindvall, H., Larsen, A. W., Fruergaard, T., Damgaard, A., Manfredi, S., Boldrin, A., Riber, C., and Hauschild, M. 2007. Experience with the use of LCA-modelling (EASEWASTE) in waste management. *Waste Management & Research*, 25(3), 257–262.

Cleary, J. 2009. Life cycle assessments of municipal solid waste management systems: a comparative analysis of selected peer-reviewed literature. *Environment International*, 35(8), 1256–1266.

Cleary, J. 2010. The incorporation of waste prevention activities into life cycle assessments of municipal solid waste management systems: methodological issues. *The International Journal of Life Cycle Assessment*, 15(6), 579–589.

Comieco. 2008. Cardboard: Introduction [Italian: Carta: Introduzione]. Comieco. Available at: http://www.matrec.it./ (accessed January 2010).

Cooper, J. S. and Vigon, B. 2001. *Life Cycle Engineering Guidelines—EPA/600/R-01/101*, USEPA, National Risk Management Research Laboratory, Office of Research and Development, Cincinnati, OH.

Corti, A. and Lombardi, L. 2004. End life tyres: alternative final disposal processes compared by LCA. *Energy*, 29(12–15), 2089–2108.

Council of the European Union. 1999. Council Directive 1999/31/EC of 26 April 1999 on the landfill of waste. *Official Journal of European Union*, L182, 1–19.

Crawford, R. H. 2011. *Life Cycle Assessment in the Built Environment*, Spon Press, New York.

Curran, M. A., Mann, M., and Norris, G. 2005. The international workshop on electricity data for life cycle inventories. *Journal of Cleaner Production*, 13(8), 853–862.

Dalemo, M., Sonesson, U., Björklund, A., Mingarini, K., Frostell, B., Nybrant, T., Jonsson, H., Sundqvist, J. O., and Thyselius, L. 1997. ORWARE—a simulation model for organic waste handling systems. Part 1: model description. *Resources, Conservation and Recycling*, 21(1), 17–37.

Damgaard, A. 2010. Implementation of life cycle assessment models in solid waste management. Doctoral dissertation, Department of Environmental Engineering, Technical University of Denmark, Lyngby, Denmark.

Dehoust, G., Gebhardt, P., and Gärtner, S. 2002. *The Contribution of the Thermal Treatment Of Waste on Climate Change, Air Pollution Control and Resource Conservation* [German: Der Beitrag der thermischen Abfallbehandlung zu Klimaschutz, Luftreinhaltung und Ressourcenschonung], Öko-Institut e.V., Darmstadt.

den Boer, E., den Boer, J., and Jager, J. (Eds.). 2005. *Waste Management Planning and Optimisation—Handbook for Municipal Waste Prognosis and Sustainability Assessment of Waste Management Systems*, Ibidem-Verlag, Stuttgart.

den Boer, J., den Boer, E., and Jager, J. 2007. LCA-IWM: a decision support tool for sustainability assessment of waste management systems. *Waste Management*, 27(8), 1032–1045.

Diaz, R. and Warith, M. 2006. Life-cycle assessment of municipal solid wastes: development of WASTED model. *Waste Management*, 26(8), 886–901.

Ecobilan. 2004. *WISARD—Waste Integrated System for Analysis of Recovery and Disposal*, Ecobilan.

Ecoinvent. 2006. Ecoinvent Data v1.2. Ecoinvent Centre, Swiss Centre for Life Cycle Inventory.

Eggels, P. and van der Ven, B. 1995. Allocation model for landfill. In: *International Workshop of Life Cycle Assessment and Treatment of Solid Waste*, AFR-Report 98 (Eds Finnveden, G. and Huppes, G.), Swedish EPA, Stockholm, Sweden, pp. 149–157.

Ekvall, T. 2000. A market-based approach to allocation at open-loop recycling. *Resources, Conservation and Recycling*, 29(1–2), 93–111.

Ekvall, T. and Tillman, A.-M. 1997. Open-loop recycling: criteria for allocation procedures. *International Journal of Life Cycle Assessment*, 2(3), 155–162.

Ekvall, T., Tillman, A.-M., and Molander, S. 2005. Normative ethics and methodology for life cycle assessment. *Journal of Cleaner Production*, 13(13–14), 1225–1234.

Ekvall, T., Assefa, G., Björklund, A., Eriksson, O., and Finnveden, G. 2007. What life-cycle assessment does and does not do in assessments of waste management. *Waste Management*, 27(8), 989–996.

ETH Zürich. 2008. Databases. ETH Zürich. Available at: http://www.ethbib.ethz.ch/index_e.html (accessed January 2010).

European Aluminium Association. 2007. *Aluminium Recycling in LCA*. European Aluminium Association.

European Commission, Joint Research Centre and Institute for Environment and Sustainability (EC-JRC-IES). 2010a. *Framework and Requirements for Life Cycle Impact Assessment Models and Indicators*, Publications Office of the European Union, Luxembourg.

European Commission, Joint Research Centre and Institute for Environment and Sustainability (EC-JRC-IES). 2010b. *Specific Guide for Life Cycle Inventory Data Sets*, Publications Office of the European Union, Luxembourg.

European Commission, Joint Research Centre and Institute for Environment and Sustainability (EC-JRC-IES). 2011. *Supporting Environmentally Sound Decisions for Waste Management: A technical guide to Life Cycle Thinking (LCT) and Life Cycle Assessment (LCA) for Waste Experts and LCA Practitioners*, European Union, Luxembourg.

European Environment Agency (EEA). 2007. *European Monitoring and Evaluation Program and Core Inventory of Air Emissions (EMEP/CORINAIR) Atmospheric Emission Inventory Guidebook*, EEA, Copenhagen.

European Environment Agency (EEA). 2009. *European Monitoring and Evaluation Program and European Environment Agency (EMEP/EEA) Air Pollutant Emission Inventory Guidebook—2009*, EEA, Copenhagen.

European Parliament and Council. 2004. Directive 2004/12/EC of the European Parliament and of the Council of 11 February 2004 amending Directive 94/62/EC on packaging and packaging waste. *Official Journal of the European Union*, L47, 26–31.

European Parliament, and Council. 2008. Directive 2008/98/EC of the European Parliament and of the Council of 19 November 2008 on waste and repealing certain Directives. *Official Journal of the European Union*, L312, 3–30.

Finnveden, G. 1999. Methodological aspects of life cycle assessment of integrated solid waste management systems. *Resources, Conservation and Recycling*, 26(3–4), 173–187.

Finnveden, G., Albertsson, A. C., Berendson, J., Eriksson, E., Hoglund, L. O., Karlsson, S., and Sundqvist, J.-O. 1995. Solid waste treatment within the framework of life-cycle assessment. *Journal of Cleaner Production*, 3(4), 189–199.

Finnveden, G., Hauschild, M. Z., Ekvall, T., Guinée, J., Heijungs, R., Hellweg, S., Koehler, A., Pennington, D., and Suh, S. 2009. Recent developments in Life Cycle Assessment. *Journal of Environmental Management*, 91(1), 1–21.

Forster, P., Ramaswamy, V., Artaxo, P., Berntsen, T., Betts, R., Fahey, D. W., Haywood, J., Lean, J., Lowe, D. C., Myhre, G., Nganga, J., Prinn, R., Raga, G., Schulz, M., and van Dorland, R. 2007. Changes in atmospheric constituents and in radiative forcing. In: *Climate Change 2007: The Physical Science Basis. Contribution of Working Group I to the Fourth Assessment Report of the Intergovernmental Panel on Climate Change* (Eds Solomon, S., Qin, D., Manning, M., Chen, Z., Marquism, M., Averyt, K. B., Tignor, M., and Miller, H. L.), Cambridge University Press, Cambridge and New York.

Fricke, K. and Müller, W. 1999. Stabilization of residual waste by mechanical-biological treatment and impact on the landfill, the joint project. In: *Mechanical Biological Treatment of Landfilled Waste* [German: Stabilisierung von Restmüll durch mechanisch-biologische Behandlung und Auswirkungen auf die Deponierung, Verbundsvorhaben "Mechanisch-Biologische Behandlung von zu deponierenden Abfällen"], Witzenhausen, Germany.

Fricke, K., Franke, H., Dichtl, N., Schmelz, K.-G., Weiland, P., and Bidlingmaier, W. 2002. Biological processes for bio-and green waste recycling [German: Biologische Verfahren zur bio- und Grünabfallverwertung]. In: *ATV Handbuch—Mechanische und Biologische Verfahren der Abfallbehandlung* (Ed. Loll, U.), Ernst and Sohn Verlag für Architektur und technische Wissenschaften, Berlin.

Fricke, K., Hake, J., Hüttner, A., Müller, W., Santen, H., Wallmann, R., and Turk, T. 2003. *Treatment Technologies for Plants in the Mechanical-Biological Treatment of Residual Waste* [German: Aufbereitungstechnologien für Anlagen der mechanisch-biologischen Restabfallbehandlung]. Müll Handbuch, Lfg. 4/03, Band 5, n° 5615, Erich Schmidt Verlag, Berlin.

Frischknecht, R., Hofstetter, P., Knoepfel, I., Dones, R., and Zollinger, E. 1996. Life Cycle Inventories of Energy Systems—Foundations for ecological comparison of energy systems and the inclusion of LCA for Switzerland [German: Ökoinventare für Energiesysteme—Grundlagen für den ökologischen Vergleich von Energiesystemen und den Einbezug von Ökobilanzen für die Schweiz]. Auflage, 3, Zürich, Schweiz.

Frischknecht, R. Jungbluth, N., Althaus, H.-J., Bauer, C., Doka, G., Dones, R., Hischier, R., Hellweg, S., Humbert, S., Kollner, T., Lőerincik, Y., Margini, M., and Nemecek, T. 2007. *Implementation of Life Cycle Impact Assessment Methods*, Ecoinvent report n° 3, v2.0, Swiss Centre for Life Cycle Inventories, Dübendorf, Switzerland.

Gentil, E. C. Damgaard, A., Hauschild, M., Finnveden, G., Eriksson, O., Thorneloe, S., Kaplan, P. O., Barlaz, M., Muller, O., Matsui, Y., Ii, R., and Christensen, T. H. 2010. Models for waste life cycle assessment: review of technical assumptions. *Waste Management*, 30(12), 2636–2648.

Gentil, E. C., Gallo, D., and Christensen, T. H. 2011. Environmental evaluation of municipal waste prevention. *Waste Management*, 31(12), 2371–2379.

Goedkoop, M. and Spriensma, R. 2000. *The Eco-Indicator 99—A Damage-Oriented Method for Life Cycle Impact Assessment*, Methodology Report, PRé Consultants B.V.

Goedkoop, M. J. Heijungs, R., Huijbregts, M., De Schryver, A., Struijs, J., and van Zelm, R. 2009. ReCiPe 2008, a life cycle impact assessment method which comprises harmonised category indicators at the midpoint and the endpoint level. Publications. Available at: http://www.lcia-recipe.net./ (accessed November 2012).

Gomes, C. 2009. Productivity indicators of selective collection of MSW considering different operational conditions [Portuguese: Análise de indicadores de produtividade de circuitos de recolha selectiva de RSU com diferentes características operacionais]. Master's Degree thesis, Faculty of Sciences and Technology, Nova University of Lisbon, Lisbon.

Google. 2010. Get directions. Google Maps. Available at: http://maps.google.com (accessed 10 April 2010).

Grosso, M., Nava, C., Testori, R., Rigamonti, L., and Viganò, F. 2012. The implementation of anaerobic digestion of food waste in a highly populated urban area: an LCA evaluation. *Waste Management & Research*, 30(9), 78–87.

Guinée, J. B., Gorree, M., Heijungs, R., Huppes, G., Kleijn, R., van Oers, L., Wegener Sleeswijk, A., Suh, S., Udo de Haes, H. A., de Bruijn, J. A., van Duin, R., and Huijbregts, M. A. J. (Eds) 2002. *Handbook on Life Cycle Assessment—Operational Guide to the ISO Standards*, Kluwer Academic Publishers, Dordrecht, The Netherlands.

Guinée, J. B., Heijungs, R., and Huppes, G. 2004. Economic allocation: examples and derived decision tree. *The International Journal of Life Cycle Assessment*, 9(1), 23–33.

Hauschild, M. Z. and Barlaz, M. A. 2011. LCA of waste management. In: *Solid Waste Technology and Management* (Ed. Christensen, T. H.), Wiley, Chichester, West Sussex, UK.

Hauschild, M. Z. and Potting, J. 2005. *Spatial Differentiation in Life Cycle Impacts Assessment—the EDIP2003 Methodology*. Environmental News n.° 80., Danish Ministry of the Environment, Environmental Protection Agency, Copenhagen, Denmark.

Hauschild, M., Olsen, S. I., Hansen, E., and Schmidt, A. 2008. Gone ... but not away— addressing the problem of long-term impacts from landfills in LCA. *The International Journal of Life Cycle Assessment*, 13(7), 547–554.

Hauschild, M. Z. Goedkoop, M., Guinée, J., Heijungs, R., Huijbregts, M., Jolliet, O., Margni, M., Schryver, A., Humbert, S., Laurent, A., Sala, S., and Pant, R. 2012. Identifying best existing practice for characterization modeling in life cycle impact assessment. *The International Journal of Life Cycle Assessment*, 18(3), 683–697.

Heijungs, R. and Guinée, J. B. 2007. Allocation and "what-if" scenarios in life cycle assessment of waste management systems. *Waste Management*, 27(8), 997–1005.

Heijungs, R. and Suh, S. 2002. *The Computational Structure of Life Cycle Assessment*, Kluwer Academic Publishers, Dordrecht, The Netherlands.

Hischier, R., Weidema, B., Althaus, H.-J., Bauer, C., Doka, G., Dones, R., Frischknecht, R., Hellweg, S., Humbert, S., Jungbluth, N., Köllner, T., Lőerincik, Y., Margini, M., and Nemecek, T. 2010. *Implementation of Life Cycle Impact Assessment Methods*, Ecoinvent report n.° 3, v2.0, Swiss Centre for Life Cycle Inventories, Dübendorf, Switzerland.

Humbert, S., De Schryver, A., Bengoa, X., Margni, M., and Jolliet, O. 2012. Impact 2002+: user guide. Draft for version Q.2.21 (version adapted by Quantis). Available at: http://www.quantis-intl.com/pdf/IMPACT2002_UserGuide_for_vQ2.21.pdf (accessed November 2014).

Institut Für Energie und Umweltforschung (IFEU). 1994. LCA for packaging. Partial Report: Energy, Transport, Disposal [German: Ökobilanz für Verpackungen, Teilbericht: Energie— Transport—Entsorgung]. Im Auftrag des Umweltbundesamtes, Berlin.

Institut für Umweltinformatik (IFU) Hamburg. 2009. Umberto version 5.5. http://www.ifu.com/en.

International Organization for Standardization (ISO). 1999. *ISO 14041 Environmental Management—Life Cycle Assessment: Goal and Scope Definition and Inventory Analysis*, ISO.

International Organization for Standardization (ISO). 2006a. *ISO 14040 Environmental Management—Life Cycle Assessment: Principles and Framework*, ISO.

International Organization for Standardization (ISO). 2006b. *ISO 14044 Environmental Management—Life cycle Assessment: Requirements and Guidelines*, ISO.

Joint Research Centre (JRC). 2013. European platform on LCA. JRC, Available at: http://ec.europa.eu/environment/ipp/lca.htm (accessed January 2013).

Jolliet, O., Margni, M., Charles, R., Humbert, S., Payet, J., Rebitzer, G., and Rosenbaum, R. 2003. IMPACT 2002+: a new life cycle impact assessment methodology. *International Journal of Life Cycle Assessment*, 8(6), 324–330.

Knörr, W., Höpfner, U., and Helms, H. 1997. *Data and Computer Model: Energy Consumption and Pollutant Emissions from Motor Traffic in Germany 1980–2020* [German: Daten- und Rechenmodell: Energieverbrauch und Schadstoffemissionen des motorisierten Verkehrs in Deutschland 1980–2020], Im Auftrag des Umweltbundesamtes. Ufoplan Nr. 10506057, Heidelberg, Deutschland.

Ligthart, T. N. and Ansems, A. M. M. 2012. Modelling of recycling in LCA. In: *Post-Consumer Waste Recycling and Optimal Production* (Ed. Damanhuri, E.), InTech, Croatia, pp. 185–210.

Lindfors, L.-G., Christiansen, K., Hoffman, L., Virtanen, Y., Juntilla, V., Leskinen, A., Hanssen, O.-J., Rønning, A., Ekvall, T., and Finnveden, G. 1995a. LCA-Nordic. Technical reports No. 1–9, TemaNord 1995:502. Nordic Council of Ministers, Copenhagen.

Lindfors, L.-G., Christiansen, K., Hoffman, L., Virtanen, Y., Juntilla, V., Hanssen, O.-J., Ronning, A., Ekvall, T., and Finnveden, G. 1995b. *Nordic Guidelines on Life Cycle Assessment*, Nord 1995:20, Nordic Council of Ministers, Arhus.

Loll, U. 1994. Treatment of wastewater from aerobic and anaerobic process through biological waste treatment [German: Behandlung von Abwässern aus aeroben und anaeroben Verfahren zur biologischen Abfallbehandlung]. In: *Verwertung Biologischer Abfälle* (Eds Wiemer, K. and Kern, M.), M.I.C. Baeza-Verlag, Witzenhausen.

Loll, U. 1998. Leachate from composting and anaerobic digestion [German: Sickerwasser aus Kompostierungs- und Anaerobanlagen]. *Entsorgungspraxis*, 7–8(98), 52–58.

Martinho, G. and Pires, A. 2009. Assessment of recovery technologies of waste lube oil in Portugal [Portuguese: Avaliação das tecnologias de valorização de óleos usados em Portugal]. Technical report for SOGILUB.

Martinho, G., Santana, F., Santos, J., Brandão, A., and Santos, I. 2008. *Management and Treatment of Leachate from Municipal Solid Waste Sanitary Landfills* [Portuguese: Gestão e tratamento de lixiviados produzidos em aterros sanitários de resíduos urbanos]. DCEA/FCT/UNL. Relatório IRAR n.° 03/2008, IRAR, Lisboa.

Mata, T. 1998. Comparison of reuse/recycling process through life cycle assessment methodology [Portuguese: Comparação de processos de reutilização/reciclagem usando a metodologia de análise de ciclo-de-vida]. Master's thesis, Faculty of Engineering, Porto University, Porto.

McDougall, F., White, P., Franke, M., and Hindle, P. 2001. *Integrated Solid Waste Management: a Life Cycle Inventory*, Blackwell Science Ltd., Oxford.

Münster, M. and Meibom, P. 2010. Long-term affected energy production of waste to energy technologies identified by use of energy system analysis. *Waste Management*, 30(12), 2510–2519.

Norgate, T. E. 2004. Metal recycling: an assessment using life cycle energy consumption as a sustainability indicator. CSIRO Minerals Report DMR-2616.

Oeko-Institut. 2001. GEMIS—Global emission model for integrated systems. Oeko-Institut. Available at: http://www.oeko.de/service/gemis (accessed June 2009).

Ozihel, H. 2012. *Avoided Burden—Life Cycle Assessment, Recycling, Reuse, Environmental Issue, Polyethylene Terephthalate*, Frac Press.

Patyk, A. and Reinhardt, G. 1997. *Fertilisers—Energy and Material Flow Balances* [German: Düngemittel—Energie- und Stoffstrombilanzen], Vieweg Umweltwissenschaften, Braunschweig.

Pesonen, H.-L. and Horn, S. 2013. Evaluating the sustainability SWOT as a streamlined tool for life cycle sustainability assessment. *The International Journal of Life Cycle Assessment*, 18(9), 1780–1792.

Pires, A. and Martinho, G. 2013. Life cycle assessment of a waste lubricant oil management system. *The International Journal of Life Cycle Assessment*, 18(1), 102–112.

Pires, A., Chang, N.-B., and Martinho, G. 2011. Reliability-based life cycle assessment for future solid waste management alternatives in Portugal. *The International Journal of Life Cycle Assessment*, 16(4), 316–337.

Potting, J. and Hauschild, M. 2005. *Background for Spatial Differentiation in LCA Impact Assessment—The EDIP2003 Methodology*, Danish Ministry of the Environment, Environmental Protection Agency, Copenhagen, Denmark.

PRé Consultants. 2004. *SimaPro 6—Introduction to LCA with SimaPro*.

ProBas. 2004. Zum ProBas-Projekt. ProBas. Available at: http://www.probas.umwelt bundesamt.de/php/index.php? (accessed January 2010).

Rettenberger, G. 1996. *Economic Leachate Treatment* [German: Wirtschaftlicthe Sickerwasserreinigung]. Trierer Berichte zur Abfallwirtschaft Bd. 10, Economia-Verlag, Bonn.

Rettenberger, G. and Schneider, R. 1997. Overview of the requirements and the state of leachate treatment technology [German: Überblick über die Anforderungen und den Stand der Sickerwasserreinigungstechnik]. In: *Wirtschaftliche Sickerwasser-Reinigung (Trierer Berichte zur Abfallwirtschaft, Band 10)* (Ed. Rettenberger, G.), Economica Verlag, Bonn.

Rettenberger, G. and Stegmann, R. 1997. *New Aspects of Landfill Gas Utilization* [German: Neue Aspekte der Deponiegasnutzung]. Trierer Berichte zur Abfallwirtschaft Bd. 11, Economia-Verlag, Bonn.

Rigamonti, L., Grosso, M., and Sunseri, M. C. 2009a. Influence of assumptions about selection and recycling efficiencies on the LCA of integrated waste management systems. *The International Journal of Life Cycle Assessment*, 14(5), 411–419.

Rigamonti, L., Grosso, M., and Giugliano, M. 2009b. Life cycle assessment for optimising the level of separated collection in integrated MSW management systems. *Waste Management*, 29(2), 934–944.

Rigamonti, L., Grosso, M., and Giugliano, M. 2010. Life cycle assessment of sub-units composing a MSW management system. *Journal of Cleaner Production*, 18(16–17), 1652–1662.

Rodrigo, J. and Castells, F. 2000. Environmental evaluation of different strategies for the management of municipal waste in the region of Catalonia (project founded by Junta de Residus, the regional Waste Agency of Catalonia).

Rodrigues, M. 2009. Sorting plants: characterization and assessment of national situation [Portuguese: Estações de triagem: caracterização e avaliação da situação nacional]. Master's degree thesis, Faculty of Sciences and Technology, Nova University of Lisbon, Lisbon.

Rotter, S. 2004. *Heavy Metals in Household Waste* [German: Schwermetalle in Haushaltsabfällen], Müll Handbuch, Lfg. 1/04, Band 4, n° 2829, Erich Schmidt Verlag, Berlin.

Ryding, S.-O. and Steen, B. 1991. The EPS system. A PC-based system for development and application of environmental priority strategies in product design—from cradle to grave. IVL-report Nr B 1022 [Swedish]. IVL, Gothenburg, Sweden.

Schäfl, A. 1995. *Mass and Energy Balances, and Disposition of Waste Incineration Plants* [German: Massen- und Energiebilanzen sowie Disposition von Müllheizkraftwerken]. Manuskripte zur Abfallwirtschaft 13.1], Verlag Abfall Now e.V., Stuttgart, Deutschland.

Schwing, E. 1999. Assessment of the emissions of the combined mechanical-biological and thermal waste treatment method in Southern Hesse [German: Bewertung der Emissionen der Kombination mechanisch-biologischer und thermischer Abfallbehandlungsverfahren in Südhessen]. Dissertation, Institut WAR, TUD Darmstadt, WAR-Schriftenreihe Bd. 111, Darmstadt.

Schmidt, M., Knörr, W., Patyk, A., and Höpfner, U. 1998. *Evaluation of Common Data Models to Identify Environmental Impacts of Road Traffic* [German: Evaluierung gängiger Datenmodelle zur Ermittlung verkehrlicher Umweltbelastungen]. Umweltinformatik 98, Marburg.

Schulz, J., Pschom, J., Kara, S., Hermann, C., Ibbotson, S., Dettmer, T., and Luger, T. 2011. Environmental footprint of single-use surgical instruments in comparison with multi-use surgical instruments. In: *Glocalized Solutions for Sustainablity in Manufacturing* (Eds. Hesselbach, J. and Herrmann, C.), Springer-Verlag, Berlin, Heidelberg, pp. 623–628.

Scipioni, A., Mazzi, A., Niero, M., and Boatto, T. 2009. LCA to choose among alternative design solutions: the case study of a new Italian incineration line. *Waste Management*, 29(9), 2462–2474.

Shen, L. Nieuwlaar, E., Worrell, E., and Patel, M. K. 2011. Life cycle energy and GHG emissions of PET recycling: change-oriented effects. *The International Journal of Life Cycle Assessment*, 16(6), 522–536.

Sonnemann, G. and Vigon, B. (Eds.) 2011. *Global Guidance Principles for Life Cycle Assessment Databases: A Basis for Greener Processes and Products*, UNEP/SETAC Life Cycle Initiative, Paris/Pensacola.

Soyez, K., Thrän, D., Koller, M., and Hermann, T. 2000. Results of research projects on mechanical-biological treatment of waste to be landfilled [in German: Ergebnisse von Forschungsvorhaben zur mechanisch-biologischen Behandlung von zu deponierenden Abfällen]. In: *Die Zukunft der Mechanisch-biologischen Restabfallbehandlung – Proceedings der Potsdamer Abfallage* (Eds Soyez, K., Hermann, T., Koller, M., and Thrän, D.), Potsdam Mai 22–23, 2000. Brandenburgische Umweltberichte, Heft 6.

Steen, B. 1999. *A Systematic Approach to Environmental Priority Strategies in Product Development (EPS)*, Chalmers University of Technology, Technical Environmental Planning. Available at: http://www.cpm.chalmers.se/document/reports/99/1999_4.pdf (accessed November 2012).

Stora Enso. 2008. EMAS Declaración Ambiental 2008. Stora Enso.

Sundberg, J. 1993. A system approach to municipal solid waste management: results from a case study of Goteborg—Part 1. In: Proceedings of International Conference on Integrated Energy and Environmental Management, Air and Waste Management Association, New Orleans.

Sundberg, J. 1995. Municipal Solid Waste Management with the MIMES/Waste model: a complementary approach to LCA studies for evaluating waste management options. In: *Proceedings of the International Workshop—Life Cycle Assessment and Treatment of Solid Waste* (Eds Finnveden, G. and Huppes, G.), Swedish Environmental Protection Agency, Stockholm.

Tabasaran, O. and Rettenberger, R. 1987. *Basis for the Planning of Biogas Extraction Manual* [German: Grundlage zur Planung von Entgasungsanlagen Müllhandbuch], Erich Schmidt Verlag, Berlin.

Thorneloe, S., Weitz, K., and Jambeck, J. 2007. Application of the US decision support tool for materials and waste management. *Waste Management*, 27(8), 1006–1020.

Thomassen, M. A., Dalgaard, R., Heijungs, R., and de Boer, I. 2008. Attributional and consequential LCA of milk production. *The International Journal of Life Cycle Assessment*, 13(4), 339–349.

Tillman, A.-M., Ekvall, T., Baumann, H., and Rydberg, T. 1994. Choice of system boundaries in the life cycle assessment. *Journal of Cleaner Production*, 2(1), 21–29.

Umweltbundesamt (UBA). 1999. *Thermal, Mechanical-Biological Treatment Plants and Landfills for Municipal Solid Waste in the Federal Republic of Germany* [German: Thermische, mechanisch-biologische Behandlungsanlagen und Deponien für Siedlungsabfälle in der Bundesrepublik Deutschland], UBA FG III.3.3, Berlin.

United States Environmental Agency (USEPA). 2006. Life cycle impact assessment. USEPA. Available at: http://www.epa.gov/nrmrl/std/lca/lca.html (accessed December 2012).

Valorsul. 2008. Sustainability report [Portuguese: Relatório de sustentabilidade]. Valorlsul. Available at: http://www.valorsul.pt (accessed October 2010).

Vaz, F. 2009. Source separated organic waste features from Lisbon metropolitan area and its influence on the performance of the anaerobic digestion process [Portuguese: As características da fracção orgânica dos RSU recolhidos selectivamente na área metropolitana de Lisboa e a sua influência no comportamento do processo de digestão anaeróbia]. Master's degree thesis, Faculty of Sciences and Technology, Nova University of Lisbon, Lisbon.

Vogt, R., Knappe, F., Giegrich, J., and Detzel, A. 2002. *LCA of Organic Waste Recycling, Studies on Environmental Impact of Systems for Biological Organic Waste Recycling* [German: Ökobilanz Bioabfallverwertung, Untersuchungen zur Umweltverträglichkeit von Systemen zur Verwertung von biologischorganischenAbfällen], Erich Schmidt Verlag, Berlin.

Wallmann, R. and Fricke, K. 2002. Energy balance in the recovery of organic and green waste and the mechanical-biological waste treatment [German: Energiebilanz bei der Verwertung von Bio- und Grünabfällen und bei der mechanisch-biologischen Restabfallbehandlung]. In: *ATV Handbuch—Mechanische und Biologische Verfahren der Abfallbehandlung* (Ed. Loll, U.), Ernst and Sohn Verlag für Architektur und technische. Wissenschaften, GmbH, Berlin.

Weber, B. 1990. *Minimizing Emissions of the Landfill* [German: Minimierung von Emissionen der Deponie], Veröffentlichungen des Instituts für Siedlungswasserwirtschaft und Abfalltechnik der Universität Hannover o. Prof. Dr.-Ing. C.F. Seyfried. H. 74, Hannover, Deutschland.

Weitz, K. Barlaz, M., Ranji, R., Brill, D., Thorneloe, S., and Ham, R. 1999. Life cycle management of municipal solid waste. *International Journal of Life Cycle Assessment*, 4(4), 195–201.

White, P., Franke, M., and Hindle, P. 1995. *Integrated Solid Waste Management: A Life-Cycle Inventory*, Blackie Academic and Professional, Glasgow.

Williams, T. G. J. L., Heidrich, O., and Sallis, P. J. 2010. A case study of the open-loop recycling of mixed plastic waste for use in a sports-field drainage system. *Resources, Conservation and Recycling*, 55(2), 118–128.

Yamada, M. and Jung, S. 2007. Application of reverse osmosis for landfill leachate treatment. APEC—Virtual Centre for Environmental Technology Exchange. Available at: http://www.apec-vc.or.jp/e/modules/tinyd00/index.php?id=35andkh_open_cid_00=41 (accessed October 2010).

Zamagni, A., Buttol, P., Porta, P. L., Buonamici, R., Masoni, P., Guinée, J., Heijungs, R., Ekvall, T., Bersani, R., Bieńkowska, A., and Pretato, U. 2008. *Critical Review of the Current Research Needs and Limitations Related to ISO-LCA Practice*, ENEA – Italian National Agency for New Technologies, Roma.

CHAPTER 11

STREAMLINED LIFE CYCLE ASSESSMENT FOR SOLID WASTE TREATMENT OPTIONS

When dealing with a complex solid waste management (SWM) system, the appropriate system boundaries embedded in a series of nested structures from small to large scale may deeply compound the decision-making process. One salient example is to choose appropriate solid waste treatment technologies, such as a waste-to-energy (WTE) facility, across differing locations as the first step of system planning based on the criteria of cost, benefit, and life cycle assessment (LCA). When technology options are finalized, system boundaries can then be expanded to tackle larger scales of system analysis. With such a hierarchical approach comes the need to develop a deeper insight of LCA. This chapter begins with a general review of LCA applications for SWM that adopt various types of system boundaries. The next effort is to conduct a comparative streamlined life cycle assessment (CSLCA) for two types of solid waste incineration technologies, fluidized bed incinerator (FBI) and mechanical grate incinerator (MGI), both of which are commonly used in East Asia and Europe for burning municipal solid waste (MSW). Different consumption levels of energy and/or materials within the streamlined process were thoroughly evaluated for comparison without the inclusion of disposal of incineration ashes and emissions from landfills. Within this context, 1 metric ton (tonne) of MSW was chosen as the functional unit to support our inventory items. Both Eco-indicator 99 and ReCiPe assessment methods were selected for environmental impact assessment based on a suite of indicators leading to the improvement of reliability of the assessment. Through the use of life cycle inventory and impact assessment tools, the environmental impacts were collectively assessed for possible damage to human health, ecosystem integrity, and resources. Finally, a generalized numeric scheme was produced for the overall assessment of

Sustainable Solid Waste Management: A Systems Engineering Approach, First Edition. Ni-Bin Chang and Ana Pires.
© 2015 The Institute of Electrical and Electronics Engineers, Inc. Published 2015 by John Wiley & Sons, Inc.

these two incinerators comparatively. Such a real-world case study recognizes and signifies the growing importance of value engineering and systems analysis toward the achievement of high level decision-making for SWM.

11.1 APPLICATION OF LIFE CYCLE ASSESSMENT FOR SOLID WASTE MANAGEMENT

Life cycle assessment (LCA) is a tool used to analyze the environmental burden of products at all stages in their "cradle to grave" life cycle, namely from the extraction of resources, through the production of materials and parts, consumption of the product itself, and the use of the product through end-of-life management after it is discarded, by either reuse, recycling, or final disposal (Guinée et al., 2002). The concept of LCA arose in the 1960s as a way to cumulatively account for energy use to project future resource supplies and demand (Bayer et al., 2010). Since the 1990s, LCA has been applied in a broad range of different fields including SWM. The International Organization for Standardization (ISO) 14040 standard (ISO, 2006a) specifies the main areas of applications of LCA, including the identification of improvement possibilities, decision-making, selection of relevant environmental performance indicators, and marketing. The most influential momentum in the context of ISO 14040 applied to SWM is the exploration of the decision-making processes in a social context to produce different final options.

The salient use of LCA for SWM was conducted in Denmark in the 1990s, aiming to properly manage the packaging waste embedded in packaging materials (Person et al., 1998; Ryberg et al., 1998; Frees and Weidema, 1998). LCA may also be applicable for conducting intercomparisons among waste treatment alternatives for specific waste streams, such as recycling of cardboard (Finnveden et al., 1994) and recycling versus incineration of scrap paper (Finnveden and Ekvall, 1998). Earlier applications of LCA for SWM focused on evaluating the waste hierarchy principle in some industrialized countries during the 1990s (Sakai et al., 1996). The literature (Klöpffer, 2000) clearly indicates that the waste hierarchy principle cannot be a substitute for thorough LCA in most cases. Once LCA is capable of promoting a holistic view of SWM, several combinations of different technologies can be meaningfully applied to support essential SWM against the contradictory suggestions based on the waste hierarchy principle. For this reason, LCA is recommended by the European Union (EU) Waste Framework Directive 2008/98/EC (European Parliament and Council, 2008) to verify whether or not the waste hierarchy principle is the best solution, particularly for the environment.

Applications of LCA for SWM appeared mainly in Europe and the United States for screening waste treatment and disposal technologies under the umbrella of the integrated solid waste management (ISWM) systems. Several LCA models were tailored specifically for ISWM systems, including but not limited to Waste Analysis Software Tool for Environmental Decisions (WASTED) (Diaz and Warith, 2006) and Environmental Assessment of Solid Waste Systems and Technologies

(EASEWASTE) (Kirkeby et al., 2005). Specifically, Integrated Waste Management (IWM) versions 1 and 2 (White et al., 1995; McDougall et al., 2001) for SWM systems provide life cycle inventory (LCI), enabling decision makers and waste managers to use an LCA to assess specific waste management configurations without in-depth knowledge of the theory and methodology and to learn how changes in the system could affect environmental impacts through scenario analysis (Winkler and Bilitewski, 2007). In addition, extended tools focusing specially on the possible impact of economic features on the decision-making process were developed. In the United States, for example, the Municipal Solid Waste Decision Support Tool (MSW-DST) developed by the Research Triangle Institute and the United States Environmental Protection Agency (USEPA) (Weitz et al., 1999; Thorneloe et al., 2007) is mainly designed for evaluating the life cycle environmental trade-offs and full costs of SWM.

LCA has been increasingly used to objectively evaluate the performances of different SWM solutions (Scipioni et al., 2009; Zhao et al., 2011; Assamoi and Lawryshyn, 2012) by comparing the environmental performance of different scenarios to manage mixed solid waste (Cherubini et al., 2009; Özeler et al., 2006; Chaya and Gheewala, 2007; De Feo and Malvano, 2009; Eriksson et al., 2005) as well as to summarize global warming potential (GWP) (Buttol et al., 2007; Khoo, 2009). The growing importance of LCA applied to assess integrated SWM systems is evidenced by many case studies worldwide (Consonni et al., 2005). LCA practices are also capable of changing packaging and packaging waste management, influencing the selection of waste treatment technologies, affecting regulation assessment, challenging waste hierarchy paradigms, increasing knowledge, and developing tools and methods for decision support. Application of LCA in SWM systems has also been promoted through combination with other systems analysis tools to reach a sustainable decision (Chang et al., 2011). In this context, LCA was combined with site-specific approaches to evaluate the taxation policy of WTE (Nilsson et al., 2005). LCA and strategic environmental assessment were integrated to assess economic and environmental impacts of weight-based taxes for waste incineration (Björklund and Finnveden, 2007). The ORganic WAste REsearch (ORWARE) model, developed by Dalemo et al. (1997), Björklund et al. (1999), and Eriksson et al. (2002, 2005), combines LCA with simulation tools and material flow analysis for ISWM. For example, Rives et al. (2010) conducted an LCA for containers in urban regions. Solano et al. (2002a, 2002b) developed a model for ISWM to obtain the best solution through integrated LCA and optimization model to balance economic and environmental considerations. Those systems analysis models can be flexibly woven to manage SWM issues with varying features (Harrison et al., 2001; Chang et al., 2011). Regardless of cases where LCA was variously applied to support SWM systems, LCA can influence the decision analysis, especially the conflict-resolution decision-making process, in many SWM systems. One common methodology capable of linking different criteria is Multiattribute Decision-Making (MADM) (Kijak and Moy, 2004; Skordilis, 2004; Contreras et al., 2008), which can be used to promote understanding in SWM decision-making.

11.2 LCA FOR SCREENING TECHNOLOGIES OF SOLID WASTE TREATMENT

Incineration (including WTE) is one of the most important processes in an ISWM system due to its ability to destroy hazardous materials, reduce mass and volume of waste streams, and recover possible material and energy contents. With varying design criteria, MSW incineration technologies may include but are not limited to MGI, modular incineration, FBI, and rotary kiln incineration. Despite the evolution of incineration technologies that have notably reduced environmental impacts to some extent in recent years, many incinerators are still perceived as major pollution sources, particularly due to their gas emissions from the stacks (Morselli et al., 2005). From a sustainability viewpoint, any SWM system having lower global and regional environmental impacts is highly desirable (Mendes et al., 2004). In other words, multiple concerns associated with environmental quality and public health and safety at different scales must be addressed along with environmental benefits accrued from effective SWM practices (Cherubini et al., 2009).

For this reason, the need for standard methods to assess the holistic environmental impact across different MSW incineration technologies is acute. One of the most useful procedures for a holistic environmental impact evaluation is the LCA, a methodology that considers the entire life cycle of products and services from cradle to grave (i.e., from raw material acquisition through production, use, and disposal) (Barton et al., 1996; Özeler et al., 2006; Liamsanguan and Gheewala, 2008; Morselli et al., 2008). LCA has proven to be a systematic tool to measure and compare the environmental impacts of human activities (Lee et al., 2007; Rebitzer et al., 2004) and, when applied to ISWM plans, constitutes a relatively new field of environmental systems analysis that introduces great potential development (Hong et al., 2006). LCA therefore has been used to evaluate air pollution control, energy recovery, and auxiliary fuel design in waste incineration (Damgaard et al., 2010; Møller et al., 2011; Zhao et al., 2012).

When choosing the most appropriate SWM alternative for a region, decision makers account for not only the technical aspects and implementation costs but also the environmental impacts produced by the treatment and disposal processes, as well as the opinion of the local communities (De Feo and Malvano, 2009). Within this context, the implementation of LCA is one of the most important environmental assessments for various MSW incineration alternatives (Mendes et al., 2004; Buttol et al., 2007; Liamsanguan and Gheewala, 2008; Scipioni et al., 2009; Morselli et al., 2007, 2008) due to different MSW characteristics, technology options, spatial and temporal factors, and the varying availability of management information (Liamsanguan and Gheewala, 2008). However, LCA cannot provide a truly comprehensive and all-inclusive assessment due to limited input and output data or industrial processes that are extensively interconnected globally, and complete consideration of these interdependencies is prohibitive (Todd and Curran, 1999). These limitations resulted in establishing a baseline of environmental performance to compare alternative product systems in the context of streamlined LCA (Todd and Curran, 1999). Yet, the key for a successful streamlined LCA is determining what can be eliminated from a full-scale LCA design and still meet the study goals.

An inevitable problem of LCA is the existence of uncertainties, discrepancies, and variations of the data collected and used for assessment, regardless of which type of LCA is chosen. The reliability of LCA is thus heavily affected by the dependence on data integrity, especially when data are collected from different countries or sources based on differing operational scenarios. This reliance becomes a crucial limitation for clear interpretation of LCA results (Sonnemann et al., 2003). Improving data reliability requires using rather extensive field and laboratory measurements to bridge the gap embedded in previous studies (Consonni et al., 2005). To date, limitations due to the data quality and heterogeneity have been recognized as unresolved problems in LCA, and without characterizing and clarifying these relevant issues of data uncertainty, the reliability associated with LCA outputs cannot be adequately addressed in the decision-making process. Sensitivity analysis therefore becomes a necessary step for realizing the most sensitive factors of the input elements leading to the improvements of system reliability.

This chapter evaluates the environmental impact associated with two different types of solid waste incineration technology. Instead of using the cradle-to-grave approach, comparative streamlined LCA, which is a routine element of defining the boundaries and data needs (SETAC, 1999), was chosen to simplify the analytical process and reduce the uncertainties driven by the data availability issues associated with larger system boundaries. During assessment, the case study in Taiwan was organized as a comparative streamlined life cycle assessment (CSLCA) for two well-known MSW incineration technologies, the FBI and the MGI, both of which are associated with differing consumption levels of energy and/or materials. Disposal of incineration ash and methane emissions from associated landfills was excluded for the reasons of simplicity. The two incineration technologies were designed for burning MSW with similar composition and properties. To improve the credibility of this case study with respect to the two types of incineration process, two LCA methods, Eco-indicator 99 (EI'99) and ReCiPe methods, were applied for intercomparisons, although the two methods are not compatible with each other. Final efforts were directed toward the sensitivity embedded in all input/output (I/O) information involved in the practice of such a unique CSLCA.

11.3 LCA ASSESSMENT METHODOLOGY

LCA is defined by the Society of Environmental Toxicology and Chemistry (SETAC) as an objective process to evaluate the environmental burdens associated with a product, process, or activity by identifying and quantifying energy and materials used and waste released to environment, and to evaluate and implement opportunities to effect environmental improvements (Barton et al., 1996; Cherubini et al., 2009; De Feo and Malvano, 2009). The development of the international standards for LCA (ISO 14040 in 1997, ISO 14041 in 1999, ISO 14042 in 2000a, ISO 14043 in

2000b, ISO 14044 in 2006b) was an important step to consolidate procedures and methods of LCA. A revised ISO 14040 (ISO, 2006b) standard (Environmental Management – Life Cycle Assessment-Principles and Framework) and a new standard 14044 (ISO, 2006b) containing all requirements (Environmental management – Life Cycle Assessment – Requirements and Guidelines) were proposed in 2006, which are the most authoritative references of LCA. With these requirements and guidelines, the methodology of CSLCA in this study can be described by four interrelated phases: (1) goal and scope definition, (2) LCI, (3) life cycle impact assessment (LCIA), and (4) life cycle interpretation (ISO, 2006a), described separately below.

11.3.1 Goal and Scope Definition

Objective   The objective of this study was to quantify and compare the environmental impacts resulting from two different types of popular waste incineration systems: the FBI and the MGI. LCI, LCIA, and sensitivity analysis were conducted to identify the advantages and limitations of these two types of incineration technology.

Functional Unit   In LCA, the functional unit is defined as the functional outputs of the product system. The main purpose of the functional unit is to provide a reference to relate inputs and outputs (Rebitzer et al., 2004); however, for waste management practices, the functional unit must be defined in terms of systems input (Cherubini et al., 2009). Because the main purpose of incineration is MSW treatment, 1 tonne of MSW treated was chosen as the functional unit for each type of incineration technology in the context of the CSLCA in this study.

System Boundary   During the comparison between the two waste incineration technologies, the system boundary considered for LCA must be sufficiently broad to account for all environmentally relevant burdens within the life cycle (Khoo, 2009). The system boundary is therefore composed of the areas from the waste streams entering the incineration plant to the residual substances being discharged and emitted into the environment. Processes of waste feed, waste treatment (incineration), flue gas treatment, and ash generation were collectively assessed for the environmental impact evaluation (Figure 11.1). The wastewater effluent was not considered because both types of incinerator were designed to meet the zero discharge requirements, and all wastewater streams were treated, recycled, and reused in the plant. Further, both MSW collection and transfer processes are all handled by public utility in Taiwan and are assumed to be identical in terms of environmental impact across different types of incineration technology. For these reasons, the environmental impacts of MSW collection and transfer were not considered in this CSLCA.

11.3.2 Life Cycle Inventory Analysis

LCI is a tool designed to investigate resource and material use as well as fuel and electricity consumption and air pollutant emissions for each type of incinerator, in

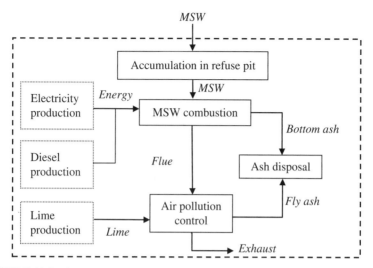

FIGURE 11.1   System Boundary of this LCA study. *Source*: From Ning et al. (2013)

which the data can be illustrated as corresponding quantities per functional unit. Data used in this CSLCA were normally adopted from the regular monitoring records and the reports published by the Environmental Protection Agency in Taiwan. Information related to fuel and electricity consumption of these two types of incinerator was collected from local sources (Yang, 2008). The remaining information was produced by the SimaPro database and the self-organized chemical mass balance analysis.

11.3.3 Life Cycle Impact Assessment

LCA aims to produce a system analysis of all the environmental burdens in every phase of a product's lifetime that can apply various tools and platforms as support. A full-scale LCA begins with a systematic inventory analysis of all possible emissions and the required resource consumption during a product's entire life cycle. The result is an inventory table with a list of emissions, consumed resources, and non-material impacts such as land use. Because inventory tables are normally lengthy, LCA tends to classify the impacts listed in an inventory table using several designated impact categories, such as greenhouse effect, ozone layer depletion, and acidification, among others. How these abstract impact categories are weighted in final decision-making is subject to detailed decision analysis. Such complexity produces LCA outcomes that cannot be unambiguously interpreted. Different impact assessment methods might create even more uncertainty in decision analysis.

For instance, the environmental impact assessment based on both the EI'99 and ReCiPe methods was used to calculate and compare the potential environmental impacts in this study in Taiwan. EI'99 is an evaluation system that uses a damage-oriented approach; in comparison, ReCiPe is evaluation software that uses problem-oriented and damage-oriented approaches that combines the EI'99 method and

Centrum voor Milieuwetenschappen Leiden (CML) model to carry out the CSLCA. CML is a database that contains characterization factors of the LCIA developed by Leiden University. Both methods have a similar assessment structure in which damages are categorized into three groups: human health, ecosystem quality, and resources. With this inclusion, the evaluation results of environmental impacts based on the EI'99 and ReCiPe methods are comparable, although not compatible.

Eco-indicator 99   EI'99, the successor of EI'95, allows designers to aggregate LCA results into easily understandable and user-friendly units. The EI'99 methodology developed by the PRé Consultants is an LCA weighting method specially developed for product design, refined by a number of Swiss and Dutch LCA experts and the Dutch National Institute of Public Health and the Environment (in Dutch: *Rijksinstituut voor Volksgezondheid en Milieu* (RIVM)) (Goedkoop and Spriensma, 1999; Goedkoop et al., 1999; PRé Consultants, 2010). This method adopts an assessment methodology that can transform the data of the inventory table into damage scores and present the relation between the impact and the damage to human health or to the ecosystem (Goedkoop and Spriensma, 1999).

The analysis procedure includes characterization, damage assessment, normalization, and weighting, leading to the generation of a single score over selected scenarios for comparative analyses. First, the impacts are aggregated to form a score card with respect to a number of environmental effects. Characterization of the impacts is in accordance with the degree to which they contribute to an effect (PRé Consultants, 2010). A panel of 365 persons from a Swiss LCA interest group was asked to assess the severity of the three damage categories (endpoints) (Goedkoop et al., 1999; PRé Consultants, 2010):

- Damage to human health is expressed as the number of years of life lost and the number of years lived disabled, which combines results as disability adjusted life years (DALY), an index also used by the World Bank and the World Health Organization.
- Damage to ecosystem quality is expressed as the loss of species over a certain area during a certain time.
- Damage to resources is expressed as the surplus energy needed for future extractions of minerals and fossil fuels.

General representation of the EI'99 methodology includes three steps. Step 1 is the inventory analysis from mining to converting, to milling, pressing, transporting, and disposal. Step 2 is designed to conduct (1) resource analysis, land-use analysis, and fate analysis; (2) exposure and effect analysis; and (3) damage analysis. Step 3 focuses on normalization and weighting across the assessment metrics in Step 2 to conclude the LCA. Normalization is performed on a damage category level. Normalization data are calculated based on a European database, mostly referring to 1993 as the base year, with some updates for the most important emissions. Finally, weighting is performed at damage category level (endpoint level in ISO). For the single score stage, the different damage categories are totaled to form a single environmental index with a unit of measurement expressed in Peta tons (Pt; 1 Pt = 10^9 tonnes) (Morselli et al., 2008).

In scientific methodology development, two types of uncertainty are considered (Goedkoop et al., 1999; PRé Consultants, 2010).

- **Fundamental uncertainty:** The choice of a concept or a method implies that the assumptions that form the basis of this concept or method are fixed. This uncertainty cannot be easily quantified.

- **Operational uncertainty:** The variation in the result of the calculations, caused by the variation of the parameters involved. This uncertainty can be quantified by using Monte Carlo simulation approach.

To address the subjective choice in the impact assessment models, the EI'99 recommends three different versions of the methodology, using the archetypes specified in Cultural Theory (Thompson et al., 1990; PRé Consultants, 2010).

- **Egalitarian perspective:** The chosen time horizon is extremely long term; substances are included if any effect is indicated. In the egalitarian perspective, damages cannot be avoided and may lead to catastrophic events.

- **Hierarchist perspective:** The chosen time horizon is long term, and substances are included if there is consensus regarding their effect. In the hierarchist perspective, damages are assumed to be avoidable through effective management.

- **Individualist perspective:** The chosen time horizon is short term. Substances are included if complete proof regarding their effect is evident. In the individualist perspective, damages are assumed to be recoverable by technological and economic development.

The hierarchist perspective in many issues occupies some score of middle ground between the two extremes of egalitarian and individualist; thus, the option chosen in this study was the hierarchist/average perspective from the EI'99.

ReCiPe Method   ReCiPe is the successor of both EI'99 and CML, developed to integrate the advantages of both the problem-oriented approach of CML and damage-oriented approach of EI'99. The problem-oriented approach defines the impact categories at a midpoint level; the uncertainty of the results at this point is relatively low. The damage oriented-approach of the EI'99 results in only three impact categories, which facilitates drawing conclusions. ReCiPe implements both strategies and has both midpoint and endpoint impact categories. The analysis procedure incorporates characterization, damage assessment, normalization, and weighting to generate a single score. At the midpoint level, 18 impact categories are addressed. The midpoint characterization factors are then multiplied with a damage factor to obtain the endpoint characterization values following the aforementioned endpoint categories: (1) human health (expressed as DALY), (2) ecosystem quality (expressed as species loss in an area for a specific timeframe), and (3) resources (expressed as surplus costs over an infinitive timeframe, considering a 3% inflation). The normalization figures used in SimaPro are recalculated per citizen; the population size of EU25+3 chosen was 464,036,294 citizens, given a world population of 6,055,000,000. In this method, weighting is performed at damage category level. A panel is needed to perform weighting generation of the three damage categories (PRé Consultants, 2010).

The characterization models are a main source of uncertainty that reflects our incomplete and uncertain knowledge of the environmental mechanisms. Like the calculation in EI'99, the ReCiPe method was designed to group different sources of uncertainty and different choices into a limited number of perspectives or scenarios.

Comparison between the Eco-indicator 99 and ReCiPe Methods In this study, the ReCiPe method with an average weighting set was chosen as the default to avoid emphasizing any particular phase. The various weighting sets reflect different phases of concern. The assessment criteria in the EI'99 and ReCiPe practices can be summarized for comparison (Table 11.1) (Goedkoop et al., 2000; PRé Consultants, 2010).

TABLE 11.1   Assessment criteria for EI'99 and ReCiPe methods

EI'99 method		ReCiPe method	
Damage types	Impact categories	Damage types	Impact categories
Human health	Carcinogens	Human health	Climate change human health
	Respiratory organics		Ozone depletion
	Respiratory inorganics		Human toxicity
	Climate change		Photochemical oxidant formation
	Radiation		Particulate matter formation
	Ozone layer		Ionizing radiation
Ecosystem quality	Ecotoxicity	Ecosystems	Climate change ecosystems
			Terrestrial acidification
			Freshwater eutrophication
	Acidification/ eutrophication		Marine eutrophication
			Terrestrial ecotoxicity
			Freshwater ecotoxicity
			Marine ecotoxicity
	Land use		Agricultural land occupation
			Urban land occupation
			Natural land transformation
Resources	Minerals	Resources cost	Metal depletion
	Fossil fuels		Fossil depletion

Source: From Ning et al. (2013).

11.3.4 Interpretation

The useful results from inventory analysis and impact assessment are discussed in accordance with the goal and scope. Comparison of environmental impacts resulting from two types of MSW incinerators is emphasized and illustrated by the categories involved in the proposed CSLCA.

11.3.5 Sensitivity Analysis

As previously described, the problem of uncertainty may limit the reliability of this CSLCA for decision-making. To overcome this barrier, we identified a few significant parameters that mostly affect the environmental impact evaluation by checking all environmental indicators via a sensitivity analysis. The most important parameters identified can then be elucidated, particularly to identify possible variations linked to environmental impacts due to the unexpected changes of their parameter values.

The different weighting sets in EI'99 model originate from the concept of cultural theory (Thompson et al., 1990). These value systems are derived examining the strength of the relation people have with their group and the degree an individual's life is circumscribed by externally imposed prescriptions (their "grid"). The viable combinations of the position of each individual in this group-grid typology and their cultural bias are called "way of life" (Thompson et al., 1990). The detailed perceptions of concern cover management style, perception of time, energy future, attitude to nature, attitude toward humans, and other similar perspectives. The hierarchist archetype is a more moderate setting compared to the other archetypes (Thompson et al., 1990) and has been more frequently selected as the evaluation model in related research. Furthermore, the weighting sets in other archetypes (Egalitarian or Individualist) are different from each other, which change the final value of evaluation but do not reverse the results in terms of the order of environmental impacts. For this reason, we did not consider the sensitivity analysis of the weighting sets in this study.

11.4 DESCRIPTION OF THE CSLCA

The FBI and MGI were compared for environmental performance in this study. A small-scale FBI located in Caotun, a small township in Central Taiwan, was selected as Case 1. The construction of this incinerator in 2003 was established through a Build-Operate-Transfer (BOT) contract to solve the waste management problem of this town. The design capacity and heating value were 95 tonne $\cdot$ per day and 2300 kcal $\cdot$ kg^{-1}, respectively (Table 11.2). The FBI consists of a refractory-lined combustion vessel partially filled with particles of sand or other inert materials. Combustion air is supplied through a distributor plate at the base of the vessel at a sufficient velocity to fluidize the sand bed. It can be operated by two modes, including bubbling and circulating bed modes. In the circulating bed mode, air velocities are greater, and the solids are blown overhead, separated in a cyclone, and then returned

TABLE 11.2 Design profile for two types of incinerators in this study

Item		Case 1 (FBI)	Case 2 (MGI)
	Designed capacity	95 t · d^{-1} (1 set)	300 t · d^{-1} (2 sets)
	Designed heat value	2300 kcal · kg^{-1}	2400 kcal · kg^{-1}
	Furnace	Fluidized bed	Mechanical grate
Exhaust	Dust collector	Baghouse filter	Baghouse filter
treatment	Scrubber	Dry scrubber	Semi-dry scrubber
	Denitrification system	N/A	Selective non-catalytic reaction equipment (SNCR)
	Dioxin removal	Activated carbon injection (dry)	Activated carbon injection (dry)

Source: From Ning et al. (2013).
Tonne (t) is metric ton (= 1000 kg).

to the combustion chamber. This type of incinerator is primarily used for sludge or shredded waste materials (i.e., refuse-derived fuel (RDF)) and is rarely used for MSW incineration (Saxena and Jotshi, 1996). In the bubbling bed mode, however, shredded MSW (i.e., fluff RDF) after recycling can be the feedstock to the sand bed for complete combustion, given a sufficient auxiliary air supply.

In the Caotun plant, the waste streams in the storage pits must first be crushed and shredded to the appropriate size. After complete stirring, the shredded waste materials is fed into the hopper at the top of furnace and then pushed over to the incineration bed by a ram feeder. The fluidized bed increases the contact between the waste, the combustion air, and the hot sand bed, thus facilitating complete combustion. Thermal destruction of a wide variety of materials is effectively accomplished in a high temperature combustion environment. The flue gas treatment process connected to the fluidized bed is mainly designed to cool down the gas temperature and remove the acid gases, particulate matters, heavy metals, and toxic organic pollutants (such as dioxins and furans) produced by the incineration process. A series of flue gas treatment units, including a gas cooler, a heat exchanger, activated carbon injection equipment, and a baghouse filter, are installed in the flue gas treatment process to minimize the air pollution impact and ensure that the emissions comply with the environmental regulations (Figure 11.2).

An MGI in Keelung, Northern Taiwan, commissioned in 2005 was selected as Case 2 for comparison. The design capacity and heating value are 600 tonne · per day (two treatment trains) and 2400 kcal · kg^{-1}, respectively (Table 11.2). The MGI feeds waste streams by using a feed chute and transport waste streams by shaking or rotating the grate within the combustor. In this incineration process, transporting and mixing are carried out by mechanical grates simultaneously. Raw waste streams without shredding are appropriately pushed over the grate by a ram feeder to form a bed. This bed, situated at the top of a mechanical grate, can perform the initial heat-up, evaporation of moisture, pyrolysis, gas-phase combustion, and oxidation of the char sequentially by means of the self-sustained combustion temperature. When

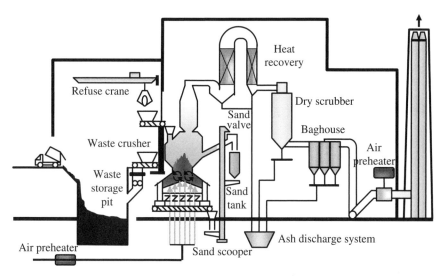

FIGURE 11.2   Schematic representation of incineration process of the FBI (Caotun plant, Case 1). *Source*: From Ning et al. (2013)

the incineration process is complete, the remaining ash is cooled by the air supply and is finally discharged into a hopper (Shina and Choi, 2000). The flue gas treatment system is integrated by a series of unit operations, including selective non-catalytic reaction equipment, a semi-dry scrubber, activated carbon injection equipment, and a baghouse filter (Figure 11.3).

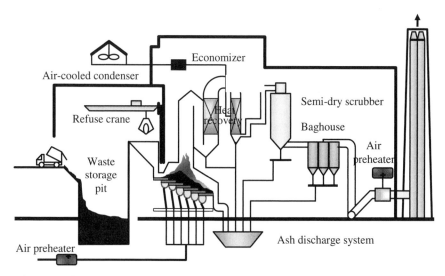

FIGURE 11.3   Schematic representation of incineration process of the MGI (Keelung plant, Case 2). *Source*: From Ning et al. (2013)

Heat recovery facilities were installed in both incinerators to supply part of the energy requirement of the plants and substantially decrease external energy input. No specific measurements of heat recovery efficiencies are available for individual evaluation, however; therefore thermal effects were disregarded in this CSLCA.

11.5 INTERPRETATION OF CSLCA RESULTS

11.5.1 Life Cycle Inventory

The LCI realm in this study conducted in Taiwan includes waste treatment, flue gas treatment, and ash disposal, which are the common processes of the FBI and MGI. Regular monitoring data were collected from operational reports of these two plants. Some essential measurements required by our LCI were estimated based on the mass balance method, simulation software, and related references when direct measurements were unavailable. The individual feedstock for waste treatment and the corresponding pollutant emissions were summarized for these two plants in Tables 11.3 and 11.4.

Because Taiwan is a small island, the waste composition of feedstock destined for incineration is deemed similar in these two case studies (i.e., the two plants); therefore, for this CSLCA we adopted the average waste composition over the past 10 years in Taiwan for both cases (Table 11.5), which decreases the uncertainty in the CSLCA. The FBI requires a substantial supply of fuel oil and sand for MSW incineration,

TABLE 11.3   The technical parameters for two types of incinerator

Inventory item	Unit	Case 1 (FBI)	Case 2 (MGI)
Waste lower heating value	$kcal \cdot t \cdot d^{-1}$	1,588,900	1886
Waste higher heating value	$kcal \cdot t \cdot d^{-1}$	2,057,100	2388.9
Electricity consumption	$kWh \cdot t \cdot d^{-1}$	176.94	3126.48
Oil consumption for hot start-up	$kg \cdot t \cdot d^{-1}$	328.4	–
Sand consumption	$kg \cdot t \cdot d^{-1}$	11.377	–
Water consumption for exhaust gas cooler	$kg \cdot t \cdot d^{-1}$	2.037.47	–
Water consumption for wetting fly ash	$kg \cdot t \cdot d^{-1}$	707.37	–
5% Urea water consumption for selective non-catalytic reaction equipment	$kg \cdot t \cdot d^{-1}$	–	4.64
Water consumption for boiler	$kg \cdot t \cdot d^{-1}$	–	1088.8
Hydrated lime consumption for attemperator	$kg \cdot t \cdot d^{-1}$	–	5.76
Cooling water consumption for attemperator	$kg \cdot t \cdot d^{-1}$	–	22.76
Activated carbon consumption for dust collector	$kg \cdot t \cdot d^{-1}$	–	0.12
Auxiliary fuel oil consumption for hot blast stove	$kg \cdot t \cdot d^{-1}$	–	10.36

Source: From Ning et al. (2013).
Tonne (t) is metric ton.

TABLE 11.4 Results of pollutant emissions for two types of incinerator

Inventory item	Unit	Case 1 (FBI)	Case 2 (MGI)
Displacement of flue gas	$Nm^3 \cdot t \cdot d^{-1}$	2877	6153
Ash from economizer	$kg \cdot t \cdot d^{-1}$	–	0.48
Ash from attemperator	$kg \cdot t \cdot d^{-1}$	–	2.52
Ash from dust collector	$kg \cdot t \cdot d^{-1}$	50.53	10.04
CO_2	$kg \cdot t \cdot d^{-1}$	656.33	741.03
NO_x	$kg \cdot t \cdot d^{-1}$	0.402	1.909
SO_x	$kg \cdot t \cdot d^{-1}$	0.226	0.137
CO	$kg \cdot t \cdot d^{-1}$	0.180	0.117
HCl	$kg \cdot t \cdot d^{-1}$	0.071	0.22
PM_{10}	$kg \cdot t \cdot d^{-1}$	0.0661	0.0636
Pb	$kg \cdot t \cdot d^{-1}$	0.0001535	0.0000225
Cd	$kg \cdot t \cdot d^{-1}$	0.0000283	0.0000006
Hg	$kg \cdot t \cdot d^{-1}$	0.0000391	0.0000254
Dioxin	$ng\ I\text{-}TEQ \cdot t^{-1}$	377.75	227.05

Source: From Ning et al. (2013).
Tonne (t) is metric ton.

whereas the MGI requires a supply of urea, lime, and activated carbon for flue gas treatment. Nevertheless, the electricity consumption of the MGI is much higher than that of the FBI. These contradictory demand agglomerates inherently compete in terms of sustainability implications across these two cases in our comparative study.

An obvious difference exists between the pollutant emissions of these two types of incinerators. The emissions of carbon dioxide (CO_2), nitrogen oxides (NO_x), hydrogen chloride (HCl), and lead (Pb) from the MGI are higher than those from the FBI. In contrast, more sulfur oxides (SO_x), carbon monoxide (CO), particulate matters (PMs), cadmium (Cd), mercury (Hg), and dioxin/furans (polychlorinated dibenzo-dioxins (PCDD) and polychlorinated dibenzo-furans (PCDF)) were emitted from the FBI. These differing levels of emission from the two incinerators firmly support the fundamental differentiation of environmental impacts in the subsequent individual assessments. Although air pollutants such as ammonia (NH_3), volatile organic compound (VOC), hydrogen fluoride (HF), arsenic (As), chromium (Cr), cobalt (Co), nickel (Ni) may occur during the incineration process, they were not included in this LCA study because these monitoring data were not available in our LCI.

11.5.2 Impact Assessment

Normalized Analysis Normalization in our CSLCA converts the actual measurements to an equivalent dimensionless quantity that indicates the relative significance of various environmental impacts during a designated period. In this study, the normalization factor provided by EI'99 and ReCiPe methods was directly used to accomplish the conversion.

TABLE 11.5 Results of chemical compositions of two cities

District	Year	Moisture	Ash	Combustibles								HHV (kcal · kg⁻¹)	LHV (kcal · kg⁻¹)
				Total	C	H	O	N	S	Cl	Others	Chemical compositions (wt%)	
Caotun	2010	60.28	3.77	35.95	19.77	3.21	12.50	0.29	0.10	0.09	–	2172.96	1637.94
	2009	57.62	4.55	37.83	20.9	2.53	13.82	0.35	0.16	0.07	–	2113.34	1631.03
	2008	55.07	4.21	40.72	21.47	2.83	15.77	0.29	0.24	0.12	–	2294.32	1811.08
	2007	49.06	8.47	42.46	20.46	3.43	17.99	0.29	0.15	0.15	–	2476.94	1997.11
	2006	53.08	5.03	41.89	19.35	3.10	18.96	0.26	0.16	0.06	–	2244.51	1758.86
	2005	57.78	6.05	36.17	17.34	2.80	15.11	0.32	0.50	0.11	–	2105.74	1628.16
	2004	56.02	6.34	37.64	17.72	2.77	16.13	0.46	0.32	0.04	0.20	1951.51	1465.98
	2003	53.55	7.20	39.25	17.90	2.72	17.48	0.43	0.41	0.05	0.26	2057.05	1588.90
	2002	45.16	13.52	41.32	22.39	3.45	14.71	0.43	0.09	0.25	–	2327.48	1870.15
	2001	57.72	8.36	33.92	18.34	2.73	11.65	0.63	0.19	0.15	0.23	1926.00	1483.00
Average		54.53	6.75	38.72	19.56	2.96	15.41	0.38	0.23	0.11	0.07	2166.99	1687.22
Keelung	2010	54.50	5.63	39.87	21.16	3.66	14.17	0.59	0.18	0.11	–	2192.21	1667.55
	2009	56.89	4.44	38.67	20.51	2.64	14.98	0.28	0.17	0.10	–	2137.01	1653.37
	2008	55.62	3.45	40.93	22.40	3.07	14.53	0.52	0.32	0.09	–	2388.93	1889.43
	2007	55.67	3.85	40.47	19.93	3.40	16.58	0.26	0.14	0.17	–	2365.84	1848.32
	2006	52.49	5.00	42.51	19.49	3.18	19.16	0.31	0.29	0.07	–	2199.75	1712.96
	2005	51.87	8.55	39.59	17.53	2.88	18.33	0.28	0.50	0.08	–	2146.68	1728.71
	2004	48.37	4.58	47.05	23.05	3.59	19.29	0.35	0.52	0.05	0.20	1836.90	1352.85
	2003	49.59	8.80	41.61	18.67	2.61	19.00	0.41	0.64	0.04	0.24	2086.66	1648.53
	2002	48.15	13.66	38.19	20.70	3.18	13.37	0.44	0.12	0.38	–	2154.06	1693.57
	2001	53.82	11.7	34.48	18.61	2.82	12.24	0.44	0.09	0.11	0.17	2058.00	1654.00
Average		52.70	6.97	40.34	20.21	3.10	16.17	0.39	0.30	0.12	0.06	2156.60	1684.93

Source: Ning et al. (2013).
HHV, high heating value (wet basis); LHV, low heating value (wet basis).

Eleven categories of environmental impact were considered in EI'99 method including carcinogens, respiratory organics, respiratory inorganics, climate change, radiation, ozone layer, ecotoxicity, acidification/eutrophication, land use, minerals, and fossil fuels. The respiratory inorganics, climate change, and fossil fuels are the significant environmental impacts in both cases (Figure 11.4a). The impact categories

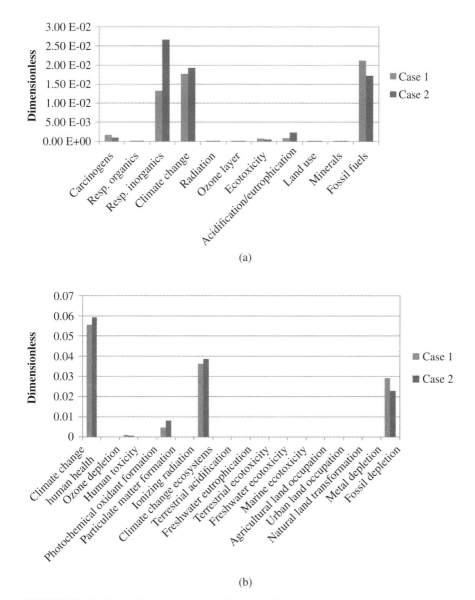

(a)

(b)

FIGURE 11.4 Normalized environmental impacts for each item in the two selected cases. (a) The EI'99 method. (b) The ReCiPe method. *Source*: From Ning et al. (2013)

TABLE 11.6   Results of normalized environmental impact assessment

Types of damage considered in each method	Case 1 (FBI)	Case 2 (MGI)
EI'99 method		
Human health	0.0326	0.0470
Ecosystem quality	0.0016	0.0028
Resources	0.0212	0.0172
ReCiPe method		
Human health	0.0614	0.0682
Ecosystems	0.0363	0.0388
Resources	0.0291	0.0228

Source: From Ning et al. (2013).

can then be used to quantify the three types of damage, including human health, ecosystem quality, and resources. The analysis clearly indicates that the human health impact is relatively higher than other impacts. Within the two cases, the normalized values of human health are 0.0326 in Case 1 and 0.047 in Case 2 (Table 11.6 and Figure 11.5a), implying that the impact of human health arising from the MGI is relatively higher than that from the FBI, partially due to the high flue gas generation rate (3.267 Nm$^3 \cdot$ kg^{-1} or 741.03 kg CO_2 per tonne waste equivalent) and nitrogen oxide emission (1.909 kg NO_x per tonne waste equivalent) from the MGI. This effect may have a larger impact on respiratory issues associated with human health and climate change impact, simultaneously.

The ReCiPe method analysis findings also indicate that the impact of human health is more significant. Note, however, that the result of ecosystem quality damage obtained from ReCiPe method is higher than its counterpart from the EI'99 method, because the damage assessment transfer factor of ecotoxicity in EI'99 is only one-tenth the order of magnitude compared to the corresponding values of other impact categories; therefore, the level of ecosystem quality damage was lower for the EI'99 method. More observations may collectively illustrate the detailed results in the evaluation (Figures 11.4b and 11.5b and Table 11.6). To have a unified combustion assessment for CO_2 emission, waste combustion in both incinerators was assumed to be complete with the aid of auxiliary fuel and air supply and contribute more than 80% of the total CO_2 production in both cases. The remaining CO_2 contributions were associated with the consumption of electricity, fuel oil delivery, and the preparation of hydrated lime in the process of incineration. Various sources of CO_2 emissions can be summarized with respect to such technical settings (Figure 11.6).

Weighted Evaluation   Weighted evaluation summed up the environmental impacts from each categorical output and followed two steps: (1) total scores of each impact category associated with each pollutant were first estimated during the environmental impact evaluation process, and (2) the scores associated with each impact category were added following the EI'99 and ReCiPe methods to represent the contribution of the overall environmental impacts regarding category. The

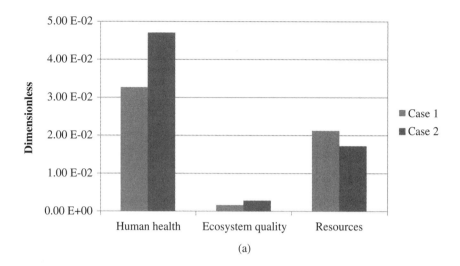

(a)

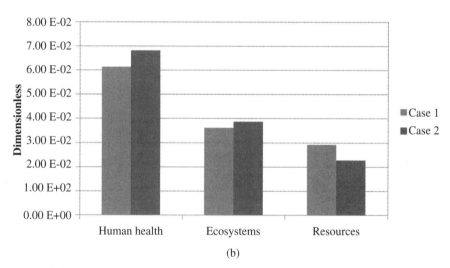

(b)

FIGURE 11.5 Normalized environmental impacts for each damage category in the two selected cases. (a) The EI'99 method. (b) The ReCiPe method. *Source*: From Ning et al. (2013)

weighted environmental impacts could then be summarized with respect to these two types of incinerator for comparative analyses (Table 11.7 and Figure 11.7).

Using the EI'99 method, the environmental impact indicator of Case 2 was 23.35 per kg waste incinerated, which is higher than 17.91 per kg waste incinerated in Case 1, confirming that the FBI is relatively environmentally benign when compared with the MGI. Similar results were also found when using ReCiPe method, in which the impact indicators of Cases 1 and 2 are 45.03 and 47.47 per kg waste

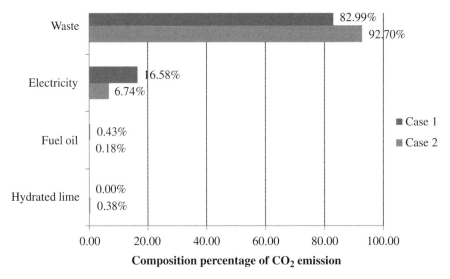

FIGURE 11.6   The composition percentage of CO_2 emission. *Source*: From Ning et al. (2013)

incinerated, respectively. The proportion of potential damage of ecosystem quality was higher in the ReCiPe method than it would be in EI'99 method, however, not only due to the higher weighting set applied for weighted evaluation, but also due to the higher damage assessment transfer factor produced through the normalization stage. Nevertheless, MGI exhibits a relatively higher environmental impact due to the potential damage of human health. Both CO_2 and NO_x emissions associated with the FBI and MGI are additional factors that merit attention in human health impacts.

TABLE 11.7   Results of weighted environmental impact assessment

Impact category considered in each method	Unit	Case 1 (FBI)	Case 2 (MGI)
EI'99 Method			
Human health	Pt	13.0	18.81
Ecosystem quality	Pt	0.66	1.11
Resources	Pt	4.25	3.43
Total	Pt	17.91	23.35
ReCiPe Method			
Human health	Pt	24.60	27.34
Ecosystems	Pt	14.60	15.56
Resources	Pt	5.83	4.57
Total	Pt	45.03	47.47

Source: Ning et al. (2013).

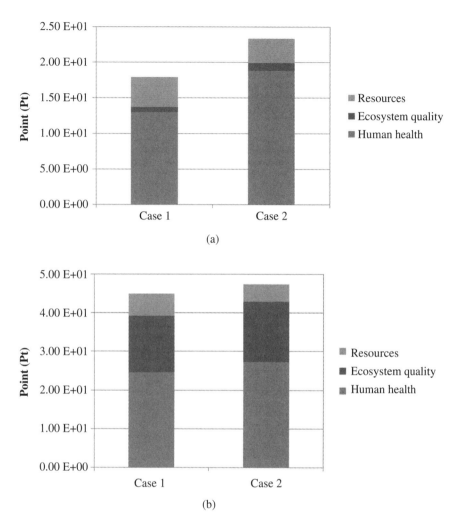

FIGURE 11.7 Weighted environmental impacts for the two selected cases. (a) The EI'99 method. (b) The ReCiPe method. *Source*: From Ning et al. (2013)

11.5.3 Sensitivity Analysis

Numerous parameters embedded in the environmental impact evaluation inevitably influence the assessment outcome. Sensitivity analysis could help reveal the uncertainty arising from varying parameter settings. Based on the value we set initially, sensitivity can be tested in a multitude of ways through the selection of the weighted environmental impact indicators. To systematically assess the sensitivity, 10% of each parameter value was selected for this analysis. The EI'99 method simulation results show that CO_2 emission exhibits the greatest influence in Case 1, caused by the presence of higher composition of carbon in the MSW. The second largest influential parameter is associated with the electricity consumption because the majority of electricity generation facilities in Taiwan depend on the use of thermal power plants such as coal-fired power plants and natural gas power plants, the major source of emissions of both greenhouse gases and heavy metals. In addition, the same impact categories were evaluated for the ReCiPe method, except for the consideration of Cd and Hg. The overall sensitivity analysis of Case 1 can then be elucidated (Figure 11.8).

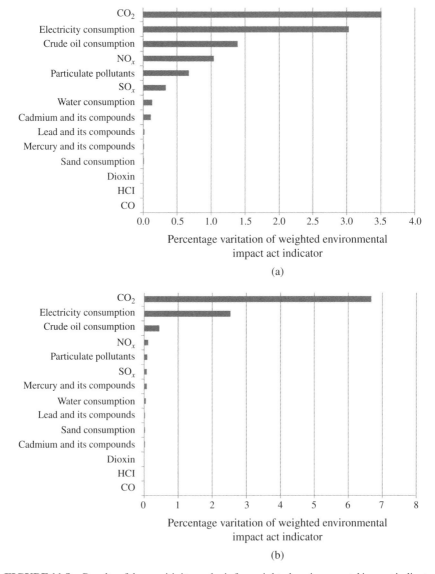

FIGURE 11.8   Results of the sensitivity analysis for weighted environmental impact indicator in Case 1. (a) The EI'99 method. (b) The ReCiPe method. *Source*: From Ning et al. (2013)

Conversely, in Case 2, CO_2 emission exhibits the greatest influence as the result of high emissions of CO_2 on a per tonne waste basis (i.e., 741.03 kg CO_2), which is approximately 1.13 times larger when compared to its counterpart measured at the Caotun plant (i.e., Case 1). The second largest impact when using the EI'99 method is NO_x emissions, which are substantially different from the ReCiPe method results because of the emphasis of NO_x in relation to acidification in the EI'99 method, which is not the case in the ReCiPe method. The overall sensitivity analysis of Case 2 can be further illustrated graphically (Figure 11.9).

11.5.4 Improvement Analysis

From an environmental management viewpoint, the incinerator must reduce the generation rate of flue gas to decrease the environmental impacts on GWP. Additionally, electricity consumption has a significant effect on environmental impact, which is mainly caused by the choice of thermal power plants commonly used in Taiwan for power supply. For improvement analysis, the diversity of power generation facilities (e.g., nuclear power generation, hydropower generation, coal power generation) is significant when examining the importance of power consumption for incinerator operation and its related environmental impacts; however, our intent was not to discuss the essential power generation structures in Taiwan. Our CSLCA cannot lead to a sound suggestion for tuning the structure of power industry in Taiwan to meet the sustainability goal. Rather, the influential level of electricity consumption and emission factors is the focus in our environmental impact assessment, which is indirectly linked with the possible improvements as the result of CSLCA. In any circumstance, such a limitation would not rule out the opportunities for investigating the power generation structure in some developed countries. A more rigorous analysis using MADM, in which the CSLCA outputs could be an integral part of the overall modeling efforts, may be formulated for advanced assessment.

The power generation structures in France (78% nuclear power), Norway (98% hydropower) and Poland (93% coal power) were selected for comparison (IEA, 2008). The environmental impacts derived from differing energy structures in various countries are based on the built-in information of the SimaPro software database (Figure 11.10). According to the simulation results from the EI'99 method, we observed that the environmental impacts of carcinogens, respiratory inorganics, climate change, and fossil fuels were relatively higher when the current structure of electricity generation in Taiwan was employed. In particular, respiratory impacts associated with the use of fossil fuels had the most severe impact. This value is several times higher than the corresponding values based on the current structure of electricity generation in France and Norway, although it is still lower than the corresponding value when choosing the current structure of electricity generation in Poland. This implies an inherent value of using clean energy for power generation (Figure 11.11). Overall, the analysis in this study clearly indicates that the electricity consumption to sustain the two prescribed incineration processes had the greatest influence on environmental impacts via the proposed CSLCA. In a country with a high reliance on nuclear power generation, such as France, or a clean-energy-oriented power generation structure,

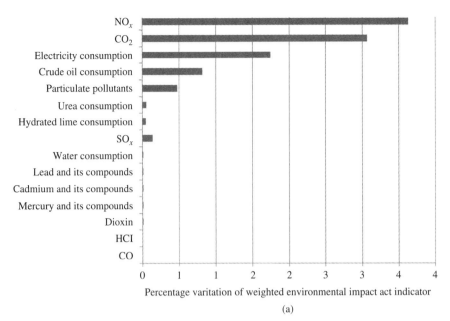

(a)

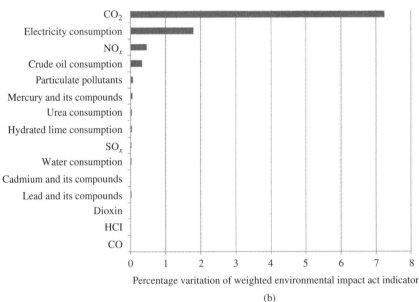

(b)

FIGURE 11.9   Results of the sensitivity analysis for weighted environmental impact indicator in Case 2. (a) The EI'99 method. (b) The ReCiPe method. *Source*: From Ning et al. (2013)

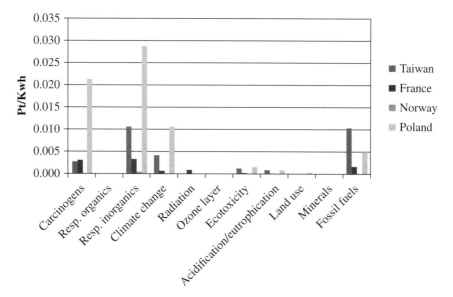

FIGURE 11.10   Single-score-based categorical environmental impact with respect to various power structures in different countries (by EI'99). *Source*: From Ning et al. (2013)

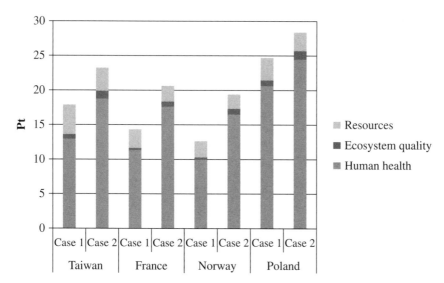

FIGURE 11.11   Evaluation results of weighted environmental impacts based on various power structures in different countries. *Source*: From Ning et al. (2013)

like Norway, the decrease of environmental impacts based on the same level of electricity consumption rate would certainly be anticipated. Quantifying and reducing uncertainty in such a CSLCA may be further carried out by using the Bayesian Monte Carlo method in the future (Lo et al., 2005).

11.6 FINAL REMARKS

This study quantifies and compares the environmental impacts of two types of MSW incinerators (fluidized bed vs. mass-burn mechanical grate systems) used for handling waste treatment. Through the processes of CSLCA in this chapter, several major findings can be summarized, as the weighted values of environmental impacts associated with mass-burn MGI and FBI are 23.35 and 17.91 Pt per kg waste incinerated, respectively, which is based on the EI'99 method in this case study. Also, the ReCiPe method for the same analysis shows that the weighted values of environmental impacts associated with mass-burn MGI and FBI are 47.47 and 45.03 Pt per kg waste incinerated, respectively. Both methods indicate that the FBI is relatively environmentally benign; however, the reliability of the CSLCA assessment results relies on the quality, precision, completeness, and representativeness of collected data. Because the distinction between these two technologies is not significant, a more comprehensive database may be required to enhance the reliability of the CSLCA.

REFERENCES

Assamoi, B. and Lawryshyn, Y. 2012. The environmental comparison of landfilling vs. incineration of MSW accounting for waste diversion. *Waste Management*, 32(5), 1019–1030.

Barton, J. R., Dalley, D., and Patel, V. S. 1996. Life cycle assessment for waste management. *Waste Management*, 16(1–3), 35–50.

Bayer, C., Gamble, M., Gentry, R., and Joshi, S. 2010. *AIA Guide to Building Life Cycle Assessment in Practice*, The American Institute of Architects, Washington, DC.

Björklund, A. and Finnveden, G. 2007. Life cycle assessment of a national policy proposal—the case of a Swedish waste incineration tax. *Waste Management*, 27(8), 1046–1058.

Björklund, A., Dalemo, M., and Sonesson, U. 1999. Evaluating a municipal waste management plan using ORWARE. *Journal of Cleaner Production*, 7(4), 271–280.

Buttol, P., Masoni, P., Bonoli, A., Goldoni. S., Belladonna, V., and Cavazzuti, C. 2007. LCA of integrated MSW management systems: case study of the Bologna District. *Waste Management*, 27(8), 1059–1070.

Chang, N. B., Pires, A., and Martinho, G. 2011. Empowering systems analysis for solid waste management: challenges, trends and perspectives. *Critical Reviews in Environmental Science and Technology*, 41(16), 1449–1530.

Chaya, W. and Gheewala, H. 2007. Life cycle assessment of MSW-to-energy schemes in Thailand. *Journal of Cleaner Production*, 15(15), 1463–1468.

Cherubini, F., Bargigli, S., and Ulgiati, S. 2009. Life cycle assessment (LCA) of waste management strategies: landfilling, sorting plant and incineration. *Energy*, 34(12), 2116–2123.

Consonni, S., Giugliano, M., and Grosso, M. 2005. Alternative strategies for energy recovery from municipal solid waste Part B: emission and cost estimates. *Waste Management*, 25(2), 137–148.

Contreras, F., Hanaki, K., Aramaki, T., and Connors, S. 2008. Application of analytical hierarchy process to analyze stakeholders preferences for municipal solid waste management plans, Boston, USA. *Resources, Conservation and Recycling*, 52(7), 979–991.

Dalemo, M., Sonesson, U., Bjorklund, A., Mingarini, K., Frostell, B., Nybrant, T., Jonsson, H., Sundqvist, J.-O., and Thyselius, L. 1997. ORWARE—a simulation model for organic waste handling systems. Part 1: model description. *Resources, Conservation and Recycling*, 21(1), 17–37.

Damgaard, A., Riber, C., Fruergaard, T., Hulgaard, T., and Christensen, T. H., 2010. Lifecycle-assessment of the historical development of air pollution control and energy recovery in waste incineration. *Waste Management*, 30(7), 1244–1250.

De Feo, G. and Malvano, C. 2009. The use of LCA in selecting the best MSW management system. *Waste Management*, 29(6), 1901–1915.

Diaz, R. and Warith, M. 2006. Life-cycle assessment of municipal solid wastes: development of the WASTED model. *Waste Management*, 26(8), 886–901.

Eriksson, O., Frostell, B., Bjorklund, A., Assefa, G., Sundwvist, J.-O., Granath, J., Carlsoon, M., Baky, A., and Thyselius, L. 2002. ORWARE – a simulation tool for waste. *Resources, Conservation and Recycling*, 36(4), 287–307.

Eriksson, O., Reich, M. C., Frostell, B., Björklund, A., Assefa, G., Sundqvist, J. O., Granath, J., Baky, A., and Thyselius, L. 2005. Municipal solid waste management from a systems perspective. *Journal of Cleaner Production*, 13(3), 241–252.

European Parliament, and Council. 2008. Directive 2008/98/EC of the European Parliament and of the Council of 19 November 2008 on waste and repealing certain Directives. *Official Journal of the European Union*, L312, 3–30.

Finnveden, G. and Ekvall, T. 1998. *Energy Recovery or Material Recycling of Paper Packaging?* [in Swedish: Energieller materialåtervinning av pappersförpakningar?]. Svensk Kartongåtervinning AB, Stockholm.

Finnveden, G., Person, L., and Steen, B. 1994. Packaging in circulation: recycling of cardboard for milk—an LCA study of the differences in environmental load [in Swedish: Förpackningar i kretsloppet: Återvinning av mjölkkartong—En LCA-studie av skillnader i miljöbelastning], Report 4301, Swedish Environmental Protection Agency, Stockholm.

Frees, N. and Weidema, B. 1998. *Life Cycle Assessment of Packaging Systems for Beer and Soft Drinks, Energy and Transport Scenarios (Environmental Project 406).* Danish Environmental Protection Agency, Copenhagen.

Goedkoop, M. and Spriensma, R. 1999. The Eco-indicator 99—a damage oriented method for life cycle impact assessment. Methodology Report. PRé Consultants, Amersfoort, Netherlands.

Goedkoop, M., Effting, S., and Collignon, M. 1999. *The Eco-indicator 99. A Damage Oriented Method for Life Cycle Impact Assessment. Manual for Designers*, PRé Consultants, Amersfoort, Netherlands.

Goedkoop, M., Effting, S., and Collignon, M. 2000. *The Eco-Indicator 99 Manual for Designers*, PRé Consultants.

Guinée, J. B., Gorree, M., Heijungs, R., Huppes, G., Kleijn, R., van Oers, L., Wegener Sleeswijk, A., Suh, S., Udo de Haes, H. A., de Bruijn, J. A., van Duin, R., and

Huijbregts, M. A. J. (Eds.) 2002. *Handbook on Life Cycle Assessment—Operational Guide to the ISO Standards*, Kluwer Academic Publisher, Dordrecht, Netherlands.

Harrison, K., Dumas, R., Solano, E., Barlaz, M., Brill, E., and Ranjithan, S. 2001. Decision support tool for life-cycle-based solid waste management. *Journal of Computing in Civil Engineering*, 15(1), 44–58.

Hong, R. J., Wang, G. F., Guo, R. Z., Cheng, X., Liu, Q., Zhang, P. J., and Qian, G. R. 2006. Life cycle assessment of BMT-based integrated municipal solid waste management: case study in Pudong, China. *Resources, Conservation and Recycling*, 49(2), 129–146.

IEA. 2008. *Energy Balance of OECD Countries*, International Energy Agency, Paris, France.

International Organization for Standardization (ISO). 1997. *Environmental Management-Life Cycle Assessment-Principles and Framework*, ISO.

International Organization for Standardization (ISO). 1999. *ISO 14041 Environmental Management-Life Cycle Assessment-Goal and Scope Definition and Inventory Analysis*, ISO.

International Organization for Standardization (ISO). 2000a. *ISO 14043 Environmental Management-Life Cycle Assessment-Life Cycle Interpretation*, ISO.

International Organization for Standardization (ISO). 2000b. *ISO 14042 Environmental Management-Life Cycle Assessment-Life Cycle Impact Assessment*, ISO.

International Organization for Standardization (ISO). 2006a. *ISO 14040 Environmental Management-Life Cycle Assessment-Principles and Framework*, ISO.

International Organization for Standardization (ISO). 2006b. *ISO 14044 Environmental Management-Life Cycle Assessment-Requirements and Guidelines*, ISO.

Khoo, H. H. 2009. Life cycle impact assessment of various waste conversion technologies. *Waste Management*, 29(6), 1892–1900.

Kijak, R. and Moy, D. 2004. A decision support framework for sustainable waste management. *Journal of Industrial Ecology*, 8(3), 33–50.

Kirkeby, J., Christensen, T., Bhander, G., Hansen, T., and Birgisdottir, H. 2005. LCA modelling of MSW management system: approach and case study. In: Proceedings Sardinia 2005, Tenth International Waste Management and Landfill Symposium, Cagliari, Italy.

Klöpffer, W. 2000. Conference reports: LCA in Brighton. *International Journal of Life Cycle Assessment*, 5(4), 249.

Lee, S. H., Choi, K. I., Osako, M., and Dong, J. I. 2007. Evaluation of environmental burdens caused by changes of food waste management systems in Seoul, Korea. *Science of Total Environment*, 387(1-3), 42–53.

Liamsanguan, C. and Gheewala, S. H. 2008. LCA: a decision support tool for environmental assessment of MSW management systems. *Journal of Environmental Management*, 87(1), 132–138.

Lo, S. C., Ma, H. W., and Lo, S. L. 2005. Quantifying and reducing uncertainty in life cycle assessment using the Bayesian Monte Carlo method. *Science of Total Environment*, 340(1-3), 23–33.

McDougall, F., White, P., Franke, M., and Hindle, P. 2001. *Integrated Solid Waste Management: A Life Cycle Inventory*, Blackwell Science Ltd., Oxford.

Mendes, M. R., Aramaki, T., and Hanaki, K. 2004. Comparison of the environmental impact of incineration and landfilling in São Paulo City as determined by LCA. *Resources, Conservation and Recycling*, 41(1), 47–63.

Møller, J., Munk, B., Crillesen, K., and Christensen, T. H. 2011. Life cycle assessment of selective non-catalytic reduction (SNCR) of nitrous oxides in a full-scale municipal solid waste incinerator. *Waste Management*, 31(6), 1184–1193.

Morselli, L., Bartoli, M., Bertaccchini, M., Brighetti, A., Luzi, J., Passarini, F., and Masoni, P. 2005. Tools for evaluation of impact associated with MSW incineration: LCA and integrated environmental monitoring system. *Waste Management*, 25(2), 191–196.

Morselli, L., Luzi, J., De Robertis, C., Vassura, I., Carrillo, V., and Passarini, F. 2007. Assessment and comparison of the environmental performances of a regional incinerator network. *Waste Management*, 27(8), S85–S91.

Morselli, L., De Robertis, C., Luzi, J., Passarini, F., and Vassura, I. 2008. Environmental impacts of waste incineration in a regional system (Emilia Romagna, Italy) evaluated from a life cycle perspective. *Journal of Hazardous Materials*, 159(2-3), 505–511.

Nilsson, M., Bjorklund, A., Finnveden, G., and Johansson, J. 2005. Testing a SEA methodology for the energy sector: a waste incineration tax proposal. *Environmental Impact Assessment Review 2005*, 25(1), 1–32.

Ning, S. K., Chang, N. B., and Hong, M. C. 2013. Comparative streamlined life cycle assessment of two types of municipal solid waste incinerator. *Journal of Cleaner Production*, 53(15), 56–66.

Özeler, D., Yetis, Ü., and Demirer, G. N. 2006. Life cycle assessment of municipal solid waste management methods: Ankara case study. *Environment International*, 32(3), 405–411.

Person, L., Ekvall, T., and Weidenma, B. P. 1998. *Life Cycle Assessment of Packaging Systems for Beer and Soft Drinks, Refillable Pet Bottles (Environmental Project 404)*, Danish Environmental Protection Agency, Copenhagen.

PRé Consultants. 2010. *SimaPro 7 Database Manual*, Methods Library, Amersfoort, Netherlands.

Rebitzer, G., Ekvall, T., Frischknecht, R., Hunkeler, D., Norris, G., Rydberg, T., Schmidt, W. P., Suhh, S., Weidema, B. P., and Pennington, D. W. 2004. Life cycle assessment Part 1: framework, goal and scope definition, inventory analysis, and applications. *Environment International*, 30(5), 701–720.

Rives, J., Rieradevall, J., and Gabarrell, X., 2010. LCA comparison of container systems in municipal solid waste management. *Waste Management*, 30(6), 949–957.

Ryberg, A., Ekvall, T., Person, L., and Weidema, B. P. 1998. *Life Cycle Assessment of Packaging Systems for Beer and Soft Drinks, Aluminum Cans (Environmental Project 402)*, Danish Environmental Protection Agency, Copenhagen.

Sakai, S., Sawell, S. E., Chandler, A. J., Eighmy, T. T., Kosson, D. S., Vehlow, J., van der Sloot, H. A., Hartlén, J., and Hjelmar, O. 1996. World trends in municipal solid waste management. *Waste Management*, 16(5–6), 341–350.

Saxena, S. C. and Jotshi, C. K. 1996. Management and combustion of hazardous wastes. *Progress in Energy and Combustion Science*, 22(5), 401–425.

Scipioni, A., Mazzi, A., Niero, M., and Boatto, T., 2009. LCA to choose among alternative design solutions: the case study of a new Italian incineration line. *Waste Management*, 29(9), 2462–2474.

Shina, D. and Choi, S. 2000. The combustion of simulated waste particles in a fixed bed. *Combustion and Flame*, 121(1–2), 167–180.

Skordilis, A. 2004. Modelling of integrated solid waste management systems in an island. *Resources, Conservation and Recycling*, 41(3), 243–254.

Society of Environmental Toxicology and Chemistry (SETAC). 1999. Streamlined life-cycle assessment. A Final Report from the SETAC North America Streamlined LCA Workgroup.

Solano, E., Dumas, R., Harrison, K., Ranjithan, S., Barlaz, M., and Brill, E. 2002a. Life-cycle-based solid waste management. I: model development. *Journal of Environmental Engineering*, 128(10), 981–992.

Solano, E., Dumas, R., Harrison, K., Ranjithan, S., Barlaz, M., and Brill, E. 2002b. Life-cycle-based solid waste management. II: illustrative applications. *Journal of Environmental Engineering*, 128(10), 993–1005.

Sonnemann, G. W., Schuhmacher, M., and Castells, F. 2003. Uncertainty assessment by a Monte Carlo simulation in a life cycle inventory of electricity produced by a waste incinerator. *Journal of Cleaner Production*, 11(3), 279–292.

Thompson, M., Ellis, R., and Wildavsky, A. 1990. *Cultural Theory*, Westview Press, Boulder, CO.

Thorneloe, S., Weitz, K., and Jambeck, J. 2007. Application of the US decision support tool for materials and waste management. *Waste Management*, 27(8), 1006–1020.

Todd, J. A. and Curran, M. A. 1999. Streamlined life-cycle assessment. A Final Report from the SETAC North America Streamlined LCA Workgroup. Published by the Society of Environmental Toxicology and Chemistry (SETAC) and SETAC Foundation for Environmental Education.

Weitz, K., Barlaz, M., Ranji, R., Brill, D., Thorneloe, S., and Ham, R. 1999. Life cycle management of municipal solid waste. *International Journal of Life Cycle Assessment*, 4(4), 195–201.

White, P., Franke, M., and Hindle, P. 1995. *Integrated Solid Waste Management: A Life-Cycle Inventory*, Blackie Academic & Professional, Glasgow, UK.

Winkler, J. and Bilitewski, B. 2007. Comparative evaluation of life cycle assessment models for solid waste management. *Waste Management*, 27(8), 1021–1031.

Yang, Y. H. 2008. Life cycle assessment and uncertainty analysis for fossil-fired power plants and fuel selection, Doctoral dissertation, Department of Environmental Engineering, National Cheng Kung University, Tainan, Taiwan.

Zhao, Y., Christensen, T. H., Lu, W., Wu, H., and Wang, H. 2011. Environmental impact assessment of solid waste management in Beijing City, China. *Waste Management*, 31(4), 793–799.

Zhao, Y, Xing, W., Lu, W., Zhang, X., and Christensen, T. H. 2012. Environmental impact assessment of the incineration of municipal solid waste with auxiliary coal in China. *Waste Management*, 32(10), 1989–1998.

CHAPTER 12

CARBON-FOOTPRINT-BASED SOLID WASTE MANAGEMENT

Life cycle assessment (LCA) provides a legitimate framework to quantify various environmental impacts for solid waste management (SWM) systems in which life cycle impact assessment (LCIA) may help reach a single comparable number with the aid of normalization and weighting factors. Yet, the complexity of impact assessment and interpretation for a myriad of environmental impact categories at different scales are still difficult for waste managers and the general public to understand. In addition, the uncertainty in the LCA procedure sometimes yields a result that cannot be easily interpreted with confidence. This uncertainty leads to a simplified LCIA that addresses a specific environmental impact with relative importance, such as the global-warming potential (GWP). This chapter provides a descriptive definition of GWP in terms of greenhouse gas (GHG) emissions, followed by a methodological discussion of SWM under carbon constraint. The case study explains a series of GWP calculations using an LCA software package to demonstrate a real-world LCIA in a typical SWM system with GHG emissions. The chapter concludes with a discussion of application potential of carbon footprint-based SWM.

12.1 THE GLOBAL-WARMING POTENTIAL IMPACT

GWP is a comprehensive estimation of the total emissions of GHG, translated as carbon dioxide equivalents (CO_2eq) through emissions factors. CO_2eq is the unit to correlate the radiative forcing of a GHG in terms of absorbing and reemitting infrared

Sustainable Solid Waste Management: A Systems Engineering Approach, First Edition. Ni-Bin Chang and Ana Pires.
© 2015 The Institute of Electrical and Electronics Engineers, Inc. Published 2015 by John Wiley & Sons, Inc.

radiation relative to CO_2 (i.e., the reference GHG). Known as GHG emissions, the major SWM system emissions are biogenic and fossil CO_2, methane (CH_4), and nitrous oxide (N_2O), with minor contribution from fluorinated gases (Polettini, 2012). To accurately estimate GWP, the sources of GHG (e.g., physical units or processes that release GHG into the atmosphere) and the sinks of GHG (e.g., physical units or processes that remove a GHG from the atmosphere) must be identified (ISO, 2006a).

The importance of GWP applied to SWM is evidenced by an independent discussion in the Fourth Assessment Report published by Intergovernmental Panel on Climate Change (IPCC) (Bogner et al., 2007). In the European Union (EU), for instance, the quantification of GWP estimates that the net GHG emissions from municipal waste in 27 EU countries plus Norway and Switzerland will decrease by about 85 million tonnes of CO_2eq between 1990 and 2020 (Bakas et al., 2011) (tonnes are metric tons). With a reliable quantification of possible climate change impacts, an accurate calculation of GWP is thus deemed necessary for risk management in many SWM projects, leading to the identification of mitigation measures and adaptive management strategies. In addition, under the Kyoto protocol, possible saving of GHG emissions can be applied to promote carbon credit exchange, resulting in cost savings. Having this common indicator of GWP shared across different SWM activities, the LCIA may ultimately reach a single comparable number with the aid of normalization and weighting factors (Barton et al., 2008). GWP associated with SWM activities would be a useful indicator for risk management across different alternatives when a carbon-regulated SWM system has to be considered.

12.2   THE QUANTIFICATION PROCESS

British Standards Institution (BSI) norm PAS 2050:2011 (BSI, 2011) and the technical specification ISO/TS 14067 (ISO, 2013) present strong cases indicating that carbon footprint has a considerable impact on environmental policies and at decision-making process. The quantification process of GWP involves GHG accounting. ISO 14040/14044 (ISO, 2006b, 2006c) describes relevant LCIA methods that include GWP calculations. Besides PAS 2050:2011 (BSI, 2011) and ISO/TS 14067 (ISO, 2013), other GHG quantification methodologies are also available including (1) Protocol for the Quantification of Greenhouse Gases Emissions from Waste Management Activities (EpE, 2010), (2) the family of ISO 14064-1 (there is also ISO 14064-2 and ISO 14064-3 related to GHG accounting) norms (ISO, 2006a), (3) Product Life Cycle Accounting and Reporting Standard of the GHG Protocol (WRI and WBCSD, 2011), (4) the upstream-operating-downstream (UOD) framework (Gentil et al., 2009), (5) the USEPA Waste Reduction Model (WARM) (USEPA, 2009), and (6) the Federation of Canadian Municipalities' Partners for Climate Change Protection (FCM, 2009). Carbon footprint assessment, a pragmatic method to calculate GWP, has been encouraged by nongovernmental organizations, companies, and various private actions (Weidema et al., 2008). This movement toward various calculation procedures has generated several published standards to help professionals calculate carbon footprint.

TABLE 12.1 Publications concerning GWP

Carbon footprint publications	1990–1999	2000–2009	2010–present
Norms and standards			ISO (2006a, 2013), BSI (2011), WRI and WBCSD (2011)
Greenhouse gas emissions inventory		Astrup et al. (2009) and Calabrò (2009)	Andersen et al. (2010)
Guidelines and manuals	IPCC (1996)	Climate Leaders (2004) and IPPC (2006)	
Reviews		Weidema et al. (2008) and Gentil et al. (2009)	Mohareb et al. (2011) and Laurent et al. (2012)
Case studies		Weitz et al. (2002), Liamsanguan and Gheewala (2008), Boldrin et al. (2009), Fruergaard et al. (2009), and Zhao et al. (2009)	Stichnothe and Azapagic (2009), Hermann et al. (2011), Vergara et al. (2011), Chang et al. (2012), Cifrian et al. (2012), Fitzgerald et al. (2012), Pires and Martinho (2012), Yoshida et al. (2012) and Jeswani et al. (2013)

For any scenario, the GWP impact quantification process for an SWM system considers the following elements: functional unit definition, waste type and composition, system boundaries and allocation, GHG selection, GHG accounting (including biogenic and fossil CO_2), GWP calculation, and interpretation of sensitivity and uncertainty. The major components and tasks required for GWP assessment are discussed in the following sections, and greater detail may be found in the literature (Table 12.1). Carbon footprint quantification is in a growing phase, especially for specific norms to be concluded in the near future by ISO.

12.2.1 Functional Unit, Waste Type, and Composition

In a GWP assessment, the GHG emissions arising from a specific system (product or service) must be calculated based on the selected functional unit capable of representing the system. As in a general LCA, the functional unit is intended to be a reference for the inputs and outputs, which can be clearly defined and measured. The functional unit for an SWM system may be an amount of solid waste stream to be managed by the prescribed SWM system (e.g., 1 tonne of waste), or the total annual amount of waste generated in a specific residence area, or the solid waste streams entering into a treatment process or to an operational unit of the prescribed SWM system. When defining the functional unit, a business-to-business approach often specifies 1 tonne of waste as the functional unit to facilitate comparisons of waste treatment technologies or SWM solutions. If the intention is to publish the results and disseminate them to the residents, a business-to-consumer approach is preferred that specifies the functional unit as the amount of municipal solid waste (MSW) generated by household or by family.

Defining the functional unit is also tied to the waste type and composition because it may influence how and which data will be collected, which GHG will be chosen, and where the boundaries of significance are set. In SWM, waste characterization campaigns usually identify organic, nonorganic, biogenic, and combustible components such as paper/cardboard, plastic, metals, glass, fines, fermentables, and others. Consistency of criteria for waste characterization campaigns is not unified among countries, however, so that data from other countries might be unusable. Data from the literature have geographic heterogeneity and temporal differences, and careful validation is thus required to reduce uncertainty. Conducting characterization campaigns at the facility level or at system level for SWM to reduce errors in GWP calculation is a typical solution if budget limitation is not an issue.

Major types of waste composition capable of causing GHG emissions, especially those with high carbon content, include MSW, industrial, hazardous, clinical, and construction and demolition wastes; however, construction and demolition waste may be disregarded if the potential source of fluorinated gases is not deemed an issue at landfill sites (Kjeldsen and Scheutz, 2003; Gentil et al., 2009). Other types of waste deemed potential sources of GHG emissions may include end-of-life vehicles (air conditioning systems), electronics waste, scrap tires, and scrap lubricating oils (during recovery).

12.2.2 System Boundaries and Allocation

When conducting a GWP assessment, the system boundary has to be first defined. Such system boundary may cover a single operation unit such as an incinerator or a landfill, part of the SWM system, or even an entire SWM system. The emissions from the system can be classified as direct emissions resulting from the waste management operation, or indirect emissions resulting from upstream emissions prior to the process or operation, or downstream emissions occurring in subsequent processes or operations. Another way to define the system is based on the traditional logic of LCA, in which foreground and background systems can be defined by cradle-to-grave, cradle-to-cradle, or streamlined approaches, as long as it can be justified.

SWM systems with by-products must be clarified in GWP assessments, similar to general practices of LCA, including ISO 14040/14044 (ISO, 2006b, 2006c), PAS 2050:2011 (BSI, 2011), and Product Life Cycle Accounting and Reporting Standard (WRI and WBCSD, 2011) procedures. These methods avoid system expansion, but if necessary, allocation is made using economic value or physical measures, depending on the criteria in relevant standards; for example, the criterion of PAS 2050:2011

(BSI, 2011) suggests that if allocation is unavoidable, economic value may be applied. A GWP study of open-loop and closed-loop recycling systems can be based on the same procedure as PAS 2050:2011 (BSI, 2011). GWP standards supported by the ISO/TS 14067 (ISO, 2013) and PAS 2050:2011 (BSI, 2011) apply attributional and consequential approaches. Attributional approach searches for all the burdens associated with the life cycle of the product, service, or process, at a specific moment. Yet, consequential approach aims to establish the environmental burdens of a decision or a planned modification at a component of the life cycle of the product or service in a study with regard to market and economic implications. More information on attributional and consequential approaches can be found in Chapter 10. If the GHG inventory follows a life cycle attributional approach, it means that average rather than marginal data are used (WRI and WBCSD, 2011). Most standards cannot cover all relevant aspects, however, and it is the practitioner's duty to search for all possibilities in real-world applications. If the system of interest has a specific geographic context, that focus reflecting the local effects on global impacts must be considered when collecting the GHG-related data to carry out the GWP assessment.

To reduce complexity, a cutoff criterion or value can be applied to a GWP assessment procedure. For example, PAS 2050:2011 (BSI, 2011) only includes GHG emissions from all inputs that contribute more than 1% of the GHG emissions from the product during its life cycle, and the final result can be scaled up to account for any excluded emissions (Sinden, 2009). Another applied cutoff criterion relates to capital goods. PAS 2050:2011 (BSI, 2011) mentions that capital goods can only be included if life cycle inventory (LCI) databases with relevant information are available. Therefore, any existing GWP studies for SWM that do not include capital goods must justify that the GWP during SWM facility operations over the life cycle of the facility, such as incineration plants or landfills, would be much higher than the GWP value of the capital goods if all construction and operation of the facility have to be considered together (Astrup et al., 2009). Whatever the decision, the study must clearly justify whether the inclusion/exclusion of capital goods is meaningful.

Time aspects influence reporting and accounting mechanisms related to a number of time-dependent variables, including reporting time frame, time lag of emissions (landfill and compost), GWP time horizon, and GHG residence time (Gentil et al., 2009). The temporal boundaries in a GWP assessment must be correctly and transparently defined to ensure that decision-making is reliable. Data collected from waste management operations are normally based on a 1-year period; however, gases released into the atmosphere have different residence times. For example, the residence time of CH_4 in the atmosphere is much shorter than that of CO_2; as a result, the relative CO_2eq impact of CH_4 emissions reduces compared to CO_2 as the time horizon increases (Sinden, 2009). In addition, future emissions resulting from the reference year must also be considered; the most well-known case is the GHG emissions from landfill and compost use. The common time frame applied is 100 years in these cases because decisions are assumed to have an impact during a human life cycle, and a shorter time period in a GWP study is not sufficient to address the inclusive impacts of such emissions. A 500-year period would be sufficient to include most GHG emission impacts, but decision-making over this length of time is obviously impracticable. Thus, selecting a reasonable time frame is necessary to ensure a transparent GWP study. In the case of carbon sequestration or storage through landfilling and composting, a 100-year time frame is reasonable, especially for stabilized residue used as daily cover or soil cover at landfills and compost and digested sludge applied as soil conditioners, resulting in a long-term carbon storage in soil.

12.2.3 GHG Selection

Although six GHG emissions, including CO_2, CH_4, N_2O, hydrofluorocarbons (HFCs), perfluorocarbons (PFCs), and sulfur hexafluoride (SF_6), are often cited, assigning priority to each substance is an important step in the GWP assessment for SWM. The six GHG emissions can be generally classified into three groups (Figure 12.1) (USEPA, 2013): Group 1 can include emissions from fossil fuels burned on site, emissions from vehicles, and other direct sources; Group 2 emissions are indirect GHG emissions resulting from the generation of electricity, heating and cooling, or steam generated inside or outside; Group 3 GHG emission sources currently required for federal GHG reporting include transmission and distribution (T&D) losses associated with purchased electricity, employee travel and commuting, contracted solid waste disposal, and contracted wastewater treatment. Substances such as CO_2, CH_4, and N_2O are most often chosen once they result directly from the SWM due to incineration and landfilling, but indirect emissions from upstream and

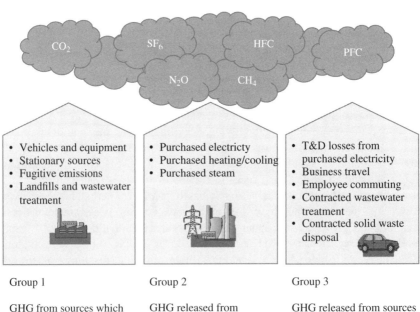

FIGURE 12.1 Classification of GHG by groups. *Source*: From USEPA (2013)

downstream units include other sources of GHG associated with different groups as described above. Although the main GHG sources must be defined, a sensitivity analysis at the end of the GWP study is suitable by using an iterative approach to identify the relative importance of all substances being considered. Final decisions of SWM alternatives can lead to the reduction of major GHG emissions by affecting one or more of the following factors: (1) energy consumption, (2) nonenergy-related emissions, (3) CH_4 emissions from landfills, (4) CO_2 and N_2O emissions from waste combustion and biological treatment processes, and (5) carbon sequestration from compost or digestate. Substances to be inventoried depend on the standards used and the purposes of the study. Rather than focusing on the six GHG included in Kyoto Protocol, other standards such as PAS 2050:2011 (BSI, 2011) and ISO/TS 14067 (ISO, 2013) include an expanded list of GHG. For instance, ISO/TS 14067 (ISO, 2013) establish 95 substances of GHG for consideration (besides CO_2 itself).

12.2.4 GHG Accounting

The goal of GHG accounting or inventory is to obtain GHG emissions and residence time occurring at each process of waste operation, including both direct and indirect emissions. To account for GHG in SWM, specific data related to facility construction and operation should be examined; however, using standard factors of various emissions to calculate GHG emissions is critical, as in the case of waste collection and transport, because the audited fuel spending data for collecting waste streams and relevant emission factors would likely be common in many SWM systems. The emission factors are generic and thus are transferrable to a number of processes representing similar characteristics, calculated per unit of activity (Gentil et al., 2009). They facilitate, in terms of time and costs, the acquisition of GHG data. To remain transparent, emissions factors must be correctly registered, including any assumptions made when selecting the factor and its features, such as date, geography, time, and technical validity. Yet, actual monitored emissions are always preferred when available. If not available, for example, audited data collected from different locations can be used to obtain the electricity consumptions and possible emissions to account for the GWP due to waste management operations. This setting is convenient when it is not possible to have local monitored emissions from the neighboring power plants to obtain the upstream data of GHG emissions. Box 12.1 presents a formula for calculating GHG to account for MSW incineration. In Table 12.2, the emissions factors used for GHG accounting are presented based on various energy sources to support the GWP calculation.

BOX 12.1   ACCOUNTING GHG EMISSIONS FROM MSW INCINERATION

Based on any MSW composition, IPCC (2006) defines the following equation to transform MSW into CO_2 during incineration:

$$E_{CO_2} = MSW \cdot \sum \left(WF_j \cdot DM_j \cdot CF_j \cdot FCF_j \cdot OF_j \right) \cdot \frac{44}{12},$$

where E_{CO_2} is the CO_2 emissions from MSW combustion (tonne), MSW the amount of MSW (as wet weight) (tonne), WF_j the fraction of waste component j in the MSW (as wet weight), DM_j the fraction of dry matter content in the component j of the MSW, CF_j the fraction of carbon in the dry matter of component j, FCF_j the fraction of fossil carbon in the total carbon of component j, OF_j the oxidation factor (assumed equal to 1), and 44/12 the conversion from C to CO_2.

Emissions avoided by using by-products from the SWM system may substitute existing products and materials. The SWM system can also achieve carbon seques- tration by landfilling of biogenic carbon materials as well as by applying compost and digestate products into soil. These processes removes CO_2 from the atmosphere because wood and paper decay slowly and accumulate in the landfills as long-term storage (EpE, 2010). By spreading compost, for instance, the product retains a por- tion of the carbon rather than the mineralized by-products in the soil. In reality, this stabilized organic matter has a turnover of 100–1000 years, thereby binding a fraction of the carbon to the soil for a longer period of time (EpE, 2010). Carbon sequestration is only related to biogenic carbon, which is recycled naturally in the carbon cycle by being removed from the atmosphere and bound to the soil. Carbon-based plastic materials, however, do not eliminate CO_2 from the atmosphere, and therefore only return carbon back into the sub-soil when landfilled.

Several methods can be used to calculate the carbon sequestration. The Waste Sector guidelines in IPCC Guidelines for National Greenhouse Gas Inventories (IPCC, 2006) estimate the biogenic carbon stored in a landfill on a long-term ($DOC_{m \text{ long-term stored } T}$) basis as:

$$DOC_{m \text{ long-term stored } T} = W_T \cdot DOC \cdot (1\text{-}DOC_f) \cdot MCF,$$

where W_T is the mass of waste disposed in year T (tonnes); DOC is the degradable organic carbon in disposal year (tonnes C per tonne of waste); DOC_f is the fraction

TABLE 12.2   Emissions factors and their CO_2eq energy provisions used for GHG accounting

Type of processes/emissions	Emissions factors
Provision of diesel oil	0.4–0.5 kg CO_2eq $\cdot$ L^{-1}
Combustion of diesel oil	2.7 kg CO_2eq $\cdot$ L^{-1}
Provision of fuel oil (heavy)	0.4–0.6 kg CO_2eq $\cdot$ L^{-1}
Combustion of fuel oil (heavy)	2.9 kg CO_2eq $\cdot$ L^{-1}
Provision of natural gas	0.2–0.3 kg CO_2eq $\cdot$ L^{-1}
Combustion of natural gas	1.9–2.2 kg CO_2eq $\cdot$ Nm^{-3}
Provision of electricity	0.1–0.9 kg CO_2eq $\cdot$ kWh^{-1}
Provision of heat	0.075 kg CO_2eq $\cdot$ MJ^{-1}

Source: From USEPA (2008) and Fruergaard et al. (2009).

of DOC that can decompose in the anaerobic conditions in the landfill (fraction); and MCF is the CH_4 correction factor for year of disposal (fraction).

Carbon sequestration can then be calculated by multiplying DOC_m long-term stored T by 44/12 (the conversion factor from CO_2 to carbon). Additional formulas are found in the literature (like is the case of SWICS (2009)) for the carbon storage factors of different waste materials in landfills (EpE, 2010). To estimate carbon sequestration in the soil after the application of compost products, the carbon sequestration factors for GHG accounting can be obtained from the United States Environmental Protection Agency (USEPA) (2006), and Prognos and IFEU (2008) (acronym of *Institut für Energie- und Umweltforschung* in German) (EpE, 2010). Avoided CO_2 emission due to carbon sequestration can be calculated by the following formula (Boldrin et al., 2009; Hermann et al., 2011):

$$CO_{2,bind} = C_{input} \cdot C_{bind} \cdot \frac{44}{12},$$

where $CO_{2,bind}$ is the sink of CO_2 (kg); C_{input} is the C content in compost (kg); and C_{bind} is the stable fraction of C.

12.2.5 GWP Calculation

GWP calculation involves the translation of the GHG inventory of different substances into a common substance, expressed in terms of CO_2eq, which is used to normalize the results. GHG emitted and removed is determined by multiplying the mass of GHG emitted or removed by the 100-year impact factor given by IPCC in a unit of kg CO_2eq kg^{-1} emission. Although the 100-year time horizon is commonly applied, other time horizons such as 20-year and 500-year can still be used with different characterization factors in IPCC. Once those characterization factors or impact factors are changed, however, the GWP would change based on the same scenario, yielding new results that are incomparable with the historical results. For example, the impact factors retrieved from IPCC (2007) are zero for biogenic CO_2 emissions, one for fossil CO_2, and 25 for biogenic CH_4 emissions, the same factor as fossil CH_4 emissions.

12.2.6 Interpretation

Interpretation of the GWP calculations for an SWM system based on a suite of alternatives leads to a discussion of effective measures to reduce the environmental impact at the global scale. Analyses based on given assumptions, prescribed system boundaries and assumed effort of data validity, are the foundation of any successful GWP study. Before reaching a conclusion, a sensitivity analysis is oftentimes conducted to identify the most sensitive parameters, which could significantly affect the outcome of the GHG emissions analysis (Yoshida et al., 2012). Sensitivity analysis conducted in an iterative way for GHG accounting and GWP calculation can highlight relevant processes that require special efforts to collect information. Scenarios like the exclusion of life cycle stages or unit processes, exclusion of inputs and outputs that lack significance to the results of the GWP, and inclusion of new unit processes that have shown to be significant in the process can also be proposed for sensitivity analysis (Guinée et al., 2002).

In SWM systems, sensitivity parameters can be waste composition, collection method, separate collection rate, incineration efficiency, credit system in a polluter-pay community, biogas recovery rate at landfills, and others. The parameters chosen for the sensitivity analysis are equal to the ones mentioned before by Guinée et al. (2002). The most sensitive parameters can be impacted by uncertainty due to measurement errors, which must be addressed. Uncertainty associated with GHG emissions have a variety of sources, including estimation errors, missing data, imprecise measurements of emissions, calculation errors, emissions based on cutoff criterion, simplification by using the average values of emissions factors, and assumptions that simplify estimation methods.

12.3 GWP ASSESSMENT FOR SOLID WASTE MANAGEMENT

Planning and conducting a GWP study for SWM involves financial, human, and technical resources to produce relevant, transparent, accurate, complete, consistent, and useful results. A preliminary identification of the GHG substances and their potential impacts on the SWM life cycle should be made early in the GWP assessment because these decisions are crucial to data collection, identification of sources of data, and calculation methodology. The planning steps involved in a GWP study (Figure 12.2) must take existing protocols into consideration. Sources such as the Product Life Cycle Accounting and Reporting Standard (WRI and WBCSD, 2011) provide a broad concept of requirements addressed for each planning phase to conduct a credible GWP assessment for SWM.

Ideally, the process begins with the scope definition listing what is mandatory and optional, followed by the determination of system boundaries. At this stage, the GHG substances to be included must be registered for both the emissions into the atmosphere and the sequestration from the atmosphere. Functional unit, unit of analysis, and reference flow are additional information to include. The presence of cofunctions and coproducts in the system requires clarification. To establish the boundaries, all cradle-to-grave life cycle processes responsible for the GHG accounting must be included, or justified if not included. A description of the processes and the system itself, including a life cycle cradle-to-grave map, is the most valuable tool for interpreting the results. Discussion items include the time period of the inventory, calculation methods for GHG and GWP, and management of coproducts. Preference is placed on avoiding allocation through system expansion when possible; otherwise, allocation can be based on physical, economic, or other relationships, sequentially (WRI and WBCSD, 2011). A successful GWP study often requires the SWM system to be divided into several principal operational units or facilities such as waste collection, waste sorting or material recovery facility (MRF (pronounced as "merf")), mechanical biological treatment plant, composting, incineration, landfill,

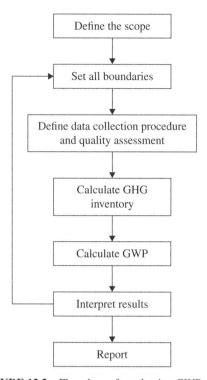

FIGURE 12.2   Flowchart of conducting GWP study

and auxiliary processes responsible for operations or facilities such as electricity and fuel production. System expansion might be needed when coproduction occurs.

A GWP study would require a prior effort to ensure that all available information and data pass the quality assurance and quality control (QA&QC) requirements during the collection stage. Data from all collection with essential QA&QC should be gathered, beginning with the waste operation activity. Data from various data sources, including waste operations data, emissions factors, and/or direct GHG emissions, should be assessed using data quality indicators to create a reliable LCI, followed by the GWP calculations.

All sources during the GHG accounting/inventory and GWP calculation, including mobile and stationary combustion process emissions, fugitive emissions, and direct and indirect sources, must be identified. The procedure to calculate the GHG also needs to be addressed and mentioned in the GWP assessment. LCI can be used as a procedure to account for GHG (Gentil et al., 2009) (Box 12.2). During GHG emissions calculation, the following items are specifically addressed (WRI and WBCSD, 2011):

- The total inventory results in CO_2eq, which includes all emissions and removals included in the boundary from biogenic sources and nonbiogenic sources;

- Percentage of total inventory results by waste operation unit (or life cycle stage);
- Biogenic and fossil emissions and removals separately when applicable;
- Presentation of GWP factors references.

BOX 12.2   UPSTREAM-OPERATION-DOWNSTREAM (UOD) PROCEDURE (Gentil et al., 2009)

According to Gentil et al. (2009), the concept of the UOD framework has emerged from the conclusion that "all the GHG accounting and reporting methodologies require similar "raw" data that are then used differently depending on the reporting scope". A generic UOD framework (Table 12.3) should be completed with metric units expressed in kg CO_2eq per tonne of waste or equivalent, the GWP reference used (e.g., GWP100), the global warming factors (GWF) of the energy mix considered (e.g., 0.1–$0.9\,kg \cdot kWh^{-1}\,CO_2$eq), and the waste considered (e.g., MSW), including indication of the cutoff rule or *de minimis* value (Gentil et al., 2009).

TABLE 12.3   UOD generic framework

	Indirect: upstream	Direct: operating	Indirect: downstream
GWF	Calculated GWF based on the sum of all accounted upstream GHG emissions	Calculated GWF based on the sum of all accounted operating GHG emissions	Calculated GWF based on the sum of all accounted downstream emissions and GHG savings
Accounted	Production of fuel, electricity, heat, ancillary materials	Collection and transport Intermediate facilities Recycling Aerobic biotreatment Anaerobic biotreatment Thermal treatment Landfill	Emissions and savings of energy substitution, material substitution, carbon binding (sequestration), fly ash transport
Not accounted	Unaccounted GHG Construction Maintenance Decommissioning Import–export Embedded energy in waste	Unaccounted GHG Unaccounted waste stream Historical waste (relevant for landfill) Staff commuting Business travel	Unaccounted GHG Decommissioning (end-of-life)

Source: From Gentil et al. (2009).

Finally, results are interpreted and the GWP report is completed describing alternative strategies with essential evaluations and courses of action. The interpretation step is expected to include a qualitative statement detailing inventory uncertainty, sensitivity analysis, and methodological choices. The methodological choices can be justified by allocation methods used, source of GWP factors applied, and calculation models adopted. The reporting period should include relevant information to ensure the integrity of the GWP study. Measures must be taken to reduce the uncertainty impact, including measurement instrumentation, conducting reliable procedures, implementing internal controls, and adopting a management validation process. Box 12.3 includes a summary of planning steps in a GWP assessment for SWM.

BOX 12.3 SUMMARY OF PLANNING STEPS IN A GWP ASSESSMENT FOR SWM

- Define the scope.
- Outline the system boundaries of SWM.
- Propose the operational boundaries.
- Reporting the period to be covered.
- Summarize the information about emissions separate of any GHG trades such as sales, acquisition, transfers, or banking of allowances.
- Collect the emissions data.
- Report the emissions data for total GHG in metric tons of CO_2eq, including direct CO_2 emissions from biologically sequestered carbon.
- Select a base year and an emissions profile over time that is consistent with policy.
- Choose a correct background for any meaningful emissions modifications that initiate base year emissions recalculation.
- Specify the exclusion of sources.
- Identify the significant issues to quantify GHG based on LCI and LCIA.
- Evaluate the completeness, sensitivity, and consistency checks.
- Describe the allocation methods, conclusions, limitations, and recommendations.

12.4 CASE STUDY

12.4.1 Structure of the SWM System

Lewisburg is a small borough in Union County, Pennsylvania, home to residential, commercial, and industrial sectors. The residential sector is the major contributor to the borough's MSW. The Municipal Waste Planning, Recycling and Waste Reduction Act of 1998 (Act 101, 2007) mandated that all towns in the state of Pennsylvania with a population of more than 5000 must have a recycling program, and at least 25% of the total generated waste stream must be recycled. In addition to the state mandate, the borough has set its own recycling goal of 30–50% per material recycled per year.

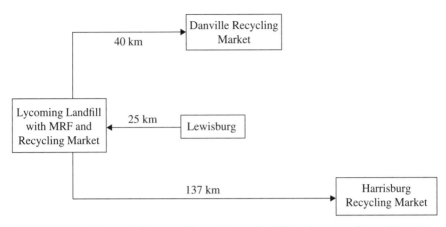

FIGURE 12.3   Schematic diagram of the study area, landfill, and recovered material markets. *Source*: From Chang et al. (2012)

It currently delivers the residential MSW to the nearby Lycoming County Landfill (Figure 12.3).

A total of 0.728 km² (179.9 ac) has been permitted to Lycoming County, which has been developing and maintaining the landfill site since 1973. This regional site serves not only Lycoming County but also Montour, Northumberland, Snyder, Columbia, and Union counties. Although the landfill has a capacity of 1600 tonnes per day, it normally receives approximately 1100 tonnes per day on average. For simplicity, we assumed that all the deposited materials at Lycoming County Landfill come from Lewisburg only.

Lycoming landfill has a MRF of about 5574 m² colocated with a recovered material market. Materials here are sorted mechanically using recycling equipment (Table 12.4) such as loaders, forklifts, balers, and trailers. The facility accepts post-consumer materials including glass containers, plastic bottles, aluminum cans, steel cans, corrugated containers, office paper, magazines, and mixed computer and office paper from businesses. The facility also accepts some postindustrial materials, such

TABLE 12.4   Summary of MRF equipment

Equipment	No. of items	Purpose
Loaders	5	Move materials, particularly unprocessed materials
Forklift	1	Move processed materials
Horizontal baler	1	Bale corrugated containers
Marathon baler with fluffer	1	Bale other fibers, steel cans, and plastics
Dens-O-Can densifier	1	Densify aluminum cans
Trailer	1	Baled material storage
Cumberland granulators	2	Grind PVC pipe and sheet

Source: From Chang et al. (2012).

TABLE 12.5 Generated waste composition

General items	Composition, mass fraction (%)
Paper	35
Corrugated	13
Newsprint	75
Mixed	12
Yard trimmings	10
Food scraps	15
Plastics	12
Metals	5
Aluminum	60
Steel cans	40
Rubber, leather, textiles	7
Glass	5
Clear glass	30
Mixed glass	70
Wood	7
Other	4
Total	100

Source: Adapted from PDEP (2004).

as polyvinyl chloride (PVC) (sheet and pipe), which they grind on-site. The facility also accepts and markets processed newspapers, magazines, corrugated containers, mixed office paper, and glass from other counties. Recyclables are directly collected from the households and sent to the MRF for sorting and recycling. The statewide composition of the generated waste stream by weight (Table 12.5) is used as a typical composition of solid waste of Lewisburg in this study.

Recycled materials are delivered to the markets according to category to meet varying product demand at different locations (Table 12.6). The Lycoming landfill itself has a potential market, and another two markets are located at Danville (40 km from the landfill) and Harrisburg (137 km from the landfill) (Figure 12.3). Because the Borough currently disposes all waste in one landfill, all information required for

TABLE 12.6 Potential recycled material markets

Material	Recycling market
Aluminum	Harrisburg, Lycoming
Steel cans	Harrisburg
Clear glass	Lycoming
Mixed glass	Harrisburg
Corrugated paper	Lycoming
Newsprint paper	Danville, Lycoming
Mixed paper	Lycoming
Plastic	Harrisburg

TABLE 12.7   Potential recycled material market prices

Materials	Danville recycling market price (US\$ · tonne^{-1})	Harrisburg recycling market (US\$ · tonne^{-1})	Lycoming recycling market (US\$ · tonne^{-1})
Aluminum	0	440	400
Steel can	0	100	0
Clear glass	0	0	10
Mixed glass	0	0	0
Corrugated paper	0	0	3
Newsprint paper	4	0	2
Mixed paper	0	0	0
Plastic	0	10	0

Source: From Chang et al. (2012).
Personal communication with recyclable materials vendors in 2010.

this study is available through a local investigation. The cost for MSW collection and for recycled material handling is approximately US\$20 per tonne in both cases. Transportation cost for MSW and recyclables is about US\$0.25 per km per tonne, and the landfill earns a tipping fee of US\$46.96 per tonne. The prices of recycled materials vary according to the location of the markets (Table 12.7). Several factors must be considered in developing an effective recycling program, including recycling rate (25% of waste generated), composition of the recyclables (30–50% per material), selection of a recycling site, the amount of waste sent to be disposed in the landfill, and cost for final disposal. In addition to the benefits from selling recycled materials, landfill gas (LFG) (e.g., CH_4) can be used to generate electricity, which may be sold at the rate of US\$0.08 Kwh^{-1} according to the comparable unit electricity price offered by Pennsylvania Public Utility (PPL) Electric Utilities Corporation.

12.4.2   Planning Background

Generally, physical, chemical, and biological processes are involved in SWM. Physical management techniques include source reduction, reuse, recycling, and recovery. Source reduction is the most desirable management system by reducing MSW generation or increasing the product life span, such as using cotton rather than paper towels in day-to-day life. Material recycling and energy recovery can be achieved using several common unit operations such as MRF, incinerators, pyrolysis-gasification process, among others. In contrast to incineration, pyrolysis is a combustion process in an oxygen-deficient chamber. Gasification is a modified pyrolysis process where a small amount of oxygen is introduced at a relatively higher temperature than pyrolysis. Landfill generally degrades the materials through a five-phase biological process of initial adjustment, transition phase, acid formation phase, CH_4 fermentation phase, and maturation phase (Vesilind et al., 2002).

CH_4 and CO_2 are the major gases emitted from SWM processes and are most responsible for high GWP in SWM systems. MSW can be decayed in an anaerobic-anoxic condition to produce end products such as CH_4, CO_2, and NH_3 for reuse, a common process for the production of biogas (mostly CO_2 and CH_4) and compost. Incineration has substantial CO_2 emissions, but GHG emissions of CH_4 and CO_2 from landfilling have enormous impacts on the environment. Although CH_4 production due to biological anaerobic process in landfill cells can be recovered from landfills for power generation, inefficiency in the CH_4 collection system exacerbates climate change. According to IPCC, GWP is 1 for CO_2 and 25 for CH_4 for a period of 100 years (Wanichpongpan and Gheewala, 2007); therefore, in terms of LCA, landfill has the lowest priority and recycling is the highest option (Moberg et al., 2005).

12.4.3 GWP Calculations

The LCA in this study was performed using GaBi 4.3 software package developed by PE International (2009), Stuttgart, Germany, to estimate GWP from the various process steps of the SWM program of Lewisburg. The functional unit of 1 tonne of the solid waste was selected. The emissions were assessed for the transport of waste from households to the landfill, from the MRF to recycling markets, and for the landfilling. Scenario settings required sorting of the entire 1 tonne solid waste stream (Table 12.8) in the MRF. Recycling ratio per material was kept constant (50%), and

TABLE 12.8   Sorting of solid waste prepared for GaBi software input

Generated items	Composition, mass fraction (%)	Functional unit (kg)	Landfilled (kg)	Recycled (kg, 50% per item)	Recycling markets
Paper	35	350	175	175	
Corrugated	13	45.5	22.75	22.75	Lycoming
Newsprint	75	262.5	131.25	131.25	Danville
Mixed	12	42	21	21	Lycoming
Yard trimmings	10	100	100	0	
Food scraps	15	150	150	0	
Plastics	12	120	60	60	Harrisburg
Metals	5	50	25	25	
Aluminum	60	30	15	15	Harrisburg
Steel cans	40	20	10	10	Harrisburg
Rubber, leather, textiles	7	70	70	0	
Glass	5	50	25	25	
Clear	30	15	7.5	7.5	Lycoming
Mixed	70	35	17.5	17.5	Harrisburg
Wood	7	70	70	0	Lycoming
Other	4	40	40	0	
Total	100	1000	715	285	
Recycling rate				28.5% > 25%	

Source: From Chang et al. (2012).

the overall recycling rate was greater than 25% to comply with the state law (Nishtala and Solano, 1997).

A screenshot of the GaBi 4.3 model (Figure 12.4) shows how the GHG emission analysis was conducted. Truck, diesel refinery, power grid mix, and landfill processes were used directly from the GaBi software database. Only MRF was created as a new process, where inputs (diesel and electricity consumption) and outputs (GHG) were calculated using Lycoming MRF equipment (Table 12.4), and corresponding data were taken from usual values (Nishtala and Solano, 1997).

The LCA was conducted based on a 1 tonne solid waste stream that was presumably carried by a truck from Lewisburg to Lycoming MRF 25 km away. More than 25% of the materials can be recycled and sent to the three recycling markets (Danville, Harrisburg, and Lycoming) according to the market demand. After mechanical sorting, 715 kg waste was disposed of at the landfill at the same place.

12.5 SYSTEMS ANALYSIS

Substantial developments in landfill technologies in the last two decades have markedly reduced emissions to the environment, both as gas and leachate. CH_4 is mostly oxidized in soil top cover or used for energy generation; CO_2 from biogenic origin is considered neutral to GWP. The reduction of GWP due to biogenic sources and oxidized CH_4 was compensated in our calculation if the leachate from the landfill was taken to the wastewater treatment plant where CH_4 and N_2O are again produced as GHG. However, to avoid model complexity it is necessary in this study to consider only CO_2 and CH_4 from a typical landfill operation for our GHG calculations. GaBi output was presented as the amount of CO_2eq emitted by each individual process (Table 12.9). According to the Lycoming Landfill authority, approximately 25% of the LFG used for electricity is sold to PPL Electric Utilities Corporation. Energy density of CH_4, which is the amount of energy per mass, is between 50 MJ $\cdot$ kg^{-1} (Thomas, 2000) and 55.6 MJ $\cdot$ kg^{-1} (Bossel and Eliason, 2002); and 50 MJ $\cdot$ kg^{-1} is assumed in this calculation. The amount of recycled CH_4 from landfill process for electricity production is 3.3 kg (or 82.53 kg CO_2eq) per 715 kg waste for disposal (Table 12.8). Assuming that the energy loss rate for the electricity production from CH_4 recovered by landfill is about 30%, the electricity produced from the LFG is thus 115.5 MJ, or 32.08 kWh per 715 kg waste for disposal. The CO_2eq emissions due to electricity production from CH_4 are 0.01015 kg $\cdot$ kg^{-1} waste for disposal, assuming that the recycled CH_4 is fully converted to CO_2 in the electricity production process.

The major findings in this LCA were that landfill processes may produce substantial amounts of CO_2. In addition, CO_2 and CH_4 are the predominant gases emitted over the entire process (Table 12.9) and are a potential threat to global climate change. With this step accomplished, GWP values from the GaBi outputs can be adopted directly in the different carbon-constrained optimization models with resource limitation to investigate the optimal patterns between recycling and landfilling (to be discussed in Chapter 17). Whereas, the linear-programming models represent single-objective

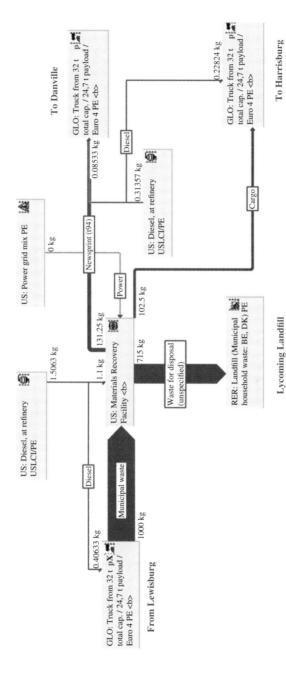

FIGURE 12.4 Screenshot of GaBi 4.3 model in conducting GHG emission analysis. *Source:* From Chang et al. (2012)

TABLE 12.9   Amount of emitted gases in the whole process calculated by GaBi software package

Source	Origin – Destination	Predominant gas	Amount of gas emitted (kg $CO_{2eq} \cdot tonne^{-1}$ MSW)
Truck	From household to landfill	CO_2	1.29
MRF	–	CO_2	253.76
MRF	–	CH_4	6.93
Disposed waste in landfill	–	CO_2	185.56
Disposed waste in landfill (75% LFG)	–	CH_4	247.58
Disposed waste in landfill for electricity production (25% LFG)	–	CH_4	82.53
Truck	From MRF to Danville Recycling Market	CO_2	0.27
Truck	From MRF to Harrisburg Recycling Market	CO_2	0.72

Source: From Chang et al. (2012).

optimization, the compromise-programming model represents multiobjective evaluation in which both GWP and economic criteria are simultaneously taken into account in Chapter 17.

12.6   FINAL REMARKS

This era has seen a unique time period with increasing emphasis on the impacts of industrial processes and products with respect to sustainability indicators, although the focus of sustainability is often only on climate change-related ecosystem and human impacts. Yet, decisions made solely on the basis of a carbon footprint in systems analysis might result in the shifting of burdens. All relevant environmental impacts must be considered when making sustainability decisions during the critical SWM decision-making process. Further studies should include features related to air pollution, noise, leachate, transportation, wastewater treatment, and water resources management.

REFERENCES

Act 101. 2007. *Pennsylvania's "Municipal Waste Planning and Recycling and Waste Reduction Act – Act of July 1988"*, Commonwealth of Pennsylvania Department of Environmental Protection, Pennsylvania. Available at: http://www.dep.state.pa.us/dep/deputate/airwaste/wm/recycle/FACTS/Act101.htm (accessed March 2011).

Andersen, J. K., Boldrin, A., Christensen, T. H., and Scheutz, C. 2010. Greenhouse gas emissions from home composting of organic household waste. *Waste Management*, 30(12), 2475–2482.

Astrup, T., Møller, J., and Fruergaard, T. 2009. Incineration and co-combustion of waste: accounting of greenhouse gases and global warming contributions. *Waste Management & Research*, 27(8), 789–799.

Bakas, I., Sieck, M., Hermann, T., Anders, F. M., and Larsen, H. V. 2011. Projections of municipal waste management and greenhouse gases. ETC/SCP working paper 4/2011. European Topic Centre on Sustainable Consumption and Production.

Barton, J. R., Issaias, I., and Stentiford, E. T. 2008. Carbon—making the right choice for waste management in developing countries. *Waste Management*, 28(4), 690–698.

Bogner, J., Abdelrafie-Ahmed, M., Diaz, C., Faaij, A., Gao, Q., Hashim-Oto, S., Mareckova, K., Pipatti, R., and Zhang, T. 2007. Waste management. In: *Climate Change 2007—Mitigation. Contribution of Working Group III to the Fourth Assessment Report of the Intergovernmental Panel on Climate Change* (Eds Metz, B., Davidson, O. R., Bosch, P. R., Dave, R., and Mayer, L. A.), Cambridge University Press, Cambridge.

Boldrin, A., Andersen, J. K., Møller, J., Christensen, T. H., and Favoino, E. 2009. Composting and compost utilization: accounting of greenhouse gases and global warming contributions. *Waste Management & Research*, 27(8), 800–812.

Bossel, U. and Eliason, B. 2002. Energy and the hydrogen economy. EV World. December 28.

British Standards Institution (BSI). 2011. PAS 2050:2011 Specification for the assessment of the life cycle greenhouse gas emissions of goods and services. BSI.

Calabrò, P. S. 2009. Greenhouse gases emission from municipal waste management: the role of separate collection. *Waste Management*, 29(7), 2178–2187.

Chang, N. B., Qi, C., Islam, K., and Hossain, F. 2012. Comparisons between global warming potential and cost–benefit criteria for optimal planning of a municipal solid waste management system. *Journal of Cleaner Production*, 20(1), 1–13.

Cifrian, E., Galan, B., Andres, A., and Viguri, J. R. 2012. Material flow indicators and carbon footprint for MSW management systems: analysis and application at regional level, Cantabria, Spain. *Resources, Conservation and Recycling*, 68, 54–66.

Climate Leaders. 2004. Direct emissions from municipal solid waste landfilling. United States Environmental Protection Agency.

EpE. 2010. Protocol for the quantification of greenhouse gases emissions from waste management activities. EpE. Available at: http://www.epe-asso.org/pdf_rapa/EpE_rapports_et_documents20.pdf (accessed November 2012).

Federation of Canadian Municipalities (FCM). 2009. Inventory quantification support spreadsheet. FCM. Available at: http://www.fcm.ca/home/programs/partners-for-climate-protection/program-resources/milestone-toolkit/milestone-1.htm (accessed December 2012).

Fitzgerald, G. C., Krones, J. S., and Themelis, N. J. 2012. Greenhouse gas impact of dual stream and single stream collection and separation of recyclables. *Resources, Conservation and Recycling*, 69, 50–56.

Fruergaard, T., Astrup, T., and Ekvall, T. 2009. Energy use and recovery in waste management and implications for accounting of greenhouse gases and global warming contributions. *Waste Management & Research,* 27(8), 724–737.

Gentil, E., Christensen, T. H., and Aoustin, E. 2009. Greenhouse gas accounting and waste management. *Waste Management & Research*, 27(8), 696–706.

Guinée, J. B., Gorree, M., Heijungs, R., Huppes, G., Kleijn, R., van Oers, L., Wegener Sleeswijk, A., Suh, S., Udo de Haes, H. A., de Bruijn, J. A., van Duin, R., and

Huijbregts, M. A. J. (Eds.). 2002. *Handbook on Life Cycle Assessment—Operational Guide to the ISO Standards*, Kluwer Academic Publisher, Dordrecht, Netherlands.

Hermann, B. G., Debeer, L., de Wilde, B., Blok, K., and Patel, M. K. 2011. To compost or not to compost: carbon and energy footprints of biodegradable materials' waste treatment. *Polymer Degradation and Stability*, 96(6), 1159–1171.

Intergovernmental Panel on Climate Change (IPCC). 1996. *Revised IPCC Guidelines for National Greenhouse Gas Inventories, Module 6–Waste*, IPCC, Geneva, Switzerland.

Intergovernmental Panel on Climate Change (IPCC). 2006. *Guidelines for National Greenhouse Gas Inventories, Volume 5–Waste*, IPCC, Geneva, Switzerland.

Intergovernmental Panel on Climate Change (IPCC). 2007. Fourth Assessment Report: Climate Change 2007 (AR4). IPCC. Available at: http://www.ipcc.ch/publications_and_data/publications_and_data_reports.shtml}.ULUAtIcz3eJ (accessed November 2012).

International Organization for Standardization (ISO). 2006a. *ISO 14064-1 Greenhouse Gases—Part 1: Specification with Guidance at the Organization Level for Quantification and Reporting of Greenhouse Gas Emissions and Removals*, ISO.

International Organization for Standardization (ISO). 2006b. *ISO 14040 Environmental Management—Life Cycle Assessment—Principles and Framework*, ISO.

International Organization for Standardization (ISO). 2006c. *ISO 14044 Environmental Management—Life Cycle Assessment—Requirements and Guidelines*, ISO.

International Organization for Standardization (ISO). 2013. *ISO/TS 14067. Carbon Footprint of Products—Requirements and Guidelines for Quantification and Communication*, ISO, Switzerland.

Jeswani, H. K., Smith, R. W., and Azapagic, A. 2013. Energy from waste: carbon footprint of incineration and landfill biogas in the UK. *The International Journal of Life Cycle Assessment*, 18(1), 218–229.

Kjeldsen, P. and Scheutz, C. 2003. Short- and long-term releases of fluorocarbons from disposal of polyurethane foam waste. *Environmental Science & Technology*, 37(21), 5071–5079.

Laurent, A., Olsen, S. I., and Hauschild, M. Z. 2012. Limitations of carbon footprint as indicator of environmental sustainability. *Environmental Science & Technology*, 46(7), 4100–4108.

Liamsanguan, C. and Gheewala, S. H. 2008. The holistic impact of integrated solid waste management on greenhouse gas emissions in Phuket. *Journal of Cleaner Production*, 16(17), 1865–1871.

Moberg, Å., Finnveden, G., Johansson, J., and Lind, P. 2005. Life cycle assessment of energy from solid waste—Part-2: landfilling compared to other treatment methods. *Journal of Cleaner Production*, 13(3), 231–240.

Mohareb, E. A., MacLean, H. L., and Kennedy, C. A. 2011. Greenhouse gas emissions from waste management—assessment of quantification methods. *Journal of the Air & Waste Management Association*, 61(5), 480–493.

Nishtala, S. R. and Solano, E. 1997. *Description of the Material Recovery Facilities Process Model: Design, Cost, and Lifecycle Inventory*. Internal Report, Department of Civil Engineering, North Carolina State University, Raleigh, NC.

Pennsylvania Department of Environmental Protection (PDEP). 2004. Pennsylvania Statewide Waste Composition. Available at: www.dep.state.pa.us/dep/deputate/airwaste/wm/recycle/waste_comp/4_State_Results.pdf (accessed June 2011).

Pires, A. and Martinho, G. 2012. Carbon footprint analysis for the waste oil management system in Portugal. *British Journal of Environment and Climate Change*, 2(3), 278–287.

Polettini, A. 2012. Waste and climate change: can appropriate management strategies contribute to mitigation? *Waste Management*, 32(8), 1501–1502.

Prognos and Institut für Energie- und Umweltforschung (IFEU). 2008. *Resource Savings and CO2 Reduction Potentials in Waste Management in Europe and the Possible Contribution to the CO2 Reduction Target in 2020*, IFEU.

Sinden, G. 2009. The contribution of PAS 2050 to the evolution of international greenhouse gas emission standards. *The International Journal of Life Cycle Assessment*, 14(3), 195–203.

Solid Waste Industry for Climate Solutions (SWICS). 2009. *Current MSW iIndustry pPosition and sState-of-the-pPractice on LFG cCollection eEfficiency, mMethane Ooxidation, and cCarbon*, SWICS.

Stichnothe, H. and Azapagic, A. 2009. Bioethanol from waste: Life cycle estimation of the greenhouse gas saving potential. *Resources, Conservation and Recycling*, 53(11), 624–630.

Thomas, G. 2000. *Overview of Storage Development DOE Hydrogen Program*, vol. 9, Sandia National Laboratories, Livermore, CA.

United States Environmental Protection Agency (USEPA). 2006. *Solid Waste Management and Greenhouse Gases: A Life Cycle Assessment of Emissions and Sinks*, USEPA.

United States Environmental Protection Agency (USEPA). 2008. Climate leaders GHG inventory protocol direct emissions from stationary combustion sources. Available at: http://www.epa.gov/climateleadership/documents/resources/stationarycombustionguidance.pdf (accessed November 2013).

United States Environmental Protection Agency (USEPA). 2009. Waste reduction model (WARM). USEPA. Available at: http://www.epa.gov/climatechange/wycd/waste (accessed November 2012).

United States Environmental Protection Agency (USEPA). 2013. EPA's greenhouse gas emission reductions. USEPA. Available at: http://www.epa.gov/oaintrnt/ghg (accessed November 2013).

Vergara, S. E., Damgaard, A., and Horvath, A. 2011. Boundaries matter: greenhouse gas emission reductions from alternative waste treatment strategies for California's municipal solid waste. *Resources, Conservation and Recycling*, 57, 87–97.

Vesilind, P. A., Worrell, W. A., and Reinhart, D. R. 2002. *Solid Waste Engineering*, Thomson Learning, USA.

Wanichpongpan, A. and Gheewala, S. H. 2007. Life cycle assessment as a decision support tool for landfill gas-to-energy projects. *Journal of Cleaner Production*, 15(18), 1819–1826.

Weidema, B. P., Thrane, M., Christensen, P., Schmidt, J., and Løkke, S. 2008. Carbon footprint *Journal of Industrial Ecology*, 12(1), 3–6.

Weitz, K. A., Thorneloe, S. A., Nishtala, S. R., Yarkosky, S., and Zannes, M. 2002. The impact of municipal solid waste management on greenhouse gas emissions in the United States. *Journal of Air & Waste Management Association*, 52(9), 1000–1011.

World Resources Institute (WRI) and World and Business Council for Sustainable Development (WBCSD). 2011. *Product Life Cycle Accounting and Reporting Standard*, World Resources Institute and World Business Council for Sustainable Development.

Yoshida, H., Gable, J. J., and Park, J. K. 2012. Evaluation of organic waste diversion alternatives for greenhouse gas reduction. *Resources, Conservation and Recycling*, 60, 1–9.

Zhao, W., van der Voet, E., Zhang, Y., and Huppes, G. 2009. Life cycle assessment of municipal solid waste management with regard to greenhouse gas emissions: case study of Tianjin, China. *The Science of the Total Environment*, 407(5), 1517–1526.

PART IV

INTEGRATED SYSTEMS PLANNING, DESIGN, AND MANAGEMENT

Considering connections across resource areas and fostering linkages across agencies require a unique means of sustainability assessment. When coping with complex sustainability issues such as SWM, which is complicated by the separated and dispersed authorities resulting from the basic legal framework, advances in environmental informatics and system analysis may provide a framework for valuable sustainability assessment.

- Multiobjective decision-making framework for SWM in a carbon regulated environment (Chapter 13)
- Integrated forecasting and optimization modeling for planning regional material recovery facilities in an SWM system (Chapter 14)
- Optimal waste collection and vehicle routing strategies (Chapter 15)
- Multiattribute decision-making framework (Chapter 16)
- Multiobjective decision-making framework for balancing waste incineration and recycling (Chapter 17)
- Environmental informatics in support of SWM (Chapter 18)

Sustainable Solid Waste Management: A Systems Engineering Approach, First Edition. Ni-Bin Chang and Ana Pires.
© 2015 The Institute of Electrical and Electronics Engineers, Inc. Published 2015 by John Wiley & Sons, Inc.

CHAPTER 13

MULTIOBJECTIVE DECISION-MAKING FOR SOLID WASTE MANAGEMENT IN A CARBON-REGULATED ENVIRONMENT

Both short-term and long-term planning of solid waste management (SWM) requires the inclusion of multiscale sustainability criteria such as cost–benefit–risk consideration, recycling effect, and global-warming potential (GWP). This chapter demonstrates a holistic integration of these criteria to optimize system planning in a typical SWM system using comparative multiobjective decision analyses. The GaBi® software package was used to estimate the possible greenhouse gas emissions throughout the scenario-based design process. Five managerial scenarios with material recycling effects, with and without the inclusion of GWP concern in the context of an optimization analysis, were carried out for possible cost–benefit–risk trade-offs toward sustainable SWM in the Borough of Lewisburg, Pennsylvania. With the aid of LINGO® software package, the multiobjective decision analyses were solved and compared sequentially to allocate different waste streams subject to the market demand to maximize net benefit and minimize GWP, simultaneously or independently. The analysis eventually led to the prioritization of the material recovery option before disposing of waste streams at the landfill site. Such a case study that considers a carbon-regulated environment bridges the large planning gap in traditional cost–benefit analyses for SWM. Major findings from this environmental systems analysis for SWM indicate that simply using the cost-effectiveness principle or cost–benefit analysis with no risk (or GWP) concern cannot compete with counterpart alternatives with GWP concerns, especially in a carbon-regulated environment. This system-of-systems engineering practice is transferable to other SWM systems in future planning, design, and operation.

Sustainable Solid Waste Management: A Systems Engineering Approach, First Edition. Ni-Bin Chang and Ana Pires.
© 2015 The Institute of Electrical and Electronics Engineers, Inc. Published 2015 by John Wiley & Sons, Inc.

13.1   CURRENT GAPS OF COST–BENEFIT ANALYSES FOR SOLID WASTE MANAGEMENT

The cost-effectiveness principle and cost–benefit analysis are modeling techniques for decision makers to assess positive and negative economic effects of a project or policy by measuring relevant impacts of monetary values over a pre-specified time frame. SWM often requires estimating the monetary values of economic inputs and outputs as direct benefits and costs over the time horizon. Sometimes, environmental and ecological impacts (i.e., indirect benefits and costs) that do not have a market price mechanism can be quantified to include the non-market value of natural resources in decision analysis for SWM (Boardman et al., 2001). Decisions can therefore be made on the basis of both environmental performance as well as technical and economic factors in a decision-making process (Azapagic and Clift, 1999). In addition to costs and benefits discussed for SWM, life cycle assessment (LCA) can also be employed to address varying concerns with different scales for risk communication.

LCA is a well-established standard method or framework that can evaluate the environmental impacts of a product, service, or project "from cradle to grave" (ISO, 2006). The functional structure of the LCA includes all life cycle stages and measures and integrates typical inputs and outputs (ISO, 2006). Given all evaluation categories (e.g., GWP, acidification, ozone depletion, lake eutrophication), a thorough LCA is inevitably complex. In an SWM system, LCA can be used to ambitiously address the energy and material flows with respect to GWP—a global-scale sustainability index. A recent study (Franchetti, 2011) in the United States revealed that the impact of solid waste disposal cost, in addition to the requirement associated with ISO 14001 certifications within an organization, is significant. Thus, a well-rounded decision analysis in simultaneously balancing both economic and environmental aspects with respect to cost–benefit–risk criteria is worthy of exploration.

The first-generation systems engineering models for SWM emphasize the use of the cost-effectiveness principle via linear programming (LP) with a single-objective optimization scheme (Anderson and Nigam, 1967; Anderson, 1968). Some models applied mixed-integer linear programming (MILP) techniques for solving real-world SWM issues related to single-network planning (Anderson and Nigam, 1967; Fuertes et al., 1974; Helms and Clark, 1974; Kuhner and Harrington, 1975) and dynamic, multi-period investment for SWM regionalization (Marks et al., 1970; Marks and Liebman, 1971). The US Environmental Protection Agency (USEPA, 1977) developed the Waste Resources Allocation Model (WRAP) model, which contains static and dynamic MILP modules. Rao (1975) also applied other types of optimization models, such as dynamic programming, for SWM planning. From the 1980s to the 1990s, major efforts in this field applied more operation research models to deal with various types of SWM problems. For example, Chapman and Yakowitz (1984) applied an LP model to size and site facilities and a cost-accounting system to incorporate economies of scale and estimate the effects of decisions. Sundberg et al. (1994) utilized a nonlinear programming model that considers energy aspects in response to the larger complexity of SWM.

In parallel with this movement of the first-generation optimization analysis after the 1980s, there was also a tendency to include more direct and indirect benefits and costs for better-informed policy decision-making. This means promoting cost-effectiveness can only guarantee limited economic management because some traditional waste management technologies are less environmentally friendly and socially acceptable. These technologies include traditional waste incinerators with no waste-to-energy consideration and landfills with no leachate treatment. Thus, systems analysis techniques were enlarged to assist in developing long-term municipal solid waste (MSW) management plans with respect to a full spectrum of costs and benefits, with or without environmental constraints (Baetz, 1990; Chang et al., 1996a, 1996b; Huhtala, 1997; Chang et al., 2005). After the 1990s, optimization analyses progressed to support decisions of short-term and long-term waste management operation by including various socioeconomic and environmental objectives and constraints to address the minimum level of sustainability concerns. These methods are deemed the second-generation optimization analyses for SWM (Chang and Wang, 1996; Chang et al., 1997; Huang et al., 2002).

LCA has much to offer in terms of selection and application of suitable MSW management techniques, technologies, and programs to achieve specific waste management objectives and goals (Özeler et al., 2006). Azapagic and Clift (1998) described how to use an LP model to perform LCA. Since then, a number of studies in the literature used LCA as a comparative tool for different SWM options (Ahluwalia and Nema, 2007; Su et al., 2007; Liamsanguan and Gheewala, 2008; Banar et al., 2009; Manfredi and Christensen, 2009; Villeneuve et al., 2009). Lu et al. (2009) integrated the emission control of greenhouse gas (GHG) in concert with optimal allocation of waste streams in a hypothetical, uncertain tricity SWM system. The inclusion of LCA as an integral part of the third-generation optimization analyses can lead to further sustainable solutions to improve the quality operation of SWM systems (Ekvall and Finnveden, 2000; Muñoz et al., 2004; Finnveden et al., 2005; Al-Salem and Lettieri, 2009; Cherubini et al., 2009; Khoo, 2009).

To avoid the insurmountable complexity involved in some interrelated aspects of LCA, environmental sustainability indicators (ESIs) were derived as a substitute in assessment (Harger and Meyer, 1996). GWP is a simpler and abstract ESI used to characterize the global climate change impact. It is a holistic estimate of the total GHG emissions, expressed as carbon dioxide equivalents (CO_2eq), as a result of a defined action over the project's life cycle or a specified period of time (Strutt et al., 2008). Thus, CO_2eq is a common metric unit used to compare the emissions from various GHG based on their GWP. Within this context, GHG of concern include, but are not limited to, CO_2, methane (CH_4), nitrous oxide (N_2O), and fluorinated gases. For GWP calculations, all GHGs are commonly converted to equivalents of CO_2 gas emissions; thus, CO_2eq becomes the base unit in many GWP analyses (United Nations, 1998). Still, the effect of including such a metric unit on decision-making is unclear in the nexus of the cost, benefit, and risk arena.

This chapter examines the GWP impact in a real-world SWM system to demonstrate that the traditional cost-effectiveness principle and cost–benefit analysis cannot permit sustainable development related to SWM. Thus, the case study in this chapter is designed to explore the potential impact to an SWM system under a global-scale, carbon-regulated environment with local-scale material/energy recycling considerations. To achieve the study goals, five management scenarios were configured and compared to demonstrate the importance of sustainability concerns in an SWM system. For this purpose, different sorting patterns in solid waste streams were developed and compared, particularly for each of the five scenarios explored in this chapter with respect to their environmental burdens, benefits, and costs, simultaneously using the compromise-programming approach. To ensure the feasibility and practicality of the final options, these scenarios were formulated as optimization analyses with technical settings from single-objective to multi-objective in the Borough of Lewisburg, Pennsylvania, with or without the inclusion of GWP. Such a wealth of mathematical formulations may uniquely demonstrate how potential costs in an SWM system could be compensated by benefits expected from recovered materials while maximizing the reduced risk of GWP. This approach helps determine the most favorable management option for optimal levels of recycling and landfilling for SWM at the Borough of Lewisburg, Pennsylvania.

This practical implementation differs from previous works by (1) applying a comparative approach to differentiate among several classical methods, from the cost-effectiveness (first-generation), to the cost–benefit (second-generation), to the latest GWP-based (third-generation) optimization analyses for SWM; (2) formulating the GWP-based optimization analysis based on a real-world example, thereby making the implementation transferable and transformative; (3) incorporating a professional database driven by Gabi® software package so that GWP estimates can be more representative with respect to a suite of flow control scenarios in the waste management networks; and (4) extending the cost–benefit analysis for SWM under a carbon-regulated environment to carry out the cost–benefit–risk trade-offs. This implementation enables us to explain why those differences are important in regard to the intrinsic comparisons among these four types of decision analyses (i.e., cost-effectiveness, cost–benefit, LCA-based optimization, carbon-regulated optimization) in the context of environmental economics, industrial ecology, and cleaner production (i.e., recycling vs. landfill).

13.2 BACKGROUND OF SYSTEM PLANNING

Generally, physical, chemical, and biological processes are involved in SWM. Physical management techniques include source reduction, reuse, recycling, recovery, and landfilling. Source reduction is the most desirable management system by reducing MSW generation or increasing the product life span, such as using cotton rather than paper towels in day-to-day life. Achieving the goal of material recycling and energy recovery involves using several common unit operations such as material recovery facilities (MRF, pronounced as 'merf'), incinerators, and pyrolysis–gasification processes. In contrast to incineration, pyrolysis is a combustion process in an oxygen-deficient chamber. Gasification is a modified pyrolysis process where a small amount of oxygen is introduced at a relatively higher temperature than pyrolysis. Landfill generally degrades the materials through a five-phase biological process including initial adjustment, transition phase, acid formation phase, fermentation phase, and maturation phase (Vesilind et al., 2002).

CH_4 and CO_2 are the major gases emitted from SWM processes and are most responsible for high GWP in SWM systems. MSW can be decayed in an anaerobic–anoxic condition by producing the end products such as CH_4, CO_2, and ammonia (NH_3) for reuse. This process is common for the production of biogas (mostly CO_2 and CH_4) and compost. Whereas incineration has substantial emissions of CO_2, landfilling has enormous impacts on the environment due to GHG emissions such as CH_4 and CO_2. Although CH_4 production due to biological anaerobic process can be recovered from landfills for power generation, inefficiency in the CH_4 collection system exacerbates the GHG situation. According to the Intergovernmental Panel on Climate Change, GWP is 1 for CO_2 and 25 for CH_4 for a period of 100 years (Wanichpongpan and Gheewala, 2007); therefore, in terms of LCA, landfill has the lowest priority and recycling has the highest (Moberg et al., 2005).

13.2.1   Structure of the Proposed Solid Waste Management System

Lewisburg, a small Borough in Union County, Pennsylvania, is home to residential, commercial, and industrial sectors. The residential sector is the major contributor to the Borough's MSW. The Municipal Waste Planning, Recycling and Waste Reduction Act of 1998 (DEP, 1988) mandated that all towns in the state of Pennsylvania with a population of more than 5000 must have a recycling program, and at least 25% of the total generated waste stream must be recycled. In addition to the state mandate, the Borough has set its own recycling goal of 30–50% per material recycled per year. It currently delivers the residential MSW to the nearby Lycoming County Landfill (Figure 13.1).

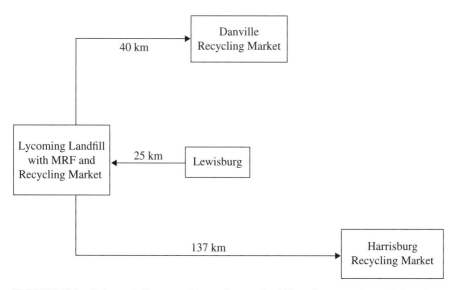

FIGURE 13.1   Schematic diagram of the study area, landfill, and recovered material markets. *Source*: From Chang et al. (2012)

A total of 0.728 km² (179.9 ac) have been permitted to Lycoming County, which has been developing and maintaining the landfill site since 1973. This regional site serves not only Lycoming County but also Montour, Northumberland, Snyder, Columbia, and Union counties. Although the landfill has a capacity of 1,600,000 kg·per day, it normally receives approximately 1,100,000 kg·per day on average. For simplicity, we assumed that all the deposited materials at Lycoming County Landfill come from Lewisburg only. The Lycoming landfill has a MRF of about 5574 m² colocated with a recovered material market. Materials here are sorted mechanically using recycling equipment (Table 12.4) such as loaders, forklifts, balers, and trailers. The facility accepts postconsumer materials including glass containers, plastic bottles, aluminum and steel cans, corrugated containers, magazines, and mixed computer and office paper from businesses. The facility also accepts some postindustrial materials, such as polyvinyl chloride (PVC) (sheet and pipe), which they grind on-site. The facility also accepts and markets processed newspapers, magazines, corrugated containers, mixed office paper, and glass from other counties. Recyclables are directly collected from the households and sent to the MRF for sorting and recycling. The statewide composition of the generated waste stream by weight (Table 12.5) is used as a typical composition of solid waste of Lewisburg in this study.

Recycled materials are delivered to the markets according to category to meet varying product demand at different locations (Table 12.6). The Lycoming landfill itself has a potential market, and two other markets are located at Danville (40 km from the landfill) and Harrisburg (137 km from the landfill) (Figure 13.1). Because the Borough currently disposes all waste in one landfill, all information required for this study is available through local investigation. The cost for MSW collection and material handling are each approximately United States dollars (US$) US$20 per metric ton (tonne), transportation cost for MSW and recyclables is about US$0.25 per km·per tonne, and the landfill earns a tipping fee of US$46.96 per tonne. The prices of recovered materials vary according to the location of the markets (Table 12.7). Several factors must be considered in developing an effective recycling program, including recycling rate (≥25% of waste generated), composition of the recyclables (30–50% per material), selection of a recycling site, the amount of waste sent to be disposed in the landfill, and cost for final disposal. In addition to the benefits from selling recycled materials, landfill gas (LFG) (e.g., CH_4) can be used to generate electricity, which may be sold at the rate of US$0.08 per kWh according to the comparable unit electricity price offered by Pennsylvania Public Utility (PPL) Electric Utilities Corporation.

13.2.2 GWP Calculations for Different Management Scenarios

The LCA in this study was performed using the GaBi® 4.3 software package developed by PE International (2009) in Stuttgart, Germany, to estimate GWP from the various process steps of the SWM program of Lewisburg. The functional unit of 1 tonne of the solid waste was selected. The emissions were assessed for the transport of waste from households to the landfill, from the MRF to recycling markets, and for landfilling. Scenario settings require that the entire solid waste stream of 1 tonne be sorted (Table 12.8) in the MRF. Recycling ratio per material is kept constant (50%), and the overall recycling is ensured to be >25% to comply with the state law.

The GHG emission analysis in the GaBi® 4.3 model (Figure 13.2) applied truck, diesel refinery, power grid mix, and landfill processes directly from the GaBi software database. Only the MRF was created as a new process where inputs (diesel and electricity consumption) and outputs (GHG) were calculated using Lycoming MRF equipment (Table 12.4), and corresponding data were standard values from Nishtala and Solano (1997).

The LCA was conducted based on 1 tonne solid waste stream (i.e., the functional unit) presumably carried by a truck from Lewisburg to Lycoming MRF 25 km away. More than 25% of the materials can be recycled and sent to the three recycling markets (Danville, Harrisburg, and Lycoming) according to the market demand. After mechanical sorting, 715 kg waste is disposed of at the landfill at the respective location.

Substantial developments have occurred in landfill technologies in the last two decades, which have brought a marked reduction of emissions to the environment, both as gas and leachate. CH_4 is mostly oxidized in soil top cover or used for energy generation, but CO_2 from biogenic origin is considered neutral to GWP. The reduction of GWP due to biogenic sources and oxidized CH_4 was compensated in our calculation if leachate from the landfill is taken to the wastewater treatment plant where, again, CH_4 and N_2O are produced as GHGs. To avoid complexity, however, we conducted this modeling work by including CO_2 and CH_4, the predominant gases in GHG emissions, as part of the typical landfill operation; thus, for simplification, only CO_2 and CH_4 are included in our GHG calculations.

GaBi® output was presented in terms of the amount of CO_2eq emitted by each individual process (Table 12.9). According to the Lycoming Landfill authority, approximately 25% of the LFG being converted into electricity is sold to PPL Electric Utilities Corporation. The production of 1 MW of electricity is enough to power approximately 1000 homes. Energy density of CH_4, which is the amount of energy per mass, is between 50 MJ·kg^{-1} (Zittel and Wurster, 1996; Thomas, 2000) and 55.6 MJ·kg^{-1} (O'Connor, 1977; Bossel and Eliason, 2002); we chose the lower boundary in our calculations. The amount of recycled CH_4 from the landfill process for electricity production is 3.3 kg (or 82.53 kg CO2eq) per 715 kg waste for disposal (Table 12.9). Assuming that the energy loss rate for the electricity production from CH_4 recovered by landfill is about 30%, the electricity produced from the LFG is thus 115.5 MJ, or 32.08 kWh per 715 kg waste for disposal. The CO_2eq emissions due to electricity production from CH_4 are 0.01015 kg·kg^{-1} waste for disposal, assuming that the recycled CH_4 is fully converted to CO_2 in the electricity production process.

The major findings in this LCA include (1) landfill processes may produce a substantial amount of CO_2, and (2) over the entire process, CO_2 and CH_4 are the predominant gases among those emitted (Table 17.6) and create a potential threat to global

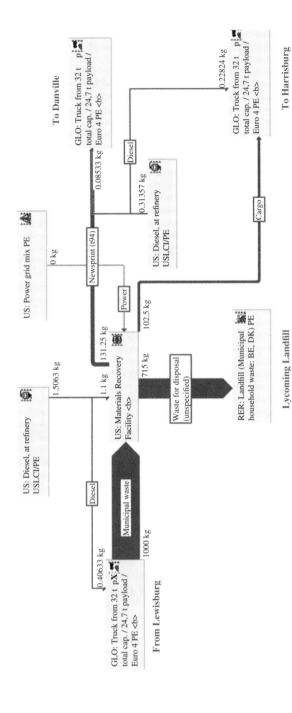

FIGURE 13.2 Screenshot of the GaBi® 4.3 model in conducting GHG emission analysis. *Source:* From Chang et al. (2012)

climate change. With this step accomplished, GWP values from the GaBi® outputs can be adopted directly into linear models (in Scenarios 3, 4, and 5) to investigate the optimal patterns between recycling and landfilling (discussed in later sections). Although the LP models (Scenarios 1, 2, and 3) represent the single-objective optimization, the compromise-programming model represents the multiobjective evaluation in which both GWP and economic criteria are simultaneously considered.

13.3 FORMULATION OF SYSTEMS ENGINEERING MODELS FOR COMPARATIVE ANALYSIS

All factors identified in the previous section were included in our modeling framework and solution procedure. LINGO® 10.0 software (LINDO Systems, 2010) was used to analyze the following scenarios: (1) cost minimization of the MSW collection and disposal, (2) net benefit maximization by collecting tipping fees and selling recovered materials and electricity by LFG, (3) minimization of GWP due to CH_4 and CO_2 gas emission from the landfill, including transportation of MSW and MRF processes, (4) trade-offs between GWP minimization and net benefit maximization, and (5) cost–benefit analysis under a carbon-regulated environment. For Scenario 4, an integrated LCA and multiple criteria decision-making (MCDM) approach was applied to find a compromise solution close to the ideal one based on the minimum distance criteria (Vatalis and Manoliadis, 2002; Shih and Lin, 2003; Chang et al., 2005; Shmelev and Powell, 2006; Xu et al., 2009; Fattahi and Fayyaz, 2010; Galante et al., 2010). For Scenario 5, a cost–benefit optimization model extended from Scenario 2 was applied under two types of different carbon-regulated settings in the waste management market.

The two different carbon-regulated settings are carbon tax and cap-and-trade (Box 13.1) approaches to addressing the policy impact. Carbon tax is a command-based approach by taxing actions that emit CO_2 or its equivalents. Denmark, along with other Scandinavian countries, implemented the world's first carbon tax on fossil fuels in the early 1990s (World Bank, 2010). Cap-and-trade is an approach to control GHG emissions that combines both market value and regulation. In a cap-and-trade scheme, the cap, an overall emissions limit, is set for a specific time range and usually declines over time by regulation. Individual parties receive permits (or allowances, either through grant or auction), giving them the legal right to emit CO_2 up to the quantity of permits they hold. The emission permits or allowances are tradable between parties who will gain from trades if they have different marginal CO_2 abatement costs. The Kyoto Protocol (United Nations, 1998), an agreement under the United Nations Framework Convention on Climate Change adopted in 1997, is an example of a cap-and-trade scheme that contains legally binding commitments to reduce GHG emissions by developed countries, also called Annex I countries in the Kyoto Protocol. The impact associated with the two different carbon-regulated settings will be discussed further in two subscenarios, 5a and 5b, in this chapter.

BOX 13.1   DEFINITION OF CAP-AND-TRADE AND ITS POTENTIAL PROBLEMS

Cap-and-trade systems are known as allowance trading, which depends on "pollution credits" for trading in the market based on lumped credits allowed by a government agency in a region. In a carbon-regulated environment, the overall air quality management goals can be set for a state, region, or the entire country. In this context, specific sources of air pollution, such as power plants and waste incineration facilities, are given a certain number of allowances (carbon credit). Trading may occur if one source wishes to buy pollution credits from others to avoid investing in hardware to remove carbon. The cap-and-trade system can achieve a higher level of overall economic efficiency, but it potentially allows certain parts of the region to become relatively more polluted, although overall air standards in the region might still be met.

The present LCA study began by collecting MSW information from residential areas, including waste transport, recycling, and landfilling of waste. Although the functional unit selected for the GWP analysis was based on the management of 1 tonne of MSW in Lewisburg, cost–benefit analysis was carried out based on 1,100,000 kg, which is the daily waste generation of Lewisburg. The landfill system was limited to the residual materials after MRF processing, assuming that steady-state solid waste streams with the same composition was maintained at all times in the SWM system. Major recyclables included clean brown glass, steel cans, newspapers, aluminum cans, and other materials. The data for the life cycle inventory were gathered from actual applications in Lewisburg, literature, and the database of the GaBi® software package, which was adjusted to the local conditions in Pennsylvania. Calculations of GWP for all the SWM processes in the study area were used as inputs for the optimization model under different scenarios. The following subsections are organized to formulate these five models sequentially.

13.3.1 Scenario-1: Total Cost Minimization

Equation (13.1) represents a cost-minimization approach in the context of the cost-effectiveness analysis. The delivery cost of recovered materials was not included in this equation because the objective of this scenario is to minimize the direct cost only as a base case for comparison:

$$\text{Minimize} \quad Z_0 = C_t + C_o, \tag{13.1}$$

where C_t = cost involved in the transportation of MSW from household to landfill (in US\$); C_o = cost involved in the operational process within the landfill (in US\$); and Z_0 = total operational cost to be minimized (in US\$). The constraints formulated as an integral part of the optimization model are the mass balance constraints, state-mandated recycling constraints, definitional cost constraints, and non-negative constraints. Constraint set can be defined below.

1. **Mass balance constraints:** The MSW streams can be conceptually divided into two parts: one stream goes to landfill and the other is recovered materials. This constraint maintains the mass balance of the generated MSW at each node in the network. Thus, the sum of all categories of recovered materials and disposed materials at landfill must be equal to the total MSW generated:

$$\sum_{j=1}^{13} X_j = R; \sum_{j=1}^{13} L_j = L; G_j = X_j + L_j \text{ and } G = L + R, \tag{13.2}$$

where G = total MSW generated (kg); G_j = MSW generated for material j (kg); R = total MSW recovered (kg); L = total MSW destined for the landfilling after passing through the MRF (kg); X_j = the amount of material j being recycled at the MRF (kg); and L_j = the amount of material j (Table 13.1) disposed of in the landfill (kg).

2. **State-mandated recycling constraints:** The state of Pennsylvania requires that recovered materials should be $\geq 25\%$ of the total MSW generated. Local government further requires recovered materials must be $>30\%$ but $\leq 50\%$:

$$R \geq sG \tag{13.3}$$

$$X_j \geq IG_j \text{ and } X_j \leq xG_j \text{ for } j = 1 \sim 8, \text{ and} \tag{13.4}$$

$$X_j = 0 \text{ for } j = 9 \sim 13, \tag{13.5}$$

where s = minimum state mandate recycling ratio (%) (i.e., 25% in this case); I = lower limit of recycle for material j (%) (i.e., 30% in this case); and x = upper limit of recycle for material j (%) (i.e., 50% in this case).

TABLE 13.1   List of materials

j	Material	Type of waste
1	Aluminum	Recyclables
2	Steel cans	Recyclables
3	Clear glass	Recyclables
4	Mixed glass	Recyclables
5	Corrugated paper	Recyclables
6	Newsprint paper	Recyclables
7	Mixed paper	Recyclables
8	Plastic	Recyclables
9	Yard trimmings	Non-recyclables
10	Food scraps	Non-recyclables
11	Rubber, leather, textiles	Non-recyclables
12	Wood	Non-recyclables
13	Other	Non-recyclables

Source: From Chang et al. (2012).

3. **Definitional constraint of cost for waste collection and handling:** The definitional constraint for waste collection and handling cost is based on the volume of MSW collected regardless of whether the waste streams are eventually destined for landfilling or MRF:

$$c \cdot G = C, \tag{13.6}$$

where c = unit collection cost of MSW (US\$·kg^{-1}) and C = total cost for waste collection (US\$).

4. **Definitional constraint of cost for waste shipping:** Total transportation cost of MSW depends on the distance shipped by a truck from a collection point to a MSW facility and the volume of MSW collected and shipped, defined by:

$$t \cdot D \cdot G = T, \tag{13.7}$$

where t = shipping cost of waste streams from a collection point to a landfill (US\$ per k·km^{-1}); D = shipping distance covered by a garbage truck (km); and T = total shipping cost of a system (US\$).

5. **Definitional constraints of total costs for waste collection and shipping:** Definitional constraints are required to summarize the shipping and operation separately:

$$C_t = T \text{ and } C_o = C. \tag{13.8}$$

6. **Non-negativity constraints:** All decision variables are non-negative.

13.3.2 Scenario 2: Net Benefit Maximization

The objective function (Equation 13.9) applied to maximize the net benefit in a system in the context of cost–benefit analysis involves selling recovered materials in the recycling market and collecting tipping fees collected from the household by landfill operators. In comparison, landfill incurs some cost due to collection, sorting, and distribution process of waste streams:

$$\text{Maximize } Z_1 = F + B_1 + B_2 - C_t - C_o - C_{t1} - C_{t2} \tag{13.9}$$

Because it is a benefit-maximization case, F = total tipping fees charged by the landfill (US\$); B_1 = total income involved with selling the recyclables (US\$); B_2 = total income involved with selling electricity from CH_4 generated by the landfill process; C_{t1} and C_{t2} = total cost involved with shipping recyclables to Market-1 (Danville) and Market-2 (Harrisburg), respectively (US\$); and Z_1 = maximized net benefit of the MSW system (US\$). In this scenario, the first four constraints are the same as constraints for Scenario-1.

1. **Definitional constraint of cost for waste shipping:** Two additional cost constraints are formulated for shipping because recovered materials must be delivered to

the recycling market:

$$t_1 D_1 \sum_{j=1}^{13} XS_{1j} = T_1 \text{ and } t_2 D_2 \sum_{j=1}^{13} XS_{2j} = T_2 \text{ and} \tag{13.10}$$

$$X_j = \sum_{k=1}^{3} XS_{kj} \text{ for } j = 1 \sim 13, \tag{13.11}$$

where t_1 and t_2 = shipping cost for recovered materials from the MRF to Market-1 and Market-2, respectively (US\$ per kg·km^{-1}); D_1 and D_2 = shipping distance covered by a truck for delivering recyclables to Market-1 and Market-2 (km), respectively; XS_{kj} = the amount of material j being recycled at MRF and shipped to Market-k ($k = 1$ Danville, $k = 2$ Harrisburg, and $k = 3$ Lycoming) (kg); T_1 and T_2 = total cost of material transported in Market-1 and Market-2 (US\$), respectively. There is no transportation cost for shipping the recycled material to Market-3 because the landfill is at the same location.

2. Definitional constraints of total costs for waste collection, recycling, and shipping: To bridge the calculation between the objective function and the constraints, four more definitional constraints are required:

$$C_t = T; C_o = C; C_{t1} = T_1; C_{t2} = T_2 \tag{13.12}$$

3. Definitional constraints for benefits: Tipping fees and income from selling recovered materials must be defined for the objective function:

$$\sum_{k=1}^{3} \sum_{j=1}^{13} XS_{kj} P_{kj} = B_1 \tag{13.13}$$

$$m \cdot L = B_2 \tag{13.14}$$

$$m = (32.08 \text{ kWh}/715 \text{ kg})0.08 \text{ US\$} \cdot \text{kWh}^{-1} = 0.003589 \text{ US\$} \cdot \text{kg}^{-1}, \text{ and} \tag{13.15}$$

$$f \cdot G = F, \tag{13.16}$$

where P_{kj} = prices of recovered material j in the Market-k (US\$·kg^{-1}); B_1 = total benefit received from recovered materials (US\$); m = the unit benefit due to sales of electricity generated by LFG (US\$·kg^{-1} LFG); B_2 = total benefit received from selling CH$_4$ (US\$); and f = tipping fees charged by the Lycoming landfill (US\$·kg^{-1} MSW).

4. Non-negative constraints: All decision variables should be non-negative.

13.3.3 Scenario-3: GWP Minimization

The GWP minimization objective function (Equation 13.17) is applied to minimize the GWP in relation to the GHG emission from the operation of the landfill, the MRF, and relevant transportation activities in an SWM system without regard to any cost–benefit concerns. Compared with the objective in Scenario-2, which only considers benefits and costs without the concern of GWP, the objective in this scenario does the opposite:

$$\text{Minimize } Z_2 = G \cdot (\text{GWP}_t + \text{GWP}_s) + L \cdot (\text{GWP}_l + \text{GWP}_e)$$

$$+ \sum_{k=1}^{2} \sum_{j=1}^{13} \text{XS}_{kj} \cdot \text{GWP}_{tk}, \tag{13.17}$$

where GWP_t = unit GWP for transportation of MSW from households to a landfill (kg CO_2 eq·kg^{-1} MSW); GWP_s = unit GWP for MRF operation (kg CO_2eq·kg^{-1} MSW); GWP_l = unit GWP for landfill operation (kg CO_2eq·kg^{-1} waste for disposal); GWP_e = unit GWP for electricity production by LFG (kg CO_2eq·kg^{-1} waste for disposal); GWP_{tk} = unit GWP for transportation of recycled materials to market-k (kg CO_2eq·kg^{-1} recycled materials); and Z_2 = total GWP of the system (kg CO_2eq). All constraints are the same as in Scenario-2, in addition to some extra definitional constraints (Figure 13.2 and Table 12.9):

$$\text{GWP}_t = 1.29 \text{ kg } CO_2\text{eq}/1{,}000 \text{ kg MSW} = 0.00129 \text{ kg } CO_2\text{eq} \cdot \text{kg}^{-1} \text{ MSW},$$
$$\tag{13.18}$$

$$\text{GWP}_s = (253.76 + 6.93)\text{kg } CO_2\text{eq}/1{,}000 \text{ kg MSW}$$
$$= 0.26069 \text{ kg } CO_2\text{eq} \cdot \text{kg}^{-1} \text{ MSW, and} \tag{13.19}$$

$$\text{GWP}_l = (185.56 + 247.58)\text{kg } CO_2\text{eq}/715 \text{ kg waste for disposal}$$
$$= 0.60579 \text{ kg } CO_2\text{eq} \cdot \text{kg}^{-1} \text{ waste for disposal.} \tag{13.20}$$

Assuming that recycled CH_4 is fully converted to CO_2 in the electricity production process, then

$$\text{GWP}_e = 3.3 \text{ kg } CH_4/715 \text{ kg waste for disposal}/20 \text{ g } CH_4 \times 44 \text{ g } CO_2$$
$$= 0.01015 \text{ kg } CO_2\text{eq} \cdot \text{kg}^{-1} \text{ waste for disposal, and} \tag{13.21}$$

$$\text{GWP}_{tk} = \begin{cases} 0.27 \text{ kg } CO_2\text{eq}/131.25 \text{ kg recycled materials} \\ = 0.002057 \text{ kg } CO_2\text{eq}/\text{kg recycled materials,} & k = 1 \\ 0.72 \text{ kg } CO_2\text{eq}/102.50 \text{ kg recycled materials} & k = 2 \\ = 0.007024 \text{ kg } CO_2\text{eq}/\text{kg recycled materials,} \end{cases} \tag{13.22}$$

13.3.4 Scenario-4: Net Benefit Maximization and GWP Minimization

This objective function (Equation 13.23) maximizes the net benefit and minimizes the GWP for the SWM system. These two objectives must be balanced through a trade-off process to apply compromise programming (Zeleny, 1973) to the search for the Pareto optimal solution set. The model formulation of this multiobjective programming can be constructed where all constraints are the same as Scenario-3:

$$\text{Maximize } Z_1 = F + B_1 + B_2 - C_t - C_o - C_{t_1} - C_{t2} \text{ and} \tag{13.23}$$

$$\text{Minimize } Z_2 = G \cdot (\text{GWP}_t + \text{GWP}_s) + L \cdot (\text{GWP}_l + \text{GWP}_e)$$

$$+ \sum_{k=1}^{2} \sum_{j=1}^{13} \text{XS}_{kj} \cdot \text{GWP}_{tk} \tag{13.24}$$

In this scenario, GWP minimization and benefit maximization are two objectives. Although these two objectives are not fully conflicting, a compromise-programming model can be formulated to determine the optimal frontier to the ideal solution (not achievable) in the solution space. The best compromise solution may be selected from the Pareto optimal solution set by a set of common criteria of shortest geometric distance to the ideal solution. A normalization function is needed before optimization analysis because a distance-based assessment function cannot be used for incommensurable units in the objectives.

13.3.5 Scenario-5: Cost–Benefit Analysis Under a Carbon-Regulated Environment

Subscenario-5a is organized to address the impact of carbon tax, and Subscenario-5b is particularly formulated to reflect a cap-and-trade environment in a waste management market. Both subscenarios are hypothetical because the United States has not yet adopted either, but with further understanding of the effects of global warming, the development of either or both carbon-regulated approaches is highly anticipated. With this condition, value assumptions must be made in this case study for a preliminary assessment. Carbon tax (Box 13.2) was assumed to be US$20 per tonne of CO_2eq emissions in Subscenario-5a. In Subscenario-5b, we also assumed that regulation requires a 2% offset in terms of CO_2eq emissions based on the outputs from Scenario-2, which is expected to be achieved purely by optimizing the operation instead of implementing new facilities and technologies. Any additional saving of carbon credit than the required level is sellable at the prevailing market carbon price, which was assumed to be US$30 per tonne of CO_2eq in this subscenario.

BOX 13.2   DEFINITION OF CARBON TAX

A carbon tax deemed as an economic instrument is a tax on emissions caused by the burning of coal, gas, and oil leading to the reduction of the production of GHG. Such a tax would be levied against high consumers of heating oil, gasoline, electricity, and other energy sources in an effort to reduce the hot spots of GHG emissions in a region.

The objective function for Scenario-5a (Equation 13.25) is applied to maximize the net benefit in a system in the context of cost–benefit analysis. Formulation of benefit terms remains the same, and cost terms are similar to those in Scenario-2, except that extra cost is incurred due to the carbon taxes for CO_2eq emissions:

$$\text{Maximize } Z_3 = F + B_1 + B_2 - C_t - C_o - C_{t1} - C_{t2} - \text{TAX}, \qquad (13.25)$$

where TAX is the carbon taxes for CO_2eq emissions (US\$). In addition to all the constraints in Scenario-2, the extra definitional constraint is

$$\text{TAX} = 0.02 \times \left[G \cdot (\text{GWP}_t + \text{GWP}_s) + L \cdot (\text{GWP}_l + \text{GWP}_e) + \sum_{k=1}^{2} \sum_{j=1}^{13} \text{XS}_{kj} \cdot \text{GWP}_{tk} \right].$$
$$(13.26)$$

The objective function for Scenario-5b (Equation 17.27) is applied to maximize the net benefit in a system in the context of cost–benefit analysis. Cost terms are the same and benefit terms are similar to those in Scenario-2, except that extra benefit is received due to the trade of surplus carbon credits:

$$\text{Maximize } Z_4 = F + B_1 + B_2 + B_{cc} - C_t - C_o - C_{t1} - C_{t2}, \qquad (13.27)$$

where B_{cc} is the benefit due to carbon credits trading (US\$). In addition to all the constraints in Scenario-2, additional constraints are

$$G \cdot (\text{GWP}_t + \text{GWP}_s) + L \cdot (\text{GWP}_l + \text{GWP}_e)$$
$$+ \sum_{k=1}^{2} \sum_{j=1}^{13} \text{XS}_{kj} \cdot \text{GWP}_{tk} \leq 0.98 Z_{2(\text{opt})}^{*}, \text{ and} \qquad (13.28)$$

$$B_{cc} = 0.03 \left[0.98 Z_{2(\text{opt})}^{*} - G \cdot (\text{GWP}_t + \text{GWP}_s) - L \cdot (\text{GWP}_l + \text{GWP}_e) \right.$$
$$\left. - \sum_{k=1}^{2} \sum_{j=1}^{13} \text{XS}_{kj} \cdot \text{GWP}_{tk} \right], \qquad (13.29)$$

where $Z_{2(\text{opt})}^{*}$ is the optimal objective function value of the following optimization problem:

$$\text{Minimize } Z_2^{*} = G \cdot (\text{GWP}_t + \text{GWP}_s) + L \cdot (\text{GWP}_l + \text{GWP}_e) + \sum_{k=1}^{2} \sum_{j=1}^{13} \text{XS}_{kj} \cdot \text{GWP}_{tk},$$
$$(13.30)$$
$$\text{subject to } F + B_1 + B_2 - C_t - C_o - C_{t1} - C_{t2} = Z_{1(\text{opt})}. \qquad (13.31)$$

All other constraints are the same as those in Scenario-2, where $Z_{1(\text{opt})}$ is the optimal objective function value in Scenario-2.

13.4 INTERPRETATION OF MODELING OUTPUT FOR DECISION ANALYSIS

Comparison of outputs from different scenarios may help explain the relative impact of GHG from landfill and power generation from the activities of recycling materials associated with their cost and benefit profiles. For clarification, the results are discussed sequentially below according to the prescribed scenarios before making overall comparisons and conclusions.

13.4.1 Interpretation of Scenario-1: Cost Minimization

Because the main objective of this scenario is to minimize total cost, all recovered materials from the MRF are kept in the local market (i.e., Lycoming) to avoid additional shipping. The minimum cost is US$52,250 when maintaining the recycling rate at 25%. The total cost would be the same for a recycling ratio >25% because the additional collected MSW streams must pass through the MRF process. To save transportation costs, the optimized sorting and allocation of MSW (Table 13.2) send nothing (i.e., all values are zero) to the other two recycling markets. Because the Lycoming market is adjacent to the landfill and the MRF, recycled materials from this market create no transportation cost.

13.4.2 Interpretation of Scenario-2: Benefit Maximization

The optimized net benefit (i.e., benefit–cost) is found to be US$10,231.38. The positive value indicates that the benefit received can balance out the cost incurred in the collection and operation process (Table 13.3). Note that all recovered aluminum and steel cans are delivered to the Harrisburg market because the prices for recovered materials are better in that market, although a higher transportation cost is incurred. For all other recovered materials, keeping them in the Lycoming market is favored to avoid transportation costs, thereby receiving more net benefits. Shipping recovered materials (e.g., newsprint paper) to Danville would not result in a positive net benefit in terms of the corresponding transportation activities.

13.4.3 Interpretation of Scenario-3: GWP Minimization

Considering gas emissions of CH_4 and CO_2, the GWP value in the SWM system was 772,614.8 kg CO_2eq, achieving a recycling ratio of 28.50%. The model output (Table 13.4) suggests that nothing should be sent to the other two markets (Danville and Harrisburg) to prevent CO_2 emissions generated by shipping the recovered materials.

13.4.4 Interpretation of Scenario-4: Benefit Maximization and GWP Minimization

The concept of compromise programming was applied to find the trade-offs between benefit maximization (Z_1) and GWP minimization (Z_2). The ideal solution was found

TABLE 13.2 Optimized sorting and allocation of MSW by LINGO to keep the cost minimum

Items	Generated MSW (kg·d⁻¹)	Amount recycled (kg·d⁻¹)	Amount disposed of in the landfill (kg·d⁻¹)	Total amount recycled at Lycoming (kg·d⁻¹)	Total amount recycled at Danville (kg·d⁻¹)	Total amount recycled at Harrisburg (kg·d⁻¹)
Total MSW	1,100,000	275,000	825,000	275,000	0	0
Aluminum	33,000	16,500	16,500	16,500	0	0
Steel cans	22,000	6,600	15,400	6,600	0	0
Clear glass	16,500	4,950	11,550	4,950	0	0
Mixed glass	38,500	11,550	26,950	11,550	0	0
Corrugated paper	50,050	15,015	35,035	15,015	0	0
Newsprint paper	288,750	131,285	157,465	131,285	0	0
Mixed paper	46,200	23,100	23,100	23,100	0	0
Plastic	132,000	66,000	66,000	66,000	0	0
Yard trimmings	110,000	0	110,000	0	0	0
Food scraps	165,000	0	165,000	0	0	0
Rubber	77,000	0	77,000	0	0	0
Wood	77,000	0	77,000	0	0	0
Other	44,000	0	44,000	0	0	0

Source: Chang et al. (2012).

TABLE 13.3 Optimized sorting and allocation of MSW by LINGO to achieve maximum benefit

Items	Generated MSW (kg·d⁻¹)	Amount recycled (kg·d⁻¹)	Amount disposed of in the landfill (kg·d⁻¹)	Total amount recycled at Lycoming (kg·d⁻¹)	Total amount recycled at Danville (kg·d⁻¹)	Total amount recycled at Harrisburg (kg·d⁻¹)
Total MSW	1,100,000	275,000	825,000	247,500	0	27,500
Aluminum	33,000	16,500	16,500	0	0	16,500
Steel cans	22,000	11,000	11,000	0	0	11,000
Clear glass	16,500	8,250	8,250	8,250	0	0
Mixed glass	38,500	11,550	26,950	11,550	0	0
Corrugated paper	50,050	25,025	25,025	25,025	0	0
Newsprint paper	288,750	144,375	144,375	144,375	0	0
Mixed paper	46,200	13,860	32,340	13,860	0	0
Plastic	132,000	44,440	87,560	44,440	0	0
Yard trimmings	110,000	0	110,000	0	0	0
Food scraps	165,000	0	165,000	0	0	0
Rubber	77,000	0	77,000	0	0	0
Wood	77,000	0	77,000	0	0	0
Other	44,000	0	44,000	0	0	0

TABLE 13.4 Optimized sorting and allocation of MSW by LINGO to achieve the minimum GWP

Items	Generated MSW (kg·d⁻¹)	Amount recycled (kg·d⁻¹)	Amount disposed of in the landfill (kg·d⁻¹)	Total amount recycled at Lycoming (kg·d⁻¹)	Total amount recycled at Danville (kg·d⁻¹)	Total amount recycled at Harrisburg (kg·d⁻¹)
Total MSW	1,100,000	313,500	786,500	313,500	0	0
Aluminum	33,000	16,500	16,500	16,500	0	0
Steel cans	22,000	11,000	11,000	11,000	0	0
Clear glass	16,500	8,250	8,250	8,250	0	0
Mixed glass	38,500	19,250	19,250	19,250	0	0
Corrugated paper	50,050	25,025	25,025	25,025	0	0
Newsprint paper	288,750	144,375	144,375	144,375	0	0
Mixed paper	46,200	23,100	23,100	23,100	0	0
Plastic	132,000	66,000	66,000	66,000	0	0
Yard trimmings	110,000	0	110,000	0	0	0
Food scraps	165,000	0	165,000	0	0	0
Rubber	77,000	0	38,500	0	0	0
Wood	77,000	0	77,000	0	0	0
Other	44,000	0	44,000	0	0	0

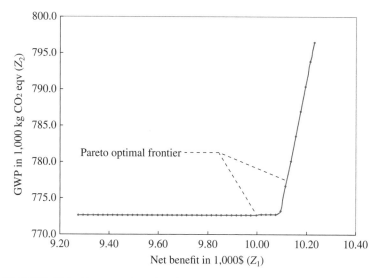

FIGURE 13.3 Pareto optimal frontiers between cost–benefit and GWP criteria

at the point with coordinates 10,231.38, 772,614.8 by solving each individual objective linear program (e.g., Scenarios 2 and 3), both of which can be integrated into a complete delineation of the compromised Pareto optimal frontier for decision analysis (Figure 13.3). Because the units for the two objectives are different, a normalized compromised Pareto optimal frontier is needed so the two objectives can be in the same scale between 0 and 1 (Figure 13.4).

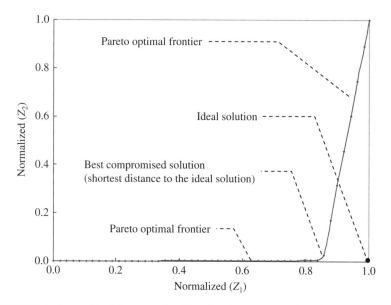

FIGURE 13.4 Normalized Pareto optimal frontiers between cost–benefit and GWP criteria

The distance from the ideal solution to each point was calculated, and the closest point to the ideal solution (coordinates 0.858, 0.027) corresponds to (Z_1, Z_2) = (10,095.9, 773,271.6). Note that GWP shows a steep increase when the net benefit is within the range of US\$10,100, US\$10,200 (Figure 13.3). Before this steep increase, the change in GWP is almost flat when the net benefit does not exceed US\$10,100. The marginal impacts for the GWP may become significant by further increasing net benefit beyond US\$10,100 (Figure 13.4). Although net benefit may increase when more GWP is generated and vice versa, the common-distance-based criteria suggest the best compromise solution (Table 13.5) is the one with the shortest geometric distance from the ideal solution.

13.4.5   Interpretation of Scenario-5: Cost–Benefit Analysis Under a Carbon-Regulated Environment

In Subscenario-5a (carbon tax approach), the optimal net benefit value is found to be −US\$5362.96, which is less than the optimal net benefit in Scenario 2 due to the extra cost of carbon tax. The negative value indicates that the benefit received cannot balance out the cost incurred in the collection, operation process, and carbon tax. In subscenario-5b (cap-and-trade approach), the optimal net benefit value is found to be US\$10,326.14, which is more than that in Scenario 2 due to the extra benefit from carbon credit trading. The model outputs (Table 13.6) suggest the same operational scheme for both subscenarios, and thus the same amount, 772,790.6 kg, of CO_2eq is emitted in both. Both subscenarios actually reduce the same percent of CO_2eq driven by carbon tax and trading benefit separately.

13.5   COMPARATIVE ANALYSIS

The overall gas emissions of CH_4 and CO_2 are higher at the landfill when compared to the emissions from the MRF processes (Table 12.9). The higher the amount of recycled materials, the lower the amount of remaining MSW destined for final disposal at the landfill, and the lower GHG emissions. Although CH_4 can be recovered from landfill to reduce the GWP and create carbon credits, the recovery of CH_4 requires a substantial financial investment, and the time lag for gas recovery is uncertain. Recycling would therefore have higher priority than landfill disposal in sustainable MSW management, an inference that is consistent with most of the literature. A summary of shipping patterns associated with each scenario links the economic and environmental impact assessment with managerial operations (Table 13.7 and Figure 13.5).

When facing a carbon-regulated environment, however, such observations may be further adjusted. By creating, calculating, and comparing all scenarios (Table 13.7 and Figure 13.5), we can identify the pros and cons of each optimized outcome. For the sorting pattern (Table 13.3) of Scenario-2, the Lewisburg MSW program can maximize benefit with a small cost increase compared with that in Scenario-1 so that the benefit is sufficient to balance out the total cost. Apart from economic considerations,

TABLE 13.5 The most suitable compromised solution of material sorting and waste allocation strategy of MSW to balance the criteria of cost–benefit and GWP simultaneously

Items	Generated MSW (kg·d⁻¹)	Total Amount recycled (kg·d⁻¹)	Amount disposed of in the landfill (kg·d⁻¹)	Total amount recycled at Lycoming (kg·d⁻¹)	Total amount recycled at Danville (kg·d⁻¹)	Total amount recycled at Harrisburg (kg·d⁻¹)
Total MSW	1,100,000	312,747	787,253	285,247	0	27,500
Aluminum	33,000	16,500	16,500	0	0	16,500
Steel cans	22,000	11,000	11,000	0	0	11,000
Clear glass	16,500	8,250	8,250	8,250	0	0
Mixed glass	38,500	18,497	20,003	18,497	0	0
Corrugated paper	50,050	25,025	25,025	25,025	0	0
Newsprint paper	288,750	144,375	144,375	144,375	0	0
Mixed paper	46,200	23,100	23,100	23,100	0	0
Plastic	132,000	66,000	66,000	66,000	0	0
Yard trimmings	110,000	0	110,000	0	0	0
Food scraps	165,000	0	165,000	0	0	0
Rubber	77,000	0	77,000	0	0	0
Wood	77,000	0	77,000	0	0	0
Other	44,000	0	44,000	0	0	0

TABLE 13.6 The compromised scheme under carbon-regulated environment

Items	Generated MSW (kg·d⁻¹)	Total Amount recycled (kg·d⁻¹)	Amount disposed of in the landfill (kg·d⁻¹)	Total amount recycled at Lycoming (kg·d⁻¹)	Total amount recycled at Danville (kg·d⁻¹)	Total amount recycled at Harrisburg (kg·d⁻¹)
Total MSW	1,100,000	313,500	786,500	286,000	0	27,500
Aluminum	33,000	16,500	16,500	0	0	16,500
Steel cans	22,000	11,000	11,000	0	0	11,000
Clear glass	16,500	8,250	8,250	8,250	0	0
Mixed glass	38,500	19,250	19,250	19,250	0	0
Corrugated paper	50,050	25,025	25,025	25,025	0	0
Newsprint paper	288,750	144,375	144,375	144,375	0	0
Mixed paper	46,200	23,100	23,100	23,100	0	0
Plastic	132,000	66,000	66,000	66,000	0	0
Yard trimmings	110,000	0	110,000	0	0	0
Food scraps	165,000	0	165,000	0	0	0
Rubber	77,000	0	77,000	0	0	0
Wood	77,000	0	77,000	0	0	0
Other	44,000	0	44,000	0	0	0

TABLE 13.7 Summary of optimized sorting and allocation of MSW for all five scenarios

Scenarios	Generated MSW (kg·d⁻¹)	Total Amount recycled (kg·d⁻¹)	Amount disposed of in the landfill (kg·d⁻¹)	Total amount recycled at Lycoming (kg·d⁻¹)	Total amount recycled at Danville (kg·d⁻¹)	Total amount recycled at Harrisburg (kg·d⁻¹)
Scenario-1: cost minimization	1,100,000	275,000	825,000	275,000	0	0
Scenario-2: benefit maximization	1,100,000	275,000	825,000	247,500	0	27,500
Scenario-3: GWP minimization	1,100,000	313,500	786,500	313,500	0	0
Scenario-4: benefit maximization and GWP minimization	1,100,000	312,747	787,253	285,247	0	27,500
Scenario-5: benefit maximization under carbon-regulated environment (5a and 5b)	1,100,000	313,500	786,500	286,000	0	27,500

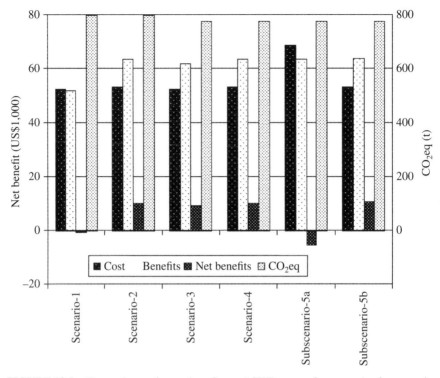

FIGURE 13.5 Comparisons of costs, benefits, and GWP among five scenarios for managing 1,100,000 kg solid waste per day in an SWM system

if the MSW program requires assessing the environmental consequences, Scenario-3 would be an appropriate case.

GWP can be minimized through the sorting and allocation of more recycled materials (Table 13.4). When considering both economic and environmental consequences before making any change to the whole SWM system, Scenario-4 is favored by its GWP (773,271.6 kg CO_2eq), which is reasonably close to the minimum possible GWP (772,614.8 kg CO_2eq) obtained in Scenario-3. A net benefit of US$10,095.9 can be achieved if the sorting and market selection follows the strategies to balance benefit and GWP simultaneously (Table 13.5). Under a carbon-regulated environment, net benefits may be affected depending on different approaches of carbon regulations, even though the operational scheme may be the same (Table 13.6).

A summary of cost and benefit distributions associated with each scenario links the economic and environmental impact assessment with managerial operations (Table 13.7 and Figure 13.6). These distributions evolve systematically over the four types of decision analyses (i.e., cost-effectiveness, cost–benefit, LCA-based optimization, carbon-regulated optimization).

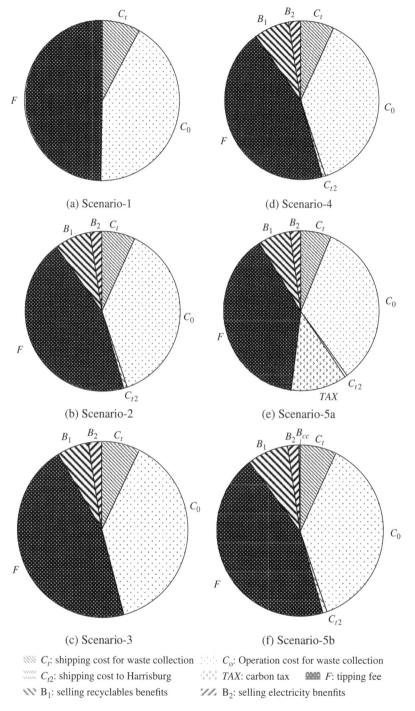

(a) Scenario-1

(d) Scenario-4

(b) Scenario-2

(e) Scenario-5a

(c) Scenario-3

(f) Scenario-5b

C_t: shipping cost for waste collection C_0: Operation cost for waste collection

C_{t2}: shipping cost to Harrisburg TAX: carbon tax F: tipping fee

B_1: selling recyclables benefits B_2: selling electricity bnenfits

FIGURE 13.6 Cost and benefit distributions associated with five scenarios for managing 1,100,000 kg solid waste per day in an SWM system

13.6 FINAL REMARKS

This chapter explored five management scenarios on a comparative basis for decision-making to determine the optimal system-wide setting. Overall, carbon regulation policy has significant impacts for sustainable management of MSW in a community. In addition to GWP, properly managed MSW in a system can help maintain the status of sustainable development in regard to several more sustainability indices, such as recycling goal and net benefit. In these scenarios, recycling gains significant importance due to the reduction of GWP, ultimately identifying the optimal operational pattern between recycling and landfilling. Results also indicate that the cost-effectiveness principle is deemed obsolete, and cost–benefit analysis needs revisions for the future. The cost–benefit–risk analysis under a carbon-regulated environment explores the potential impacts of carbon regulations. With the aid of the comparative study, findings in the LCA-based compromise-programming analysis might lead to the most sustainable solution.

Increased demand for recycled materials would act as an incentive for landfill operators to recover more materials from MSW and further reduce the GHG emissions. Consequently, more campaigns could be conducted to improve future collection and separation of recycled materials. Additional CH_4 recovered at landfill could be delivered for electricity generation through gas turbines, partially outweighing the advantage of recycling activities based on GWP concerns; however, burning of CH_4 may produce more CO_2, so source reduction for CH_4 might be a better solution. Gaining additional benefits would be difficult due to the diminishing rate of return while concurrently reducing GWP from both technical and managerial points of view, even when carbon credit trades are taken into account. Such preliminary insights gained in this chapter can be extended to other MSW systems with MRF, landfills, and even incinerators.

REFERENCES

Ahluwalia, P. K. and Nema, A. K. 2007. A life cycle based multi-objective optimization model for management for computer waste. *Resources Conservation and Recycling*, 51(4), 792–826.

Al-Salem, S. M. and Lettieri, P. 2009. Life cycle assessment (LCA) of municipal solid waste management in the State of Kuwait. *European Journal of Scientific Research*, 34(3), 395–405.

Anderson, L. 1968. A mathematical model for the optimization of a waste management system. SERL Report. Sanitary Engineering Research Laboratory, University of California, Berkeley, CA.

Anderson, L. E. and Nigam, A. K. 1967. A mathematical model for the optimization of a waste management system, ORC 67–25, Operations Research Center, University of California, Berkeley, CA.

Azapagic, A. and Clift, R. 1998. Linear programming as a tool in life cycle assessment. *The International Journal of Life Cycle Assessment*, 3(6), 305–316.

Azapagic, A. and Clift, R. 1999. Life cycle assessment and multiobjective optimization. *Journal of Cleaner Production*, 7(2), 135–143.

Baetz, B. 1990. Optimization/simulation modeling for waste management capacity planning. *Journal of Urban Planning and Development*, 116(2), 59–79.

Banar, M., Cokaygil, Z., and Ozkan, A. 2009. Life cycle assessment of solid waste management options for Eskisehir, Turkey. *Waste Management*, 29(1), 54–62.

Beck, R. W. 2003. Statewide waste composition study – Final report. Report elaborated for the Pennsylvania Department of Environmental Protection Pennsylvania, US.

Boardman, A., Greenberg, D., Vining, A., and Weimer, D. 2001. *Cost-Benefit Analysis: Concepts and Practice*, Prentice Hall, Inc., Upper Saddle River, NJ.

Bossel, U. and Eliason, B. 2002. Energy and the hydrogen economy. EV World, December 28.

Chang, N. B. and Wang, S. F. 1996. Solid waste management system analysis by multiobjective mixed integer programming model. *Journal of Environmental Management*, 48(1), 17–43.

Chang, N. B., Shoemaker, C. A., and Schuler, R. E. 1996a. Solid waste management system analysis with air pollution control and leachate impact limitations. *Waste Management & Research*, 14(5), 463–481.

Chang, N. B., Yang, Y., and Wang, S. F. 1996b. Solid waste management system analysis with noise control and traffic congestion limitations. *Journal of Environmental Engineering*, 122(2), 122–131.

Chang, N. B., Chen, Y. L, and Wang, S. F. 1997. A fuzzy interval multiobjective mixed integer programming approach for the optimal planning of solid waste management systems. *Fuzzy Sets and Systems*, 89(1), 35–60.

Chang, N. B., Davila, E., Dyson, B., and Brown, R. 2005. Optimal design for sustainable development of a material recovery facility in a fast-growing urban setting. *Waste Management*, 25(8), 833–846.

Chang, N. B., Qi, C., Islam, K., and Hossain, F. 2012. Comparisons between global warming potential and cost-benefit criteria for optimal planning of a municipal solid waste management system. *Journal of Cleaner Production*, 20(1), 1–13.

Chapman, R. and Yakowitz, H. 1984. Evaluating the risks of solid waste management programs: a suggested approach. *Resources and Conservation*, 11(2), 189–199.

Cherubini, F., Bargigil, S., and Ulgiati, S. 2009. Life cycle assessment (LCA) of waste management strategies: landfilling, sorting plant and incinerator. *Energy*, 34(12), 2116–2123.

Department of Environmental Protection (DEP). 1988. *Act 101: Pennsylvania's "Municipal Waste Planning and Recycling and Waste Reduction Act – Act of July 1988,"* Commonwealth of Pennsylvania Department of Environmental Protection, PA, USA. Available at: http://www.dep.state.pa.us/dep/deputate/airwaste/wm/recycle/FACTS/Act101.htm (accessed March 2011).

Ekvall, T. and Finnveden, G. 2000. The application of life cycle assessment to integrated solid waste management, part-2, perspectives on energy and material recovery from paper. *Process Safety and Environmental Protection*, 78(4), 288–294.

Fattahi, P. and Fayyaz, S. 2010. A compromise programming model to integrated urban water management. *Water Resources Management*, 24(6), 1211–1227.

Finnveden, G., Johansson, J., Lind, P., and Moberg, Å. 2005. Life cycle assessment of energy from solid waste- part-1: general methodology and results. *Journal of Cleaner Production*, 13(3), 213–229.

Franchetti, M. 2011. ISO 14001 and solid waste generation rates in US manufacturing organizations: an analysis of relationship. *Journal of Cleaner Production*, 19(9-10), 1104–1109.

Fuertes, L., Hudson, J., and Mark, D. 1974. Solid waste management: equity trade-off-models. *Journal of Urban Planning and Development*, 100(2), 155–171.

Galante, G., Aiello, G., Enea, M., and Panascia, E. 2010. A multiobjective approach to solid waste management. *Waste Management*, 30(8-9), 1720–1728.

Harger, J. R. E. and Meyer, F. M. 1996. Definition of indicators for environmentally sustainable development. *Chemosphere*, 33(9), 1749–1775.

Helms, B. and Clark, R. 1974. Locational models for solid waste management. *Journal of Urban Planning and Development ASCE*, 97(1), 1–13.

Huang, Y., Baetz, B., Huang, G., and Liu, L. 2002. Violation analysis for solid waste management systems: an interval fuzzy programming approach. *Journal of Environmental Management*, 65(4), 431–446.

Huhtala, A. 1997. A Post-consumer waste management model for determining optimal levels of recycling and landfilling. *Environmental and Resource Economics*, 10(3), 301–314.

International Organization for Standardization (ISO). 2006. *ISO 14040 Environmental Management-Life Cycle Assessment-Principles and Framework*, ISO.

Khoo, H. H. 2009. Life cycle impact assessment of various waste conversion technologies. *Waste Management*, 29(6), 1892–1900.

Kuhner, J. and Harrington, J. J. 1975. Mathematical models for developing regional solid waste management policies. *Engineering Optimization*, 1(4), 237–256.

Liamsanguan, C. and Gheewala, S. H. 2008. LCA: a decision support tool for environmental assessment of MSW management systems. *Journal of Environmental Management*, 87(1), 132–138.

LINDO Systems. 2010. *LINGO 10.0—Optimization Modeling Software for Linear, Nonlinear, and Integer Programming*, LINDO Systems.

Lu, H. W., Huang, G. H., He, L., and Zeng, G. M. 2009. An inexact dynamic optimization model for municipal solid waste management in association with greenhouse gas emission control. *Journal of Environmental Management*, 90(1), 396–409.

Manfredi, S. and Christensen, T. H. 2009. Environmental assesment of solid waste landfilling technologies by means of LCA-modeling. *Waste Management*, 29(1), 32–43.

Marks, D. and Liebman, J. 1971. Location models: solid waste collection example. *Journal of the Urban Planning and Development Division ASCE*, 97(1), 15–30.

Marks, D. H., ReVelle, C. S., and Liebman, J. C. 1970. Mathematical models of location: a review. *Journal of the Urban Planning and Development Division ASCE*, 96(1), 81–93.

Moberg, Å., Finnveden, G., Johansson, J., and Lind, P. 2005. Life cycle assessment of energy from solid waste- part-2: landfilling compared to other treatment methods. *Journal of Cleaner Production*, 13(3), 231–240.

Muñoz, I., Rieradevall, J., Doménech, X., and Milà, L. 2004. LCA aplication to integrated waste management planning in Gipuzkoa. *The International Journal of Life Cycle Assessment*, 9(4), 272–280.

Nishtala, S. R. and Solano, E. 1997. Description of the material recovery facilities process model: design, cost, and lifecycle inventory. Internal Report, Deptartment of Civil Engineering, North Carolina State University, Raleigh, NC.

O'Connor, R. 1977. *Fundamentals of Chemistry*, Harper and Row Publishers, New York.

Özeler, D., Yetis, Ü., and Demirer, G. N. 2006. Life cycle assessment of municipal solid waste management methods: Ankara case study. *Environment International*, 32(3), 405–411.

PE International. 2009. Gabi Software. Available at: http://www.gabi-software.com/

Rao, D. 1975. A dynamic model for optimal planning of regional solid waste management. Ph.D. Thesis, Clarkson College of Technology, Potsdam, NY.

Shih, L. H. and Lin, Y. T. 2003. Multicriteria optimization for infectious medical waste collection system planning. Practice Periodical of Hazardous. *Toxic and Radioactive Waste Management (ASCE)*, 7(2), 78–85.

Shmelev, S. E. and Powell, J. R. 2006. Ecological-economic modeling for strategic regional waste management systems. *Ecological Economics*, 59(1), 115–130.

Strutt, J., Wilson, S., Shorney-Darby, H., Shaw, A., and Byers, A. 2008. Assessing the carbon footprint of water production. *American Water Works Association*, 100(6), 80–91.

Su, J. P., Chiueh, P. T., Hung, M. L., and Ma, H. W. 2007. Analyzing policy impact potential for municipal solid waste management decision-making: a case study of Taiwan. *Resources Conservation and Recycling*, 51(2), 418–434.

Sundberg, J., Gipperth, P., and Wene, C. 1994. A systems approach to municipal solid waste management: a pilot study of Goteborg. *Waste Management & Research*, 12(1), 73–91.

Thomas, G. 2000. *Overview of Storage Development DOE Hydrogen Program*, Sandia National Laboratories, 9 May 2000, Livermore, CA. Available at: http://www1.eere.energy.gov/hydrogenandfuelcells/pdfs/storage.pdf (accessed January 2011).

United Nations. 1998. *Kyoto protocol to the United Nations Framework Convention on Climate Change*.

United States Environmental Protection Agency (USEPA). 1977. *WRAP: A Model for Solid Waste Management Planning User's Guide. US*, Environmental Protection Agency, Washington, DC.

Vatalis, K. and Manoliadis, O. 2002. A two-level multi-criteria DSS for landfill site selection using GIS: case study in Western Macedonia, Greece. *Journal of Geographic Information and Decision Analysis*, 6(1), 49–56.

Vesilind, P. A., Worrell, W. A., and Reinhart, D. R. 2002. *Solid Waste Engineering*, Brooks/Cole Thomson Learning, Pacific Grove, CA.

Villeneuve, J., Michel, P., Fournet, D., Lafon, C., Ménard, Y., Wavrer, P., and Guyonnet,D. 2009. Process-based analysis of waste management systems: a case study. *Waste Management*, 29(1), 2–11.

Wanichpongpan, A. and Gheewala, S. H. 2007. Life cycle assessment as a decision support tool for landfill gas-to-energy projects. *Journal of Cleaner Production*, 15(18), 1819–1826.

World Bank. 2010. World Development Report 2010: development and climate change. International Energy Agency Working Paper Series. Available at: http://www.siteresources.worldbank.org/INTWDR2010/Resources/5287678–1226014527953/ WDR10-Full-Text. pdf (accessed January 2011).

Xu, Y., Huang, G. H., Qin, X. S., and Cao, M. F. 2009. SRCCP: a stochastic robust chance-constrained programming model for municipal solid waste management under uncertainty. *Resources, Conservation and Recycling*, 53(6), 352–363.

Zeleny, M. 1973. Compromise programming. In: *Multiple Criteria Decision Making— Proceedings South Carolina 1972* (Eds. Cochrane, J.-L. and Zeleny, M.), University of South Carolina Press, Columbia, SC, pp. 262–301.

Zittel, W. and Wurster, R. 1996. Hydrogen in the energy sector. Ludwig-Bölkow Systemtechnik GmbH Report.

CHAPTER 14

PLANNING REGIONAL MATERIAL RECOVERY FACILITIES IN A FAST-GROWING URBAN REGION

In a fast-growing urban region, installing a material recovery facility (MRF, pronounced "merf") in a solid waste management (SWM) system could be a feasible alternative to achieving sustainable development goals if current household and curbside recycling programs do not prove successful. Yet both planning and design of MRF require accurate prediction of solid waste generation, and achieving the anticipated generation trends in fast-growing regions is challenging. The lack of complete historical records of solid waste quantity and quality due to insufficient budget and unavailable management capacity makes long-term system planning and/or short-term expansion programs intangible. To effectively handle problems based on limited data samples, this chapter begins with a system dynamics model capable of addressing socioeconomic and environmental situations leading to accurate prediction analysis of solid waste generation. This forecasting approach evaluates a variety of possible causative models and tracks inevitable uncertainties that traditional statistical least-squares regression methods cannot. Because the prediction of solid waste generation is a useful input, the second part of this chapter focuses on the optimal site selection and capacity planning of a MRF in conjunction with an optimal shipping strategy of solid waste streams in a multidistrict urban region. To address the waste management impact on city-wide sustainable development, screening of MRF capacity in the city of San Antonio, Texas, can be achieved in terms of economic feasibility, technology limitation, recycling potential, and site availability with the aid of a mixed-integer programming model.

Sustainable Solid Waste Management: A Systems Engineering Approach, First Edition. Ni-Bin Chang and Ana Pires.
© 2015 The Institute of Electrical and Electronics Engineers, Inc. Published 2015 by John Wiley & Sons, Inc.

14.1 FORECASTING MUNICIPAL SOLID WASTE GENERATION AND OPTIMAL SITING OF MRF IN A FAST-GROWING URBAN REGION

Many cities in North America rely on landfills as their main option for waste disposal in connection with a voluntary or mandated curbside recycling program. Experience gained in San Antonio, Texas, in the early 2000s indicates that budget cuts, population growth, rising waste streams, low curbside recycling participation, and taxpayer demands for lower taxes collide with a need for quality curbside garbage pickup service for commingled recyclables. These social movements motivate possible installation of material recovery facilities (MRF, pronounced as "merf") in urban regions, which can be linked with existing waste collection programs. How to create this linkage is relevant to decision-making but requires an accurate prediction of solid waste generation.

The prediction of municipal solid waste (MSW) generation plays an important role for planning SWM systems. In addition to population growth and migration, factors that interact to influence solid waste generation include underlying economic development, household size, employment changes, and the impact of waste recycling, which have confounding effect on the prediction of solid waste generation. The development of a reliable model for predicting the aggregate impact of economic trends, demographic changes, and household recycling on overall solid waste generation would be a useful advance in the practice of SWM.

Traditional forecasting methods for solid waste generation frequently depend on demographic and socioeconomic factors on a per capita basis. The per capita coefficients may be considered fixed over time or they may be projected to change with time. Grossman et al. (1974) extended per capita coefficients by including the effects of population, income level, and the dwelling unit size in a linear regression model, and Niessen and Alsobrook (1972) conducted similar estimates by providing other extensive variables characterizing waste generation; however, dynamic properties in the process of solid waste generation cannot be fully characterized in those model formulations.

Econometric forecasting, an alternative to static models, is an approach in which future forecasts are derived from current forecasts of the independent variables themselves (Chang et al., 1993). This forecasting method encompasses some of the dynamic features in forecasting analysis. When the recycling impact is phenomenal, intervention analysis may detect varying trends of solid waste generation under uncertainty (Chang and Lin, 1997a) and profoundly impact the possible structural change of solid waste generation trends in metropolitan regions. Implementing those traditional statistical forecasting methods, however, would first require collecting thorough socioeconomic and environmental information. In many cases, municipalities might not have sufficient budget and management capacity to maintain a sufficient database of solid waste quantity and quality to support needs on a long-term basis.

Most traditional statistical forecasting models, such as the geometry average method, saturation curve method, least-squares regression method, and the curve extension method are designed based on the configuration of semiempirical mathematical models. The structure of these models is simply an expression of causal effect or an illustration of trend extension to verify the inherent systematic features related to the observed database. In light of the difficulty embedded in formulating structured or semi-structured forecasting models, the synergy of fuzzy forecasting and grey dynamic modeling is viewed as a promising alternative to address forecasting issues under uncertainty.

The grey dynamic model was developed earlier to resolve the data scarcity issue (Deng, 1982). It is particularly designed to address situations in which only limited data are available for forecasting practice and in which system environments are not well defined or fully understood. In conjunction with fuzzy regression analysis, a revised dynamic forecasting method, the grey fuzzy dynamic modeling suitable only for situations with limited available samples for forecasting practice was demonstrated to dynamically predict MSW generation with reasonable accuracy (Chen and Chang, 2000).

When the database is not sufficient to support traditional statistical forecasting analyses yet ample to run several grey dynamic models with different natures, the separate dynamic efforts must be integrated to account for the interrelationships among relevant dynamic features that influence MSW generation. Such concatenation allows us to explore the interactions among a variety of socioeconomic, environmental, and managerial factors while simultaneously addressing data scarcity. This chapter presents a new approach, system dynamics modeling, for predicting MSW generation in an urban area based on a set of limited samples. To address the city-wide impact on sustainable development, the practical implementation was assessed with a case study in the City of San Antonio, South Texas, one of the fastest-growing regions in North America (Dyson and Chang, 2005). The San Antonio case presents various trends of MSW generation associated with five different solid waste generation models using Stella®, a system dynamics simulation tool (Dyson and Chang, 2005).

The optimal site selection and capacity planning of MRF in conjunction with an optimal shipping strategy of solid waste streams in a multidistrict, fast-growing urban region require accurate prediction of solid waste generation as inputs in optimization models. The optimization objectives may include economic impacts characterized by recycling income and cost components for waste management, while the constraint set consists of mass balance, capacity limitation, recycling limitation, scale economy, conditionality, and relevant screening constraints (Chang et al., 2005).

With the aid of forecasting analysis of solid waste generation, the case study of San Antonio, Texas, in the second part of this chapter presents a dynamic example where scenario planning using optimization analysis demonstrates the robustness and flexibility of site screening and selection of MRF. It proves especially useful when determining the MRF ownership structure from environmental policy viewpoints. Each scenario explores two case settings: (1) two MRF site locations proposed for selection, and (2) a single MRF site based on the anticipated solid waste generation rate. Cost analysis confirms that shipping costs, not processing fees, are the driving force in the city's operation. Sensitivity analysis confirms that significant public participation plays the most important role in minimizing SWM expenses (Chang et al., 2005).

14.2 MODELING PHILOSOPHY

The method of system thinking for system dynamics modeling has been promoted for more than 50 years (Forrester, 1961) and provides effective tools to better understand large-scale, complex management problems. System dynamics, designed based on system thinking, is a well-established methodology for studying and managing complex feedback systems as opposed to traditional statistical models. It requires constructing a unique "causal loop diagrams" or "stock and flow diagram" to form a system dynamics model for applications. Relevant studies on developing system dynamics models can be found in the literature (Forrester, 1961, 1968; Randers, 1980; Mohapatra et al., 1994).

Building a system dynamics model requires identifying a problem and developing a dynamic hypothesis explaining the cause of the problem. The mode formulation is normally designed to test a computer simulation model for alternative policies for the problem. Simulation runs in system dynamics models are governed entirely by time. This time-step simulation analysis requires a number of simulation steps along the timeframe to update the status of system variables of concern as a result of system activities. When the initial conditions are assigned for those variables, which denote the state of the system, the model may begin producing the related consequences for those system variables based on the initiation of action and the flow of information.

System dynamics modeling has been used to address almost every type of feedback system, including business systems (Sterman, 2000), ecological systems (Grant et al., 1997), socioeconomic systems (Forrester, 1969, 1971; Meadows and Meadows, 1973), agricultural systems (Qu and Barney, 1998; Saysel et al., 2002), political decision-making systems (Nail et al., 1992), and environmental systems (Vizayakumar and Mohapatra, 1992, 1993; Vezjak et al., 1998; Abbott and Stanley, 1999; Ford, 1999; Wood and Shelley, 1999; Deaton and Winebrake, 2000; Guo et al., 2001; Dyson and Chang, 2005; Xuan et al., 2010, 2012; Qi and Chang, 2011). The application matrix has covered several issues of environmental concerns, including environmental impact analysis of coalfields (Vizayakumar and Mohapatra, 1992, 1993), lake eutrophication assessment (Vezjak et al., 1998), pesticide control (Ford, 1999), wetland metal balance (Wood and Shelley, 1999), groundwater recharge (Abbott and Stanley, 1999), lake watershed management (Guo et al., 2001), river pollution control (Deaton and Winebrake, 2000), and SWM (Mashayekhi, 1993; Sudhir et al., 1997; Karavezyris et al., 2002). Within the SWM regime, Mashayekhi (1993) explored a dynamic analysis for the transition in New York State's solid waste system. Sudhir et al. (1997) further employed a system dynamics model to capture the dynamic nature of interactions among the various components in the urban SWM system, and Karavezyris et al. (2002) developed a methodology to incorporate qualitative variables such as voluntary recycling participation and regulation impacts. The model provides a platform to examine various structural and policy alternatives for sustainable SWM.

Most computer simulation applications using system dynamics models rely on Vensim® (Ventana Systems, 2012) and Stella® (High Performance System, 2012) software packages, in which the mechanisms of system dynamics can be handled by a user-friendly interface. These model development procedures are designed based on a visualization process that allows model builders to conceptualize, document, simulate, and analyze models of dynamic systems. They offer a flexible way to build a variety of simulation models from causal loops or stock and flow. The dynamic relationships between the elements, including variables, parameters, and their linkages can be created onto the interface using user-friendly visual tools. The feedback loops associated with these employed variables can be visualized at every step throughout the modeling process. Simulation runs are carried out entirely along the prescribed timeline. At the end of the process, some designated system variables of interest are updated for demonstration and policy evaluation.

While dynamic systems models may be necessarily complex, their complexity is managed through combinations of simpler submodels linked to simulate the system in question. These submodels are themselves dynamic systems models exhibiting specific system behaviors such as linear, exponential, and logistic growth or decay, overshoot and collapse, and oscillation (Deaton and Winebrake, 2000). In the present study, the system dynamics models characterize solid waste generation as a behavior of linear growth. In these models the concept of feedback within the system is not explored due to the difficulty of linking waste generation directly back to consumption activities; nevertheless, the prediction models provide sufficient insight in support of the subsequent systems engineering models to achieve higher levels of optimization analyses.

A number of systems engineering models are available for optimization analyses to exclusively address MSW management issues for regional decision-makers (Anex et al., 1996; Everett and Shahi, 1996a, 1996b; Eisenstein and Iyer, 1997; Everett and Riley, 1997; Everett and Shahi, 1997; Everett et al., 1998a, 1998b; Timms and Baetz, 1998; Wilson and Baetz, 2001a, 2001b; Solano et al., 2002a, 2002b). To reach mass throughput requirements and material/energy conservation goals, the application of systems engineering models with respect to a cost-effectiveness approach has received extensive attention in the last three decades (Fuertes et al., 1974; Helms and Clark, 1974; Walker et al., 1974; Kühner and Harrington, 1975; Male and Liebman, 1978; Hasit and Warner, 1981; Chiplunkar et al., 1982; Jenkins, 1982; Gottinger, 1986; Kirca and Erkip, 1988; Lund, 1990; Zhu and ReVelle, 1990; Huang et al., 1993, 1994, 1996; Lund et al., 1994; Chang and Wang, 1996a, 1996b; Chang et al., 1997, 2005). The spectrum of SWM systems planning in later stages covers a variety of complex issues, including the assessment of workload balancing, vehicle routing, site selection of transfer station, recycling drop-off stations, presorting facilities, incinerators and landfills, and comparative risk assessment for proposed planning alternatives (Kirca and Erkip, 1988; Chang and Wang, 1996a, 1996b; Chang and Lin, 1997b; Chang et al., 1997; Chang and Wei, 2000). Models for parts of the system as well as models covering the overall system with respect to various temporal and spatial criteria are of interest to system analysts and planners (Chang and Chang, 2003; Yeomans and Huang, 2003; Chang et al., 2005; Pires et al., 2011). In most instances, optimization approaches demonstrating a synergistic concern for economic, ecological, and environmental factors substantially improve the quality of decision-making for waste collection, separation, recycling, waste treatment, and disposal.

14.3   STUDY REGION AND SYSTEM ANALYSIS FRAMEWORK

The system dynamics model for the case study in San Antonio, South Texas, is formulated to predict solid waste generation in a fast-growing urban area based on a set of limited samples. This area is one of the fastest-growing regions in North America due to the economic impact of the North American Free Trade Agreement (NAFTA). The analysis presents various trends of solid waste generation associated with five different solid waste generation models using Stella®, a system dynamics simulation tool.

NAFTA is a comprehensive trade agreement implemented on January 1, 1994, that facilitates virtually all aspects of business within North America, resulting in a boom of economic and population growth along the United States (US)–Mexico border region, especially in the Maquiladoras and the Rio Grande/Rio Bravo river corridor along Laredo, McAllen, and Brownsville, Texas. Active economic activities due to the impact of NAFTA have rapidly extended from the US–Mexico border region to areas of Harlingen, San Antonio, and Corpus Christi, Texas. In 1995, due to the increasing trend of MSW generation, San Antonio implemented a voluntary recycling program to help reduce the amount of waste being landfilled, thus extending the life of the landfills. Revenue generated from the sale of recycled material was expected to offset the operating costs, but poor participation from the city residents produced insufficient income to justify the continued operation of the recycling program. Because ending the program will shorten the lifespan of the landfills, the city is considering building a MRF to continue the recycling program without requiring the support of the city residents.

For solid waste collection purposes, the City of San Antonio is divided into four service areas: Northloop (or Northeast), Northwest, Southcentral (or Southwets), and Southeast (Figure 14.1). Solid waste is collected in each service area and sent to a service center in that area. From the service center, the waste is shipped to any or all of the three landfills/transfer stations. With the implementation of a MRF, some or all of the collected solid waste would be first routed from a service center or community drop-off station to the MRF, where mechanical separation of recyclable material would ensue; remaining unrecyclable material would then be routed to the landfills. Two potential locations for siting the proposed MRF have been proposed (Figure 14.1). To achieve the capacity planning of MRF, an accurate amount of solid waste generation must be estimated for the City of San Antonio by 2010 (i.e., the target year in this planning conducted in the early 2000s).

San Antonio previously maintained its own landfills until the Nelson Gardens landfill site reached its permitted limit in the early 1990s. Before then, the city had routinely applied for variances from the state regulating authority to continue using the landfill site. Increasing environmental legislation led to the promulgation of regulations that required increasing accountability to the regulatory agencies. In addition, the cost of employee benefits and the liability posed by potential accidents, along with other hazards and the increasing cost of insuring city personnel and payments of workers compensations, made it economically unfeasible for the city

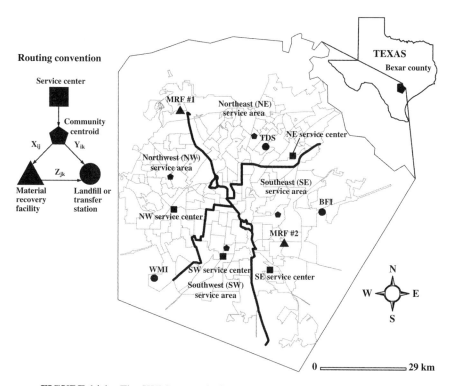

FIGURE 14.1   The SWM system in San Antonio. *Source*: Chang et al. (2005)

to continue operating a landfill. Other cities across the nation were facing the same crisis and had sought relief through privatizing the operation of their landfills.

Anticipating the closure of the Nelson Gardens landfill, Waste Management, Inc. (WMI) purchased land adjacent to the extant landfill at 8611 Covel Road and began developing a new site for the disposal of the city's refuse. At that time, San Antonio moved to fully privatize all of its landfill operations, and through competitive bidding, awarded the contract to WMI. Once the initial operation was privatized, the City of San Antonio solicited bids for another landfill site to be located at 7000 Interstate Highway (IH) 10 (Tessman Road) on the east side of the city. A long-term lease was awarded to Browning-Ferris Industries, Inc. (BFI) for the operation of this landfill site, including excavation, placement of impermeable liners, and installation of a drainage network to remove leachate. Eventually, the city awarded long-term 30-year contracts to both landfill contractors for landfill services that guaranteed minimums of 45,454 metric tons (tonnes) per year and a gradual reduction in the fee rate for exceeding the minimum with no maximum limit.

In parallel with the above movement, another contract that guarantees a minimum of 90,909 tonnes per year (TPY) was awarded to Transfer Disposal Systems (TDS) for disposal of more refuse. The transfer station, located at 11601 Starcrest Drive, north of the San Antonio International Airport, transfers the garbage approximately 112 km to a site near Buda, Texas, just outside of Austin. To date, the San Antonio Department of Environmental Services maintains its four service centers (Northloop, Northwest, Southcentral and Southeast) for routine solid waste collection (Figure 14.1). In addition to routine solid waste collection, Environmental Services performs the following services: (1) brush collection, (2) hazardous household waste collection at the Northwest Service Center on Culebra Road, (3) dead animal collection (road kill), and (4) recyclable paper pickup by Abitibi (i.e., a local contractor).

In the voluntary recycling program, revenue generated from the sale of recycled material by the city was expected to offset the operating costs. Unfortunately, poor citizen participation with curbside recycling resulted in insufficient income to justify the continued operation of the recycling program. The San Antonio administration recently proposed terminating the recycling portion of its collection operation due to a lack of participation by its citizenry. The recyclable material remains in the waste stream, however, shortening landfill life expectancy due to the space occupied by these materials. The city is contemplating a MRF combined with waste routing optimization to continue the recycling program while requiring less curbside sorting from the general public.

For solid waste collection purposes, the City of San Antonio is divided into four service areas housing trash and recyclable collection of vehicles (Figure 14.1). Recycling routes would first ensue from a service center into the service area where commingled recyclables mixed with solid waste are collected. At the MRF (two potential locations; Figure 14.1), the waste would undergo mechanical separation for recyclable material; unrecoverable material is finally routed to the landfills. The "trash only" routes would follow a similar course except the final destination is a transfer station or landfill. The first MRF location (MRF_1) is on the Northwest corner and the second (MRF_2) is near the BFI landfill in the Southeast service area.

The City of San Antonio has the option to become the primary stakeholder or to outsource the MRF installation. From this vantage point, the city must carefully evaluate system boundaries and enumerate MSW components to minimize costs. This effort might bear a Decision Circle that graphically represents objectives, waste flows, and system pressures (Figure 14.2). Three concentric circles provide a general impression of the interplay between cost minimization and system realities. The nucleus represents material flows that emerge from a fundamental need for waste disposal. Given the city's drive to minimize MSW costs, the middle ring forms an economic gauge that responds to system pressures and waste routing needs. The outer ring imposes a system boundary from a material and decision standpoint. For example, system planners must ensure that new system designs honor disposal obligations if a MRF is selected. Waste flows take on decision significance because they simultaneously impart influence on costs and constraints. Forming a Decision Circle may help municipal planners focus on key targets to complete conceptual planning. The subsequent stage involves intricate mathematical modeling to achieve material recovery that optimizes environmental quality and urban development.

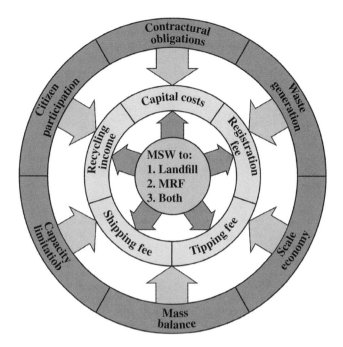

FIGURE 14.2   Decision Circle outlining the modeling components for optimization analysis.
Source: Chang et al. (2005)

14.4 PREDICTION OF SOLID WASTE GENERATION

14.4.1 Prediction Analysis

Simulation of MSW generation is predicated based on the contributing factors of
population growth, household income, people per household, and economic activity.
Dennison et al. (1995) found that waste generation on a per capita basis is inversely
related to household size. Sudhir et al. (1997) described the links among these parameters and MSW generation as follows: economic activity and population growth affect
household income, which in turn impacts per capita waste generation, and higher
income households tend to produce higher amounts of waste (Sudhir et al., 1997).
Yet higher income households are believed to achieve higher participation rates of
recycling. For example, Saltzman et al. (1993) found a positive relationship with rising income and newspaper recycling, and Schultz et al. (1995) in a review article cite
numerous studies reporting a significant positive relationship between rising income
and increased recycling effort. Hence, the situation could be further complicated if a
recycling program underway is mandated by law or regulation. Interactions among
a number of related social, economic, environmental, managerial, geographical, and
regulatory factors may further compound this understanding. Recycling is a complex

issue, not only because all factors are simultaneously involved and affected each other, but because they are dynamic over time.

To develop a predictive statistical regression model for the four service center areas we must rely on the historical population and income data provided by the US Census for 1980, 1990, and 2000 only. However, statistical issues, specifically a small degree of freedom, prohibits including more than two explanatory variables in the model due to the scarcity of datasets. With no other reliable databases available

TABLE 14.1   US Census data for San Antonio allocated into solid waste collection service centers

1980 Census SF-3 by Census tract	Population count	Median income household ^{a}(US$)	CPI^b 2000 (US$)	Average capita/H^c	Waste (tonne)
Northloop Service Center	232,177	18,734	39,150	2.71	38,745
Northwest Service Center	144,791	16,461	34,400	3.13	45,665
Southcentral Service Center	232,550	9,436	19,719	3.61	37,362
Southeast Service Center	176,505	10,194	21,303	3.10	33,211
Total	786,023				154,983

1990 Census SF-3 by Census tract	Population count	Median income household^a (US$)	CPI^b 2000 (US$)	Average capita/H^c	Waste (tonne)
Northloop Service Center	308,511	32,589	42,937	2.31	56,756
Northwest Service Center	210,494	28,445	37,477	2.60	66,890
Southcentral Service Center	239,102	15,921	20,976	3.14	54,728
Southeast Service Center	177,827	17,381	22,900	2.57	48,648
Total	935,934				227,022

2000 Census SF-3 by Census tract	Population count	Median income household^a (US$)	CPI^b 2000 (US$)	Average capita/H^c	Waste (tonne)
Northloop Service Center	411,417	48,791	48,791	2.55	74,620
Northwest Service Center	290,135	43,156	43,156	2.83	87,945
Southcentral Service Center	259,969	25,477	25,477	3.45	71,955
Southeast Service Center	183,125	27,428	27,428	2.92	63,960
Total	1,144,646				298,480

Source: Dyson and Chang (2005).
Note: US$ are United States dollars.
^{a}Normalized income level based on US Department of Labor Bureau of Labor Statistics Washington, DC 20212 in 2000.
^{b}Consumer Price Index (CPI)—2000, Average.
^{c}Capita/H is the average capita per each household in 2000.

in this case, a system dynamics model must be employed to project the amount of waste generated up to 2010 to analyze the implications of different MRF site selection alternatives in the near future.

14.4.2 Data Collection for Prediction Analysis

Population and average people per household data for 1980, 1990, and 2000 were collected by census tract from the US Census Bureau (Census, 2003), and the City of San Antonio allocated the census tracts to the appropriate service center. Data for median household income were likewise allocated, yielding values for population and people per household according to each of the four service center regions (BLS, 2003). Per capita income and population growth were further predicted based on the increasing trend of Texas economic growth due to the impact of NAFTA (NAFTA, 2000) and related employment growth (BLS, 2003).

Plotting the data, a linear relationship relating waste generation and some socioeconomic data with time was indicated. Hence, the values of population, median income per household, population per household, and tonnes of waste collected were regressed individually versus time for the three US Census periods of 1980, 1990, and 2000 (Table 14.1). The calculated relationships of income per service center, tonnes of waste generated normalized to income, tonnes generated per capita, tonnes generated per household income, and tonnes generated per household population were also tested. With only three data points, normality is neither assumed nor is causality sought but rather an indication of trend by goodness of fit. In all cases R^2 values ranged from 0.89 to 0.99 except for the population per household relationship (R^2 range 0.11–0.31). The slope of the regressed line was used as the annual growth rate in the model. NAFTA-related rate converters were developed by determining (1) the average rate of increase in employment growth factored into the population growth rate (Table 14.2) and (2) the average rate of increase in sales from exports to NAFTA partners factored into the income per sector growth rate (Table 14.3).

TABLE 14.2   Employment growth in San Antonio after NAFTA implementation

Year	Employment (jobs)	Growth (%)
1993	640,537	
1994	668,235	4.32
1995	668,646	0.06
1996	696,192	4.12
1997	712,224	2.30
1998	728,379	2.27
1999	743,056	2.02
Average growth (%)	2.52	

Source: BLS (2003).

TABLE 14.3   Increase in exports to NAFTA partners

Year	NAFTA exports (US$ million)	Growth (%)
1993	24,675	
1994	29,379	19.06
1995	28,794	−1.99
1996	34,545	19.97
1997	40,706	17.83
1998	46,654	14.61
1999	52,096	11.66
Average growth (%)	13.53	

Source: NAFTA (2000).

14.4.3   System Dynamics Modeling

The simulations were performed using the software package Stella®, an iconographic software using basic building blocks such as stocks, flows, and converters (Figure 14.3) that are intuitively assembled to simulate the dynamic processes of a system. Stocks represent the accounting of a system component, either spatially or temporally (e.g., population, waste generated); flows are the rate at which the component flows in or out of the stock; and converters modify rates of change and unit conversions.

Solid waste generation simulation was performed for each service sector in this study. The predicted amount of waste generated was simulated using five models, each of which simulated solid waste generation (tonnes·per year) as a function of the various socioeconomic factors (Table 14.4). For this simulation, tonnes generated were represented as a product of two stocks, rate of generation as a flow and economic effect of NAFTA as a converter (Figures 14.4–14.8). Each model was run for all service centers with a time step of 1 year.

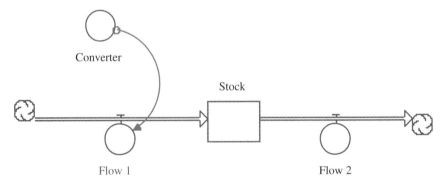

FIGURE 14.3   Stella® diagram showing stocks, flows, and converters. *Source*: Dyson and Chang (2005)

TABLE 14.4   Prediction models used to estimate the solid waste generation in 2010

Models	Driving factor in generation/service center
1	Total income per service center
2	People per household
3	Historical amount generated
4	Income per household
5	Population

Source: Dyson and Chang (2005).

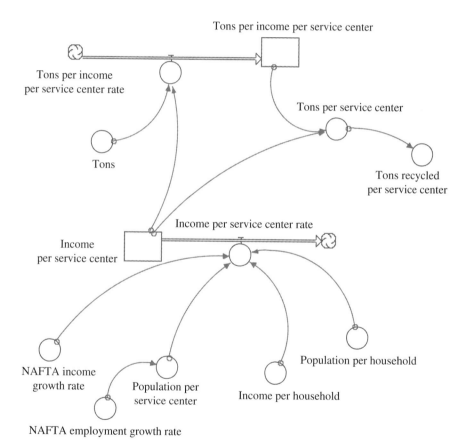

FIGURE 14.4   Generalized form of model 1 used to simulate tonnes generated and recycled per service center based on socioeconomic data. *Source*: Dyson and Chang (2005)

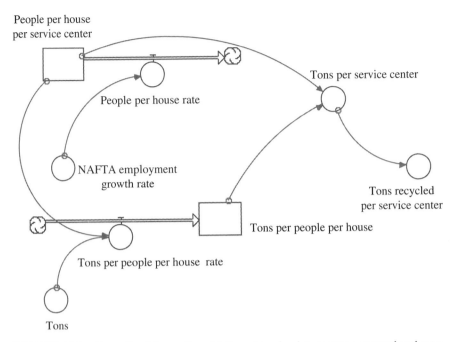

FIGURE 14.5 Generalized form of model 2 used to simulate tonnes generated and recycled by people per household based on socioeconomic data. *Source*: Dyson and Chang (2005)

For each model, the ratio of tonnes generated to the driving factor (i.e., population) was established for each period of record (1980, 1990, and 2000) and regressed to find the growth rate for the ratio to establish the effect on waste generation due to the factor in question. Year 2000 values were taken for initial values in the stocks and used to project results yearly until 2010. Total tonnes generated were calculated by multiplying tonnes per population by population. A yearly recycling rate was developed by comparing tonnes recycled versus tonnes of waste collected from 1995 to 2002 and multiplied by the simulated amount of tonnes generated to estimate the amount of waste recycled under current conditions.

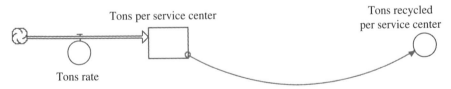

FIGURE 14.6 Generalized form of model 3 used to simulate tonnes generated and recycled by historical amount generated. *Source*: Dyson and Chang (2005)

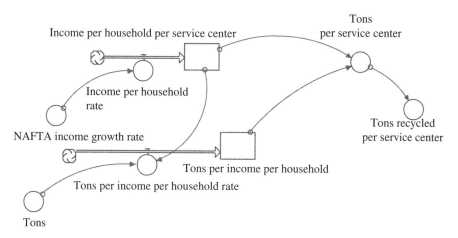

FIGURE 14.7   Generalized form of model 4 used to simulate tonnes generated by income per household. *Source*: Dyson and Chang (2005)

14.4.4 Prediction of Solid Waste Generation with System Dynamic Modeling

The simulated results for waste generation in 2010 for the four service centers (Table 14.5) and the estimated amount of waste generation from regression analysis in terms of population were provided as a base case for comparison. The estimates were recorded in order of increasing tonnage, and the range of values were projected based on the five models proposed in system dynamics modeling analysis. These five interval numbers constitute a prediction matrix that potentially reflects the uncertainties in decision-making. Model 1 or 2 tends to generate lower bounds, and model 4 or 5 tends to produce upper bounds. The reliability of predicted values obtained

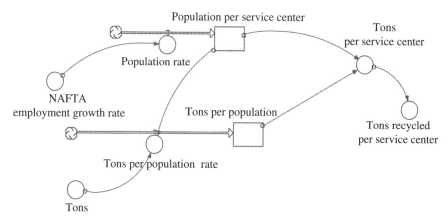

FIGURE 14.8   Generalized form of model 5 used to simulate tonnes generated and recycled by population based on socioeconomic data. *Source*: Dyson and Chang (2005)

TABLE 14.5   Simulation results for MSW generation in San Antonio

	MSW generation (tonne·y^{-1})				
Model	Northeast	Southeast	Southcentral	Northwest	Total
Base case	93,833	65,355	75,058	113,449	347,675
1	88,238	93,483	89,319	95,085	366,125
2	90,761	77,825	87,984	105,379	361,949
3	92,550	79,333	89,251	109,077	370,210
4	96,481	82,875	93,224	113,748	386,328
5	127,912	80,208	91,773	108,316	408,209
Range	88,238–	77,825–	75,058–	95,085–	361,949–
	127,912	93,483	93,224	113,748	408,209
Disparity	39,674	15,658	18,139	18,663	46,260

Source: Dyson and Chang (2005).

from regression analyses varies over different service areas and exhibits a potential underestimate in Southeast and Southcentral service areas compared with those with interval numbers.

The disparity in the extremes of the estimates indicates the importance of choosing the most appropriate model to prevent over- or underestimation of tonnage generated. Models 2, 3, 4, and 5 are functions of single factors and exhibit similar linear growth patterns (Figure 14.9). Model 1 exhibits slightly different behavior (tonnes per income), (Figure 14.10). While all service center regions exhibit waste generation increases, the behavior of the increase is markedly different. The change in average number of people per household is different than the other factors that affect waste generation. Average number of people per household is decreasing over time while all

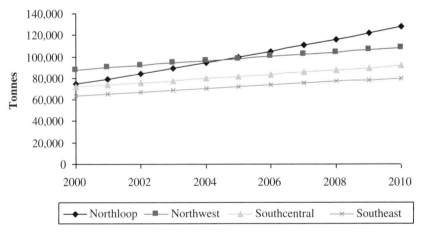

FIGURE 14.9   Prediction of solid waste generation as a function of population (Model 5). *Source*: Dyson and Chang (2005)

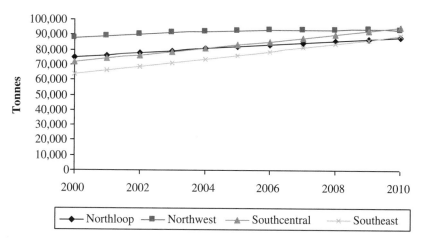

FIGURE 14.10   Prediction of tonnes generated based on income per service center region (Model 1). *Source*: Dyson and Chang (2005)

other factors are increasing, an effect accounted for in the tonnes per income model. The only model that incorporates all the driving factors of population, people per household, income per household, and economic activity is model 1; therefore, it best reflects the dynamics of the system.

This analysis thus concluded that model 1 is the most appropriate model to reflect the system dynamics in solid waste generation for the simulated amount of waste generated in 2010 per service center and the amount of simulated waste recycled under the current recycling system (Table 14.6). The recycled waste in this simulation is viewed as the basic amount as long as the private recycling market is still active. The increasing trend of historical records and predictive results of San Antonio population, income, and waste generation is phenomenal in all four relevant factors including population, median household income, household size, and waste generation (Figure 14.11).

The greatest increase in population by service center has and will continue to occur in Northloop and Northwest (Figure 14.11). Median income shows a steady increase over time, with the ratio relative to service center remaining the same for all service centers. The average persons per household exhibits some fluctuation over the past 20 years, with a slight decreasing trend evident in the Northloop and Northwest service

TABLE 14.6   Income per service center model simulation results at 2010

Income per service center	Northloop	Southeast	Southcentral	Northwest	Total
Total tonnes generated	88,238	93,483	89,319	95,085	366,125
Total population	501,037	186,435	273,679	362,805	1,323,956
Median household income (US$)	53,611	30,486	28,357	47,536	159,990
Total tonnes recycled	6,177	6,252	6,656	6,544	25,629

Source: Dyson and Chang (2005).

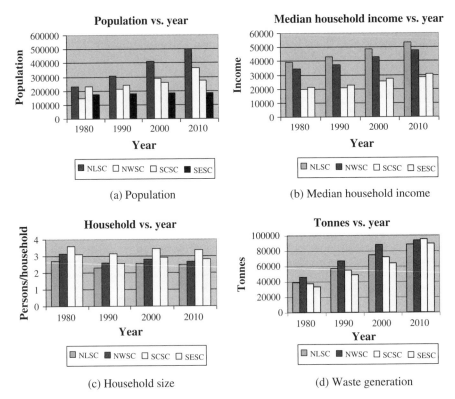

FIGURE 14.11 Summary of San Antonio population, household size and income, and waste generation. (a) Population. (b) Median household income. (c) Household size. (d) Waste generation. *Source*: Dyson and Chang (2005)

centers. This decrease accounts for the relative increase in solid waste generation in the Southcentral and Southeast service centers in the 2010 projections.

14.5 REGIONAL PLANNING OF MATERIAL RECOVERY FACILITIES

14.5.1 Model Formulation

The following integer programming model formulation details the methods used to choose the best location and optimal design capacity for the MRF with respect to associated cost and income components. Two types of management models include (Chang et al., 2005)

- Scenario A: City of San Antonio constructs and operates the MRF.
- Scenario B: the MRF is privatized, and thus a private company constructs and operates the facility while the city pays a processing fee on a per ton basis.

Both scenarios provide MRF site selection and capacity planning integration with the existing collection program and its disposal capability. To test system planning robustness and flexibility, both scenarios were run for two different case settings (Chang et al., 2005)

- Case 1: both MRF candidate locations compete for solid waste processing using integer programming (i.e., Scenarios A_1 and B_1 denote this practice hereafter).
- Case 2: the models are slightly modified to enforce selection of one MRF using integer programming (i.e., Scenarios A_2 and B_2 denote this practice hereafter).

The formulation outcome is a nonlinear integer programming model in both scenarios. The information incorporated into the optimization objectives includes economic impacts characterized by recycling income and cost components for waste management. These components remain steady over time. The constraint set therefore consists of mass balance, capacity limitation, recycling limitation, scale economy, conditionality, and relevant screening constraints.

Scenario A_1　In this scenario, the City of San Antonio is the investor and in charge of construction and operation during the lifecycle. Integer programming allows both MRF candidate sites to be considered simultaneously or independently to provide the greatest flexibility in decision-making. Objective function is to

Minimize net system cost Z = Total shipping cost ($C1$) + Total disposal cost ($C2$) − Income from recycling ($C3$) + Total operation cost of MRF ($C4$) + Total construction cost of MRF ($C5$) + Registration cost ($C6$)

$$C1 = \sum_i \sum_j X_{ij} \cdot D1_{ij} \cdot P_1 + \sum_i \sum_k Y_{ik} \cdot P_1 \cdot D2_{ik}$$
$$+ \sum_j \sum_k Z_{jk} \cdot P_1 \cdot D3_{jk}, \tag{14.1}$$

$$C2 = \sum_i Y_{i1} \cdot P_2 + \sum_j Z_{j1} \cdot P_2 + \sum_i Y_{i2} \cdot P_3 + \sum_j Z_{j2} \cdot P_3$$
$$+ \sum_i Y_{i3} \cdot P_4 + \sum_j Z_{j3} \cdot P_4, \tag{14.2}$$

$$C3 = \sum_j \sum_l R_{jl} \cdot I_{jl} \cdot \sum_i X_{ij}, \tag{14.3}$$

$$C4 = \sum_j (a \cdot DC_j + b \cdot I_j)UCF, \tag{14.4}$$

$$C5 = \sum_j (c \cdot DC_j + d \cdot I_j)UCF \cdot CPIR \cdot CRF^*, \tag{14.5}$$

$$C6 = \sum_j e \cdot I_j \cdot CRF^* \tag{14.6}$$

in which $C1$ is total shipping cost (US$·per year); $C2$ is total disposal cost (US$·per year); $C3$ = income from recycling (US$·per year); $C4$ is the annual total MRF operating cost (US$·per year); $C5$ is amortized MRF construction cost; $C6$ is amortized MRF registration cost, which is lumped with the construction cost in the cost/benefit analysis. All cash flows follow the present worth conversion via a government bond financial structure based on a capital recovery factor (CRF) that has a driving force in cash flow patterns for long-term waste management

$$
{}^{*}\text{Capital Recovery Factor (CRF)} = \left[m \cdot i \left(y \cdot \frac{\left(1 + \frac{y}{m}\right)^{n \times m}}{\left(1 + \frac{y}{m}\right)^{n \times m} - 1} \right) + \left(y \cdot \frac{(1 + y)^{n}}{(1 + y)^{n} - 1} \right) \right]
$$

(14.7)

in which CRF is a capital recovery factor with i as interest rate, m as number of intervals the interest is compounded, y as yield to maturity on the bond, and n as the number of years to repay the municipal bond; P_1 is unitary shipping cost (US$ per tonne per km); P_2 is unitary disposal cost at Landfill 1 (WMI) (US$·per tonne); P_3 is unitary disposal cost at Landfill 2 (BFI) (US$·per tonne); P_4 is unitary disposal cost at Transfer Station (TDS) (US$·per tonne); X_{ij} is commingled recyclable waste stream shipped from i to j in TPY (i = 1, 2, 3, 4 service centers) and (j = 1, 2 MRF); Y_{ik} is waste stream shipped from i to k (TPY) (k = 1, 2, 3 Landfills and Transfer Station); Z_{jk} is the residual waste stream shipped from j to k for disposal (TPY); DC_j is design capacity of the MRF in tonnes per day (TPD) (j = 1, 2, MRF); I_{jl} is average income from selling recyclable l associated with site j (US$·per tonne) ($l$ = 1 for paper; l = 2 for glass, l = 3 for plastics, and l = 4 for metal); R_{jl} is the recycling ratio of recyclable l in the waste stream at MRF site j (%) (l = 1, 2, 3, 4); UCF is a unit conversion factor that makes the unit of cost in the derived cost functions to be consistent with those in the other cost components; a, b, c, and d are defined regression coefficients in the cost functions, and e is the cost that is amortized; $D1_{ij}$ is the average shipping distance from i to j (km); $D2_{ik}$ is the average shipping distance from i to k (km); $D3_{jk}$ is the average shipping distance from j to k (km); and $CPIR$ is a consumer price index ratio that updates the construction an operation cost information from the baseline in the past to the present along the timeline (unitless). Constraint set is defined as

1. Mass balance constraint for source location: propels all solid waste generated in a service area ships to other treatment or disposal components in the network.

$$
G_i = \sum_j X_{ij} + \sum_k Y_{ik} \quad \forall i.
$$

(14.8)

2. Commingled recyclable waste stream: delineates recyclable waste stream as a product of waste generation in a service area and overall citizen participation.

$$
\sum_j X_{ij} \le (C_p \cdot P_r) G_i \quad \forall i.
$$

(14.9)

3. Mass balance constraint for MRF: confirms that rate of incoming waste equals the rate of outgoing waste plus the amount removed in the material recovery process.

$$\sum_i X_{ij} = \sum_k Z_{jk} + \left(\sum_l R_{jl}\right)\left(\sum_i X_{ij}\right) \quad \forall j. \tag{14.10}$$

4. Capacity limitation constraint for landfill/transfer station: guarantees waste inflow destined for final disposal matches or exceeds contract-based minimums.

$$\sum_i Y_{ik} + \sum_k Z_{jk} \geq CAP_k \quad \forall k. \tag{14.11}$$

5. Capacity limitation constraint for MRF: maintains the waste inflow destined for material recovery at or below the design capacity.

$$\left(\sum_i X_{ij}\right)/365 \leq DC_j \quad \forall j. \tag{14.12}$$

6. Scale economy constraint: ensures the MRF investment abide economies of scale (Chang and Wang, 1995). It also justifies the use of linear construction and operation cost functions.

$$DC_j \leq DC_{max} \cdot I_j \quad \forall j, \tag{14.13a}$$
$$DC_j \geq DC_{min} \cdot I_j \quad \forall j. \tag{14.13b}$$

7. Recycling limitation constraint: characterizes the recyclables in the commingled waste stream destined for the MRF.

$$L_{jl} \leq R_{jl} \leq U_{jl} \quad \forall jl. \tag{14.14}$$

8. Nonnegativity constraint: eliminates infeasibilities by filtering only positive waste streams for consideration in the optimal solution.

$$I_j \in \{0, 1\} \quad \forall j,$$
$$X_{ij}, Y_{ik}, Z_{jk} \geq 0 \quad \forall i, j, k, \tag{14.15}$$

where C_p is the percentage of citizen participation in curbside collection; P_r is the percentage of the overall waste stream that is comprised of commingled recyclables; CAP_k is the minimum of waste destined for disposal site k (TPY); I_j is a binary integer variable, equal to one when the MRF facility j is included for consideration in the SWM system, otherwise zero (unitless); U_{jl} and L_{jl} are the upper and lower bounds of the recyclables l in the commingled waste stream destined for MRF site j, and

DC_{min} and DC_{max} are correspondingly the minimum and maximum MRF capacity justified by economies of scale (TPD).

Scenario A_2   In this scenario only one of the two MRF location candidates is chosen in any alternative. It requires Equations (14.1)–(14.15) with an additional constraint.

9. Conditionality constraint: assures the one-time initialization of a new MRF site in the system.

$$\sum_j I_j = 1. \tag{14.16}$$

Scenario B_1   In this scenario the City of San Antonio chooses to privatize the MRF so that the city pays a processing fee on a per ton basis. Both candidate sites can be simultaneously included in the system optimization alternative. Information incorporated into the objective functions includes economic impacts characterized by all relevant cost components for waste management. The constraint set contains mass balance, capacity limitation, recycling limitation, recycling summary, and related screening constraints. All constraints, except for the MRF tipping fee screening, are similar to those in Scenario A_1. Objective function is defined as

Minimize total cost Z = Total shipping cost ($C1$) + Total disposal cost ($C2$) + Total processing cost at MRF ($C3$)

$$C1 = \sum_i \sum_j X_{ij} \cdot D1_{ij} \cdot P_1 + \sum_i \sum_k Y_{ik} \cdot P_1 \cdot D2_{ik},$$
$$+ \sum_j \sum_k Z_{jk} \cdot P_1 \cdot D3_{jk}, \tag{14.17}$$

$$C2 = \sum_i Y_{i1} \cdot P_2 + \sum_j Z_{j1} \cdot P_2 + \sum_i Y_{i2} \cdot P_3 + \sum_j Z_{j2} \cdot P_3,$$
$$+ \sum_i Y_{i3} \cdot P_4 + \sum_j Z_{j3} \cdot P_4, \tag{14.18}$$

$$C3 = \sum_j DC_j \cdot P_{MRF_j} \cdot 365, \tag{14.19}$$

where $C1$ is total shipping cost (US\$·per year); $C2$ is total disposal cost (US\$·per year); $C3$ is redefined as MRF processing fee (US\$·per year); and P_{MRF} is the processing fee governmental agency needs to pay for recycling at the MRF (US\$·per tonne). Constraint set is defined as

1. Equations (14.8)–(14.15): classify waste streams, enforce mass balance across MSW components, meet disposal obligations, and purport economies of scale that also define Scenario B_1.

2. Screening constraint: screens possible processing fees in the nonlinear domain.

$$P_{MRF_j} \leq PUB_j \cdot I_j \quad \forall j, \qquad (14.20a)$$

$$P_{MRF_j} \geq PLB_j \cdot I_j \quad \forall j, \qquad (14.20b)$$

in which PLB_j and PUB_j are the lowest and highest MRF tipping fees (US\$·per tonne) in Texas, respectively.

Scenario B₂ In this scenario a private entity invests, constructs, and operates the MRF, and a single MRF candidate site is selected in any solution. This requires amending Scenario B_1 to include Equation (14.16). Equations (14.7)–(14.16) and (14.20) allow decision-makers and planners to model the optimal cost of privatizing a single MRF operation.

14.5.2 Data Collection for Optimization Analysis

The selection of data analysis methods may affect the quality and comparability of optimal solutions produced in different scenarios. Given the wide array of data available in SWM, site-specific information in the form of locations, waste generation, and waste composition were first investigated. To conduct the simulation, the role of monetary flows in waste management and municipal finance must be understood.

Political and demographic boundaries undergo frequent changes in fast-growing urban areas, and San Antonio is no exception. While the current administrative map might have slight changes to service area borders, the overall effect on the macro analysis of the solid waste analysis will be negligible, and the global shipping patterns will remain. The service areas for waste collection were analyzed from select Environmental Services geographic information system layers. Once a service area centroid is calculated, the distance traveled in the service area during solid waste pickup in the absence of micro routing distances can be simulated (J. Perez, personal communication with City of San Antonio Environmental Services Department, January 16, 2004). The total shipping distance includes distance traveled from the service center to the community plus the distance traveled from the community to the MRF or the landfills/transfer station.

A typical waste stream composition in Texas (Figure 14.12) is characterized in concert with the public's recycling participation rate to define the maximum waste stream eligible for commingled recyclable pickup with the city. The average tipping fee for a processing facility ranges from US\$20.07 to US\$103.17 per tonne, averaging around US\$41.21 per tonne (Chartwell Solid Waste Group, 2001). Waste generation rate in the target year of 2010 was applied directly according to the results in a companion study (Dyson and Chang, 2005). Using operational records for the 2001–2002 fiscal year (Tables 14.7 and 14.8) and the projected values of waste generation rate by service centers (Table 14.5), cost–benefit components, technological parameters, and economic index were assessed and reorganized for appropriate use in the model.

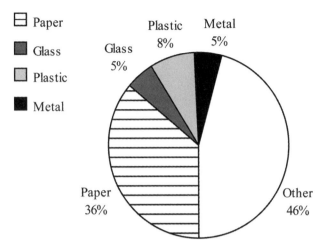

FIGURE 14.12   Composition of Texas waste stream. *Source*: TNRCC (2000)

TABLE 14.7   Environmental Services disposal cost in 2003

Company	Tonnage	Amount (US$)
BFI	66,893	1,210,670
WMI	158,306	2,641,665
TDS	91,598	2,257,995
Total	316,797	6,110,330

Source: Chang et al. (2005).

Municipal projects of this magnitude are often financed by municipal bonds with an average interest rate of 5% (Merrill Lynch, 2003). To reflect that standard, part of the system cost goes into amortized yearly payments on a 10-year bond, which in this system was simulated at 4.75% interest compounded semiannually (Berk, 2001). The interest rate was based on San Antonio's credit rating of "AA+" by the largest bond rating agencies Fitch Ratings and Standard & Poor's (American City Business Journals Inc., 2002).

An investigation into average operation and construction costs for publicly or privately owned MRF concludes that design capacity exhibits an economy of scale

TABLE 14.8   Environmental Services revenue in 2003

Company	Tonnage	Amount (US$)
Abitibi Paper	13,167	490,730
Bitters Brush Site	8,409	213,761
Total	21,576	704,491

Source: Chang et al. (2005).

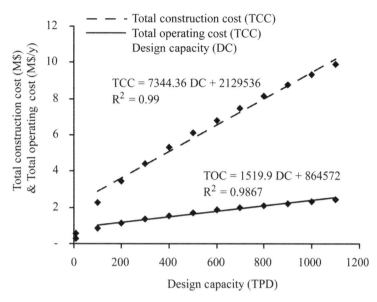

FIGURE 14.13   Total construction cost and total operating cost in million dollars versus design capacity. *Source*: Chang and Wang (1995)

(Chang and Wang, 1995). Concentrating on the linear section of the nonlinear Total Construction Cost and Total Operation Cost functions allows one to stay within the range that exhibits the best scale economy (100–1000 TPD) (Figure 14.13). Operating a MRF at 50 TPD will cost more on average compared to a facility at 500 TPD because the unit cost of providing the service decreases with a rise in magnitude of the service.

Consequently, the construction and operating costs in objective function in Scenario A include linearized cost curves built from the linear tail of the operation and construction cost curves over the best scale economy range. These are subsequently corrected by an Engineering News Record (ENR, 2003) building cost index to adjust dollars to present value and location differences in applications. Scenarios A and B include the scale economy, a constraint to assure the best return on the MRF investment. Thus, the adoption of linearized construction and operation cost functions of MRF that were summarized and updated based on the previous study eases the efforts of the cost–benefit matrix (Table 14.9) (Chang and Wang, 1995).

14.5.3 Optimal Siting of MRF by Optimization Analysis

To provide the decision maker with the best comparison, the scenario outcomes are compared in a cost-competitive manner. Depending on what the City of San Antonio will find valuable and deem viable for a specific scenario, it is the analyst's duty to clearly communicate the results. A cost–benefit analysis together with sensitivity and policy analyses will sort out the implications of the model simulation.

TABLE 14.9 Parameter values used in LINGO program

	Definition	Value		Definition	Value
P_1	Unit shipping cost (US$ per tonne·km^{-1})	4.81	D142	Northwest SC to MRF$_2$	40.70
P_2	Tipping fee at BFI (US$·tonne^{-1})	18.22	D2$_{ik}$	Average shipping distance	(km)
P_3	Tipping fee at WMI (US$·tonne^{-1})	16.72	D211	Northeast SC to BFI	43.40
P_4	Tipping fee at TDS (US$·tonne^{-1})	24.89	D212	Northeast SC to WMI	45.55
G_i	Waste generation rate	(TPY)	D213	Northeast SC to TDS	13.05
	i = 1: Northeast service area	92,550	D221	Southeast SC to BFI	28.65
	i = 2: Southeast service area	79,333	D222	Southeast SC to WMI	45.20
	i = 3: Southwest Service Area	89,251	D223	Southeast SC to TDS	32.85
	i = 4: Northwest service area	109,077	D231	Southwest SC to BFI	27.05
I_{jl}	Average recycling income	(US$·per tonne)	D232	Southwest SC to WMI	18.45
	l = 1 paper[a]	27.50	D233	Southwest SC to TDS	28.80
	l = 2 glass[b]	0.55	D241	Northwest SC to BFI	50.55
	l = 3 plastic[b]	264.00	D242	Northwest SC to WMI	32.40
	l = 4 metal[c]	704.00	D243	Northwest SC to TDS	32.10
R_{jl}	Recyclables in waste stream	(%)	D3$_{jk}$	Average shipping distance	(km)
	l = 1 for paper	[0.54, 0.58]	D311	MRF$_1$ to BFI	43.35
	l = 2 for glass	[0.04, 0.08]	D312	MRF$_1$ to WMI	41.10
	l = 3 for plastic	[0.08, 0.13]	D313	MRF$_1$ to TDS	17.00
	l = 4 for metal	[0.04, 0.08]	D321	MRF$_2$ to BFI	11.35
CAP_k		(TPY)	D322	MRF$_2$ to WMI	32.30
CAP1	Minimum waste to BFI	45,454	D323	MRF$_2$ to TDS	18.00
CAP2	Minimum waste to WMI	45,454	CPIR	Other parameters: ENR (2003)	1.29
CAP3	Minimum waste to TDS	90,909			
D1$_{ij}$	Avg. shipping distance	(km)			
D111	Northeast SC^c to MRF$_1$	23.75	a	Variable construction cost ($ per ton)	1519.90
D112	Northeast SC to MRF$_2$	33.60	b	Fixed construction cost ($)	864,572
D121	Southeast SC to MRF$_1$	45.90	c	Variable operating cost ($ per ton)	7344.36
D122	Southeast SC to MRF$_2$	16.95	d	Fixed operating cost ($)	2,129,536
D131	Southwest SC to MRF$_1$	32.25	e	MRF Registration cost ($)	482,500
D132	Southwest SC to MRF$_2$	16.65	P_{MRF}	Avg. MRF tipping fee ($ per ton)	42.21
D141	Northwest SC to MRF$_1$	26.15	$DC_{min/max}$	Scale Economy Constraint (TPD)	[100, 1,000]

Source: Chang et al. (2005).

[a] Acco-Bfi (2003).
[b] RecycleNet (2003).
[c] Newell (2003).

TABLE 14.10   Overview of recycling results per scenario in TPD

Scenario	DC_1	DC_2	Total
A_1	172	112	284
A_2	0	284	284
B_1 & B_2	0	100	100

Source: Chang et al. (2005).

The location of the better MRF and its optimal capacity design are determined by a series of LINGO® optimization programs (LINDO Systems, 2003). The base case scenario optimizes waste shipment without a MRF present in the system. Northwest and Southwest service areas ship waste to the WMI landfill in the east, the Northeast region opts for the transfer station (TDS) in the west, and the Southeast corner should principally contract with the BFI landfill. This routing pattern adheres to current contract obligations. The optimal MRF capacities for scenarios with material recovery indicate that Scenarios A_1 and B_1 allow both recycling facilities to compete, although alone A_1 elects dual facilities (Table 14.10). In the single MRF site selection scenarios, MRF_2 near the BFI landfill is preferred.

The recycling effort in Scenario A is 284 TPD. When the system is forced to choose one recovery facility, Scenario A_2 realizes the effect of scale economy when it opts for MRF_2 (Figure 14.14). The 97 TPD of waste routed from the Northeast to MRF_1 in Scenario A_1 is simply shifted to MRF_2 in Scenario A_2. No surplus waste is routed to the transfer aside from capacity condition, likely due to the combination of a high capacity requirement, the largest tipping fees for final disposal, and distance from southern service areas.

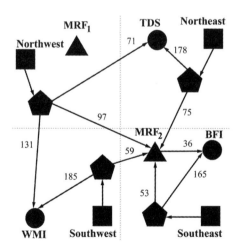

FIGURE 14.14   Scenario A_2 optimal waste routing within each route in tonne per day (TPD). *Source*: Chang et al. (2005)

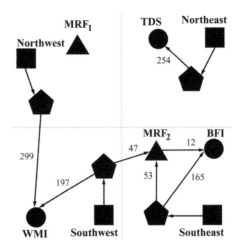

FIGURE 14.15 Scenario B_1 and B_2 optimal waste routing within each route in tonne per day (TPD). *Source*: Chang et al. (2005)

Scenario B suggests that a MRF does little to minimize the system costs (Figure 14.15). Paying a tipping fee for private processing proves marginally beneficial in terms of system costs compared to the base case while achieving only half the material recovery of Scenario A. Scenario B's low material recovery level indicates a strain to meet the MRF requirement with an optimal design capacity of 100 TPD. Analogous to Scenario A, the transfer station receives minimal waste excess from its contract from the Northeast service area.

Direct Cost–benefit Analysis   The present value net cost breakdown for all scenarios under consideration (Table 14.11) indicates that the shipping cost for the base case is the major cost contributor, accounting for 86% of the total cost in the MSW system at a price of US$48.7 million per year. If a publicly or privately owned MRF could alleviate MSW expenses, then the MRF system could realize cost savings for the City of San Antonio.

Meeting recycling mandates can be difficult as a result of increased complexities in the shipping strategy. In A_2, 86% of system costs go toward shipping. In contrast, A_1 has site selection flexibility to distribute across two MRFs and three final disposal facilities, which explains the decrease in shipping expenditures and the smaller system net cost. The first model tested (Scenario A_1) is to illustrate the considerable cost for building and operating MRFs; nevertheless, the investment is justified based on potential recycling revenue, which is estimated at US$11.1 million per year in Scenario A. Unfortunately, to maintain this level of recycling effort with a single MRF requires an additional US$3.5 million per year in shipping. On average, the single public MRF realizes a net savings around US$7 million per year compared to the optimal base case.

TABLE 14.11 Cost distribution on a yearly basis for the base case, private, and public MRF ownership

Scenario	Shipping cost	Disposal cost	Operation cost	Construction cost	Potential income	Net cost
A_1	US$41,636,899 83%	US$5,511,909 11%	US$2,161,355 4%	US$1,089,340 2%	US$11,139,864	US$39,259,639 –
A_2	US$45,187,798 86%	US$5,523,689 10%	US$1,296,783 2%	US$700,230 1%	US$11,139,864 –	US$41,568,636 –

Scenario	Shipping cost	Disposal cost	MRF disposal cost	Net cost
B_1 & B_2	US$40,676,615 83%	US$6,508,993 13%	US$1,540,665 3%	US$48,726,272 –

Scenario	Shipping cost	Disposal cost	Net Cost
Base	US$41,661,544 86%	US$7,065,061 14%	US$48,726,605 –

Source: Chang et al. (2005).
Note: Percentages represent fractions of the total cost per scenario and does not include benefits.

At a net cost of US$48.7 million per year, Scenario B recycles 100 TPD, which means that the cost of privatization is on a par with the base case. Lacking a source of income in this scenario, privatization offers little relief from the base costs regardless of the MRF tipping fee. Sensitivity analysis serves an imperative evaluation function for any positive outlook that might exist for privatization. The intent is to test tipping fee sensitivity and citizen participation to identify a range of prices that present privatization as cost-competitive, or bolster public ownership support.

Sensitivity Analysis   Testing the privatization scenario for sensitivity across the lower and upper average tipping fees in Texas did not produce compelling results until the fee fell below US$41.21 per tonne. Although Scenario B lacks income to make the system cost-competitive with Scenario A, privatization would lower shipping costs when compared to the base or the material recovery case.

Examining citizen participation (C_p) across a spectrum of values produces diverse consequences across scenarios (Figure 14.16). Varying participation from 32% to 90% and measuring MRF capacity and net cost indicates that as participation increases, the system cost in Scenario A_2 decreases with an increase in design capacity. In A_2, a boost in citizen participation clearly diminishes net costs with the increased potential for recycling income. Scenario B_2, in comparison, does not exhibit a drastic cut in system costs; the difference between the lowest and highest simulated participation rate reveals a maximum US$2.4 million per year savings. A_2 yields a more pronounced US$9.2 million per year differential; therefore, participation plays a more pivotal role in cost reduction for public MRF ownership connected to larger income potential. While Scenario A_2 capitalizes on available recyclables in the waste stream, Scenario B_2 does not. The design capacity in B_2 remains stagnant despite an increase in recycling participation, and a lack of income provides no incentive for the formulation to increase MRF capacity above 100 TPD.

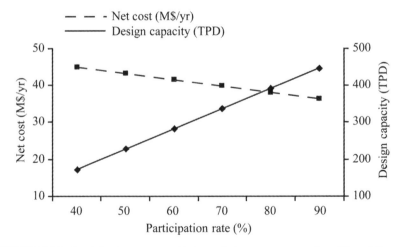

FIGURE 14.16   Participation rate sensitivity as it impacts net cost in million dollars and design capacity of MRF$_2$ in scenario A$_2$. *Source*: Chang et al. (2005)

Modeling robustness is evident in that design capacity grows with increased citizen participation. Scenario A_2 is particularly affected by the participation range, and extraordinary decreases in system costs accompany an increased potential to capitalize from recyclables. This opens the possibility to advance household participation in curbside recycling by educating the public to achieve a high level of participation (~70%), although the expenses to achieve such a lofty public relations goal are not easily quantified.

Indirect Cost–benefit Analysis   Indirect costs and benefits allow closer scrutiny of scenario planning. Although sensitivity analysis shows the importance of public participation of the recycling program, the cost associated with promotion and cultivation of public support cannot be easily modeled. If the promotion of recycling consists of a short-term investment, then cost savings from material recovery may leave capital to invest in an ambitious marketing strategy directed at increasing citizen participation and program sustainability.

Solid waste permits present an important hidden cost to San Antonio. Landfill permits for new construction or expansion present opportunities for landfill operators to pass increased surcharges. Although MRF are not yet common in Texas, they present planners with an opportunity to explore future regulations at the state level. Material recovery facilities in Texas currently lie in a grey area but are beginning to manifest themselves as "Type VI facilit[ies]... involving a new or unproven method of managing or utilizing MSW, including resource and energy recovery projects (Texas Administrative Code RULE §330.41)." MRF undoubtedly qualify for that criterion, meaning a potential site can register rather than undergo expensive permitting (J. Bard, personal communication with TCEQ Region 13 Solid Waste Inspector, October 31, 2003).

Pending changes to the Texas Administrative Code showcase how regulators plan to cope with municipality (and private industry) demands for recycling permit flexibility. These changes are meant to eliminate cumbersome permit requirements that deter legitimate recycling (J. Bard, personal communication with TCEQ Region 13 Solid Waste Inspector, October 31, 2003). A Title V permit could range from US$225,000 to US$1,000,000 and registration costs from US$215,000 to US$750,000 according to analysts with the Texas Commission for Environmental Quality (TCEQ) Strategic Planning and Appropriations in the early 2000s. Facilities may register as MRF if they recycle more than 10% of the incoming waste stream that would normally go to a landfill (Texas Administration Code, TAC Rule Log No. 2003-028-330-WS). By design, MRF can easily meet that requirement by maintaining a minimum level of curbside sorting. As registration becomes viable, MSW permits will cease to present a hurdle to private industry or cities pursuing construction and operation of a MRF.

Identifying sustainable markets for the recycled material throughout the lifetime of the operation presents a critical hidden cost in material recovery. The costs of evaluating and sustaining these secondary material markets would be similar in each scenario, except that private waste handlers have greater industry reach. This exception presents a key advantage toward privatization if San Antonio management cannot engage the recyclable material markets in the near- or long-term future.

Policy Analysis The MRF registration issue emphasizes the legal ramifications of the decision to recover recyclables. SWM can quickly galvanize strong social and political unrest. Policy analysis may help managers anticipate social dynamics and widen the system planners' awareness for solid waste issues as they promote centralized material recovery options.

A city's interest is best served by saving landfill space to avoid unpleasantness with the citizenry over complaints over landfill odor, sound, and health concerns (Fields, 2002). Amid global increases in landfill activity, rifts between stakeholders and solid waste decision-makers grow with actual and perceived health risks (Ishizaka and Tanaka, 2003). "Top-down" decision-making in the public or private sectors often fails to adequately engage community involvement. A lack of risk communication between local partners and service providers consequently exposes waste management projects to condemnation, irrespective of the underlying common good (Ishizaka and Tanaka, 2003).

As an example, San Antonio's southeastern residents perceive recent BFI permit amendments as undesirable because the changes would increase the facility size and extend landfill lifespan for an additional 50 years, even though the changes would provide the city with a much-needed disposal option. Community members sued BFI in Texas courts, largely due to a perceived lack of environmental stewardship amid residents' health and noise concerns (Sorg, 2003). The permit case is currently before the Texas Supreme Court after BFI fought a lower appellate court's decision to deny the new permit (Needham, 2003).

The lesson from BFI is that contentious solid waste decisions in this urban area require new forms of outreach and disposal options to balance environmental concerns with genuine public participation. Including a MRF option may help the City avoid potential debacles as they face impending disposal pressures that accompany a rapidly increasing population.

Landfill space savings can be expressed in terms of volume, compaction, and remaining years for the base case and the single public/private MRF options (Table 14.12). Upon close inspection, Scenario A_2 prolongs landfill life in San Antonio by 25 years, whereas Scenario B adds only about 8 years; therefore, a publicly owned MRF will allow the city to develop an integrated waste disposal network with disposal contractors to extend the life expectancy of landfill.

14.6 FINAL REMARKS

Poor participation from the city residents in curbside recycling programs results in a need to install MRF in an urban setting. To predict solid waste generation, system dynamics models were developed to predict solid waste generation in an urban setting having a higher economic growth potential. A case study for the City of San Antonio, Texas, presented unique forecasting solutions based on system dynamics outputs of solid waste generation. Five planning models were considered based on different types of socioeconomic conditions, while the base case was designed according to a traditional regression analysis. All five planning models were based

TABLE 14.12   Impacts of Scenario A$_2$ and Scenario B$_2$ compared to present landfill characteristics

	Acres (km^2)	Depth (m)	Remaining volume (m^3)	Compaction rate (kg·m^{-3})	Remaining space (tonnes)	Landfilled (TPY)	Remaining years (y)
			Present Consumption [a]				
WMI	2.82	32	90,384,602	739	66,760,701	612,776	108.9
BFI	1.07	32	34,306,092	866	29,715,373	1,114,801	26.7
			Scenario A$_2$				
WMI	2.82	32	90,384,602	739	66,760,701	508,982	131.2
BFI	1.07	32	34,306,092	866	29,715,373	1,011,007	29.4
			Scenario B$_2$				
WMI	2.82	32	90,384,602	739	66,760,701	576,276	115.8
BFI	1.07	32	34,306,092	866	29,715,373	1,078,301	27.6

Source: Chang et al. (2005).
[a]TCEQ (2004).

on an assumption that the existing recycling pattern through the private sector will remain in the entire planning horizon; thus, the system will constantly maintain a minimum level of recycling, although participation in public recycling programs remains inactive. Interactions among several system components within a prescribed timeframe were examined dynamically using Stella® software.

These models simulated five combinations of essential socioeconomic factors that influence solid waste generation. The disparity in the extremes of the estimates indicates the importance of choosing the most appropriate model to prevent over- or underestimation of tonnage generated. Model 1 was chosen as the representative estimation for this reason as well as its incorporation of all possible driving factors; however, the disparity in the extremes of the estimates indicates that systematic uncertainty embedded in the estimation is influential. Based on the historical records and predictive results of San Antonio's population, income, and waste generation, we concluded that the increasing trend is phenomenal for all relevant factors. The modeling results are directly useful for associated system planning for site selection and capacity planning of MRFs in the near future.

These estimates of solid waste generation support a series of optimization models to address the optimal site selection and capacity planning for MRFs in conjunction with an optimal shipping strategy for waste streams in a multidistrict urban region. The macro analysis presented in this chapter organizes two possible scenarios based on whether MRFs are run by a private or public agency. The modeling results provide decision-makers opportunity to weigh MRF options alongside important solid waste policies for the City of San Antonio. Cost–benefit analysis confirms that shipping costs, not processing fees, are the driving expense of the city's operation. Sensitivity analysis confirms that public participation plays the most important role in minimizing MSW expenses when the system embraces MRF options. With limited municipal budgets in mind, this analysis points to employ an integrated forecasting and optimization analysis for seeking optimal solution that uses a public policy outlook to reason that a MRF may alleviate solid waste costs.

Recent developments in state environmental policy, citizen unrest with landfill expansion, and city efforts to overhaul waste routes compel system planners to examine MRF as a viable alternative to improve the municipal bottom line and deliver quality environmental services. The most favorable design of a MRF complements system optimization techniques with cost–benefit, sensitivity, and policy analysis. The result is supported by a holistic modeling approach that forms a nexus between environmental sustainability and cost minimization. The next phase for San Antonio should examine collection vehicle routing to achieve a more global cost minimization within an integrated disposal solution. The next chapter will discuss the details of optimal collection vehicle routing with respect to household recycling impacts.

REFERENCES

Abbott, M. D. and Stanley, R. S. 1999. Modeling groundwater recharge and flow in an upland fracture bedrock aquifer. *System Dynamics Review*, 15(2), 163–184.

Acco-Bfi. 2003. *Redemption Price of Recycling Materials*, Acco-Bfi, San Antonio, TX.

American City Business Journals Inc. 2002. San Antonio gets high marks for debt management. San Antonio Business Journal, November 26. Available at: http://www.bizjournals.com (accessed October 2003).

Anex, R. P., Lawver, R. A., Lund, J. R., and Tchobanoglous, G. 1996. GIGO spreadsheet-based simulation for MSW systems. *Journal of Environmental Engineering, ASCE*, 122(4), 259–263.

Berk, J. 2001. Corporate Finance, Chapter 6. "What is a Bond?" Available at: http://faculty.haas.berkeley.edu/berk/teaching (accessed November 2003).

Bureau of Labor Statistics (BLS). 2003. Employment Data. Bureau of Labor Statistics. Available at: http://www.bls.gov/bls/employment.htm (accessed November 2003).

Census. 2003. Census. United States Census Bureau. Available at: http://www.census.gov (accessed November 2003).

Chang, Y. H. and Chang, N. B. 2003. Compatibility analysis of material and energy recovery in a regional solid waste management system. *Journal of Air and Waste Management Association*, 53(1), 32–40.

Chang, N. B. and Lin, Y. T. 1997a. An analysis of recycling impacts on solid waste generation by time series intervention modeling. *Resource Conservation and Recycling*, 19(3), 165–186.

Chang, N. B. and Lin, Y. T. 1997b. Optimal siting of transfer station locations in a metropolitan solid waste management system. *Journal of Environmental Science and Health*, A32(8), 2379–2401.

Chang, N. B. and Wang, S. F. 1995. The development of material recovery facilities in the United States: status and cost structure analysis. *Resources Conservation and Recycling*, 13(2), 115–128.

Chang, N. B. and Wang, S. F. 1996a. Solid waste management system analysis by multi-objective mixed integer programming model. *Journal of Environmental Management*, 48(1), 17–43.

Chang, N. B. and Wang, S. F. 1996b. Comparative risk analysis of solid waste management alternatives in a metropolitan region. *Environmental Management*, 20(1), 65–80.

Chang, N. B. and Wei, Y. L. 2000. Siting recycling drop-off stations in an urban area by genetic algorithm-based fuzzy multi-objective nonlinear programming modeling. *Fuzzy Sets and Systems*, 114(1), 133–149.

Chang, N. B., Pan, Y. C., and Huang, S. D. 1993. Time series forecasting of solid waste generation. *Journal of Resources Management Technology*, 21(1), 1–10.

Chang, N. B., Chang, Y. H., and Chen, Y. L. 1997. Cost-effective and workload balancing operation in solid waste management systems. *Journal of Environmental Engineering, ASCE*, 123(2), 178–190.

Chang, N. B., Davila, E., Dyson, B., and Brown, R. 2005. Optimal site selection and capacity planning of a municipal solid waste material recovery facility in an urban setting. *Waste Management*, 25(8), 833–846.

Chartwell Solid Waste Group. 2001. *Solid Waste Digest*, Chartwell Information Publishers, Alexandria, VA.

Chen, H. W. and Chang, N. B. 2000. Prediction of solid waste generation via grey fuzzy dynamic modeling. *Resources, Conservation and Recycling*, 29(1–2), 1–18.

Chiplunkar, A. V., Mehndiratta, S. L., and Khanna, P. 1982. Optimization of refuse collection systems. *Journal of Environmental Engineering, ASCE*, 107(EE6), 1203–1210.

Deaton, M. L. and Winebrake, J. J. 2000. *Dynamic Modeling of Environmental Systems*, Springer-Verlag, New York.

Deng, J. L. 1982. Control problems of grey systems. *Systems and Control Letters*, 1(5), 288–294.

Dennison, G. J., Dodd, V. A., and Whelan, B. 1995. A socio-economic based survey of household waste characteristics in the city of Dublin, Ireland – II. Waste quantities. *Resources, Conservation and Recycling*, 17(3), 245–257.

Dyson, B. and Chang, N. B. 2005. Forecasting of solid waste generation in an urban region by system dynamics modeling. *Waste Management*, 25(7), 669–679.

Eisenstein, D. D. and Iyer, A. V. 1997. Garbage collection in Chicago: a dynamic scheduling model. *Management Science*, 43(7), 922–933.

Engineering News-Record (ENR). 2003. Construction economics. ENR. Available at: http://enr.construction.com (accessed February 2014).

Everett, J. W. and Riley, P. 1997. Curbside collection of recyclable materials: simulation of collection activities and estimation of vehicle and labour needs. *Journal of Air & Waste Management Association*, 47(10), 1061–1069.

Everett, J. W. and Shahi, S. 1996a. Curbside collection of yard waste: I. Estimating route time. *Journal of Environmental Engineering, ASCE*, 122(2), 107–114.

Everett, J. W. and Shahi, S. 1996b. Curbside collection of yard waste: II Simulation and Application. *Journal of Environmental Engineering, ASCE*, 122(2), 115–121.

Everett, J. W. and Shahi, S. 1997. Vehicle and labour requirements for yard waste collection. *Waste Management & Research*, 15(6), 627–640.

Everett, J. W., Dorairaj, R., Maratha, S., and Riley, P. 1998a. Curbside collection of recyclables II: simulation and economic analysis. *Resources, Conservation and Recycling*, 22(3–4), 217–240.

Everett, J. W., Maratha, S., Dorairaj, R., and Riley, P. 1998b. Curbside collection of recyclables I: route time estimation model. *Resources, Conservation and Recycling*, 22(3–4), 177–192.

Fields, J. 2002. Letter to BFI from the San Antonio City Manager. Available at: http://stopbfi.com (accessed December 2002).

Ford, A. 1999. *Modeling the Environment*, Island Press, Washington, DC.

Forrester, J. W. 1961. *Industrial Dynamics*, The MIT Press, Cambridge, MA.

Forrester, J. W. 1968. *Principles of System*, Productivity Press, Cambridge, MA.

Forrester, J. W. 1969. *Urban Dynamics*, The MIT Press, Cambridge, MA.

Forrester, J. W. 1971. *World Dynamics*, Wright-Allen Press, Cambridge, MA.

Fuertes, L. A., Hudson, J. F., and Mark, D. H. 1974. Solid waste management: equity trade-off models. *Journal of Urban Planning and Development, ASCE*, 100(2), 155–171.

Gottinger, H. W. 1986. A computational model for solid waste management with applications. *Applied Mathematical Modelling*, 10(5), 330–338.

Grant, W. E., Pedersen, E. K., and Marin, S. L. 1997. *Ecology & Natural Resource Management: Systems Analysis and Simulation*, John Wiley & Sons, New York.

Grossman, D., Hudson, J. F., and Mark, D. H. 1974. Waste generation methods for solid waste collection. *Journal of Environmental Engineering, ASCE*, 6, 1219–1230.

Guo, H. C., Liu, L., Huang, G. H., Fuller, G. A., Zou, R., and Yin, Y. Y. 2001. A system dynamics approach for regional environmental planning and management: a study for Lake Erhai Basin. *Journal of Environmental Management*, 61(1), 93–111.

Hasit, Y. and Warner, D. B. 1981. Regional solid waste planning with WRAP. *Journal of Environmental Engineering, ASCE*, 107(3), 511–525.

Helms, B. P. and Clark, R. M. 1974. Locational models for SWM. *Journal of Urban Planning and Development, ASCE*, 97(1), 1–13.

High Performance System. 2012. Stella®. Available at: http://www.hps-inc.com/stellavpsr.htm (accessed January 2012).

Huang, G. H., Baetz, B. W., and Patry, G. G. 1993. A grey fuzzy linear programming approach for waste management and planning under uncertainty. *Civil Engineering Systems*, 10(2), 123–146.

Huang, G. H., Baetz, B. W., and Patry, G. G. 1994. Grey dynamic programming for SWM planning under uncertainty. *Journal of Urban Planning and Development, ASCE*, 120(3), 132–156.

Huang, G. H., Baetz, B. W., and Patry, G. G. 1996. A grey hop, skip, and jump method for generating decision alternatives: planning for the expansion/utilization of waste management facilities. *Canadian Journal of Civil Engineering*, 23(6), 1207–1209.

Ishizaka, K. and Tanaka, M. 2003. Resolving public conflict in site selection process – a risk communication approach. *Waste Management*, 23(5), 385–396.

Jenkins, L. 1982. Parametric mixed integer programming: an application to solid waste management. *Management Science*, 28(11), 1271–1284.

Karavezyris, V., Timpe, K., and Marzi, R. 2002. Application of system dynamics and fuzzy logic to forecasting of municipal solid waste. *Mathematics and Computers in Simulation*, 60(3–5), 149–158.

Kirca, Ö. and Erkip, N. 1988. Selecting transfer station locations for large solid waste systems. *European Journal of Operational Research*, 35(3), 339–349.

Kühner, J. and Harrington, J. J. 1975. Mathematical models for developing regional solid waste management policies. *Engineering Optimization*, 1(4), 237–256.

LINDO Systems. 2003. *LINGO – Optimization Modeling Software for Linear, Nonlinear, and Integer Programming*.

Lund, J. R. 1990. Least-cost scheduling of solid waste recycling. *Journal of Environmental Engineering, ASCE*, 116(1), 182–197.

Lund, J. R., Tchobanoglous, G., Anex, R. P., and Lawver, R. A. 1994. Linear programming for analysis of material recovery facilities. *Journal of Environmental Engineering, ASCE*, 120(5), 1082–1094.

Male, J. W. and Liebman, J. C. 1978. Districting and routing for solid waste collection. *Journal of Environmental Engineering, ASCE*, 104(EE1), 1–14.

Mashayekhi, A. N. 1993. Transition in New York State solid waste system: a dynamic analysis. *System Dynamics Review*, 9(1), 23–47.

Meadows, D. L. and Meadows, D. H. (Eds.). 1973. *Toward Global Equilibrium: Collected Papers*, Wright-Allen Press, MA.

Merrill Lynch. 2003. Municipal Bonds. Merril Lynch. Available at: http://askmerrill.ml. com/ (accessed November 2003).

Mohapatra, P. K. J., Mandal, P., and Bora, M. C. 1994. *Introduction to System Dynamics Modeling*, Orient Longman Ltd., Hyderabad, India, New York.

Nail, R. F., Gelanger, S., Klinger, A., and Peterson, E. 1992. An analysis of cost effectiveness of US energy policies to mitigate global warming. *System Dynamics Review*, 8(2), 111–128.

North American Free Trade Agreement (NAFTA). 2000. NAFTA delivers for Texas – 1999 report, The Trade Partnership, Washington, DC.

Needham, J. 2003. City official expects BFI to keep on top of trash. San Antonio Express News. Available at: http://news.mysanantonio.com (accessed October 2003).

Newell. 2003. *Redemption Price of Recyclable Materials*, Newell Recycling Co., San Antonio, TX.

Niessen, W. R. and Alsobrook, A. F. 1972. Municipal and industrial refuse: composition and rates. In: Proceedings of National Waste Processing Conference, New York, N.Y., USA, pp. 112–117.

Pires, A., Chang, N. B., and Martinho, G. 2011. Reliability-based life cycle assessment for future solid waste management alternatives in Portugal. *The International Journal of Life Cycle Assessment*, 16(4), 316–337.

Qi, C. and Chang, N. B. 2011. System dynamics modeling for municipal water demand forecasting in a fast growing region under uncertain economic impacts. *Journal of Environmental Management*, 92(6), 1628–1641.

Qu, W. and Barney, G. O. 1998. *Projecting China's Grain Supply and Demand Using a New Computer Simulation Model*, Millennium Institute, Arlington, VA.

Randers, J. (Ed.). 1980. *Elements of the System Dynamics Method*, Productivity Press, Cambridge, MA.

RecycleNet. 2003. Daily spot market prices. RecycleNet. Available at: http://www .recycle.net/price (accessed November 2013).

Saltzman, C., Duggal, V. G., and Williams, M. L. 1993. Income and the recycling effort: a maximization problem. *Energy Economics*, 15(1), 33–38.

Saysel, A. K., Barlas, Y., and Yenigun, O. 2002. Environmental sustainability in an agricultural development project: a system dynamics approach. *Journal of Environmental Management*, 64(3), 247–260.

Schultz, P. W., Oskamp, S., Mainieri, T. 1995. Who recycles and when? A review of personal and situational factors. *Journal of Environmental Psychology*, 15(2) 105–121.

Solano, E., Dumas, R. D., Harrison, K. W., Ranjithan, S., Barlaz, M. A., and Brill, E. D. 2002a. Life cycle-based solid waste management – 2. Illustrative applications. *Journal of Environmental Engineering, ASCE*, 128 (10), 981–992.

Solano, E., Ranjithan, S., Barlaz, M. A., and Brill, E. D. 2002b. Life cycle-based solid waste management – 1. Model development. *Journal of Environmental Engineering, ASCE*, 128(10), 993–1005.

Sorg, L. 2003. *The Wasteland*, San Antonio Current. Available at: http://www.sacurrent.com (accessed November 2003).

Sterman, J. D. 2000. *Business Dynamics: Systems Thinking and Modeling for a Complex World*, McGraw-Hill, Irwin, Boston, MA.

Sudhir, V., Srinivasan, G., and Muraleedharan, V. R. 1997. Planning for sustainable solid waste in urban India. *System Dynamics Review*, 13(3), 223–246.

Texas Commission on Environmental Quality (TCEQ, formerly TNRCC). 2004. Present consumption of solid waste management facilities.

Texas Natural Resource Conservation Commission (TNRCC). 2000. Composition of Texas waste stream.

Timms, C. E. and Baetz, B. W. 1998. Sizing of centralized leaf and yard waste composting facilities. *Canadian Journal of Civil Engineering*, 25(5), 967–973.

Ventana Systems. 2012. Vensim®. Ventana Systems, Inc. Available at: http://www.vensim.com (accessed January 2012).

Vezjak, M., Savsek, T., and Stuhler, E. A., 1998. System dynamics of eutrophication processes in lakes. *European Journal of Operational Research*, 109(2), 442–451.

Vizayakumar, K. and Mohapatra, P. K. J. 1992. Environmental impact analysis of a coalfield. *Journal of Environmental Management*, 34(2), 73–103.

Vizayakumar, K. and Mohapatra, P. K. J. 1993. Modeling and simulation of environmental impacts of a coalfield: system dynamic approach. *Journal of Environmental Systems*, 22(1), 59–79.

Walker, W., Aquilina, M., and Schur, D. 1974. Development and use of a fixed charge programming model for regional solid waste planning. Presented at the 46th joint meeting of the Operation Research Society of America and the Institute of Management Sciences, Puerto Rico.

Wilson, B. G. and Baetz, B. W. 2001b. Modelling municipal solid waste collection systems using derived probability distributions: II. Applications. *Journal of Environmental Engineering, ASCE*, 127(11), 1039–1047.

Wilson, B. G. and Baetz, B. W. 2001a. Modelling municipal solid waste collection systems using derived probability distributions: I. Model Development. *Journal of Environmental Engineering, ASCE*, 127(11), 1031–1038.

Wood, T. S. and Shelley, M. L. 1999. A dynamic model of bioavailability of metals in constructed wetland sediments. *Ecological Engineering*, 12(3–4), 231–252.

Xuan, Z. M., Chang, N. B., Daranpob, A., and Wanielista, M. 2010. Modeling the Subsurface Upflow Wetlands (SUW) systems for wastewater effluent treatment. *Environmental Engineering Science*, 27(10), 879–888.

Xuan, Z. M., Chang, N. B., and Wanielista, M. 2012. Modeling the system dynamics for nutrient removal in an innovative septic tank media filter. *Bioprocess and Biosystems Engineering*, 35(4), 545–552.

Yeomans, J. S. and Huang, G. H. 2003. An evolutionary grey, hop, skip, and jump approach: generating alternative policies for the expansion of waste management facilities. *Journal of Environmental Informatics*, 1(1), 37–51.

Zhu, Z. and ReVelle, C. 1990. A cost allocation method for facilities siting with fixed-charge cost functions. *Civil Engineering Systems*, 7(1), 29–35.

CHAPTER 15

OPTIMAL PLANNING FOR SOLID WASTE COLLECTION, RECYCLING, AND VEHICLE ROUTING

Effective planning of solid waste collection and recycling programs is challenging for current solid waste management (SWM) systems in urban regions. The process usually requires evaluating many assignment alternatives of collection vehicles, bins and recycling drop-off stations, as well as appropriate scheduling of vehicles and labor that can be optimally allocated or dispatched with respect to a suite of physical, technical, and economic constraints. This type of system planning by no means is an easy job. Large-scale system planning, design and operation models for vehicle routing and scheduling can encounter heavy computational loading, which makes the computational time rapidly increase to infinity within a few computational steps as the total number of network nodes increases. Both simulation models with heuristic algorithms and optimization models with explicit constraints for solving large-scale, complex solid waste collection problems would encounter the same type of computational issue: non-deterministic polynomial-time hard (NP-hard). How to overcome this issue is still an ongoing research task nowadays. This chapter begins with simulation-driven approach to support optimization analysis. Using a Geographical Information System (GIS) for siting the recycling drop-off stations in a fast-growing urban district in the City of Kaohsiung, Taiwan. A heuristic algorithm was employed with respect to the dynamics of population growth and shipping distance required by collection vehicles in the beginning. For comparison, the second part of this chapter formulates a multiobjective, nonlinear mixed-integer programming model to replace the heuristic algorithm to achieve the same goal by applying genetic algorithms in the same GIS environment. Sampling and analysis of waste generation rate and

Sustainable Solid Waste Management: A Systems Engineering Approach, First Edition. Ni-Bin Chang and Ana Pires.
© 2015 The Institute of Electrical and Electronics Engineers, Inc. Published 2015 by John Wiley & Sons, Inc.

composition were carried out as a prior study to provide an essential database for such a complex system analysis.

15.1 SYSTEMS ENGINEERING APPROACHES FOR SOLID WASTE COLLECTION

15.1.1 Vehicle Routing and Scheduling Programs for Handling Solid Waste Streams

Studies of vehicle routing and scheduling for SWM can be traced back to the mid-1970s (Beltrami and Bodin, 1974). This type of research became widespread in the 1990s, when SWM problems caused by the fast growth of urban population were highlighted (Golden et al., 2002). If the collection of waste has no time restrictions and no precedence relationships exist, then the problem is a pure routing problem. If there is a specified time for the waste collection service to take place, then a scheduling problem exists. Otherwise, we have to deal with a combined routing and scheduling problem for waste collection. Solid waste collection system planning consists of fine-tuning the allocation strategies of waste stream processing on a regional scale and formulating vehicle routing strategies on a local scale to match the regional demand (Lu et al., 2013). Optimization analyses aims to minimize the fleet operational burden in terms of vehicle numbers and labor when fixed costs of such fleet capacity are taken into account (Kim et al., 2006; Li et al., 2008; Arribas et al., 2010) for the reduction of collection costs (Li et al., 2008; Arribas et al., 2010) and total shipping distances (Bautista et al., 2008).

Waste collection vehicle routing problem (VRP) can be divided into three categories of problems: (1) arc routing, (2) node routing, and (3) roll-on-roll-off (Bodin et al., 2000; Golden et al., 2002; Kim et al., 2006). Service elements are streets in arc routing problem but points in node routing problem. Roll-on-roll-off problem, also named skip collection problem, is a restricted version of node routing problem where only one large container can be served per trip. Arc routing problem is less favorable than node routing problem in modeling (Kulcar, 1996) and can be appropriately transformed into node routing (Bautista et al., 2008). The VRP and its variants actually have time restrictions, because we know when a vehicle can departure the depot and when it must come back. Time restrictions are obvious in the periodical VRP or the VRP with time windows. So, routing and scheduling are indeed made simultaneously. Since capacitated VRP and other restricted variants are known to be NP-hard (Box 15.1) (Solomon, 1987), it is not easy to obtain an exact solution within an acceptable time frame. Consequently, fast approximation and enumerative optimization are often used (Desrochers et al., 1987). Approximation algorithms such as dynamic programming and column generation (Desrochers et al., 1992) and advanced bounding methods such as Lagrangian relaxation (Fisher, 1981) were employed to solve large-scale integer programming problems including waste collection VRP. Lower-bounding based on transportation model and upper-bounding based on the Clark–Wright algorithm were effective to reduce solution space in arc routing and roll-on-roll-off problems (De Meulemeester et al., 1997; Bodin et al., 2000; Mourão and Almeida, 2000). Multiple bounding methods can also be combined to solve a roll-on-roll-off problem with multiple disposal sites and time constraints (Baldacci et al., 2006).

BOX 15.1   NON-DETERMINISTIC POLYNOMIAL-TIME HARD

NP-hard is a term found in computational complexity theory. A problem is called NP-hard if the algorithm for solving it can be translated into one for solving any problem requiring non-deterministic polynomial time. In other words, the computational time will go up exponentially toward infinity as the total number of decision variables continuously increase. NP-hard is a class of problems that are, informally, "at least as hard as the hardest" for determining an optimal solution numerically, although it might, in fact, be even harder.

To avoid NP-hard issue encountered oftentimes in optimization analyses, heuristic algorithms are preferable and the more flexible Solomon's insertion algorithm, which outperforms several other construction heuristics such as nearest neighbor and sweep (Solomon, 1987), was modified to construct initial solutions in node routing problems with time windows (Tung and Pinnoi, 2000; Kim et al., 2006). After the seed routes are constructed, improvement techniques and metaheuristics, such as ant colonies algorithm (Bautista et al., 2008), tabu search, and variable neighborhood search (Benjamin and Beasley, 2010), were employed to improve solutions. In addition, the collection network can be partitioned when real situation is complicated (Mourão et al., 2009). Sometimes, the node routing problem can be directly solved by optimization software packages after partition (Chang et al., 1997a).

Unlike ordinary VRPs that only include depots and customers, waste collection VRPs also involve disposal sites (i.e., intermediate facilities). Although some simplified problems do not distinguish disposal sites and depots, optimal visiting sequence with the distinction is quite different from the sequence without the distinction (Kim et al., 2006). The disposal sites were dealt with by the modifications of Solomon's algorithm (Kim et al., 2006; Benjamin and Beasley, 2010) and a nearest insertion algorithm different from the sweep method (Angelelli and Speranza, 2002). Roll-on-roll-off problems including disposal sites were transformed into combination problems of decomposed elementary trips (Bodin et al., 2000; Baldacci et al., 2006). Furthermore, the route compactness of solutions (Kim et al., 2006) and balanced trip assignments to disposal sites (Li et al., 2008) should also be noticed besides the costs.

Optimization of waste collection VRP can be enhanced by informatics techniques. Using a geographical information system (GIS) is efficient to display optimization strategies and simplify processing of large sets of data such as distance matrices (Chang et al., 1997a; Teixeira et al., 2004; Kim et al., 2006). Optimization models can be coupled with GIS as an integral part of a decision support system in support of proper human-machine interactions (Lu et al., 2013). One method to streamline the cost-effectiveness or cost–benefit assessment of large-scale collection vehicle routing and curbside/community recycling schemes is to employ a sequential hybrid simulation and optimization technology at different decision scales and decision sequences with the aid of GIS. In particular, a hierarchical approach that reflects a system of systems engineering (SOSE) philosophy may be designed to handle situations with highly uncertain solid waste generation rates due to the impact of floating fee collection policies and rapid expansion of residential areas in fast-growing urban regions. Within such complex decision analyses, a suite of urban-scale recycling programs, such as regional material recovery facilities (MRF, pronounced as "merf"), are often designed to improve urban sustainability, which requires evaluating the solid waste collection programs at the local scale from time to time. The simulation and optimization analyses in this chapter is deemed as an integral part of modeling hierarchies in connection with these regional-scale models in previous two chapters.

15.1.2   Recycling Programs with Optimal Vehicle Routing and Scheduling Approaches

SWM system planning has received wide attention because of its complex integration of various management strategies. Due to the variations over social, economic, and regional factors, SWM programs must be frequently reorganized to overcome barriers. One of the salient issues is how to effectively distribute the collection crew size and vehicles in a growing metropolitan region. Recent studies have focused on the use of heuristic algorithms or mathematical models to evaluate the capability of GIS spatial analysis and decision support because optimizing waste collection in municipal solid waste (MSW) management can yield large savings.

Recycling is often viewed as a partially desirable management option before landfill and incineration (Table 15.1) in the context of waste management hierarchy (Gertsakis and Lewis, 2003). Fundamental simulation modeling analyses are commonly applied to solve various operational problems of recycling programs. Most previous recycling programs have implemented a separate collection and processing system for several individual recyclables operated by nonprofit, independent foundations in parallel with the existing MSW collection, processing, and disposal system. As a result, studies on short-term planning of vehicle routing and scheduling problems would be supportive of long-term regional planning for integrated solid waste management (ISWM).

With the rapid depletion of landfill space and the continuing concerns of air pollution emissions from incineration, ISWM strategies must be reorganized to improve the success of various types of recycling programs. One strategy that could aid household, curbside, and centralized recycling programs is the use of MRF. Some curbside

TABLE 15.1   The environmental attributes and outcomes of the waste management hierarchy

Goals	Attributes	Outcomes
Waste reduction	Preventative	Most desirable
Reuse	Relatively ameliorative; partially preventative	⇕
Recycling	Relatively ameliorative; partially preventative	Partially desirable
Treatment and recovery	Relatively assimilative; partially ameliorative	⇕
Disposal	Assimilative	Least desirable

and centralized recycling programs, however, have utilized a voluntary separation of recyclables by the residents, being collected in parallel with existing raw solid waste collection, processing, and disposal systems. In some European countries, this recycling initiative was promoted by extended producer responsibility, which is known as the Green Dot system, for managing packaging waste (see Chapters 2 and 4). The Green Dot system finances all the costs of collection and sorting at MRF through the fee paid by packaging producers and importers. Besides, to make recycling programs successful, voluntary separation via consumers should count on a strong economic instrument that promotes citizens participation (i.e., more information on economic instruments applied to solid waste can be seen in Chapters 3 and 4).

A different approach has been proposed to reduce recycling costs by analyzing the entire system rather than individual recycling elements to generate a holistic view of cost versus benefit profiles (Chang and Wang, 1996a, 1996b; Chang and Chang, 1998; Chang et al., 2005). Some management practices for both material and energy recovery have developed engineering facilities that recover recyclables from MSW and generate refuse-derived fuels prior to incineration (Chang et al., 1997b; Chang and Chang, 1998). Other management practices have opted to co-collect recyclables using the same routes of collection vehicles for processing at either a common transfer station or a centralized MRF (Chang and Wang, 1996a, 1996b; Chang et al., 2005). Yet multichannel recycling programs conducted by communities, MSW collection teams, schools, and independent foundations can be coordinated through a common organization supervised by a government authority (Chang and Wei, 1999). Transfer stations may be used to aid in recycling, balancing a cost-effective and workload operation in an ISWM system (Chang et al., 1997c; Chang and Lin, 1997). This type of complexity may be further improved by using various SOSE approaches.

The concept of building a MRF to support separate curbside recycling programs has been supported by many studies (Chang and Wang, 1996a, 1996b; Chang et al., 2005), but the latest practical implementation employs a very different arrangement. Curbside recycling programs collect commingled recyclables every few days while concurrently collecting other household waste flows on a regular schedule, a structure that can be analyzed using various SOSE approaches. To improve curbside collection schemes, the use of dedicated recycling vehicles and containers can be examined based on differing cost-and-benefit criteria. For example, efforts of collecting recyclables from all channels could have a quick financial "payback" from the recycling income supported by a nonprofit organization. Many ISWM systems could be greatly enhanced by such economic recycling incentives through either publicly owned waste collection teams or privately invested enterprises. By evaluating different financial schemes to sustain the recycling programs, systems engineering models can be formulated to assess the trade-offs among cost-benefit-risk criteria at differing scales.

Within a fast-growing district, for instance, the first effort of a curbside recycling program is the distribution of separate recycling containers designed to store plastic, metal, glass, and paper in more flexible and dynamic ways. The recycling problem, therefore, focuses on selecting the most appropriate community drop-off station locations periodically for users to bring their recyclables to the containers at any time. Public education is required to support these recycling activities, with a possible condition that part of the recycling income can be returned to a community foundation or used to reduce the charges for cleaning up the daily solid waste. The associated MRF projects in an ISWM system should be aimed at attracting a private enterprise to perform the recycling activities. Collection vehicles must be designed to withstand the rigors of the transportation network based on the frequency of emptying at various drop-off stations, which could eventually affect the operation of regional MRF. To address these changing conditions, GIS planning tools along with heuristic algorithms or systems engineering models may be applied to generate waste management strategies when screening and sizing recycling alternatives, which could in turn affect local-scale collection vehicle routing and labor scheduling.

It would be interesting to begin by describing the use of simulation-driven design optimization and GIS modeling to site recycling drop-off stations in a fast-growing urban district. In the first part of this chapter, a heuristic algorithm was employed to evaluate the dynamics of population growth and the shipping distances required by collection vehicles. In contrast, the simulation analysis with the aid of GIS and heuristic algorithm is followed by formulating a multiobjective, nonlinear mixed-integer programming model to replace the heuristic algorithm and achieve the same goal, which is solved by the genetic algorithms in the same GIS environment in the second part of this chapter.

15.2 SIMULATION FOR PLANNING SOLID WASTE RECYCLING DROP-OFF STATIONS

15.2.1 Planning Philosophy

If only one vehicle is available to route the nodes in a collection network, we have a traveling salesman problem (TSP) (Box 15.2), for which both exact and approximate solution procedures exist. If only one vehicle is available to route over the branches of a network, we have a Chinese postman problem (CPP) (Box 15.3). A VRP arises when attempting to design optimal collection routes from one or several depots to a number of scattered destinations that may follow either TSP or CPP logistics. Recent optimization models used to perform large-scale collection and vehicle routing practices are derived from many well-established studies dating back to the 1970s. Most involve an integration of multiple vehicle routes and varying service frequencies based on capacity and time limitations (Liebman et al., 1975; Bodin and Kursh, 1978; Chiplunkar et al., 1981; Schrage, 1981; Madsen, 1983; Current et al., 1987; Brodie and Waters, 1988; Feiring, 1990; Ong et al., 1990; Achuthan and Caccetta, 1991; ReVelle et al., 1991; Lysgaard, 1992; Dror, 1993; Thangiah, 1995). Recent applications of various vehicle routing models to a myriad of waste-collection problems can be found in the literature (Tung and Pinnoi, 2000; Angelelli and Speranza, 2002; Kim et al., 2006; Benjamin and Beasley, 2010; Benjamin, 2011). Methodological complexity has been variously addressed by the memetic algorithm and tabu search (Box 15.4) for the multi-compartment vehicle routing (El Fallahi et al., 2008) and capacitated arc routing problems to distribute the shipping loads in a shipping network with capacity limitation associated with each section of the network (Muyldermans and Pang, 2010a, 2010b).

BOX 15.2   TRAVELING SALESMAN PROBLEM

The TSP is a class of transportation systems engineering problems, which entails: "what is the shortest possible route that enables the salesman to visit each city exactly one time and then return to the origin city given a list of cities and the distances between each pair of cities?" It is an NP-hard problem oftentimes (See Box 15.1).

BOX 15.3   CHINESE POSTMAN PROBLEM

The CPP is also a class of systems engineering problems. It is also known as route inspection problem in graph theory. Suppose there is a postman whose job is to deliver mails to a certain neighborhood and wants to find the shortest route through the neighborhood, which meets the following two criteria:

- the postman needs to go through every street at least once;
- it ends at the same point it starts within a closed circuit.

BOX 15.4   TABU SEARCH

Tabu search is a metaheuristic search method which provides a set of neighborhood search rules to trigger a more effective local search for the optimal solution by avoiding a tendency to become stuck in suboptimal regions or on plateaus where many solutions are equally fit. If a potential solution has been previously evaluated or identified as violation of a rule, it is marked as "tabu" (forbidden) so that the algorithm does not consider that option repeatedly. In short, Tabu search enhances the holistic performance of local search techniques by using memory structures.

Heavy computational loading frequently increases computational time to infinity within a few steps when using those optimization models or heuristic algorithms to solve large-scale solid waste collection problems, however. Finding an appropriate method to improve the cost-effectiveness for the "co-collect" system in which raw waste and recyclables are collected separately could become a research focus. Improving the cost-effectiveness and feasibility of curbside recycling and collection schemes using an advanced simulation technology aided by GIS was examined in the first part of this chapter. Due to the rapid depletion of landfill space and the continuing debate of the public health impacts of MSW incinerators, MSW collection strategies must be frequently reorganized for successful recycling. Failure to utilize modern environmental informatics tools such as GIS to empower systems engineering models to meet changing managerial goals in a changing environment would impede public decision-making.

In recent decades, significant advances have been made toward applying GIS to solve a variety of issues in environmental planning and management (Dedic et al., 1992; Downer et al., 1992; Hromadka et al., 1992). Research efforts have integrated GIS with various analytical models through system architectures and schematic representation of the network, covering the spectrum of groundwater, surface water, water distribution, soil erosion, water resources management, SWM, and chemical emergency response systems planning (Johnston, 1987; Lupien et al., 1987; Hass et al., 1992; Kilborn et al., 1992; Tsakiris and Salahoris, 1993; Muzik, 1994; Summer, 1994; Zhang and Parks, 1994; Cargin and Dwyer, 1995; Massie, 1995; Chang et al., 1998; Bartlett, 2004). In particular, several computer packages have been designed for spatial decision support in various types of SWM systems (Chang and Wang, 1996c; MacDonald, 1996). The GIS technology for vehicle routing and scheduling using a multiobjective programming model was applied by dividing the service area into 14 subregions to reduce the computational loading (Chang et al., 1997a). With the GIS capacity to manage large amounts of network information and integrate with environmental models, complicated SWM practices become feasible for different planning scenarios. These types of simulation analysis particularly illustrate the spatial analytical capability of GIS to integrate simulation modeling techniques when planning solid waste collection systems.

15.2.2 GIS-Based Simulation Analysis for Siting Recycling Drop-Off Stations

GIS offers the functionality and tools to collect, store, retrieve, analyze, and display geographical information. GIS also allows the user to create and store as many layers of data or maps as needed and provides various possibilities to integrate tremendous amounts of data and large numbers of map overlays into a single output to aid in decision-making. GIS and environmental modeling, however, are synergistic. Various applications of GIS for environmental planning and management have been presented since the 1990s (Downer et al., 1992; Cargin and Dwyer, 1995). The rapid development of GIS has opened the possibility to not only serve as a common data and analysis framework for external environmental models or software components, but also allow GIS itself to become an integral part of a model by providing both the mapped variables and the processing environment.

The recent interest of environmental modelers in GIS is stimulating the development of its internal spatial and network analysis modules as an advanced modeling tool that moves GIS progressively away from the simple role of data presentation into cartographic output. With the advent of GIS technology, complex analysis has become possible for large-scale environmental planning programs. This section explores the simulation modules in GIS to examine the recycling and collection practices in an SWM system employing a co-collect approach. An investigation was first conducted to evaluate the impact of waste reduction due to the allocation of recycling drop-off containers in the MSW collection network. The routine collection program of solid waste streams was then assessed by a spatial analysis approach in a GIS environment. This management strategy is particularly important in a privatized solid waste collection system in which recycling, waste collection, and MRF are owned by a single private or public agency.

The network analysis module of ARC/INFO® GIS, which provides network-based spatial analysis developed by the Environmental Systems Research Institute (ESRI), was used for the demonstration purpose in this study (ESRI, 1992). ARC/INFO® GIS (i.e., later version is called ArcGIS Network Analyst) is a vector-based system that stores a network as a set of line features with associated attributes. These line features consist of coordinate pairs with an explicit identification number. Due to the complexity of the attribute tables, vector data are usually organized and stored in a relational database management system (RDBMS). Using a topological structured vector, reality is modeled through a series of GIS map layers of homogeneous information, and different combinations of these associated attributes can be dynamically organized from various viewpoints to fulfill specific query requirements for spatial analysis.

Records in the database tables also contain explicit identification numbers that relate them to a particular feature. When the coordinate information for a network is modified, perhaps by the addition of new features, the attribute file in the RDBMS is automatically updated. An approach has been adopted in the study whereby environmental models were used in conjunction with GIS to generate a set of systematic vehicle routing and scheduling program for SWM. GIS thus has a dual role: (1) in data integration and quantification as an input to environmental models for recycling analysis, and (2) in data interpolation, visualization, and assessment of assignment alternatives for vehicle routing using spatial analysis modules.

The curbside recycling program with an "ideal" weekly collection proposal, one that would perform the curbside commingled collection each Sunday while remaining household waste flows are collected from Monday through Saturday, was assumed in a fast-growing district. Building such analytical capabilities for developing effective recycling strategies is essential for creating a more efficient SWM practice. First stage modeling analysis requires GIS to link with a recycling investigation, which is a promising way to enhance the effectiveness of the co-collect system because the initial participation rate of residents is relatively low.

One important feature of GIS that supports the analysis of environmental model for recycling analysis is its capability to calculate population densities, waste generation rates, percentage of recyclables, and waste distributions in the network. Hence, the essential map digitizing and editing procedures for building both spatial and nonspatial database support the creation of the cartographic portion of a network database within the service area. The ARC/INFO® GIS system allows assignment of the attributes of recyclables distribution to each collection point in the network cartography so that the management scenarios for siting recycling drop-off stations can be performed using the optimization model as an explicit tool in the model base. A self-designed interface is required for performing the communication duty in the modeling analysis.

Projection of solid waste sources requires an explicit linkage between solid waste generation patterns and recycling processes that provides an unambiguous estimate of spatial heterogeneity of solid waste distribution essential for subsequent analysis in the vehicle dispatch program. A spatial analysis module in GIS can work together with the environmental models to perform the vehicle routing and scheduling for solid waste collection in the second stage (Figure 15.1). Performing a system analysis for collection vehicle routing and scheduling in a street network must consider economic and equitable objectives, which can be a complex task.

Such a task can be performed by the space filling curve (SFC) algorithm embedded in ARC/INFO® GIS that uses a strategy of "cluster first-route second" (Lu, 1996). The efficiency of these commands in ARC/INFO® GIS to achieve near-optimal vehicle routing and scheduling is sufficient to support a large-scale SWM system. SFC can be applied with a few computational resources and may achieve the dispatch program based on several types of mixture of vehicles, a task not possible using optimization schemes. As a result, a GIS framework with a simulation model applied in the first stage and a valid vehicle routing program applied in the second stage form an integrated tool to achieve the planning goal for SWM in an urban region.

15.2.3 Results of Practical Implementation

Simulation Results for Recycling Program   The best siting pattern of recycling drop-off containers in one district of the City of Kaohsiung, Taiwan, in the simulation scheme can be presented as a network output (Figure 15.2) that shows a set of nodes and links selected by the simulation scheme. The circles associated with each container (i.e., each selected node) represent the service radius of each container, and the routing orders prepared for the dedicated collection vehicle are marked by Arabic numerals near those drop-off stations on the map. Five performance indexes were defined in the assessment metrics for the comparative evaluation of the efficiency and effectiveness of recycling programs in this study. The first index describes the service ratio that represents the percentage of population serviced by recycling drop-off stations; the second index describes the utilization rate of those drop-off stations; the third index presents the average walking distance of residents from their household to collection tanks; the fourth index demonstrates the percentage of recyclables to be collected by the curbside recycling program; and the fifth index depicts the routing ratio of total routing distance to total distance in the network links (denoted as a normalized achievement ratio) (Chang and Wei, 1999).

- Service ratio $= \dfrac{\text{The population serviced by recycling drop-off stations}}{\text{Total population in the district}}$

- Utilization rate $= \dfrac{\text{Recyclables colected by drop-off stations}}{\text{Total capacity provided by drop-off staions in the district}}$

- Average walking distance $= \dfrac{\text{Total service distance between node } i \text{ and node } k \text{ of network links}}{\text{Total number of drop-off stations in the district}}$

- Recycling rate $= \dfrac{\text{Recyclable collected by drop-off stations}}{\text{Total recyclables in the district}}$

- Routing ratio $= \dfrac{\text{Routing distance by collection vechicle}}{\text{Total distance of network links in the district}}$

Assuming 90 m as the average distance from a household to a drop-off station, the suggested number of recycling drop-off stations is 23. This calculation accounts

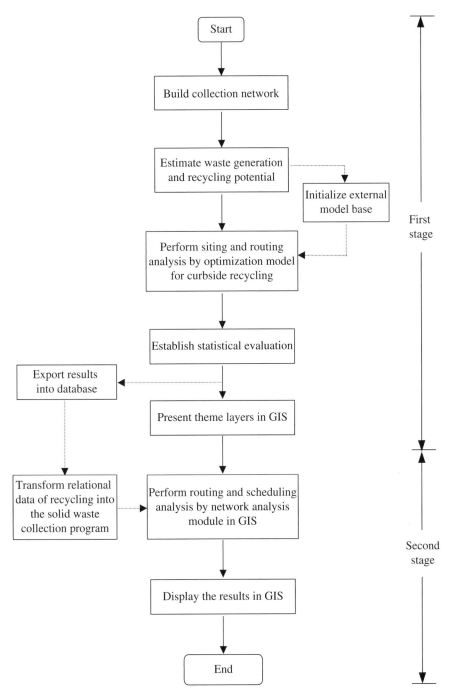

FIGURE 15.1   Analytical framework of this GIS study for allocation of recycling drop-off containers

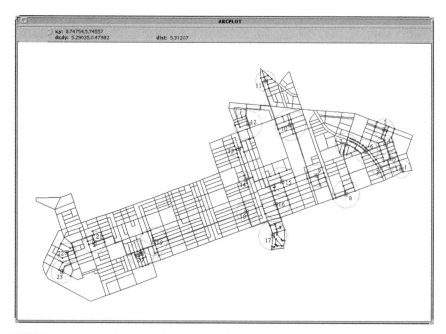

FIGURE 15.2 Curbside recycling program in the district. The red lines are the optimal pathway in the collection network and each of the green circle is the service radius of each collection node

for a 10% reduction of the solid waste streams and 41.9% of the population that can be serviced by this recycling network in the initial stage. The actual average walking distance for those residents to access the containers is 89.31 m. Net system income is anticipated in such a recycling system, which implies the existence of economic incentives in the privatization process (Table 15.2).

The allocation of drop-off stations, which were favored over individual household collection, is relatively effective in highly populated cities where households cannot be easily identified from a residential door-to-door configuration, and the participation rate is difficult to promote by other methods. Once the recyclables are removed from the co-collect recycling network, a revised program for regular collection of solid waste streams in the highly populated district may be performed. The analytical results of the collection vehicle routing, considering the impact associated with such a curbside recycling program performed by GIS simulation technology, can then be presented by a heuristic algorithm based on the minimum distance principle.

The result indicates that 39 vehicles are required to achieve the regular task of solid waste collection. The amount of solid waste collected, the corresponding collection time and distance, the number of nodes encountered in each service tour, and the utilization efficiency of each vehicle can be summarized for decision-making (Table 15.3). The solid waste collection network of the proposed district is thus delineated from A to E to facilitate presentation (Figure 15.3), although GIS is capable of representing the entire region. The scheduling information can be further summarized through the presentation of routing order for those vehicles (Table 15.4).

TABLE 15.2   Results of optimization analysis for curbside recycling

Planning parameters	Value
The ratio of recyclables (%)	10.00
Optimal number of drop-off stations	23.00
Population serviced by drop-off stations in the district (capita)	55,015
Service ratio (%)	41.90
Collected recyclables (kg)	7,426.99
Recycling rate (%)	37.71
Utilization rate of tanks (%)	35.88
Average walking distance (m)	89.31
Routing distance by collection vehicle (m)	9,819.10
Routing ratio (%)	9.03
Price of recyclables (US\$·kg^{-1})	0.07
Recycling cost (US\$·d^{-1})	221.00
Recycling income (US\$·d^{-1})	563.88
Net system income (US\$·d^{-1})	342.70

US\$ is United States dollars.

The equity principles of the entire dispatch program based on the proposed routing scheme for each of the 39 collection vehicles can be examined as well. All the node numbers in support of a full routing illustration as described in Table 15.4 can be seen in Figure 15.4. It clearly shows the diversity of application areas made by GIS for an SWM program.

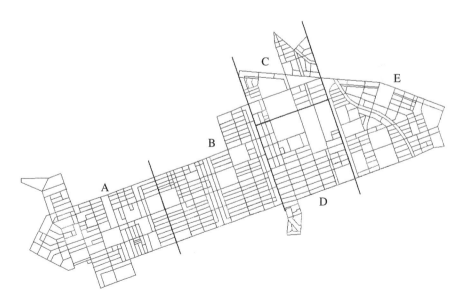

FIGURE 15.3   Subregions of collection network from A to E defined for the presentation of vehicle routing. The letters A to E represent the subregions divided for the convenience of presentation. Different colors of routing pathway in each subdistrict represent the truck fleet dispatched for waste collection

TABLE 15.3 Results of GIS analysis for solid waste collection after recycling

No. of vehicle	Amount of collected waste (kg)	Collection time for each service tour (min)	Distance of collection for each service tour (m)	Total collection points for each service tour	Utilization of collection capacity (%)
1	4,460.32	70.87	4,615.32	36	0.99
2	3,971.32	50.85	1,048.71	10	0.88
3	4,445.40	64.05	2,955.01	36	0.98
4	4,269.76	53.72	889.38	18	0.94
5	4,419.56	63.05	2,781.66	33	0.98
6	4,316.33	53.84	781.58	9	0.95
7	4,404.36	62.73	2,745.84	41	0.97
8	4,406.15	53.88	528.89	9	0.97
9	4,273.39	61.69	2,871.61	19	0.94
10	4,288.34	54.11	931.92	15	0.95
11	4,459.03	61.60	2,302.52	16	0.99
12	4,251.03	54.59	1,161.15	18	0.94
13	4,488.02	60.62	1,973.76	23	0.99
14	4,485.26	55.72	756.28	9	0.99
15	4,441.11	60.38	2,049.26	24	0.98
16	4,386.52	56.19	1,163.80	18	0.97
17	4,463.10	60.25	1,952.35	27	0.99
18	4,253.08	56.20	1,556.61	27	0.94
19	4,384.63	59.93	2,104.76	27	0.97
20	4,391.61	56.25	1,163.10	18	0.97
21	4,317.50	59.64	2,229.28	26	0.95
22	4,365.28	56.38	1,273.18	18	0.97
23	4,453.88	59.41	1,769.26	27	0.98
24	4,430.35	56.66	1,151.93	12	0.98
25	4,486.38	59.35	1,659.34	19	0.99
26	4,364.91	56.67	1,347.81	20	0.96
27	4,434.46	59.32	1,805.93	31	0.98
28	4,454.98	56.96	1,154.19	12	0.98
29	4,473.76	59.22	1,663.40	28	0.99
30	4,335.17	57.65	1,679.74	18	0.96
31	4,204.87	59.19	2,445.89	41	0.93
32	4,348.32	57.71	1,655.55	29	0.96
33	4,403.29	58.99	1,814.44	23	0.97
34	4,347.83	57.72	1,659.08	25	0.96
35	4,470.73	58.81	1,570.81	23	0.99
36	4,452.57	58.00	1,422.31	19	0.98
37	4,484.80	58.46	1,442.05	17	0.99
38	4,489.25	58.28	1,385.09	18	0.99
39	4,257.53	58.40	2,094.14	15	0.94

TABLE 15.4   Detailed vehicle routing scheme of solid waste collection program from GIS analysis

No. of vehicle	Routing order
1	E126→E120→E132→E141→E153→E162→E181→E189→E195→E196→ E194→E187→E179→E169→E159→E150→E152→E143→E160→E170→ E180→E186→D46→D36→D30→D20→D14→D6→E151→E116→E105→ E74→E66→E82→E95→E118
2	D19→D28→D41→D49→D52→D42→D31→D37→D29→D22
3	C63→C56→C61→C60→C58→C64→C52→C57→C42→C47→C49→C62→ C65→C80→C76→C72→C69→C68→C91→C50→C30→C26→C19→C23→ C28→C36→C34→C41→C51→C43→C53→C54→C55→C46→C45→C44
4	B130→B147→B145→B141→B134→B152→B159→B168→B178→B191→ B182→B187→B190→B194→B183→B174→B166→B154
5	C17→C14→C8→C3→C13→C21→C16→C22→C25→C31→C33→C35→ C40→C38→C37→C32→C27→C20→C28→C30→C24→C18→C11→C7→ C4→C5→C2→C1→C6→C12→C10→C9→C15
6	D87→D93→D96→D92→D84→D79→D74→D71→D85
7	E173→E182→E190→E191→E185→E175→E165→E192→E184→E177→ E165→E155→E146→E136→E122→E112→E106→E94→E99→E102→ E107→E109→E114→E127→E138→E145→E156→E154→E148→E144→ E135→E124→E110→E142→E134→E123→E111→E103→E101→E79→E68
8	B41→B48→D45→D43→D40→D47→D35→D25→D27
9	D116→D119→D124→D120→D121→D118→D114→D123→D122→D125→ D113→D111→B181→B170→B192→B200→B179→B169→B160
10	B150→B161→B158→B167→B131→B138→B156→B165→B176→D103→ D105→D102→D101→D100→D98
11	C79→C88→C98→C102→C104→C99→C92→C81→C85→C98→C103→ C106→D5→D7→D12→D17
12	D21→D23→D28→B29→B30→B35→B38→B42→B39→B45→B43→B53→ B50→B52→B61→B69→B77→B94
13	A174→A154→A148→A141→A139→A134→A121→A126→A132→A135→ A118→A105→A102→A96→A111→A123→A115→A128→A145→A165→ A179→A171→A155
14	D78→D73→D68→D60→D51→D58→D67→D72→D59
15	A125→A158→A186→A196→A177→A159→A150→A176→A167→A151→ A181→A188→A208→A213→A199→A191→A209→A222→A226→A224→ A218→A201→A194→A185
16	A20→A34→A29→A39→A49→A57→A68→A73→A78→A80→A83→ A67→A79→A93→A97→A119→A136→A130
17	A204→A206→A216→A210→A214→A203→A193→A184→A182→A189→ A178→A173→A168→A162→A122→A116→A86→A87→A108→A117→ A101→A92→A82→A129→A142→A156→A172
18	A137→A127→A98→A89→A77→A85→A91→A99→A112→A109→A106→ A100→A138→A140→A133→A144→A149→A153→A157→A163→A170→ A166→A160→A152→A146→A175→A169
19	E92→E104→E115→E128→E140→E131→E117→E108→E98→E84→E70→ E64→E76→E121→E137→E172→E174→E167→E158→E147→E161→ E171→E176→E183→E168→E157→E163

TABLE 15.4 *(Continued)*

No. of vehicle	Routing order
20	B24→B22→B19→B21→B18→B14→B10→B6→B5→B2→B4→B3→B8→ B11→B15→B13→B7→C109
21	A6→A4→A3→A1→A7→A16→A22→A30→A41→A35→A46→A55→A66→ A61→A53→A42→A27→A19→A25→A36→A47→A56→A38→A28→ A21→A15
22	D68→D76→D80→D88→D82→D86→D83→D90→D94→D104→D109→ D106→D107→D108→D110→D112→D115→D117
23	A9→A2→B80→B76→B72→B83→B97→B108→B122→B127→B116→ B114→B109→B121→B115→B103→B101→B88→B91→A5→A8→A12→ A14→A17→A26→A18→A10
24	B9→B18→B20→B23→B26→B31→B33→B37→B34→B32→B28→B25
25	A230→A229→A200→A197→A195→A190→A187→A180→A164→A147→ A131→A120→A107→A143→A161→B226→B229→B221→B216
26	A24→B132→B144→B140→B155→B151→B157→B162→B175→B188→ B185→B171→B172→B164→B149→B148→B123→B135→B112→B126
27	B12→B17→C118→C117→C116→C113→C114→C112→C111→C105→ C107→C108→C100→C95→C89→C83→C74→C71→C73→C67→C75→ C66→C70→C84→C90→C94→C93→C101→C86→C82
28	D48→D38→D32→D33→D24→D10→D8→D13→D18→D9→D11→D13
29	E19→E15→E7→E17→E25→E27→E40→E54→E37→E32→E22→E51→ E63→E71→E61→E49→E45→E31→E34→E42→E57→E48→E53→E65→ E75→E81→E88→E100
30	A13→A23→A31→A50→A65→A82→A48→A45→A40→A11→A37→A74→ A63→A75→A84→A95→A110→A124
31	E133→E130→E125→E119→E113→E97→E93→E89→E96→E91→E87→ E85→E83→E78→E69→E72→E73→E77→E80→E86→E90→E56→E33→ E24→E16→E12→E5→E39→ E44→E62→E58→E52→E47→E41→E35→ E26→E20→E36→E46→E55→E59
32	B113→B99→B87→B75→B65→B60→B68→B79→B71→B85→B98→B92→ B106→B102→B89→B86→B81→B67→B58→B46→B40→B51→B44→ B47→B57→B59→B64→B50→B54
33	B202→B207→B204→B213→B217→B220→B214→B212→206→B208→ B199→B219→B224→B215→B209→B201→B210→B203→B196→B186→ B195→B177→B180
34	A58→A64→A54→A43→A33→A32→A44→A51→A60→A71→A81→ A69→A59→A52→A70→A72→A66→A90→A88→A76→A103→A114→ A104→A94→A113
35	E4→E10→E13→E11→E9→E3→E2→E6→E14→E23→E18→E21→E30→ E33→E50→E60→E67→C97→C87→C78→C70→C59→C48
36	A220→A211→A207→A205→A198→A183→A192→A202→A212→A221→ A225→A228→A231→A232→A227→A223→A217→A215→A219
37	B124→B55→B63→B73→B84→B93→B110→B128→B142→B119→A136→ A148→B153→B163→B173→B184→B193
38	B62→B66→B82→B78→B96→B100→B107→B120→B130→B125→B118→ B104→B90→B74→B70→D63→D61→D56
39	E196→E201→E199→E200→E202→E204→E203→D77→D55→D62→D64→ D57→D53→D44→D34

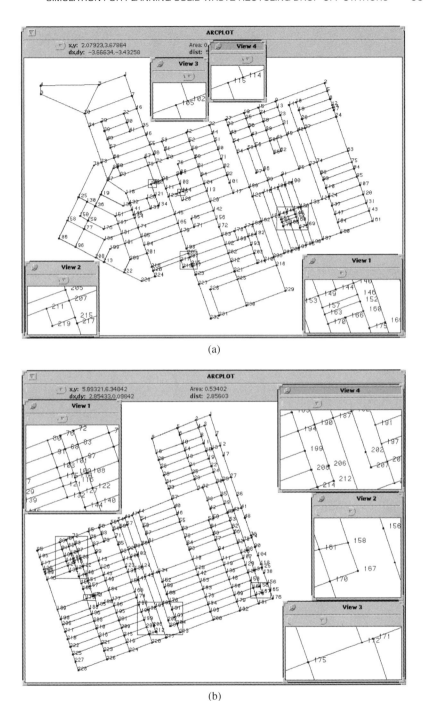

(a)

(b)

FIGURE 15.4 (a) Subregion A of solid waste collection network; (b) subregion B of solid waste collection network; (c) subregion C of solid waste collection network; (d) subregion D of solid waste collection network; and (e) subregion E of solid waste collection network

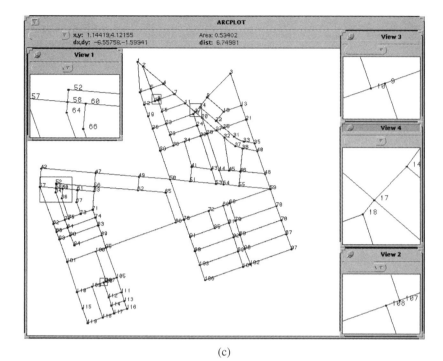

(c)

(d)

FIGURE 15.4 (*Continued*)

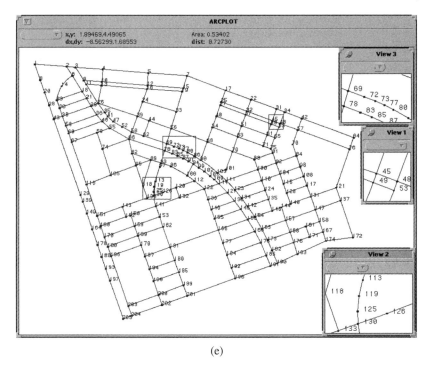

(e)

FIGURE 15.4   (*Continued*)

15.3   MULTIOBJECTIVE PROGRAMMING FOR PLANNING SOLID WASTE RECYCLING DROP-OFF STATIONS

This section introduces the use of multiobjective programming model to enhance system planning of siting the recycling drop-off stations in the same district. To ensure maximum consensus and minimum disruption, three broad-based planning goals were proposed in this analysis (Chang and Wei, 1999, 2000) to (1) consolidate routes of dedicated collection vehicles, saving haulers time and costs, (2) provide as many recycling containers as possible within the collection network, and (3) distribute recycling containers as equitably as possible.

15.3.1   Objective Function and Constraints

In general, three objectives can be considered for siting recycling drop-off stations and collection vehicle routing: (1) maximization of population serviced by recycling drop-off stations, (2) minimization of total walking distance from household to recycling drop-off stations, and (3) minimization of total driving distance during vehicle routing. Maximizing the population serviced is equivalent to maximizing waste or recyclables collected because uniform production rate per capita was chosen for this analysis; however, GIS should be capable of managing a nonuniform production rate of solid waste in a dynamic environment in which the dynamic information can be transformed into the total yield of waste at each node. The application of the first objective does not mean that some subgroups of the population cannot be serviced by the recycling containers; rather, some residents may encounter a longer walking distance to access the recycling containers, a component of the second objective. Population density varies over subregions, however, and to improve access to recycling containers, a surrogate parameter (i.e., walking distance) was applied to evaluate siting feasibility based on population density. The third objective to minimize operational costs encompasses various cost factors, but routing distance of the dedicated vehicles is the only factor intimately tied to the siting pattern of recycling containers; hence, if other physical conditions are fixed, the shorter the routing distance, the smaller the operational cost.

Objective Function   The three objectives for siting recycling drop-off stations and collection vehicle routing can be further quantified (Chang and Wei, 1999, 2000) in the following way:

- **Maximization of population serviced by those recycling drop-off stations**. This objective depends on the distribution of population density in the service area. Hence, the first objective is to maximize the service ratio based on total residents in the service area.

$$\text{Max} \sum_{k=1}^{K'} \sum_{i=1}^{N'} C_i P_{ik}, \qquad (15.1)$$

where C_i is the representative population allocated to node i in the collection network (Capita) and P_{ik} is the binary integer variable for the selection of node k as a drop-off station for those residents living around node i. P_{ik} is equal to 1 if node i is serviced by the recycling drop-off tank installed at node k in the network; zero otherwise, N' is the number of nodes in the network; and K' is the number of candidate sites that could be picked up as the recycling drop-off stations in the network.

- **Minimization of total walking distance from household to recycling drop-off stations.** Although incorporating environmental awareness into daily lives might be a key factor, the convenience of depositing recyclables for collection, achieved by minimizing the possible walking distance, is also a key factor for the success of a recycling program. Hence, the second objective is formulated to minimize the total walking distance for those residents.

$$\text{Min} \sum_{k=1}^{K'} \sum_{i=1}^{N'} d_{ik} P_{ik}, \qquad (15.2)$$

where d_{ik} is the distance from node i to node k (m), and if the value of P_{ik} is equal to 1, the distance d_{ik} would be counted in the objective function, zero otherwise.

- **Minimization of total driving distance during vehicle routing.** The distribution of recycling drop-off stations is also dependent on the efficiency of collection vehicle routing; however, the collection time is equivalent to the sum of the travel time in the links and the pick-up time at the collection nodes in the network. To facilitate the competitive application, the third objective only minimizes the summation of routing distance required for the collection task. Although the routing distance is defined as vector distance instead of actual distance, the difference can be minimized if the digitized network with higher accuracy in the GIS system is used. Even so, a nonlinear description in this objective function cannot be fully avoided. Higher computational loading within large-scale collection network inevitably requires the use of advanced solution techniques, such as genetic algorithm, to determine the global optimal strategies. This objective function can be formulated as

$$\text{Min} \sum_{t=1}^{M'} \sum_{k=1}^{K'} - \left[\sum_{j=1}^{K'} O_{kt}(X_k - X_j)^2 + \sum_{j=1}^{K'} O_{kt}(Y_k Y_j)^2 \right]^{1/2}, \qquad (15.3)$$

where O_{kt} is the binary integer variable for the description of the order of node k to be visited in sequence by the collection vehicle, which is equal to 1 if the node k with recycling drop-off tanks in the network is collected at the order t, zero otherwise; and (X_k, Y_k) and (X_j, Y_j) represent the coordinate of nodes k and j in the network, respectively. Overall, this objective would result in a nonlinear formulation that could increase the computational complexity in the solution procedure.

Constraint Set   The basic constraint set consists of the node selection condition, capacity limitation, walking distance limit, routing and scheduling, and binary constraints. The configuration of the proposed constraint set in this model can be described as follows (Chang and Wei, 1999, 2000).

1. Upper bound constraint. This constraint ensures that only a limited number of recycling drop-off stations can be sited in the network. The upper bound might be determined in terms of economic, environmental, social, and even political factors as

$$\sum_{k=1}^{K'} S_k \leq M', \qquad (15.4)$$

where S_k is the binary integer variable for the possible selection of node k as a candidate site of recycling drop-off station, which is equal to 1 if the node k is picked up as a recycling drop-off station in the network, zero otherwise; and M' is the selected upper limit of the number of recycling drop-off stations. This constraint ensures that, at most, M' recycling drop-off stations can be picked up from K' candidate sites in the network.

2. Service efficiency constraint. This constraint may avoid the possible overlap of service areas of individual recycling drop-off stations.

$$\sum_{k=1}^{K} P_{ik} \leq 1 \qquad \forall i \in N, \tag{15.5}$$

$$S_k - P_{ik} \geq 0 \qquad \forall i \in N; \forall k \in K, \tag{15.6}$$

where K is the set of all possible numbers of candidate sites for the selection as recycling drop-off stations; and N is the set of all possible numbers of nodes in the network. In general, the value of N is equivalent to the value of K. Equation (15.5) ensures that, at most, one recycling drop-off station at node k can be arranged by an analytical scheme for the recycling service around node i. Equation (15.6) further ensures that if any node i is serviced by the recycling drop-off station installed at node k, then node k should be picked up as the candidate site of recycling drop-off station (i.e., S_k should be equal to 1).

3. Capacity limitation constraint. This constraint limits the total recyclables collected at node k, in case it is picked as a candidate site, which should not exceed the storage capacity provided by those recycling tanks. If post planning of recycling tanks is allowed, however, this constraint may become optional.

$$\sum_{i=1}^{N'} Q_i P_{ik} \leq C_a \qquad \forall k \in k, \tag{15.7}$$

where Q_i is the upper limit of the recyclables that could be collected at node i (kg); and C_a is the largest capacity of collection tanks arranged at each node (kg).

4. Routing and scheduling constraint. This constraint, coordinated with the third objective function, ensures that every node picked up as a recycling drop-off station is visited once in the vehicle routing process.

$$S_k - \sum_{t=1}^{M'} O_{kt} = 0 \qquad \forall k \in k \tag{15.8}$$

$$\sum_{k=1}^{K'} O_{kt} = 0 \qquad \forall t \in M' \tag{15.9}$$

Equation (15.8) ensures that once the recycling drop-off station at node k has been assigned for recycling service, node k should be included in the collection vehicle routing program. Equation (15.9) guarantees that any two recycling drop-off stations cannot have the same routing order t in the collection sequence.

5. Service area constraint. This constraint guarantees that the service radius of each recycling drop-off station is limited to a specified distance. Without this constraint, even if the walking distance can be minimized in the objective function, many centralized drop-off stations would result without regard to the equity principles.

$$P_{ik}d_{ik} \leq r_k \qquad \forall i \in N; \forall k \in K \qquad (15.10)$$

$$a \leq r_k \leq b \qquad \forall k \in k, \qquad (15.11)$$

where r_k is the service radius of drop-off tanks at node k (m); and a and b are the upper and lower bounds of service radius (m).

6. Non-negativity and binary constraint. All the decision variables (i.e., O_{kt}, S_k, and P_{ik}) are non-negative and binary variables.

15.3.2 Solution Procedure

Genetic algorithm (GA) has to be employed to solve this multiobjective, nonlinear mixed integer programming model. GA was originally developed in the field of artificial intelligence by Holland (1975) and was designed to simply find the best possible solution instead of the optimal solution when dealing with nonlinear optimization problems. The major assumption of GA is that the process for searching a global optimal solution in a multidimensional domain is similar to an evolutionary process in a biological organism. Similar to the selection of the most advantageous chromosomes over generations, numerical solutions are continuously improved by a conventional steep-ascent approach. Once the decision variables are coded, as chromosomes would be in a genetic structure, the evolutionary process can then be designed as a global search process for finding the optimal solution.

GA has recently received wide attention in environmental planning and management (Ritzel and Eheart, 1994; Cieniawski et al., 1995; Dandy et al., 1996). The merits of GA include, among others, (1) GA is an efficient global method for nonlinear optimization problems and is able to search for a global optimal solution with a simple algorithm in which linearization assumptions and the calculation of partial derivatives are not required; (2) GA may avoid numerical instabilities associated with essential matrix inversion, a problem frequently encountered in conventional mathematical programming algorithms; and (3) GA is much more efficient and robust in search of the global optimal solution compared with conventional Monte Carlo simulation and previous optimization algorithms used for solving nonlinear programming models.

The use of GA technology to solve TSP and VRP is actually a relatively new topic of system analysis but applies the criteria of genetic evolution in an iterative procedure in search of the global optimal solution (Goldberg and Lingle, 1985; Grefenstette et al., 1985; Current et al., 1987; Oliver et al., 1987; Thangiah, 1995). For dealing with the complexity of a multiobjective, nonlinear mixed-integer programming model like the one used in this research, GA is an advanced tool for finding the best possible solution.

In the GA solution procedure for solving the multiobjective, nonlinear mixed-integer programming models (Figure 15.5), the traditional operators of crossover and mutation are designed as tools in the optimization process. More sophisticated operators for crossover and mutation can be found in the literature (Michalewicz and Janikow, 1991; Starkweather et al., 1991; Smith and Tate, 1993; Homaifar et al.,

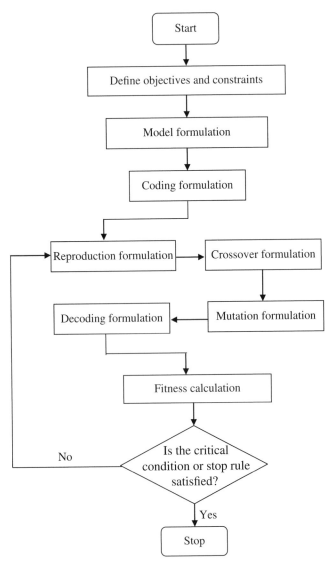

FIGURE 15.5 The GA-based multiobjective evaluation procedure. *Source:* Chang and Wei (1999)

1994). The possible alteration of the mutation rate and the selection of different types of crossover operations would result in a different convergence speed in the computational process; however, one of the most salient weaknesses of GA as a solution procedure is that it is a problem-based solution procedure, and different optimization formulas may require different arrangements of coding structure in the programming effort (Chang and Wei, 1999). A special technique was therefore applied in this analysis in which the violation of constraint in the fitness function is balanced by choosing a reference value (Homaifar et al., 1994).

The design for the location strategy of recycling drop-off stations and collection vehicle routing and scheduling involves both implicit and explicit data integration in the GIS environment (Chang and Wei, 1999, 2000). To support the analysis, GIS is capable of calculating realistic population densities, waste generation rates, waste distributions, and recycling potentials. Hence, map digitizing and editing procedures support the cartographic creations of a network database for these service areas (Chang and Wei, 1999, 2000). Characteristics of the population living within these service areas must be investigated and stored in the GIS database to predict spatially distributed sources of waste generation. The ARC/INFO® system used in this analysis assigns waste distribution attributes to each collection point in the network cartography so that the management scenarios can then be optimally generated using an external multiobjective programming model base (Chang and Wei, 1999, 2000). The true cost of building a topologically and directionally correct road network and associated population and waste attributes in a GIS environment should be justified by the inherent cost-saving through the use of such a multiobjective evaluation analysis (Chang and Wei, 1999, 2000).

15.3.3 Planning Scenarios, Assessment Metrics, and Planning Outcome

The multiobjective programming model used for planning curbside recycling programs can answer three broad classes of questions regarding the curbside recycling programs: (1) What is the best siting pattern of recycling drop-off stations? (2) How should a recycling coordinator allocate collection tanks in the network? and (3) What is the best routing pattern using the dedicated collection vehicles? Each of these questions is discussed below in relation to our study of solid waste recycling in the same proposed district applied in Section 15.2 (Chang and Wei, 1999, 2000).

Planning Scenarios   Five planning scenarios were presented for this case study. Cases 1–3 examine the optimal distribution strategies with respect to the fixed-service radius, capacity of each recycling container, and the equality condition of upper bound constraint determining the maximum allowable number of recycling drop-off stations. Case 4 was designed to generate the optimal distribution strategies for different crossover rates and inequality conditions of the upper bound constraint. Case 5 specifically applied varying mutation rates with different crossover operations (i.e., single- and double-point mutations) to investigate the sensitivity of the parameters in GA (Chang and Wei, 1999, 2000).

Assessment Metrics   Five performance indices similar to those used in Section 15.2 were defined for the comparative evaluation of management alternatives. The first index describes the service ratio that represents the percentage of population serviced by recycling drop-off stations; the second index defines the predicted usage rate of collection tanks with fixed capacity in the management scenarios; the third index presents the average walking distance of residents from their households to recycling containers to evaluate social feasibility; the fourth index helps estimate the percentage of recyclables to be collected by the curbside recycling program; and the

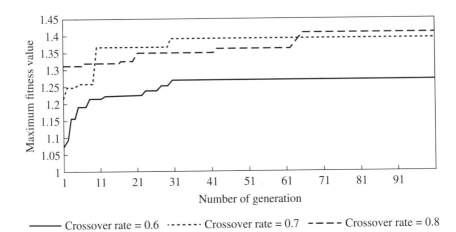

FIGURE 15.6 Evolutionary process of GA with varying crossover rate in Case 4. *Source: Chang and Wei (1999)*

fifth index depicts the routing ratio of the total routing distance to the total distance in the network links (Chang and Wei, 1999, 2000).

Planning Outcome The evolutionary process of maximum fitness values corresponding to Case 4 (Figure 15.6) indicates that the speed of convergence in both cases is reasonably good due to the appropriate arrangement of coding structure for both objective functions and constraints (Chang and Wei, 1999, 2000). All cases converge as expected. The planning outputs for different ranges of service radius and container capacity (Tables 15.5–15.9) indicate that in Cases 1–3, the allowable number of

TABLE 15.5 Analytical results of Case 1

	Service radius of each drop-off station:100–200 m			
	Capacity/tank: 1,250 kg allowable drop-off stations (N)			
Condition	5	10	30	50
Population served by drop-off stations in district (N)	4,958	13,891	56,780	65,413
Service ratio (%)	3.78	10.58	43.24	49.82
Collected recyclables (kg)	2,633	6,876	26,749	32,744
Recycling rate (%)	4.45	11.63	45.27	55.42
Utilization rate of tanks (m)	42.13	55.01	71.32	52.39
Average walking distance (m)	66.92	81.44	85.06	81.08
Routing distance per collection vehicle (m)	5,208	7,357	12,595	16,439
Routing ratio (%)	4.79	6.76	11.58	15.12

Source: Chang and Wei (1999).

TABLE 15.6   Analytical results of Case 2

Condition	Service radius of each drop-off station: 50–250 m			
	Capacity/tank: 2,000 kg allowable drop-off stations (N)			
	5	10	30	50
Population served by drop-off stations in district (N)	8,757	15,914	63,667	76,154
Service ratio (%)	6.67	12.12	48.49	58.00
Collected recyclables (kg)	5,091	6,976	34,113	39,410
Recycling rate (%)	8.61	11.81	57.74	66.70
Utilization rate of tanks (m)	50.91	44.88	56.85	39.41
Average walking distance (m)	91.88	57.46	102.13	77.90
Routing distance per collection vehicle (m)	5,695	6,881	13,216	18,998
Routing ratio (%)	5.24	6.33	12.15	17.47

Source: Chang and Wei (1999).

recycling drop-off stations was fixed so that the sensitivity of various types of planning scenarios for these five performance indices become comparable (Chang and Wei, 1999, 2000). As a consequence, the service ratio, recycling rate, and routing ratio are roughly proportional to the number of drop-off stations selected in the network; however, the highest usage rate of containers and average walking distance exists when 30 recycling drop-off stations are used in the network (Chang and Wei, 1999, 2000).

TABLE 15.7   Analytical results of Case 3

Condition	Service radius of each drop-off station: 100–300 m			
	Capacity/tank: 1,250 kg allowable drop-off stations (N)			
	5	10	30	50
Population served by drop-off stations in district (N)	8,757	15,914	63,667	76,154
Service ratio (%)	6.67	12.12	48.49	58.00
Collected recyclables (kg)	5,091	6,976	34,113	39,410
Recycling rate (%)	8.61	11.81	57.74	66.70
Utilization rate of tanks (m)	50.91	44.88	56.85	39.41
Average walking distance (m)	91.88	57.46	102.13	77.90
Routing distance by collection vehicle (m)	5,695	6,881	13,216	18,998
Routing ratio (%)	5.24	6.33	12.15	17.47

Source: Chang and Wei (1999).

TABLE 15.8   Analytical results of Case 4

Condition	Service radius of each drop-off station: 100–200 m		
	Capacity/tank: 1,250 kg Upper limit of number of drop-off stations allowable: 50 Crossover rate		
	0.6	0.7	0.8
Optimal number of drop-off stations	26	31	37
Population served by drop-off stations in district (N)	42,277	53,721	63,404
Service ratio (%)	32.20	40.91	48.29
Collected recyclables (kg)	21,342.75	27,066.88	32,116.00
Recycling rate (%)	36.12	45.81	54.36
Utilization rate of tanks (m)	65.67	69.85	69.44
Average walking distance (m)	79.83	87.25	82.69
Routing distance by collection vehicle (m)	10,228	11,548	15,188
Routing ratio (%)	9.40	10.62	13.97

Source: Chang and Wei (1999).

TABLE 15.9   Analytical results of Case 5

Condition	Service radius of each drop-off station: 100–200 m			
	Capacity/tank: 1,250 kg Upper limit of no. of drop-off stations allowable: 50 Crossover rate			
	0.8	0.8	0.8	0.8
Mutation rate crossover	0.01	0.005	0.01	0.01
Operation	One point	One point	One point	Two point
Optimal no. of drop-off stations	37	34	37	42
Population served by drop-off stations in district (N)	42,404	55,576	63,404	71,922
Service ratio (%)	48.29	42.33	48.29	54.78
Collected recyclables (kg)	32,116.00	27,509.90	32,116.00	34,951.73
Recycling rate (%)	54.36	46.56	54.36	59.16
Utilization rate of tanks (m)	69.44	64.73	69.44	66.57
Average walking distance (m)	82.69	86.97	82.69	80.75
Routing distance by collection vehicle (m)	15,188	13,170	15,188	17,000
Routing ratio (%)	13.97	12.11	13.97	15.64

Source: Chang and Wei (1999).

Case 2 allows a larger service radius and container capacity, and the higher the recycling rate, the larger the service and routing ratios (Table 15.5) (Chang and Wei, 1999, 2000). The results of the sensitivity of service radius (Table 15.6) indicates that the use of 30 recycling drop-off stations may require the longest average walking distance, resulting in lower social feasibility (Chang and Wei, 1999, 2000). The change of crossover rate under different conditions (Table 15.8), however, indicates only a slight influence on the optimal number of selected drop-off stations. Yet other indices, such as the service ratio, recycling rate, and routing ratio, present relatively different degrees of sensitivity (Chang and Wei, 1999, 2000). The mutation rate and crossover operation are relatively less sensitive to the optimal distribution pattern of recycling containers (Table 15.9) (Chang and Wei, 1999, 2000).

According to Chang and Wei (1999), the suggested network for recycling drop-off stations for the Case 4 scenario (Figure 15.7) depicts the optimal pattern along with the service radius and detailed collection routing in the proposed district, which can be systematically updated with GIS should the parameter values change in the multiobjective evaluation scheme (Chang and Wei, 1999). Each Arabic number marked beside the recycling drop-off station represents the order of collection sequence in the routing procedure (Chang and Wei, 1999). The zig-zag collection route that starts at node 1 in the northern part of the map and terminates at a node 31 in the southwestern part of the map represents the optimal collection vehicle routing in the network (Chang and Wei, 1999). The number of collection vehicles required to support this collection operation depends on the normal loading capacity of each truck, the amount of recyclables distributed in the network, and the frequency of collection. A higher percentage of the population (based on the population inside the shaded circles in Figure 15.7) and shorter walking distances can be achieved in the network with respect to both economic, social and technical feasibilities when compared to the counterpart of the simulation analysis. Decision makers can use the magnitude of the four performance indices (Figure 15.8) comparing Cases 1–3 to make final recommendations (Chang and Wei, 1999). Solving large-scale VRPs with time windows may be a focal point in relation to supply chain management and demand market of recyclables in future research (Gendreau and Tarantilis, 2010).

15.4 FINAL REMARKS

Various recycling strategies have been evaluated in this chapter. The key issues discussed and analyzed were presented using a local-level, generic facility location problem of locational strategies of routes and collection nodes for curbside recycling bins, managed by either a privatized or a public program. Although the simulation approach using only the GIS tool (Section 15.2) can arrive at a systematic planning outcome, a multiobjective nonlinear mixed-integer programming model (Section 15.3) was formulated for three objectives to maximize population served, minimize walking distance, and minimize total routing distance for the collection vehicle. These objectives were subject to several physical constraints providing an even more effective planning for the distribution of recycling containers (Chang and Wei, 1999).

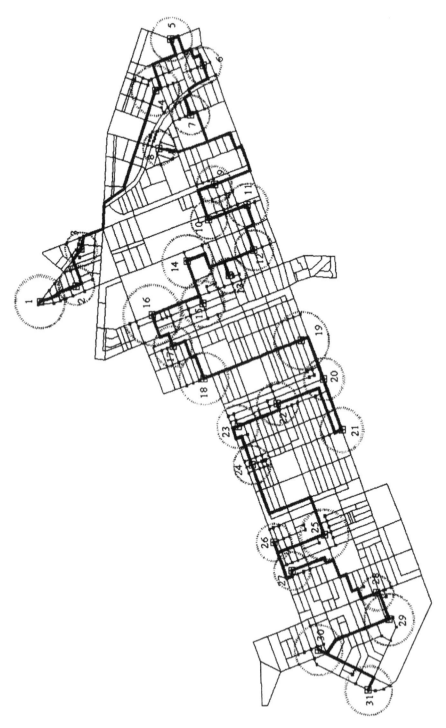

FIGURE 15.7 GIS outputs of location/allocation information of recycling drop-off stations in Case 4 (crossover rate = 0.7). *Source:* Chang and Wei (1999)

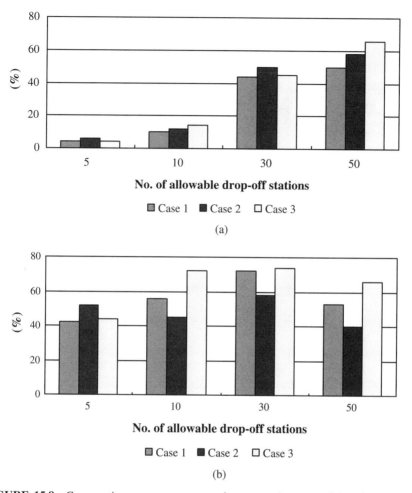

FIGURE 15.8 Comparative assessment across three cases in terms of four indexes: (a) service ratio, (b) utilization rate, (c) average walking distance, and (d) routing ratio. *Source:* Chang and Wei (1999)

GA for determining recycling routes as well as allocating and sizing those drop-off stations were employed to solve the model. As part of the user interface to support such an analytical framework, GIS is used to perform data entry, integration, analysis, and display, for receiving data from modeling systems or data base management systems, and for depicting model results by generating cartographic products (Section 15.3). This GIS environment associated with the GA-based optimization technique constitutes a novel and convenient computerized platform to achieve the strategic planning goal. With the aid of this environmental informatics platform, the managerial trend can be linked to various integrated simulation and optimization models. The planning scheme would contribute to operational cost savings and carbon footprint

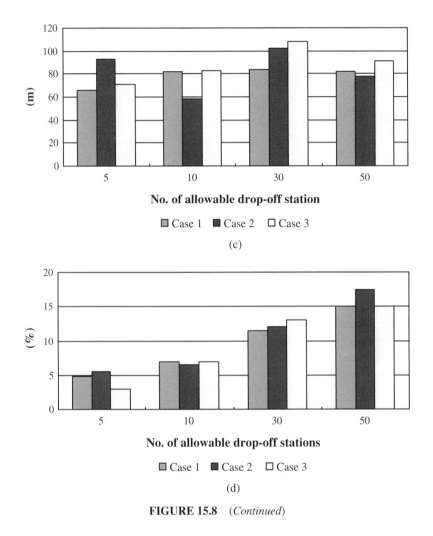

FIGURE 15.8 *(Continued)*

reduction under resource limitation in an SWM system. The advances of Internet of Things technologies may empower such a platform further for achieving some high-end network operational missions. Overall, sufficient insight gained in the application provides additional system thinking in decision-making.

REFERENCES

Achuthan, N. R. and Caccetta, L. 1991. Integer linear programming formulation for a vehicle routing problem. *European Journal of Operational Research*, 52(1), 86–89.

Angelelli, E. and Speranza, M. G. 2002. The application of a vehicle routing model to a waste-collection problem: two case studies. *Journal of the Operational Research Society*, 53(9), 944–952.

Arribas, C. A., Blazquez, C. A., and Lamas, A. 2010. Urban solid waste collection system using mathematical modelling and tools of geographic information systems. *Waste Management & Research*, 28(4), 355–363.

Baldacci, R., Bodin, L., and Mingozzi, A. 2006. The multiple disposal facilities and multiple inventory locations rollon–rolloff vehicle routing problem. *Computers & Operations Research*, 33(9), 2667–2702.

Bartlett, D. 2004. *GIS for Coastal Zone Management*, Taylor & Francis, Boca Raton, FL.

Bautista, J., Fernández, E., and Pereira, J. 2008. Solving an urban waste collection problem using ants heuristics. *Computers & Operations Research*, 35(9), 3020–3033.

Beltrami, E. J. and Bodin, L. D. 1974. Networks and vehicle routing for municipal waste collection. *Networks*, 4, 65–94.

Benjamin, A. M. 2011. Metaheuristics for the waste collection vehicle routing problem with time windows. PhD thesis, Department of Mathematical Sciences, Brunel University.

Benjamin, A. M. and Beasley, J. E. 2010. Metaheuristics for the waste collection vehicle routing problem with time windows, driver rest period and multiple disposal facilities. *Computers & Operations Research*, 37(12), 2270–2280.

Bodin, L. D. and Kursh, S. J. 1978. A computer-assisted system for the routing and scheduling of street sweepers. *Operation Research*, 26(4), 525–537.

Bodin, L., Mingozzi, A., Baldacci, R., and Ball, M. 2000. The Rollon–Rolloff vehicle routing problem. *Transportation Science*, 34(3), 271–288.

Brodie, G. R. and Waters, C. D. J. 1988. Integer linear programming formulation for vehicle routing problems. *European Journal of Operational Research*, 34(3), 403–404.

Cargin, J. and Dwyer, J. 1995. Pennsylvania's low-level radioactive waste disposal facility siting project: special GIS operations. In: Proceedings of 1995 ESRI User Conference, Red Wood City, CA pp. 162.

Chang, Y. H. and Chang, N. B. 1998. Optimization analysis for the development of short-term solid waste management strategies using presorting process prior to incinerator. *Resources Conservation and Recycling*, 24(1), 7–32.

Chang, N. B. and Lin, Y. T. 1997. Optimal siting of transfer station locations in a metropolitan solid waste management system. *Journal of Environmental Science and Health*, 32(8), 2379–2401.

Chang, N. B. and Wang, S. F. 1996a. Solid waste management system analysis by multiobjective mixed integer programming model. *Journal of Environmental Management*, 48(1), 17–43.

Chang, N. B. and Wang, S. F. 1996b. Comparative risk analysis of solid waste management alternatives in a metropolitan region. *Environmental Management*, 20(1), 65–80.

Chang, N. B. and Wang, S. F. 1996c. The development of an environmental decision support system for municipal solid waste management. *Computers, Environment, and Urban System*, 20(3), 201–212.

Chang, N. B. and Wei, Y. L. 1999. Strategic planning of recycling drop-off stations by multiobjective programming. *Environmental Management*, 24(2), 247–263.

Chang, N. B. and Wei, Y. L. 2000. Siting recycling drop-off stations in an urban area by genetic algorithm-based fuzzy multiobjective nonlinear programming. *Fuzzy Sets and Systems*, 114(1), 133–149.

Chang, N. B., Lu, H. Y., and Wei, Y. L. 1997a. GIS technology for vehicle routing and scheduling in solid waste collection systems. *Journal of Environmental Engineering-ASCE*, 123(9), 901–910.

Chang, N. B., Chang, Y. H., and Chen, W. C. 1997b. Evaluation of heat values and its prediction for refuse-derived fuel. *The Science of the Total Environment*, 197(1–3), 139–148.

Chang, N. B., Chang, Y. H., and Chen, Y. L. 1997c. Cost-effective and workload balancing operation in solid waste management systems. *Journal of Environmental Engineering, ASCE*, 123(2), 178–190.

Chang, N. B., Wei, Y. L., Tseng, C. C., and Kao, C. Y. 1998. The design of a GIS-based decision support system for chemical emergency preparedness and response in an urban environment. *Computers, Environment and Urban System*, 21(1), 1–28.

Chang, N. B., Davila, E., Dyson, B., and Brown, R. 2005. Optimal site selection and capacity planning of a municipal solid waste material recovery facility in an urban setting. *Waste Management*, 25(8), 833–846.

Chiplunkar, A. V., Mehndiratta, S. L., and Khanna, P. 1981. Optimization of refuse collection systems. *Journal of Environmental Engineering ASCE*, 107(6), 1203–1210.

Cieniawski, S. E., Eheart, W., and Ranjithan, S. 1995. Using genetic algorithms to solve a multiobjective groundwater monitoring problem. *Water Resources Research*, 31(2), 399–409.

Current, J. R., ReVelle, C. S., and Cohon, J. L. 1987. The median shortest path problem: a multiobjective approach to analyze cost vs. accessibility in the design of transportation networks. *Transportation Science*, 21(3), 188–197.

Dandy, G. C., Simpson, A. R., and Murphy, L. J. 1996. An improved genetic algorithm for pipe network optimization. *Water Resources Research*, 32(2), 449–458.

De Meulemeester, L., Laporte, G., Louveaux, F. V., and Semet, F. 1997. Optimal sequencing of skip collections and deliveries. *Journal of the Operational Research Society*, 48(1), 57–64.

Dedic, A., Murn, R., and Pecek, D. 1992. Map data processing in a geographic information system environment. In: *Computer Techniques in Environmental Studies IV* (Ed. Zannetti, P.), Computation Mechanics Publication, Portsmouth, UK, pp. 819–825.

Desrochers, M., Lenstra, J. K., Savelsbergh, M. W. P., and Soumis, F. 1987. Vehicle routing with time windows: optimization and approximation. Department of Operations Research and System Theory, Centrum voor Wiskunde en Informatica.

Desrochers, M., Desrosiers, J., and Solomon, M. 1992. A new optimization algorithm for the vehicle routing problem with time windows. *Operations Research*, 40(2), 342–354.

Downer, R., Kurtz, C., and Ferson, S. 1992. Integration of environmental models in geographical spreadsheet. In: *Computer Techniques in Environmental Studies IV* (Ed. Zannetti, P.), Computation Mechanics Publication, Portsmouth, UK, pp. 797–804.

Dror, M. 1993. Modeling vehicle routing with uncertain demands as a stochastic program: properties of the corresponding solution. *European Journal of Operational Research*, 64(3), 432–441.

El Fallahi, A., Prins, C., and Calvo, R. W. 2008. A memetic algorithm and a tabu search for the multi-compartment vehicle routing problem. *Computers & Operations Research*, 35(5), 1725–1741.

Environmental Systems Research Institute (ESRI). 1992. Network analysis: modeling network system. *ARC/INFO∗ User's Guide*, Redlands, CA.

Feiring, B. R. 1990. An efficient procedure for obtaining feasible solutions to the n-city traveling salesman problem. *Mathematical Computer Modelling*, 13(3), 67–71.

Fisher, M. L. 1981. The Lagrangian relaxation method for solving integer programming problems. *Management Science*, 27(1), 1–18.

Gendreau, M. and Tarantilis, C. D. 2010. Solving large-scale vehicle routing problems with time windows: the state-of-the-art. CIRRELT. Available at: https://www.cirrelt.ca/DocumentsTravail/CIRRELT-2010–04.pdf (accessed March 2010).

Gertsakis, J. and Lewis, H. 2003. Sustainability and the waste management hierarchy. Available at: http://ecorecycle.vic.gov.au/resources/documents/TZW_-_Sustainability_and_the_Waste_Hierarchy_%282003%29.pdf (accessed March 2009).

Goldberg, D. E. and Lingle, R., Jr. 1985. Alleles, loci, and the traveling salesman problem. In: *Proceedings of the First International Conference on Genetic Algorithms and other Applications*, Carnegie-Mellon University, Pittsburgh, PA0, pp. 154–159.

Golden, B. L., Assad, A. A., and Wasil, E. A. 2002. Routing vehicles in the real world: applications in the solid waste, beverage, food, dairy, and newspaper industries. In: *The Vehicle Routing Problem* (Eds. Toth, P. and Vigo, D.), Society for Industrial Mathematics, pp. 245–286.

Grefenstette, J., Gopal, R. Rosmaita, B., and Gucht, D. V. 1985. Genetic algorithm for the traveling salesman problem. In: Proceedings of the First International Conference on Genetic Algorithms and other Applications, Carnegie-Mellon University, Pittsburgh, PA, pp. 160–168.

Hass, W., Schewig, D., and Resch, M. M. 1992. Numerical simulation of ground water flow and ground water pollution in a graphical software environment. In: *Computer Techniques in Environmental Studies IV* (Ed. Zannetti, P.), Computation Mechanics Publication, Portsmouth, UK, pp. 827–841.

Holland, J. H. 1975. *Adaptation in Natural and Artificial Systems*, University of Michigan Press, Ann Arbor, MI.

Homaifar, A., Qi, C. X., and Lai, S. H. 1994. Constrained optimization via genetic algorithm. *Simulation*, 62(4), 242–254.

Hromadka, T. V., Whitley, R. J., Harryman, R. R., and Braksator, M. J. 1992. Application of a graphics data base management system: computerized master plan of drainage. In: *Computer Techniques in Environmental Studies IV* (Ed. Zannetti, P.), Computation Mechanics Publication, Portsmouth, UK, pp. 805–817.

Johnston, E. M. 1987. Natural resource modeling in the geographical information system environment. *Photogrammetric Engineering and Remote Sensing*, 53(10), 1411–1415.

Kilborn, K., Rifai, H. S., and Bedient, P. B. 1992. Connecting groundwater models and GIS. Geo Info Systems, February, 26–30.

Kim, B. I., Kim, S., and Sahoo, S. 2006. Waste collection vehicle routing problem with time windows. *Computers & Operations Research*, 33(12), 3624–3642.

Kulcar, T. 1996. Optimizing solid waste collection in Brussels. *European Journal of Operational Research*, 90(1), 71–77.

Li, J. Q., Borenstein, D. and Mirchandani, P. B. 2008. Truck scheduling for solid waste collection in the City of Porto Alegre, Brazil. *Omega*, 36(6), 1133–1149.

Liebman, J. C., Male, J. W., and Wathne, M. 1975. Minimum cost in residential refuse vehicle routes. *Journal of Environmental Engineering ASCE*, 101(3), 399–411.

Lu, G. Y. 1996. System planning for medical waste collection in the metropolitan region. Master Thesis, Department of Resources Engineering, National Cheng-Kung University, Tainan, Taiwan, R.O.C.

Lu, J. W., Chang, N. B., and Liao, L. 2013. Environmental informatics for solid and hazardous waste management: advances, challenges, and perspectives. *Critical Reviews in Environmental Science and Technology*, 43(15), 1557–1656.

Lupien, Y. E., Moreland, W. H., and Dangermond, J. 1987. Network analysis in geographical information systems. *Photogrammetric Engineering and Remote Sensing*, 53(10), 1417–1421.

Lysgaard, J. 1992. Dynamic transportation networks in vehicle routing and schedule. *Interfaces*, 22(3), 45–55.

MacDonald, M. L. 1996. A multi-attribute spatial decision support system for solid waste planning. *Computers, Environment and Urban Systems*, 20(1), 1–17.

Madsen, O. B. G. 1983. Methods for solving combined two level location-routing problems of realistic dimensions. *European Journal of Operational Research*, 12(3), 295–301.

Massie, K. 1995. Using GIS to improve solid waste management and recycling programs. In: Proceedings of 1995 ESRI User Conference, California, pp. 18.

Michalewicz, Z. and Janikow, C. Z. 1991. Handling constraints in genetic algorithm. In: Proceedings of the Fourth International Conference on Genetic Algorithms, University of California, San Diego, CA, pp. 151–157.

Mourão, M. C. and Almeida, M. T. 2000. Lower-bounding and heuristic methods for a refuse collection vehicle routing problem. *European Journal of Operational Research*, 121(2), 420–434.

Mourão, M. C., Nunes, A. C., and Prins, C. 2009. Heuristic methods for the sectoring arc routing problem. *European Journal of Operational Research*, 196(3), 856–868.

Muyldermans, L. and Pang, G. 2010a. A guided local search procedure for the multi-compartment capacitated arc routing problem. *Computers & Operations Research*, 37(9), 1662–1673.

Muyldermans, L. and Pang, G. 2010b. On the benefits of co-collection: experiments with a multi-compartment vehicle routing algorithm. *European Journal of Operational Research*, 206(1), 93–103.

Muzik, I. 1994. HYDROGGISS - hydrograph generating GIS software. In: *Environmental Systems* (Ed. Zannetti, P.), Vol. 2. Computation Mechanics Publication, Portsmouth, UK, pp. 311–318.

Oliver, I. M., Smith, D. J., and Holland, J. R. C. 1987. A study of permutation crossover operators on the traveling salesman problem. In: Proceedings of the Second International Conference on Genetic Algorithms and Other Applications, Massachusetts Institute of Technology, Cambridge, MA, pp. 224–230.

Ong, H. L., Goh, T. N., and Poh, K. L. 1990. A computerised vehicle routing system for refuse collection. *Advances in Engineering Software*, 12(2), 54–58.

ReVelle, C., Cohon, J., and Shobrys, D. 1991. Simultaneous siting and routing in the disposal of hazardous wastes. *Transportation Science*, 25(2), 138–145.

Ritzel, B. J. and Eheart, J. W. 1994. Using genetic algorithms to solve a multiple objective groundwater pollution containment problem. *Water Resources Research*, 30(5), 1589–1603.

Schrage, L. 1981. Formulation and structure of more complex/realistic routing and scheduling problems. *Network*, 11(2), 229–232.

Smith, A. E. and Tate, D. M. 1993. Genetic optimization using a penalty function. In: Proceedings of the Fifth International Conference on Genetic Algorithms, University of Illinois at Urbana-Champaign, USA, pp. 499–503.

Solomon, M. M. 1987. Algorithms for the vehicle routing and scheduling problems with time window constraints. *Operations Research*, 35(2)254–265.

Starkweather, T., McDaniel, T., and Whitley, C. 1991. A comparison of genetic sequencing operators. In: Proceedings of the Fourth International Conference on Genetic Algorithms, University of California, San Diego, CA, pp. 69–76.

Summer, W. 1994. GIS and soil erosion models as tools for the development of soil conservation strategies. In: *Environmental Systems* (Ed. Zannetti, P.), Vol. 2. Computation Mechanics Publication, Portsmouth, UK, pp. 303–310.

Teixeira, J., Antunes, A. P., and de Sousa, J. P. 2004. Recyclable waste collection planning—a case study. *European Journal of Operational Research*, 158(3), 543–554.

Thangiah, S. R. 1995. An adaptive clustering method using a geometric shape for vehicle routing problems with time window. In: Proceedings of the Sixth International Conference on Genetic Algorithms, University of Pittsburgh, USA, pp. 536–543.

Tsakiris, G. and Salahoris, M. 1993. GIS technology for management of water distribution networks. In: *Water Supply Systems* (Eds. Cabrera, E. and Martinez, K.), Computation Mechanics Publication, Portsmouth, UK, pp. 361–378.

Tung, D. V. and Pinnoi, A. 2000. Vehicle routing-scheduling for waste collection in Hanoi. *European Journal of Operational Research*, 125(3), 449–468.

Zhang, J. and Parks, Y. 1994. Dynamic linking between GIS and surface water database. In: *Environmental Systems* (Ed. Zannetti, P.), Vol. 2. Computation Mechanics Publication, Portsmouth, UK, pp. 319–327.

CHAPTER 16

MULTIATTRIBUTE DECISION-MAKING WITH SUSTAINABILITY CONSIDERATIONS

Decision analysis is an essential tool in support of many solid waste management (SWM) practices. Planning scenarios for SWM, such as how to choose the best treatment technology or organize a technology portfolio out of a contemporary technology hub, are some of the typical practices in which decision makers are key players. Decision-making based on a single criterion, usually the cost and benefit factors, is not sufficient anymore although cost-benefit analysis is still the core tool in decision-making arena to date. Factors, such as emission standards, recycling targets, and diversion requirements of biodegradable organic waste streams from landfills, must be taken into account in many occasions. In addition, the selection of the best treatment technology requires social acceptance; social syndromes, such as "not in my back yard" and "build absolutely nothing anywhere near anyone", can be even more difficult to deal with than cost limitation and environmental legislation. Decision-making criteria might be conflicting with one another so that no single alternative can be unanimously accepted among stakeholders. This chapter demonstrates how multiattribute decision analysis can be formalized and applied to align the disparity of decision-making criteria and alternatives for SWM. Two case studies are introduced; one devoted to construction and demolition waste (CDW) planning in Germany, and the other conducted a comparison of waste collection systems in Portugal. Both cases present the application potential of multiattribute decision analysis.

Sustainable Solid Waste Management: A Systems Engineering Approach, First Edition. Ni-Bin Chang and Ana Pires.
© 2015 The Institute of Electrical and Electronics Engineers, Inc. Published 2015 by John Wiley & Sons, Inc.

16.1 DETERMINISTIC MULTIPLE ATTRIBUTE DECISION-MAKING PROCESS

Multiple criteria decision-making (MCDM) can solve two different types of problems: those that select/evaluate a finite number of alternatives or those that find the "best" alternative. The method used to solve the first type of problems is multiple attribute decision-making (MADM), and the method used to solve the second type of problems is multiple objective decision-making (MODM). MCDM can be viewed as a decision support system (DSS) capable of dealing with complex problems featuring high uncertainty, conflicting objectives, and different forms of data (Wang et al., 2009a). Whereas MADM methods analyze explicit alternatives' attributes, MODM considers various interactions within the design constraints that best satisfy the decision maker by attaining acceptable levels of a set of quantifiable objectives (Hwang and Yoon, 1981). In a broader generalization, a DSS can also provide a wealth of sociotechnical inputs for tackling complexity embedded in biophysical and socioeconomic systems. According to Wang et al. (2009a), sustainable SWM systems can be possibly formulated to be a well-structured model to promote an integrated evaluation via MCDM.

Recent decision analyses in SWM systems are intimately tied to the concept of sustainable development (Pires et al., 2011; Chang et al., 2011, 2012); more indicators being developed and used in the context of sustainable development can be factored into MADM. The growth of synoptic applications also leads to the development of various algorithms with complexities in the interfaces across several dimensions in MADM (Figure 16.1). Nevertheless, the MADM process is always composed of some common elements (Hwang and Yoon, 1981): (1) alternatives (a limited group of possible solutions), (2) criteria (each problem has a group of independent attributes that must be fulfilled by the alternatives), (3) units (each criterion can be quantified

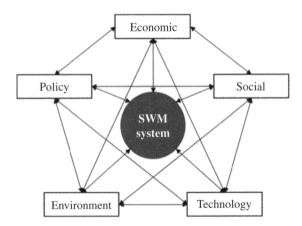

FIGURE 16.1 The complex interactions in SWM systems

in different units), and (4) weights (the relative importance of each criterion). Based on Wang et al. (2009a), the MADM process is designed to integrate all the elements into a single metric to rank alternatives that comply with the goals. The multiattribute decision analysis can be divided into four main stages: (1) criteria selection, (2) criteria weighting, (3) evaluation, and (4) interpretation.

The four main stages have been applied sequentially for managing municipal solid waste (MSW), with special focus on MADM methodologies applied to SWM systems:

- for versatile assessment of waste management strategies (Hokkanen and Salminen, 1997; Karagiannidis and Moussiopoulos, 1997; Skordilis, 2004; Contreras et al., 2008; Pires et al., 2011; Madadian et al., 2013);
- for the selection of SWM technologies and operations (Courcelle et al., 1998; Generowicz et al., 2011); and
- for characterization of different types of waste, like medical waste (Dursun et al., 2011).

Practical implementation includes ranking treatment alternatives (Herva and Roca, 2013), sorting planning alternatives (Karmperis et al., 2012), managing waste streams (Hung et al., 2007; Hanan et al., 2013), and compensatory fund (i.e., fair fund) distribution due to waste treatment plant (Chiueh et al., 2008; Chang et al., 2009). The two case studies we present in this chapter will demonstrate how sustainable management and technical options to collect CDW can be assessed with the aid of MADM.

16.1.1 Criteria Selection

Several criteria can be used to evaluate alternatives for managing sustainable SWM systems. Usually, the method or assumptions used to select criteria are not only reported in publications, but also described as an emerging knowledge base as our understanding of sustainability evolves over time. Criteria may be chosen by decision makers and/or stakeholders involved in the decision-making process; however, the greater the more criteria, the higher the complexity. The following generic principles are used to select the "major" criteria for SWM decision-making (Ye et al., 2006; Jin and Wei, 2008):

- **Systemic principle.** The relevant criteria need to thoroughly mirror the principal features and the entire SWM performance. An inclusive evaluation function of multiple attributes should obtain improved results than the sum of single criteria evaluations.
- **Consistency principle.** The relevant criteria should be compatible with the decision makers' goals.

- **Independency principle.** The relevant criteria should not have equal all-inclusive relationship as other criteria at the same level. The relevant criteria should mirror the alternatives performance from versatile aspects.
- **Measurability principle.** The relevant criteria should be quantifiable as quantitative values if possible or qualitatively expressed.
- **Comparability principle.** The more the rationale, the higher the comparability with the relevant criteria. The relevant criteria should be normalized to compare or operate immediately with both benefit criteria and cost criteria.

Although these fundamental principles are considered, criteria to be selected must also reflect the concept of sustainability science and engineering. Selection criteria can be aided by mathematical methods such as the Delphi method, least mean square, minimax deviation, correlation coefficient, grey relational method, analytical hierarchy process (AHP), clustering method, principal component analysis, and rough set method, all of which can be applied to eliminate the relevance among criteria and help select the independent criteria (Wang et al., 2009a).

Developing evaluation criteria and matrices which could measure sustainability is a prerequisite for selecting the best alternative, identifying non-sustainable SWM options, informing decision makers and stakeholders of the integrated performance of the alternatives, and monitoring impacts on the natural and social environments (Wang et al., 2009a). Yet, measuring sustainability in an SWM system is a major challenge when discussing sustainable resources management strategies and consumption with no externality effect. Most evaluation criteria are divided into five categories representing the pillars of sustainability: technical, regulatory, economic, environmental, and social; however, criteria associated with each category are diversified due to the required parameters in MADM related to reliability, comprehensiveness, comprehension, and limitations such as data availability. Some criteria with overlapped implications might be tied to socio-economic, socio-environmental, and economic-environmental aspects simultaneously in final interpretation.

Technical Criteria   A significant number of technical criteria are used to achieve goals and define alternatives to be assessed. The common technical criteria are capacity of pollution prevention, capacity of research and development (R&D), and technology level (Kaya, 2012). Other technical criteria from organizations regarding service quality may be applied to evaluate different solutions when considering options to outsourcing waste stream management (Kaya, 2012). Some examples of technical criteria include:

- **Installed capacity and production capacity**. This criterion reflects the total capacity of the alternative (e.g., equipment, infrastructure) and the function of the alternative.
- **Feasibility**. This criterion is related to the possibility of success of a specific alternative, also known as the "applicability" criterion (Su et al., 2007).

- **Existing resources**. Resources are related to waste treatment technologies available in the specific location of the SWM system; for example, if an SWM alternative intends to produce refuse-derived fuel, an existing cement plant nearby would be an advantage.

- **Landfill space saving**. This criterion refers to the ability to save landfill space in an SWM system (Pires et al., 2011). Rather than focus on the environmental impact of using land resources for landfilling, it extends the lifetime of facility operation in the SWM system.

- **Land demand or required land for waste management**. This criterion focuses on the land resources needed to deposit and manage waste, as well as the location and land required to build the relevant infrastructure. This criterion is also considered an economic criterion when it reflects constraints of high land prices (Rousis et al., 2008) or, conversely, of declining land values in the surrounding area, or even the compatibility of the potential site with the city development plan (Cheng et al., 2003).

- **Adaptability to local conditions**. The efficacy of each SWM system is defined by the specific features of the region (e.g., available quantities of waste for management, minimum required capacity for the system to keep viable (Rousis et al., 2008)).

- **Flexibility**. This criterion is also known as "flexibility technology" (Kaya, 2012) or "independence" (Tseng, 2009). It is related to the ability to adapt each alternative of the SWM system to potential changes in the amount of waste (increasing or decreasing) or "total operating time" (Gomes et al., 2008). This criterion is particularly important to landfills because of its extensibility (Cheng et al., 2003), which relates directly to the flexibility of the landfill's life span.

- **Technology maturity**. This criterion is related to the use of a specific technology over a long enough time so that most of its inherent problems can be fixed during its development. For example, a sanitary landfill is a technology with maturity; however, co-gasification of waste with other fuels like coal is not yet proven, and it is inherently unknown if the reliability of this technology is sustainable. This criterion is also known as "existing experience-reliability" (Rousis et al., 2008) or "reliability" (Cheng et al., 2003).

- **Schedule.** This criterion is characterized by progress in schedule and implementation procedures of the alternative (Su et al., 2007). For managerial purposes, a clear timeline must be identified for the specific waste technologies that affect the SWM.

- **Functionality**. The parameters examined for the proposed SWM system may include "the potential of constant and smooth operation, the requirements in specialized personnel maintenance, the simplicity of operation, the resistance of equipment to time and natural deteriorations, and the expected lifetime of the installation and safety of the installation" (Rousis et al., 2008). Some cases focus on the complexity of the process and operation only (Madadian et al., 2013).

Economic Criteria Economic criteria are devoted to costs and revenues from the waste management operation. Some examples include:

- **Investment costs**. This criterion comprises costs related to the construction of waste management infrastructures, including landfills, purchase of equipment, and all capital goods. It also includes the cost of engineering consultancy services. Personal costs or maintenance costs are not included. Investment costs can also be divided into fixed and variable costs.

- **Operational costs**. This criterion includes the costs of employee wages, consumables like fuels and electric energy, as well as other products and services for the SWM operation. Operational costs also incorporate costs related to maintenance, which ensure SWM operation without errors and failures that could lead to operation suspension. Operational costs can also be divided into fixed and variable costs. Such a breakdown is a commonly applied criterion. Yet, the more detailed breakdown of collection, transportation, treatment, recycling, and disposal costs are preferred in the assessment of some alternatives.

- **Net present value**. This criterion is defined as the present value of a time series (present to future) of cash flows, which is deemed as a standard method for calculating the time value of money to appraise long-term SWM projects (Wang et al., 2009a).

- **Operational revenues**. This criterion refers to system benefits from the overall gains from products resulting from SWM systems, such as recyclables, compost, and electric energy; the waste service charged to the population is not included. Operational revenues are equivalent to the "marketing potential of the byproducts" (Hung et al., 2007), "resource recycling" (Su et al., 2007), or "income to cost ratio" (Madadian et al., 2013); disposal/treatment costs can be considered separately from operational costs (Gomes et al., 2008). Environmental credits from Kyoto Protocol mechanisms that allow greenhouse gas (GHG) emissions to be traded in the carbon market to reduce or limit GHG emissions can also be included in this criterion. For example, a cement plant that uses refuse-derived fuel rich in biogenic carbon to replace fossil fuels receives a credit on carbon dioxide emissions (because fossil carbon dioxide was substituted by biogenic carbon dioxide) to sell in the carbon market.

- **Total net cost**. Investment cost, operational costs, and operational revenues can all be applied to calculate the total net cost for comparing all the alternatives on the same basis. This criterion is sometimes called "net production cost" (Generowicz et al., 2011).

- **Full cost accounting**. This is a method of quantifying total monetary costs of resources utilized or consigned for the SWM system, including costs from direct and indirect operating costs from the system, and past and future expenses. If possible, it should also include environmental and social externalities, which today are usually assessed in monetary terms.

Regulatory Criteria   Decision makers need to know if the SWM alternatives meet the regulatory requirements. Regulatory criteria applied to SWM case studies include:

- **Harmonization with the existing institutional/legislative framework**. Regulatory requirements can be set up at global, regional, national, and local scales in connection to regulatory directives, waste management strategies, and laws. A regulatory criterion can also be considered a social criterion in specific cases when they are highly tied to cost and benefit impacts.
- **Application priorities of legislation**. This criterion is related to any described priority defined in the legislation for the types of waste considered, including SWM targets. This criterion can be specific: for example, waste diversion of organic waste streams from landfill, landfilled waste processing and recycling, net energy savings in waste-to-energy facilities, and GHG emissions reduction (El Hanandeh and El-Zein, 2009).
- **Adaptability to environmental policy**. This criterion assesses alternatives to determine if they can withstand changes in regulation while simultaneously maintaining operability and functionality. In other words, it determines if the alternative is able to comply with the new changes in regulatory requirements.

Environmental Criteria   Environmental criteria are related to pollutant releases, their impacts on the environment, and environmental risk. In addition, issues related to energy consumption or energy efficiency are also addressed by these criteria. Some examples include:

- **Life cycle impact assessment categories**. The impact assessment categories like acidification, eutrophication, global warming, human toxicity, and photochemical oxidation are all relevant environmental impacts considered in life cycle assessment, which can be used as environmental criteria to evaluate alternatives in SWM. The lower the environmental impact, the better the evaluation results attributed to the criteria. Environmental criteria can be used individually and flexibly as needed (Pires et al., 2011).
- **Ecological footprint**. This criterion defines the required space to bear an activity, determined by the area necessary to supply the resources consumed and to assimilate the waste generated in an alternative (Wackernagel and Rees, 1996).
- **Potential environmental impacts**. This criterion is devoted to the environmental consequences caused by an alternative, including installation requirements and antipollution/prevention systems needed for the proposed technology. Impacts can be classified into air emissions, water pollution, and soil contamination: for example, environmental impacts in the proposed alternative can be related to siting distance from the transfer stations to the waste treatment center.
- **Human health**. This criterion is focused on the release of hazardous substances, typically heavy metals and dioxins, detrimental to human health, also described

as release with health effects and use of harmful materials. This criterion can be combined with environmental impacts to form environmental and health impacts.

- **Potential environmental risk**. This criterion derives almost exclusively from hazardous waste management which may further cause various types of environmental and public health risk.

- **Resources**. This criterion refers to the consumption or the use of natural resources by each alternative.

- **Air emissions**. This is a sensitive criterion to deal with because emissions are specific to each type of treatment technology and dispersion of air pollutants are faster than others. This criterion is also known as "air emissions of organic compounds" (Herva and Roca, 2013), "air emissions of dusts" (Herva and Roca, 2013), and "air residuals and environmental impacts" (Dursun et al., 2011). Specific pollutants of concern include volatile organic compounds, dioxin, particulate matter, nitrogen oxides, and sulfur dioxide. In addition, GHG emissions (Karagiannidis and Perkoulidis, 2009) as well as acidification gases and smog precursors (El Hanandeh and El-Zein, 2010) are sometimes considered in specific conditions.

- **Wastewater**. This criterion is quantitatively applied as the generation of wastewater (Rousis et al., 2008) in some cases and as "water emissions of suspended solids" (Herva and Roca, 2013) and "water residuals and environmental impacts" (Dursun et al., 2011) in other cases. The criterion can also be referred to as the "cumulate hazard in released wastewater" (Generowicz et al., 2011). The presence of suspended solids reduces photosynthetic activity of aquatic vegetation, which may cause ecosystem problems (Herva and Roca, 2013).

- **Water consumption**. This criterion is related to the amount of water consumed in each alternative, which is site specific. The more scarce the water resource, the more significant this criterion.

- **Generation of solid waste**. This criterion is referred to as "solid wastes" (Kaya, 2012) or "solid residuals and associated environmental impacts" (Dursun et al., 2011). The harmful consequences to ecosystem and human health are considered in this criterion, such as the case related to accumulated hazard in waste disposal (Generowicz et al., 2011).

- **Noise pollution**. Although not commonly considered, this attribute is significant to the environment and human health. Noise pollution may result from not only waste collection and shipping, but also from all the operational units involved in each SWM alternative. This criterion can also be considered a social criterion.

- **Aesthetic nuisance**. This criterion considers changes to the natural landscape from the installation and operation of the unit considered in the alternative (Rousis et al., 2008). Although not a common criterion, it can influence the equipment variety utilized and the conditions for the additional work in the infrastructure (Rousis et al., 2008). This criterion can be also considered a social criterion because aesthetics are based on human perception.

- **Energy**. This criterion is related to the consumption of gross energy, energy consumption, energy recovery (Perkoulidis et al., 2010; Madadian et al., 2013), energy cumulated, lost energy, net energy savings, green electricity production potential, and green energy recovery. The energy balance depends on waste content energy and process efficiency in using and producing energy.
- **Material recovered**. This criterion is related to the amount of waste that can be recycled, the energy recovered, or even materials biologically recovered; it can also be called "rate of recycling" (Madadian et al., 2013). If associated with a regulatory target, this criterion can be a regulatory criterion, such as that associated with diversion of biodegradable waste streams from MSW landfill in European Union (EU) countries.
- **Occupied landfill volume**. This criterion measures the capability of an alternative to reduce the volume of waste, and therefore the land required to waste disposal (Herva and Roca, 2013).

Social Criteria   Aspects concerning labor, welfare, and public acceptance are all possible social criteria. A short list of social criteria used in MADM for SWM includes:

- **Employment**. This criterion considers the number of jobs created due to SWM (Roussat et al., 2009); it is also known as the "potential for creation of new jobs" (Rousis et al., 2008). Not only is the amount of work considered in this criterion, but also the quality of the work.
- **Quality of life**. This criterion includes aspects such as area destroyed or preserved by SWM system as well as traffic flows linked with waste collection vehicles.
- **Social acceptance/acceptability**. This criterion examines the degree of social acceptance of the proposed solution, dependent on many factors such as existing management practices, environmental repercussions, prevention/reduction of environmental repercussions, the cognizance level of citizens on environmental field, the level of environmental sensitization of citizens, and the education system (Rousis et al., 2008). It can also be known as "public acceptance" (Madadian et al., 2013), "public acceptance obstacles" (Dursun et al., 2011), or "stigma perceived by affected community" (Kijak and Moy, 2004).
- **Social justice**. This criterion evaluates equity of alternatives, such as waste disposal fees assessed to users, which can be applied as a single criterion related to both social justice and economics.
- **Social welfare**. This criterion considers the welfare of populations around the SWM system. One of the main factors affecting population welfare is odor.
- **Tourism**. This criterion is related to the effects of SWM on tourism, especially on islands, where tourism is a dominant economic activity (Hanan et al., 2013).

16.1.2   Criteria Weighting Methods

Each criterion has a different level of importance or relevance to decision makers who evaluate the alternatives, or in a broader view, the stakeholders assigned to the decision-making process. In an MADM process, this relevance is demonstrated by weighting the attributes/criteria.

The method used to determine the weights must ensure trust and confidence because weighting influences the final results. Weights are needed to consider the varying degrees of criteria, the independency of criteria, and the subjective preference of the decision makers (Wang et al., 2009a). The methods applied in SWM decision-making include (Wang et al., 2009a; Xiao and Guo, 2010) equal weights methods, subjective weighting methods, objective weighting methods, and combination or integrative weighting methods.

Equal Weights Methods   In equal weights methods all attributes are treated as equally important. The criteria weight in equal weights method is defined as

$$w_i = \frac{1}{n}, \quad i = 1, 2, \ldots, n, \tag{16.1}$$

where w_i is the equal weights vector, the domain of $0 \le w_i \le 1$, and $\sum_i^n w_i = 1$.

Because all attributes have the same weight, stakeholder and decision maker participation is not necessary to obtain weights. Dawes and Corrigan (1974) argued that this method often produces results similar to optimal weighting methods. If extended, each sustainability aspect would have the same importance, and therefore given the same weight. The application of such an equity rule might complicate gaining consensus from heterogonous groups of stakeholders and decision makers.

Subjective Weighting Methods   Subjective weighting methods consider the preferences of stakeholder (including experts) and decision maker regardless of whether the quantitative data of SWM can be available. These methods contain expert survey method, AHP (Saaty, 1980), simple multiattribute rating technique (SMART) (Edwards, 1977), swing (von Winterfeldt and Edwards, 1986), trade-off (Keeney and Raiffa, 1976), Simos' method (Simos, 1990), least-square method (Chu et al., 1979), eigenvector method (Gao et al., 2010), Delphi method (Dalkey and Helmer, 1963), and so on. Popular methods in SWM include AHP (explained in Chapter 8), expert survey method, SMART, and eigenvector method:

- **Expert survey method**. This method consists of conducting questionnaires to collect opinions from stakeholders. A stakeholder panel, usually composed of government staff, experts, nongovernmental organizations, and waste managers, then classifies the criteria and ranks the alternatives according to scores in a specific range.

- **SMART**. This method proposed by Edwards (1977) is a 10-step procedure, the first five of which are common to a generic MADM procedure. The participants/stakeholders are asked to rank the importance of the criteria from the worst to best levels, with the least important assigned an importance of 10; the next-least-important criterion is assigned a number reflecting the ratio of relative importance to the least important dimension (Olson, 1996). A raising number of points (without an explicit upper limit) is allocated to the other criteria to address their relevance respective to the least important criterion (Wang et al., 2009b). Consequently, weights are normalized and then will be used to help assess alternative being considered for each criterion. The improved version, named simple multiattribute rating technique extended to ranking (SMARTER) (Edwards and Barron, 1994), was proposed to solve calculation difficulties.
- **Eigenvector method**. This method is based on pairwise comparisons, consisting of basing weights on the components of the eigenvector of the judgment matrix (Gao et al., 2010). Eigenvector is defined by

$$AW = \lambda_{max}W, \tag{16.2}$$

where λ_{max} is the largest eigenvalue of the judgment square matrix A, and W is a non-zero vector. This eigenvector solution is normalized additively, so that $\sum_{i=1}^{n} w_i = 1$.

Objective Weighting Methods   In these methods, the weights associated with objectives are obtained by mathematical methods based on the analysis of initial data (Wang et al., 2009a). Objective weighting includes principal component analysis, entropy technology, and maximum deviation (Zhang et al., 2008). These methods are not commonly used in SWM decision-making.

Integrative/Integrated Weighting Methods   Subjective and objective weighting have inherent drawbacks. Subjective weighting approach depends on the decision maker's knowledge of the subject, whereas objective weighting approach requires clear justification. Such dilemmas have promoted an integrated approach between both weighting approaches to overcome their shortcomings (Ma et al., 1999).

16.1.3 Evaluation

The evaluation step in MADM is related to the methods described in Chapter 8, also described systematically by Tzeng and Huang (2011). To improve accessibility to nonexpert users, interactive solution tools have been developed, including a range of MADM software packages (Morrissey and Browne, 2004; Weistroffer et al., 2005). The most applicable MADM software packages for SWM issues are Criterium Decision Plus (CDP) (InfoHarvest), Decision Lab 2000 (Visual Decision Inc.), *Elimination Et Choix Traduisant la Realité* (or Elimination and Choice

Expressing Reality—ELECTRE) III-IV (Laboratory for Analysis and Modelling of Decision Support Systems, Université Paris-Dauphine—LAMSADE), Expert Choice (Expert Choice©), Hiview (Catalyze, Ltd and London Scholl of Economics), and Super Decisions (Creative Decisions Foundation). Generic software packages such as Microsoft Excel® (Microsoft Corporation) and MATLAB® (MathWorks, Inc.) can also be used to implement MADM methods and weighting criteria methods.

CDP provides users a choice between a simple multiattributive rating technique and AHP (Weistroffer et al., 2005). Some of the CDP's primary strengths (Haerer, 2000) include a transparent structured decision-making framework, easy assessment of sensitivities of alternative rankings to weights/trade-offs to help groups focus on key aspects, and the ability to incorporate uncertainties in performance scores. CDP has been applied to select remedial strategies and technologies for hazardous waste sites (Haerer, 2000) and to analyze SWM system alternatives (Kijak and Moy, 2004).

Decision Lab 2000 was developed by Visual Decision Company and is based on preference ranking organization method for enrichment of evaluations (PROMETHEE) and geometrical analysis for interactive aid (GAIA) methods. Sensitivity analysis is made by applying techniques of walking weights, intervals of stability, and the graphical axis of decision presented by GAIA (Weistroffer et al., 2005). Some applications can be found in literature, like the CDW management schemes comparison (Kourmpanis et al., 2008), waste of electrical and electronic equipment (WEEE) management (Rousis et al., 2008), and MSW management (Vego et al., 2008). PROMETHEE CALCulations (PROMCALC) software package was the previous version of Decision Lab 2000, and replaced later on by PROMETHEE and GAIA software. A free academic version is available at http://www.promethee-gaia.net/software.html.

ELECTRE III–IV is a software package that implements ELECTRE III and ELECTRE IV methods. Developed by LAMSADE of the University Paris-Dauphine, ELECTRE is used to implement the ranking procedure. ELECTRE IV builds various nonfuzzy outranking relations for the case when weighting criteria is not possible (Weistroffer et al., 2005). ELECTRE III software has been applied for demotion waste management strategies studies (Roussat et al., 2009), which is more commonly used in waste treatment infrastructure location assessments. A free full version exists in LAMSADE website: http://www.lamsade.dauphine.fr/.

Expert Choice 2000, developed by Expert Choice Inc., is the best known software package that applies AHP. It enables the decision maker solve the problem with a visualized hierarchy approach and then conduct pairwise judgments in an interactive, verbal mode (Alidi, 1996). A group decision support software application exists. Expert Choice application has been more devoted to weighting criteria applied to petrochemical waste management (Alidi, 1996), solid waste planning (MacDonald, 1996), replication of community-based decentralized composting (Yedla, 2012), evaluation of possibilities to promote a country's performance in recycling (Lin et al., 2010), and identification of the best future technology to waste-to-energy (Liang et al., 2013). Several licenses exist: limited academic license, web-based software, and desktop-based software.

Hiview, a software developed by London School of Economics, uses multiattribute value theory in a weighted linear combination approach to justify decision-making where various reciprocally exclusive alternatives are available (Bastin and Longden, 2009). This software has been applied to assess waste transportation options (Bastin and Longden, 2009) and to assess management options of radioactive waste (Morton et al., 2009). A 20-day trial version exists, as well standard licenses and education licenses available from http://www.catalyze.co.uk.

The Super Decisions software package implements the analytic network process (ANP) method, produced by the Creative Decisions Foundation. The process begins by dividing the problem criteria into different tribes, then setting up numerous nodes with decision criteria under these tribes and conducting networked analysis on the decision problem; thus, a complex problem is divided into several element groups to clarify the problem (He, 2010). This process has mostly been used for cases of waste treatment infrastructures, such as comparing WEEE management scenarios (Üçüncüoğlu and Ulukan, 2010), and evaluating reverse logistics operations to manage waste appliances (He, 2010). A free download is available from http://www.superdecisions.com.

16.1.4 Interpretation

After ranking the criteria the results are interpreted. The ranking outcome provides a basis for decision-making practitioners to define the optimal solution, the best second and third, and even the worst solution. Reliability can be enhanced by many methods, especially sensitivity analysis because it shows how the scores and rankings of the proposed SWM system are influenced by fluctuations in the weighted coefficients of the criteria (Rousis et al., 2008). Uncertainty analysis conducted in MADM can be addressed by using fuzzy membership values, established at the beginning of the MADM process.

Not only sensitivity analysis, but also uncertainty analysis provides credibility to the MADM processes, especially when considerable budgeted resources and public support are available, which is usually the case for SWM systems. Sustainability can be driven by decision-making practitioners by applying different MADM methods together to increase final decision accuracy. If ranking results from the different MADM methods are not the same, then is difficult to choose which MADM ranking results should be considered (Wang et al., 2009a). The solution for this issue is to aggregate the results. Hwang and Yoon (1981) proposed three sets of aggregation: average ranking, the Borda method based on majority rule binary relation, and the Copeland method, which is a modification of Borda method that accounts for "losses" as well as "wins." Other methods suggested by Wang et al. (2009a) are the vertical and horizontal aggregation method and the singular value decomposition aggregation method. Wang et al. (2009a) proposes singular value decomposition aggregation method.

Average ranking is the simplest procedure, even if two alternatives have the same average rank because the alternative with the smallest standard deviation can be selected (Hwang and Yoon, 1981). However, this process has no guarantee of acquiring best results when substantial differences exist between the rankings of alternatives (Jahan et al., 2011).

The Borda method (named by the French scientist Jean Charles de Borda (1733–1799) who formulated this preferential voting system) (Pomerol and Barba-Romero, 2000) is a voting procedure which selects the highest option in the voter's ranking. According to the description made by Wang et al. (2009a) and Kumar (2009), Borda method works as follows: each alternative is assigned with points according to the individual preferences, namely N for the top ranked alternative, the second with $N-1$ points, and so on; then all votes from all individuals are added, resulting the Borda score $B(a)$, being the preferred alternative the one with the highest score.

The Copeland method, which is also a voting procedure, starts where the Borda method stops by considering not only how many "wins" an alternative has, but also explicitly including the "losses." First, calculate the alternative's number it beats by a majority and the number of alternatives it loses in opposition to each alternative; then calculate the difference between the two numbers; in the end, the alternative in the social preference is the larger number of the higher ranked (Wang et al., 2009a).

Singular value decomposition aggregation method is introduced according to Wang et al. (2009b). Assume that $p\,(p \le \min{(m,n)})$ is the rank of the real matrix $G_{m \times n}$, along with the orthogonal matrixes $U_{m \times m}$ and $V_{n \times n}$. The following expression can be obtained (Wang et al., 2009b):

$$U^T GV = \begin{bmatrix} \Gamma & 0 \\ 0 & 0 \end{bmatrix} = Q \in R^{m \times n}, \tag{16.3}$$

where $\Gamma = \mathrm{diag}(\sigma_1, \sigma_2, \ldots, \sigma_p)\sigma_1 \ge \sigma_2 \ge \cdots \ge \sigma_p$, and $\sigma_1, \sigma_2, \ldots, \sigma_p$ is the non-zero singular values of matrix G. When $p = 1$, the sequencing results are identical in n different evaluation methods; otherwise, the sequencing results are not the same, and $p > 1$.

Non-zero singular values, $\sigma_1 (i = 1, 2, \ldots, p)$, describe the quantitative comparison of the p characteristics in the sequence value matrix G. Larger singular values describe more characteristics of G than smaller ones; therefore, by maintaining larger anterior k singular values in G and other singular values at zero, the corresponding approximate matrix G_k of G can be calculated as (Wang et al., 2009b):

$$\hat{G}_k = U \hat{Q} V^T, \tag{16.4}$$

where $\hat{Q} = \begin{bmatrix} \Gamma & 0 \\ 0 & 0 \end{bmatrix}, \hat{G}_k = \left(g_{ij}^{(k)}\right) \in R^{m \times n}, i = 1, 2, \ldots, m, j = 1, 2, \ldots, n$.

Thus, the commonness of evaluation information is kept and non-commonness (or it can be thought as noise) is eliminated. Then, a logical k value should be determined to make $\hat{G}_k$ meet the following two points as closely as possible (Wang et al., 2009b): (1) the error between $\hat{G}_k$ and G is small, and the consistency of evaluation results is enhanced and (2) the mass information and characteristics are maintained to avoid a greater deviation of $\hat{G}_k$ from G. Here, "consistent degree" and "reliability" are defined to reflect the two points. Consistent degree is the closeness between $\hat{G}_k$ and $\hat{G}_1$, expressed as (Wang et al., 2009b):

$$n_k = \frac{\|G\|_F - \|\hat{G}_k\|_F}{\|G\|_F - \|\hat{G}_1\|_F}, k = 1, 2, \ldots, p, \tag{16.5}$$

where $\|.\|_F$ is the Frobenius norm of matrix ".", $\|G\|_F = \|\hat{G}_p\|_F$, $\hat{G}_1$ is the approximate matrix that is kept as the only largest singular value, and $n_k \in [0, 1]$ and n_k increases with the increase of k. Reliability is the closeness between $\hat{G}_k$ and G, defined as (Wang et al., 2009b):

$$\varepsilon_k = \sum_{i=1}^{k} \varphi_i, \tag{16.6}$$

where $\varphi_i = \varphi_i / \sum_{j=1}^{p} \sigma_j$, $i = 1, 2, \ldots, k$, $1 \leq k \leq p$, $\varepsilon_k \in [0, 1]$, and ε_k decreases with the increase of k. To meet these two aspects as consistently as possible, a parameter, "consistent and reliability degree" is defined and calculated as (Wang et al., 2009b):

$$\pi_k = \alpha_1(\beta_1\eta_k + \beta_2\varepsilon_k) + \alpha_2(\eta_k\varepsilon_k), \tag{16.7}$$

where $\alpha_i, \beta_i \in [0, 1]$ $(i = 1, 2)$, $\alpha_1 + \alpha_2 = 1$, and $\beta_1 + \beta_2 = 1$, $\pi_k \in [0, 1]$. The linear combination term $(\beta_1\eta_k + \beta_2\varepsilon_k)$ shows the complementary functionality between consistency and reliability and the nonlinear term $(\eta_k\varepsilon_k)$ indicates their proportionality. Based on the relative importance between consistency and reliability, decision makers set the β_1 and β_2 values. Let $v_k = \beta_1\eta_k + \beta_2\varepsilon_k$, $\tau_k = \eta_k\varepsilon_k$, $k = 1, 2, \ldots, p$, and then $v = (v_1, v_2, \ldots, v_p)^T$ and $\tau = (\tau_1, \tau_2, \ldots, \tau_p)^T$ can be obtained (Wang et al., 2009b).

Next, select the appropriate α_1 and α_2 values to maximize the holistic discretization of $\{\pi_k | k = 1, 2, \ldots, p\}$ in the previous equation, which expresses the difference in different values of k. To obtain the suitable α_1 and α_2, the optimal model can be constructed as (Wang et al., 2009b):

$$\max \quad \sum_{k=1}^{p} \left[\alpha_1 v_k + \alpha_2 \tau_k - \frac{1}{p} \sum_{k=1}^{p} (\alpha_1 v_k + \alpha_2 \tau_k) \right]^2.$$

$$\text{s.t.} \quad \alpha_1^2 + \alpha_2^2 = 1, \quad \alpha_1, \alpha_2 \geq 0 \tag{16.8}$$

The optimal model can be solved in computation software such as MATLAB and LINGO (LINDO Systems, 2003). The solution $(\alpha_1, \alpha_2)^T$ is the eigenvector, which then is normalized to obtain α_1 and α_2 values; then, select the corresponding k value to the maximum of π_k to obtain the final sequence result in $\hat{G}_k = U\hat{Q}V^T$ (Wang et al., 2009b).

16.2 MADM FOR SOLID WASTE MANAGEMENT

MADM has been applied to solve SWM problems in various planning and design cases. In this section, two case studies are presented to demonstrate; the first case assesses screening waste management technology options, and the second case ranks choices of waste collection schemes for collecting MSW.

16.2.1 Case 1—Selecting Construction and Demolition Waste Management

CDW is one of the most prevalent solid waste streams generated in the EU and China over the recent past. It accounts for approximately 25–30% of all waste generated and consists of several types of materials with recycling potential, such as concrete, bricks, gypsum, wood, glass, plastics, solvents, asbestos, and excavated soil (European Commission, 2012). The EU Waste Framework Directive (European Parliament and Council, 2008) established a minimum target of 70% of CDW to be reused, recycled, or other material recovery by 2020 in EU Member States. To achieve this target, solutions that ensure technical, environmental, and economic affordability must be explored.

A survey was conducted in the federal state of Baden-Württemberg in south-west Germany (Hiete et al., 2011). The focus of the study was to plan a CDW (only for demolition waste, mostly inert materials) recycling network that minimizes costs while also defining and considering environmental impacts. The network considers CDW supply and recycled material demand and selects policy measures that foster recycling (Figure 16.2). The CDW study was conducted during two non-consecutive years (2010 and 2050) to determine how projected demographic changes in the region could affect the CDW chain.

During the study, 19 scenarios were developed for managing CDW (Table 16.1), considering factors such as the type of CDW, the demand for recycled materials, the processing technology focusing on sorting performance and capacity, the natural aggregates supply, CDW disposal fees, and the transport costs. For processing technology, the option of retrofitting existing recycling plants was considered in some scenarios (Hiete et al., 2011).

The optimization model developed by Hiete et al. (2011) was applied in this case study of MADM techniques to choose the best scenario to manage CDW now and in

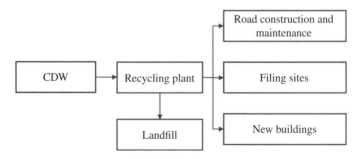

FIGURE 16.2 CDW management system

TABLE 16.1 CDW scenarios considered in the study

Scenarios	Year	Recycling plants retrofit	Scenario parameters in planning stage
S.Base	2010	Yes	Existing disposal fee, where transport cost is US$0.02 per kilometer per tonne
S.DF2	2010	Yes	Disposal fee + US$2.78 per tonne
S.DF5	2010	Yes	Disposal fee + US$6.76 per tonne
S.TC7	2010	Yes	Transport costs of US$0.10 per kilometer per tonne
S.TC30	2010	Yes	Transport costs of US$0.42 per kilometer per tonne
S.Free	2010	No—free design	Existing disposal fee
S.Free.DF2	2010	No—free design	Disposal fee + US$2.78 per tonne
S.Free.DF5	2010	No—free design	Disposal fee + US$6.76 per tonne
S.Free.TC7	2010	No—free design	Transport costs of US$0.10 per kilometer per tonne
S.Free.TC30	2010	No—free design	Transport costs of US$0.42 per kilometer per tonne
S.2050.Base	2050	No	Existing disposal fee
S.2050.S20	2050	No	Residential building demolition rate: 20%
S.2050.S50	2050	No	Residential building demolition rate: 50%
S.2050.D40	2050	No	Recycled concrete demand: 40%
S.2050.D40.S20	2050	No	Residential building demolition rate: 20%Recycled concrete demand: 40%
S.2050.D.40.S50	2050	No	Residential building demolition rate: 50%Recycled concrete demand: 40%
S.2050.D80	2050	No	Recycled concrete demand: 80%
S.2050.D80.S20	2050	No	Residential building demolition rate: 20%Recycled concrete demand: 80%
S.2050.D80.S50	2050	No	Residential building demolition rate: 50%Recycled concrete demand: 80%

Note: US$ is United States dollars.
Source: Adapted from Hiete et al. (2011).

the future (2010 and 2050). The 19 scenarios were separated in two groups: the first group is focused on the actual CDW generated in the region, and the second group with the amount of CDW production for year 2050.

The first of the four stages of MADM (criteria selection, criteria weighting, evaluation and interpretation) is to choose the criteria to be used. In this case study, economic, technical, and environmental criteria were chosen (Table 16.2). Economic criteria considered were total costs (TC), total costs (without disposal taxes (TCo)); technical criteria considered were direct disposal rate (DDR), recycling rate (RR), high quality recycled material (HQRC), and number of plants (NP). Abiotic depletion (AD) and global warming potential (GWP) were the environmental criteria chosen. The equal weights method was often assumed for the weighting criteria, and the technique for order preference by similarity to ideal solution (TOPSIS), a well-known

TABLE 16.2 Criteria used to characterize the scenarios

Groups	Scenario	TC (million US$ per 10 years)	TCo (%)	DDR (%)	RR (%)	HQRC (%)	AD (%)	GWP (%)	NP
2010	S.Base	66.82	100	7	82	51	−100	100	13
	S.DF2	69.60	100	2	86	51	−114	100	13
	S.DF5	72.39	101	0	89	51	−126	99	14
	S.TC7	65.43	98	2	87	52	−122	98	13
	S.TC30	64.04	96	15	75	49	−73	101	14
	S.Free	55.68	83	22	72	56	−54	103	4
	S.Free.DF2	58.47	84	12	82	56	−88	103	4
	S.Free.DF5	62.64	86	1	88	52	−118	101	7
	S.Free.TC7	50.11	74	11	83	55	−101	101	3
	S.Free.TC30	55.68	82	34	59	53	−2	105	5
2050	S.2050.Base	192.11	287	67	30	47	376	353	4
	S.2050.S20	161.48	240	63	34	49	273	299	4
	S.2050.S50	115.54	172	52	44	49	121	217	4
	S.2050.D40	174.01	260	46	50	61	98	341	8
	S.2050.D40.S20	143.38	214	36	60	61	−25	286	8
	S.2050.D40.S50	100.23	150	17	78	59	−165	204	9
	S.2050.D80	162.87	243	37	60	67	−38	334	10
	S.2050.D80.S20	135.03	202	31	65	64	−86	282	9
	S.2050.D80.S50	94.66	142	15	80	61	−181	203	8

Source: Hiete et al. (2011).

MADM method, was applied to evaluate the alternatives. A brief description of TOP-SIS was provided in Chapter 8, but a more in-depth description of the steps required to conduct the practice is provided here.

TOPSIS, developed by Hwang and Yoon (1981), was selected based on the concept that the chosen alternative should have the shortest distance from the ideal solution and the farthest from the negative-ideal solution. A utility value $D(i)$ for each alternative i is obtained by calculating the relative distance for i to the ideal solution, which can be described as follows (Jahanshahloo et al., 2006).

Step 1. Calculate the normalized decision matrix. The normalized value n_{ij} is calculated as

$$n_{ij} = \frac{x_{ij}}{\sqrt{\sum_{j=1}^{m} x_{ij}^2}}, \quad j = 1, \ldots, m, i = 1, \ldots, n. \tag{16.9}$$

Step 2. Calculate the weighted normalized decision matrix. The weighted normalized value v_{ij} is calculated as

$$v_{ij} = w_i n_{ij}, \quad j = 1, \ldots, m, i = 1, \ldots, n, \tag{16.10}$$

TABLE 16.3   Normalized scenarios for the year 2010 (step 2)

Scenario	TC (US$ million per 10 years)	TCo (%)	DDR (%)	RR (%)	HQRC (%)	AD (%)	GWP (%)	NP
S.Base	0.042	0.044	0.019	0.040	0.038	−0.041	0.039	0.051
S.DF2	0.044	0.044	0.005	0.042	0.038	−0.047	0.039	0.051
S.DF5	0.046	0.044	0.000	0.044	0.038	−0.051	0.039	0.055
S.TC.7	0.041	0.043	0.005	0.043	0.039	−0.050	0.038	0.051
S.TC.30	0.041	0.042	0.040	0.037	0.037	−0.030	0.039	0.055
S.Free	0.035	0.036	0.059	0.035	0.042	−0.022	0.040	0.016
S.Free.DF2	0.037	0.037	0.032	0.040	0.042	−0.036	0.040	0.016
S.Free.DF5	0.040	0.037	0.003	0.043	0.039	−0.048	0.039	0.027
S.Free.TC7	0.032	0.032	0.029	0.041	0.041	−0.041	0.039	0.012
S.Free.TC30	0.035	0.036	0.091	0.029	0.040	−0.001	0.041	0.020

where w_i is the weight of the ith attribute or criterion, and $\sum_{i=1}^{n} w_i = 1$ (Tables 16.3 and 16.4). Because the equal weight method was used, every criteria has the same importance (i.e., 0.125).

Step 3. Determine the positive ideal and negative ideal solutions.

$$A^+ = \left\{ v_1^+, \ldots, v_n^+ \right\} = \left\{ \left(\max_j v_{ij} | i \in I \right) . \left(\min_j v_{ij} | i \in J \right) \right\}$$
$$A^- = \left\{ v_1^-, \ldots, v_n^- \right\} = \left\{ \left(\min_j v_{ij} | i \in I \right) . \left(\max_j v_{ij} | i \in J \right) \right\},$$
(16.11)

where I is associated with benefit criteria, and J is associated with cost criteria.

TABLE 16.4   Normalized scenarios for the year 2050 (step 2)

Scenario	TC (US$ million per 10 years)	TCo (%)	DDR (%)	RR (%)	HQRC (%)	AD (%)	GWP (%)	NP
S.2050.Base	0.055	0.055	0.063	0.022	0.034	0.084	0.052	0.022
S.2050.S.20	0.046	0.046	0.060	0.024	0.035	0.061	0.044	0.022
S.2050.S.50	0.033	0.033	0.049	0.032	0.035	0.027	0.032	0.022
S.2050.D.40	0.050	0.050	0.044	0.036	0.044	0.022	0.050	0.045
S.2050.D.40 S.20	0.041	0.041	0.034	0.043	0.044	−0.006	0.042	0.045
S.2050.D.40 S.50	0.029	0.029	0.016	0.056	0.042	−0.037	0.030	0.050
S.2050.D.80	0.047	0.047	0.035	0.043	0.048	−0.009	0.049	0.056
S.2050.D.80 S.20	0.039	0.039	0.029	0.047	0.046	−0.019	0.041	0.050
S.2050.D.80 S.50	0.027	0.027	0.014	0.057	0.044	−0.041	0.030	0.045

The positive and negative ideal solutions were both obtained (Table 16.3), described as

- Positive ideal solution $A^+ = \{0.032, 0.032, 0.000, 0.044, 0.042, -0.051, 0.038, 0.012\}$.
- Negative ideal solution $A^- = \{0.046, 0.044, 0.091, 0.029, 0.037, -0.001, 0.041, 0.055\}$.

Step 4. Calculate the separation measures using the n-dimensional Euclidean distance. The separation of each alternative for the ideal solution and for the negative ideal solution is given as, respectively,

$$
d_j^+ = \left\{ \sum_{i=1}^{n} \left(v_{ij} - v_i^+ \right)^2 \right\}^{1/2}
$$
$$
,j = 1, \ldots, m. \qquad (16.12)
$$
$$
d_j^- = \left\{ \sum_{i=1}^{n} \left(v_{ij} - v_i^- \right)^2 \right\}^{1/2}
$$

The calculated distance to ideal solutions is presented for various scenarios (Tables 16.5 and 16.6), which can be used in the step 5.

Step 5. Calculate the relative closeness to the ideal solution, R_j, defined for alternative A_j with respect to A^+ as

$$
R_j = \frac{d_j^-}{\left(d_j^+ + d_j^- \right)}, j = 1, \ldots, m \qquad (16.13)
$$

because $d_j^- \geq 0$ and $d_j^+ \geq 0$, then $R_j \in [0, 1]$.

Step 6. Rank the preference decreasing order.

The best scenario to manage CDW according to the ranking outcome (Table 16.7) is through the free planning of recycling plants plus the addition of US\$ 6.96 per tonne to the current disposal fee. The second best option is the application of transportation cost from the CDW site to the recycling plant, which is lower by 10% of US\$ per km per tonne required. The current situation is shown to be one of the worst solutions according to MADM results.

The same procedure was applied for the scenarios projected for 2050. The best result according to the rankings (Table 16.8) is the one with the highest demand for recycled material and the highest residential demolition rate. The second best scenario is the one in which the demand for recycled material is 40%, but the residential demolition rate remains at 50%. Maintaining today's situation until 2050 is the worst scenario.

In the interpretation stage of MADM process (the last stage), sensitivity analysis can improve interpretation of the ranking outcome. Sensitivity analysis can change the weights applied to characterize criteria. Criteria TC, TCo, and DDR have higher

TABLE 16.5   Euclidean distance to positive and negative ideal solutions (step 4) for 2010 alternatives

	Criteria								
Scenario	TC (US$ million per 10 years)	TCo (%)	DDR (%)	RR (%)	HQRC (%)	AD (%)	GWP (%)	NP	d_j^+
S.Base	0.0001	0.0001	0.0003	0.0000	0.0000	0.0001	0.0000	0.0015	0.0476
S.DF2	0.0002	0.0001	0.0000	0.0000	0.0000	0.0000	0.0000	0.0015	0.0435
S.DF5	0.0002	0.0001	0.0000	0.0000	0.0000	0.0000	0.0000	0.0019	0.0471
S.TC.7	0.0001	0.0001	0.0000	0.0000	0.0000	0.0000	0.0000	0.0015	0.0423
S.TC.30	0.0001	0.0001	0.0016	0.0000	0.0000	0.0005	0.0000	0.0019	0.0647
S.Free	0.0000	0.0000	0.0035	0.0001	0.0000	0.0009	0.0000	0.0000	0.0666
S.Free.DF2	0.0000	0.0000	0.0010	0.0000	0.0000	0.0002	0.0000	0.0000	0.0367
S.Free.DF5	0.0001	0.0000	0.0000	0.0000	0.0000	0.0000	0.0000	0.0002	0.0191
S.Free.TC7	0.0000	0.0000	0.0009	0.0000	0.0000	0.0001	0.0000	0.0000	0.0313
S.Free.TC30	0.0000	0.0000	0.0083	0.0002	0.0000	0.0026	0.0000	0.0001	0.1055

	Criteria								
Scenario	TC (US$ million per 10 years)	TCo (%)	DDR (%)	RR (%)	HQRC (%)	AD (%)	GWP (%)	NP	d_j^-
S.Base	0.0000	0.0000	0.0052	0.0001	0.0000	0.0016	0.0000	0.0000	0.0835
S.DF2	0.0000	0.0000	0.0073	0.0002	0.0000	0.0021	0.0000	0.0000	0.0980
S.DF5	0.0000	0.0000	0.0083	0.0002	0.0000	0.0026	0.0000	0.0000	0.1051
S.TC.7	0.0000	0.0000	0.0073	0.0002	0.0000	0.0024	0.0000	0.0000	0.0997
S.TC.30	0.0000	0.0000	0.0026	0.0001	0.0000	0.0008	0.0000	0.0000	0.0593
S.Free	0.0001	0.0001	0.0010	0.0000	0.0000	0.0005	0.0000	0.0015	0.0571
S.Free.DF2	0.0001	0.0001	0.0035	0.0001	0.0000	0.0012	0.0000	0.0015	0.0807
S.Free.DF5	0.0000	0.0000	0.0078	0.0002	0.0000	0.0022	0.0000	0.0008	0.1052
S.Free.TC7	0.0002	0.0001	0.0038	0.0001	0.0000	0.0016	0.0000	0.0019	0.0882
S.Free.TC30	0.0001	0.0001	0.0000	0.0000	0.0000	0.0000	0.0000	0.0012	0.0379

importance, and the remaining are lower. In this case study, the ranking outcome of R_j was also calculated for environmental criteria including AD, GWP, and DDR, the most relevant criteria considered. The ranking outcome with sensitivity analysis for scenarios in 2010 (Table 16.9) indicates that the first and second best options oscillate between scenarios S.Free.TC7 and S.Free.DF5. When only environmental criteria are highlighted, the second best option is to add US$6.96 per tonne to the current disposal fee and retrofitting the recycling plants.

The results for the ranking outcomes in 2050 (Table 16.10) are consistent with those from 2010. The best solutions are still S.2050.D80.S50 and S.2050.D40.S50.

TABLE 16.6   Euclidean distance to positive and negative ideal solutions (step 4) for 2050 alternatives

	Criteria								
Scenario	TC (US$ million per 10 years)	TCo (%)	DDR (%)	RR (%)	HQRC (%)	AD (%)	GWP (%)	NP	d_j^+
S.2050.Base	0.001	0.001	0.002	0.001	0.000	0.016	0.000	0.000	0.1470
S.2050.S.20	0.000	0.000	0.002	0.001	0.000	0.010	0.000	0.000	0.1210
S.2050.S.50	0.000	0.000	0.001	0.001	0.000	0.005	0.000	0.000	0.0821
S.2050.D.40	0.001	0.001	0.001	0.000	0.000	0.004	0.000	0.000	0.0849
S.2050.D.40 S.20	0.000	0.000	0.000	0.000	0.000	0.001	0.000	0.000	0.0536
S.2050.D.40 S.50	0.000	0.000	0.000	0.000	0.000	0.000	0.000	0.001	0.0289
S.2050.D.80	0.000	0.000	0.000	0.000	0.000	0.001	0.000	0.001	0.0626
S.2050.D.80 S.20	0.000	0.000	0.000	0.000	0.000	0.000	0.000	0.001	0.0445
S.2050.D.80 S.50	0.000	0.000	0.000	0.000	0.000	0.000	0.000	0.000	0.0227

	Criteria								
Scenario	TC (US$ million per 10 years)	TCo (%)	DDR (%)	RR (%)	HQRC (%)	AD (%)	GWP (%)	NP	d_j^-
S.2050.Base	0.000	0.000	0.000	0.000	0.000	0.000	0.000	0.001	0.0335
S.2050.S.20	0.000	0.000	0.000	0.000	0.000	0.001	0.000	0.001	0.0436
S.2050.S.50	0.000	0.000	0.000	0.000	0.000	0.003	0.000	0.001	0.0779
S.2050.D.40	0.000	0.000	0.000	0.000	0.000	0.004	0.000	0.000	0.0692
S.2050.D.40 S.20	0.000	0.000	0.001	0.000	0.000	0.008	0.000	0.000	0.1008
S.2050.D.40 S.50	0.001	0.001	0.002	0.001	0.000	0.015	0.000	0.000	0.1420
S.2050.D.80	0.000	0.000	0.001	0.000	0.000	0.009	0.000	0.000	0.1014
S.2050.D.80 S.20	0.000	0.000	0.001	0.001	0.000	0.011	0.000	0.000	0.1157
S.2050.D.80 S.50	0.001	0.001	0.002	0.001	0.000	0.016	0.000	0.000	0.1470

TABLE 16.7   Ranking outcome of R_j for the year 2010

	Scenarios ranking (best to worse scenarios)				
Scenario	S.Free.DF5	S.Free.TC7	S.TC7	S.DF2	S.DF5
TOPSIS results	0.8462	0.7381	0.7024	0.6926	0.6906
Scenario	S.Free.DF2	S.Base	S.TC30	S.Free	S.Free.TC30
TOPSIS results	0.6875	0.6366	0.4782	0.4616	0.2644

16.2.2   Case 2—Choosing Waste Collection System

The second case study examines the problem of waste collection schemes for commingled waste in Lisbon, Portugal (Figure 16.3). Waste collection systems can be divided in two main types: curbside collection and neighborhood collection. In

TABLE 16.8   Ranking outcome of R_j for the year 2050

	Scenarios ranking (best to worse scenarios)				
Scenario	S.2050.D80.S50	S.2050.D40.S50	S.2050.D80.S20	S.2050.D40.S20	S.2050.D80
TOPSIS results	0.8661	0.8309	0.7221	0.6526	0.6184
Scenario	S.2050.S50	S.2050.D40	S.2050.S20	S.2050.Base	
TOPSIS results	0.4869	0.4490	0.2649	0.1855	

TABLE 16.9   Sensitivity analysis – ranking outcome of R_j for the year 2010

	Ranking based on weights: TC, TCo, and DDR = 0.25; other criteria = 0.05				
Scenario	S.Free.TC7	S.Free.DF5	S.Free.DF2	S.TC7	S.DF2
TOPSIS results	0.8008	0.7262	0.6826	0.5994	0.5585
Scenario	S.DF2	S.DF2	S.Free	S.TC30	S.Free.TC30
TOPSIS results	0.5565	0.5270	0.5269	0.4396	0.3687
	Ranking based on weights: AD, WP, and DDR = 0.25; other criteria = 0.05				
Scenario	S.Free.DF5	S.DF5	S.TC7	S.DF2	S.Base
TOPSIS results	0.946	0.918	0.908	0.896	0.781
Scenario	S.Free.TC7	S.Free.DF2	S.TC30	S.Free	S.Free.TC30
TOPSIS results	0.706	0.661	0.557	0.376	0.067

curbside collection, each family has a single container that is collected by the vehicle at dwellings or buildings up to three floors. In neighborhood collection, a single container serves several waste producers. The container's location can be near an apartment complex or building, where each family is responsible for dropping their waste into the container. These areas also provide specific containers (called *Ecopontos* in Portuguese) for packaging waste to be sorted by the user—drop-off or bring system. Another option, zone containers (called *Ecoilhas* in Portuguese) system, consists in the location of the commingled recyclables or residual/mixed MSW container near to source separated packaging waste containers all together at central areas, for several dwellings, apartments, commercial and services.

Waste collection and shipping are expensive relative to other SWM treatment operations due to labor cost, fuel costs, and vehicles and container maintenance expenses. Currently, a wide range of technical options exists for MSW collection systems, but MADM can be applied to help decision makers select the best solution for each case by evaluating indicators that represent each waste collection system.

TABLE 16.10 Sensitivity analysis–ranking outcome R_j for the year 2050

Ranking based on weights: TC, TCo and DDR = 0.25; other criteria = 0.05

Scenario	S.2050.D80.S50	S.2050.D40.S50	S.2050.D80.S20	S.2050.D40.S20	S.2050.S50
TOPSIS results	0.9294	0.9003	0.6582	0.5727	0.5326
Scenario	S.2050.D80	S.2050.D40	S.2050.S20	S.2050.Base	
TOPSIS results	0.4893	0.3361	0.2471	0.1006	

Ranking based on weights: Weight of AD, WP and DDR = 0.25; other criteria = 0.05

Scenario	S.2050.D80.S50	S.2050.D40.S50	S.2050.D80.S20	S.2050.D80	S.2050.D40.S20
TOPSIS results	0.9669	0.9490	0.7966	0.6992	0.6981
Scenario	S.2050.D40	S.2050.S50	S.2050.S20	S.2050.Base	
TOPSIS results	0.4790	0.4600	0.1910	0.0481	

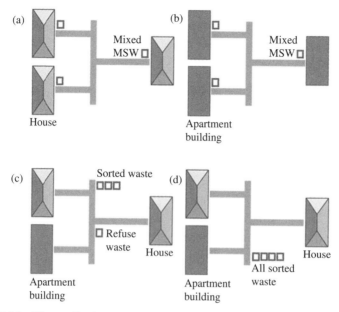

FIGURE 16.3 Waste collection systems: (a) dwelling curbside (DC), (b) apartments curbside (AC), (c) refuse waste neighborhood collection with drop-off containers for recyclables (RNC), and (d) zone containers (ZC)

The case study is based on indicators to assess the best waste collection system for dwelling areas and for apartment building areas in Lisbon, Portugal's capital, to collect commingled or refuse MSW. To do so, several indicators from Santos (2011) can be used in the MADM. Following the traditional MADM procedure, the criteria defined to compare waste collection options are technical, technical/economic, and technical/environmental criteria. Technical/economic and technical/environmental criteria are technical aspects with economic or environmental impacts; however, due to missing cost data these considerations were not included. The relevant attributes of waste collection systems (Table 16.11) compare apartments curbside (AC), zone containers (ZC), refuse neighborhood collection (RNC), and dwelling curbside (DC).

TABLE 16.11 Criteria selected and relevant attributes

Criteria	AC	DC	ZC	RNC
Distance per ton collected (km·tonne⁻¹)	0.84	3.11	2.04	1.13
Effective time per ton collected (hour·tonne⁻¹)	0.24	0.54	0.3	0.47
Fuel consumption per ton (L·tonne⁻¹)	4.52	7.7	7.07	5.27
Number of containers collected per tonne of waste (n.°·tonne⁻¹)	38.22	32.7	12.33	55.66
Amount of waste collected per effective time (tonne·h⁻¹)	4.35	1.93	3.84	2.32
Collection speed (km·h⁻¹)	3.41	5.74	6.5	2.41
Effective time work per total circuit collection time (%)	64.88	72.52	54.31	67.94

Source: Santos (2011).

Equal weights were applied to all criteria in this case study. Simple additive weighting (SAW) is an intuitive and simple MADM method common in many other applications, as described in Chapter 8. The steps involved in SAW are (adapted from Setiawan and Sadiq (2011):

Step 1. Normalization of the matrix

SAW method requires a process of normalization which is provided by

$$r_{ij} = \begin{cases} \dfrac{x_{ij}}{\max x_{ij}} \\ \dfrac{\min x_{ij}}{x_{ij}} \end{cases} . \tag{16.14}$$

$\dfrac{x_{ij}}{\max x_{ij}}$ is used if the attribute/criteria is benefit,

$\dfrac{\min x_{ij}}{x_{ij}}$ is used if the attribute/criteria is costs,

where r_{ij} is the normalized performance ratings of alternatives A_i on attributes C_j, $i = 1, 2, \ldots, m$ and $j = 1, 2, \ldots, n$.

Step 2. Weight calculation

An equal weighting criteria of 1/7 for each weight will be applied.

Step 3. Construct the final matrix

Preference value for each alternative (V_i) is given as

$$V_i = \sum_{j=1}^{n} w_j r_{ij}. \tag{16.15}$$

Step 4. Rank the results

From comparing AC with ZC and RNC, the obtained ranking outcome (Table 16.12) indicates that the best waste collection system for apartment buildings areas is the AC, and the worst is the RNC.

TABLE 16.12   SAW results for apartment building areas

Criteria	AC	ZC	RNC
Distance per ton collected (km·tonne^{-1})	1.000	0.412	0.743
Effective time per ton collected (h·tonne^{-1})	1.000	0.800	0.511
Fuel consumption per ton (L·tonne^{-1})	1.000	0.639	0.858
Number of containers collected per tonne of solid waste (n.°·tonne^{-1})	0.687	0.222	1.000
Amount of waste collected per effective time (tonne·h^{-1})	1.000	0.883	0.533
Collection speed (km·h^{-1})	0.525	1.000	0.371
Effective time work per total circuit collection time (%)	0.837	1.000	0.799
SAW results	0.756	0.619	0.602

TABLE 16.13   SAW results for dwelling area

Criteria	DC	ZC	RNC
Distance per ton collected (km·tonne^{-1})	0.363	0.554	1.000
Effective time per ton collected (h·tonne^{-1})	0.556	1.000	0.638
Fuel consumption per ton (L·tonne^{-1})	0.684	0.745	1.000
Number of containers collected per ton of waste (n.°·tonne^{-1})	0.922	0.222	1.000
Amount of waste collected per effective time (tonne·h^{-1})	0.503	1.000	0.604
Collection speed (km·h^{-1})	0.883	1.000	0.371
Effective time work per total circuit collection time (%)	0.749	1.000	0.799
SAW results	0.583	0.690	0.677

TABLE 16.14   Sensitivity analysis

Apartment buildings area			
Collection alternatives	AC	ZC	RNC
SAW results	0.756	0.619	0.602
Collection alternatives	AC	ZC	RNC
TOPSIS results	0.635	0.438	0.500
Dwelling area			
Collection alternatives	DC	ZC	RNC
SAW results	0.583	0.690	0.677
Collection alternatives	DC	ZC	RNC
TOPSIS results	0.451	0.543	0.569

Comparing DC with ZC and RNC, the obtained ranking outcome (Table 16.13) indicates that the best waste collection system for dwelling area is the ZC, and the worst result is DC.

As in the first case study, sensitivity analysis can be conducted based on different MADM methods, in this case between SAW and TOPSIS. The results obtained from comparative sensitivity analysis (Table 16.14) show that for apartment building areas, the best waste collection system is still the curbside collection. In the case of dwelling areas, the results now show that refuse collective collection is the best solution; however, the MADM is consistent in that curbside collection for houses is not the best solution, considering the criteria selected and the weight applied. The results are indicative of the strength and weakness of indicators used to compare waste collection systems. Even if these indicators can be obtained, the choice of a waste collection system is also dependent on other factors also as important as those used in the analysis, such as investment costs, noise, and pollution.

16.3   FINAL REMARKS

Making decisions with multiple attributes is by no means a trivial exercise, especially for cases involving sustainability criteria for SWM planning, design, and operation.

Identifying alternatives, determining weighting factors, evaluating the alternatives, and interpreting the final outcome must be seamlessly woven together. The project alternatives to be considered may be highly technically oriented and complex in relation to all proven and emerging technologies, and the economic, environmental, and social impacts might be intertwined with one another. System integration with varying levels of technology complexity could result in more intricacy. In particular, decisions involving sustainability issues tend to include an array of objective attributes based on highly subjective value judgments. In those cases, SWM have to find a process to lead qualitative attributes including social and environmental impacts into the quantitative decision-making process. In all circumstances, today's solid waste managers confront difficult decisions on a daily basis and must consider an increasingly wide range of criteria in making those decisions.

REFERENCES

Alidi, A. S. 1996. A multiobjective optimization model for the waste management of the petrochemical industry. *Applied Mathematical Modelling*, 20(12), 925–933.

Bastin, L. and Longden, D. M. 2009. Comparing transport emissions and impacts for energy recovery from domestic waste (EfW): centralised and distributed disposal options for two UK Counties. *Computers, Environment and Urban Systems*, 33(6), 492–503.

Chang, N.-B., Chang, Y.-H., and Chen, H.-W. 2009. Fair fund distribution for a municipal incinerator using GIS-based fuzzy analytical hierarchy process. *Journal of Environmental Management*, 90(1), 441–454.

Chang, N.-B., Pires, A., and Martinho, G. 2011. Empowering Systems Analysis for Solid Waste Management: Challenges, Trends, and Perspectives. *Critical Reviews in Environmental Science and Technology*, 41(16), 1449–1530.

Chang, N.-B., Qi, C., Islam, K., and Hossain, F. 2012. Comparisons between global warming potential and cost–benefit criteria for optimal planning of a municipal solid waste management system. *Journal of Cleaner Production*, 20(1), 1–13.

Cheng, S., Chan, C. W., and Huang, G. H. 2003. An integrated multicriteria decision analysis and inexact mixed integer linear programming approach for solid waste management. *Engineering Applications of Artificial Intelligence*, 16(5–6), 543–554.

Chiueh, P.-T., Lo, S.-L., and Chang, C.-L. 2008. A GIS-based system for allocating municipal solid waste incinerator compensatory fund. *Waste Management*, 28(12), 2690–2701.

Chu, A. T. W., Kalaba, R. E., and Spingarn, K. 1979. A comparison of two methods for determining the weights of belonging o fuzzy sets. *Journal of Optimisation Theory and Application*, 27, 531–538.

Contreras, F., Hanaki, K., Aramaki, T., and Connors, S. 2008. Application of analytical hierarchy process to analyze stakeholder's preferences for municipal solid waste management plans, Boston, USA. *Resources, Conservation and Recycling*, 52(7), 979–991.

Courcelle, C., Kesmont, M., and Tyteca, D. 1998. Assessing the economic and environmental performance of municipal solid waste collection and sorting programmes. *Waste Management & Research*, 16, 253–263.

Dalkey, N. and Helmer, O. 1963. An experimental application of the Delphi method to the use of experts. *Management Science*, 9, 458–467.

Dawes, R. M. and Corrigan, B. 1974. Linear models in decision making. *Psychological Bulletin*, 81, 95–106.

Dursun, M., Karsak, E. E., and Almula, M. 2011. Expert systems with applications a fuzzy multi-criteria group decision making framework for evaluating health-care waste disposal alternatives. *Expert Systems with Applications*, 38(9), 11453–11462.

Edwards, W. 1977. How to use multiattributive utility measurement for social decision making. *IEEE Transactions on Systems, Man, and Cybernetics SMC*, 7(5), 326–340.

Edwards, W. and Barron, F. H. 1994. SMARTS and SMART: improved simple methods for multiattributive utility measurement. *Organizational Behavior and Human Decision Processes*, 60(3), 306–325.

El Hanandeh, A. and El-Zein, A. 2009. Strategies for the municipal waste management system to take advantage of carbon trading under competing policies: the role of energy from waste in Sydney. *Waste Management*, 29(7), 2188–2194.

El Hanandeh, A. and El-Zein, A. 2010. The development and application of multi-criteria decision-making tool with consideration of uncertainty: the selection of a management strategy for the bio-degradable fraction in the municipal solid waste. *Bioresource Technology*, 101(2), 555–561.

European Commission. 2012. Construction and demolition waste (CDW). Waste. Available at: http://ec.europa.eu/environment/waste/construction_demolition.htm (accessed July 3, 2013).

European Parliament and Council. 2008. Directive 2008/98/EC of the European Parliament and of the Council of 19 November 2008 on waste and repealing certain directives. *Official Journal of the European Union*, L312, 3–30.

Gao, S., Zhang, Z., and Cao, C. 2010. Calculating weights methods in complete matrices and incomplete matrices. *Journal of Software*, 5(3), 304–311.

Generowicz, A., Kulczycka, J., Kowalski, Z., and Banach, M. 2011. Assessment of waste management technology using BATNEEC options, technology quality method and multi-criteria analysis. *Journal of Environmental Management*, 92(4), 1314–1320.

Gomes, C. F. S., Nunes, K. R. A., Xavier, L. H., Cardoso, R., and Valle, R. 2008. Multicriteria decision making applied to waste recycling in Brazil. *Omega*, 36(3), 395–404.

Haerer, W. 2000. Criterium decision plus 3.0—versatile multi-criteria tool excels in its ability to support decision-making. ORMS. Available at: http://www.orms-today.org/orms-2-00/cdpreview.html (accessed December 29, 2012).

Hanan, D., Burnley, S., and Cooke, D. 2013. A multi-criteria decision analysis assessment of waste paper management options. *Waste Management*, 33(3), 566–573.

He, B. 2010. Selection of the reverse logistics mode of the waste appliances based on ANP. In: *International Conference on Future Information Technology and Management Engineering* (Ed. Zhou, Q.), Institute of Electrical and Electronics Engineers, Inc., Changzhou, China, pp. 134–138.

Herva, M. and Roca, E. 2013. Ranking municipal solid waste treatment alternatives based on ecological footprint and multi-criteria analysis. *Ecological Indicators*, 25, 77–84.

Hiete, M., Stengel, J., Ludwig, J., and Schultmann, F. 2011. Matching construction and demolition waste supply to recycling demand: a regional management chain model. *Building Research & Information*, 39(4), 333–351.

Hokkanen, J. and Salminen, P. 1997. Choosing a solid waste management system using multicriteria decision analysis. *European Journal of Operational Research*, 98(1), 19–36.

Hung, M., Ma, H., and Yang, W. 2007. A novel sustainable decision making model for municipal solid waste management. *Waste Management*, 27(2), 209–219.

Hwang, C.-H. and Yoon, K. 1981. *Multiple Attribute Decision Making: Methods and Applications*, Springer, New York.

Jahan, A., Ismail, M. Y., Shuib, S., Norfazidah, D., and Edwards, K. L. 2011. An aggregation technique for optimal decision-making in materials selection. *Materials & Design*, 32(10), 4918–4924.

Jahanshahloo, G. R., Lotfi, F. H., and Izadikhah, M. 2006. An algorithmic method to extend TOP- SIS for decision-making problems with interval data. *Applied Mathematics and Computation*, 175(2), 1375–1384.

Jin, J. and Wei, Y. M. 2008. *Generalized Intelligent Assessment Methods for Complex Systems and Applications*, Science Press, Beijing.

Karagiannidis, A. and Moussiopoulos, N. 1997. Application of ELECTRE III for the integrated management of municipal solid wastes in the Greater Athens area. *European Journal of Operational Research*, 97(3), 439–449.

Karagiannidis. A. and Perkoulidis, G. 2009. A multi-criteria ranking of different technologies for the anaerobic digestion for energy recovery of the organic fraction of municipal solid wastes. *Bioresource Technology*, 100(8), 2355–2360.

Karmperis, A. C., Sotirchos, A., Aravossis, K., and Tatsiopoulos, I. P. 2012. Waste management project's alternatives: a risk-based multi-criteria assessment (RBMCA) approach. *Waste Management*, 32(1), 194–212.

Kaya, İ. 2012. Evaluation of outsourcing alternatives under fuzzy environment for waste management. *Resources, Conservation and Recycling*, 60, 107–118.

Keeney, R. L. and Raiffa, H. 1976. *Decisions with Multiple Objectives: Preferences and Value Tradeoffs*, John Wiley & Sons, New York.

Kijak, R. and Moy, D. 2004. A decision support framework for sustainable waste management. *Journal of Industrial Ecology*, 8(3), 33–50.

Kourmpanis, B., Papadopoulos, A., Moustakas, K., Kourmoussis, F., Stylianou, M., and Loizidou, M. 2008. An integrated approach for the management of demolition waste in Cyprus. *Waste Management & Research*, 26(6), 573–581.

Kumar, A. 2009. Fusion, rank-level. In: *Encyclopedia of Biometrics, I–Z Vol.2* (Eds Li, S. Z. and Jain, A. K.), Springer, New York, pp. 607–611.

Liang, X., Sun, X., Shu, G., Sun, K., Wang, X., and Wang, X. 2013. Using the analytic network process (ANP) to determine method of waste energy recovery from engine. *Energy Conversion and Management*, 66, 304–311.

Lin, C.-H., Wen, L., and Tsai, Y.-M. 2010. Applying decision-making tools to national e-waste recycling policy: an example of analytic hierarchy process. *Waste Management*, 30(5), 863–869.

LINDO Systems. 2003. LINGO – Optimization Modeling Software for Linear, Nonlinear, and Integer Programming. http://www.lindo.com/ (accessed January, 2013).

Ma, J., Fan, Z.-P., and Huang, L.-H. 1999. A subjective and objective integrated approach to determine attribute weights. *European Journal of Operational Research*, 112(2), 397–404.

MacDonald, M. L. 1996. A multi-attribute spatial decision support system for solid waste planning. *Computers, Environment and Urban Systems*, 20(1), 1–17.

Madadian, E., Amiri, L., and Abdoli, M. A. 2013. Application of analytic hierarchy process and multicriteria decision analysis on waste management: a case study in Iran. *Environmental Progress & Sustainable Energy*, 32(3), 810–817.

Morrissey, A. J. and Browne, J. 2004. Waste management models and their application to sustainable waste management. *Waste Management*, 24(3), 297–308.

Morton, A., Airoldi, M., and Phillips, L. D. 2009. Nuclear risk management on stage: a decision analysis perspective on the UK's Committee on Radioactive Waste Management. *Risk Analysis*, 29(5), 764–779.

Olson, D. L. 1996. *Decision Aids for Selection*, Springer, New York.

Perkoulidis, G., Papageorgiou, A., Karagiannidis, A., and Kalogirou, S. 2010. Integrated assessment of a new waste-to-energy facility in Central Greece in the context of regional perspectives. *Waste Management*, 30(7), 1395–1406.

Pires, A., Chang, N.-B., and Martinho, G. 2011. An AHP-based fuzzy interval TOPSIS assessment for sustainable expansion of the solid waste management system in Setúbal Peninsula, Portugal. *Resources, Conservation and Recycling*, 56(1), 7–21.

Pomerol, J. and Barba-Romero, S. 2000. *Multicriterion Decision in Management: Principles and Practices*, Kluwer Academic Publishers, Boston, MA.

Rousis, K., Moustakas, K., Malamis, S., Papadopoulos, A., and Loizidou, M. 2008. Multi-criteria analysis for the determination of the best WEEE management scenario in Cyprus. *Waste Management*, 28(10), 1941–1954.

Roussat, N., Dujet, C., and Méhu, J. 2009. Choosing a sustainable demolition waste management strategy using multicriteria decision analysis. *Waste Management*, 29(1), 12–20.

Saaty, T. 1980. *The Analytical Hierarchy Process*, McGraw-Hill, New York.

Santos, P. H. M. R. 2011. Waste collection circuits evaluation: operational indicators [Portuguese: Avaliação de circuitos de recolha de resíduos urbanos: indicadores operacionais]. MSc. Dissertation. Faculdade de Ciências e Tecnoogia da Universidade Nova de Lisboa. Lisbon.

Setiawan, M. A. and Sadiq, S. 2011. Experience driven process improvement. In: *Enterprise, Business-Process and Information Systems Modeling – 12th International Conference, BPMDS 2011 and 16th International Conference, EMMSAD 2011 Held at CAiSE 2011* (Eds Halpin, T., Nurcan, S., Krogstie, J., Soffer, P., Proper, E., Schmidt, R., and Bider, I.), London, UK, June 2011, Proceedings. Springer, Berlin, Heidelberg, pp. 75–87.

Simos, H. A. 1990. L'Evaluation Environnementale: Un Processus Cognitive Négocié. Thèse de doctoral, DGF-EPFL, Lausanne.

Skordilis, A. 2004. Modelling of integrated solid waste management systems in an island. *Resources, Conservation and Recycling*, 41, 243–254.

Su, J.-P., Chiueh, P.-T., Hung, M.-L., and Ma, H.-W. 2007. Analyzing policy impact potential for municipal solid waste management decision-making: a case study of Taiwan. *Resources, Conservation and Recycling*, 51(2), 418–434.

Tseng, M. L. 2009. Application of ANP and DEMATEL to evaluate the decision-making of municipal solid waste management in Metro Manila. *Environmental Monitoring Assessment*, 156(1–4), 181–197.

Tzeng, G.-H. and Huang, J.-J. 2011. *Multiple Attribute Decision Making: Methods and Applications*, CRC Press, Boca Raton, FL.

Üçüncüoğlu, C. and Ulukan, H. Z. 2010. A fuzzy multi-attribute method for the evaluation of WEEE management scenarios. In: *Computational Intelligence: Foundations and*

Applications: Proceedings of the 9th International FLINS Conference (Eds Ruan, D., Li, T., Xu, Y., Chen, G., and Kerre, E. E.), World Scientific Publishing, Singapore, pp. 348–354.

Vego, G., Kucar-Dragicevic, S., and Koprivanac, N. 2008. Application of multi-criteria decision-making on strategic municipal solid waste management in Dalmatia, Croatia. *Waste Management*, 28(11), 2192–2201.

von Winterfeldt, D. and Edwards, W. 1986. *Decision Analysis and Behavioral Research*, Cambridge University Press, Cambridge, UK.

Wackernagel, M. and Rees, W. 1996. *Our Ecological Footprint: Reducing Human Impact on the Earth*, New Society Publishers, Gabriola Island.

Wang, J.-J., Jing, Y.-Y., Zhang, C.-F., and Zhao, J.-H. 2009a. Review on multi-criteria decision analysis aid in sustainable energy decision-making. *Renewable and Sustainable Energy Reviews*, 13(9), 2263–2278.

Wang, J.-J., Jing, Y.-Y., and Zhang, C.-F. 2009b. Weighting methodologies in multi-criteria evaluations of combined heat and power systems. *International Journal of Energy Research*, 33(12), 1023–1039.

Weistroffer, H. R., Smith, C. H., and Narula, S. C. 2005. Multiple criteria decision support software. In: *Multiple Criteria Decision Analysis – State of the Art Surveys* (Eds Figueira, J., Greco, S., and Ehrgott, M.), Springer, New York, pp. 989–1018.

Xiao, X. and Guo, H. 2010. Optimization method of grey relation analysis based on the minimum sensitivity of attribute weights. In: *Advances in Grey Systems Research* (Eds Liu, S. and Forrest, J. Y.-L.), Springer, Berlin, Heidelberg, pp. 177–190.

Ye, Y. C., Ke, L. H., and Huang, D. Y. 2006. *System Synthetical Evaluation Technology and Its Application*, Metallurgical Industry Press, Beijing.

Yedla, S. 2012. Replication of urban innovations—prioritization of strategies for the replication of Dhaka's community-based decentralized composting model. *Waste Management & Research*, 30(1), 20–31.

Zhang, F. T., Su, W. C., and Zhou, J. X. 2008. Assessment of urban ecological security based on entropy-weighted grey correlation analysis. *Chinese Journal of Ecology*, 28(7), 1249–1254.

CHAPTER 17

DECISION ANALYSIS FOR OPTIMAL BALANCE BETWEEN SOLID WASTE INCINERATION AND RECYCLING PROGRAMS

Rising prices of raw materials and concerns of energy conservation have resulted in two goals: the simultaneous reuse of recyclables and recovery of energy from the waste streams. Concerns of compatibility between these two goals exist due to several economic, environmental, and managerial reasons. Because heating values of waste streams may deviate from the design levels for waste incineration, one way to balance both goals is to install an on-site or off-site facility to presort waste before it reaches the incinerator. The other feasible alternative is to achieve successful household recycling programs to fine-tune the heating values of waste stream. In some urban regions, both successful household recycling programs and material recovery facilities may become essential components in their integrated solid waste management (SWM) systems. If the household recycling program cannot succeed in local communities, the regional impacts of presorting solid waste prior to the waste-to-energy facility must be systematically assessed due to the inherent complexity of solid waste composition and quantity over different service areas. This chapter begins with a series of regression analyses of heating values based on the products generated by a typical refuse-derived fuel (RDF) system designed to coordinate subsequent waste-to-energy facilities for improving the balance between reuse of recyclables and recovery of energy system-wide. The second part of this chapter applies a system engineering model to assess the impact of installing an RDF incineration system to ensure that the optimal size of the RDF process and the associated shipping patterns correspond to the regional demand. The integrated effort will demonstrate how municipal solid waste (MSW) with different rates of generation, physical and chemical compositions, and heating values collected from various administrative

Sustainable Solid Waste Management: A Systems Engineering Approach, First Edition. Ni-Bin Chang and Ana Pires.
© 2015 The Institute of Electrical and Electronics Engineers, Inc. Published 2015 by John Wiley & Sons, Inc.

districts can be properly handled within a harmonized framework. A case study conducted in Taipei County, Taiwan, one of the most densely populated metropolitan areas in the world, demonstrates the application potential of such a methodology.

17.1 SYSTEMS ANALYSIS FOR INTEGRATED MATERIAL RECYCLING AND WASTE-TO-ENERGY PROGRAMS

Systems analysis for SWM has received wide attention from both economic and environmental planners because of the complex nature of these multifaceted linkages between source generation, collection, recycling, treatment, and disposal (Chang and Wang, 1997). The focal point of research is usually linked with the holistic SWM system with the aid of system of systems engineering approaches from waste collection to recycling of separate materials, and to waste treatment processes such as incineration, composting, and landfilling. Models for parts of the system, as well as models covering the overall system for various temporal and spatial criteria, are of interest to environmental planners in many parts of the world (Chang and Wang, 1996a,1996b; Chang et al., 1997).

Several studies were conducted in European countries and the United States (Bunsow and Dobberstein, 1987; Jackson, 1987; Sommer et al., 1989) that explored the efficiency of resource recovery from solid waste (Wilson, 1979); yet the impacts of material recovery on heating value remain unclear when processing local MSW with relatively complex composition. A number of surveys in the literature, however, have shown that recycling and waste-to-energy seem to work well together (Jackson, 1987). While the rising prices of raw materials and benefits from waste recycling have increased concern for material recovery technologies and reuse potential, thermal treatment using incineration technologies has become an attractive method prior to waste disposal based on hygienic control, volume reduction, and energy recovery (Chang and Chang, 2003).

Due to fast urbanization in developing countries, common SWM problems facing many metropolitan regions around the world include but are not limited to insufficient design capacity for waste treatment and disposal, residents' reluctance to recycle, and the rapid increase of solid waste generation within the designated service area. Proper integration of a presorting process to be associated with a target incinerator could possibly alleviate the pressure of solid waste disposal in the short-run. Previous experience in SWM indicates that solid waste presorting prior to incineration is solely a function of material recycling, but the economic feasibility of presorting facilities is difficult to justify due to unstable prices of recyclable goods in the secondary material market. In recent years, the focus has been changing in response to increasing public health and environmental concerns of increased heating values, incinerator emissions, and ash properties. Recognizing the value of integrating solid waste presorting before the incineration process would present a new perspective in SWM that could end up coordinating environmental benefits from solid waste presorting via waste minimization and pollution prevention, improving incinerator performance, and realizing direct market revenues from recycled materials.

Despite the potential benefits, this type of regionalization program among several districts in an SWM system is sometimes difficult to analyze because of the managerial complexity of multiple private–public partnerships (Chang and Chang, 1998, 2001). Considering only one incineration facility associated with its service region at a time could be more achievable and applicable when assessing real-world SWM systems based on the "system of systems engineering approach." Although empirical studies evaluated the calorific value of RDF in the United States (Kirklin et al., 1982; Buckley and Domalski, 1988), reexamining them based on the local MSW in a case-based engineering practice is worthwhile. The RDF production process introduced in the next section was proven functionally effective for subsequent systems analysis of the integrative potential between a presorting process and a waste-to-energy facility (Chang et al., 1998a). With this philosophy, technological evolutions, such as RDF processes and material recovery facilities, can be flexibly factored into a systems engineering model to simultaneously achieve higher levels of economic and environmental goals.

Possible interactions between solid waste presorting processes and waste-to-energy facilities were fully investigated by Chang et al. (1997, 1998a, 1998b) from a regional perspective. From these studies (Chang et al., 1997, 1998a, 1998b), a short-term operating policy may be optimally arranged and confirmed with respect to not only the energy recovery targets and throughput requirements in an incineration plant, but also a cost-effective shipping strategy in terms of an emergent presorting process. A nonlinear programming model for assessing the optimal size of a presorting process prior to an incinerator became essential for seeking the near-optimal solution via a suite of cost-effective shipping patterns. These optimal shipping patterns are designed to balance or reconcile the impacts resulting from different solid waste inflows with varying heating values and contents of recyclables to meet the energy recovery, material recycling, and throughput requirements in an incineration program. Such a thrust may provide insight into bridging the gaps among technological evolution, recycling efforts, and waste management efficiency from time to time. The practical implementation of this nonlinear programming model was eventually assessed by a case study in Taipei County, Taiwan, a typical metropolitan area with rapid economic development, population growth, and rapid urbanization. This chapter presents and evaluates a typical solid waste presorting plant (Section 17.2) developed in Taiwan at first. The operations included in the presorting pilot plant consist of several standard units, such as mechanical shredding, magnetic separation, trommel screening, and air classification. It is followed by the formulation of a regional optimization model (Section 17.3) to achieve the balance between energy recovery and waste recycling goals simultaneously in a typical SWM system in Taipei, Taiwan.

17.2   REFUSE-DERIVED FUEL PROCESS FOR SOLID WASTE MANAGEMENT

17.2.1   The Refuse-Derived Fuel Process

The design of the proposed presorting process consists of three major subsystems: shredding, air classification, and screening. The facility can process 30 tonnes·per

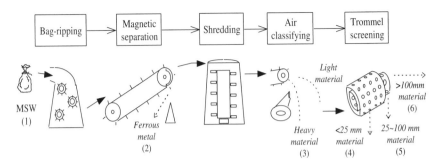

FIGURE 17.1 Flowchart of the solid waste presorting process. *Source*: From Chang et al. (1997)

hour at maximum capacity per line (Figure 17.1) (tonnes are metric tons) (Chang et al., 1997, 1998a, 1998b). The MSW is packed in plastic bags and delivered by packer trucks to the facility, where a bag-ripping unit initializes the sorting process. Following the bag-ripping unit, ferrous metal is extracted from the MSW stream by magnets and is conveyed to a ferrous storage bin for recycling (Chang et al., 1997, 1998a, 1998b). The remaining MSW is then moved on belt-type conveyor into a vertical hammer mill, followed by an air classifier (Chang et al., 1997, 1998a, 1998b). Nonferrous materials, such as aluminum cans, as well as other MSW, are crushed by the vertical hammer mill (Chang et al., 1997, 1998a, 1998b). The air classifier, blowing a regular air stream of 200 $m^3 \cdot min^{-1}$ from the vertical hammer mill, further isolates and separates the inert materials such as glass and ceramics to reduce the content of heavy material in the residual MSW streams (Chang et al., 1997, 1998a, 1998b).

Light materials passing through the air classifier are sent into the trommel screen for advanced separation (Chang et al., 1997, 1998a, 1998b). The dimensions of the openings on the surface of trommel screen can be varied to fine-tune the processing function and assure maximum combustibles recovery (Chang et al., 1997, 1998a, 1998b). The trommel is designed as two concentric shells; the outer shell, 2.33 m in diameter and 4.3 m in length, has many circular holes on the surface designed to remove the shredded materials smaller than 25 mm (trommel underflow); the inner shell, 1.9 m in diameter and 4.56 m in length, separates partial waste stream sized 25–100 mm (trommel middle flow) (Chang et al., 1997, 1998a, 1998b). A third waste stream sized >100 mm (trommel overflow) passes through the trommel screen (Chang et al., 1997, 1998a, 1998b). The lightest portion in the MSW (trommel underflow) is the end product of the sorting process, identified as fluff-RDF (Chang et al., 1997, 1998a, 1998b) (Box 17.1). Overflow and middle flow can be used as alternative fuels in the waste-to-energy facilities (Chang et al., 1997, 1998a, 1998b). Every unit is fully enclosed for noise and dust control, but odor control is not currently considered (Chang et al., 1997, 1998a, 1998b). A series of sampling and analysis programs were conducted, and the performance of a pilot plant was characterized so that the regression studies for the prediction of RDF heating values could be used for the subsequent system engineering analysis later in this chapter (Chang et al., 1997, 1998a, 1998b).

BOX 17.1   CLASSIFICATION OF RDF

The main difference between MSW incineration in a mass burn incinerator versus an RDF incinerator is that in RDF incineration, the MSW is processed prior to burning. Processing the MSW stream can vary from simple removal of bulky items and shredding to extensive processing into various types of RDF suitable for incineration or cofiring. These RDF production facilities can sometimes be combined with transfer stations rather than on-site facilities in front of an incinerator. RDF is classified by the American Society for Testing and Materials (ASTM) (Table 17.1).

TABLE 17.1   ASTM classification of RDF

Class	Form	Description
RDF-1 (MSW)	Raw	MSW with minimal processing to remove oversized bulky waste
RDF-2 (C-RDF)	Coarse	MSW processed to coarse particle size with or without ferrous metal separation such that 95% by weight passes through a 6 in^2 mesh screen
RDF-3 (F-RDF)	Fluff	Shredded fuel derived from MSW processed for the removal of metal, glass, and other entrained inorganics; 95% by weight of this particle size passes through a 2 in^2 mesh screen
RDF-4 (P-RDF)	Powder	Combustible waste fraction is processed into powdered form such that 95% by weight passes through a 10-mm mesh screen (0.035 in^2)
RDF-5 (D-RDF)	Densified	Combustible waste fraction is densified (compressed) into pellets, slugs, cubettes, briquettes, or similar forms
RDF-6	Liquid	Combustible waste fraction is processed into a liquid fuel
RDF-7	Gas	Combustible waste fraction is processed into a gas fuel

---

17.2.2 Experimental Results

A sample of more than 10 tonnes of MSW was collected from various locations (Figure 17.1) to conduct performance test of this presorting process during 1995 to 1996. The database is composed of some sample and analyses runs (Table 17.2). The mean, the variance, and the range of the measurements (Tables 17.3 and 17.4) show that while the plastics content dramatically increased from 26.33% in the MSW to 57.81% in the trommel overflow on a dry basis, food waste, metal, glass, and ceramics were reduced close to zero. The water content decreased from 50.65% in the MSW to 40.28% in the trommel overflow due to the evaporization effect during the air classification process (Chang et al., 1997, 1998a, 1998b). The combustible content increased from 37.15% in the MSW to 49.76% in the trommel overflow (Chang et al., 1997, 1998a, 1998b). In addition, chlorine and sulfur became the major elements for the examination of the impact on air pollution control (Chang et al., 1997, 1998a,

TABLE 17.2 Sampling background of RDF

Sampling date	Sampling location	Amount of samples
Period I: July 1995	(1) MSW	3
	(1) Heavy material	5
	(4) <25 mm	5
	(5) 25~100 mm	5
	(6) >100 mm	5
Period II: March 1996	(1) MSW	10
	(2) Heavy material	10
	(4) <25 mm	10
	(5) 25~100 mm	10
	(6) >100 mm	8

Source: From Chang et al. (1997).

1998b). The sulfur content decreased from 0.8% in the MSW to 0.05% in the trommel overflow, and the chlorine content was slightly increased from 0.18% in the MSW to 0.23%, probably due to the increase of plastics; however, the high heating value was increased from 2277 kcal·kg^{-1} in the MSW to 3715 kcal·kg^{-1} (Chang et al., 1997, 1998a, 1998b). Based on this study, an increase of almost 66% of high heating value in the RDF stream would result in a higher degree of energy recovery if a waste-to-energy facility is installed together with such a presorting process (Chang et al., 1997, 1998a, 1998b).

17.2.3 Regression Analysis to Predict Heating Value

A series of regression analysis conducted to specifically predict heating values found that the higher and lower heating values are strongly related to moisture content. The regression equations, in terms of the moisture content, fit well in both *t*-ratios and R^2. The numbers in the parentheses below each regression coefficient from Equations (17.1)–(17.6) represent the *t*-ratios in statistics in each regression model from which statistical inferences may be performed to test whether or not the corresponding regression coefficient is statically significant. A 5% level of significance was chosen as the critical value in the statistical testing procedure to indicate whether the coefficient is statistically significant. Both R^2 and adjusted R^2 values were presented for the adjustment of the degree of freedom.

- Trommel overflow (>100 mm):

$$\text{HHV} = 6295.7 - 64.05\, M_{(>100\text{ mm})}$$
$$(15.737)(-6.546)$$
$$R^2 = 0.7812 \qquad \text{adjusted } R^2 = 0.7630, \tag{17.1}$$

$$\text{LHV} = 5780.1 - 61.67 M_{(>100\text{ mm})}, \text{ and}$$
$$(15.372)(-6.707)$$
$$R^2 = 0.7894 \qquad \text{adjusted } R^2 = 0.7719. \tag{17.2}$$

TABLE 17.3   The mean and variance of the samples corresponding to each property at different sampling locations

	MSW		Heavy material		<25 mm		25~100 mm		>100 mm	
	Mean	Variance	Mean	Variance	Mean	Variance	Mean	Variance	Mean	Variance
Physical composition (on dry basis, wt%)										
Paper	28.62	8.57	2.50	1.73	0.10	0.24	8.08	3.29	5.70	3.09
Plastics	26.33	7.03	9.00	7.19	1.00	1.07	29.15	7.19	57.81	9.49
Garden trimmings	4.05	2.13	5.40	2.97	0.20	0.21	4.60	3.96	4.21	2.37
Textiles	9.03	6.90	9.30	7.96	0.00	0.00	7.43	3.38	18.23	6.71
Food waste	14.04	10.01	0.20	0.00	0.00	0.00	0.01	0.00	0.00	0.00
Leather/rubber	0.58	1.39	0.50	0.76	0.10	0.16	1.13	1.54	1.48	2.73
Metal	6.99	4.56	0.30	0.81	0.00	0.00	1.09	1.44	0.03	0.11
Glass	7.26	8.83	0.00	0.00	0.00	0.00	0.00	0.00	0.00	0.00
Ceramics and china	0.47	0.88	0.00	0.00	0.00	0.00	0.00	0.00	0.00	0.00
<5 mm	1.59	1.96	35.00	22.73	54.80	10.67	16.15	5.52	8.89	5.09
>5 mm	1.04	1.97	37.80	11.61	43.80	11.99	32.36	15.96	3.65	3.59
Heat value										
HHV (kcal·kg^{-1})	2277.8	477.2	1643.7	455.03	1376.8	310.6	2554.5	342.0	3715.9	551.6
LHV (kcal·kg^{-1})	1816.3	489.2	1183.2	484.28	936.7	331.7	2095.7	342.2	3296.0	528.4
Chemical composition (on wet basis, wt%)										
C	20.11	5.00	16.89	3.47	20.55	3.51	24.45	4.34	29.24	4.58
H	2.92	0.29	2.17	0.47	2.30	0.44	3.21	1.09	3.30	1.54
N	0.55	0.41	0.76	0.48	1.05	0.51	1.09	0.55	1.04	0.85
Cl	0.18	0.14	0.08	0.01	0.11	0.08	0.16	0.15	0.23	0.22
S	0.80	2.69	0.05	0.05	0.06	0.05	0.10	0.11	0.05	0.05
O	12.58	3.19	11.27	4.46	11.27	3.82	11.69	3.53	15.90	9.54
Proximate analysis (on wet basis, wt%)										
Moisture	50.65	6.71	59.20	4.90	52.65	5.43	47.55	4.87	40.28	7.61
Ash	12.21	5.90	11.02	4.18	12.00	2.99	11.75	5.00	9.96	4.78
Combustibles	37.14	10.96	29.78	5.57	35.35	3.60	40.70	5.35	49.76	9.96

Source: From Chang et al. (1997).

TABLE 17.4 The ranges of measurement of the samples corresponding to each property at different sampling locations

	MSW range	Heavy material range	<25 mm range	25~100 mm range	>100 mm range
Physical composition (on dry basis, wt%)					
Paper	16.9–47.5	0.3–5.3	0.0–0.7	3.7–14.2	2.7–11.0
Plastics	15.4–39.1	1.0–18.9	0.0–3.2	24.0–50.2	40.0–76.2
Garden trimmings	1.0–8.3	2.1–9.9	0.0–1.4	0.0–12.2	0.7–7.8
Textiles	0.0–22.1	0.0–20.0	0.0–0.0	3.1–15.6	6.0–26.1
Food waste	2.2–31.0	0.0–0.0	0.0–0.0	0.0–0.0	0.0–0.0
Leather/rubber	0.0–4.9	0.0–2.4	0.0–0.5	0.0–4.6	0.0–10.0
Metal	0.0–16.2	0.0–3.1	0.0–0.0	0.0–4.5	0.0–0.4
Glass	0.0–26.9	0.0–0.0	0.0–0.0	0.0–0.0	0.0–0.0
Ceramics and china	0.0–3.2	0.0–0.0	0.0–0.0	0.0–0.0	0.0–0.0
<5 mm	0.0–5.6	0.9–86.4	40.3–68.3	9.1–25.4	4.0–20.3
>5 mm	0.0–5.9	3.9–53.5	26.8–59.7	10.2–52.3	0.0–11.4
Heating value					
HHV (kcal·kg^{-1})	1708–3372	111–2432	864–1961	2120–3304	2986–4749
LHV (kcal·kg^{-1})	1234–2950	627–2042	364–1512	1719–2883	2599–4238
Chemical Composition (on wet basis, %)					
C	14.6–31.1	23.7–12.1	13.1–24.7	19.0–36.0	24.4–42.5
H	2.51–3.43	1.16–2.83	1.65–3.27	1.88–5.90	0.94–6.32
N	0.08–1.32	0.19–1.89	0.20–1.63	0.21–1.89	0.11–2.75
Cl	0.06–0.54	0.03–0.16	0.06–0.37	0.00–0.47	0.01–0.80
S	0.01–9.74	0.00–0.16	0.00–0.13	0.00–0.34	0.00–0.16
O	6.03–19.65	0.86–17.96	3.89–17.24	5.13–18.25	2.62–33.01
Proximate analysis (on wet basis, wt%)					
Moisture	42.6	50.5–68.2	47.9–64.8	37.4–54.6	28.2–48.7
Ash	3.9–20.4	4.7–21.1	2.4–15.0	1.9–18.4	3.4–18.3
Combustibles	11.4–45.2	22.4–39.2	28.4–40.6	32.9–51.9	38.9–67.0

Source: From Chang et al. (1997).

- Trommel middle flow (25~100 mm):

$$\text{HHV} = 5112.0 - 52.08\, M_{(25\sim100\text{ mm})}$$
$$(8.042)\ (-3.964)$$
$$R^2 = 0.5473 \qquad \text{adjusted } R^2 = 0.5125, \text{ and} \tag{17.3}$$

$$\text{LHV} = 4851.0 - 56.39\, M_{(25\sim100\text{ mm})}$$
$$(8.563)\ (-4.816)$$
$$R^2 = 0.6409 \qquad \text{adjusted } R^2 = 0.6132 \tag{17.4}$$

However, due to the existing of large proportion of plastics in the waste stream, plastics are also identified as a valid explanatory variable in the regression:

$$\text{HHV} = 5441.4 + 33.10\, P - 80.45\, M_{(25\sim100\text{ mm})}$$
$$(15.085)(5.436)(-8.926)$$
$$R^2 = 0.8692 \qquad \text{adjusted } R^2 = 0.8475, \text{ and} \tag{17.5}$$

$$\text{LHV} = 5150.0 + 30.04\, P - 82.14\, M_{(25\sim100\text{ mm})}$$
$$(16.808)(5.808)\quad(-10.728)$$
$$R^2 = 0.9058 \qquad \text{adjusted } R^2 = 0.8901. \tag{17.6}$$

- Integrated equation for both middle flow and overflow:
  Because both waste streams from trommel middle flow and overflow can be used as RDFs, an integrated prediction equation of HHV based on Equations (17.1) and (17.3) can be generated as (i.e., note that t-ratios and R^2 value are not available in this case.):

$$\text{HHV} = 5850.9 - 30.70 M_{(>100\text{ mm})} + 17.23 P_{(25\sim100\text{ mm})}$$
$$- 41.89 M_{(25\sim100\text{ mm})} \tag{17.7a}$$

In the same way, an integrated prediction equation of LHV based on Equations (17.2) and (17.4) can be generated as (i.e., note that t-ratios and R^2 value are not available in this case):

$$\text{LHV} = 5452.19 - 29.57 M_{(>100\text{ mm})} + 15.643 P_{(25\sim100\text{ mm})}$$
$$- 42.75 M_{(25\sim100\text{ mm})} \tag{17.7b}$$

where HHV = high heating value (kcal·kg^{-1}); LHV = low heating value (kcal·kg^{-1}); $M_{(25\sim100\text{ mm})}$ = the moisture content in the trommel middle flow; $M_{(>100\text{ mm})}$ = the moisture content in the trommel overflow; and $P_{(25\sim100\text{ mm})}$ = the plastics content in the trommel middle flow.

17.3 REGIONAL SHIPPING STRATEGIES

17.3.1 Formulation of Mathematical Programming Model

Although energy recovery is an important goal in most incineration projects, the variations of heating values of the solid waste streams and the rising prices of virgin materials in the market have resulted in a renewed interest in assessing the installation of a presorting facility prior to an existing incinerator. To handle the varying contents of solid waste streams over time, the focus of this cost–benefit analysis in an SWM system is to minimize the net value of total cost minus total benefit with respect to a multitude of technical, environmental, and economic criteria. The information incorporated into the objective function in this optimization model includes economic impacts associated with transportation costs, construction costs, and operating costs for waste treatment and disposal, as well as possible operational income from the recovery of recyclables and the sales of electricity/steam. The constraint set thereby consists of treatment capacity constraint, energy recovery constraint, mass balance constraint, and non-negativity requirements. The output of this model may be used to generate a set of management strategies to identify the optimal operational option in an SWM system with respect to the changing situation of waste characteristics and material/energy recovery goals. The generic expressions of the objective function and constraints are separately described (below). A yearly or daily basis is normally chosen as the time frame when the cash flow, solid waste stream, and the shipping strategy are considered in the planning scenarios simultaneously.

Objective Function

$$\text{Minimize } Z = \text{TC} + \text{CC} + \text{OC} - \text{IE} - \text{IR}, \tag{17.8}$$

where Z is the net value of total cost minus total benefit. The cost terms included in the objective function consist of

1. **Total transportation cost:** $\text{TC} = (\text{TC}_1 + \text{TC}_2 + \text{TC}_3 + \text{TC}_4 + \text{TC}_5)$, in which

$$\text{TC}_1 = \sum_i \sum_n (X_{in} T_{in} D_{in}) \text{ (from service area to landfills)};$$

$$\text{TC}_2 = \sum_i \sum_k (X_{ik} T_{ik} D_{ik}) \text{ (from service area to RDF plants)};$$

$$\text{TC}_3 = \sum_i \sum_j (X_{ij} T_{ij} D_{ij}) \text{ (from service area to incineration plants)}$$

$$\text{TC}_4 = \sum_k \sum_j (Y_{kj} T_{kj} D_{kj}) \text{ (from RDF plants to incineration plants)}; \text{ and}$$

$$\text{TC}_5 = \sum_j \sum_m (Z_{jm} T_{jm} D_{jm}) \text{ (from incineration plants to ash landfills)}.$$

2. **Total construction cost:** $CC = CC_1 + CC_2 + CC_3 + CC_4$, which includes construction costs for landfill(s), incineration plant(s), RDF plant(s), and ash landfill(s), respectively.

3. **Total operating cost:** $OC = OC_1 + OC_2 + OC_3 + OC_4$, in which

$$OC_1 = \sum_n \sum_i (X_{in}O_n) \text{ (for landfills)};$$

$$OC_2 = \sum_j \sum_i (X_{ij}O_j) \text{ (for incineration plants)};$$

$$OC_3 = \sum_k \sum_i (X_{ik}O_k) \text{ (for RDF plants); and}$$

$$OC_4 = \sum_m \sum_j (X_{jm}O_m) \text{ (for ash landfills)}.$$

4. **Total income from selling electricity:** $IE = IE_1 + IE_2$, in which

$$IE_1 = \sum_j \sum_i X_{ij}E_{ij}P_j \text{ (generated from incinerating MSW); and}$$

$$IE_2 = \sum_j \sum_k Y_{kj}E_{kj}P_j \text{ (generated from incinerating RDF)}.$$

5. **Total income from selling recyclables:** $IR = \sum_k \sum_i (r_i X_{ik} P_k)$, in which subscripts $i, j, k, n,$ and m represent waste collection service area, waste incinerator, RDF plant, landfill site, and ash landfill site, respectively, in an SWM system. $X_{ij}, X_{ik},$ and X_{in} are defined as the waste inflow (tonnes per year, TPY) from service area i destined directly for waste incinerator j, RDF plant k, and landfill site n, respectively; X_{jm} is the amount of ashes going to landfill (TPY); Y_{kj} is defined as the RDF flow from RDF plant k to incinerator j (TPY); Z_{jm} is defined as the ash amount from incinerator j to ash landfill site m (TPY); T_{ij} is the transportation cost function for hauling MSW from service area i to incinerator j ($ per year); T_{ik} is the transportation cost function for hauling MSW from service area i to RDF plant k ($ per year); T_{in} is the transportation cost function for hauling MSW from service area i to landfill site n ($ per year); T_{kj} is the transportation cost function for hauling RDF from RDF plant k to incinerator j ($ per year); T_{jm} is the transportation cost function for hauling ash from incinerator j to ash landfill site m ($ per year); D_{ij} is the average hauling distance from service area i to incinerator j (km); D_{ik} is the average hauling distance from service area i to RDF plant k (km); D_{in} is the average hauling distance from service area i to landfill site n (km); D_{kj} is the average hauling distance from RDF plant k to incinerator j (km); D_{jm} is the average hauling distance from incinerator j to ash landfill site m (km); O_n is the operating cost function of landfill site n ($ per year); O_j is the operating cost function of incinerator j ($ per year); O_k is the operating cost function of RDF plant k ($ per year); O_m

is the operating cost function of ash landfill site m ($ per year); E_{ij} is a function in terms of heating value (H_i) of MSW, which is the electricity converted from MSW at incinerator j; H_i is the heating value of MSW from district i within the service area (kJ·kg^{-1} or kJ·tonne^{-1}); E_{kj} is a function in terms of RDF (H_{ik}), which is the electricity converted from RDF at incinerator j (KWhr·tonne^{-1}); H_{ik} is the resultant heating value of RDF that is made by the MSW being hauled from district i, and being processed at RDF plant k (kJ·kg^{-1} or kJ·tonne^{-1}); r_i is the percentage of recyclables in the MSW stream in district i (%); P_j is the average selling price of electricity generated from incinerating MSW or RDF ($·kWh^{-1}); and P_k is the average selling price of recyclables collected at RDF plant k ($ per tonne).

Constraint Set

1. **Treatment capacity constraints:** These constraints describe the integrated waste inflows from either a presorting facility or service area destined for incineration and should be as close as possible to the design capacity of incinerator j. Such an expression should be valid for all incinerators j (i.e., $\forall j$) in a system.

$$\sum_i X_{ij} + \sum_k Y_{kj} \leq Q_j \quad \forall j, \tag{17.9}$$

$$\sum_i X_{ij} + \sum_k Y_{kj} \geq F1_j \cdot Q \quad \forall j, \tag{17.10}$$

where Q_j is the design capacity of incinerator j (TPD); and $F1_j$ is the minimum throughput requirement for incinerator j.

2. **Energy recovery constraints:** These constraints describe the target value of energy recovery and should be as close as possible to the design capacity at incinerator j. Such an expression should be valid for all incinerators j (i.e., $\forall j$) in a system. The estimation of H_{ik} may be carried out based on Equations (17.1)–(17.6) depending on the type of RDF process adopted:

$$\sum_i X_{ij} \cdot H_i + \sum_k \frac{\sum_i \alpha_{ik} \cdot X_{ik} \cdot H_{ik}}{\sum_i \alpha_{ik} \cdot X_{ik}} Y_{kj} \leq Q_j \cdot H_j \quad \forall j, \tag{17.11}$$

$$\sum_i X_{ij} \cdot H_i + \sum_k \frac{\sum_i \alpha_{ik} \cdot X_{ik} \cdot H_{ik}}{\sum_i \alpha_{ik} \cdot X_{ik}} Y_{kj} \geq F2_j \cdot Q_j \cdot H_j \quad \forall j, \tag{17.12}$$

where H_j is the design heating value of solid waste at incinerator j (kcal·kg^{-1}); α_{ik} is the production rate of presorting facility k, in which the MSW is collected from district i; $F2_j$ is the minimum energy recovery efficiency for incinerator j.

3. **Mass balance constraints:** All solid waste generated in each service area should be shipped to treatment or disposal facilities in a system. Equation (17.13) implies that all the sources generated in each collection district are shipped to management facilities and no accumulation is allowed. Equations (17.14) and (17.15) are the mass balance equations corresponding to each type of facility:

$$B_i = \sum_j X_{ij} + \sum_k X_{ik} + \sum_n X_{in} \quad \forall i, \tag{17.13}$$

$$B_k = \sum_i X_{ik} \quad \forall k, \tag{17.14}$$

$$\sum_j Y_{kj} = \sum_i \alpha_{ik} \cdot X_i \quad \forall k, \tag{17.15}$$

$$B_n = \sum_i X_{in} \quad \forall n, \tag{17.16}$$

$$B_m = \sum_j Z_{jm} \quad \forall m, \text{ and} \tag{17.17}$$

$$\sum_m Z_{jm} = \sum_i X_{ij} \cdot Rl_j + \sum_k Y_{kj} \cdot R2_j \sum_m \quad \forall j, \tag{17.18}$$

where B_i is the waste generated in collection district i (TPY); B_k is the waste presorted at the presorting facility k (TPY); B_n is the amount of solid waste disposed directly at landfill site n; B_m is the ash disposed in ash landfill site m; Z_{jm} is the ash generated from the incinerator j (TPY); $R1_j$ and $R2_j$ are the ash production ratio (%) based on one unit of raw waste stream and presorted waste stream, respectively.

4. **Non-negativity constraints:** All decision variables are required to be non-negative.

$$X_{ij} \geq 0 \quad \forall i, j, \tag{17.19}$$

$$X_{ik} \geq 0 \quad \forall i, k, \tag{17.20}$$

$$X_{in} \geq 0 \quad \forall i, n, \tag{17.21}$$

$$Y_{kj} \geq 0 \quad \forall i, n, \tag{17.22}$$

$$Y_{kn} \geq 0 \quad \forall k, n, \text{ and} \tag{17.23}$$

$$Z_{jm} \geq 0 \quad \forall j, m. \tag{17.24}$$

17.3.2 Application of the Mathematical Programming Model for Decision Analysis

The following case study illustrates the proposed model only. To implement such an analysis, site, source, process, and transportation information are required for various purposes.

System Environment of Taipei Metropolitan Region   In the Taipei metropolitan region, two administrative systems—Taipei City and Taipei County—form a unique urban region. Taipei City, located in the central part of Taipei County, is the largest city in Taiwan (Chang and Chang, 2003). The Taipei metropolitan region generated more than 6600 tonnes per day (TPD) of solid waste streams in the late 1990s; however, Taipei City and the County Government Agencies handle their solid waste streams independently at that time (Chang and Chang, 2003). Such a large metropolitan region, with a total population of more than 6.2 million and an area >2000 km^2 in the late 1990s, requires a regional system analysis for SWM. Taipei County has 29 organized administrative districts, and each has its own garbage collection team in charge of waste shipping and disposal duty (Chang and Chang, 2003). MSW streams generated in Taipei County are shipped to two existing incinerators (Shu-Lin and Hsin-Tein) or to the regional sanitary landfills (San-Hsia and Pa-Li) for waste treatment and disposal (Chang and Chang, 2003). Although local, public, or private agencies in several administrative districts of Taipei County operate a number of small landfills, they will soon reach capacity. Rapid economic development in this region and the uncertain need for solid waste treatment in the long term have resulted in a new incinerator located in the Pa-Li area (Chang and Chang, 2003). This incinerator was added to the MSW management system of Taipei County in the early 2000s (Chang and Chang, 2003). No large-scale material recovery facility had previously existed in the Pa-Li SWM system up to the present, however. The following systems analysis is simply presented for the purpose of demonstration (Chang and Chang, 2003).

The geographical information of the SWM system in Taipei County (Figure 17.2) denotes each type of existing service areas for each incinerator (Chang and Chang, 2003). Districts other than Taipei City and Keelung City belong to Taipei County, in which six SWM facilities (marked from A to E, Figure 17.2), are in service (Chang and Chang, 2003). However, the actual mass throughput is much less than the design capacity in both Shu-Lin and Hsin-Tein incinerators due to unexpected rise in heating values of the solid waste streams in each service area over the last few years (Chang and Chang, 2003). In response to this, the design capacity heating value of solid waste streams destined for the Pa-Li incinerator is currently up to 9600 kJ·kg^{-1}, which is much higher than the heating value designed for Shu-Lin and Hsin-Tein incinerators (i.e., 6500 kJ·kg^{-1}) (Chang and Chang, 2003). The capacity of incineration and disposal is still regionally insufficient, however, which could result in the need to install presorting facilities prior to incinerators in the SWM system. In view of the relatively higher heating value design for Pa-Li incinerator with relatively lower heating value of the solid waste streams in its corresponding service area, the Pa-Li incinerator may be a good candidate for the possible installation of a presorting process (Chang and Chang, 2003). The pilot study of an RDF process (Section 17.2) is applicable for fine-tuning the solid waste streams destined for the Pa-Li incinerator.

The questions of interest to the planners and decision makers are (1) what is the optimal size of the presorting process to be installed prior to Pa-Li incinerator for assessing the optimal shipping strategy in the service area?, (2) what is the short-term shipping pattern of solid waste stream in the designated service area, given

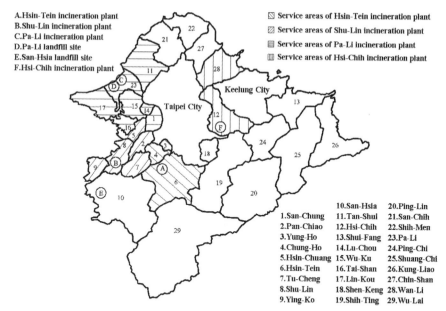

A.Hsin-Tein incineration plant
B.Shu-Lin incineration plant
C.Pa-Li incineration plant
D.Pa-Li landfill site
E.San-Hsia landfill site
F.Hsi-Chih incineration plant

▨ Service areas of Hsin-Tein incineration plant
▨ Service areas of Shu-Lin incineration plant
▤ Service areas of Pa-Li incineration plant
▥ Service areas of Hsi-Chih incineration plant

1.San-Chung	10.San-Hsia	20.Ping-Lin
2.Pan-Chiao	11.Tan-Shui	21.San-Chih
3.Yung-Ho	12.Hsi-Chih	22.Shih-Men
4.Chung-Ho	13.Shui-Fang	23.Pa-Li
5.Hsin-Chuang	14.Lu-Chou	24.Ping-Chi
6.Hsin-Tein	15.Wu-Ku	25.Shuang-Chi
7.Tu-Cheng	16.Tai-Shan	26.Kung-Liao
8.Shu-Lin	17.Lin-Kou	27.Chin-Shan
9.Ying-Ko	18.Shen-Keng	28.Wan-Li
	19.Shih-Ting	29.Wu-Lai

FIGURE 17.2   The service areas of incineration plants in Taipei County, Taiwan. *Source*:
From Chang and Chang (2003)

that the installation of such a presorting processes prior to Pa-Li may become a
reality in Taipei County?, and 3) would it be possible to enlarge the service area if
the installation of an RDF process provides additional capacity? (Chang and Chang,
2003). These questions can be analyzed using the nonlinear programming model
(formulated below) as a tool, but solid waste generation rates and heating values
in each district within the service area must be investigated before performing the
optimization analysis.

Data Investigation and Technical Settings   The functional parameter values
of RDF process gained (Section 17.2) can be utilized directly to support this analysis.
Many of the service districts associated with designated SWM facilities in Taipei
County (Table 17.2) share municipal incinerators over regions (Chang and Chang,
2001). However, the actual throughput is much less than the design capacity in both
Shu-Lin and Hsin-Tein incinerators due to unexpected rise of heating values in the
solid waste streams over the last few years (Chang and Chang, 2001). In response, the
designed heating value of solid waste streams destined for the Pa-Li incinerator was
increased up to 9600 kcal·kg^{-1} by the consultant firm, which is much higher than the
heating value designed for Shu-Lin and Hsin-Tein incinerators (i.e., 6500 kcal·kg^{-1})
(Chang and Chang, 2001).

In addition to these three large-scale incinerators, two administrative districts,
His-Chih and Wan-Li, have proposed a joint management program by installing a
small-scale municipal incinerator with a capacity of 300 TPD through a privatization

TABLE 17.5   The base information for incineration plant in Taipei County

Plant name	Design capacity (TPD)	Design heating value	Service areas
Hsin-Tein	900	650 kcal·kg^{-1}	Yung-Ho, Chung-Ho, Hsin-Tein, Wu-Lai
Shu-Lin	1350	650 kcal·kg^{-1}	Shu-Lin, Ying-Ko, Pan-Chiao, Hsin-Chuan, Tu-Cheng
Pa-Li	1350	9600 kcal·kg^{-1}	Pa-Li, Wu-Ku, Lin-Kou, Tai-Shan, San-Chung, Lu-Chou, Tan-Shui
Hsi-Chih	300	In planning	Hsi-Shih, Wan-Li

Source: From Chang and Chang (2001).

process (i.e., the Built-Operate-Transfer, or BOT, process) (Chang and Chang, 2001). The upward trend of solid waste generation in Taipei County (Table 17.3) with limited household recycling efforts has increased (Chang and Chang, 2001). Insufficient incineration and disposal capacity over the last decade have resulted in a need to consider installing presorting facilities prior to several municipal incinerators in this region (Chang and Chang, 2001). In view of the relatively higher heating value design for the Pa-Li incinerator and relatively lower heating value of the solid waste streams, the Pa-Li incinerator could be a valuable candidate for assessing the feasibility of installing a presorting process.

To confirm the application potential of the presorting facility, the recyclables in the solid waste streams were investigated in several service districts in Taipei County (Table 17.4) to help formulate the material balance in the RDF process (Section 17.2). Knowing the historical record of heating values in the solid waste streams in Taipei County (Table 17.5) is also important (Chang and Chang, 2001). Various forecasting methods, such as the regression analysis methods, may be applied to estimate the quantity and quality of solid waste streams in Taipei County during the target year.

A summary of the relevant parameter values in Tables 17.6 and 17.7 indicates that the original service area assigned to the Pa-Li incinerator could be expanded to cover more administrative districts if the presorting process can be installed (Chang and Chang, 2003). To fulfill this assessment, historical data of low heating value (Table 17.8), the related operational parameters (Table 17.9), and the information of waste management related to each district (Table 17.10) are required as well. The original service area of Pa-Li incinerator only covers seven districts, consisting of Wu-Ku, Lin-Kou, Pa-Li, Tai-Shan, San-Chung, Tan-Shui, and Lu-Chou (Figure 17.2); thus, two cases (i.e., defined as Case A and Case B below) can account for the impact due to the enlargement of service area (Chang and Chang, 2003). To enhance the operational efficiency of facility utilization and energy recovery without overburdening the economic system, the flexibility of shipping patterns of solid waste streams should be limited to some extent. While Case A does not impose any upper limit of shipping distance, Case B limits the maximum shipping distance to no farther than 50 km (Chang and Chang, 2003).

TABLE 17.6   The rate of waste generation in Taipei County

Area	1992	1993	1994	1995	1996	1997
San-Chung	418	454	477	474	471	481
Pan-Chiao	560	594	618	565	449	428
Yung-Ho	210	245	250	250	247	231
Chung-Ho	325	380	383	392	400	400
Hsin-Chuang	312	326	330	323	280	271
Hsin-Tein	200	217	228	258	267	283
Tu-Cheng	221	250	215	236	250	250
Shu-Lin	186	183	231	158	128	102
Ying-Ko	59	61	64	82	70	75
San-Hsia	62	64	60	65	66	70
Tan-Shui	144	146	150	152	150	149
Hsi-Chih	111	151	106	163	150	153
Shui-Fang	42	60	60	60	60	100
Lu-Chou	120	127	143	150	151	167
Wu-Ku	77	83	80	83	85	84
Tai-Shan	91	112	120	130	129	130
Lin-Kou	30	30	31	34	39	51
Shen-Keng	13	15	18	25	29	32
Shih-Ting	3	5	9	10	9	9
Ping-Lin	20	20	16	17	20	20
San-Chih	15	47	30	39	46	45
Shih-Men	10	10	10	10	19	20
Pa-Li	27	35	51	61	49	38
Ping-Chi	3	4	15	21	21	21
Shuang-Chi	11	13	13	13	20	20
Kung-Liao	15	31	21	20	17	13
Chin-Shan	18	29	23	23	23	24
Wan-Li	33	34	44	42	46	52
Wu-Lai	8	9	9	10	13	10
Total	3386	3736	3807	3864	3703	3749

Source: From Chang and Chang (2001).
Data are in TPD.

In response to additional processing capacity due to the installation of a presorting facility prior to the Pa-Li incinerator, Case A is designed to cover up to 12 districts, consisting of San-Chung, Tan-Shui, Hsi-Chih, Lu-Chou, Wu-Ku, Tai-Shan, Lin-Kou, San-Chih, Shih-Men, Pa-Li, Chin-Shan, and Wan-Li (Chang and Chang, 2003). Case B excludes the districts of Wa-Li and Chin-Shan due to shipping distance limitations. To account for the privatization process, two different planning scenarios can be classified, depending on whether the construction cost for building the RDF process is included or not (Chang and Chang, 2003).

Planning and Analysis   The nonlinear programming model was eventually simplified as a linear programming model because only one incinerator is involved in the

TABLE 17.7 The content of recyclables in MSW (on wet basis)

Composition (wt%)	Lin-Kou	Lu-Chou	Pa-Li	Tan-Shui	Pan-Chiao	San-Hsia	Hsin-Chuang	Ying-Ko	Wu-Ku	Hsi-Chih
Glass content	3.29	4.34	3.48	3.10	3.33	1.89	8.73	3.71	6.79	0.24
White	1.61	3.11	1.56	0.65	0.91	1.27	5.12	1.85	0.73	0.24
Green	0.44	0.00	0.00	0.00	0.00	0.00	1.62	0.19	0.00	0.00
Brown	1.24	1.23	1.92	2.45	2.42	0.62	1.99	1.67	6.06	0.00
Plastic content	24.23	23.62	18.84	17.21	17.63	15.89	11.87	19.52	18.51	13.80
PET	1.48	1.31	0.93	0.80	1.52	0.74	0.45	0.91	0.65	0.52
HDPE	3.76	3.61	2.49	2.19	2.19	2.27	0.79	1.19	0.88	0.53
PP	0.70	1.78	0.48	0.31	0.00	0.00	0.00	0.00	0.13	0.00
PVC	1.81	1.50	1.62	0.00	0.00	0.00	0.76	0.00	0.00	0.37
Plastic foam	1.71	1.50	0.75	0.41	0.47	0.91	0.21	0.16	0.20	0.43
Plastic bag	7.18	5.82	5.79	4.93	8.25	7.81	3.51	6.73	7.92	
Others	7.59	8.10	6.78	8.57	5.20	4.16	6.15	10.53	8.73	11.95
Paper content	25.95	15.77	21.86	25.41	18.93	29.01	19.18	26.82	17.81	28.09
Paper cardboard	8.63	3.09	3.12	4.90	2.70	1.03	7.84	8.49	10.01	
Magazine				1.96	1.96	8.11	0.72	1.48	0.63	2.36
Newspaper	8.79	4.18	6.93	8.05	4.41	5.57	3.61	6.60	2.59	5.75
Others	8.53	8.50	11.81	10.50	9.86	14.30	7.01	10.25	4.58	19.98
Metal content	1.31	5.86	2.46	3.25	1.43	3.44	2.03	5.72	4.88	0.90
Ferrous	0.81	2.73	1.74	0.95	1.03	0.85	1.07	5.41	3.95	0.66
Aluminum	0.50	3.13	0.72	1.06	0.40	0.97	0.96	0.31	0.93	0.24
Others				1.24		1.62				
Total	54.79	49.59	46.64	48.97	41.32	50.23	41.81	55.77	47.36	43.03

Source: From Chang and Chang (2001).

TABLE 17.8   Historical data of LHV for Taipei County

Service area	1994	1995	1996	1997	Max	Min
San-Chung	8,984.22	9,492.84	7,889.7	6,696.48	9,492.84	6,696.48
Pan-Chiao	–	4,504.08	6,720.00	5,398.26	6,720.00	4,504.08
Yung-Ho	–	7,165.20	7,533.96	6,656.16	7,533.96	6,656.16
Chung-Ho	–	4,567.08	4,623.36	5,698.56	5,698.56	4,567.08
Hsin-Chuang	–	10,775.94	6,612.48	7,468.44	10,775.94	6,612.48
Hsin-Tein	–	7,947.24	7,182.42	6,995.94	7,947.24	6,995.94
Tu-Cheng	–	–	–	6,485.22	6,485.22	6,485.22
Shu-Lin	–	5,336.52	4,618.74	6,183.66	6,183.66	4,618.74
Ying-Ko	–	5,387.34	–	5,946.78	5,946.78	5,387.34
San-Hsia	–	7,186.62	5,809.02	6,727.98	7,186.62	5,809.02
Tan-Shui	6,798.54			6,712.02	6,798.54	6,712.02
Hsi-Chih	–	–		7,446.60	7,446.60	7,446.60
Shui-Fang	–	–		5,733.03	5,733.03	5,733.03
Lu-Chou	–	7,640.64	5,518.8	5,207.58	7,640.64	5,207.58
Wu-Ku	–	–	5,613.3	10,880.52	10,880.52	5,613.3
Tai-Shan	–	8,517.18	6,861.12	8,659.56	8,659.56	6,861.12
Lin-Kou	–	–	8,050.14	7,512.12	8,050.14	7,512.12
Shen-Keng	–	4,930.38	4,937.94	6,383.58	6,383.58	4,930.38
Shih-Ting	–	5,027.40	4,503.24	6,995.94	6,995.94	4,503.24
Ping-Lin	–	4,391.94	4,008.48	5,309.64	5,309.64	4,008.48
San-Chih	–	–	–	5,443.20	5,443.20	5,443.20
Shih-Men	–	–	–	6,887.16	6,887.16	6,887.16
Pa-Li	8,901.48	7,156.80	4,998.84	4,641.00	8,901.48	4,641.00
Ping-Chi	–	–	–	5,524.26	5,524.26	5,524.26
Shuang-Chi	–	–	–	6,538.14	6,538.14	6,538.14
Kung-Liao	–	–	–	5,226.06	5,226.06	5,226.06
Chin-Shan	–	–	–	5,073.18	5,073.18	5,073.18
Wan-Li	–	–	–	6,260.94	6,260.94	6,260.94
Wu-Lai	–	–	3,976.56	5,990.88	5,990.88	3,976.56

Source: From Chang and Chang (2001).
Data are in kJ·kg^{-1}.

systems analysis, and the problem was solved using the software package LINDO® as a computer solver (Chang and Chang, 2003). Findings clearly indicate that the shipping pattern is the same whether the construction cost for building the RDF process is included or not. The optimal shipping pattern of solid waste streams in relation to all districts involved in both cases (Table 17.11) shows that the optimal size of such an RDF process is equivalent to 705 TPD and 372 TPD in Cases A and B, respectively (Chang and Chang, 2003). Case B qualifies for a much smaller RDF facility size because the solid waste streams collected in Wa-Li and Chin-Shan districts, which have much lower heating values, are eliminated in the planning scenario, and part or all of the solid waste streams collected in Tan-Shui and His-Chih districts, which have relatively higher heating values, are sent to the Pa-Li incinerator directly rather than to the RDF process, achieving a higher throughput requirement (Chang and Chang, 2003).

TABLE 17.9 The related operational parameters used in the case study (2001 basis)

Parameter	Value
Average unit transportation cost	US$0.7 per km·tonne^{-1}
Average operating cost for landfill	US$5.7 per t·d^{-1}
Average operating cost for incinerator	US$28.57 per tonne·d^{-1}
Average operating cost for RDF plant	US$4.6 per tonne·d^{-1}
Average operating cost for ash disposal	US$14.3 per tonne·d^{-1}
Construction cost for RDF plant	US$4.0 per tonne·d^{-1}
Price of selling electricity	US$0.4 per 10 kwh
Average price of recyclables	US$285.7 per tonne·d^{-1}
Conversion rate from heat to electricity	23.9%
Production rate of RDF in RDF plant	46.4%
Percentage of ash produced in incinerating RDF	5.0%
Percentage of ash produced in incinerating MSW	20.0%
Minimum percentage of throughput requirement for incinerator	90.0%
Minimum percentage of energy recovery goal for incinerator	90.0%

Source: From Chang and Chang (2003).

Two statistical indices can be applied to evaluate system performance in both cases. One is the percentage of capacity utilization defined as the ratio between the actual throughput and the design capacity (Chang and Chang, 2003). The other is the percentage of energy recovery defined as the ratio between the actual electricity generation and the design capacity (Chang and Chang, 2003). The statistical summary (Table 17.12) shows that the percentage of capacity utilization and energy recovery for the Pa-Li incinerator is 90.2% and 90% in Case A, and 95.5% and 90% in

TABLE 17.10 The information of waste management related to each district

District	Shipping distance (km)	Heating value of MSW (kJ·kg^{-1})	Recyclables content (%)
San-Chung	17.6	9,448.0	2.9
Tan-Shui	16.1	6,766.0	2.1
Hsi-Chih	31.8	7,411.0	4.4
Lu-Chou	14.4	7,604.0	6.7
Wu-Ku	9.7	10,829.0	4.8
Tai-Shan	11.4	8,618.0	4.4
Lin-Kou	7.1	8,012.0	6.2
San-Chih	37.6	5,417.0	6.1
Shih-Men	44.1	6,854.0	9.9
Pa-Li	1.1	8,859.0	4.3
Chin-Shan	57.0	5,049.0	5.4
Wan-Li	68.4	6,231.0	3.2

Source: From Chang and Chang (2003).

TABLE 17.11 Summary of optimal shipping pattern in the case study

	Case A			Case B		
District	Pa-Li incineration plant	Pa-Li landfill site	Pa-Li RDF plant	Pa-Li incineration plant	Pa-Li landfill site	Pa-Li RDF plant
San-Chung	530.8	0.0	0.0	530.8	0.0	0.0
Tan-Shui	0.0	0.0	123.2	94.4	0.0	28.8
Hsi-Chih	49.7	0.0	132.4	182.1	0.0	0.0
Lu-Chou	0.0	0.0	257.1	0.0	0.0	257.1
Wu-Ku	76.4	0.0	0.0	76.4	0.0	0.0
Tai-Shan	104.6	0.0	0.0	104.6	0.0	0.0
Lin-Kou	84.9	0.0	0.0	84.9	0.0	0.0
San-Chih	0.0	0.0	45.5	0.0	0.0	45.5
Shih-Men	0.0	0.0	40.8	0.0	0.0	40.8
Pa-Li	43.8	0.0	0.0	43.8	0.0	0.0
Chin-Shan	0.0	0.0	39.5	—	—	—
Wan-Li	0.0	0.0	66.8	—	—	—
Total	890.2	0.0	705.3	1117.0	0.0	372.2
		1595.5			1489.2	

Source: From Chang and Chang (2003).
Data are in TPD.

Case B, respectively (Chang and Chang, 2003). Overall, all four planning scenarios meet the minimum percentage of throughput requirement and energy recovery for incinerator. Meeting the throughput requirement is preferred over meeting the energy recovery goal, however, because the actual heating values of solid waste streams in the optimization process are not high enough and the construction and operating costs

TABLE 17.12 The statistical assessment of planning results in the case study

	Case A			Case B		
Item	Pa-Li incineration plant	Pa-Li landfill site	Pa-Li RDF plant	Pa-Li incineration plant	Pa-Li landfill site	Pa-Li RDF plant
Total throughput (TPD)	1217.3		705.3	1289.6		372.2
Ash generation (TPD)	213.3	—	—	242.0	—	—
Percentage of capacity utilization (%)	90.2	—	—	95.5	—	—
Percentage of energy recovery (%)	90.0	—	—	90.0	—	—
Tipping fee (US$·tonne^{-1})	24.5 [a]/22.7[b]			24.7 [a]/23.7[b]		

Source: From Chang and Chang (2003).
[a]Including construction cost of RDF plant;
[b]Not including construction cost of RDF plant.

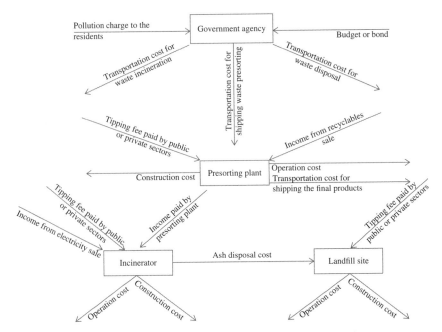

FIGURE 17.3   Cost–benefit structure in systems analysis

for handling the RDF facility are not low enough to allow the trade-off process to achieve better energy recovery (Chang and Chang, 2003).

For various private and public partnerships, the cost–benefit structure applied in this system analysis can be elaborated (Figure 17.3). Several sets of pricing levels for handling waste streams in the case study were described by using tipping fees to connect the managerial components in operation, but the proposed fee ranges were not sufficient to drive a different waste distribution pattern in this study. With the current settings for all planning scenarios, the cost–benefit distribution for Cases A and B in this case study (Figure 17.4a and 17.4b, respectively) indicates that more than half the income is gained from selling the electricity generated by burning the MSW rather than the RDF, and close to a quarter is obtained by selling the recyclables (Chang and Chang, 2003). Transportation cost always constitutes more than half of the total expenditure, however, implying that the shipping distance limitation is regionally essential in a regional SWM system (Chang and Chang, 2003). The need to collect tipping fees (last row in Table 17.12) reflects the inability to justify the cost–benefit analysis without imposing user charges in this SWM system.

17.4   FINAL REMARKS

Recycling raw materials and energy recovery become essential in SWM systems during the era of building intelligent and sustainable infrastructure systems in urban

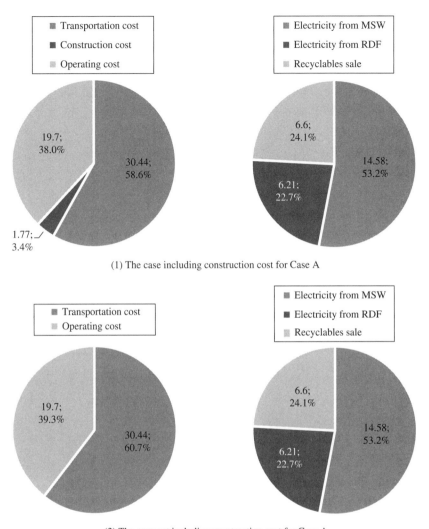

(1) The case including construction cost for Case A

(2) The case not including construction cost for Case A

FIGURE 17.4 (a) The cost–benefit distribution for Case A in the case study. *Source*: Adapted from Chang and Chang (2003). (b) The cost–benefit distribution for Case B in the case study. *Source*: Adapted from Chang and Chang (2003)

settings. Meeting mass throughput requirement and energy recovery goals in an SWM system is significant from both environmental and economic aspects. This chapter specifically addresses two problems of energy recovery: the first problem is inherent to the coordination between RDF processing and waste-to-energy facilities to promote its efficiency, and second problem is related to the shipping strategy of solid waste streams in a multidistrict analytical framework that includes a new presorting process prior to an incinerator. For the first problem, it is possible to observe that regression

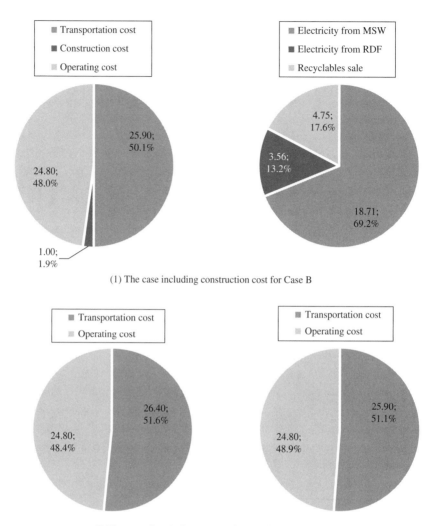

(1) The case including construction cost for Case B

(2) The case of excluding construction cost for Case B

FIGURE 17.5 (*Continued*)

models developed maybe used to control the production quality of the RDF presorting process. For the second case, the inclusion of presorting prior to incineration may help achieve the goals of energy recovery and capacity utilization that can be coordinated or reconciled through an optimization process. Such a systems analysis explores the compatibility issues between the energy recovery and material recycling goals in an incinerator in conjunction with a region-wide, on-site, or off-site presorting facility from a system of systems engineering perspective.

The case study presented in this chapter covers a broader situation that applies an RDF facility (a presorting process) to aid in material and energy conservation. The planning scenarios bridge possible gaps between technology evolution and manage-rial efficiency by predicting minimum tipping fees in an SWM system. The gaps were eventually minimized by using a centralized presorting facility prior to incineration to meet both the energy recovery and throughput requirements. The cost–benefit analysis at the conclusion of the case study further illuminates the potential of using a system-based optimization approach to enhance the efforts of resource conserva-tion with an engineered infrastructure system. Even if space limitations prevent the installation of a presorting facility at an incineration site, the promotion of household recycling programs can be in place to refine the characteristics of the SWM system. The main findings within this series of companion studies finally lead to the evalu-ation of new managerial strategies for various types of SWM scenarios in terms of cost-effectiveness, environmental impacts, and material/energy recovery. While the task of waste collection, presorting, incineration, and landfill may be currently per-formed by different management agencies in an SWM system, trade-offs among those private and public sectors could still be evaluated within different decision-making processes, given varying private–public partnerships.

REFERENCES

Buckley, T. J. and Domalski, E. S. 1988. Evaluation of data on higher heating values and elemental analysis for refuse-derived fuels. In: Proceedings of the 1988 National Waste Processing Conference, Pennsylvania, American Society of Mechanical Engineers, pp. 77–84.

Bunsow, W. and Dobberstein, J. 1987. Refuse-derived fuel: composition and emissions from combustion. *Resources and Conservation*, 14, 249–256.

Chang, Y. H. and Chang, N. B. 1998. Optimization analysis for the development of short-term solid waste management strategies using presorting process prior to incinerators. *Resources Conservation and Recycling*, 24(1), 7–32.

Chang, Y. H. and Chang, N. B. 2001. Regional shipping strategy assessment based on installing a refuse-derived-fuel process in a municipal incinerator municipal incinerator. *Waste Management & Research*, 19(6), 504–517.

Chang, Y. H. and Chang, N. B. 2003. Compatibility analysis of material and energy recovery in a regional solid waste management system. *Journal of Air & Waste Management Association*, 53(1), 32–40.

Chang, N. B. and Wang, S. F. 1996a. Solid waste management system analysis by multi-objective mixed integer programming model. *Journal of Environmental Management*, 48(1), 17–43.

Chang, N. B. and Wang, S. F. 1996b. Managerial fuzzy optimal planning for solid waste management systems. *Journal of Environmental Engineering, ASCE*, 122(7), 649–658.

Chang, N. B. and Wang, S. F. 1997. Integrated analysis of recycling and incineration pro-grams by goal programming techniques. *Waste Management & Research*, 15(2), 121–136.

Chang, N. B., Chang, Y. H., and Chen, W. C. 1997. Evaluation of heat values and its prediction for refuse-derived fuel. *The Science of the Total Environment*, 197(1–3), 139–148.

Chang, N. B., Chang, Y. H., and Chen, W. C. 1998a. Systematic evaluation and uncertainty analysis of refuse-derived fuel process in Taiwan. *Journal of Air & Waste Management Association*, 48(6), 174–185.

Chang, N. B., Chen, W. C., and Chang, Y. H. 1998b. Comparative evaluation of RDF and MSW incineration. *Journal of Hazardous Materials*, 58(1–3), 33–45.

Jackson, D. V. 1987. Advances in thermal treatment and RDF. *Resources and Conservation*, 14, 1–13.

Kirklin, D. R., Colbert, J. C., and Decker, P. H. 1982. The variability of municipal solid waste and its relationship to the determination of the calorific value of refuse-derived fuels. *Resources and Conservation*, 9, 281–300.

Sommer, E. J., Kenny, G. R., and Roos, C. E. 1989. Mass burn incineration with a presorted MSW fuel. *Journal of Air & Waste Management Association*, 39(4), 511–516.

Wilson, D. C. 1979. The efficiency of resource recovery from solid waste. *Resource Recovery and Conservation*, 4, 161–188.

CHAPTER 18

ENVIRONMENTAL INFORMATICS FOR INTEGRATED SOLID WASTE MANAGEMENT

Environmental informatics is a highly interdisciplinary science, where synergistic efforts among environmental sciences, electronic engineering, and computer sciences can be signified and magnified in regard to data collection, data analysis, data evaluation, and data visualization using specific computational intelligence methodologies, networking and sensing, cyberinfrastructure platforms, virtual reality and computer vision, and data science tools for solving environmental problems. Environmental informatics technology can further improve the efficiency of integrated solid waste management (ISWM). This chapter elaborates the major categories of the environmental informatics methods and tools applied in ISWM—from data collection through sensors and sensor networks, to data management through database design, and to data evaluation and assessment through geographical information systems (GIS), global positioning systems (GPS) and associated spatial analysis methods.

18.1 HOW DOES ENVIRONMENTAL INFORMATICS HELP SOLID WASTE MANAGEMENT?

In the 1980s, environmental informatics began to orchestrate various informatics tools for decision makers, allowing them to link the domain knowledge with conceptualized social, economic, and environmental objectives (Lu et al., 2013). This accumulation of environmental research findings and comprehensive measurements of environmental pollution have induced media reports elaboration, increasing environmental awareness, resulting in political responses (Pillmann et al., 2006).

Sustainable Solid Waste Management: A Systems Engineering Approach, First Edition. Ni-Bin Chang and Ana Pires.
© 2015 The Institute of Electrical and Electronics Engineers, Inc. Published 2015 by John Wiley & Sons, Inc.

Informatics permits analysis of real-world problems in a considered environmental field and defines functional requirements in regard to information processing, but it also "introduces the problem solving potential of informatics methodology and tools into the environmental field" (Page and Wohlgemuth, 2010).

Environmental informatics "is becoming more important for solid waste management (SWM) due to the increasing need for large-scale complex data storage, communication, analysis, and applications in concert with" distributed and parallel computational capability (Lu et al., 2013). Applying informatics methods and tools that could help solve the acute needs for SWM is useful at various levels of decision-making. These methods and tools include database systems (DBS), GIS, GPS, decision support systems (DSS), expert systems (ES), integrated environmental information systems (IEIS), and management science/operational research (MS/OR), all of which have been applied for management control (e.g., risk analysis, site management, process optimization, public participation), strategic planning (e.g., optimal siting of locations for waste treatment facilities, short- and long-term planning), and operational control (e.g., site monitoring with the aid of sensing and sensor networks).

18.2 SENSORS AND SENSOR NETWORKS FOR SOLID WASTE MANAGEMENT

A sensor is a specific device that perceives and measures real-world properties and converts them into signals that can be directly utilized by another device. The spectrum of these properties include but are not limited to acoustic, biological, chemical, electrical, magnetic, mechanical, optical, radiation, and thermal properties (Madou, 2011). In the networking, sensing and control framework, a sensor is the interface that allows communication between the controlled process and the controlling agent. A sensor may at least have one sensing device and a transducer. The transducer converts the information from the sensing device to an electrical or a pneumatic (air pressure) signal. The transduction techniques used to quantify the property measured can be piezoelectric, piezoresistive, capacitive, optical, electrical, electrochemical, thermoelectric, and photosensitive (Table 18.1). Advanced skills of sensor synergy and telecommunication make large-scale, ground-based sampling schemes feasible (Huang and Chang, 2003). These advancements have stimulated the creation of sensor networks, which usually consist of several sensing devices that communicate over wired or wireless media, and "have as intrinsic properties limitations in computational capability, communication, or energy reserve" (Iyengar et al., 2010). These sensors can all be applicable for SWM at different levels of management control, operational control, and strategic planning.

The advantages of applying sensors, albeit expensive, to SWM are salient for meeting the need to find fast, nondestructive, and time-saving solutions. In addition to these cost factors, sensor features must be considered as well, which may include but are not limited to: (1) static characteristics (i.e., if, after the stabilization, the sensor signal represents the amount measured) in terms of sensitivity, resolution, linearity, zero drift and full-scale drift, accuracy, precision, range, repeatability, and

TABLE 18.1 Sensors applied for SWM problems

Transduction technique	Type of sensor	Application	Reference
Optical and radiation	Fiber-optic	Remote monitoring of waste storage tanks in radioactive waste repository	Greenwell et al. (1992)
	NIR	To sort post-consumer plastics	Huth-Fehre et al. (1995)
	Two-color infrared color sensor	To sort waste plastic efficiently and quickly, for examples PET and PVC	Scott (1995)
	Raman	To monitor nitrate in the nuclear waste tank	Khijwania et al. (2003)
	Image recognition	To carry out an indirect sorting process based on integration of optical sensor and air separator	Huang et al. (2010)
Piezoresistive	Strain gauge	Load cell used to weigh waste containers	Chowdhury and Chowdhury (2007)
Piezoelectric	Semiconductor sensor/tin oxide sensor	To investigate the possibility of sensor array for odor assessment in the neighborhood of a landfill or landfill gas collection network	Micone and Guy (2007)
Electrical	Capacitance	To measure the moisture content of MSW	Fuchs et al. (2008)
	Resistive	To measure the moisture content of MSW	Gawande et al. (2003)
Electrochemical	Screen-printed carbon strip	For detection of compost maturity	Chikae et al. (2007)
	Electronic nose	To monitor biofilter emissions	López et al. (2011)
Thermoelectric	Online calorific value sensor	Optimization of MSW combustion processes	van Kessel et al. (2002)

MSW, municipal solid waste; NIR, near infrared; PET, polyethylene terephthalate; PVC, Polyvinyl chloride.

reproducibility (Olsson and Piani, 1992; Bagad and Godse, 2009); and (2) dynamic characteristics in terms of rise time, delay time, peak time, settling time percentage error, and steady-state error (Olsson and Piani, 1992; Bagad and Godse, 2009) that can characterize the time response of the sensor system. Accuracy and precision are also important features to describe the system behavior of sensors.

For SWM applications, wireless sensor networks have obtained a significant relevance due to their flexibility and lightweight nature. Wireless sensor networks that have been developed for SWM include monitoring temperature in composting heaps (Neehaarika and Sindhura, 2011), solid waste collection (Longhi et al., 2012), and monitoring landfills (Nasipuri et al., 2006; Beirne et al., 2009; Mitra et al., 2012). Due to the popularity of Radio Frequency Identification (RFID) wireless networks (Box 18.1), their application has a great potential to be used in concert with other communication devices for SWM.

BOX 18.1 SMART SWM SYSTEMS (Chowdhury and Chowdhury, 2007)

Intelligent systems can be applied to pay-as-you-throw systems to promote waste reduction, reduce waste collection costs, and identify misplaced containers or bins. The smart waste management system, represented in Figure 18.1, is composed of

- RFID waste tag;
- smart tag reader, such as a smartphone or Personal Digital Assistant (PDA);
- wireless sensor network;
- waste management information technology system (WMITS) composed of a load cell sensor.

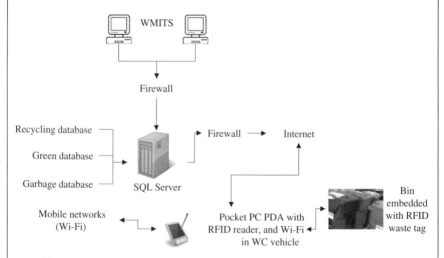

FIGURE 18.1 Main components of RFID and sensor-based waste management system. *Source*: Adapted from Chowdhury and Chowdhury (2007)

The system works by first placing RFID tags in waste bins. The antenna gets radio waves or electromagnetic energy beams from a reader device fixed to a smartphone/PDA located in the collection vehicle. The container can be identified through the use of chip for distinctive identification (Shepard, 2005). Using a wireless network, information related to the waste and the waste producer is sent from a smartphone/PDA to the database server. When the robotic/lifting arms load a container onto the vehicle, the weighing system (load cell sensor) measures the weight of each container. After emptying each container, waste disposal charges are calculated using the data and are sent to the smartphone/PDA. At the end of the shift, the smartphone/PDA sends the information to the Sequential Query Language (SQL) server, through the wireless fidelity (Wi-Fi) connection and Internet for relevant Waste Container (WC) vehicles.

18.3   DATABASE DESIGN FOR SOLID WASTE MANAGEMENT

If data constitute any amount of information, the database is where such information is collected and stored as electronic files that can be easily accessed by end users. The end user intention (i.e., how the user applies the data) defines how the database must be elaborated. For storage purposes, databases can be analytical, operational, data warehoused, distributed, end-user, external, and hypermedia. This section is devoted to the characterization of general-purpose and analytical databases for SWM.

According to Elangovan (2006), four types of database concepts are important— hierarchical, network, relational, and object orientations. Hierarchical databases organize data at different levels with a one-to-one relationship. In a network database, a connection exists between tables. In relational databases, connections are related to each other by keys. The object-oriented model applies functions for spatial and non-spatial modeling of the relationships and the characteristics; the object consists of an enclosed unit composed of attributes with a series of guidelines and rules (Lindsjørn and Sjøberg, 1988). All of these concepts can be applied to SWM.

The components of a database system are (1) a database, an organized collection of data for different purposes; (2) a database management system (DBMS) software for user-database interactions; and (3) a data model, which gives rise to essential principles that support both a database and its DBMS (Lu et al., 2013). The type of data in a database component can be technical, economic, environmental, and even social information; for example, a specific plastic waste infrared sorting machine that requires a unique database of infrared absorbencies to identify the plastic waste to be sorted.

To produce annual environmental reports, an SWM facility must collect continuous monitoring data related to air and water emissions due to waste treatment; hence, a DBMS is used to construct, use, and maintain a database system meeting various needs throughout the SWM system. Different software exists for DBMS, including Oracle®, IBM DB2® (acronym of International Business Machines Database), Sybase®, IBM Informix®, PostgreSQL™ (SQL is acronym for Sequential Query Language), and MySQL™. "A data model is characterized by the inherent structure and a set of tools and techniques used in the process of designing, constructing, and manipulating model systems (e.g., databases)" (Lindsjørn and Sjøberg, 1988). Designing a customized database system includes (Lu et al., 2013): "1) demand analysis for clarifying user intent, 2) conceptual modeling for abstracting the realistic relationships, 3) logical modeling that transfers the entity relation-to-relation schema with selected DBMS and data models, and 4) physical modeling that specifies physical configurations of hardware." In SWM, databases are used to support sensing, monitoring, and modeling for management operation. Databases allow us to collect and archive specific data so they are easily and rapidly accessible and retrievable by the end users. Several existing databases for SWM are used in different countries (Table 18.2).

18.4 SPATIAL ANALYSIS WITH GIS AND GPS FOR SOLID WASTE MANAGEMENT

Several definitions for GIS exist in recent literature.

- GIS is applied to stock, operate, analyze, and present data that are needed to sustain accurate modeling of Earth's environmental processes, and environmental risk assessment (RA) (Lukasheh et al., 2001).
- GIS, or a geospatial information system, "is any system for capturing, storing, analyzing, and managing data and associated attributes which are spatially referenced to the Earth" (Walia, 2010).
- GIS is a computer-based information system (CBIS) "that can gather, store, manipulate, manage, analyze, display, and even share all kinds of spatial data" (Lu et al., 2013).

These GIS definitions are mostly devoted to the technological approach and less to the problem-solving and decision-making process (Foote and Lynch, 1995; Malczewski, 1999).

- GIS can be regarded as a special-purpose digital database where a spatial coordinate system is the first method of storing and obtaining data. GIS systems have the capacity to conduct several tasks using spatial and attributive data stored inside them.
- GIS is an integrated technology that permits integration of several geographical technologies, including remote sensing (RS), GPS, computer-aided design (CAD), and automated mapping. GIS can be integrated with analytical and decision-making techniques.
- GIS can be considered as "a decision support system involving the integration of spatially referenced data in a problem solving environment" (Cowen, 1988). The way in which data are introduced, saved, and examined within a GIS should reflect the way that information will be utilized for a particular analysis or decision-making tasks.

TABLE 18.2 Current databases for SWM

Database system	Description
Eurostat/New Cronos (2013)	Data on the generation of waste by economic sector, waste streams, and waste treatment
Basel Convention (2011)	National reporting concerning generation of hazardous and other waste, also the transboundary movement of hazardous waste
European Pollutant Release and Transfer Register (E-PRTR) (EEA, 2013a)	Previously named European Pollutant Emission Register (EPER); data from individual polluting industrial sources and activities, including information on off-site transfers of waste
WasteInfo (Dialog, 2013)	Several types of waste like solid waste, household, commercial, and industrial wastes, hazardous waste; several types of waste management options like minimization, recovery, reuse, recycling; several treatment options including separation, composting; elimination options like incineration and landfill; other waste management aspects like waste policy, legislation, economics are also included in the database
WasteBase (EIONET, 2009)	Historical information on waste in the European countries
Phyllis2 database (ECN, 2012)	Composition of waste (also biomass)
Healthcare Waste Management (WHO, 2011)	Healthcare waste management, including waste treatment technologies, country information
International Solid Waste Association (ISWA) knowledge base (ISWA, 2013)	Searchable database concerning SWM, including reports, training materials, web links, papers, books, conferences, and workshops
Australian Waste Database (CSIRO and Department of Environment Heritage, 2004)	Solid waste and hazardous waste, waste composition, and waste generation
Biennial Reporting System (USEPA, 2013)	Hazardous waste, including waste generation, and waste received
Organisation for Economic Co-operation and Development/European Environmental Agency Database (OECD and EEA, 2012)	Instruments applied to several environmental policies and natural resources management aspects, including waste management
Database on Transboundary Movement of Wastes destined for Recovery Operations (OECD, 2013)	Waste transboundary movement for recovery operations within OECD countries
EEA Data and Maps (EEA, 2013b)	Data and maps format concerning environmental themes, including waste management issues such as waste generation

Beginning with McHarg (1969), who expressed the basic mapping ideas for site suitability analysis, the use of GIS focuses on tasks that delineate the best route to connect two points or identify the best location for a specific function. McHarg (1969) utilized a technique named "overlap maps" in which each feature (criterion) is plotted on a map of the region in colors; different color intensities across the map denote variation in fulfilling the criterion (Christensen et al., 2011). For instance, by superimposing all the thematic maps, the most suitable areas for landfills can be identified as well as sites that should be avoided (Christensen et al., 2011).

According to Lukasheh et al. (2001), in the 1960s, GIS was mentioned as the application of computers in managing considerable mapping information for the Canada land inventory. GIS is established as computer system for introducing, storing, manipulating, analyzing, and presenting geo-referenced data, being data represented as points, lines, and polygons (Lukasheh et al., 2001). For example, a point translating a groundwater monitoring well at a landfill site can stand for related the groundwater chemistry data as its characteristics; lines can represent roads, rivers, or any other linear feature; polygons can represent a lake or any closed-boundary region with uniform characteristics near the landfill (Lukasheh et al., 2001).

GIS has the ability to manage large amounts of spatial and non-spatial data and statistical information. It can merge several demographic, geological, land use, and census tract maps to apply landfill criteria to locate adequate sites to locate landfills (Michaelis, 1991). GIS can integrate data from different times and scales and use various methods of capturing data. The data sources to create a GIS spatial database can be digitized paper maps, scanning materials, traditional surveying data, paper records and field notes, photogrammetry, remote sensing, and GPS. Maps can be combined by means of Boolean functions that add or subtract thematic features or search for particular patterns (Christensen et al., 2011). The main output of this procedure is a map (an image) produced as a result of a query.

This structure of GIS was defined by several authors. Aronoff (1989) identified such a structure in terms of data input, data management, data manipulation and analysis, and data presentation. Malczewski (1999) proposed four head components in a GIS: data input, data storage and management, data manipulation and analysis, and data output. Lu et al. (2013) defined spatial data production, data management, display and cartography, and various analysis tools. Regardless of the method used to define the structure, GIS has aided many SWM applications, such as siting potential waste disposal facilities (Chang et al., 2008), collection vehicle routing (Chang et al., 1997; Chang and Wei, 2002), production rate estimation of municipal solid waste (MSW) in urban areas (Purcell and Magette, 2009), RA for fair fund distribution (Chang et al., 2009), and system planning (Chang and Wei, 1999). How GIS can be configured to help reach these goals of SWM is discussed in subsequent sections.

18.4.1 Data Input

Data input module comprises procedures and methods for entering geospatial data into the computer hardware system. Spatial data are collected from various sources and entered in the computer system through manual digitizing, keyboard entry, scanning, and use of existing digital files. Spatial data for SWM are usually provided by the agencies so that the data-producing module is optional, but end users occasionally must produce the spatial data for SWM, such as locations of collection points and waste treatment facilities. In cases where data are not available, spatial data must be created, through data capture, quality inspection, and format conversion.

Depending on the purpose of SWM studies, data must be collected, provided, and/or produced for inclusion in a GIS. To site landfills, GIS data inputs include residential areas, population, surface water bodies, ground water, land use and land cover, price of land, land slope, elevation, roads and railway networks, waste recycling centers, geology, soil permeability, natural hazards, hydrology, forest inventory, ecological sensitivity, protected areas, settlements, and air quality (Chang et al., 2008; Sumathi et al., 2008). To site a hazardous waste landfill, the data input for GIS used by Sharifi et al. (2009) were lithological, land use, slope and aspect, surface water, human settlement patterns, climatology maps, protected regions, hot spots, karst areas, aquifers, cultural heritage sites, springs, ghanats and wells, and infrastructure. For sound system planning using GIS, specific data were considered to develop a fair fund distribution strategy (see Chapter 20) for a municipal incinerator with respect to environmental impact categories of interest, such as air pollution from trucks and stacks, suspension of ash monofill, noise from incinerators and trucks, wastewater from monofill and incinerators, incinerator ash and sludge from wastewater, traffic impact, income in local communities, real estate fluctuations, and demographic variations (Chang et al., 2009).

For management control, GIS can be used to collect information about production factors associated with biodegradable municipal waste (BMW) to help managers identify various types of producers, calculate the total amount of BMW, and locate those producers geographically (Purcell and Magette, 2009). To help forecast MSW generation, spatial data such as geographic coordinates of waste producers and generation factors can be linked with non-spatial data such as waste statistics, socioeconomic features, and demographic features to achieve management goals.

To route vehicles for operational control of waste collection, a data input model can be linked to road networks, vehicle speed, vehicle load, road gradient, bin capacity, vehicle collection capacity and configurations, and collection methodology type (Ghose et al., 2006; Tavares et al., 2009). To optimize the location and type of containers used for the collection of light packaging waste material, Alvarez et al. (2009) applied a GIS in which data input was related to location of collection points, inventory of containers, accessibility for pedestrians and vehicles, visibility, and density.

18.4.2 Data Management

Data management module stores and retrieves spatial data on a regular basis to update and make changes to ensure data integrity (Sugumaran, 1999). This component determines the model of spatial data representation as well as the structure for data storage (Sugumaran, 1999). According to Malczewski (1999), the GIS database "can be defined as a collection of non-redundant data in a computer, organized so that it can be expanded, updated, retrieved and shared by various users". Spatial data represented at the physical level are arranged in one of two formats available in GIS for data representation (Malczewski, 1999): raster (or grid), where each layer consists of a number of equally sized square cells forms a grid; and vector (or polygon), where points are represented as pairs of x, y coordinates, lines as strings of points, and polygons as lines that form closed areas. In the raster, for example, satellite images, the size of the grid determines the spatial resolution of the raster; thus, increasing the resolution results in a corresponding increase in computer storage requirements (Phadke, 2006). In a vector representation, data require less computer storage and are preferred for display purposes because they reflect the exact shape of the object of interest through visualization; however, their geometric calculations are complex and time-consuming (Phadke, 2006). Data management is required for strategic planning, management control, and operational control of ISWM with differing types of representation and scales in real-world applications.

18.4.3 Data Analysis

Spatial statistics analysis are needed to draw inferences from empirical data. Statistical methods are employed "to interrogate spatial data to determine whether or not the data are "typical" or "unexpected" relative to a statistical model" (O'Sullivan and Unwin, 2003). Spatial analysis use queries to reach the appropriate geographic zone. To manipulate and analyze GIS data, analytical tools can be applied for various purposes. All queries such as "what if," "what is on," and "what is closest" can be obtained from this component. The five main types of spatial analyses are retrieval, measurement, overlay, neighborhood, and connectivity.

According to Singhal and Gupta (2010), "retrieval operations include selective search of spatial and attribute data in such a way that the geographical locations of features are not changed (i.e., the outputs shows selectively retrieved data in their original geographical positions)". Measurement operations are distances from a feature, between points, lengths of lines (areas), perimeters of polygons, volume, areas of polygons, number of points falling in a polygon, and number of raster cells in each class (Engel and Navulut, 1998; Singhal and Gupta, 2010). Overlay operations allow the user to combine and overlay multiple thematic information, at different scales and formats, involving both arithmetic functions (addition, subtraction, division, and multiplication) as well as logical types of functions (AND, OR, and NOT) (Engel and Navulur, 1998; Singhal and Gupta, 2010). Neighborhood operations deal with local characteristics, or characteristics surrounding a specific location or neighborhood cells (Engel and Navulur, 1998; Johnson, 2009; Singhal and Gupta, 2010) and are useful in finding local variability and neighboring/adjoining information (Singhal and Gupta, 2010). Functions included in neighborhood operations are topography, Thiessen polygons, interpolation, and contour generation. Connectivity operations are grouped into contiguity, proximity, network, spread, perspective-view and classification functions (Singhal and Gupta, 2010). Two-dimensional (2D) analysis is the basis of queries in most cases, but three-dimensional (3D) analysis can also be performed for database queries and spatial analysis. As a supplement, time efficient computation of morphometric analysis and 3D modeling of topographic and thematic variables have been treated as a typical example of 3D analysis (Hurni et al., 1999).

Network analysis is conducted in GIS in the context of spatial infrastructure, communication, and transportation. Network graphs are popular analysis tools as they allow users to visualize relationships between multiple objects so that the human eye can detect patterns, and also because they provide a mathematical structure that allows algorithmic analysis (Sierra and Stephens, 2012). In summary, four classes of GIS analysis are applicable to ISWM (Lu et al., 2013).

- **Spatial analysis**: including measuring and calculating of spatial distance an area, buffer analysis, analysis, distance mapping, raster interpolation, and surface analysis
- **3D analysis:** such as 3D roaming of vehicle routing and sight line analysis
- **Spatial statistical analysis**: which complements spatial analysis with statistical models
- **Network analysis:** which makes use of the graph datasets

18.4.4 Data Output

Data output consists of the display and cartography components that allow outputs to be generated in the form of maps, tables, or text in both hard and soft copies. A GIS organizes all the analyzed information in a format that best communicates the results of analysis to end users. Outputs can be in soft or hard copy form and are usually maps and tables accompanied by charts (Figures 18.2 and 18.3). As examples, a final map can present appropriate places for siting a landfill (Figure 18.2) or the location of recycling containers (Figure 18.3). The advantage of GIS outputs such as maps is to ease the visualization and interpretation of the results. This feature has been used in concert with environmental impact assessment (EIA) when siting and operating waste management infrastructures, such as landfills and incineration plants. GIS may be used collaboratively to support decision-making among stakeholders, such as fair fund distribution (Chang et al., 2009) (see Chapter 20). Visualization is considered the best method for stakeholders to view and discuss plans to reevaluate the problems, make changes, and propose improvements.

18.4.5 GIS Software

Open-source and licensed software packages are used to conduct GIS analyses. The most popular software packages in the SWM community are the Intergraph GIS system (Intergraph, 2014), ArcView and ArcGIS (ESRI, 2012), MapInfo (MapInfo, 2014), and Integrated Land and Water Information System (ILWIS) (IIGISEO, 2011). The ArcGIS family of software products by ESRI, aimed at both end users and

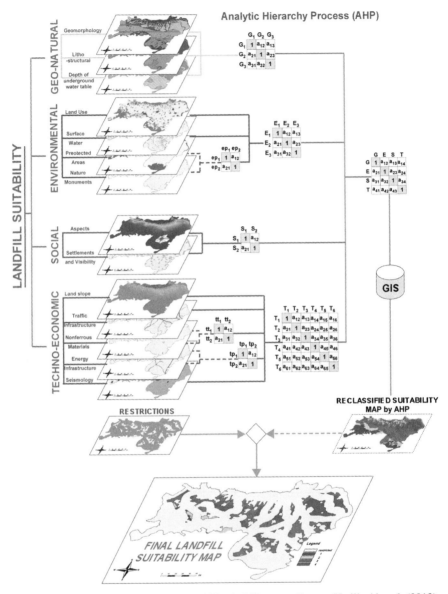

FIGURE 18.2   GIS process to find landfill suitability map. *Source*: Vasiljević et al. (2012)

technical developers, provides data visualization, query, analyses, and integration capabilities, and the ability to create and edit geographic data (Duggal, 2004). The ESRI suite has extensions for special purposes, such as 3D analyst, data interoperability, geostatistical analyst, job tracking, network analyst, spatial analyst, and survey analyst (Wright and Yoon, 2006). The primary GIS product from Intergraph is

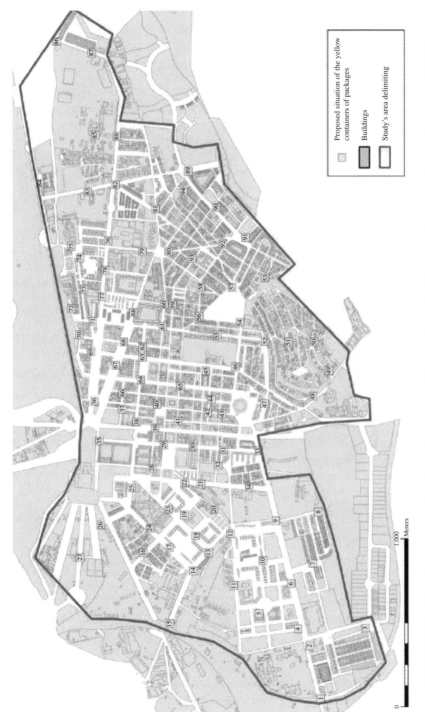

FIGURE 18.3 Map of the proposed location of the new recycling containers for light-packaging waste. *Source*: Alvarez et al. (2009)

Proposed situation of the yellow containers of packages

Buildings

Study's area delimiting

Meters

623

Intergraph Modular GIS Environment, which facilitates the capture, storage, retrieval, and analysis of geographic data (Wright and Yoon, 2006). Another GIS product is GeoMedia, which can be used to manage spatial data that resides in various databases, to be brought into a single GIS environment, turning it into valuable information (Longley et al., 2005; Wright and Yoon, 2006). MapInfo is a global software company that developed MapInfo GIS software, with several existing applications, most notably MapInfo Professional with a built-in geocoding ability which permits fast and accurate placement of address data for maps, combining and displaying in a single map data from several sources existing in different formats and projections (Wright and Yoon, 2006; Siddiqui et al., 2012). Via interpolation, it is capable to relate raster or vector layers, being quite popular in business and public sector (Wright and Yoon, 2006; Siddiqui et al., 2012). ILWIS software allows us to construct a GIS through input, manage, analyze and output geographic data (Ahmed et al., 2006a). According to Hengl et al. (2009), the true advantage of ILWIS is the ability to combine vector, raster, and database operations with geomorphometric analysis. Some of these software programs have specific features that can solve specific SWM problems (Table 18.3). For example, Arc/Info can be used to select possible locations; Route View Pro can be used to optimize waste collection; and ArcGIS Network Analyst can be applied for siting waste bins and optimizing waste collection routes.

18.5 EXPERT SYSTEMS, DECISION SUPPORT SYSTEMS, AND COMPUTATIONAL INTELLIGENCE TECHNIQUES

Although not typical for all countries, solid waste managers in developed countries rely on a variety of information systems to support their decision-making processes, including management information systems, DSS, executive support systems, and ES. Among these systems, DSS and ES have been the most successful types of applications in several areas, including SWM. Artificial intelligence (AI) has also helped decision makers and waste managers control waste processes and planning. The features, abilities, and contributions of DSS and ES to SWM are discussed in the following sections.

18.5.1 Decision Support System

Decision-making requires an understanding of the various processes involved in facilitating the design of the computer-based system support and to increase efficiency (Lukasheh et al., 2001). A DSS is computer-based information system designed to affect and improve the process of decision-making (Chang et al., 2011). It attempts to automate decision-making tasks rather than operate individual models (Turban, 2005) by addressing all decision-making phases, including intelligence, design, choice, and implementation. DSS consist of (Chang et al., 2011) (a) an interactive graphic display capacity for managing the interface between the decision makers and the system, (b) a data management system (DMS), and (c) a model base management system (MBMS), which aggregates different models, such as optimization, forecasting, and simulation

TABLE 18.3 A summary of GIS applications in SWM

Cat.	Scope	Methodology	Reference
Suitable site selection	Low-level waste disposal site selection	Remote sensing-based spatial analysis and 10 Code of Federal Regulations 61 low-level waste sitting criteria	Stewart (1988)
	To carry out solid and hazardous waste disposal site selection	Remote sensing-based spatial analysis and industrial location constraint criteria	Jensen and Christensen (1986)
	To screen low-level waste sites	GIS and MCE	Scott et al. (1989)
	To screen low-level waste sites	GIS and MCE	Judd et al. (1990)
	To reduce time and costs in site selection under public and technical demands	GIS and ES	Davies and Lein (1991)
	To reduce execution time of site selection	Markov chain-based simulated annealing, using GRASS GIS	Muttiah et al. (1996)
	To screen waste disposal sites considering storage capacity	GIS screening, on-site investigation and 3D analysis, using Arc/Info	Basagaoglu et al. (1997)
	Preliminary landfill site screening	Raster-based MCE combining fuzzy logic and AHP, using Arc/Info	Charnpratheep et al. (1997)
	To analysis the waste generation and to find locations of in-vessel composting	Basic GIS functions and screening	McLeod et al. (1997)
	To select the most suitable radioactive waste disposal site	3D GIS, using ERMA 3D GIS	Flinn (1998)
	To locate a landfill site considering land compactness	GIS and mixed-integer compactness model, using GRASS GIS	Kao (1996)
	To select a site for low-level waste disposal	GIS	Veitch (2000)
	To select a for animal waste treatment	GIS-based weighted summation, and AHP, using Arc/Info	Basnet et al. (2001)
	To locate high-level waste repository	GIS and groundwater flow modeling	Sheng and Almasi (2002)

(continued)

625

TABLE 18.3 (*Continued*)

Cat.	Scope	Methodology	Reference
	To select a landfill	GIS and geotechnical evaluation, using ArcView	Allen et al. (2003)
	To select a waste disposal site	GIS and DRASTIC, using Arc/Info	Lee (2003)
	To select a grid raster-based heuristic method for multi-factor landfill site	GIS, weighted summation, MIP	Lin and Kao (2005)
	To select a site for a hazardous waste landfill	GIS, MCE and global sensitivity analysis, using Idrisi	Gómez-Delgado and Tarantola, (2006)
	To enhance public participation in facility sitting	GIS and MCE with public participation	Higgs (2006)
	To select a site for solid waste disposal sites, considering vulnerability to groundwater pollution	GIS and weighted summation	Simsek et al. (2006)
	To select a site for solid waste landfill	GIS, raster-based fuzzy MCE, using MapInfo, Vertical Mapper and Idrisi	Gemitzi et al. (2007)
	To select a site for landfill, considering environmental impacts	GIS and simple MCE, using ArcGIS	Zamorano et al. (2008)
	To plan CCA-treated wood waste recommendation units	GIS and clustering methods (self-organizing maps and k-means), using ArcGIS, MATLAB and SOM toolbox	Gomes et al. (2007)
	To select a site for solid waste landfill	GIS and fuzzy logic, using Arc/Info	Lotfi et al. (2007)
	To select a site for solid waste landfill	GIS, fuzzy MCE, Monte Carlo simulation-based sensitivity analysis, using ArcGIS	Chang et al. (2008)
	To select a site for composting plant of vegetable waste	GIS and contingent valuation method, using Idrisi	Parra et al. (2008)
	To select a site for hazardous waste landfill	GIS and MCE	Sharifi et al. (2009)
	To select a site for solid waste landfill	GIS, AHP and LCA, using GIS and SimaPro	Sumiani et al. (2009)
	To select a site for solid waste landfill	GIS and AHP, using ArcGIS	Wang et al. (2009)
	To select a site for waste storage locations	GIS, fuzzy logic and AHP	Ahmadi et al. (2010)
	To select and rank inert landfill sites	GIS, MCE and stakeholder analysis, using ILWIS	Geneletti (2010)

	To select a site for solid waste landfill	GIS and MCE, using ArcGIS	Nas et al. (2010)
	To select a site for solid waste disposal	GIS and MCE, using ArcGIS	Nishanth et al. (2010)
	To select a site for solid waste disposal	GIS and MCE, using ArcView	Sfakianaki and Kasis (2010)
	To select a site for hazardous waste treatment and disposal	GIS and MCE	Sauri-Riancho et al. (2011)
	To select a site for solid waste disposal	GIS and AHP, using ArcGIS	Şener et al. (2011)
	To select a site for industrial waste landfill	GIS and MCE	Taghizadeh (2011)
Optimal location	To select the optimal locations for transfer stations on a large-scale metropolitan region	GIS screening and MIP optimization, using ArcView	Chang and Lin (1997)
	To select the optimal locations for waste pickup stations	GIS and customized shortest service location model, using ArcView	Kao and Lin (2002)
	Optimal location for collection points considering the operational convenience	GIS and p-median constrained model, using Arc/Info	Vijay et al. (2008)
	To select the optimal locations for packaging waste collection points	GIS, using ArcGIS	Alvarez et al. (2009)
	To select a site for solid waste storage containers	GIS, GPS and investigation	Eshkevari and Panahande (2010)
	To select a site for multi-compartment containers for urban sorted waste	GIS and multiobjective MIP, using ArcGIS Server	Tralhão et al. (2010)
Vehicle routing and scheduling	To carry out the optimal vehicle routing and scheduling in solid waste collection systems	GIS and multiobjective MIP, using Arc/Info	Chang et al. (1997)
	To ensure security dominated hazardous waste transportation	GIS network analysis and AHP	Huang (2006)
	To carry out route scheduling of waste collection	GIS network analysis, using Arc/Info	Ghose et al. (2006)

(continued)

TABLE 18.3 *(Continued)*

Cat.	Scope	Methodology	Reference
	To optimize waste collection	GIS network analysis, suing Route View Pro	Apaydin and Gonullu (2008)
	To comparing ACA with GIS network analysis for waste collection optimization	GIS network analysis and ACA, using Arc/Info	Karadimas et al. (2007a)
	To optimize waste collection	GA	Karadimas et al. (2007b)
	To optimize bulky waste collection	GIS network analysis, using ArcGIS Network Analyst	Karadimas et al. (2007c)
	To optimize waste collection	ACA	Karadimas et al. (2007d)
	To reduce total distance of waste collection	Using TransCAD, Routine Arc Routing	Brasileiro and Lacerda (2008)
	To comparing GIS network analysis with ACA for optimal routing	GIS network analysis and ACA, using ArcGIS Network Analyst	Karadimas et al. (2008)
	To comparing optimization of waste collection in different scenarios	GIS network analysis and scenario development, using ArcGIS Network Analyst	Chalkias and Lasaridi (2009a, 2009b)
	To optimize hazardous waste transport	GIS network analysis and AHP-based MCE, using ArcGIS Network Analyst	Monprapussorn et al. (2009)
	To optimize waste collection routes for minimum fuel consumption	3D-based GIS network analysis, using ArcGIS	Tavares et al. (2009)
	To optimize waste transport considering the lowest fuel cost route	3D-based GIS network analysis	Zsigraiová et al. (2009)
	To find the shortest routes after adding new collection points for separated waste collection and allocation	GIS spatial analysis and network analysis, using ArcGIS	Zamorano et al. (2009)
	To optimize routing and vehicle scheduling for waste collection and transport	GIS network analysis, combinatorial optimization and integer programming, using ArcGIS Network Analyst	Arribas et al. (2010)

	Objective	Method	Reference
	To optimize routing for solid waste collection	GA and simulate algorithm integration	Fan et al. (2010)
	To optimize routing and monitoring	Single-objective network analysis and on-board GPS, using ArcGIS	Jovičić et al. (2010)
	To carry out real-time routing management of hazardous waste transportation	GIS, on-board GPS tracking and GPRS communication, using customized GIS	Zhu et al. (2010)
	To carry out optimal routing for waste collection and transportation to composting plant	GIS and vehicle tracking system, using MapInfo	Kanchanabhan et al. (2011)
Regional planning	To make decision for solid waste planning considering multiattribute	GIS, scenario development and MCE, using TransCAD	MacDonald (1996a)
	To assess the demand and supply of land for waste disposal	GIS and scenario development	Leao et al. (2001)
	To carry out ecological–economic modeling for strategic regional SWM systems	GIS, LCA and MCE	Shmelev and Powell (2006)
	To maximize the recovery and utilization of solid waste generated and to address harmful treatment	GIS and scenario development	Guzman et al. (2010)
	To construct a complete graph of waste allocation in Czech Republic	GIS network analysis, using GIS Network Analyst	Hřebíček and Soukopová (2010)
	To quantify the relationship between the demand and supply of suitable land for waste disposal over time	GIS and cellular automata based prediction, using ArcView	Leao et al. (2004)
Estimation of waste generation	To develop a waste production model for identification of the optimal number and location of the waste bins	GIS and spatial clustering, using ArcGIS	Karadimas et al. (2005)
	To carry out waste production estimation and containers location	GIS slope analysis, using Arc/Info	Vijay et al. (2005)

(continued)

TABLE 18.3 *(Continued)*

Cat.	Scope	Methodology	Reference
	To estimate allocation and to propose relocation for waste containers	GIS, using ArcGIS	Ahmed et al. (2006b)
	To estimate waste productivity	GIS and fuzzy logic	Karadimas et al. (2006)
	To estimate waste productivity considering commercial activities	GIS and spatial clustering, using ArcGIS	Karadimas and Loumos (2008)
	To investigate current MSW management in an urban-rural fringe area	GIS, questionnaire survey and GPS tracking, using ArcGIS	Hiramatsu et al. (2009)
	To investigate relationship between hazardous waste generation and socioeconomic characteristics	GIS for geostatistics, using ArcGIS	Lara-Valencia et al. (2009)
	To develop a partial relationship between land use classes and solid waste characteristics and generating indexes	GIS and linear regression analysis, using ArcGIS	Katpatal and Rao (2011)
ISWM activities	To develop a general architecture of GIS for design of SWM systems	GIS, dynamic calculation and interactive analyses	Caputo and Pelagagge (2000)
	To develop an SWM framework for calculating total cost and identifying the most cost-effective alternative	GIS, financial analysis and scenario development	Karadimas et al. (2004)
	To display results for a compost-dominated ISWM framework	GIS	Bilitewski et al. (2005)
	To reduce construction waste and efficiently manage material and equipment	GIS, GPS and IRP barcode system, using ArcGIS, Trimble and GeoExplorer	Li et al. (2005)
	To carry out emission control with route optimization in solid waste collection	GIS and emission calculation, using Route View Pro	Apaydin and Gonullu (2008)
	To allocate waste incinerator compensatory fund	GIS and AHP, using ArcView	Chiueh et al. (2008)

	Objective	GIS approach	Reference
	To compare transport emissions and impacts of different alternatives	GIS and MCE, using ArcGIS	Bastin and Longden (2009)
	To assess the fair fund distribution for a municipal incinerator	GIS and fuzzy AHP, using ArcGIS	Chang et al. (2009)
	To build an Internet-based reverse logistics monitoring system of e-waste	WebGIS	Tao (2010)
	To develop an EIS for urban industrial SWM and exchanges	GIS and Web integration	Zheng and Pan (2010)
	To carry out energy and environmental assessment for construction and demolition waste recycling chain	GIS, scenario development and LCA, using ArcGIS	Blengini and Garbarino (2010)
	To develop a framework for waste disposal sites management	Data management and interaction	Garcia-Frias et al. (1993)
	To discuss general design scheme of GIS in SWM, data and operations management	Basic GIS functions, using Arc/Info	Zaheer et al. (2002)
	To develop an SDSS for hazardous waste sites monitoring	Recursive, flexible and integrative design, using ArcGIS	Jensen et al. (2009)
Assessment	To implement risk analysis for aquifer contamination from hazardous materials transporter spills	GIS and DRASTIC assessment model	Padgett (1991)
	To implement risk analysis hazardous waste transport	GIS, scenario development and RA	Brainard et al. (1996)
	To display and communicate the results to interested parties	GIS and modular RA approach	Whelan et al. (1996)
	To develop regional strategies for RA of hazardous waste transport	GIS, investigation and RA	Lazar et al. (2001)

(continued)

TABLE 18.3 (*Continued*)

Cat.	Scope	Methodology	Reference
	To implement risk analysis for high-level waste repository closure	GIS and one-dimensional ground waste transport model, using ArcView	Bollinger et al. (2004)
	To carry out reassessment of racial inequality in the distribution of hazardous waste facilities	GIS and environmental inequality research methods	Mohai and Saha (2007)
	To evaluate construction and operation cost for a waste landfill site	GIS and energy synthesis	Franzese et al. (2008)
	To assess environmental racism and injustice caused by sitting hazardous facilities	GIS and qualitative methods	Sicotte (2008)
	To carry out environmental impact assessment for waste transport	GIS, using ArcGIS	Bastin and Longden (2009)
	To assess groundwater contamination caused by solid waste disposal	Field sampling, statistics and GIS	Rajkumar et al. (2010)
	To assess soil pollution caused e-waste recycling	Field sampling, statistics using SPSS and GIS, using ArcView	Tang et al. (2010)
	To develop an environment RA of organic waste reuse in agriculture	GIS and a long-term dynamic modeling	Rio et al. (2011)

Source: Adapted from Lu et al. (2013).

ACA, ant colony algorithm; AHP, analytical hierarchical process; DRASTIC, Depth to water, net Recharge, Aquifer media, Soil media, Topography, Impact of the vadose zone media and hydraulic Conductivity of the aquifer; EIS, environmental information system; ERMA, Environmental Response Management Application; GA, genetic algorithm; GPRS, general packet radio service; GRASS, Geographic Resources Analysis Support System; IRP, incentive reward program; RA, risk assessment; LCA, life cycle assessment; LP, linear programming; MATLAB, MATrix LABoratory; MIP, mixed integer programming; MCE, multicriteria evaluation; SDSS, spatial DSS; SOM, self-organizing map.

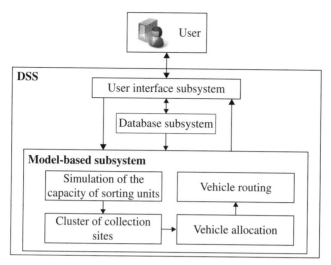

FIGURE 18.4 Decision support systems for a waste collection system. *Source*: Adapted from Simonetto and Borenstein (2007)

models. The DMS can collect and manage internal and external data used by the MBMS, resulting in information for decision makers to analyze and interpret in their decision process (Figure 18.4).

DSS can be classified into five generic types (Power, 2004): "Communications-driven, Data-driven, Document-driven, Knowledge-driven, and Model-driven DSS". "Data-driven DSS help managers organize, retrieve, and analyze large volumes of relevant data using database queries and online analytical processing techniques. Model-driven DSS use formal representations of decision models and provide analytical support using the tools of decision analysis, optimization, stochastic modeling, simulation, statistics, and logic modeling" (Bhargava et al., 2007). A knowledge-driven (or based) DSS (KBDSS) is the recommendation component of a DSS, with suggestions to managers; a communication-driven DSS relies on electronic communication technologies to link multiple decision makers over different location or at different time with relevant information and tools; a document-driven DSS integrate a variety of storage and processing technologies to provide decision makers with essential document retrieval and perform necessary analysis (Bhargava et al., 2007). These DSS have all been promoted by the Internet, increasing their influence and application in web-based DSS. Web-based DSS can be accessed anywhere to improve public participation (Carver et al., 2000).

A DSS is basically a long questionnaire that includes solution procedures, and the results are influenced by human decision makers' preferences (Karmakar et al., 2007). A DSS can be constructed using different approaches, including spreadsheet software such as QuattroPro® and Excel® (Karmakar et al., 2007), special software packages such as SAS® (Statistical Analysis System) (Chang and Wang, 1996), and self-developed systems (Bhargava and Tettelbach, 1997; Huang and Sheng, 2006). Cortés et al. (2000) presented a sophisticated framework for environmental DSS, including data interpretation, diagnosis, and decision support.

Practical DSS for SWM are relatively uncomplicated (Table 18.4). The application spectrum includes regional planning, vehicle routing and scheduling (Li et al., 2007; Simonetto and Borenstein, 2007), and site rehabilitation (Carlon et al., 2007) with the aid of models. Most of these applications are composed of two stages: first, the incorporation of systems engineering models such as linear programming (LP) and mixed integer programming (MIP) models for the waste flow assessment; and second, the implementation of multicriteria decision-making (MCDM) with regard to multiple factors or criteria such as economic, environmental, and social impacts. Cases in the second stage include a multicriteria decision matrix (John, 2010) and a fuzzy multicriteria evaluation (Zeng and Trauth, 2005; Alves et al., 2009). For example, a web-based DSS was developed to help the public understand radioactive waste management and determine educated opinions of proper disposal sites (Carver et al., 2000). Further, with GIS support, a web-based spatial DSS (i.e., web-based SDSS) is more helpful for public participation because of the map-based interactions. The history and categories of web-based SDSS can be found in the literature (Rinner, 2003); most DSSs applied to SWM are model-driven (Table 18.4).

18.5.2 Expert Systems

Other types of DSS models may include a fourth component related to a knowledge-based system that estimates input parameters and helps interpret modeling results (Lukasheh et al., 2001). These knowledge-based systems, also known as ES, are computer programs designed to imitate the advice of a human expert and draw conclusions from information where no precise, unambiguous answer exists (AEA Technology, 1998). ES is a branch of AI, appropriate for a specific domain of judgment problems that make no attempt at structured representation or no traceable paths from inputs to conclusions (de Kock, 2004; Sun et al., 2009). Thus, an ES consists of "(a) a knowledge base, (b) an inference engine that applies built-in rules (often rather rough estimates) to the knowledge base to draw conclusions, and (c) a user interface that enables the user to ask questions and understand the answers" (Chang et al., 2011). A case-based ES in SWM can be developed through the acquisition of relevant data and information, providing the planner with technical information that could not be available. For example, an ES database was used to characterize a waste stream and estimate implications concerning transport, processing, and disposing of materials and waste (MacDonald, 1996b).

Development of an ES involves the human expert, the knowledge engineer, and the knowledge-base of ES. Human knowledge can be obtained from textual sources including books, manuals, technical reports, and research publications (Basri, 1998; Basri, 2000; Alani et al., 2009); and domain expertise via interviews or phone interviews (Coursey et al., 1993; Manamperi et al., 2005). Knowledge engineering (or knowledge representation) develops the process for the ES specifically. It encodes

TABLE 18.4   A summary of DSS applied to SWM

Cat.	Scope	Methodology	Reference
Regional planning	To develop SWAP, a computer package for aiding in regional planning management when considering waste recovery	Model-driven: LP	Ossenbruggen and Ossenbruggen (1992)
	To develop a web-based DSS for optimization of delivery of waste to transfer stations	Model-driven: LP and web integration	Bhargava and Tettelbach (1997)
	To review the development of DSS for SWM in recent years and give a demonstration in Taiwan, China	Model-driven: LP by SAS®	Chang and Wang (1996)
	To develop a DSS for ISWM at a regional level in Italy	Model-driven: multiobjective dynamic MIP and *ad hoc* user interface	Bazzani (2000)
	To show how urban information systems can be shift SWM knowledge system	Knowledge-based	Rubenstein-Montano (2000)
	To plan the optimal number of landfills and treatment plants, and then to determine the optimal waste allocation	Model-driven: constrained nonlinear optimization	Fiorucci et al. (2003)
	To model BestCity for intelligent management of MSW	Knowledge-based	Jayawardhana et al. (2003)
	To select the optimal waste treatment combination from eight alternatives	Model-driven: fuzzy MCE and web-based user interaction	Zeng and Trauth (2005)
	To find a cost-effective model solution in municipal SWM	Model-driven: web-based net present value evaluation	Xiangyun et al. (2007)
	To find a sustainable alternative for waste allocation to landfill	Model-driven: multicriteria decision matrix	John (2010)
Suitable site selection	To generate alternatives for the sitting of new waste treatment or disposal facilities and then to allocate waste flow	Model-driven: MCE	Haastrup et al. (1998)
	To develop a DSS to select landfill site	Model-driven: MCE and GIS	Vatalis and Manoliadis (2002)
	To develop a fuzzy DSS for landfill site selection	Model-driven: fuzzy MCE	Alves et al. (2009)

(continued)

635

TABLE 18.4 (Continued)

Cat.	Scope	Methodology	Reference
Vehicle routing and scheduling	To aid human schedulers to obtain optimal vehicle assignments and reassignments for minimized operation and delay costs	Model-driven: single objective quasi-assignment model (linear integer programming); web-based	Li et al. (2007)
	To optimize waste allocation from collection sites to sorting units and then to assign vehicles for each sorting unit	Model-driven: single objective LP for two stages, Arena® software	Simonetto and Borenstein (2007)
	To included optimization methods in a DSS to manage reverse logistical waste lubricating oils	Model-driven: web-based	Repoussis et al. (2009)
Site management	To select the best scheme for contaminated mega-sites rehabilitation	Model-driven: MCE	Carlon et al. (2007)
	To assess the revegetation potentiality, to choose trees/plant species and to select land consolidation approach	Model-driven: fuzzy MCE, grey relationship analysis, cluster analysis and rule-based reasoning, ArcIMS and ArcObjects	Chen and Li (2008)
	To achieve satisfaction of multiple objectives of economy and health and environmental risk for computer waste management	Model-driven: LP-based multi-time step optimal material flow analysis	Ahluwalia and Nema (2007)
	To select optimal landfill design alternatives	Model-driven: MCE	Celik et al. (2010)
	To develop a prototype model to determine the potentials for reuse and recycling of waste materials, to select the treatments needed to recycle waste materials or for treatment before disposal, and to determine potentials for co-treatment of wastes	Knowledge-based	Boyle and Baetz (1998a, 1998b)
	To identify the optimal construction and demolition waste management strategy that minimizes end-of-life costs and maximizes the recovery of salvaged building materials	Model-driven: MILP, waste generation and web-based	Banias et al. (2011)
	To develop a DSS for a waste company that buys waste and sells it to recyclers	Communication via model-driven	Derigs and Friederichs (2009)

Source: Adapted from Lu et al. (2013).

MILP, mixed-integer linear programming; SWAP, Solid Waste Allocation Package.

acquired knowledge and is crucial to the ease and speed of reasoning. The forms of knowledge representation (or the knowledge-base) can be

- unstructured styles such as rules and cases;
- structured styles such as frames, semantic nets and objects; and
- implicit styles such as artificial neural network (ANN) models.

Rule-based representation, often termed a production rule system (de Kock, 2004), is the most commonly employed system, in which knowledge reasoning is key to effectively and sufficiently capitalizing on expertise to solve problems. The inference engine is activated by user consultation and then draws conclusions or infers new knowledge according to a certain reasoning mechanism, deductive or inductive. Knowledge reasoning has three directions: forward, backward, and hybrid. Forward reasoning is generally suitable for searching all potential conclusions according to a set of facts; backward reasoning is efficient for proving whether the goal is true; and hybrid reasoning is suitable for integrated problems. Forward reasoning for diagnosis has been the most popular, as indicated in the literature (Sun et al., 2009).

An ES is developed either by AI languages such as Prolog, Lisp, CLIPS (C Language Integrated Production System), and Erlang; or by high level languages such as Fortran and C; or by an ES shell, the inference and interpretation mechanism of ES (Turban, 2005). Using a shell can reduce the development time of the ES. Basri and Stentiford (1995) reviewed ES applied for SWM before the mid-1990s, and Lukasheh et al. (2001) reviewed ES application and its combination with GIS and DSS in landfill design and management. ES were developed to solve unstructured SWM problems (Table 18.5), including hazardous waste management in accordance with issued regulations, evaluation and recommendation of potential waste treatment alternatives, ranking or selection of waste treatment sites, waste identification and classification, facility design, site management, and automatic fault diagnosis and control. Various kinds of ES shells improve development efficiency, such as Level5 Object^{TM}, Insight 2+^{TM} (Rouhani and Kangari, 1987), HYSYS® (Halim and Srinivasan, 2006), Kappa-PC^{TM} (Basri, 2000), Design++^{TM}, AcquireTM (Manamperi et al., 2005), and Visual Rule Studio^{TM} (Chau, 2005); MATLAB® is also used for reasoning (Alani et al., 2009). These tools reduce complexity and allow developers to focus attention on substantive content rather than form.

Rule-based representation (Barrow, 1988; Kim et al., 1993) and object-oriented representation (Staudinger et al., 1997; Basri, 2000; Alani et al., 2009) are most commonly used for knowledge representation, including heuristic knowledge represented in a rule base with fuzzy logic (Chau, 2005, 2006). Systems using object-oriented representation still capitalize on the rule-based reasoning, however. Fuzzy logic, Bayesian networks, and ANN have been used to manage uncertainty (Hodges et al., 2001). A heuristic approach such as genetic algorithm (GA) has been employed to accelerate reasoning searches (Hirokane et al., 1995). Other ES applications, such as solving odor problems in a wastewater treatment plant (Kordon et al., 1996), can be found in the literature.

TABLE 18.5 A summary of ES applied to SWM

Cat.	Scope	Methodology	References
Regulation of hazardous waste	To assist the USEPA in regulating hazardous waste	Providing alternatives arranged according to likelihood	Anandalingam (1987)
	The regulation of hazardous waste	Rules derived from current regulations and algorithm from literature	Barrow (1988)
	To estimate and control the costs of the initial phases of hazardous waste cleanup		Ketterer et al. (1991)
Regional planning	To analyze the possibility of applying ES and GA to solid waste disposal services		Hirokane et al. (1995)
	To select the best waste treatment alternatives	Knowledge bases covering economic, technical, social and political aspects, Turbo Prolog	Wei and Weber (1996)
Suitable site selection	To rank the candidate sites according to five rules including ground water routes	USEPA documents for ranking controlled hazardous waste sites, Insight 2+TM ES shell	Rouhani and Kangari (1987)
	To develop a prototype ES for landfill site selection	Hybrid application with ES, ANN, and fuzzy inference, Visual Rule StudioTM shell	Chau (2005, 2006)
	To select the optimal waste incinerator sites	Multicriteria decision analysis that combined MS/OR with ES	Wey (2005)
Waste generation	To estimate inventories of hardware waste in greater-than-class-C waste and to suggest packing and disposal options	Local-scale prototype to gain familiarity for the full-scale system that would follow	Williamson (1990)
ISWM activities	To develop a ES for industrial waste management and minimization	The expertise gained from Industrial Assessment Center research	Gopalakrishnan et al. (2003)
	To minimize waste in the chemical industry	CAPE-OPEN capability of HYSYS® simulator and XML data exchange	Halim and Srinivasan (2006)
	To develop a highway construction ES for giving advices on waste minimization in highway construction	OO knowledge representation and rule-based reasoning through MATLAB®	Alani et al. (2009)
	To detect, classify, squeeze, and recycle metal cans automatically	A heuristic feedback control theory, micro-chip processor, sensors, etc.	Lin et al. (2009)

Site management	To rank hazardous waste sites for remedial action priority under the Department of Defense's Installation Restoration Program in the United States	Evaluation of multiple factors associated with environmental and ecological contamination	Hushon and Read (1991)
	To plan an ES for the landfill restoration	Acquiring knowledge from multiple expertise sources such as textbooks, research publications	Basri (1998)
Fault diagnosis and automatic control	To aid operation in a rod consolidation process, including tracking the transition, diagnosing the process status and giving operators advises	A knowledge base including three database groups and 60 rules with production and techniques	Kim et al. (1993)
	To control the use of a gas chromatography system in the analysis of hazardous wastes, in both a more rigorous and systematic manner	Design and validation with assistance of analytical chemists	Matek and Luger (1997)
	To recognize changes in the sign of process gain and implement appropriate control laws for continuous anaerobic digesters	A T-test for determination	Pullammanappallil et al. (1998)
	To develop an ES for better management of waste composting in Sri Lanka	Giving accurate and real feel to the user via different channels of knowledge browsing	Jayawardhana et al. (2003)
	To develop an ES for proper implementation of landfill technology in Sri Lanka	OO ES shell - AcquireTM 2.1, and various kinds of knowledge source	Manamperi et al. (2005)
	To develop a fault diagnosis of garbage crusher	Knowledge base in the domain, forward reasoning	Sun et al. (2009)
Other applications	To evaluate six types of renewable energy technology fuelled by biogas from MSW landfills	ES through fuzzy multi-rules and fuzzy multi-sets	Barin et al. (2013)
	To develop a real-time ES with a multi-analyzer including two electronic noses for odor control	ANN used to estimate variables from electronic nose signal responses	Bachinger and Mandenius (2001)

Source: Lu et al. (2013).
CAPE-OPEN, Computer Aided Process Engineering (OPEN is for free available); HYSYS, Hyprotech System Simulation; LRPA, Landfill Restoration Plan Advisor; OO, object-oriented; USEPA, United States Environmental Protection Agency; XML, Extensible Markup Language.

18.5.3   Artificial Neural Networks and Genetic Algorithms

An ANN is a reasoning model based on the human brain (Youssef, 2007). An ANN (or alternatively NN) is a massive, parallel computational system composed of simple nonlinear processing elements with adaptable interconnections (Dong et al., 2003). ANN simulates human functions like learning from experience and abstracting crucial attributes from inputs having irrelevant data for analysis (Greenman, 1998). Neurons usually operate in parallel and are configured in regular architectures (Antanasijević et al., 2013). Each connection strength is indicated by a weight (numerical value), which can be updated (Jahandideh et al., 2009). ANN models are often organized in three layers: input, output, and hidden layers (Figure 18.5).

The most important feature of ANN is the ability to learn based on the surrounding situations, improving its performance due to learning or training. The three learning paradigms are reinforced, supervised, and unsupervised learning. In terms of modeling structure, there are two types of ANN including feedback (recurrent) model and feedforward model. Whereas feedback (recurrent) model has connections from a direct cycle to convey the modeling outputs back for adjustment, feedforward model has no such cycle, moving in one direction and never going backward. In supervised

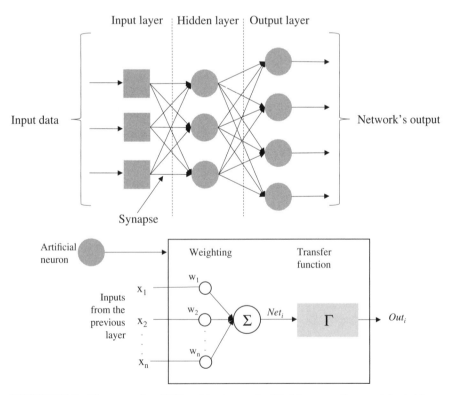

FIGURE 18.5   Elements of an ANN architecture and artificial neuron. *Source*: Adapted from Hernández-Caraballo et al. (2003)

learning, the network produces its own answer and presents it to the supervisor for validation. The most common models applying supervised learning are perceptron, the Widrow-Hoff learning rule, the delta rule, and error back-propagation (and its derivations). In unsupervised learning, there are no initial output values, so outputs are not validated. The main models are competitive learning and Kohonen's learning. Reinforcement learning consists of making agent actions based on the ambience. In this learning method, there is no teacher to give the correct answer, except a hint to indicate the correct path. Models that reinforce the learning method are the Hebb learning rule and Hopfield learning rule.

GA is a stochastic search technique which mimics the natural evolution theorem, being applied to many problems with success, like NP-hard (non-deterministic polynomial-time hard) problems (see Box 15.1), and has proved superior to many other heuristics (Youssef, 2007). GA can often reach the optimal or close-to-optimal solutions, and thus are used in many optimization techniques that assist in the design of different DSS, such as ANN systems (Youssef, 2007). GA searches for solutions by simulating Darwinian survival-of-the-fittest theory to evolve solutions over a series of generations (Holland, 1975 in Rubenstein-Montano, 2000). A solution set is generated to contain the population from which better solutions evolve over a sequence of generations; the initial population of solutions is created at random and uniformly distributed over the solution space; prospering populations are produced from the foregoing ones via search operators, which work to preserve and merge the desirable traits of the better members of the antecedent population of solutions (Rubenstein-Montano et al., 2000).

GA has contributed significantly to SWM, specifically in solid waste collection problems. The merits of GA include (Chang and Wei, 1999, 2000): (1) efficiency as a global method for nonlinear optimization problems, and the ability to search for global optimal solutions with a simple algorithm that does not require linearization assumptions and the calculation of partial derivatives; (2) avoidance of numerical instabilities associated with essential matrix inversion, a problem frequently encountered in the conventional mathematical programming algorithms; and (3) efficiency and robustness when searching for a global optimal solution, as compared with conventional Monte Carlo simulations or previous optimization algorithms used for solving nonlinear programming models.

The easiest to use, most common software to implement ANN and GA is MATLAB, although others are available, such as NeuroSolutions, NeuroDimension, Neuro Laboratory Solution, and Learning and Intelligent OptimizatioN Solver (known as LIONSolver). Open-source software is also available, including OpenNN (known as open source neural network), Encog, and Neuroph. The literature provides examples of GA and ANN implemented for SWM (Table 18.6).

18.6 INTEGRATED ENVIRONMENTAL INFORMATION SYSTEMS

Each kind of EIS has its own strengths and limitations. A DBS is the basic component in environmental informatics applications, which are limited to processes that

TABLE 18.6 Examples of ANN and GA applied to SWM

Cat.	Scope	Methodology	References
Solid waste characteristics	To predict MSW low heating value	ANN by feedforward network	Dong et al. (2003)
	To predict MSW low heating value	Multilayer perception	Shu et al. (2006)
	To predict MSW higher heating value, water, carbon, hydrogen, nitrogen, oxygen, sulfur, and ash	ANN based on Marquardt back-propagation learning algorithm	Akkaya and Demir (2010)
	Gasification properties from MSW	Back-propagation	Xiao et al. (2009)
	To simulate and optimize biogas production process from organic waste anaerobic digestion	ANN (back-propagation) and GA	Qdais et al. (2010)
Solid waste emissions	To forecast dioxin emissions from waste incineration	Back-propagation neural network	Bunsan et al. (2013)
	To predict methane fraction in landfill gas from landfill bioreactors	Back-propagation neural network	Ozkayam et al. (2007)
	For modeling leachate flow-rate in MSW landfill site	Back-propagation algorithm	Karaca and Özkayab (2006)
	To predict PCDD/PCDF emissions from MSW incineration	GA and ANN	Chang and Chen (2000)
Solid waste generation	To forecast MSW generation	Back-propagation and general regression NN	Antanasijević et al. (2013)
	To forecast weekly MSW generation	Several algorithms were used: resilient back-propagation, scale conjugate gradient, one step secant, and Levenberg-Marquardt	Noori et al. (2010)
	To forecast medical waste generation totally and by sharp, infectious and general types	Feedforward with back-propagation algorithm	Jahandideh et al. (2009)
Solid waste planning	To generate policy alternatives for evaluation by human decision makers	GA	Rubenstein-Montano et al. (2000)
	To optimize a sustainable recovery network of scrap tires	Multiobjective optimization with GA	Dehghanian and Mansour (2009)
	To determine the optimal siting and routing drop-off collection	GA with multiobjective programming model	Chang and Wei (1999)
Products eco-design with end-of-life stage	To solve the reverse logistic problem	GA with nonlinear mixed-integer programming model	Min et al. (2006)
	To optimize the multiobjectives of material selection to obtain more sustainable products	ANN (back-propagation) and GA	Zhou et al. (2009)
	To optimize construction steel bars waste, minimizing cutting losses	GA	Khalifa et al. (2006)

PCDD/PCDF, Polychlorinated dibenzodioxins/ Polychlorinated dibenzofurans.

create, read, update, and delete. GIS is dedicated to information visualization and spatial analysis and generally falls short in utility models. DSS excels in extracting implicated knowledge from multidimensional data comparison and organizing the interactions among users and models, but it is susceptible to decision maker preferences. ES utilizes knowledge to imitate expert judgment, but it is limited to knowledge acquisition. An IEIS combines two or more kinds of tools and/or platforms to develop unique advantages tailored for applications. For example, Lukasheh et al. (2001) assessed the integration of ES and GIS in connection with several simulation models to form a DSS framework for landfill design and management. In addition, the computer networking capability not only reduces the cost of information dissemination to end users, but also avoids the difficulty that each user may encounter when installing and managing tools on the computer (Chen et al., 1996). Huang and Sheng (2006) noted that webGIS allows more stakeholders to participate in the decision-making loops. SDSS, KBDSS, and a spatial expert system (SES) can be representative components in IEIS, in which advanced integration with internet technologies and decision analysis models can be anticipated for specific research solutions. The following discussion provides supporting evidence.

Spatial Decision Support Systems. A GIS integrated with a decision analysis model can be defined as an SDSS (Table 18.3). Ohri and Singh (2009) compared SDSS with DSS for SWM and noted that segmented DSS could not provide a holistic view of the interactions and effects among all functional elements in the complex system; however, traditional decision support techniques without GIS lack the ability to simultaneously address all aspects of site selection and landfill design (Baban and Flannagan, 1998; Allen et al., 2003). While low level SDSS provides visualization via GIS, high level SDSS offers spatial data mining to perform advanced spatial analysis as a data-driven DSS. A salient example for the latter may be a DSS linking a rule-based ES within a GIS environment to screen candidate locations for landfill (Davies and Lein, 1991).

Spatial Expert Systems. An SES addresses knowledge acquisition and reasoning. In theory, GIS analysis is theoretically capable of discovering new knowledge to enhance the knowledge acquisition of an ES, but it has rarely been used for SWM to explore the relationship between the GIS analysis and knowledge acquisition. GIS was typically integrated into SES simply to improve visualized interactions, or integrated with both DSS and ES to enhance routine frameworks. Chen et al. (1996) developed a prototypical SES in which the knowledge base was composed of forward chaining with related rules extracted from various literatures to facilitate landfill siting.

Knowledge-Based Decision Support Systems. A KBDSS integrates traditional DSS with the advances of ES, embracing symbolic reasoning and explanation capabilities (Klein and Methlie, 1995). It may have an additional knowledge base in support of specific domain knowledge and a separate reasoning mechanism for analyses which offers expert assistance to support decision-making and interpret modeling outputs. A small number of KBDSS were developed to assist in selecting regional waste management planning alternatives (Smith et al., 1997; Boyle and Baetz, 1998a, 1998b; Karmakar et al., 2010).

TABLE 18.7   Model-based DSS with the aid of GIS

Purpose		Mathematical models
Siting landfills	Spatial analysis	AHP, multicriteria evaluation, spatial model
Siting waste treatment facilities	Spatial analysis	AHP, Euclidean distance
Forecasting of solid waste generation	Spatial analysis	Addition (overlay) Spatial auto-regression Geographically weighted regression
	Statistical analysis	Multiple regression analysis, ordinary least squares regression
Routes	3D analysis	3D modeling of terrain and road network
	Network analysis	Traveling salesman, heuristic and genetic algorithm
Waste management planning	Spatial analysis	AHP, multicriteria evaluation

Model-Based Decision Support Systems. Spatial analysis can be combined with a variety of models to carry out decision support. These models are related to decision science regimes and may include spatial statistical models, optimization models, and computational intelligence models (Table 18.7).

18.7   FINAL REMARKS

In the 1990s, various EIS applications defined the basic milestones for environmental informatics in SWM. Since the 2000s, more techniques for automatic data acquisition have been enriching environmental informatics. These informatics solutions can address a series of problems that ISWM faces today, including the lack of source separation and reduction, poor efficiency in waste collection and transport, and imperfect monitoring. In the future, the development of environmental informatics for SWM systems will require greater depth and scope in all aspects. Given that the entire waste management life cycle from source separation and reduction to final disposal must be cohesively managed, high-level integrated, intelligent, informatics systems such as Internet of Things (IoT) should be developed in practice in the nexus of industrial ecology and sustainable engineering. Based on the search frequency from Google Trend over recent years, the 2010s are predicted to become the decade of IoT, prefiguring a smart internet with ubiquitous tentacles detecting and connecting people and/or things through techniques such as RFID and sensors. IoT reflects the developmental tendency of informatics over the next few years but is different from supervisory control and data acquisition (SCADA) because it is "smart" enough to improve both efficiency and effectiveness of waste management.

Although IoT is a new concept, it is actually derived from current internet techniques. To apply IoT to SWM systems (Figure 18.6), the whole framework may

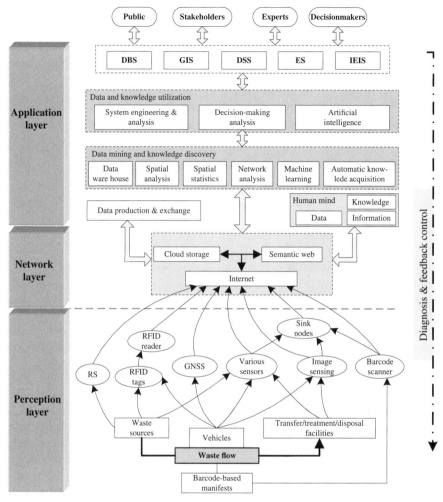

FIGURE 18.6 The framework of IoT-based SWM systems. *Source*: Lu et al. (2013). GNSS, global navigation satellite system

be divided into three layers according to IBM's proposition. The bottom layer is a perception layer that collects data about waste amounts, source producers, vehicle positions and states, facility states, environmental impacts, images, and surveillance videos. Barcode-based manifests may be used to record waste flow and exchange data among different operation companies. Those data collection devices are all abstracted as data acquisition nodes, which have their own intercommunications and can access the Internet directly or via sink nodes.

The intermediate layer is a network layer that couples the Internet, cloud storage, and semantic web. Data acquisition nodes access the network layer similar to the

Internet. Heterogeneous data may be exchanged in response to demand. Users can access the network layer to find what they want, regardless of the storage site. The semantic web transforms the network layer into a network of data, information, and knowledge and teaches machines to understand the human mind.

The top layer is an application layer that is actually a super-integrated EIS that adapts to heterogeneous data environments. This type of EIS, which combines the advantages of all types of EIS, is able to solve well-structured or poorly-structured problems of SWM and maintain itself independently. It manages input of data and models, storage, human-machine interactions, computing processes, and information output and representation. It understands the human mind, uses data mining and GIS analysis to discover new knowledge from existing non-spatial and spatial data, and uses inference engines to learn new knowledge from existing rules. Moreover, this integrated EIS can create wisdom, consult with experts, and accumulate domain expertise. It informs the public and accommodates partnerships, considering the public a part of the stakeholders. The public and stakeholders who are not domain experts can learn from the system and give their opinions in a share-vision mode. These opinions will influence the decision-making with different design weights, however, which may be assigned and adjusted dynamically according to the human–machine interactive learning and progressive improvements. Moreover, the application layer gives other layers instructions for diagnosis and feedback control.

Besides, big data analytics, which is an emerging area in computer science, is the examination procedure of large amounts and different types data (the big data), to expose unseen patterns, undisclosed correlations that may be applied to ISWM. The gaps between perception and decision-making will be minimized with the development of big data analytics and IoT. It can be envisioned that next-generation sensing technologies will be applied to collect large-scale and accurate data, identification techniques will be applied to promote waste separation and source reduction, and more sophisticated systems combining various advantages of different EIS and models will be developed to support spatiotemporal analysis. These combined efforts will push forward the intelligent promotion of ISWM in the future.

REFERENCES

AEA Technology. 1998. Computer-based models in integrated environmental assessment. Technical report 14. European Environment Agency.

Ahluwalia, P. K. and Nema, A. K. 2007. A goal programming based multi-time step optimal material flow analysis model for integrated computer waste management. *Journal of Environmental Informatics*, 10(2), 82–98.

Ahmadi, M., Zade, P. S., Pouryani, S. B. M., and Gilak, S. 2010. Site selection of waste storage locations with fuzzy logic and analytic hierarchy process in a GIS framework, case study: Si-sakht town in Dena city, Kohgilouye O Boyer Ahmad province, Iran. In: Proceedings of the 10th International Multidisciplinary Scientific Geoconference, Vol. 1, pp. 1143–1150.

Ahmed, M. H., Donia, N., and Fahmy, M. A. 2006a. Eutrophication assessment of Lake Mazala, Egypt using geographical information systems (GIS) techniques. *Journal of Hydroinformatics*, 8(2), 101–109.

Ahmed, S. M., Muhammad, H., and Sivertun, A. 2006b. Solid waste management planning using GIS and remote sensing technologies: case study Aurangabad City, India. In: Proceedings of the 2006 International Conference on Advances in Space Technologies (IEEE Cat No. 06EX1482 C), pp. 196–200.

Akkaya, E. and Demir, A. 2010. Predicting the heating value of municipal solid waste-base materials: an artificial neural network model. *Energy Sources, Part A: Recovery, Utilization, and Environmental Effects*, 32(19), 1777–1783.

Alani, I. A. R., Basri, N. E. A., Abdullah, R. A., and Ismail, A. 2009. Artificial intelligence expert system for minimizing solid waste during highway construction activities. In: Proceedings of the IMECS 2009: International Multi-Conference of Engineers and Computer Scientists, vol. 1–2, pp. 54–57.

Allen, A. R., Brito, G., Caetano, P., Costa, C., Cummins, V. A., Donnelly, J., Koukoulas, S., O'Donnell, V. A, Robalo, C., and Vendas, D. 2003. A landfill site selection process incorporating GIS modelling. In: Proceedings Sardinia 2003, Ninth International Waste Management and Landfill Symposium, S. Margherita di Pula, Cagliari, Italy.

Alvarez, J. V. L., Larrucea, M. A., Santandreu, F. S., and De Fuentes, A. F. 2009. Containerisation of the selective collection of light packaging waste material: the case of small cities in advanced economies. *Cities*, 26(6), 339–348.

Alves, M. C., Lima, B. S., Evsukoff, A. G., and Vieira, I. N. 2009. Developing a fuzzy decision support system to determine the location of a landfill site. *Waste Management & Research*, 27(7), 641–651.

Anandalingam, G. 1987. REGWASTE: an expert system for regulating hazardous wastes. In: Proceedings of the 1987 International Conference on Systems, Man, and Cybernetics (Cat. No. 87CH2503-1).

Antanasijević, D., Pocajt, V., Popović, I., Redžić, N., and Ristić, M. 2013. The forecasting of municipal waste generation using artificial neural networks and sustainability indicators. *Sustainability Science*, 8(1), 37–46.

Apaydin, O. and Gonullu, M. T. 2008. Emission control with route optimization in solid waste collection process: a case study. *Sadhana-Academy Proceedings in Engineering Sciences*, 33(2), 71–82.

Aronoff, S. 1989. *Geographic Information Systems: A Management Perspective*, WDL Publications, Ottawa, Canada.

Arribas, C. A., Blazquez, C. A., and Lamas, A. 2010. Urban solid waste collection system using mathematical modelling and tools of geographic information systems. *Waste Management & Research*, 28(4), 355–363.

Baban, S. M. J. and Flannagan, J. 1998. Developing and implementing GIS- assisted constraints criteria for planning landfill sites in the UK. *Planning Practice & Research*, 13(2), 139–151.

Bachinger, T. and Mandenius, C. G. 2001. Physiologically motivated monitoring of fermentation processes by means of an electronic nose. *Engineering in Life Sciences*, 1(1), 33–42.

Bagad, V. S. and Godse, A. P. 2009. *Mechatronics and Microprocessor*, Technical Publications, Pune, India.

Banias, G., Achillas, C., Vlachokostas, C., Moussiopoulos, N., and Papaioannou, I. 2011. A web-based decision support system for the optimal management of construction and demolition waste. *Waste Management*, 31(12), 2497–2502.

Barin, A., Canha, L. N., Abaide, A. R., Magnago, K. F., Matos, M. A., and Orling, R. B. 2013. A novel fuzzy-based expert system for RET selection. *Journal of Intelligent and Fuzzy Systems*, 25(2), 325–333.

Barrow, P. A. 1988. An expert system applied to the regulation of hazardous waste. Simulators V. In: Proceedings of the SCS Simulators Conference.

Basagaoglu, H., Celen, E., Mariulo, M. A., and Usul, N. 1997. Selection of waste disposal sites using GIS. *Journal of the American Water Resources Association*, 33(2), 455–464.

Basel Convention. 2011. Basel Convention – controlling transboundary movements of hazardous wastes and their disposal. Basel Convention. Available at: http://www.basel.int/Home/tabid/2202/Default.aspx (accessed February 2013).

Basnet, B. B., Apan, A. A., and Raine, S. R. 2001. Selecting suitable sites for animal waste application using a raster GIS. *Environmental Management*, 28(4), 519–531.

Basri, H. 1998. An expert system for planning landfill restoration. *Water Science and Technology*, 37(8), 211–217.

Basri, H. B. 2000. An expert system for landfill leachate management. *Environmental Technology*, 21(2), 157–166.

Basri, H. B. and Stentiford, E. I. 1995. Expert systems in solid waste management. *Waste Management & Research*, 13(1), 67–89.

Bastin, L. and Longden, D. M. 2009. Comparing transport emissions and impacts for energy recovery from domestic waste (EfW): centralised and distributed disposal options for two UK Counties. *Computers Environment and Urban Systems*, 33(6), 492–503.

Bazzani, G. M. 2000. PRUS: a DSS for multicriteria integrated solid waste management. In: Proceedings of the Development and Application of Computer Techniques to Environmental Studies Eighth International Conference, ENVIROSOFT 2000, pp. 227–236|544.

Beirne, S., Kiernan, B. M., Fay, C., Foley, C., Corcoran, B., Smeaton, A. F., and Diamond, D. 2009. Autonomous greenhouse gas measurement system for analysis of gas migration on landfill sites. In: Proceedings of the Sensors Applications Symposium.

Bhargava, H. K. and Tettelbach, C. 1997. A web-based decision support system for waste disposal and recycling. *Computers, Environment and Urban Systems*, 21(1), 47–65.

Bhargava, H. K., Power, D. J., and Sun, D. 2007. Progress in web-based decision support technologies. *Decision Support Systems*, 43(4), 1083–1095.

Bilitewski, B., Karagiannidis, A., Theodoseli, M., Malamakis, A., Reichenbach, J., and Janz, A. 2005. Composting as an integrated solid waste management tool lessons learned from Asean-European co-operation: the example of Pieria, Greece. In: Proceedings of the 9th International Conference on Environmental Science and Technology Vol B – Poster Presentations, B86–B91.

Blengini, G. A. and Garbarino, E. 2010. Resources and waste management in Turin (Italy): the role of recycled aggregates in the sustainable supply mix. *Journal of Cleaner Production*, 18(10–11), 1021–1030.

Bollinger, J. S., Newman, J., and Collard, L. 2004. High level waste tank closure modeling with geographic information systems (GIS). *Transactions of the American Nuclear Society*, 91, 94–95.

Boyle, C. A. and Baetz, B. W. 1998a. A prototype knowledge-based decision support system for industrial waste management: Part I. The decision support system. *Waste Management*, 18(2), 87–97.

Boyle, C. A. and Baetz, B. W. 1998b. A prototype knowledge-based decision support system for industrial waste management: Part II. Application to a Trinidadian industrial estate case study. *Waste Management*, 17(7), 411–428.

Brainard, J., Lovett, A., and Parfitt, J. 1996. Assessing hazardous waste transport risks using a GIS. *International Journal of Geographical Information Systems*, 10(7), 831–849.

Brasileiro, L. A. and Lacerda, M. G. 2008. Analysis of the use of GIS in the collecting vehicle routing of domestic solid waste. *Engenharia Sanitaria e Ambiental*, 13(4), 356–360 (in Portuguese).

Bunsan, S., Chen, W. Y., Chen, H. W., Chuang, Y. H., and Grisdanurak, N. 2013. Modeling the dioxin emission of a municipal solid waste incinerator using neural networks. *Chemosphere*, 92(3), 258–264.

Caputo, A. C. and Pelagagge, P. M. 2000. Integrated geographical information system for urban solid waste management. In: *The Sustainable City* (Eds. Brebbia, C. A., Ferrante, A., Rodriguez, M., and Terra, B.), WitPress, pp. 159–170.

Carlon, C., Critto, A., Ramieri, E., and Marcomini, A. 2007. DESYRE: decision support system for the rehabilitation of contaminated megasites. *Integrated Environmental Assessment and Management*, 3(2), 211–222.

Carver, S., Evans, A., Kingston, R., and Turton, I. 2000. Accessing geographical information systems over the World Wide Web: improving public participation in environmental decision-making. *Information Infrastructure and Policy*, 6(3), 157–170.

Celik, B., Girgin, S., Yazici, A., and Unlú, K. 2010. A decision support system for assessing landfill performance. *Waste Management*, 30(1), 72–81.

Chalkias, C. and Lasaridi, K. 2009a. A GIS based model for the optimisation of municipal solid waste collection: the case study of Nikea, Athens, Greece. In: Proceedings of the WSEAS Transactions on Environment and Development, pp. 640–650.

Chalkias, C. and Lasaridi, K. 2009b. Optimizing municipal solid waste collection using GIS. In: Proceedings of the conference Energy, Environment, Ecosystems, Development and Landscape Architecture, Athens, Greece, pp. 45–50.

Chang, N. B. and Chen, W. C. 2000. Prediction of PCDDs/PCDFs emissions from municipal incinerators by genetic programming and neural network modeling. *Waste Management and Research*, 18(4), 341–351.

Chang, N. B. and Lin, Y. T. 1997. Optimal siting of transfer station locations in a metropolitan solid waste management system. *Journal of Environmental Science and Health Part A-Environmental Science and Engineering and Toxic and Hazardous Substance Control*, 32(8), 2379–2401.

Chang, N. B. and Wang, S. F. 1996. The development of an environmental decision support system for municipal solid waste management. *Computers, Environment and Urban Systems*, 20(3), 201–212.

Chang, N. B. and Wei, Y. L. 1999. Strategic planning of recycling drop-off stations and collection network by multiobjective programming. *Environmental Management*, 24(2), 247–263.

Chang, N. B. and Wei, Y. L. 2000. Siting recycling drop-off stations in an urban area by genetic algorithm-based fuzzy multi-objective nonlinear programming modeling. *Fuzzy Sets and Systems*, 114(1), 133–149.

Chang, N. B. and Wei, Y. L. 2002. Comparative study between heuristic algorithm and optimization technique for vehicle routing and scheduling in the solid waste management system. *Civil Engineering and Environmental Systems*, 19(1), 41–65.

Chang, N. B., Lu, H. Y., and Wei, Y. L. 1997. GIS technology for vehicle routing and scheduling in solid waste collection systems. *Journal of Environmental Engineering, ASCE*, 123(9), 901–910.

Chang, N. B., Parvathinathan, G., and Breeden, J. B. 2008. Combining GIS with fuzzy multiple attribute decision making for landfill siting in a fast growing urban region. *Journal of Environmental Management*, 87(1), 139–153.

Chang, N. B., Chang, Y. H., and Chen, H. W. 2009. Fair fund distribution for a municipal incinerator using a GIS-based fuzzy analytic hierarchy process. *Journal of Environmental Management*, 90(1), 441–454.

Chang, N. B., Pires, A., and Martinho, G. 2011. Empowering systems analysis for solid waste management: challenges, trends, and perspectives. *Critical Reviews in Environmental Science and Technology*, 41(16), 1449–1530.

Charnpratheep, K., Zhou, Q. M., and Garner, B. 1997. Preliminary landfill site screening using fuzzy geographical information systems. *Waste Management & Research*, 15(2), 197–215.

Chau, K. W. 2005. Prototype expert system for site selection of a sanitary landfill. *Civil Engineering and Environmental Systems*, 22(4), 205–215.

Chau, K. W. 2006. An expert system on site selection of sanitary landfill. *International Journal of Environment and Pollution*, 28(3–4), 402–411.

Chen, Y. Y. and Li, D. L. 2008. A web-GIS based decision support system for revegetation in coal mine waste land. In: Proceedings of the WSEAS: Advances on Applied Computer and Applied Computational Science, pp. 579–584.

Chen, W. Y., Lin, H. Y., and Guo, S. J. 1996. Network expert geographic information system for landfill siting. *Journal of Computing in Civil Engineering*, 10(4), 307–317.

Chikae, M., Kerman, K., Nagatani, N., Takamura, Y., and Tamiya, E. 2007. An electrochemical on-field sensor system for the detection of compost maturity. *Analytica Chimica Acta*, 581(2), 364–369.

Chiueh, P. T., Lo, S. L., and Chang, C. L. 2008. A GIS-based system for allocating municipal solid waste incinerator compensatory fund. *Waste Management*, 28(12), 2690–2701.

Chowdhury, B. and Chowdhury, M. U. 2007. RFID-based real-time smart waste management system. In: Proceedings of the 2007 Australasian Telecommunication Networks and Applications Conference, IEEE, Christchurch, New Zealand, pp. 175–180.

Christensen, T. H., Scharff, H., and Hjelmar, O. 2011. Landfilling: concepts and challenges. In: *Solid Waste Technology and Management* (Ed. Christensen, T. H.), Blackwell Publishing, Ltd., Chichester, West Sussex, pp. 685–694.

Commonwealth Scientific and Industrial Research Organisation (CSIRO) and the Australian Department of Environment Heritage. 2004. Australian Waste Database (AWD). CSIRO and the Department of Environment and Heritage. Available at: http://awd.csiro.au/about.aspx (accessed February 2013).

Cortés, U., Sànchez-Marrè, M., Ceccaroni, L., R-Roda, I., and Poch, M. 2000. Artificial intelligence and environmental decision support systems. *Applied Intelligence*, 13(1), 77–91.

Coursey, D., Bretschneider, S., and Blair, J. 1993. IWSAS: expert system phone survey assistance for collecting data on hazardous waste generation. *Interfaces*, 23(3), 79–90.

Cowen, D. J. 1988. GIS versus CAD versus DBMS: what are the differences? *Photogrammetric Engineering and Remote Sensing*, 54(11), 1551–1555.

Davies, R. E. and Lein, J. K. 1991. Applying an expert system methodology for solid waste landfill site selection. In: Proceedings of the Annual Conference of the Urban and Regional Information Systems Association. San Francisco, CA, pp. 40–53.

de Kock, E. 2004. Decentralising the codification of rules in a decision support expert knowledge base. M.Sc. Dissertation on Computer Science. University of Pretoria, Pretoria, South Africa.

Dehghanian, F. and Mansour, S. 2009. Designing sustainable recovery network of end-of-life products using genetic algorithm. *Resources, Conservation and Recycling*, 53(10), 559–570.

Derigs, U. and Friederichs, S. 2009. On the application of a transportation model for revenue optimization in waste management: a case study. *Central European Journal of Operations Research*, 17(1), 81–93.

Dialog, ˙2013. WasteInfo. Dialog. Available at: http://library.dialog.com/bluesheets/html/bl0110.html (accessed February 2013).

Dong, C., Jin, B., and Li, D. 2003. Predicting the heating value of MSW with a feed forward neural network. *Waste Management*, 23(2), 103–106.

Duggal, S. K. 2004. *Surveying*, 2nd edition, Tata McGraw-Hill Publishing Company, Delhi, India.

Elangovan, K. 2006. *GIS: Fundamentals, Applications and Implementations*, New India Publishing Agency, New Delhi, India.

Energy Research Centre of the Netherlands (ECN). 2012. Phyllis2, database for biomass and waste. ECN. Available at: http://www.ecn.nl/phyllis (accessed February 2013).

Engel, B. A. and Navulur, K. C. S. 1998. The role of geographical information systems in groundwater engineering. In: *The Handbook of Groundwater Engineering* (Ed. Delleur, J. W.), Vol. 1, CRC Press, pp. 16–21.

Eshkevari, M. M. and Panahande, M. 2010. Site selection of exiting and proposed containers for solid waste storing, case study: Semnan, Iran. In: Proceedings of International Conference on Environmental Science and Development, pp. 299–303.

ESRI. 2012. ArcGIS - Mapping and analysis for understanding our world. ESRI. Available at: http://www.esri.com/software/arcgis (accessed February 2013).

European Environment Agency (EEA). 2013a. Welcome to E-PRTR. The European Pollutant Release and Transfer Register. Available at: http://prtr.ec.europa.eu/ (accessed February 2013).

European Environment Agency (EEA). 2013b. Data and maps. European Environment Agency. Available at: http://www.eea.europa.eu/data-and-maps (accessed February 2013).

European Environment Information and Observation Network (EIONET). 2009. Wastebase. European Topic Centre on SCP. Available at: http://scp.eionet.europa.eu/facts/wastebase (accessed February 2013).

Eurostat/New Cronos. 2013. Eurostat – Your key to European statistics. Eurostat. Available at: http://epp.eurostat.ec.europa.eu/portal/page/portal/eurostat/home/ (accessed February 2013).

Fan, X., Zhu, M., Zhang, X., He, Q., and Rovetta, A. 2010. Solid waste collection optimization considering energy utilization for large city area. In: Proceedings of the International

Conference on Logistics Systems and Intelligent Management (ICLSIM 2010), Harbin, China. Vol. 1–3, pp. 1905–1909.

Fiorucci, P., Minciardi, R., Robba, M., and Sacile, R. 2003. Solid waste management in urban areas development and application of a decision support system. *Resources, Conservation and Recycling*, 37(4), 301–328.

Flinn, J. C. 1998. 3-D GIS helps dispose of radioactive waste. *GIS World*, 11(4), 46–49.

Foote, K. E. and Lynch, M. 1995. Geographic information systems as an integrating technology: context, concepts, and definitions. The geographer's craft project, Department of Geography, The University of Colorado at Boulder. Available at: http://www.colorado.edu/geography/gcraft/notes/intro/intro.html (accessed February 2013).

Franzese, P. P., Russo, G. F., and Ulgiati, S. 2008. Geographical information system (GIS) and energy synthesis evaluation of urban waste management. In: *Sustainable Energy Production and Consumption: Benefits, Strategies and Environmental Costing* (Eds. Barbir, F. and Ulgiati, S.), Springer, pp. 339–352.

Fuchs, A., Zangl, H., Holler, G., and Brasseur, G. 2008. Design and analysis of a capacitive moisture sensor for municipal solid waste. *Measurement Science and Technology*, 19(2), 025201.

Garcia-Frias, B., Harrop, J., Gifford, M., and Stephens, J. 1993. EPA/DOD GIS pilot for the identification, characterization, evaluation, and prioritization of sites contaminated with radioactive/mixed waste. In: Proceedings of Waste Management Technology and Programs for Radioactive Waste Management and Environmental Restoration, Vol. 1–2, pp. 697–700.

Gawande, N. A, Reinhart, D. R., Thomas, P. A, McCreanor, P. T., and Townsend, T. G. 2003. Municipal solid waste in situ moisture content measurement using an electrical resistance sensor. *Waste Management*, 23(7), 667–674.

Gemitzi, A., Tsihrintzis, V. A., Voudrias, E., Petalas, C., and Stravodimos, G. 2007. Combining geographic information system, multicriteria evaluation techniques and fuzzy logic in siting MSW landfills. *Environmental Geology*, 51(5), 797–811.

Geneletti, D. 2010. Combining stakeholder analysis and spatial multicriteria evaluation to select and rank inert landfill sites. *Waste Management*, 30(2), 328–337.

Ghose, M. K., Dikshit, A. K., and Sharma, S. K. 2006. A GIS based transportation model for solid waste disposal–a case study on Asansol municipality. *Waste Management*, 26(11), 1287–1293.

Gomes, H., Ribeiro, A. B., and Lobo, V. 2007. Location model for CCA-treated wood waste remediation units using GIS and clustering methods. *Environmental Modelling and Software*, 22(12), 1788–1795.

Gómez-Delgado, M. and Tarantola, S. 2006. Global sensitivity analysis, GIS and multi-criteria evaluation for a sustainable planning of a hazardous waste disposal site in Spain. *International Journal of Geographical Information Science*, 20(4), 449–466.

Gopalakrishnan, B., Plummer, R. W., Kulkarni, R., and Mangalampalli, P. 2003. Solvent and paint waste reduction with an expert system for industrial waste minimization. *Journal of Environmental Systems*, 29(1), 39–53.

Greenman, R. M. 1998. 2-D high-lift aerodynamic optimization using neural networks. NASA TM-1998-112233. Available at http://ntrs.nasa.gov/archive/nasa/casi.ntrs.nasa.gov/19980218795.pdf (accessed April, 2014).

Greenwell, R. A., Addlemanb, R. S., Crawfordb, B. A., Mechb, S. J., and Troyerb, G. L. 1992. A survey of fiber optic sensor technology for nuclear waste tank applications. *Fiber and Integrated Optics*, 11(2), 141–150.

Guzman, J. B., Paningbatan, E. P., and Alcantara, A. J. 2010. A geographic information systems-based decision support system for solid waste recovery and utilization in Tuguegarao City, Cagayan, Philippines. *Journal of Environmental Science and Management*, 13(1), 52–66.

Haastrup, P., Maniezzo, V., Mattarelli, M., Rinaldi, F. M., Mendes, I., and Paruccini, M. 1998. A decision support system for urban waste management. *European Journal of Operational Research*, 109(2), 330–341.

Halim, I. and Srinivasan, R. 2006. Supporting waste minimization studies by integrating expert system with process simulators. In: 16th European Symposium on Computer Aided Process Engineering and 9th International Symposium on Process Systems Engineering (Ed. Marquardt, W. P. C.), Garmisch-Partenkirchen, Germany, pp. 1003–1007.

Hengl, T., Maathuis, B. H. P., and Wang, L. 2009. Geomorphometry in ILWIS. In: *Geomorphometry: Concepts, Software, Applications – Developments in Soil Science* (Eds. Hengl, T. and Reuter, H. I.), Vol. 53, Elsevier, pp. 309–332.

Hernández-Caraballo, E. A., Avila-Gómez, R. M., Capote, T., Rivas, F., and Pérez, A. G. 2003. Classification of Venezuelan spirituous beverages by means of discriminant analysis and artificial neural networks based on their Zn, Cu and Fe concentrations. *Talanta*, 60(6), 1259–1267.

Higgs, G. 2006. Integrating multi-criteria techniques with geographical information systems in waste facility location to enhance public participation. *Waste Management & Research*, 24(2), 105–117.

Hiramatsu, A., Hara, Y., Sekiyama, M., Hinda, R., and Chiemchaisri, C. 2009. Municipal solid waste flow and waste generation characteristics in an urban-rural fringe area in Thailand. *Waste Management & Research*, 27(10), 951–960.

Hirokane, M., Furuta, H., and Habara, H. 1995. Application of genetic algorithms and expert system to design and planning of solid waste disposal. In: *Developments in Neural Networks and Evolutionary Computing for Civil and Structural Engineering* (Ed. Topping, B. H. V.), Civil-Comp Press, Edinburgh, UK, pp. 205–212.

Hodges, J. E., Bridges, S., Sparrow, C., Weerakkody, G., Tang, B., Jun, C., and Luo, J. 2001. Assessing the performance of a waste characterization expert system. *Applied Artificial Intelligence*, 15(4), 385–396.

Holland, J. H. 1975. *Adaptation in Natural and Artificial Systems*, The University of Michigan Press, Ann Harbor, MI.

Hřebíček, J. and Soukopová, J. 2010. Modeling integrated waste management system of the Czech Republic. In: Proceedings of the Latest Trends on Systems: 14th WSEAS International Conference on Systems (Part of the 14th WSEAS CSCC Multiconference), Corfu, Greece, pp. 510–515.

Huang, B. 2006. GIS-based route planning for hazardous material transportation. *Journal of Environmental Informatics*, 8(1), 49–57.

Huang, G. H., and Chang, N. B. 2003. The perspectives of environmental informatics and systems analysis. *Journal of Environmental Informatics*, 1(1), 1–7.

Huang, L. X. and Sheng, G. 2006. Web-services-based spatial decision support system to facilitate nuclear waste siting. In: Proceedings of 14th International Conference on Geoinformatics, Geospatial Information Technology (Proc. SPIE 6421, 642115).

Huang, J., Pretz, T., and Bian, Z. 2010. Intelligent solid waste processing using optical sensor based sorting technology. In: Proceedings of the 2010 3rd International Congress on Image and Signal Processing, IEEE, pp. 1657–1661.

Hurni, L., Bar, H. R., and Sieber, R. 1999. The atlas of Switzerland as an interactive multimedai atlas information system. In: *Multimedia Cartography* (Eds. Cartwright, W., Peterson, M. P., and Gartner, G.), Springer-Verlag, Berlin, Heidelberg, pp. 99–112.

Hushon, J. M. and Read, M. W. 1991. Defense priority model: experience of developing an environmental expert system for remedial site ranking. In: Proceedings of International Symposium on Artificial Intelligence, Cancun, Mexico, pp. 252–258.

Huth-Fehre, T., Feldhoff, R., Kantimm, T., Quick, L., Winter, F., Cammann, K., van der Broek, W., Wienke, D., Melssen, W., and Buydens, L. 1995. NIR – Remote sensing and artificial neural networks for rapid identification of post consumer plastics. *Journal of Molecular Structure*, 348, 143–146.

Intergraph. 2014. GIS software. Intergraph. Available at: http://www.intergraph.com/ (accessed March 2014).

International Institute for Geo-Information Science and Earth Observation (IIGISEO). 2011. ILWIS – Remote Sensing and GIS software. IIGISEO. Available at: http://www.itc.nl/Pub/research_programme/Research_output/ILWIS_-_Remote_Sensing_and_GIS_software.html (accessed February 2013).

International Solid Waste Association (ISWA). 2013. Welcome to the ISWA knowledge base. ISWA. Available at: http://www.iswa.org/en/525/knowledge_base.html (accessed February 2013).

Iyengar, S. S., Parameshwaran, N., Phona, V. V., Balakrishnan, N., and Okoye, C. D. 2010. *Fundamentals of Sensor Network Programming – Applications and Technology*, Wiley-IEEE Press.

Jahandideh, S., Jahandideh, S., Asadabadi, E. B., Askarian, M., Movahedi, M. M., Hosseini, S., and Jahandideh, M. 2009. The use of artificial neural networks and multiple linear regression to predict rate of medical waste generation. *Waste Management*, 29(11), 2874–2879.

Jayawardhana, L. C., Manipura, A., Alwis, A. D., Ranasinghe, M., Pilapitiya, S., and Abeygunawardena, I. 2003. BESTCOMP: expert system for Sri Lankan solid waste composting. *Expert Systems with Applications*, 24(3), 281–286.

Jensen, J. R. and Christensen, E. J. 1986. Solid and hazardous waste disposal site selection using digital geographic information system techniques. *Science of The Total Environment*, 56, 265–276.

Jensen, J. R., Hodgson, M. E., Garcia-Quijano, M., Im, J., and Tullis, J. A. 2009. A remote sensing and GIS-assisted spatial decision support system for hazardous waste site monitoring. *Photogrammetric Engineering and Remote Sensing*, 75(2), 169–177.

John, S. 2010. Sustainability-based decision-support system for solid waste management. *International Journal of Environment and Waste Management*, 6(1–2), 41–50.

Johnson, L. E. 2009. *Geographic Information Systems in Water Resources Engineering*, CRC Press, Boca Raton, FL.

Jovičić, N. M., Bošković, G., B., Vujić, G., Jovičić, G., Despotović, M. Z., Milovanović, D. M., and Gordić, D. R. 2010. Route optimization to increase energy efficiency and reduce fuel consumption of communal vehicles. *Thermal Science*, 14, S67-S78.

Judd, D. D., Haak, A., and Johnson, K. 1990. Using a GIS for sting a low-level radioactive waste disposal facility. *Transactions of the American Nuclear Society*, 62, 39–40.

Kanchanabhan, T. E., Mohaideen, J. A., Srinivasan, S., and Sundaram, V. L. 2011. Optimum municipal solid waste collection using geographical information system (GIS) and vehicle tracking for Pallavapuram municipality. *Waste Management & Research*, 29(3), 323–339.

Kao, J. J. 1996. A raster-based C program for siting a landfill with optimal compactness. *Computers and Geosciences*, 22(8), 837–847.

Kao, J. J. and Lin, T. I. 2002. Shortest service location model for planning waste pickup locations. *Journal of the Air and Waste Management Association*, 52(5), 585–592.

Karaca, F. and Özkayab, B. 2006. NN-LEAP: a neural network-based model for controlling leachate flow-rate in a municipal solid waste landfill site. *Environmental Modelling and Software*, 21(8), 1190–1197.

Karadimas, N. V. and Loumos, V. G. 2008. GIS-based modelling for the estimation of municipal solid waste generation and collection. *Waste Management & Research*, 26(4), 337–346.

Karadimas, N. V., Loumos, V. G., and Mavrantza, O. D. 2004. Quality of service ensuring in urban solid waste management. In: Proceedings of the 2nd International IEEE Conference Intelligent Systems, Vol. 1–2, pp. 288–292.

Karadimas, N. V., Mavrantza, O. D., and Loumos, V. G. 2005. GIS integrated waste production modeling. In: Proceedings of the Eurocon 2005: The International Conference on Computer as a Tool, Belgrade, Serbia. Vol. 1–2, pp. 1279–1282.

Karadimas, N. V., Loumos, V., and Orsoni, A. 2006. Municipal solid waste generation modelling based on fuzzy logic. In: Proceedings of the 20th European Conference on Modelling and Simulation ECMS 2006, Bonn, Germany, pp. 309–314.

Karadimas, N. V., Kolokathi, M., Defteraiou, G., and Loumos, V. 2007a. Ant colony system VS ArcGIS network analyst: the case of municipal solid waste collection. In: Proceedings of the 5th WSEAS International Conference on Environment, Ecosystems and Development, pp. 133–139.

Karadimas, N. V., Papatzelou, K., and Loumos, V. G. 2007b. Genetic algorithms for municipal solid waste collection and routing optimization. In: Proceedings of the 4th IFIP International Conference on Artificial Intelligence Applications and Innovations: From Theory to Applications, Peania, Greece, pp. 223–231.

Karadimas, N. V., Kolokathi, M., Defteraiou, G., and Loumos, V. 2007c. Municipal waste collection of large items optimized with ArcGIS network analyst. In: Proceedings of the 21st European Conference on Modeling and Simulation, Prague, Czech, pp. 80–85.

Karadimas, N. V., Papatzelou, K., and Loumos, V. G. 2007d. Optimal solid waste collection routes identified by the ant colony system algorithm. *Waste Management & Research*, 25(2), 139–147.

Karadimas, N. V., Doukas, N., Kolokathi, M., and Defteraiou, G. 2008. Routing optimization heuristics algorithms for urban solid waste transportation management. In: Proceedings of the WSEAS Transactions on Computers, 7(12), 2022–2031.

Karmakar, S., Laguë, C., Agnew, J., and Landry, H. 2007. Integrated decision support system (DSS) for manure management: a review and perspective. *Computers and Electronics in Agriculture*, 57(2), 190–201.

Karmakar, S., Nketia, M., Laguë, C., and Agnew, J. 2010. Development of expert system modeling based decision support system for swine manure management. *Computers and Electronics in Agriculture*, 71(1), 88–95.

Katpatal, Y. B. and Rao, B. 2011. Urban spatial decision support system for municipal solid waste management of Nagpur urban area using high-resolution satellite data and geographic information system. *Journal of Urban Planning and Development, ASCE*, 137(1), 65–76.

Ketterer, N., Schmuller, J., and Buehler, R. 1991. The superfund cost estimating expert system. In: Proceedings of the 1991 Symposium on Applied Computing, pp. 483–487.

Khalifa, Y., Salem, O., and Shahin, A. 2006. Cutting Stock Waste Reduction using genetic algorithms. In: Proceedings of the GECCO '06 Proceedings of the 8th Annual Conference on Genetic and Evolutionary Computation, Seattle, WA, pp. 1675–1680.

Khijwania, S. K., Kumar, A., Yueh, F. Y., and Singh, J. P. 2003. Raman sensor to monitor the nitrate and nitrite in the nuclear waste tank. In: Proceedings of the SPIE Conference on Chemical and Biological Standoff Detection, Providence, RI, pp. 349–354.

Kim, H. D., Kim, K. J., and Yoon, W. K. 1993. Development of a prototype expert system for intelligent operation aids in rod consolidation process nuclear materials. *Journal of the Korean Nuclear Society*, 25(1), 1–7.

Klein, M. and Methlie, L. 1995. Knowledge-based decision support systems, Annual reports in Organic Synthesis, Elsevier 1997, pp. 458.

Kordon, A. K., Dhurjati, P. S., and Bockrath, B. J. 1996. On-line expert system for odor complaints in a refinery. *Computers and Chemical Engineering*, 20(Supplement 2), S1449–S1454.

Lara-Valencia, F., Harlow, S. D., Lemos, M. C., and Denman, C. A. 2009. Equity dimensions of hazardous waste generation in rapidly industrialising cities along the United States-Mexico border. *Journal of Environmental Planning and Management*, 52(2), 195–216.

Lazar, R. E., Dumitrescu, M., and Stefanescu, I. 2001. Risk assessment of hazardous waste transport-perspectives of GIS application. In: Proceedings of International Conference Nuclear Energy in Central Europe (CD-ROM), Portoroz, Slovenia, pp. 737–744.

Leao, S., Bishop, I., and Evans, D. 2001. Assessing the demand of solid waste disposal in urban region by urban dynamics modelling in a GIS environment. *Resources Conservation and Recycling*, 33(4), 289–313.

Leao, S., Bishop, I., and Evans, D. 2004. Spatial-temporal model for demand and allocation of waste landfills in growing urban regions. *Computers, Environment and Urban Systems*, 28(4), 353–385.

Lee, S. 2003. Evaluation of waste disposal site using the DRASTIC system in Southern Korea. *Environmental Geology*, 44(6), 654–664.

Li, H., Chen, Z., Yong, L., and Kong, S. C. W. 2005. Application of integrated GPS and GIS technology for reducing construction waste and improving construction efficiency. *Automation in Construction*, 14(3), 323–331.

Li, J. Q., Borenstein, D., and Mirchandani, P. B. 2007. A decision support system for the single-depot vehicle rescheduling problem. *Computers and Operations Research*, 34(4), 1008–1032.

Lin, H. Y. and Kao, J. J. 2005. Grid-based heuristic method for multifactor landfill siting. *Journal of Computing in Civil Engineering*, 19(4), 369–376.

Lin, S. S., Horng, S. C., and Lin, C. I. H. 2009. A heuristic feedback control theory based interactive expert system – RSCD. *Expert Systems with Applications*, 36(9), 11907–11917.

Lindsjørn, Y. and Sjøberg, D. 1988. Database concepts discussed in an object-oriented perspective. In: *ECOOP'88* (Eds. Gjessing, S. and Nygaard, K.), Springer-Verlag, Heidelberg, Berlin, Germany, pp. 300–318.

Longhi, S., Marzioni, D., Alidori, E., di Buo, G., Prist, M., Grisostomi, M., and Pirro, M. 2012. Solid waste management architecture using wireless sensor network technology. In: Proceedings of the 5th International Conference on New Technologies, Mobility and Security (NTMS), pp. 1–5.

Longley, P. A., Goodchild, M. F., Maguire, D. J., Rhind, D. W. 2005. *Geographic Information Systems and Science*, 2nd edition, John Wiley & Sons, New York.

López, R., Cabeza, I. O., Giráldez, I., and Díaz, M. J. 2011. Biofiltration of composting gases using different municipal solid waste-pruning residue composts: monitoring by using an electronic nose. *Bioresource Technology*, 102(17), 7984–7993.

Lotfi, S., Habibi, K., and Koohsari, M. J. 2007. Integrating GIS and fuzzy logic for urban solid waste management (a case study of Sanandaj city, Iran). *Pakistan Journal of Biological Sciences*, 10(22), 4000–4007.

Lu, J. W., Chang, N. B., and Liao, L. 2013. Environmental informatics for solid and hazardous waste management: advances, challenges, and perspectives. *Critical Reviews in Environmental Science and Technology*, 43(15), 1557–1656.

Lukasheh, A. F., Droste, R. L., and Warith, M. A. 2001. Review of expert system (ES), geographic information system (GIS), decision support system (DSS), and their applications in landfill design and management. *Waste Management & Research*, 19(2), 177–185.

MacDonald, M. L. 1996a. A multi-attribute spatial decision support system for solid waste planning. *Computers Environment and Urban Systems*, 20(1), 1–17.

MacDonald, M. 1996b. Solid waste management models: a state of the art review. *Journal of Solid Waste Technology and Management*, 23(2), 73–83.

Madou, M. J. 2011. *From MEMS to Bio-MEMS and Bio-NEMS: Manufacturing Techniques and Applications*, CRC Press, Boca Raton, FL.

Malczewski, J. 1999. *GIS and Multicriteria Decision Analysis*, John Wiley & Sons, New York.

MapInfo. 2014. Introducing the mapInfo GIS suite. MapInfo. Available at: http://www.mapinfo.com/products/desktop (accessed March 2014).

Manamperi, A., Jayawardhana, L. C., de Alwis, A., and Pilapitiya, S. 2005. Development of an expert system for landfilling applications in Sri Lanka. In: *Artificial Intelligence Applications and Innovations II* (Eds. Li, D and Wang, B), Springer, pp. 643–653.

Matek, J. E. and Luger, G. F. 1997. An expert system controller for gas chromatography automation. *Instrumentation Science and Technology*, 25(2), 107–120.

McHarg, I. L. 1969. *Design with Nature*, Doubleday and Company, New York.

McLeod, C. A., Terazawa, M., and Yamamura, E. 1997. Using geographical information systems to evaluate decentralized management of municipal food waste. *Compost Science and Utilization*, 5(1), 49–61.

Michaelis, M. 1991. *GIS Expected to Make Landfill Siting Easier*, Geographical Information Systems, 30–35.

Micone, P. G. and Guy, C. 2007. Odour quantification by a sensor array: an application to landfill gas odours from two different municipal waste treatment works. *Sensors and Actuators B-Chemical*, 120(2), 628–637.

Min, H., Ko, H. J., and Ko, C. S. 2006. A genetic algorithm approach to developing the multi-echelon reverse logistics network for product returns. *Omega*, 34(1), 56–69.

Mitra, S., Duttagupta, S. P., Tuckley, K., and Ekram, S. 2012. Wireless sensor network based localization and threat estimation of hazardous landfill gas source. In: Proceedings of the 2012 IEEE International Conference on Industrial Technology, IEEE, pp. 349–355.

Mohai, P. and Saha, R. 2007. Racial inequality in the distribution of hazardous waste: a national-level reassessment. *Social Problems*, 54(3), 343–370.

Monprapussorn, S., Thaitakoo, D., Watts, D. J., and Banomyong, R. 2009. Multi criteria decision analysis and geographic information system framework for hazardous waste transport sustainability. *Journal of Applied Sciences*, 9(2), 268–277.

Muttiah, R. S., Engel, B. A., and Jones, D. D. 1996. Waste disposal site selection using GIS-based simulated annealing. *Computers and Geosciences*, 22(9), 1013–1017.

Nas, B., Cay, T., Iscan, F., and Berklay, A. 2010. Selection of MSW landfill site for Konya, Turkey using GIS and multi-criteria evaluation. *Environmental Monitoring and Assessment*, 160(1–4), 491–500.

Nasipuri, A., Subramanian, K. R., Ogunro, V., Daniels, J. L., and Hilger, H. 2006. Development of a wireless sensor network for monitoring a bioreactor landfill. In: *GeoCongress 2006: Geotechnical Engineering in the Information Technology Age* (Eds. DeGroot, D. J., DeLong, J. T., Frost, D., and Baise, L. G.), American Society of Civil Engineers, pp. 1–6.

Neehaarika, V. and Sindhura, S. 2011. Evaluation of routing protocols used in wireless sensor networks monitoring temperature in composting heaps. In: Proceedings of the 2011 Annual IEEE India Conference, IEEE, pp. 1–4.

Nishanth, T., Prakash, M. N., and Vijith, H. 2010. Suitable site determination for urban solid waste disposal using GIS and remote sensing techniques in Kottayam Municipality, India. *International Journal of Geomatics and Geosciences*, 1(2), 197–210.

Noori, R., Karbassi, A., and Sabahi, M. S. 2010. Evaluation of PCA and Gamma test techniques on ANN operation for weekly solid waste prediction. *Journal of Environmental Management*, 91(3), 767–771.

Ohri, A. and Singh, P. K. 2009. Landfill site selection using site sensitivity index: a case study of Varanasi city in India. In: Proceedings of 1 st International Conference on Solid Waste Management. Khudiram Anushilan Kendra (KAK) & NetajiIndoor Stadium (NIS), Kolkata, India, pp. 31–38.

Olsson, G. and Piani, G. 1992. *Computer Systems for Automotation and Control*, Prentice Hall International, London.

Organisation for Economic Co-operation and Development (OECD). 2013. Database on transboundary movement of wastes destined for recovery operations. OECD. Available at: http://www2.oecd.org/waste (accessed February 2013).

Organisation for Economic Co-operation and Development (OECD) and European Environment Agency (EEA). 2012. Welcome to the OECD/EEA database on instruments used for environmental policy and natural resources management. OECD. Available at: https://infoeuropa.eurocid.pt/registo/000041025/ (accessed February 2013).

Ossenbruggen, P. J. and Ossenbruggen, P. C. 1992. SWAP: a computer package for solid waste management. *Computers, Environment and Urban Systems*, 16(2), 83–100.

O'Sullivan, D. and Unwin, D. J. 2003. *Geographic Information Analysis*, John Wiley & Sons, Hoboken, NJ.

Ozkayam, B., Demir, A., and Bilgili, M. S. 2007. Neural network prediction model for the methane fraction in biogas from field-scale landfill bioreactors. *Environmental Modelling and Software*, 22(6), 815–822.

Padgett, D. A. 1991. An application of GIS in the risk assessment of aquifer contamination from hazardous materials transporter spills. In: Proceedings of the Annual Conference of the Urban and Regional Information Systems Association, San Francisco, CA, pp. 250–261.

Page, B. and Wohlgemuth, V. 2010. Advances in environmental informatics: integration of discrete event simulation methodology with ecological material flow analysis for modelling eco-efficient systems. *Procedia Environmental Sciences*, 2, 696–705.

Parra, S., Aguilar, F. J., and Calatrava, J. 2008. Decision modelling for environmental protection: the contingent valuation method applied to greenhouse waste management. *Biosystems Engineering*, 99(4), 469–477.

Phadke, D. N. 2006. *Georaphic Information Systems (GIS) in Library and Information Services*, Concept Publishing Company, New Delhi, India.

Pillmann, W., Geiger, W., and Voigt, K. 2006. Survey of environmental informatics in Europe. *Environmental Modelling and Software*, 21(11), 1519–1527.

Power, D. J. 2004. Specifying an expanded framework for classifying and describing decision support systems. *Communications of the Association for Information Systems*, 13(13), 158–166.

Pullammanappallil, P. C. Svoronos, S. A., Chynoweth, D. P., and Lyberatos, G. 1998. Expert system for control of anaerobic digesters. *Biotechnology and Bioengineering*, 58(1), 13–22.

Purcell, M. and Magette, W. 2009. Prediction of household and commercial BMW generation according to socio-economic and other factors for the Dublin region. *Waste Management*, 29(4), 1237–1250.

Qdais, H. A., Hani, K. B., and Shatnawi, N. 2010. Modeling and optimization of biogas production from a waste digester using artificial neural network and genetic algorithm. *Resources, Conservation and Recycling*, 54(6), 359–363.

Rajkumar, N., Subramani, T., and Elango, L. 2010. Groundwater contamination due to municipal solid waste disposal GIS based - A study in Erode City. *International Journal of Environmental Sciences*, 1(1), 39–55.

Repoussis, P. P., Tarantilis, C. D., and Ioannou, G. 2009. An Arc-guided evolutionary algorithm for the vehicle routing problem with time windows. *IEEE Transactions on Evolutionary Computing*, 13(3), 624–647.

Rinner, C. 2003. Web-based spatial decision support: status and research directions. *Journal of Geographic Information and Decision Analysis*, 7(1), 14–31.

Rio, M., Franco-Uría, A., Abad, E., and Roca, E. 2011. A risk-based decision tool for the management of organic waste in agriculture and farming activities (FARMERS). *Journal of Hazardous Materials*, 185(2–3), 792–800.

Rouhani, S. and Kangari, R. 1987. Landfill site selection, a microcomputer expert system. *Microcomputers in Civil Engineering*, 2(1), 47–54.

Rubenstein-Montano, B. 2000. A survey of knowledge-based information systems for urban planning: moving towards knowledge management. *Computers, Environment and Urban Systems*, 24(3), 155–172.

Rubenstein-Montano, B., Anandalingam, G., and Zandi, I. 2000. A genetic algorithm approach to policy design for consequence minimization. *European Journal of Operational Research*, 124(1), 43–54.

Sauri-Riancho, M. R., Cabañas-Vargas, D. D., Echeverría-Victoria, M., Gamboa-Marrufo, M., Centeno-Lara, R., and Méndez-Novelo, R. I. 2011. Locating hazardous waste treatment facilities and disposal sites in the State of Yucatan, Mexico. *Environmental Earth Sciences*, 63(2), 351–362.

Scott, D. M. 1995. A 2-color near-infrared sensor for sorting recycled plastic waste. *Measurement Science and Technology*, 6(2), 156–159.

Scott, M., Thompson, S. N., Anderson, W. A., and Williams, J. S. 1989. Status of Maine's low-level radioactive waste program. In: Proceedings of the Waste Management Symposium Waste Processing, Transportation, Storage and Disposal, Technical Programs and Public Education, Vol. 2, pp. 39–42.

Şener, Ş., Sener, E., and Karaguzel, R. 2011. Solid waste disposal site selection with GIS and AHP methodology: a case study in Senirkent-Uluborlu (Isparta) Basin, Turkey. *Environmental Monitoring and Assessment*, 173(1–4), 533–554.

Sfakianaki, E. and Kasis, G. 2010. Identification and evaluation of suitable locations for waste management sites. *International Journal of Decision Sciences, Risk and Management*, 2(1), 66–84.

Sharifi, M., Hadidi, M., Vessali, E., Mosstafakhani, P., Taheri, K., Shahoie, S., and Khodamoradpour, M. 2009. Integrating multi-criteria decision analysis for a GIS-based hazardous waste landfill sitting in Kurdistan Province, western Iran. *Waste Management*, 29(10), 2740–2758.

Sheng, G. and Almasi, I. 2002. A siting approach for nuclear waste disposal using GIS. In: Proceedings of 6th International Conference on Probabilistic Assessment and Management, Vol. 1–2, pp. 1147–1152.

Shepard, S. 2005. *RFID Radio Frequency Identification*, The McGraw-Hall Companies, Inc,.

Shmelev, S. E. and Powell, J. R. 2006. Ecological-economic modelling for strategic regional waste management systems. *Ecological Economics*, 59(1), 115–130.

Shu, H., Lu, H., Fan, H., Chang, M., and Chen, J. 2006. Prediction for energy content of Taiwan municipal solid waste using multilayer perceptron neural networks. *Journal of the Air and Waste Management Association*, 56(6), 852–858.

Sicotte, D. 2008. Dealing in toxins on the wrong side of the tracks: lessons from a hazardous waste controversy in Phoenix. *Social Science Quarterly*, 89(5), 1136–1152.

Siddiqui, S. T., Alam, M., S., and Bokhari, M. U., 2012. Software tools required to develop GIS applications: an overview. In: Proceedings of the Secton International Conference on Advanced Computing & Communication Technologies, IEEE Computer Society, Rohtak, Haryana, India, pp. 51–56.

Sierra, R. and Stephens, C. R. 2012. Exploratory analysis of the interrelations between co-located boolean spatial features using network graphs. *International Journal of Geographical Exploratory Information Science*, 26(3), 441–468.

Simonetto, E. D. O. and Borenstein, D. 2007. A decision support system for the operational planning of solid waste collection. *Waste Management*, 27(10), 1286–1297.

Simsek, C., Kincal, C., and Gunduz, O. 2006. A solid waste disposal site selection procedure based on groundwater vulnerability mapping. *Environmental Geology*, 49(4), 620–633.

Singhal, B. B. S. and Gupta, R. 2010. *Applied Hydrogeology of Fractured Rocks*, 2nd edition, Springer.

Smith, E. G., Lindwall, C. W., Green, M., and Pavlik, C. K. 1997. PARMS: a decision support system for planting and residue management. *Computers and Electronics in Agriculture*, 16(3), 219–229.

Staudinger, J., Oralkan, G. A., Levitt, R. E., and Roberts, P. V. 1997. The Haztimator knowledge-based (expert) system: providing design and time/cost estimates for hazardous waste remediation. *Environmental Progress*, 16(2), 82–87.

Stewart, J. C. 1988. The application of low-level waste siting criteria to geographic information systems. *Transactions of the American Nuclear Society*, 56, 52.

Sugumaran, V. 1999. A web based intelligent interface to geographic information systems using agent technology. In: *Information Resources Management Association International Conference* (Eds. Khosrowpour, M. and Travers, J.), Idea Group Publishing, Hershey, PA, pp. 569–575.

Sumathi, V. R., Natesan, U., and Sarkar, C. 2008. GIS-based approach for optimized siting of municipal solid waste landfill. *Waste Management*, 28(11), 2146–2160.

Sumiani, Y., Onn, C. C., Din, M. A. M., and Jaafar, W. Z. W. 2009. Environmental planning strategies for optimum solid waste landfill siting. *Sains Malaysiana*, 38(4), 457–462.

Sun, Y., Li, C., and Huang, M. 2009. A fault diagnosis expert system of the municipal solid garbage crusher. In: Proceedings of Fifth International Workshop on Energy and Environment of Residential Buildings and Third International Conference on Built Environment and Public Health, Guilin, China, Vol. 1–2, pp. 1716–1723.

Taghizadeh, F. 2011. Evaluation and site selection of petrochemical industrial waste land filling using SMCE method. *Journal of Food Agriculture and Environment*, 9(1), 684–688.

Tang, X., Shen, C., Chen, L., Xiao, X., Wu, J., Khan, M. I., Dou, C., and Chen, Y., 2010. Inorganic and organic pollution in agricultural soil from an emerging e-waste recycling town in Taizhou area, China. *Journal of Soils and Sediments*, 10(5), 895–906.

Tao, J. 2010. Reverse logistics information system of e-waste based on internet. In: Proceedings of International Conference on Challenges in Environmental Science and Computer Engineering, pp. 447–450.

Tavares, G., Zsigraiova, Z., Semião, V., and Carvalho, M. G. 2009. Optimisation of MSW collection routes for minimum fuel consumption using 3D GIS modelling. *Waste Management*, 29(3), 1176–1185.

Tralhão, L., Coutinho-Rodrigues, J., and Alcada-Almeida, L. 2010. A multiobjective modeling approach to locate multi-compartment containers for urban-sorted waste. *Waste Management*, 30(12), 2418–2429.

Turban, E., Leidner, D., McLean, E., and Wetherbe, J. 2005. *Information Technology for Management: Transforming Organizations in the Digital Economy*, John Wiley & Sons, Inc., Hoboken.

United States Environmental Protection Agency (USEPA). 2013. National biennial RCRA hazardous waste report: documents and data. United States Environmental Protection Agency. Available at: http://www.epa.gov/epawaste/inforesources/data/biennialreport/index.htm (accessed February 2013).

van Kessel, L. B. M., Leskens, M., and Brem, G. 2002. On-line calorific value sensor and validation of dynamic models applied to municipal solid waste combustion. *Process Safety and Environmental Protection*, 80(B5), 245–255.

Vasiljević, T. Z., Srdjević, Z., Bajčetić, R., and Miloradov, M. V. 2012. GIS and the analytic hierarchy process for regional landfill site selection in transitional countries: a case study from Serbia. *Environmental Management*, 49(2), 445–458.

Vatalis, K. and Manoliadis, O. 2002. A two-level multicriteria DSS for landfill site selection using GIS: case study in Western Macedonia, Greece. *Journal of Geographic Information and Decision Analysis*, 6(1), 49–56.

Veitch, S. M. 2000. Lessons from an environmental information system developed to select a radioactive waste disposal site. In: Proceedings of 3rd International Symposium on Environmental Software Systems, pp. 177–186.

Vijay, R., Gupta, A., Kalamdhad, A. S., and Devolta, S. 2005. Estimation and allocation of solid waste to bin through geographical information systems. *Waste Management & Research*, 23(5), 479–484.

Vijay, R., Gautam, A., Kalamdhad, A., Gupta, A., and Devotta, S. 2008. GIS-based locational analysis of collection bins in municipal solid waste management systems. *Journal of Environmental Engineering and Science*, 7(1), 39–43.

Walia, A. 2010. Approaches to solid waste management: a case study of Delhi. In: *Solid Waste Management: Present and Future* (Eds. Singh,J. and Ramanathan, A. L.), International Publishing House Pvt. Ltd., New Delhi, India, pp. 212–221.

Wang, G., Qin, L., Li, G., and Chen, L. 2009. Landfill site selection using spatial information technologies and AHP: a case study in Beijing, China. *Journal of Environmental Management*, 90(8), 2414–2421.

Wei, M. S. and Weber, F. 1996. An expert system for waste management. *Journal of Environmental Management*, 46(4), 345–358.

Wey, W. M. 2005. An integrated expert system/operations research approach for the optimization of waste incinerator siting problems. *Knowledge-Based Systems*, 18(6), 267–278.

Whelan, G., Buck, J. W., Castleton, K., and Nazarali, A. 1996. Assessing multiple waste sites using decision-support tools. In: Proceedings of International Conference on the Application of GIS in Hydrology and Water Resources, Vienna, Austria, pp. 373–381.

Williamson, D. A. 1990. An expert system for greater-than-class-C waste classification. *Transactions of the American Nuclear Society*, 61, 79.

World Health Organization (WHO). 2011. Health care waste management – resources. WHO. Available at: http://www.who.int/mediacentre/factsheets/fs281/en/ (accessed February 2013).

Wright, N. T. and Yoon, J. (Eds.), 2006. Application of GIS technologies in port facilities and operations management. American Society of Civil Engineers.

Xiangyun, G., Kulczycka, J., Koneczny, K., and Daoliang, L., 2007. Web-based DSS for economic evaluation of municipal solid waste management. In: Proceedings of the 3rd WSEAS/IASME International Conference on Energy, Environment Ecosystems and Sustainable Development (EEESD'07), pp. 30–35.

Xiao, G., Ni, M. J., Chi, Y., Jin, B.-S., Xiao, R., Zhong, Z.-P., and Haung, Y.-J. 2009. Gasification characteristics of MSW and an ANN prediction model. *Waste Management*, 29(1), 240–244.

Youssef, W. A., 2007. *Intelligent Gateways Placements Techniques for Improved Dependability of Wireless Sensor Networks*, ProQuest, Ann Arbor, MI.

Zaheer, I., Cui, G., and Khalid, M. S. 2002. General design scheme of urban solid waste management using Geographical Information System. *Journal of Natural Science Nanjing Normal University*, 4(2), 66–72.

Zamorano, M., Molero, E., Hurtado, A., Grindlay, A., and Ramos, A. 2008. Evaluation of a municipal landfill site in Southern Spain with GIS-aided methodology. *Journal of Hazardous Materials*, 160(2–3), 473–481.

Zamorano, M., Molero, E., Grindlay, A., Rodríguez, M. L., Hurtado, A., and Calvo, F. J. 2009. A planning scenario for the application of geographical information systems in municipal waste collection: a case of Churriana de la Vega (Granada, Spain). *Resources Conservation and Recycling*, 54(2), 123–133.

Zeng, Y. and Trauth, K. M. 2005. Internet-based fuzzy multicriteria decision support system for planning integrated solid waste management. *Journal of Environmental Informatics*, 6(1), 1–15.

Zheng, G. L. and Pan, W. B. 2010. A GIS based mode research for urban industrial solid waste management and exchanges. In: Conference on Environmental Pollution and Public Health, Vol. 1–2, pp. 393–397.

Zhou, C. C., Yin, G. F., and Hu, X. B. 2009. Multi-objective optimization of material selection for sustainable products: artificial neural networks and genetic algorithm approach. *Materials and Design*, 30(4), 1209–1215.

Zhu, Y., Lin, C. J., Zhong, Y., Zhou, Q., Lin, C. J., and Chen, C. 2010. Cost optimization of a real-time GIS-based management system for hazardous waste transportation. *Waste Management & Research*, 28(8), 723–730.

Zsigraiová, Z., Tavares, G., Semião, V., and Carvalho, M. G. 2009. Integrated waste-to-energy conversion and waste transportation within island communities. *Energy*, 34(5), 623–635.

PART V

UNCERTAINTY ANALYSES AND FUTURE PERSPECTIVES

Risk analysis that fails to account for measurement uncertainties may produce misleading and sometimes dangerous results. Quantitative uncertainty analyses might be useful in systematically evaluating the possible or plausible changes in decision analysis outcomes due to changes in measurement accuracy, sources of data, communication, and social behavior.

- Evaluating the significance of uncertainty with random phenomenon and game theory for SWM in decision-making (Chapter 19)
- Considering linguistic uncertainty related to institutional settings and social behavior by fuzzy multiattribute analysis for SWM in decision-making (Chapter 20)
- Considering linguistic uncertainty related to institutional settings and technological implications by fuzzy multiattribute analysis for SWM in decision-making (Chapter 21)
- Assessing linguistic uncertainty by fuzzy multiobjective programming for SWM in decision-making (Chapter 22)
- Formalizing grey uncertainty by interval programming for SWM in decision-making (Chapter 23)
- Current challenges and future perspectives (Chapter 24)

Sustainable Solid Waste Management: A Systems Engineering Approach, First Edition. Ni-Bin Chang and Ana Pires.
© 2015 The Institute of Electrical and Electronics Engineers, Inc. Published 2015 by John Wiley & Sons, Inc.

CHAPTER 19

STOCHASTIC PROGRAMMING AND GAME THEORY FOR SOLID WASTE MANAGEMENT DECISION-MAKING

Uncertainty plays a critical role in decision-making. Challenges in characterizing relevant sources of uncertainty in system engineering analyses and validating predictions are problems that permeate systems science in general and decision-making in particular. Not only the uncertainty in socioeconomic dynamics, but also uncertainties from model parameters, type of models, inherent process uncertainties, uncertainties due to lack of knowledge about specific processes, risk, and uncertainties embedded in decision-making have to be considered. Holistic uncertainty analysis leads to the creation of a new spectrum of uncertainty quantification (UQ) that has been recognized as a critical element necessary for continued advancement in decision analysis for environmental sustainability. Overall, the core domains that illuminate decision analysis are theories of stochastic processes, game theory (GT), fuzzy set theory, and gray system theory. This chapter describes the rationale and application of stochastic programming and GT to start a series of discussions concerning UQ as applied to waste management.

19.1 BACKGROUND OF STOCHASTIC PROGRAMMING

Stochastic programming can be viewed as an extension of mathematical programming models for decision analysis whose coefficients (input data) could be expressed by random variables under an uncertain environment. Some pioneering studies include Dantzig (1955), Charnes et al. (1958), and Charnes and Cooper (1959). Stochastic linear programming, two-stage programming, chance-constraint programming

Sustainable Solid Waste Management: A Systems Engineering Approach, First Edition. Ni-Bin Chang and Ana Pires.
© 2015 The Institute of Electrical and Electronics Engineers, Inc. Published 2015 by John Wiley & Sons, Inc.

(CCP), and stochastic dynamic programming are typical methods for handling single objective stochastic programming models. Thus, stochastic goal programming and stochastic programming with utility function are suitable for a variety of applications in terms of multiple objectives under uncertainty.

Within a spectrum of formulations, CCP techniques are often applied to find cost-effective environmental quality management strategies (Guldmann, 1988). In the 1980s, acid rain control strategies in North America were a focus. CCP techniques proved useful for assessing the management policy relating to binational (United States/Canada) acid rain control issues (Fortin and McBean, 1983; Ellis et al., 1985, 1986; Ellis, 1988, 1990). To provide an overview of the applicability of CCP techniques for environmental management, Ellis (1987) summarized various types of stochastic programming models, including stochastic linear programming, two-stage programming under uncertainty, and stochastic goal programming, that can be employed in dealing with regional acid rain control issues.

19.2 MODEL FORMULATIONS OF STOCHASTIC PROGRAMMING

19.2.1 Stochastic Linear Programming

Focusing on stochastic linear programming, the objective function is a function in terms of decision variables X_j. If some or all of the coefficients C_j are random variables, then

$$Z = \sum_{j=1}^{n} C_j \cdot X_j \tag{19.1}$$

is also a random variable for any given solution. Since it is meaningless to maximize a random variable, Z must be replaced by some deterministic forms. The most common choice of this form is the expected value of Z:

$$E(Z) = \sum_{j=1}^{n} E(C_j) \cdot X_j. \tag{19.2}$$

Similarly, the functional constraints

$$\sum_{j=1}^{n} a_{ij} \cdot X_j \leq b_i \quad \forall i \tag{19.3}$$

must be reinterpreted if any of the a_{ij} and b_i are random variables. If they are mutually independent, then each of these a_{ij} and b_i with multiple possible values will be replaced by the most restrictive value for its constraint; and the functional constraint i could become

$$\sum_{j=1}^{n} (\max a_{ij})X_j \leq \min b_i, \tag{19.4}$$

where $\max a_{ij}$ is the largest value that the random variable a_{ij} can take on and $\min b_i$ is the smallest value that the random variable b_i can take on. By replacing the random variables with deterministic values in these constraints, the new model fitted with a suite of crisp equivalent forms can ensure that the original constraint will be satisfied for every possible combination of values for the random variable parameters.

Stochastic linear programming as formulated above requires all constraints to hold with the probability constraint, whereas CCP, which will be described in Section 19.2.2, permits a small violation of probability. The general approach deals with both types of models by reformulating them as new equivalent deterministic linear programming problems where the certainty assumption is satisfied. Then, the model can be solved by the traditional simplex method.

19.2.2 Chance-Constrained Programming Model

Fundamental Theory of CCP   The CCP problem was first introduced by Charnes and Cooper (1959). The problem is formulated as

$$\text{Max/min } Z = \boldsymbol{CX} \tag{19.5}$$

$$\text{s.t.} \quad P(A_i\boldsymbol{X} \leq b_i) \geq \beta_i, \quad i = 1, 2, \ldots, m \tag{19.6}$$

$$\beta_i \in [0, 1], \quad \forall i, \quad \text{and } \boldsymbol{X} \geq 0, \tag{19.7}$$

in which P stands for probability expression and constraint $A_i\boldsymbol{X}{\leq}b_i$ becomes a random variable. β_i is the reliability assigned in this optimization analysis. $\boldsymbol{X}$ or $\tilde{\boldsymbol{X}}$ are decision variables in vector form. Thus, an ordinary linear programming model can be defined as an equivalent CCP model if its linear constraints are expressed based on a set of probability measures indicating the extent of violation of the constraints. The CCP model may be viewed as a technique for providing appropriate "reliability," for it allows partial violation of the constraints. A complementary probability $1 - \beta_i$ represents the "admissible risk" that random variables have such values when considering the compliance with $A_i\boldsymbol{X}{\leq}b_i, \forall i$. Chance constraints can also appear as

$$P(A_i\boldsymbol{X} \leq b_i) \geq \beta, \quad i = 1, 2, \ldots, m. \tag{19.8}$$

If threshold β imposed can apply to all constraints, the unique risk level provides a unified risk assessment in decision-making no matter how many uncertain factors are included.

The key to the solution procedure in CCP is the transformation of chance constraints into crisp equivalent constraints. Then the traditional simplex method is applicable when seeking an optimal solution. Thus, the objective is to select the best optimal solution that "probably" will in turn satisfy each of the original constraints with risk or reliability implications when the random variables (a_{ij}, b_i, and c_j) take on values. Several important cases in the area of the general chance-constrained linear programming problem, solved by using the crisp equivalent constraints approach, include the two common cases described as follows (Hiller and Lieberman, 1991):

CASE 1

- All the a_{ij} parameters are constants and only some or all of the c_j and b_i are random variables.
- The probability distribution of the b_i is a known multivariate normal distribution.
- c_j is statistically independent of b_i for all i and j.
- There is no statistical linkage between different b_i that would imply that the chance constraints themselves are interrelated with each other.

In the way to deal with crisp equivalent constraints, let us consider

$$P\left\{ \sum_{j=1}^{n} a_{ij} \cdot X_j \leq b_i \right\} \geq \alpha_i \quad \forall i. \tag{19.9}$$

The goal is to convert these constraints into legitimate linear programming constraints so as to apply the simplex method. This idea can be implemented by the following:

$$P\left\{ \sum_{j=1}^{n} a_{ij} \cdot X_j \leq b_i \right\} = P\left\{ \frac{\sum_{j=1}^{n} a_{ij} \cdot X_j - E(b_i)}{\sigma_{b_i}} \leq \frac{b_i - E(b_i)}{\sigma_{b_i}} \right\} \geq \alpha_i \quad \forall i, \tag{19.10}$$

where $E(b_i)$ and σ_{b_i} are the mean and standard deviation of b_i, respectively. Since b_i is assumed to have a normal distribution, $[b_i - E(b_i)]/\sigma_{b_i}$ must also be normal with a mean of zero and standard deviation of one. In any statistical textbook, we can find out a cut-off value k_α such that

$$Y = \frac{b_i - E(b_i)}{\sigma_{b_i}} \tag{19.11}$$

$$P\{Y \geq K_\alpha\} = \alpha \tag{19.12}$$

where α is any given number between zero and one, and where Y is the random variable whose probability distribution is normal with mean of zero and standard

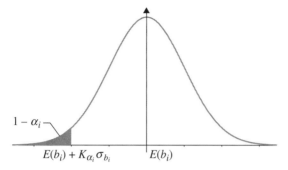

FIGURE 19.1   Probability density function of b_i

deviation of one. Almost every statistical book gives k_α for various values of α. For example,

$$K_{0.9} = -1.28 \quad K_{0.95} = -1.645 \quad K_{0.99} = -2.33.$$

Therefore, it now follows that

$$P\left\{ K_{\alpha_i} \leq \frac{b_i - E(b_i)}{\sigma_{b_i}} \right\} = \alpha_i \quad \forall i. \tag{19.13}$$

Note that this probability would be increased if k_{α_i} is replaced by a number that is less than or equal to k_{α_i}. Based on Figure 19.1, the constraint could be transformed by the following inequality:

$$P\left\{ \frac{\sum_{j=1}^{n} a_{ij} \cdot X_j - E(b_i)}{\sigma_{b_i}} \leq \frac{b_i - E(b_i)}{\sigma_{b_i}} \right\} \geq \alpha_i \quad \forall i. \tag{19.14}$$

The cross-hatched area is equal to $1 - \alpha_i$ and the cut-off value is $E(b_i) + K_{\alpha_i}\sigma_{b_i}$. For a given solution if and only if

$$\frac{\sum_{j=1}^{n} a_{ij} \cdot X_j - E(b_i)}{\sigma_{b_i}} \leq K_{\alpha_i} \quad \forall i, \tag{19.15}$$

then the crisp equivalent constraint is

$$\sum_{j=1}^{n} a_{ij} \cdot X_j \leq E(b_i) + K_{\alpha_i} \cdot \sigma_{b_i} \quad \forall i. \tag{19.16}$$

Therefore, the CCP problem can be reduced to the following equivalent linear programming problem:

$$\max \quad E(Z) = \sum_{j=1}^{n} E(c_j) \cdot X_j \tag{19.17}$$

$$\text{s.t.} \quad \sum_{j=1}^{n} a_{ij} \cdot X_j \leq E(b_i) + K_{\alpha_i} \cdot \sigma_{b_i} \quad \forall i \tag{19.18}$$

$$X_j \geq 0 \quad \forall j. \tag{19.19}$$

CASE 2

- All the b_i parameters are constants and only some or all of the c_j and a_{ij} are random variables.
- The probability distribution of the a_{ij} is a known multivariate normal distribution.
- c_j is statistically independent of a_{ij} for all i and j.
- There is no statistical linkage between different a_{ij} that would imply chance constraints themselves are interrelated with each other.

Let us consider the chance constraint

$$\left\{ \sum_{j=1}^{n} a_{ij} \cdot X_j \leq b_i \right\} \geq \alpha_i \quad \forall i. \tag{19.20}$$

Again, the goal is to convert these constraints into legitimate linear programming constraints so as to apply the simplex method. This idea can be implemented by the following:

$$P\left\{ \sum_{j=1}^{n} a_{ij} \cdot X_j \leq b_i \right\} = P\left\{ \frac{\sum_{j=1}^{n} (a_{ij} \cdot X_j - E(a_{ij}) \cdot X_j)}{\sigma_{a_{ij}}} \leq \frac{b_i - \sum_{j=1}^{n} E(a_{ij}) \cdot X_j}{\sigma_{a_{ij}}} \right\} \geq \alpha_i \quad \forall i. \tag{19.21}$$

Similarly, Y can be defined as

$$Y = \frac{\sum_{j=1}^{n} (a_{ij} X_j - E(a_{ij}) X_j)}{\sigma_{a_{ij}}} \quad \forall i \tag{19.22}$$

$$P\{Y \leq K_{\alpha_i}\} \geq \alpha_i \quad \forall i. \tag{19.23}$$

Different density functions with different degrees of freedom might generate different corresponding critical values. If t-distribution is chosen, which is the most

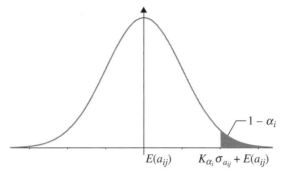

FIGURE 19.2   Probability density function of a_{ij}

practical one in application, given that the degree of freedom is sufficiently large, the values of k_α for various values of α may include

$$K_{0.9} = 1.28 \quad K_{0.95} = 1.645 \quad K_{0.99} = 2.33.$$

Based on Figure 19.2, the cross-hatched area is equal to $1 - \alpha_i$ and the cut-off value is $K_{\alpha_i}\sigma_{a_{ij}} + E(a_{ij})$. For a given solution if and only if

$$\frac{b_i - \sum\limits_{j=1}^{n} E(a_{ij}) \cdot X_j}{\sigma_{a_{ij}}} \geq K_\alpha \quad \forall i, \tag{19.24}$$

then the crisp equivalent constraint is

$$\sum_{j=1}^{n} E(a_{ij}) \cdot X_j + K_{\alpha_i} \cdot \sigma_{a_{ij}} \leq b_i \quad \forall i. \tag{19.25}$$

Therefore, the CCP problem can be reduced to the following equivalent linear programming problem:

$$\max \quad E(Z) = \sum_{j=1}^{n} E(c_j) \cdot X_j \tag{19.26}$$

$$\text{s.t.} \quad \sum_{j=1}^{n} E(a_{ij}) \cdot X_j + K_{\alpha_i} \cdot \sigma_{a_{ij}} \leq b_i \quad \forall i \tag{19.27}$$

$$X_j \geq 0 \quad \forall j. \tag{19.28}$$

Or, the model above can be expressed by the matrix form in some literature as below:

$$\text{Max} \quad Z = E\left(\tilde{c}^T \tilde{X}\right) \tag{19.29}$$

$$\text{s.t.} \quad \sum_{j=1}^{n} E(a_{ij}) \cdot X_j + F^{-1}(\alpha_i) \left(\tilde{X}^T \tilde{V} \tilde{X}\right)^{1/2} \le b_i \quad \forall i \tag{19.30}$$

$$X_j \ge 0 \quad \forall j, \tag{19.31}$$

where $\tilde{V}$ is the covariance matrix of a_{ij} and $F^{-1}(\cdot)$ is the inverse value of accumulative probability density function with mean of zero and standard deviation of one. $\tilde{c}^T$ is the transposed vector of coefficients associated with the objective function.

Applications of CCP

Example 19.1   Consider the following stochastic programming and assume that there is no correlation between these two constraints. Given that $E(b_1) = 3$ and $E(b_2) = 4$ and standard deviations are 0.02 and 0.2, respectively, to solve this problem using CCP:

$$\text{Max} \quad 5x_1 + 6x_2 \tag{19.32}$$
$$\text{s.t.} \quad P\{4x_1 + 3x_2 \ge 3\} \ge 0.95 \tag{19.33}$$
$$P\{5x_1 - 2x_2 \le 4\} \ge 0.9 \tag{19.34}$$
$$3x_1 - 5x_2 \ge 13 \tag{19.35}$$
$$x_i \ge 0, \quad \forall i. \tag{19.36}$$

Solution: Assume Z is defined as

$$Z = \sum_{j=1}^{n} (c_j \cdot x_j). \tag{19.37}$$

Chance constraint is defined by

$$P\left\{\sum_{j=1}^{n} (a_{ij} \cdot x_j) \le b_i\right\} \ge \alpha_i, \tag{19.38}$$

in which α_i is a number between 0 and 1.

Given that no correlation between these two chance constraints, c_j and b_i, are random variables and they are statistically independent, by using the crisp equivalent form:

$$\sum_{j=1}^{n} a_{ij} \cdot x_j \leq E(b_i) + K_{\alpha i} \cdot \sigma_{bi} \tag{19.39}$$

$$K_{0.95} = -1.645, K_{0.9} = 1.28. \tag{19.40}$$

We have

$$4x_1 + 3x_2 \geq 3 + (-1.645) \cdot (0.02) \tag{19.41}$$
$$5x_1 - 2x_2 \leq 4 + (1.28) \cdot (0.2). \tag{19.42}$$

By rearranging the two crisp constraints, we have

$$4x_1 + 3x_2 \geq 2.9671 \tag{19.43}$$
$$5x_1 - 2x_2 \leq 4.256. \tag{19.44}$$

Then, the crisp equivalent form of the linear programming model is:

$$\text{Max} \quad Z = 5x_1 + 6x_2 \tag{19.45}$$
$$\text{s.t.} \quad 4x_1 + 3x_2 \geq 2.9671 \tag{19.46}$$
$$5x_1 - 2x_2 \leq 4.256 \tag{19.47}$$
$$3x_1 - 5x_2 \geq 1 \tag{19.48}$$
$$x_1 \geq 0 \tag{19.49}$$
$$x_2 \geq 0. \tag{19.50}$$

The modeling outputs based on the software package LINDO or LINGO (LINDO Inc., 2012) are $x_1 = 1.0417$, $x_2 = 0.4088$, $Z = 7.5267$.

Example 19.2   CCP may be applied for solid waste management (SWM). In many urban regions, SWM has gained considerable attention due to limited landfill space, the emphasis of material conservation, and the concern for environmental quality. The SWM process normally consists of waste generation/storage, collection, transportation, treatment, recycling, and landfill. In this case, households in two different cities (i.e., city A and city B) generate substantial waste streams, which require building at least one incineration facility and one landfill to meet disposal needs (Figure 19.3). The waste generation rate is 900 and 1350 tonnes · per day in cities A and B, respectively (tonnes are metric tons). Due to the space limitation and the constraint of environmental assimilative capacity, the maximum treatment

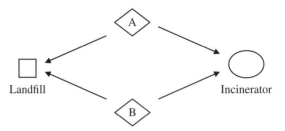

FIGURE 19.3 The proposed SWM system

capacities of incineration and landfill are 1000 tonnes per day and 1250 tonnes per day, respectively. Shipping the waste streams around the network could add more costs to the budget. The transportation costs for shipping waste streams from A to incinerator and landfill are US (US$) US$35 per tonne and US$75 per tonne, respectively. Meanwhile, the transportation costs for shipping waste streams from B to incinerator and landfill are US$75 per tonne and US$25 per tonne, respectively.

The constructions cost functions for these two facilities are

for incinerator: CC_1 (US$million) $= 9.45 + 0.215 DCP_1$
for landfill: CC_2 (US$million) $= 4.21 + 0.05 DCP_2$,

in which CC stands for construction cost and DCP is the design capacity. The installation costs of facilities have to be factored into the cost–benefit framework by using capital recovery factor (CRF) to convert the investment into annual cash flow. The definition of CRF in engineering economics is $\{[r(1 + r)^n]/[(1 + r)^n - 1]\}$. The planning time period is 30 years and the social discount rate is 9%. Each tonne of waste burned at incineration facility will generate energy recovery income at the rate of US$60. In contrast, each tonne of waste buried at landfill could generate methane gas recovery income at the rate of US$35. However, the operating costs of incineration and landfilling are US$45 and US$60 per tonne of waste. The following two problems address typical system engineering practices for SWM.

(a) Formulate a mix integer-programming model and solve it by LINDO to search for the best development plan. In this model formulation, mass balance, capacity limitation, and those necessary for performing fixed charge functionality must be included in the constraint sets.

(b) If the decision variables DCP_1 and DCP_2 are random variables and they are not interrelated, the above model will be modified by the CCP framework in which two capacity limitation constraints are handled as chance constraints with a reliability level of 95%. The DCP_1 has a mean value of 1000 tonnes and a standard variation of 80 tonnes; on the other hand, DCP_2 has a mean value of 1250 tonnes and a standard variation of 150 tonnes. Using the crisp equivalent form will transform this CCP into LP and therefore solvable by LINDO.

Solution (a):
Decision Variables

WAL = Waste from town A to landfill (tonnes per day)

WAI = Waste from town A to incinerator (tonnes per day)

WBL = Waste from town B to landfill (tonnes per day)

WBI = Waste from town B to incinerator (tonnes per day)

WI = Waste incinerated (tonnes per day)

WL = Waste landfilled (tonnes per day)

I_1, I_2 = Binary integer variables for the screening of treatment plant and disposal site.

Parameter Values

The maximum treatment capacities of incineration = 1000 tonnes per day

The maximum treatment capacities of landfill = 1250 tonnes per day

The transportation costs for shipping waste streams from A to incinerator = US$35 per tonne

The transportation costs for shipping waste streams from A to landfill = US$75 per tonne

The transportation costs for shipping waste streams from B to incinerator = US$75 per tonne

The transportation costs for shipping waste streams from B to landfill = US$25 per tonne.

In general, the investment of SWM can be recovered in a period of 30 years and assume the social discount rate is 9%. Therefore,

$$CRF = \left\{ [r(1 + r)^n] / [(1 + r)^n - 1] \right\} = 0.0973.$$

Objective Function

Minimize Total Cost = Incineration cost + Transportation cost + Disposal cost − Recovery cost

Yearly investment for incinerator = $(9.45 + 0.215\ WI) \cdot (1000000) \cdot (0.0973)$
$$= 919485 + 20919.5\ WI$$

Yearly investment for landfill = $(4.21 + 0.05\ WL) \cdot (1000000) \cdot (0.0973)$
$$= 409633 + 4865\ WL$$

Transportation cost (per day): $75WAL + 35WAI + 75WBI + 25WBL$

Operating cost (per day): $45WI + 60\ WL$

Recovery income (per day): $60WI + 35\ WL$

Total cost (per year) = $[919485 + 20919.5\ WI + 409633 + 4865\ WL] +$
$$365[75\ WAL = 35\ WAI = 75\ WBI + 25\ WBL = 45\ WI$$
$$+60\ WL - (60\ WI = 35\ WL)].$$

The whole model can be reorganized as below:

$$Z = 919485\ I_1 + 409633\ I_2 + 15444.5\ WI + 13990\ WL$$
$$+12775\ WAI + 27375\ WBI + 9125\ WBL + 27375\ WAL.$$

Subject to:

1. Mass balance constraint: The first two are defined for waste generation and the second two are defined for treatment and disposal.

   $WAL + WAI = 900$

   $WBL + WBI = 1350$

   $WAL + WBL \leq WL$

   $WAI + WBI \leq WI.$

2. Capacity limitation constraint:

   $WAL + WBL \leq 1250$

   $WAI + WBI \leq 1000.$

3. Site screening constraint:

   $WI - 1000I_1 \leq 0$

   $WL - 1250I_2 \leq 0.$

Modeling outputs with interpretation based on LINDO or LINGO are

Total cost = US\$59,902,370

Waste transported from Town A → Landfill = 0 tonnes per year

Waste transported from Town B → Landfill = 1250 tonnes per year

Waste transported from Town A → Incinerator = 900 tonnes per year

Waste transported from Town B → Incinerator = 100 tonnes per year.

Solution (b): Mean design capacity (incinerator) = 1000 tonnes

Mean design capacity (landfill) = 1250 tonnes

Standard variation (incinerator) = 80 tonnes

Standard variation (landfill) = 150 tonnes

Reliability level = 95%

$P(WI < 900) > 0.95$

$P(WL < 1100) > 0.95$

$K_{0.95} = -1.645.$

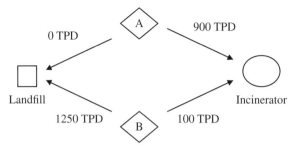

FIGURE 19.4   The flow control strategy in proposed SWM system (Case I)

The equivalent linear form is

$WI \leq 1000 + (-1.645)\cdot(80) \leq 868.4$
$WL \leq 1250 + (-1.645)\cdot(150) \leq 1003.25.$

Modeling outputs with interpretation based on LINDO or LINGO are as follows.

Case I: No feasible solution was obtained as the design capacity of the landfill and incinerator combined after the consideration of randomness is less than the amount of waste generated from towns A and B.

Case II: Therefore, for a systems engineer, the design capacity of landfill and incinerator had better be increased from 1000 and 1250 to 1200 and 1550 respectively and the problem is reformulated as follows.

The equivalent deterministic linear constraints are

$$WI = 1200 + (-1.645) \cdot 80 \leq 1068.4.$$
$$WL = 1550 + (-1.645) \cdot 150 \leq 1303.25.$$

Modeling outputs with interpretation based on LINDO or LINGO are

Total cost = US\$58,853,100
Waste transported from Town A → Landfill = 0 tonnes per year
Waste transported from Town B → Landfill = 1303.250 tonnes per year
Waste transported from Town A → Incinerator = 900 tonnes per year
Waste transported from Town B → Incinerator = 46.75 tonnes per year.

In summary (Figure 19.4), the Case I representing the situation before the inclusion of reliability analysis is described below:

Waste transported from Town A → Landfill = 0 tonnes per year
Waste transported from Town B → Landfill = 1250 tonnes per year[1]
Waste transported from Town A → Incinerator = 900 tonnes per year
Waste transported from Town B → Incinerator = 100 tonnes per year.

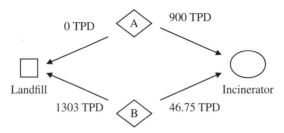

FIGURE 19.5 The flow control strategy in proposed SWM system (Case II)

In summary (Figure 19.5), the Case II representing the situation after the inclusion of reliability analysis is described below:

In Case II, the total cost decreased from US$59,902,370 to US$58,853,100.

Joint Chance Constraints Once in a while, correlations among some parameters themselves, such as a_{ij} or $\underline{b}_i$, crossing different constraints could result in a need to employ joint chance constraints for addressing possible interactions during the optimization step. The solution procedure when dealing with joint chance constraints in the CCP framework has been investigated and formalized as follows:

$$\text{Max } Z = CX \tag{19.51}$$

$$\text{s.t.} \quad AX \leq b \tag{19.52}$$

$$\sum_j \alpha_{ij} X_i \leq b_i \quad \forall i \in S, b_i \text{ unrestricted in sign} \tag{19.53}$$

$$\prod_{i \in S} g_i(\beta_i) \geq \beta \tag{19.54}$$

$$X \geq 0, \ \beta \geq 0. \tag{19.55}$$

In this case it is assumed that the elements b_i ($i = 1, 2, \ldots , m$) only are random variables. A is a coefficient matrix defined for these constraints without having stochastic implication; α_{ij} is an n-element row vector in the stochastic constraints; Z is assumed as a concave function with continuous derivatives satisfying certain regularity conditions; and the set of S is chance constraint set. The decision variable is the n-element column vector $\tilde{x}$. β is a specified probability value. Note that a nonlinear programming model can be defined as a convex programming problem if $f(x)$ is convex and the function h_i are all convex in the following model:

$$\text{min } Z = f(x)$$

$$\text{s.t.:} \quad h_i(x) \leq b_i, \quad i = 1, 2, \ldots , m. \tag{19.56}$$

By examining the following properties, a local optimal can be confirmed to be the same as a global optimum:

- A local minimum of a convex function on a convex feasible region is also a global minimum.

- A local maximum of a concave function on a convex feasible region is also a global maximum.
- A local minimum of a strictly convex function on a convex feasible region is the unique global minimum.
- A local maximum of a strictly concave function on a convex feasible region is also a global maximum.

In the stochastic case (19.56), $(b_1, b_2, \ldots, b_m)$ are random variables that are assumed to have a continuous joint distribution with continuous first-order derivatives with respect to all variables in any point of the m-dimensional space of the form

$$[g_1(x), g_2(x), \ldots, g_m(x)] \tag{19.57a}$$

and accordingly the stochastic constraints can be reformulated by

$$G(x) = P\left[g_1(x) \geq \beta_1, g_2(x) \geq \beta_2, \ldots, g_m(x) \geq \beta_m\right] \geq \beta \tag{19.57b}$$

$$\prod_{i \in S} g_i(\beta_i) \geq \beta, \tag{19.57c}$$

where the scalar β is such that $0 < \beta < 1$, and only those values of β are considered for which the set of feasible solutions is not empty.

$$g_i(b_i) = P\left(\sum_j a_{ij} X_i \leq b_i\right) \geq \beta_i, \quad 0 \leq \beta_i \leq 1. \tag{19.57d}$$

The joint form is expressed by (19.57d), where $g_i(b_i) = 1 - F_i(Y)$, $F_i(Y)$ being the cumulative distribution of the random element b_i is assumed to be continuous. For each $i \in S$ the marginal distribution of the random variable b_i is denoted by $F_i(Y) = P(b_i < Y)$.

It is known that the left-hand side of (19.57d) is a concave function; hence, a global optimum with respect to a nonlinear programming problem can be found with any existing type of nonlinear programming algorithm. Sometimes, however, it may not be the case unless additional conditions are imposed, such as when a probability density function is a normal distribution, a gamma distribution, or a uniform distribution. In cases where the product $\prod_{i \in S} g_i(\beta_i)$ is not concave, a suitable transformation will lead to a concave constraint.

If the occurrence of upstream events is independent, the joint probability density function (PDF) can be formed by multiplying the marginal PDF. Otherwise, the joint probability must be estimated by an appropriate multivariate statistical method. Singh et al. (2007) summarized empirical formulas for bivariate distributions, including the bivariate normal distribution, bivariate exponential distribution, bivariate Gumbel mixed distribution, bivariate Gumbel logistic distribution, bivariate gamma distribution, and bivariate lognormal distribution. However, most of the proposed empirical bivariate distribution models require the marginal distributions to be of the same type, whereas in reality this is not always the case. With the introduction of copulas, multivariate probability distributions with arbitrary marginal distributions can be constructed in a flexible manner.

In this context of stochastic programming, the "copula" method may play a critical role to harmonize the gap and provide a convenient avenue to assimilate the various sources of uncertainty. The term "copula" was first employed by Sklar (1959), and then developed and addressed by many researchers (Galambos, 1978; Genest and Mackay, 1986; Schweizer, 1991; Genest and Rivest, 1993; Joe, 1997; Shih and Louis, 1995; Nelsen, 2006). Copulas were defined by Nelsen (2006) as "functions that join or "copula" multivariate distribution functions to their one-dimensional marginal distribution functions." The biggest advantage of the copula method is that it is capable of determining the multivariate distribution in an easy way regardless of the marginal distributions. Unlike the other methods, copulas may not involve either very complicated derivation or have some strict requirements, that is, the margins are the same type of distribution.

19.3 STOCHASTIC PROGRAMMING WITH MULTIPLE OBJECTIVE FUNCTIONS

The topology of multiple criteria decision-making models consists of two main dimensions (Karpak and Zionts, 1989; Huang and Chang, 2003): (1) the nature of outcomes—fuzzy, grey, game, or stochastic versus deterministic and (2) the nature of alternative generating mechanism—whether the constraints limiting the alternatives are explicit or implicit. The metrics in Table 19.1 improve our understanding of these dimensions. When constraints are implicit, the alternatives must be explicit. One of the alternatives is then selected. When the constraints are explicit, then the

TABLE 19.1 A topology of multiple criteria decision methods

	Implicit constraints (Explicit solutions)	Explicit constraints (Implicit solutions)
Deterministic outcomes	Choosing among deterministic discrete alternatives or deterministic decision analysis	Deterministic mathematical programming
Stochastic outcomes	Stochastic decision analysis	Stochastic mathematical programming
Game outcomes	Conflict resolution matrixes	Multiobjective programming
Fuzzy outcomes	Fuzzy multicriteria decision analysis	Fuzzy mathematical programming
Gray outcomes	Gray decision analysis	Gray mathematical programming
Hybrid outcomes	Fuzzy stochastic multicriteria decision analysis	Fuzzy stochastic mathematical programming

alternative solutions are implicit and may be indefinite in number if the solution space is continuous. Problems in the explicit constraint category are generally regarded as mathematical programming problems. More dimensions may be added to the topology. We may consider each decision maker as a dimension to formalize the group decision-making process. If the feedback in the group decision-making process can be used to redirect the decision actions, it can be recognized as an interactive group decision-making process. In these cases of interactive decision-making processes involving multiple decision makers or stakeholders, GT would be an appropriate framework to solve the problem.

As stated in Chapter 8, a multiobjective programming technique may provide a set of more realistic alternatives in which stakeholders play a more active role in decision-making. The inclusion of uncertainty in a multiobjective programming model can be deemed as a progress to model a complex system. A stochastic programming model with multiple objectives can be defined as

$$\begin{aligned}
\text{maximize} \quad & Z(x) = [Z_1(x), Z_2(x), \ldots, Z_k(x)] \\
\text{subject to:} \quad & g(x) \leq 0, \\
& x \geq 0
\end{aligned}$$

in which either objective functions or constraints are random. The solution procedure can be obtained based on the following two approaches.

- Replace the stochastic constraints by crisp equivalent forms, and then solve the model by any multiobjective decision-making (MODM) techniques suitable to be applied.
- Transform this MODM problem into a single-objective stochastic programming model that can be solved by those algorithms available in the literature.

In terms of the goal programming model, assume that the elements of C in the objective function $f(x) = CX$ are random variables and that $f(x) = (f_1', f_2', \ldots, f_r')$ are prescribed target values to be reached by r objective functions. For each x, there exist possible deviations of CX from all target values $f(x)$ denoted by $d^+(x)$ and $d^-(x)$. The solution can be obtained by minimizing the mean value of norm vector $f(x) - CX$. One realization is

$$\begin{aligned}
\text{min} \quad & E\left[\sum_k (d_k^-(x) + d_k^+(x)) \right] \\
\text{s.t.} \quad & f_k(x) + d_k^- - d_k^+ = f_k' \quad k = 1, \ldots, r \\
& x \in X \text{ and } d_k^-, d_k^+ \geq 0, \quad \forall k.
\end{aligned}$$

Example 19.3   One municipality is launching a bold environmental agenda to achieve a sustainable development plan in terms of resource conservation and

TABLE 19.2   Treatment train production capacity and costs

	Compost (tonne)	Energy production (tonne)	Material recovery (tonne)	Cost (US$·tonne^{-1})
Integrated technology I	100	300	200	540,000
Integrated technology II	80	200	300	440,000

pollution prevention using benign environmental technologies. There are three goals during a year.

> *Goal 1:* Recycling 2000 tonnes of organic wastes to produce compost for landscape design, storm water management, and other uses.
>
> *Goal 2:* Collect 2000 tonnes of residential and commercial waste and recover the heat content through an incineration process for energy conservation.
>
> *Goal 3:* Recycling 3000 tonnes of residential and commercial wastes for material recovery using material recovery facilities, curbside recycling program, etc.

The total budget is 6 million that can be spent for achieving any one of these goals. There are two types of integrated technologies that may accommodate all three goals at the same time. The potential production rate on a per-treatment-train basis is listed in Table 19.2.

The decision analysis is to determine the possible combination of both technologies that could make the plan better off for this municipality. Both questions listed below need to include a budget constraint that is formulated as a chance constraint with the reliability level of 95% and a standard deviation is equal to 15% of its current budget level (i.e., 6 million).

(a) It is known that these three goals are non-preemptive goals. Solving the problem by a "once-through" approach assumes that both deviational variables are equally important and that all goals are also equally important.

(b) It is known that these three goals are preemptive goals (P1>>P2>>P3). Solving the problem based on a "streamline" approach assumes that both deviational variables are equally important at the same priority level.

Solutions:
Definition of decision variable

X_1 = number of treatment trains of technology I required;

X_2 = number of treatment trains of technology II required;

S_i^+ = amount by which we numerically exceed the ith goal;

S_i^- = amount by which we are numerically under the ith goal.

(a) Model formulation: non-preemptive goal programming with CCP

Assume that weighting factors associated with both deviational variables are the same.

$$\text{Min} \quad \sum_{i=1}^{3} (S_i^- + S_i^+)$$

s.t.:
$$100X_1 + 80X_2 + S_1^- - S_1^+ = 2000$$
$$300X_1 + 200X_2 + S_2^- - S_1^+ = 2000$$
$$200X_1 + 300X_2 + S_3^- - S_3^+ = 3000$$
$$540000X_1 + 440000X_1 \le 6000000.$$

Consider CCP with budget constraint
Reliability level $= 95\%$
Standard variation $= 15\% = \text{US\$6 million} \cdot (0.15) = \text{US\$900,000}$
$P[540000 \times 1 + 440000 \times 2 \le 6000000] > 0.95$
$K_{0.95} = -1.645.$
The equivalent linear form is
$540000 \times 1 + 440000 \times 2 \le 6000000 + (-1.645)(900000) \le 4519500.$
LINDO Program:

Min $S + S12 + S21 + S22 + S31 + S32$

s.t.:
$100X1 + 80X2 + S11 - S12 = 2000$
$300X1 + 200X2 + S21 - S22 = 2000$
$200X1 + 300X2 + S31 - S31 = 3000$
$540000X1 + 440000X2 <= 4519500$

END

Interpretation of outputs: Finding suggests that investing 10 treatment trains of technology II to be the most desirable option that will generate 800 tonnes of compost, recover high quality steam using 2000 tonnes of waste, and recycle materials of 3000 tonnes per year.

(b) Model formulation: preemptive goal programming with CCP

Assume that weighting factors associated with both deviational variables are the same. Yet priority of the first goal (P1) >> priority of the second goal (P2) >> priority of the third goal (P3).

LINDO program for the first priority with CCP:

Min $S11+S12$

s.t.:
$100X1 + 80X2 + S11 - S12 = 2000$
$300X1 + 200X2 + S21 - S22 = 2000$
$200X1 + 300X2 + S31 - S31 = 3000$
$540000X1 + 440000X2 <= 4519500$

END

TABLE 19.3 Comparative analysis of sustainable solid waste management plan

	(a): Without CCP	(b): With CCP	Difference between two cases
Selected technology	Technology I	Technology I	Same
No. of treatment trains	11	8	27% decrease
Compost	1,112 tonnes	837 tonnes	24.7% decrease
Energy production	3,333 tonnes	2,511 tonnes	24.7% decrease
Material recovery	2,223 tonnes	1,674 tonnes	24.7% decrease
Total cost	US$3,600,720	US$2,711,880	24.7% decrease

LINDO program for the second priority with CCP:

Min $S21 + S22$

s.t.: $100X1 + 80X2 + S11 - S12 = 2000$
$300X1 + 200X2 + S21 - S22 = 2000$
$200X1 + 300X2 + S31 - S31 = 3000$
$540000X1 + 440000X2 <= 4519500$
$S11 = 1163.056$

END

LINDO program for the third priority with CCP:

Min $S31 + S32$

s.t.: $100X1 + 80X2 + S11 - S12 = 2000$
$300X1 + 200X2 + S21 - S22 = 2000$
$200X1 + 300X2 + S31 - S31 = 3000$
$540000X1 + 440000X2 <= 4519500$
$S11 = 1163.056$
$S22 = 510.8191$

END

Interpretation of outputs: Preemptive goal programming with CCP—These findings suggest that investing in eight treatment trains of technology I may be the most desirable option that will generate approximately 837 tonnes of compost, while recovering high quality streams of solid waste from 2511 tonnes of raw waste, and recover 1674 tonnes of recyclable materials per year. However, with uncertainty analysis, the extent of activities would be smaller than the case without uncertainty analysis. A summary of all possible options can be seen in Table 19.3; it shows that the conservative option is the most attractive.

19.4 STOCHASTIC DYNAMIC PROGRAMMING

Continuous time stochastic optimization methods are a very powerful means to account for all time-varying events, but are not used widely in decision science because of the computational burden. Stochastic dynamic programming is equivalent to a dynamic control problem in which the element to be optimized is a function. Discrete-time stochastic models are popular considering that dynamic systems stochastically evolve over discrete stages in time, depending upon decisions and assuming that time is discrete and has many finite time steps. Assuming that the total cost of a decision is the sum of instantaneous costs where cost is to be minimized, and the strategy is to optimize the decision function.

$$\text{Cost} = c_1 + c_2 + \cdots + c_T, \; c_i = c(i, x_i, d_i),$$

where x_i is the state at time step i, the w_i are a random process, $x_i = f(x_{i-1}, d_{i-1}, w_i)$, and $d_{i-1} = $ strategy (x_{i-1}, w_i).

Stochastic dynamic programming is based on the following Bellman's Optimality Principle (Bellman, 1957):

"Take the decision at time step t such that the sum "cost at time step t due to your decision" plus "expected cost from time steps $t+1$ to T from the state resulting from your decision" is minimal."

Nevertheless, it can only be applied if the expected cost from time steps $t+1$ to T can be guessed, depending on the current state of the system and the decision. With this philosophy, Bellman's Optimality Principle reduces the control problem to the computation of this function (Bellman, 1957).

If x_t can be computed from x_{t-1} and d_{t-1} (i.e., if f is known) then this is reduced to the computation of $V(t, x_t) = Ec(t, x_t, d_t) + c(t + 1, x_{t+1}, d_{t+1}) + \cdots + c(T, x_T, d_T)$ depending on the chosen strategy. Thus, this expectation can be applied for any optimal strategy. Even if many strategies are optimal, V is uniquely determined.

In fact, stochastic dynamic programming is equivalent to the computation of V backwards in time, based on the following equation:

$$V(t, x_t) = \inf_{d_t} [c(t, x_t, d_t) + E_{w_{t+1}} V(t + 1, f(x_t, d_t, w_{t+1}))],$$

where inf is the infimum of a set of V. With Bellman's Optimality Principle, the computation of V is sufficient to define an optimal strategy by $d_{i-1} = \arg \min c_{i-1} + V(i, x_i) = \arg \min[c(i - 1, x_{i-1}, d_{i-1}) + E_{w_i} V(i, f(X_{i-1}, d_{i-1}, w_i))]$. arg *min* stands for the argument of the minimum, which is the set of points of the given argument for which the given function attains its minimum value. d_{i-1} is a robust solution applied in many areas to account for uncertainty impact in multistage decision-making.

Example 19.4   An enterprise needs to propose an inventory plan to go into the recycling market during the first season of a year—January, February, and March. Based on their investigation and projection, the local waste stream can produce 400, 300, 300 tonnes of comingled recyclables (i.e., glass, metal, paper, and plastics), respectively, in these 3 months. But the demand of secondary materials market does

TABLE 19.4   Income from recyclables sales

Sale (tonnes)	January income (US$1000)	February income (US$1000)	March income (US$1000)
0	0	0	0
100	1	2	2
200	2	3	3
300	3	4	4
400		5	4
500		4	5
600		3	7

not exactly meet their production during each month from January to March. A market analysis reveals the net income from recyclables sales each month (Table 19.4). The maximum sales are 300, 400, and 600 tonnes in the 3 months. Since prospective sales do not match production, the manager decided to build a storage warehouse to store excess production from early months for later sale when demand is higher. Storage cost is given in Table 19.5. Note that each parenthesis in the following Table 19.5 includes the probability that space for storage is required for the corresponding case. The warehouse cannot hold more than 200 tonnes. The assumption here is there will be no storage at the beginning of January and April. The problem is to determine how much to sell and store in each month using a dynamic programming model.

Solution: Definition of decision variables

X_t: amount of recyclables sold during month t (tonne)
S_t: amount of recyclables in storage at the beginning of month t (tonne).

Mass balance relationships are required for the storage facility. In general,

$$S_{t+1} = S_t + I_t - X_t,$$

where I_t is the amount of recyclables produced in month t. Since the storage contents must be zero at the beginning and end of the third month, we have

$$S_1 = 0$$

TABLE 19.5   Monthly direct cost of recyclables storage

Amount in storage at the beginning of month (tonnes)	January storage cost (US$1000)	February storage cost (US$1000)	March storage cost (US$1000)
0	1	1	2
100	2 (50%) or 3 (50%)	2 (60%) or 3 (40%)	2 (30%) or 3 (70%)
200	3 (30%) or 4 (70%)	2 (40%) or 3 (60%)	3 (30%) or 4 (70%)

and

$$S_3 + I_3 - X_3 = 0.$$

Profits from selling recyclables are equal to net sales income minus storage costs. Sales income in month t, $B_t(X_t)$ are given in Table 19.4 and storage cost in month t, $C_t(S_t)$, is given in Table 19.5. Profits for the month are thus $R(S_t, X_t) = B_t(X_t) - C_t(S_t)$,

$$\max \quad Z = \sum_{t=1}^{3} [B_t(X_t) - C_t(S_t)]$$

$$\begin{aligned}
\text{s.t.} \quad & S_1 = 0 \\
& S_2 - S_1 + X_1 = 400 \\
& S_3 - S_2 + X_2 = 300 \\
& -S_3 + X_3 = 300 \\
& S_t \in \{0, 100, 200\} \quad \forall t \\
& X_1 \in \Omega_1 \\
& X_2 \in \Omega_2 \\
& X_3 \in \Omega_3 \\
& X_4 \in \Omega_4.
\end{aligned}$$

A clearer representation is accomplished by a flow diagram for final presentation of the optimum sales/storage policy due to backward induction (Figure 19.6). Setting up calculation sub-tables in Table 19.6, we may begin the sorting process to achieve the optimum policy toward maximum net income overall. With all the information sorted, we can construct the final network diagram and follow optimum policy using backward induction toward maximum net income. The maximum net income US$6500 is achieved when storage in February is 100 tonnes in this quarter and with such a small return on the investment (when one considers storage) it is advisable to inform this company to do a yearly projection in order to maximize profits. Uncertainty in storage costs did not seem to affect the overall global solution.

19.5 GAME THEORY

Conflict resolution problems arise from social and political aspects of the design, operation, and management of complex environmental systems in which socioeconomic, ecological, environmental, water, and energy factors are intertwined. It usually affects a wide range of stakeholders that perceive reality from various vantage points, thereby finding themselves in serious conflicts (Hipel et al., 1997). GT has been demonstrated as one useful method to analyze situations of conflict and competition.

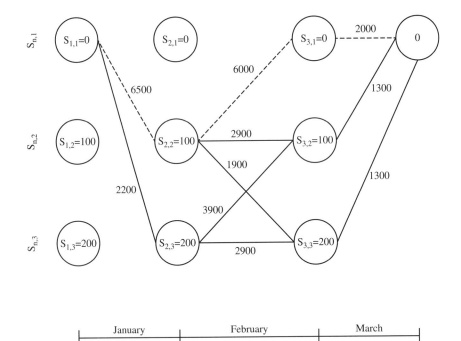

FIGURE 19.6 The multistage flow control strategies

Conflict resolution requires stakeholders to be placed in an analytical framework with varying semantic properties, subjective preferences, and uncertainty considerations due to knowledge gaps or insufficient information. It is believed that the conflicts in environmental resources allocation can be solved in a collaborative way through a participatory process. In this context, GT may be used to predict how stakeholders behave, following their own interests in a game. Typically, it includes players, rules, strategies, and payoffs in a study of strategic decision-making to analyze and resolve the problems with respect to multiple criteria and/or objectives related to conflicting interests in the field of politics, economic management, and so on (Buckley, 1984; Wu and Parlar, 2011; Li, 2012). In a game, decision makers or individuals can be modeled as players, with their own goals, trying to outsmart one another by anticipating each other's decision. In other words, GT is developed to illustrate how strategic interactions among players would result in ultimate outcomes in an equilibrium with respect to the preferences of those players (Madani, 2010). Basically, the payoffs among players for single-objective play a key role in determining the analytical framework and final outcomes of GT because it enables players to ponder the socioeconomic aspects of conflicts and conduct planning, design, and policy analysis when quantitative information is not readily available (Madani, 2010).

When analyzing, operating, or designing a complex environmental project, a decision maker must ensure that the undertaking is not only physically, financially, and economically achievable, but also socially, environmentally, and politically

TABLE 19.6   Outputs of the dynamic programming model

Mar	S_3		$D_3 = S_3 + 300$		$R_3(S_3) = B_3(D_3) - C_3(S_3)$	$*R_3(S_3)$
	$S_{3,1}$	0	300		$4000 - 2000 = 2000$	2000
	$S_{3,2}$	100	400		$4000 - (.3*2000 + .7*3000) = 1300$	
	$S_{3,3}$	200	500		$5000 - (.3*3000 + .7*4000) = 1300$	

Feb	S_2		$S_2 + 300$	D_2	$S_3 = S_2 + 300 - D_2$	$R_2(S_2,D_2) + R_3(S_3)$	$*R_2(S_2)$
	$S_{2,2}$	100	400	400	0	$5000 - 1000 + 2000 = 6000$	6000
				300	100	$4000 - (.6*2000 + .4*3000) + 1300 = 2900$	
				200	200	$3000 - (.6*2000 + .4*3000) + 1300 = 1900$	
	$S_{2,3}$	200	500	400	100	$5000 - (.6*2000 + .4*3000) + 1300 = 3900$	3900
				300	200	$4000 - (.6*2000 + .4*3000) + 1300 = 2900$	

Jan	S_1		$S_1 + 400$	D_1	$S_2 = S_1 + 300 - D_1$	$R_1(S_1,D_1) + R_2(S_2)$	$*R_1(S_1)$
	$S_{1,1}$	0	400	300	100	$3000 - (.5*2000 + .5*3000) + 6000 = 6500$	6500
				200	200	$2000 - (.3*3000 + .7*4000) + 3900 = 2200$	

feasible. GT provides a more realistic simulation of players/stakeholders' interest-based behavior in a game, in which each decision maker plays the game to optimize his own objective, knowing that other players' decisions would affect his payoffs and that his decision affects others' payoffs and decisions. Stable outcomes of the game predicted by GT are not necessarily the "Pareto optima." Note that the Pareto solution (or named non-dominant solution) is defined if none of the objective functions can be improved in value without degrading some of the other objective values without involving a gaming process. The main concern of players is to maximize their own benefit in the game knowing that the final outcome is the product of all the decisions to be made. In addition, GT allows the players to select applicable alternatives from a set of available strategies or options according to their respective preferences after an interactive decision-making procedure (Wu and Parlar, 2011), in which every choice from one player will affect and limit the choices of others. Another advantage of GT is its ability to predict the possible resolutions in the absence of quantitative payoff information.

However, strategies, styles, and techniques used in managing conflicts between individuals or between groups in relation to multiple participants with multiobjective decision-making issues deserve further attention, whenever humans interact through their decisions. For example, two or more individuals or groups may have (1) opposing objectives, as when a seller tries to get a high price while the buyer aims for a low price, or (2) differing strategies, as when one political party wants to remove the current ruler through a peaceful protest while another would like a revolution. These analyses of conflicting objectives generally consist of two categories: (1) methods that generate Pareto solution sets and (2) those that incorporate multiobjective trade-off preferences to select the best alternative (Cohon, 2004). To handle conflict resolution problems, a number of formal methodologies have been proposed, such as the theory of moves (Brams, 1994) and the graph model for conflict resolution (Hipel et al., 1997; Hamouda et al., 2004; Xu et al., 2009), all of which are related to GT (Kilgour, 1995) and provide a means to represent and analyze conflict situations with at least two decision makers or stakeholders, each of whom has multiple options (i.e., strategies) and multiple objectives.

19.5.1 Stochastic versus Deterministic Game Theory

As we discussed in Chapter 8, GT could be divided into two branches: cooperative game theory (CGT) and noncooperative game theory (NCGT). If these players are assumed to contribute to optimizing the common objectives without giving priority to their own payoffs, such game may be titled "CGT." Compared to CGT, the NCGT focuses on the self-optimizing attitude of players and stakeholders, and it often results in noncooperative stakeholder behaviors even when cooperative behavior is more beneficial to all parties. Note that "cooperative" and "noncooperative" are technical terms and are not an assessment of the degree of cooperation among players in the game (Chatain, 2014). For example, CGT models may be formulated as how agents compete and cooperate as coalitions in unstructured interactions to create and capture value. In NCGT models, when maximizing their utility in a defined procedure, the actions of agents actually rely on a detailed description of the moves when information becomes available to each agent (Chatain, 2014). Every game in the NCGT is shown to have at least one Nash equilibrium in pure strategies (Rosenthal, 1973) and it sometimes is referred to as a "self-enforcing equilibrium" (Kacher and Larbani, 2008). Hence, the main difference among CGT and NCGT is that NCGT players make decisions independently, whereas the basic unit of analysis is based on groups of players in CGT (Maskin, 2011). CGT focuses on how much players can appropriately be given the value each coalition of player can create, while NCGT concentrates on which moves players should rationally make (Chatain, 2014), and thus, the Nash equilibrium is a solution concept of a noncooperative game (Kacher and Larbani, 2008).

In CGT, Nash solution and the Shapley value were the most widely used solutions via axiomatic approaches, which can be used in a model developed within the waste management game framework (Karmperis et al., 2013). It is known that CGT game has the advantages over NCGT game in fulfilling polytropic win–win benefit collectively; while the CGT model is the worst from the computational efficiency view, and is due to that cooperation spends more time negotiating so as to obtain

TABLE 19.7   **Distinction between deterministic and stochastic CGT and NCGT**

Nash equilibrium	Deterministic	Cooperative game	Nash bargaining solution can be used for these games
		Noncooperative game	Traditional Nash solution
	Stochastic	Cooperative game	Whenever there have any uncertainties in parameters or even variables
		Noncooperative game	Mixed strategy equilibrium

win–win options (Meng et al., 2011). Due to the fact that the essential and sufficient information in CGT is hard to obtain, NCGT is more appropriate to describe the actual game in many occasions. To some extent, NCGT methods can still help resolve the conflict based on "qualitative" knowledge about the players' payoffs (i.e., how the players order (rank) different outcomes (ordinal payoffs)) (Madani, 2010). Hence, Karmperis et al. (2013) presents how cooperative and noncooperative game-theoretic approaches can be used for the purpose of modeling and analyzing decision-making in situations with multiple stakeholders (Karmperis et al., 2013). For mixed strategies, the payoffs are the expected value of the players, and the analytical problem then becomes polylinear functions in the probabilities with which all kinds of players play their different pure strategies associated with probability distributions and stochastic games may come to play. For comparative purpose, both CGT and NCGT can be defined as models with pure or mixed strategies (Table 19.7).

19.5.2 Case Study

Municipal solid waste (MSW) management systems are often deemed to be complex environmental systems. For instance, insufficient landfill space motivated some countries to propose a bold agenda for constructing a series of waste incinerators, in parallel with the full promotion of material recycling and reuse. These efforts would result in redundant incineration capacity at the later stages and end up creating competition among municipal incinerators and even recyclers. Multiple participants (i.e., operators of incinerators demanding more waste inflows) with multiple objectives in decision-making seek more MSW using different strategies, each of which has a probability to be chosen. As a result, interactions between recyclers and incineration operators, when dealing with changing rate of solid waste generation, may need to be reevaluated from time to time in some urban regions due to changing waste management policies. Such an issue may be deemed as "conflict resolution" in waste management.

Multiobjective programming is an indispensable tool to find the decision variables in a decision support system leading to the individual or group decision-making. However, if not formulated correctly, they would not provide insights into the strategic behaviors to reach an optimal outcome from the real world status (Madani, 2010). In dealing with the conflicting objective problems, uncertain characters are often involved with the multiobjective optimization issues. Some mathematical tools based on GT for multiagent decision-making, such as covariance biplot technique (Losa

TABLE 19.8a Components and framework of a NCGT in this case study

Player	Strategies (decision variables)	Weighting factor of two objectives	Lower and upper bounds of payoff of gth objective
(A, B)	(A_1, A_2) for player A, (B_1, B_2) for player B	(3/4, 1/4) for player A, (1/6, 5/6) for player B	(70, 500) for 1st objective and player A, (2, 130) for 2nd objective and player A, (60, 400) for 1st objective and player B, (0, 1) for 2nd objective and player B.

et al., 2001), fuzzy cognitive maps (Giordano et al., 2005), the gray fuzzy control algorithm (Li and Li, 2007), fuzzy bimatrix games (Li, 2012), graph theory-based fuzzy preferences (Bashar et al., 2012), and the stochastic conflict resolution model (Kerachian and Karamouz, 2007) were developed to evaluate a specific objective as well as to compromise the possible trade-offs among stakeholders.

In this case study, a decision analysis model to support a two-person multiobjective bimatrix game for the Nash equilibrium solution (NES) with mixed strategies is presented for the purpose of illustration only. The two players, A and B, stand for two operators of municipal incinerators, facing a dilemma of redundant incineration capacity, both of which have two strategies, respectively, to compete for more waste throughput and thus approach the maximum payoff for satisfying their two objectives, where the weighting factor method is used to combine the two objectives for evaluation. The two objectives are defined for maximizing the income and minimizing the shipping distance when seeking additional waste streams. The probabilities for the selection of different strategies are defined as decision variables. Assume that the weighting factors and the possible range of both objectives in this GT model can be summarized in Table 19.8a, and the crisp payoff matrices of these two decision makers are shown in Table 19.8b. Obviously, these payoffs have different units associated with different objectives.

TABLE 19.8b Payoffs matrices with crisp payoff value

Player	A			
Objective	1st objective (US$)		2nd objective (km)	
Strategy	SB_1	SB_2	SB_1	SB_2
SA_1	100	400	31	6
SA_2	250	250	100	100

Player	B			
Objective	1st objective (US$)		2nd objective (km)	
Strategy	SB_1	SB_2	SB_1	SB_2
SA_1	350	300	0.5	0.3
SA_2	100	300	0.7	0.9

To solve the NES as defined in Table 19.8a, the probability set is defined as the decision variables (A_1, A_2, B_1, B_2) for two players that adopt these two strategies to approach the Nash equilibrium. Based on the algorithm of GT, such crisp payoffs may then be first combined with the decision variables to form the expected payoffs. Because these payoffs have different units for different objectives, they must be normalized to the satisfaction level for each player making them comparable. Then, players may reorganize the payoff matrices in Table 19.8b by expected satisfaction level as shown in Table 19.9 after normalization according to the lower and upper bounds associated with each objective. Following these steps, both objectives are transformed into a set of satisfaction functions expressed in terms of corresponding decision variables. Such expected satisfaction functions (ESF) in Table 19.9 for two players can be used as the basis for conflict resolution in the proposed multiobjective programming model.

The weighting factor method may be chosen to combine the multiple objectives for each player. The weighting factors, as shown in Table 19.8a, for player A and B are equal to (3/4, 1/4) and (1/6, 5/6), respectively. Finally, the 2-person mathematical game expressions as a conflict resolution problem with known crisp payoffs will become numerically available for decision analysis. This 2-person GT model may

TABLE 19.9   ESF for two players

Player	A	
ESF	1st objective	
Strategy	SB_1	SB_2
SA_1	$1/430(100A_1B_1 - 70)$	$1/430(400A_2B_1 - 70)$
SA_2	$1/430(250A_1B_1 - 70)$	$1/430(250A_2B_1 - 70)$
ESF	2nd objective	
Strategy	SB_1	SB_2
SA_1	$1/128(31A_1 B_2 - 2)$	$1/128(6A_2 B_2 - 2)$
SA_2	$1/128(100A_1 B_2 - 2)$	$1/128(100A_2 B_2 - 2)$
Player	B	
ESF	1st objective	
Strategy	SB_1	SB_2
SA_1	$1/340(350A_1B_1 - 60)$	$1/340(300A_1 B_2 - 60)$
SA_2	$1/340(100A_2B_1 - 60)$	$1/340(300A_2 B_2 - 60)$
ESF	2nd objective	
Strategy	SB_1	SB_2
SA_1	$1/1((0.4, 0.5, 0.6)A_1B_1 - 0)$	$1/1((0.2, 0.3, 0.4) A_1 B_2 - 0)$
SA_2	$1/1((0.6, 0.7, 0.8)A_2B_1 - 0)$	$1/1((0.8, 0.9, 1.0)A_2 B_2 - 0)$

be formulated by the following paired nonlinear programming models as shown in Equations (19.58)–(19.62) in which all of the n players may eventually approach the maximal satisfaction degree Aro_A and Aro_B.

$$\text{Maximize} \quad Aro_A \text{ and } Aro_B \tag{19.58}$$

$$
\begin{aligned}
Aro_A = {}& 3/4[(100A_1B_1 - 70)/430 + (400A_1B_2 - 70)/430 + (250A_2B_1 - 70)/430 \\
& + (250A_2B_2 - 70)/430] + 1/4[(31A_1B_1 - 2)/128 + (6A_1B_2 - 2)/128 \\
& + (100A_2B_1 - 2)/128 + (100A_2B_2 - 2)/128] \tag{19.59}
\end{aligned}
$$

$$
\begin{aligned}
Aro_B = {}& 1/6[(350A_1B_1 - 60)/340 + (300A_2B_1 - 60)/340 + (100A_1B_2 - 60)/340 \\
& + (300A_2B_2 - 60)/340] + 5/6[(0.5A_1B_1) + (0.3A_2B_1) + (0.7A_1B_2) \\
& + (0.9A_2B_2)] \tag{19.60}
\end{aligned}
$$

subject to

$$A_1 + A_2 = 1, \ B_1 + B_2 = 1, \tag{19.61}$$
$$A_1 \geq 0, A_2 \geq 0, B_1 \geq 0, B_2 \geq 0. \tag{19.62}$$

Based on the above Equations (19.58)–(19.62), the problem may be directly solved by the bounded objective method (BOM) (Hwang and Masud, 1979), involving the concept of MAX/MIN. The general algorithm of BOM for an n-person game is described as (19.63) and (19.64):

$$\text{Maximize} \quad Z_A = f(x, y) \tag{19.63a}$$
$$\text{Maximize} \quad Z_B = f(x, y) \tag{19.63b}$$
$$\cdots$$
$$\text{Maximize} \quad Z_N = f(x, y) \tag{19.63c}$$

subject to

$$x, y \geq 0$$

where the Z_A, Z_B, $\ldots$, and Z_N are nonlinear equations with crisp coefficient. Moreover, each objective in Equation (19.56) can be transformed into a goal programming model that is identical to the Equation (19.57).

$$\text{Maximize} \quad TCAro \tag{19.64a}$$

subject to

$$tmf_A \geq TCAro \tag{19.64b}$$
$$tmf_B \geq TCAro \tag{19.64c}$$
$$\cdots$$
$$tmf_n \geq TCAro \tag{19.64d}$$

where

$$tmf_A = \frac{Z_{1A} - Z_A^-}{Z_A^+ - Z_A^-} \tag{19.64e}$$

$$tmf_B = \frac{Z_B - Z_B^-}{Z_B^+ - Z_B^-} \tag{19.64f}$$

$$\cdots$$

$$tmf_N = \frac{Z_N - Z_N^-}{Z_N^+ - Z_N^-}, \tag{19.64g}$$

in which $TCAro$ is an intermediate variable, tmf_A, tmf_B, ... , tmf_N are the definitional constraints addressing the degree of satisfaction for n-persons in the game. In this case with 2persons involved, only tmf_A and tmf_B are required. Z_A^+ and Z_A^- belong to a set of ideal solutions of Z_A representing the maximum and minimum values of Z_A; by the same token, Z_B^+ and Z_B^- belong to a set of ideal solutions of Z_B representing the maximum and minimum values of Z_B, ... , and Z_N^+ and Z_N^- belong to a set of ideal solutions of Z_N representing the maximum and minimum values of Z_N. The ideal solution is easily identified using software packages such as LINDO/LINGO. Multiple objectives can be transformed into single objectives with the aid of weighting factors and solved by LINDO/LINGO software package as well. Hence, the Equations (19.58) and (19.59) can be transformed into the following Equation (19.65) and (19.66).

$$\text{Maximize } TCAro \tag{19.65}$$

Subject to

$$tmf_A \geq TCAro \tag{19.66a}$$

$$tmf_B \geq TCAro \tag{19.66b}$$

$$A_1 + A_2 = 1, \ B_1 + B_2 = 1, \ A_1 \geq 0, \ A_2 \geq 0, \ B_1 \geq 0, \ B_2 \geq 0 \tag{19.66c}$$

where

$$tmf_A = \frac{Z_A - Z_A^-}{0.205 - (-0.269)}$$

$$tmf_B = \frac{Z_B - 0}{0.780 - 0.280}$$

$$Z_A = 0.235A_1B_1 + 0.631A_2B_1 + 0.709A_1B_2 + 0.631A_2B_2 - 0.504$$

$$Z_B = 0.588A_1B_1 + 0.397A_2B_1 + 0.632A_1B_2 + 0.897A_2B_2 - 0.118,$$

in which the Z_A^- and Z_B^- are the minimum values of Z_A and Z_B, respectively. Their values as solved using the LINGO software package are 0.269 and 0.280 in this

TABLE 19.10 Conflict resolution strategies based on the proposed GT model

Player	A	B
The probability of mixed strategy	$(A_1, A_2) = (0.00, 1.00)$	$(B_1, B_2) = (0.16, 0.84)$
(Total satisfaction)	0.110	0.687
Sequence of satisfaction		B > A

case. Similarly, Z_A^- and Z_B^- are the maximum values of Z_A and Z_B, respectively. Their values as solved with the LINGO software package are 0.205 and 0.780. Finally, the Equations (19.65) and (19.66) can be solved using the LINGO software package and the conflict resolution of such a GT is presented in Table 19.10. The probability sets associated with both players who may adopt the prescribed strategies can eventually be obtained. The sequence of satisfaction confirms that player B would have higher satisfaction in this game by favoring B_2 strategies with higher odds of success.

Given the crisp payoffs matrices seen in Table 19.8b, NES in Table 19.10 can be entailed by the probabilities for player A to adopt strategies 1 and 2 that are about 0% and 100%, respectively. In other words, it would be beneficial for player A to direct all efforts on implementing strategy 2 only. Meanwhile, as seen in the NES shown in Table 19.10, the best strategy for player B is that it would be better to spend 84% effort to implement strategy 2 and 16% effort to implement strategy 1 in order to obtain the highest satisfaction in this competitive and conflicting game. Overall, player B can obtain a higher satisfaction level than player A in this game. Nevertheless, it does not imply that the actual profit that player A may earn will be smaller than that of player B during the whole game.

19.6 FINAL REMARKS

In the context of decision analysis with random characteristics, various sources of uncertainty having probability distributions must be characterized and quantified by tailored methods to tackle the problem of natural variability. In this chapter, two important interrelated decision analysis sub-disciplines were discussed in which the stakeholders were not involved in stochastic programming models but were called on for decision-making in GT for conflict resolution in relation to differing criteria. The involvement of stakeholders highlights the difficulties when conflicting objectives with a set of different strategies are positioned against each other, making final decision analysis challenging. When incorporating the GT into consideration, a series of nonlinear programming models are oftentimes needed to carry out the essential decision analysis and comprehend the implications of the interactions associated with relevant constraints. Nevertheless, a consideration of random characteristics is not enough to cover all sources of uncertainty. More uncertainty quantification techniques such as fuzzy set theory may be factored into such a framework to minimize the gap of decision analysis in real world SWM systems.

REFERENCES

Bashar, M. A., Kilgour, D. M., and Hipel, K. W. 2012. Fuzzy preferences in the graph model for conflict resolution. *IEEE Transactions on Fuzzy Systems*, 20(4), 760–770.

Bellman, R. 1957. *Dynamic Programming*, Princeton University Press, Princeton, NJ.

Brams, S. J. 1994. *In Theory of Moves*, Cambridge University Press, Cambridge, MA.

Buckley, J. J. 1984. Multiple goal non-cooperative conflicts under uncertainty: a fuzzy set approach. *Fuzzy Sets and Systems*, 13(2), 107–124.

Charnes, A. and Cooper, W. W. 1959. Chance-constrained programming. *Management Science*, 6(1), 73–79.

Charnes, A., Cooper, W. W., and Symonds, G. H. 1958. Cost horizons and certainty equivalents: an approach to stochastic programming of heating oil. *Management Science*, 4(3), 235–263.

Chatain, O. 2014. Cooperative and non-cooperative game theory. Available at: http://works. bepress.com/cgi/viewcontent.cgi?article=1017&context=olivier_chatain (accessed January 2014).

Cohon, J. L. 2004. *Multiobjective Programming and Planning*, Dover Publications, New York.

Dantzig, G. B. 1955. Linear programming under uncertainty. *Management Science*, 1, 197–206.

Ellis, J. H. 1987. Optimization models for development of acid rain abatement strategies. *Civil Engineering Systems*, 4(2), 58–66.

Ellis, J. H. 1988. Multi-objective mathematical programming models for acid rain control. *European Journal of Operational Research*, 35(3), 365–377.

Ellis, J. H. 1990. Integrating multiple long-range transport models into optimization methodologies for acid rain policy analysis. *European Journal of Operational Research*, 46(3), 313–321.

Ellis, J., McBean, E., and Farquhar, G. 1985. Chance-constrained/stochastic linear programming model for acid rain abatement-I complete collinearity and noncollinearity. *Atmospheric Environment*, 19(6), 925–937.

Ellis, J., McBean, E., and Farquhar, G. 1986. Chance-constrained/stochastic linear programming model for acid rain abatement-II limited collinearity. *Atmospheric Environment*, 20(3), 501–511.

Fortin, M. and McBean, E. A. 1983. A management model for acid rain abatement. *Atmospheric Environment*, 17(11), 2331–2336.

Galambos, J. 1978. *The Asymptotic Theory of Extreme Order Statistics*, John Wiley & Sons, New York.

Genest, C. and MacKay, J. 1986. The joy of copulas: bivariate distributions with uniform marginals. *American Statisticians Association*, 40(4), 280–283.

Genest, C. and Rivest, L. P. 1993. Statistical inference procedures for bivariate Archimedean copulas. *Journal of American Statisticians Association*, 88(423), 1034–1043.

Giordano, R., Passarella, G., Uricchio, V. F., and Vurro, M. 2005. Fuzzy cognitive maps for issue identification in a water resources conflict resolution system. *Physics and Chemistry of the Earth, Parts A/B/C*, 30(6–7), 463–469.

Guldmann, J. M. 1988. Chance-constrained dynamic model of air quality management. *Journal of Environmental Engineering*, 114(5), 1116–1135.

Hamouda, L., Kilgour, D. M., and Hipel, K. 2004. Strength of preference in the graph model for conflict resolution. *Group Decision and Negotiation*, 13(5), 449–462.

Hiller, F. S. and Lieberman, G. J. 1991. *Introduction to Mathematical Programming*, McGraw-Hill, New York.

Hipel, K. W., Kilgour, M. D., Fang, L., and Peng, X. 1997. The decision support system GMCR in environmental conflict management. *Applied Mathematics and Computation*, 83(2–3), 117–152.

Huang, G. H. and Chang, N. B. 2003. The perspectives of environmental informatics and systems analysis. *Journal of Environmental Informatics*, 1(1), 1–7.

Hwang, C. L. and Masud, A. S. M. 1979. *Multiple Objective Decision Making, Methods and Applications: A State-of-the-Art Survey*, Springer-Verlag, Berlin.

Joe, H. 1997. *Multivariate Mold and Dependence Concepts*, Chapman & Hall, London.

Kacher, F. and Larbani, M. 2008. Existence of equilibrium solution for a non-cooperative game with fuzzy goals and parameters. *Fuzzy Sets and Systems*, 159(2), 164–176.

Karmperis, A. C., Aravossis, K., Tatsiopoulos, I. P., and Sotirchos, A. 2013. Decision support models for solid waste management: review and game-theoretic approaches. *Waste Management*, 33(5), 1290–1301.

Karpak, B. and Zionts, S. 1989. *Multiple Criteria Decision Making and Risk Analysis using Microcomputers*, Springer-Verlag, Berlin.

Kerachian, R. and Karamouz, M. 2007. A stochastic conflict resolution model for water quality management in reservoir–river systems. *Advances in Water Resources*, 30(4), 866–882.

Kilgour, D. M. 1995. Book review: theory of moves. *Group Decision and Negotiation*, 4(3), 287–288.

Li, C. L., 2012. Characterization of the equilibrium strategy of fuzzy bimatrix games based on L-R fuzzy variables. *Journal of Applied Mathematics*, 15, 1–15.

Li, J. and Li, J. 2007. Study on dynamic environmental supervision based on gray fuzzy control. In: Proceedings of the IEEE International Conference on Grey Systems and Intelligent Services,Nanjing, pp. 978–982.

LINDO Inc. 2012. Optimization Software. Available at: http://www.lindo.com/ (accessed May 2012).

Losa, F. B., van den Honert, R., and Joubert, A. 2001. The multivariate analysis biplot as tool for conflict analysis in MCDA. *Journal of Multi-Criteria Decision Analysis*, 10(5), 273–284.

Madani, K. 2010. Game theory and water resources. *Journal of Hydrology*, 381(3–4), 225–238.

Maskin, E. 2011. Commentary: Nash equilibrium and mechanism design. *Games and Economic Behavior*, 71(1), 9–11.

Meng, D., Zhang, X., and Qin, K. 2011. Soft rough fuzzy sets and soft fuzzy rough sets. *Computers & Mathematics with Applications*, 62(12), 4635–4645.

Nelsen, R. B. 2006. *An Introduction to Copulas*, 2nd edition, Springer-Verlag, New York.

Rosenthal, R. 1973. A class of games possessing pure-strategy Nash equilibria. *International Journal of Game Theory*, 2(1), 65–67.

Schweizer, B. 1991. Thirty years of copulas. In: *Advances in Probability Distributions with Given Marginals* (Eds Dall's Aglio, G., Kotz, S., and Salinetti, G.), Kluwer, Dordrecht, The Netherlands, pp. 13–50.

Shih, J. H. and Louis, T. A. 1995. Inferences on the association parameter in copulas models for bivariate survival data. *Biometrics*, 51(4), 1384–1399.

Singh, V. P., Jain, S. K., and Tyagi, A. 2007. *Risk and Reliability Analysis: A Handbook for Civil and Environmental Engineers*, American Society of Civil Engineers, Reston, VA.

Sklar, M. 1959. Distribution functions in n dimensions and margins [in French: Fonctions de répartition à n dimensions et leurs marges]. *Publications de l'Institute de Statistique de L'Université de Paris*, 8, 229–231.

Wu, H. and Parlar, M. 2011. Games with incomplete information: a simplified exposition with inventory management applications. *International Journal of Production Economics*, 133(2), 562–577.

Xu, H., Li, K. W., Hipel, K. W., and Kilgour, D. M. 2009. A matrix approach to status quo analysis in the graph model for conflict resolution. *Applied Mathematics and Computation*, 212(2), 470–480.

CHAPTER 20

FUZZY MULTIATTRIBUTE DECISION-MAKING FOR SOLID WASTE MANAGEMENT WITH SOCIETAL COMPLICATIONS

Delineating the social, economic, ecological, and environmental objectives into a succinct yet representative management model for decision-making is by no means an easy task. While classical mathematics and probability theory is dichotomous in character, fuzzy set theory may consider the situations involving human factors with all the vagueness of perception, flexibility of personality trait, mentality-oriented subjectivity, culture-based attitudes, individual goals, and ethical conceptions. By introducing such linguistics into the crisp set theory, fuzzy set theory with such a unique orientation becomes more robust, flexible, general, and applicable than the classical set theory. It is the aim of this chapter to provide readers with a capsule look into the fundamentals of fuzzy set theory and existing fuzzy multiattribute decision-making problems with possible extensions and applications in dealing with two different types of solid waste management (SWM) issues. It starts with the introduction of basic concept of fuzzy sets, followed by two case studies for SWM in sequence for addressing landfill siting issues in Texas, and fair fund redistribution issues in Taiwan, all of which collectively illuminate the essence of fuzzy multiattribute decision-making.

20.1 FUNDAMENTALS OF FUZZY SET THEORY

Delineating the social, economic, ecological, and environmental objectives into a succinct yet representative management model for decision-making is oftentimes prone to considerable subjectivity, imprecision, and uncertainty. To reflect such

Sustainable Solid Waste Management: A Systems Engineering Approach, First Edition. Ni-Bin Chang and Ana Pires.
© 2015 The Institute of Electrical and Electronics Engineers, Inc. Published 2015 by John Wiley & Sons, Inc.

subjectivity, imprecision, and uncertainty issues, deterministic decision-making models are deemed incapable of tackling those situations from a practical sense. Before the 1990s, the concepts of probability theory were usually employed as a form of stochastic programming to supplement the decision-making capacity of traditional deterministic mathematical programming models. As chaos theory was developed to handle nonlinear dynamic systems in physics and mathematics in the 1980s, fuzzy set theory invented by Zadeh (1962, 1965) (Box 20.1) became widely applied about the same time for coping with linguistic uncertainty in human society. Since then, fuzzy set theory has become an important part of the techniques used in conjunction with traditional deterministic mathematical programming models. Fuzzy set theory is a theory of a matter of degree of feeling rather than a description of chance as proposed by traditional probability theory. Fuzzy set theory uses fuzzy numbers to quantify subjective fuzzy observations or estimates through the realization of fuzzy membership functions. It provides a synergistic effect in the traditional fields of operation research, management sciences, artificial intelligence, control theory, expert and decision support systems.

BOX 20.1   THE INVENTION OF FUZZY SET THEORY

In 1962, Zadeh first proposed the term "fuzzy" in a conference. In 1965, he formally published the concept of "Fuzzy Sets" in a journal paper, presenting a more robust and flexible model to delineate real-world decision-making problems involving the human dimension (Zadeh, 1965). Traditionally, the probability theory is certainly the prevailing approach to represent some knowledge about uncertain information whose boundaries can be clearly defined and where large numbers of samples can be collected. The reason that fuzzy sets are defined independently is that precise mathematics are not sufficient to model a complex system involving human aspects based on incomplete knowledge and precise information. The example of throwing coin into air and guessing either tails or heads will be up or down can easily be fitted into the paradigm of probability theory. However, there are a lot of problems satisfying the law of the excluded middle with vague contradictions. Evidence in favor of a particular hypothesis to some degree does not disconfirm it to any degree, or the contrary, at the same time (Zadeh, 1962). Salient examples in our daily life may include but are not limited to the feeling of "clean water," "beautiful scenery," "reasonable price," "good personality," or "smart people." In these situations, the probability theory is not capable of modeling these aspects in relation to incompleteness, imprecision, or vagueness when involving human judgment. The fuzzy set theory is developed to define these problems without sharp boundaries. It has also been widely applied to aid in the new formulations of system engineering models such as linear programming, nonlinear programming, integer programming, dynamic programming, and multiobjective programming.

20.1.1 Basic Concept of Fuzzy Sets

The Fuzzy Set Theory   To clearly differentiate the fuzzy set from the classical crisp set, a description of membership is necessary. Consider a basketball team as a universe that is composed of five players, including Chang, Lee, Hans, Kim, and Michael. That is, U = {Chang, Lee, Hans, Kim, Michael}. The heights for the five players are given as follows:

Chang = 180 cm, Lee = 190 cm, Hans = 205 cm, Kim = 160 cm, Michael = 220 cm

Now, let us consider the linguistic proposition "tall players." The players who belong to "tall players" then constitute a fuzzy set, A. Is Chang $\in A$ or Michael $\in A$?

One plots the heights on a real line (see Figure 20.1) in order to present the relative differences in height. According to common sense, a player who is taller than 210 cm in height in this team is absolutely considered a tall player. On the other hand, a player who is shorter than 160 cm in height is not considered a tall player at all. Therefore, it is obvious that Michael is absolutely tall and Kim is absolutely not tall. How about the other three? It is true that the taller the player, the greater the degree to which he or she belongs to a fuzzy set A in analysis. Thus, a degree-scaled line (see Figure 20.1) can be drawn corresponding to the previous height-scaled line in order to represent the degree of membership indicating that a player belongs to A.

The scale on the degree-scaled line is linearly proportional to the height-scaled line when the height belongs to the interval [160, 210]. As a result, the following grades of membership are available:

Degree (Chang $\in A$) = $\mu(x = $ Chang$)$ = 0.4
Degree (Lee $\in A$) = $\mu(x = $ Lee$)$ = 0.6
Degree (Hans $\in A$) = $\mu(x = $ Hans$)$ = 0.9
Degree (Kim $\in A$) = $\mu(x = $ Kim$)$ = 0
Degree (Michael $\in A$) = $\mu(x = $ Michael$)$ = 1.0

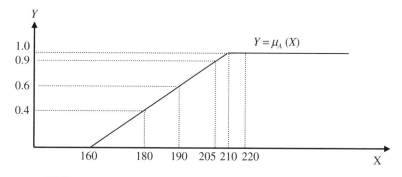

FIGURE 20.1   Derived degree of membership for the fuzzy set A

Kim	Chang	Lee	Hans	Michael		X = height
160	180	190	205	220		
0	0.4	0.6	0.9	1.0		Y = degree

FIGURE 20.2   Illustration of degree of membership for the fuzzy set A

Alternatively, we can map the fuzzy relationship as in the following illustration (Figure 20.2).

In general, a fuzzy set as initiated by Zadeh (1965) is defined as follows: Let x denote a universal set. Then a membership in a crisp subset of X is often viewed as a characteristic function $\mu_A(x)$, which is the degree of membership of x in A, such that the closer the value of $\mu_A(x)$ is to 1, the more x belongs to A. Therefore, A is completely characterized by the set of ordered pairs

$$A = \left\{ (x, \mu_A(x)) \,|\, x \in X \right\}. \tag{20.1}$$

Obviously, characteristic function can be either linear or nonlinear function. When X is not a finite set, A can be expressed as

$$A = \int_x \mu_A(x)/x. \tag{20.2}$$

The most important concept in the application of fuzzy sets is the α-cut. The α-cut of a fuzzy set A is a crisp subset of X that can be denoted by

$$A_\alpha = \left\{ x \,|\, \mu_A(x) \geq \alpha \text{ and } x \in X \right\}. \tag{20.3}$$

In the example described above, $A_{0.4} = \{$Chang, Lee, Hans, Michael$\}$. It is clear that the following property holds for the α-level sets. This relationship is illustrated in Figure 20.3.

$$\alpha_1 \geq \alpha_2 \Leftrightarrow A_{\alpha_1} \supseteq A_{\alpha_2} \tag{20.4}$$

Generation of Membership Functions   There are two approaches for generating membership functions: the axiomatic and semantic approaches (Giles, 1988). An axiomatic approach, similar to the approaches used in utility theory, is centered on the mathematical structures involved in human feeling. Yet, the semantic approach is concentrated on practical interpretation rather than the mathematical structures of human feeling. In comparison, the semantic approach actually follows the perceptions of pragmatism in its insistence that all conclusions should be firmly based on the practical meaning of the concepts involved, with no axiom or law laid down in

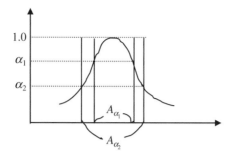

FIGURE 20.3   Examples of α-level sets

advance (Giles, 1988). In real-world applications, however, a semantic approach that is applicable when interpreting the implications of membership is generally preferred over the axiomatic approach.

To construct meaningful membership functions, there are five classical scale levels, including the nominal, ordinal, interval, ratio, and absolute scale (Box 20.2). According to Zimmermann and Zysno (1985), the interval scale level seems to be most adequate for the construction of membership functions among the five classical scale levels given that the requirements of ratio and absolute scales seem to be too strict for fuzzy sets and ordinal scale cannot delineate the specific amount of difference. Within the context of interval scale level, the distance approach, true-value approach, and payoff function are three major ways used to elicit the membership function (Zimmermann and Zysno, 1985). They will be described below sequentially.

BOX 20.2   THE FIVE CLASSICAL SCALE LEVELS IN FUZZY SET THEORY

The conversion from numerical values to linguistic representation in fuzzy set theory is called a linguistic description. To perform a linguistic measurement, it is necessary to clearly specify the relation between linguistic terms in fuzzy set theory and numbers in the general scale formalism. To construct meaningful membership functions, there are five classical scale levels which include the nominal, ordinal, interval, ratio, and absolute scale. They are defined as follows:

- The nominal scale: A nominal scale is a method for placing of data into categories, without any order or structure. A physical example of a nominal scale is the terms we use for color classification. Although the underlying spectrum is ordered, the names are nominal in subjective classification. In this case, it is necessary to employ the concept of fuzzy partition for the definition of the fuzzy nominal scale and the proximity relation on the lexical set can then be defined. For example, a fuzzy partition based on an interpolation of human knowledge can be defined on this chrominance plane (Figure 20.4).

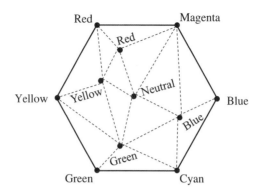

FIGURE 20.4 Examples of the fuzzy nominal scale

- The ordinal scale: When items are classified according to whether they have more or less of a characteristic, the scale used is referred to as an ordinal scale permitting the measurement of degrees of difference instead of the specific amount of difference.
- The interval scale: This scale or level of measurement has the characteristics of rank order with equal or unequal intervals (i.e., the distance between adjacent points might or might not be the same).
- The ratio scale: A ratio scale is an interval scale in which distances are stated with respect to a rational zero. In other words, it consists of not only equidistant points but also a meaningful zero point.
- The absolute scale: An absolute scale has absolute zero point. The scale of temperature with absolute zero as the minimum is a typical example.

Distance approach utilizes a distance $d(x)$ from a reference as an evaluation criterion. It entails that if an object possesses all the ideal features, the distance should be zero. On the contrary, if no similarity exists between an object and the ideal features, the distance should be infinity. Membership can thus be defined as a function in terms of distance between a given object x and an ideal standard (Zimmermann and Zysno, 1985). The same idea is applicable in multiattribute and multiobjective decision-making. However, distance was found to be quite context dependent. While the relationship between physical units and perceptions is generally exponential, Zimmermann and Zysno (1985) proposed the following distance function and membership function in terms of the context-dependent parameters a and b as follows:

$$D(x) = \frac{1}{e^{a(x-b)}}, \qquad (20.5)$$

and

$$\mu(x) = \frac{1}{1 + e^{-a(x-b)}}, \qquad (20.6)$$

where a and b can be considered as semantic parameters from a linguistic point of view. The determination of the parameters from empirical data can be established by a proper transformation prior to a linear regression analysis. That is

$$\text{Ln}\left[\frac{\mu(x)}{(1-\mu(x))}\right] = a\,(x-b)). \tag{20.7}$$

$$\text{Suppose } y = \ln\left[\frac{\mu(x)}{(1-\mu(x))}\right].$$

The linear relationship between x and y in (20.7) is obvious.

Fuzzy logic, however, has two characteristics, including the true domain is the whole [0, 1] interval and the true value can be a fuzzy subset of [0, 1]. Smets and Magrez (1988) developed a true-value approach to reduce fuzzy logic to its multivalued logic components in order to help create a membership function. In the identification procedure, a canonical scale for the true values is defined and a set of fuzzy propositions with well-defined intermediate true values is assigned. This aids in the postulation of the relation between degree of truth and grade of membership so that the degree of membership of an element x to a fuzzy set A is numerically equal to the degree of truth.

In the later period of evolution, membership functions were classified as two groups. One is the preference-based membership function and the other is possibility-based membership function. The former is constructed by describing the preference information from the decision-makers whereas the latter is created by considering the possible occurrence of the event of interest which is not equivalent to the probability of the occurrence of the same event. In any circumstance, how to generate the preference-based membership and possibility-based membership functions becomes a fundamental issue in both fuzzy and possibilistic mathematical programming problems. To define both types of membership functions, prior information has to be obtained via conducting an independent investigation, relying on either the distance approach or the true-value approach. To ease the applications, the following survey of functional forms of membership helps create relatively representative membership functions for fuzzy decision-making under uncertainty (Dombi, 1990).

(I) Membership functions based on heuristic determination
- Zadeh's unimodal functions (Zadeh, 1965, 1971, 1972, 1976)

$$\mu_{\text{young}}(x) = \begin{cases} 1/\{1+[(x-25)/5]^2\}, & \text{if } x > 25 \\ 1, & \text{if } x \le 25 \end{cases} \tag{20.8}$$

$$\mu_{\text{old}}(x) = \begin{cases} 1/\{1+[(x-50)/5]^{-2}\}, & \text{if } x \ge 50 \\ 1, & \text{if } x < 50 \end{cases} \tag{20.9}$$

- Dimitru and Luban's power functions (Dimitru and Luban, 1986)

$$\mu(x) = x^2/a^2 + 1, x \in [0, a] \tag{20.10}$$

$$\mu(x) = -x^2/a^2 - 2x/a + 1, x \in [0, a] \tag{20.11}$$

- Svarovski's sin function (Svarovski, 1987)

$$\mu(x) = 1/2 + (1/2)[\sin\{\pi/(b-a)[x-(a+b)/2]\}], x \in [a, b]. \tag{20.12}$$

(II) Membership functions based on reliability concerns
- Zimmermann's linear function (Zimmermann, 1978)

$$\mu(x) = 1 - x/a, \quad x \in [0, a] \tag{20.13}$$

- Tanaka, Uejima, and Asai's symmetric triangular function (Tanaka et al., 1982)

$$\mu(x) = \begin{cases} 1 - \dfrac{|b-x|}{a}, & \text{if } b - a \leq x \leq b + a \\ 0, & \text{otherwise} \end{cases} \tag{20.14}$$

- Hannan's piecewise linear function (Hannan, 1981)

$$\mu(x) = \sum_j \alpha_j |x - a_j| + \beta x + r, \quad j = 1, 2, \ldots, N, \tag{20.15}$$

$$\alpha_j = (t_{j+1} - t_j)/2, \tag{20.16}$$

$$\beta = (t_{N+1} + t_1)/2, \tag{20.17}$$

$$r = (s_{N+1} + s_1)/2, \tag{20.18}$$

where $\mu(x) = t_i x + s_i$ for each segment i, $a_{j-1} \leq x \leq a_j$ for all i and j is the slope and s_i is the y-intercept for the section of the curve initiated at a_{j-1} and terminated at a_j.
- Leberling's hyperbolic function (Leberling, 1981)

$$\mu(x) = 1/2 + (1/2)\tanh[a(x-b)], \quad -\infty \leq x \leq \infty \tag{20.19}$$

where a is a parameter.
- Sakawa and Yumine's exponential and hyperbolic inverse function (Sakawa and Yumine, 1983)

$$\mu(x) = c\left[1 - e^{(b-x)/(b-a)}\right], \quad x \in [a, b], \tag{20.20}$$

$$\mu(x) = 1/2 + c\tanh^{-1}[d(x-b)] \tag{20.21}$$

where c and d are parameters.

- Dimitru and Luban's function (Dimitru and Luban, 1982)

$$\mu(x) = 1/(1 + x/a) \tag{20.22}$$

where a is a parameter.
- Dubois and Prade's L-R fuzzy number (Dubois and Prade, 1988)

$$\mu(x) = \begin{cases} L\left(\dfrac{a-x}{\alpha}\right) & \text{if } x < a \\ R\left(\dfrac{x-b}{\beta}\right) & \text{if } x > a \\ 1 & \text{if } a \le x \le b \end{cases} \tag{20.23}$$

where $L(.)$ and $R(.)$ are reference functions.

Besides, payoff function is a mathematical function describing the award given to a single player at the outcome of a game. Payoff function in relation to game theory has been discussed briefly in Chapter 19 (i.e., Section 19.5). Fuzzy payoff function is normally used to aid in the delineation of traditional payoff function in economics. For example, as fuzzy payoff function may be used to replace the payoffs in a game that includes players, rules, strategies, and payoffs in a study, it would fall into the context of fuzzy game theory.

Basic Set-theoretic Operators   Fuzzy logic is derived from fuzzy set theory in order to deal with reasoning that is approximate rather than precisely deduced from classical predicate logic via the use of specific set-theoretic operators. Several set-theoretic basic operations involving fuzzy sets originally proposed by Zadeh (1965) are summarized to reason the operation of adopted membership functions as follows:

- **Equality**: The fuzzy sets A and B on X are equal, denoted by $A = B$, if and only if their membership functions are equal everywhere on X

$$A = B \Leftrightarrow \mu_B(x) \ \forall x \in X. \tag{20.24}$$

- **Containment**: The fuzzy set A is contained in B (or a subset of B), denoted by $A \subseteq B$, if and only if their membership function is less than or equal to that of B everywhere on X

$$A \subseteq B \Leftrightarrow \mu_A(x) \le \mu_B(x) \quad \forall x \in X \tag{20.25}$$

- **Complementation**: The complement of a fuzzy set A on X, denoted by $\bar{A}$ is defined by

$$\mu_{\bar{A}}(x) = 1 - \mu_A(x) \quad \forall x \in X. \tag{20.26}$$

- **Intersection**: The intersection of two fuzzy sets A and B on X, denoted by $A \cap B$, is defined by

$$\mu_{A \cap B}(x) = \min[\mu_A(x), \mu_B(x)] \quad \forall x \in X \qquad (20.27)$$

- **Union**: The union of two fuzzy sets A and B on X, denoted by $A \cup B$, is defined by

$$\mu_{A \cup B}(x) = \max[\mu_A(x), \mu_B(x)] \quad \forall x \in X \qquad (20.28)$$

These set-theoretic operations for fuzzy sets above can be viewed as a natural extension of those for ordinary sets. The intersection and union of two fuzzy sets A and B, and the complement of a fuzzy set A are illustrated in Figure 20.5.

Based on the definitions of the set-theoretic operations for fuzzy sets, Zadeh (1965) pointed out that it is impossible to extend many of the basic identities that hold for ordinary sets to fuzzy sets. With this said, the following properties for union, intersection, and complementation can still hold for fuzzy sets in a way that is similar to those properties in ordinary sets.

- Commutative laws

$$A \cup B = B \cup A, A \cap B = B \cap A \qquad (20.29)$$

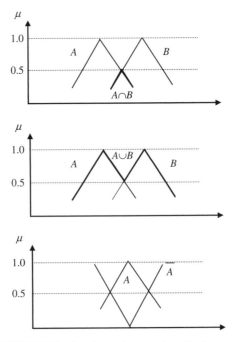

FIGURE 20.5   Set-theoretic operations for fuzzy sets

- Associative Law

$$(A \cup B) \cup C = A \cup (B \cup C), \tag{20.30}$$

$$(A \cap B) \cap C = A \cap (B \cap C). \tag{20.31}$$

- Distributive Law

$$(A \cup B) \cap C = (A \cap C) \cup (B \cap C) \tag{20.32}$$

$$(A \cap B) \cup C = (A \cup C) \cap (B \cup C) \tag{20.33}$$

- De Morgan's rule

$$\overline{(A \cup B)} = (\overline{A}) \cap (\overline{B}), \tag{20.34}$$

$$\overline{(A \cap B)} = (\overline{A}) \cup (\overline{B}). \tag{20.35}$$

- Involution

$$\overline{\overline{A}} = A. \tag{20.36}$$

It should be noted that the only law that is no longer valid for fuzzy sets is the excluded-middle law

$$A \cup (\overline{A}) = 1, \tag{20.37}$$

$$A \cap (\overline{A}) = \phi. \tag{20.38}$$

20.2 SITING A REGIONAL LANDFILL WITH FUZZY MULTIATTRIBUTE DECISION-MAKING AND GIS TECHNIQUES

Landfill siting is a difficult, complex, tedious, and protracted process requiring evaluation of many different criteria (Charnpratheep et al., 1997). This section presents a fuzzy multicriteria decision analysis alongside a geospatial analysis for the screening and selection of a set of landfill sites. It employs a two-stage analysis synergistically to form a spatial decision support system (SDSS) for SWM in a fast growing urban region, South Texas. The first-stage analysis makes use of thematic mapping tools found in a geographical information system (GIS). In conjunction with environmental, biophysical, ecological, and socioeconomic variables, a geospatial analysis can lead to the provision of essential spatial information for the second-stage analysis using the fuzzy multicriteria decision-making (FMCDM) method. A case study was made for the city of Harlingen in South Texas, which is rapidly evolving into a large urban area due to its advantageous position near the United States (US)–Mexico borderlands under the direct impact of North America Free Trade Agreement (NAFTA). Overall, the purpose of GIS analysis in the first stage was to perform an initial screening process to eliminate unsuitable land for a landfill site and it was followed by an implementation of the FMCDM method to identify the most suitable location using the information provided by a group of regional experts with reference to five chosen criteria.

20.2.1 Landfill Siting Strategies

Landfill site selection can generally be divided into two main steps: the identification of potential sites through preliminary screening, and the evaluation of their suitability based on environmental impact assessment (EIA), economic feasibility, and engineering design, and cost comparison. Many siting factors and criteria must be carefully organized and analyzed. The "not in my backyard" and "not in anyone's backyard" phenomena is creating a tremendous pressure on decision-makers involved in the selection of a landfill site. Issues related to availability of land, public acceptance, increasing amounts of waste generation complicate the process of selection for a suitable site for landfill (Chang et al., 2008). An initially chosen candidate site may be later abandoned because opposition arises due to previously neglected but important factors (Chang et al., 2008). Such delay increases costs and postpones the final decision on a landfill site (Chang et al., 2008). Of course, inappropriate siting of landfill may adversely affect the surrounding environment and other economic and sociocultural aspects.

The criteria used for preliminary screening primarily examine the proximity of potential sites with respect to geographic objects that may be affected by the landfill siting (e.g., groundwater wells) or that may affect landfill operations (e.g., areas with steep slopes) (Chang et al., 2008). The methodologies used are normally based on a composite suitability analysis using thematic map overlays (O'Leary et al., 1986) and their extensions to include statistical analysis (Anderson and Greenberg, 1982). With the development of GIS, the landfill siting process is increasingly based on more sophisticated spatial analysis and modeling (Chang et al., 2008).

Jensen and Christensen (1986) first demonstrated the use of a raster-based GIS and associated Boolean logic map algebra to identify potential waste sites based on the suitability of topography and proximity with respect to key geographic features, whereas Keir et al. (1993) discussed the use of both raster-based and vector-based GIS for the full-scale site selection process. Şener et al., (2006) started integrating GIS and multicriteria decision analysis to solve the landfill site selection problem and developed a ranking for potential landfill areas based on a variety of criteria. A GIS is used for a preliminary screening and is normally carried out by classifying an individual map, based on selected criteria, into exactly defined classes or by creating buffer zones around geographic features to be protected. All map layers are then intersected so that the resulting composite map contains two distinct areas—possible candidate sites and unsuitable sites (Chang et al., 2008). For example, if screening criteria involve the provision of a protective buffer around certain types of spatial objects, the area outside the intersected boundary is considered suitable and that inside is unsuitable. The two distinct classes separated by a sharp boundary reflect the representation of geo-referenced data based on a binary true or false Boolean logic in GIS. With the aid of this functionality, GIS have been used to facilitate the process of site selection for building sanitary landfills for about two decades (Siddiqui et al., 1996; Kao et al., 1997).

Advanced algorithms, however, may further help justify the uncertainty in siting new landfills (Chang et al., 2008). Several approaches were proposed for multicriteria decision-making (MCDM) (Chang et al., 2008). Relevant methods were developed and applied with more or less success depending on the specific problem. In the past, the analytic hierarchy process (AHP) introduced by Saaty (1980a), was one of the useful methodologies, and plays an important role in selecting alternatives (Fanti et al., 1998; Labib et al., 1998; Chan et al., 2000) based on a suite of derived weights associated with a predetermined set of criteria. Specifically, an AHP is an analytical tool enabling people to explicitly rank both "tangible" and "intangible" criteria against each other when selecting priorities. The process involves structuring a problem from a primary objective to secondary levels of criteria and alternatives in a layered diagram. Once the hierarchy has been established as a decision-making framework, a pairwise comparison matrix of each element within each level is constructed by a systematic sequence toward the final retrieval of the weights, which are a set of unknowns in the beginning.

The AHP also allows group decision-making, where group members can use their experience, values, and knowledge to collectively break down a problem into a hierarchy and solve it using the AHP steps. Participants can weigh each element against each other within each level, while each level is related to the levels above and below it, and the entire scheme is tied together mathematically. For evaluating numerous criteria, AHP has become one of the most widely used methods for the practical solution of MCDM problems (Cheng, 1997; Akash et al., 1999; Chan et al., 2000). Yet a major difficulty arises from the estimation of the required input data that address qualitative rather than quantitative observations and preferences. In early stages, the AHP is mainly used for nearly crisp decision analyses, since it does not take into account the uncertainty associated with the mapping of people's judgment to an evaluation scale (Chen, 1996; Hauser and Tadikamalla, 1996; Cheng, 1997). In order to overcome the shortcomings of the AHP, fuzzy set theory was used later on to aid AHP in making a more robust decision when selecting the best alternative (Chen, 1996; Hauser and Tadikamalla, 1996; Levary and Ke, 1998).

The practical applications of FMCDM reported in the literature have shown advantages in handling unquantifiable/qualitative criteria and obtained quite reliable results (Altrock and Krause, 1994; Teng and Tzeng, 1996; Baas and Kwakernaak, 1997; Mcintyre and Parfitt, 1998; Tang et al. 1999). Fuzzy linguistic models permit the translation of verbal expressions into numerical ones, thereby dealing quantitatively with imprecision in the expression of the importance of each criterion (Chang et al., 2008). For solving a siting issue, FMCDM utilizes linguistic variables and fuzzy numbers to aggregate the decision-makers' subjective assessment about criteria weightings and appropriateness of alternative candidate sites versus selection criteria to obtain the final scores—fuzzy appropriateness indices for quantitative comparative analysis using fuzzy sets.

Overall, Figure 20.6 illustrates the typical procedure applying a GIS practice for initial landfill siting. As mentioned above, such a landfill site selection process

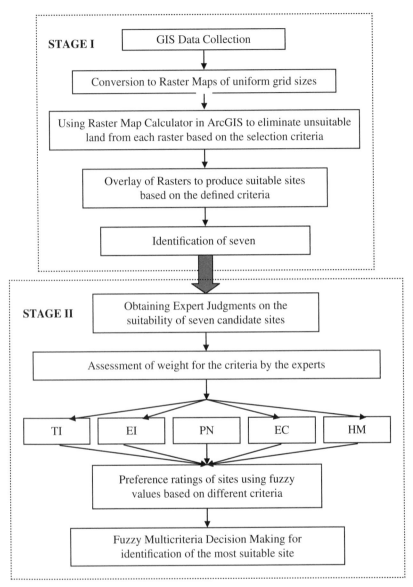

FIGURE 20.6 Flowchart of the methodology. *Source*: Chang et al. (2008). TI, transportation issues; EI, environmental and ecological impact; PN, public nuisance; EC, economical impact; HM, impact on historical markers

comprised of two stages with the first stage utilizing a GIS practice to identify a few candidate sites that were ranked later on in the second stage using FMCDM method. There are several different criteria involved in the selection of a landfill site in the first stage. A literature review may be conducted to identify the most important criteria. For example, a landfill site must be situated at a fair distance away from biophysical elements such as water, wetlands, critical habitats, and wells to reduce the risk of contamination from landfill (Dikshit et al., 2000). Different studies used different buffer distances from stream and rivers based on the size of the watershed, such as buffer of 0.8 km (Siddiqui et al., 1996), 180 m (Zeiss and Lefsrud, 1995) and 2–3 km (Lin and Kao, 1999). Considering the size of Harlingen city, a buffer distance of 1 km was used to protect the river system in this study. These suitability criteria are defined with the focus to minimize any potential health risks from direct or indirect contamination due to the proximity of a landfill site with respect to key geographic features. Thus, the first-stage analysis using a GIS is essential for the initial identification of a couple suitable landfill sites prior to undertaking further analyses or field investigations. Although the initial screening is based on criteria related to environmental and ecological factors involved in the site selection process, there are certain criteria, such as impact on historical markers, public comfort, and economic factors for which data are not always readily available, which cannot be included in the first stage. A second-stage analysis based on a handful of suitable sites obtained from the initial GIS screening was performed with the objective of including the opinions of domain experts in the region through a FMCDM approach. FMCDM was useful in addressing the issue of lack of availability of data for certain important criteria to incorporate human judgment into the selection process that can prove useful in solving political debates in the future. The second-stage analysis using FMCDM was then applied to rank the proposed candidate sites and to summarize the final selection.

20.2.2 The Study Site

The Lower Rio Grande Valley (LRGV or Valley), comprised of Cameron, Willacy, Hidalgo, and Starr counties, is located at the southernmost tip of Texas along the US–Mexico border (Figure 20.7) (Chang et al., 2008). The Office of Management

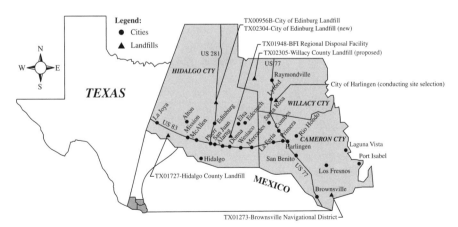

FIGURE 20.7   Map of South Texas indicating the facility location of the study area. *Source*: Chang et al. (2008)

and Budget at the White House ranks Metropolitan Statistical Areas (MSA) according to their population and economic growth. Cameron County, at the tip of Texas, comprises 3266 km^2 (1276 square miles) and includes the 28th MSA, Brownsville–Harlingen–San Benito in the early 2000s (Chang et al., 2008). Hidalgo County, the largest of the three LRGV counties, covers the western half of the region with an area of 3963 km^2 (1548 square miles). This county is mostly urbanized, containing the McAllen–Edinburg–Mission MSA, the fourth fastest growing areas in Texas. Both of the LRGV's MSA are experiencing a developmental change due to their strategic location and economic ties with the US–Mexico borderland (Chang et al., 2008). The NAFTA that was enacted in 1994 has increased trade throughout America. The Valley, with a total area of 9216 km^2 (3600 square miles), has emerged as a warehouse and transportation center between Central America and the US (TSHA, 2003). The increasing number of *maquiladoras*, or twin plants, having manufacturing industries both in the MSA of the Valley and in nearby Reynosa and Matamoros, Mexico, are positively influencing the economic development in the region. This has been a catalyst for further growth in other Valley cities located in between these two MSA (Chang et al., 2008). As a result, the population of the LRGV is growing at a tremendous pace and yard waste, food waste, and biosolid waste production is increasing over time (Chang et al., 2008). Figure 20.6 indicates the study area along with the waste disposal sites. The area's population has increased by 39.8% in the last 10 years due to the NAFTA's economic impact. It is expected to continue growing at an estimated rate of 4% per year in the coming years. The population is projected to be over 1.7 million people in 2022 (LRGVDC, 2002).

SWM is at the forefront of environmental concerns in the LRGV, South Texas. The complexity in SWM drives area decision-makers to look for innovative and forward-looking solutions to address various waste management options. The LRGV is facing the difficult reality of siting new landfills due to their large capital costs and local protests, like those seen as a result of Willacy County's intentions to site a new landfill. The hotly contested landfill permit process culminated in a hearing on August 1, 2005 with the decision pending whether to allow the process to continue amid community resistance (del Valle, 2005). Adding to the complexity of the issue, the realization by local residents of the economic value of their ecosystem from tourism dollars generated from bird watching enthusiasts means siting future landfills could become more contested.

Development of a landfill in Harlingen can possibly cause environmental impacts on the soil, groundwater, surface water, regional air quality, atmosphere, biodiversity, and landscape (Chang et al., 2008). Besides these environmental impacts, there are those related to the economy, employment, attainability and valuation of different areas, services, safety, and health (Chang et al., 2008). A landfill in this region can also affect many of the endangered and threatened species that occur at their northernmost limit in the LRGV. In light of such circumstances, there is acute necessity for a careful selection of a landfill site in order to preserve the ecological and environmental quality that is unique to the LRGV (Chang et al., 2008).

TABLE 20.1   GIS map layers used in the study

Data	Scale	Data source
Rivers	1:500,000	USEPA
Lakes	1:250,000	USEPA
Wetland	1:250,000	USEPA
Land use/land cover	1:250,000	USEPA
Roads	1:100,000	USEPA
Ground water wells		TWDB
Urban areas	1:24,000	USEPA
Soil map STATSGO[a]	1:250,000	USGS
Digital elevation model	1:250,000	EPA basins
County census data	1:2000,000	Tiger data

Source: Chang et al. (2008).
[a]STATSGO: The US General Soil Map was developed by the National Cooperative Soil Survey and supersedes the State Soil Geographic (STATSGO) data set.

20.2.3 Data Collection and Analysis

GIS data sets for land-use, rivers, wetlands, roads, demography, wildlife parks, airports, soil types, groundwater wells, and digital elevation models (DEM) were collected for the Cameron, Hidalgo, and Willacy counties from different sources, such as Texas Natural Resources Information Systems, Texas Department of Transportation, Texas Water Development Board (TWDB), US Geological Survey (USGS), and US Environmental Protection Agency (USEPA) (Chang et al., 2008). They are summarized as shown in Table 20.1. Geographical features required for the first-stage analysis were extracted with ArcGIS® software (Chang et al., 2008). For example, to obtain GIS data sets for a buffer zone, the land in the LRGV was classified by creating buffer zones around geographic features to be protected using values widely reported in the literature for landfill selection processes (Chang et al., 2008). The buffer maps were then converted into raster maps of uniform grid sizes using the raster calculator available as a spatial analyst tool in ArcGIS® (Chang et al., 2008). Then, these outputs were used to eliminate unsuitable land parcels based on the different criteria leading to identification of seven potential landfill sites in the first stage (Chang et al., 2008).

20.2.4 Application of GIS in Landfill Candidate Site Selection

The proximity of a landfill to a groundwater well is an important environmental criterion in the landfill site selection. Wells must therefore, be protected from the runoff and leaching of the landfill (Chang et al., 2008). For this study, groundwater well data were obtained from TWDB, and a buffer distance of 50 m from the wells was used to prevent contamination from landfill due to leaching of pollutants (Chang et al., 2008). Slope is also an important factor when siting a landfill since higher slopes increase runoff of pollutants from the landfill, and thereby increasing the contamination zone area (Lin and Kao, 1999). Lin and Kao's study (1999) suggested that a slope of less than 12% would be suitable for the prevention of contaminant runoff. Based on this study, regions with slope greater than 12% were defined as unsuitable for a landfill site (Chang et al., 2008). DEM data sets with a 30 m spatial resolution obtained from USEPA basins' data source were used to calculate the slope percentage area wide. In addition, a landfill should be situated at a significant distance away from urban residential areas due to public concerns, such as aesthetics, odor (Tagaris et al., 2003), noise, decrease in property value (Zeiss and Lefsrud, 1995), and health concerns, and to avoid contamination of freshwater aquifers through leaching (Nagar and Mizra, 2002). Urban buffers may range from 150 m (Lin and Kao, 1999) to 5 km (Zeiss and Lefsrud, 1995). A buffer distance of 3 km was chosen for the study area.

Economic considerations include finding the most cost-effective route for transporting wastes and locating the most suitable land for the candidate sites based on land values (Siddiqui et al., 1996). Developments on or too close to existing road and rail networks would hinder transportation and may have an impact on tourism in the region (Zeiss and Lefsrud, 1995). Baban and Flanagan (1998) used a 50-m buffer for roads, while Dikshit et al. (2000) used a 1-km buffer in his study. However, a study done by Lin and Kao (1999) stated that a 1-km buffer was too far from roadways, and would incur greater economic costs to the project over the long term since new roads would need to be constructed. Considering the huge cost of transportation, a 75-m buffer for roads was finally selected for this study (Chang et al., 2008).

The different constraint maps developed in this study include an environmental constraint map, a stream constraint map, a well constraint map, a slope constraint map, an urban constraint map, a water body constraint map, and a transportation constraint map (Chang et al., 2008). The constrained map layers are overlaid as shown in Figure 20.8 (Chang et al., 2008). Whereas the final constraint maps were developed with the candidate sites, as shown in Figure 20.9a, the seven candidate

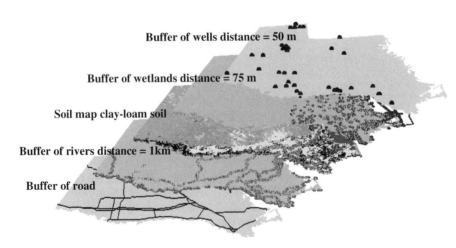

FIGURE 20.8 Overlay of different constrained maps. *Source*: Chang et al. (2008)

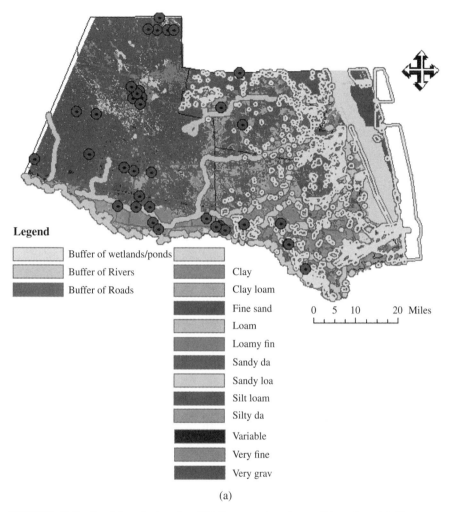

Legend

Buffer of wetlands/ponds	
Buffer of Rivers	Clay
Buffer of Roads	Clay loam
	Fine sand
	Loam
	Loamy fin
	Sandy da
	Sandy loa
	Silt loam
	Silty da
	Variable
	Very fine
	Very grav

0   5   10   20 Miles

(a)

FIGURE 20.9   Spatial analysis using GIS for screening the candidate sites. (a) Final map showing different constrained maps. (b) Map showing the candidate sites for landfill with different constraints. *Source*: Chang et al. (2008)

sites in a GIS were subjected to advanced assessment in the second-stage analysis (Figure 20.9b) (Chang et al., 2008). An ecological assessment study states that the region is divided into several ecoregions based on topographic, climatic, and edaphic factors, and plant community similarities (Chang et al., 2008). These ecoregions are characterized by high summer temperatures, high evaporation rates, and periodic droughts. The seven candidate sites are currently in use as agricultural crop land and have been cleared of native vegetation (Chang et al., 2008). Soils have a direct effect on the types of vegetation and ultimately the animal species that will occur in an

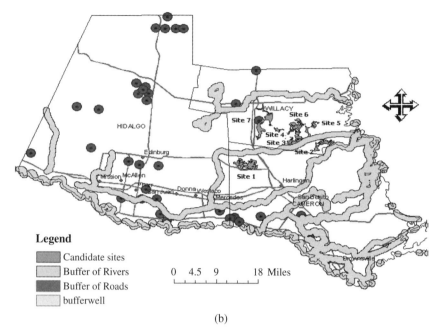

(b)

FIGURE 20.9   (*Continued*)

area (Chang et al., 2008). The US Department of Agriculture (USDA) (1977, 1982) rated the potential for soil types throughout Cameron and Willacy counties to provide elements of habitat for various species of wildlife. The soils are also rated on their potential to support wildlife species. Based on the criteria of the USDA (1977 and 1982), a rating of good indicates that the habitat is easily established and maintained. A rating of fair indicates that the habitat can be established with moderately intensive management. A poor rating indicates that the habitat type can be established, but with intensive and difficult management (Chang et al., 2008). A very poor rating indicates that creating or maintaining a habitat type is impractical or impossible (Chang et al., 2008). The terrain in South Texas is quite flat and all candidate sites are managed as agricultural land at present except site 7 (Chang et al., 2008). Future landfill to be built in this area will be designed as a plain-type rather than a gully-type landfill so that soil thickness was not an obvious issue on site selection (Chang et al., 2008). Thus, the proposed criteria did not include soil thickness and depth to bedrock, which may hamper the excavatability of the site in some cases (Chang et al., 2008).

The potential for elements of wildlife habitats to occur and their ratings are compared across the seven candidate sites in Table 20.2. The list in Table 20.3 compares the potential for types of wildlife species to occur at the seven candidate sites. Based on the ecological assessment study, all candidate sites are similar in soil type and similar in the potential for wildlife habitat and wildlife species. Because of the similarity between all sites, the potential effects on endangered and threatened species are the same for all candidate sites. Candidate sites 1–6 would result in the

TABLE 20.2 **Comparison of seven candidate sites using potential for elements of wildlife habitat**

	Site 1	Site 2	Site 3	Site 4	Site 5	Site 6	Site 7	
Soil type[a]	1	1	1	1	1	1	1	2
Coverage%	95	99	99	100	100	100	79	17
Grains	Good	Good	Good	Good	Good	Good	Good	Fair
Grasses	Good	Good	Good	Good	Good	Good	Good	Fair
Herbaceous plants	Fair	Fair	Fair	Fair	Fair	Fair	Fair	Fair
Shrubs	Good	Good	Good	Good	Good	Good	Good	Fair
Wetland plants	Poor	Poor	Poor	Poor	Poor	Poor	Poor	Good
Shallow wetlands	Very poor	Very poor	Very poor	Very poor	Very poor	Very poor	Very poor	Good

Source: Chang et al. (2008).
[a] 1, Raymondville clay loam; 2, Mercedes clay.

same ecological effects regardless of any action. Candidate site 7 is slightly different than the other six sites because an additional soil type occurs there. This soil type is better suited for the potential occurrence of wetlands and wetland wildlife species than the predominant soil types found at the other candidate sites. It is commonly known that wetlands are an important component of the ecosystem and are diminishing across the country. Therefore, candidate site 7 would be the most ecologically sensitive site because of the potential impact on wetland habitats.

20.2.5 Fuzzy Multicriteria Decision-Making

The second-stage analysis for landfill site selection requires a careful evaluation of the advantages and disadvantages of different candidate sites with respect to different predetermined criteria. Landfill siting is a complicated process that leads to different impacts in the area (Chang et al., 2008). Due to lack of crisp data, the evaluation of different alternatives against different criteria requires assessment using fuzzy numbers (Chang et al., 2008). FMCDM method is therefore chosen for ranking

TABLE 20.3 **Comparison of potential types of wildlife species occur in the seven candidate sites**

	Site 1	Site 2	Site 3	Site 4	Site 5	Site 6	Site 7	
Soil type[a]	1	1	1	1	1	1	1	2
Coverage%	95	99	99	100	100	100	79	17
Rangeland wildlife	Fair	Fair	Fair	Fair	Fair	Fair	Fair	Fair
Open land wildlife	Good	Good	Good	Good	Good	Good	Good	Fair
Wetland wildlife	Very poor	Very poor	Very poor	Very poor	Very poor	Very poor	Very poor	Good

Source: Chang et al. (2008).
Notes: [a] 1 – Raymondville clay loam, 2 – Mercedes clay.

different landfill sites for Harlingen city based on decisions given by a group of experts (Chang et al., 2008). Experts or planners were called on to participate in a questionnaire survey using linguistic variables or fuzzy numbers to give the preference ratings for each individual candidate sites (Chang et al., 2008).

A typical MCDM method was employed to solve the selection problem of distribution center location under a fuzzy environment (Chang and Chen, 1994), in which the ratings of each alternative and the weight of criterion are described by linguistic variables that can be expressed in triangular fuzzy numbers (TFN). The evaluation value of each facility site is also expressed in a TFN. By calculating the difference of evaluation values between each pair of candidate sites, a fuzzy preference relation matrix is constructed to represent the intensity of the preferences of one plant location over another (Chang et al., 2008). Then, a stepwise ranking procedure is proposed to determine the ranking order of all candidate locations (Chang et al., 2008). When conducting the inference, TFN are commonly used by the experts to describe vagueness and ambiguity in the real-world systems (Chang et al., 2008). Many methods, such as max, min, median, addition, multiplication, and mixed set operators, are available to aggregate TFN. Related literature can be found in (Kaufmann and Gupta, 1988; Paek et al., 1992).

In such decision analysis, the experts can employ an assumed weighting set W = {Very Poor, Poor, Fair, Good, and Very Good} to evaluate the appropriateness of the alternatives versus various criteria (Chang et al., 2008). The membership functions of the linguistic values in the weighting set W represented by the approximate reasoning of TFN, are shown in Figure 20.10 (Chang et al., 2008). If one does not agree with

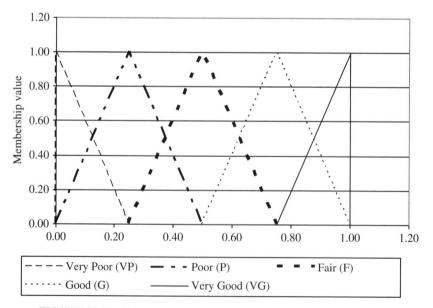

FIGURE 20.10 Fuzzy membership functions. *Source*: Chang et al. (2008)

the assumed preference rating system, one can give his own rating by using the TFN, showing perception of the linguistic variables, "importance" and "appropriateness" (Chang et al., 2008).

By the same token, the criteria that were selected in this study in Texas for evaluating the merits of the different landfill sites are: (1) environmental and ecological impact, (2) transportation issues, (3) impacts on historical markers, (4) economic impacts of the landfill, and (5) public nuisance. These criteria are described below. Transportation of waste loads from the hauling station to the landfill causes disruption of traffic within the city limits that cannot be clearly quantified in the decision-making process, thereby requiring fuzzy description of the criteria. Similarly, the possible impacts that can be caused by landfill on historical markers in terms of aesthetical impairment; bad odors, etc. are critical and vague and hence, require fuzzy concepts to represent the importance of historical makers on the landfill selection process. The criterion of economical impact reflects the possibility of decrease in land value in the neighborhood and also in the farming productivity of the region, thereby affecting the economy of the city directly, also vague in many other ways. Public nuisance is another vague but important factor that refers to the feeling of discomfort caused to the public due to the construction and operation of a landfill in the middle of a populous place.

The decision objective was to select the most appropriate landfill from seven different candidate sites. The different alternatives are defined as L = {L1, L2, L3, L4, L5, L6, L7} and the decision criteria are defined as C = {TI, EI, PN, EC, HM}, where TI = transportation issues, EI = environmental and ecological impact, PN = public nuisance, EC = economical impact, HM = historical markers. Linkage between different alternatives with different criteria is shown in Figure 20.11 (Chang et al., 2008). There is a committee of two experts (E1 and E2) who are called on for assessing the appropriateness of "m" alternatives ({L1, L2, L3, L4, L5, L6, L7}) under each of the "k" criteria ({TI, EI, PN, EC, HM}) as well as the importance weights for the criteria (Chang et al., 2008).

Let S_{itj} ($i = 1, 2 \ldots m; t = 1, 2 \ldots k; j = 1, 2 \ldots n$) be the rating assigned to alternative A_i by expert E_j under criterion C_t. Let W_{tj} be the weight given to C_t by decision maker E_j. The rating S_{itj} of n experts for each alternative versus each criterion is aggregated (Chang et al., 2008). Each pooled rating is further weighted by weight W_t according to the relative importance of the k criteria (Chang et al., 2008). Then the final score F_t, fuzzy appropriate index, of alternative A_i is obtained by aggregating S_{itj} and W_t, which is finally ranked to obtain the most suitable alternative (Chang and Chen, 1994). The experts give their own preference rating for the different alternatives and weights for different criteria by using the TFN. To present the rating, the two domain experts were invited to compare the seven alternatives (i.e., candidate sites) against the five criteria (Tables 20.4 and 20.5) (Chang et al., 2008). The weights assigned to the different criteria for decision-making are presented in Table 20.6 (Chang et al., 2008).

Following the method developed by Chang and Chen (1994), this case study utilizes a mean fuzzy operator to aggregate the expert assessment. Let $\oplus$ and $\otimes$ be the fuzzy addition and fuzzy multiplication operators, respectively. The aggregation

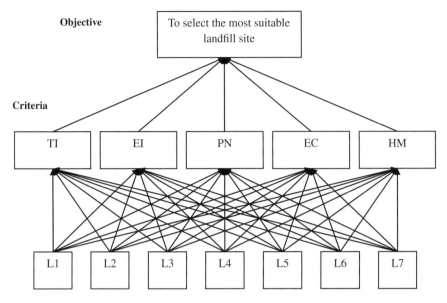

Objective

Criteria

Alternatives

L = Candidate sites (Alternatives) = {L1, L2, L3, L4, L5, L6, L7}

C = Criteria {TI, EI, PN, EC, HM}

EI = Environmental and ecological impact, TI = Transportation issues, PN = Public nuisance, EC = Economical Impact, HM = Impact on historical markers.

FIGURE 20.11 Description of decision-making process. *Source*: Chang et al. (2008). L, candidate sites (alternatives) = {L1, L2, L3, L4, L5, L6, L7}; C, criteria {TI, EI, PN, EC, HM}; EI, environmental and ecological impact; TI, transportation issues; PN, public nuisance; EC, economical impact; HM, impact on historical markers

of the different ratings is given by

$$S_{iti} = (S_{it1} \oplus S_{it2} \oplus \ldots \oplus S_{itn}) \otimes (1/n) \tag{20.39}$$

$$W_t = (W_{t1} \oplus W_{t2} \oplus \ldots \oplus W_{tn}) \otimes (1/n) \tag{20.40}$$

where S_{itj} is the average fuzzy appropriateness index rating of alternative A_i under criterion C_j, and W_t is the average importance weight of criterion C_j. Thus, the fuzzy appropriateness index F_i of the ith alternative can be obtained by aggregating S_{itj} and W_t, expressed as

$$F_i = [(S_{i1} \otimes W_1) \oplus (S_{i2} \otimes W_2) \oplus \ldots \oplus (S_{itk} \otimes W_k] \otimes (1/k) \tag{20.41}$$

TABLE 20.4 Evaluation of different alternative against all criteria by expert E1

				Alternatives			
Criteria	Site 1	Site 2	Site 3	Site 4	Site 5	Site 6	Site 7
TI	(0.6, 0.7, 0.8)	(0.6, 0.7, 0.8)	(0.6, 0.7, 0.8)	(0.6, 0.7, 0.8)	(0.6, 0.7, 0.8)	(0.6, 0.7, 0.8)	(0.6, 0.7, 0.8)
PN	(0.9, 0.95, 1.0)	(0.4, 0.5, 0.6)	(0.4, 0.5, 0.6)	(0.4, 0.5, 0.6)	(0.4, 0.5, 0.6)	(0.4, 0.5, 0.6)	(0.4, 0.5, 0.6)
EI	(0.7, 0.85, 0.9)	(0.6, 0.65, 0.75)	(0.4, 0.45, 0.55)	(0.50, 0.55, 0.60)	(0.4, 0.45, 0.55)	(0.5, 0.55, 0.65)	(0.5, 0.55, 0.65)
EC	(0.65, 0.75, 0.8)	(0.40, 0.45, 0.5)	(0.5, 0.55, 0.65)	(0.55, 0.6, 0.65)	(0.55, 0.6, 0.65)	(0.50, 0.60, 0.75)	(0.50, 0.60, 0.75)
HM	(0.50, 0.60, 0.75)	(0.55, 0.6, 0.65)	(0.55, 0.6, 0.65)	(0.55, 0.6, 0.65)	(0.55, 0.6, 0.65)	(0.55, 0.6, 0.65)	(0.55, 0.6, 0.65)

Source: Chang et al. (2008).

TABLE 20.5 Evaluation of different alternative against all criteria by expert E2

| | | | | Alternatives | | | |
Criteria	Site 1	Site 2	Site 3	Site 4	Site 5	Site 6	Site 7
TI	(0.55, 0.6, 0.70)	(0.35, 0.4, 0.45)	(0.4, 0.45, 0.5)	(0.45, 0.5, 0.55)	(0.4, 0.45, 0.5)	(0.5, 0.55, 0.6)	(0.3, 0.35, 0.4)
PN	(0.4, 0.45, 0.5)	(0.5, 0.55, 0.6)	(0.4, 0.45, 0.5)	(0.35, 0.4, 0.45)	(0.5, 0.55, 0.60)	(0.5, 0.55, 0.6)	(0.35, 0.4, 0.45)
EI	(0.75, 0.8, 0.85)	(0.5, 0.55, 0.60)	(0.55, 0.60, 0.65)	(0.30, 0.35, 0.40)	(0.50, 0.55, 0.60)	(0.30, 0.35, 0.40)	(0.55, 0.60, 0.65)
EC	(0.7, 0.75, 0.8)	(0.4, 0.45, 0.5)	(0.6, 0.65, 0.7)	(0.5, 0.55, 0.60)	(0.6, 0.65, 0.7)	(0.5, 0.55, 0.60)	(0.5, 0.55, 0.6)
HM	(0.45, 0.5, 0.55)	(0.45, 0.5, 0.55)	(0.45, 0.5, 0.55)	(0.45, 0.5, 0.55)	(0.45, 0.5, 0.55)	(0.45, 0.5, 0.55)	(0.45, 0.5, 0.55)

Source: Chang et al. (2008).

TABLE 20.6   Weights of different criteria by two experts

	Experts	
Criteria	E1	E2
TI	(0.8, 0.9, 0.95)	(0.8, 0.9, 0.95)
PN	(0.9, 0.9, 5, 1)	(0.75, 0.8, 0.9)
EI	(0.7, 0.75, 0.8)	(0.85, 0.9, 0.99)
EC	(0.8, 0.9, 0.95)	(0.7, 0.75, 0.8)
HM	(0.45, 0.55, 0.6)	(0.45, 0.55, 0.6)

Source: Chang et al. (2008).

Let $S_{itj} = (q_{itj}, o_{itj}, p_{itj})$ and $W_{itj} = (c_{tj}, a_{tj}, b_{tj})$ be TFN. Then F_i can be expressed as

$$F_i = (Y_i, Q_i, Z_i) \tag{20.42}$$

where

$$Y_i = \sum_{i=1}^{k} (q_{it} c_t / k) \qquad Q_i = \sum_{i=1}^{k} (o_{it} a_t / k) \qquad Z_i = \sum_{i=1}^{k} (p_{it} b_t / k)$$

$$q_{it} = \sum_{j=1}^{n} (q_{itj} / n) \qquad o_{it} = \sum_{j=1}^{n} (o_{itj} / n) \qquad p_{it} = \sum_{j=1}^{n} (p_{itj} / n)$$

$$c_t = \sum_{j=1}^{n} (c_{tj} / n) \qquad a_t = \sum_{j=1}^{n} (a_{tj} / n) \qquad b_t = \sum_{j=1}^{n} (b_{tj} / n)$$

for $i = 1, 2, \ldots, m; t = 1, 2, \ldots, k; j = 1, 2, \ldots, n$.

Based on the aggregation functions, the fuzzy appropriate indices are obtained and presented in Table 20.7 (Chang et al., 2008). This information may help justify the final ranking among these seven candidate sites. Therefore, the ranking values of

TABLE 20.7   Fuzzy appropriateness indices for the seven alternatives

Alternatives	Fuzzy appropriateness index
Site 1	(0.45563, 0.55988, 0.65963)
Site 2	(0.3405, 0.42463, 0.51308)
Site 3	(0.34713, 0.43275, 0.52553)
Site 4	(0.53163, 0.41638, 0.49888)
Site 5	(0.35525, 0.4415, 0.53055)
Site 6	(0.34425, 0.434, 0.5311)
Site 7	(0.33525, 0.4235, 0.52023)

Source: Chang et al. (2008).

fuzzy appropriate indices for the alternatives were computed based on the method developed in Chang and Chen (1994).

Let F_i $(i = 1, 2, \ldots, m)$ be the fuzzy appropriate indices of m alternatives. The maximizing set $M = \{[x, f_m(x)] | x \in R\}$, with

$$f_m(x) = \begin{cases} (x - x_1)/(x_2 - x_1), & x_1 \le x < x_2 \\ 0, & \text{otherwise} \end{cases} \qquad (20.43)$$

and minimizing set $G = \{[x, f_g(x)] | x \in R\}$ with

$$f_g(x) = \begin{cases} (x - x_2)/(x_1 - x_2), & x_1 \le x < x_2 \\ 0, & \text{otherwise} \end{cases} \qquad (20.44)$$

where $x_1 = \inf S$, $x_2 = \sup S$, $S = \cup_{t=1,m} F_i$, $F_i = [x | f_{F_i}(x) > 0]$, for $i = 1, 2, \ldots, m$.

Defining the optimistic utility $U_M(F_i)$ and pessimistic utility $U_G(F_i)$ for each appropriate index F_i as

$$U_M(F_i) = \sup(f_{F_i}(x) \wedge f_M(x)) \qquad (20.45)$$
$$U_G(F_i) = 1 - \sup(f_{F_i}(x) \wedge f_G(x)) \qquad (20.46)$$

for $i = 1, 2 \ldots, m$ where $\wedge$ means min. Ranking value $U_T(Fi)$ of fuzzy appropriate indices is defined as

$$U_T(F_i) = \alpha U_M(F_M) + (1 - \alpha)U_G(F_i), \quad 0 \le \alpha \le 1. \qquad (20.47)$$

The value α is an index of rating attitude (Chang et al., 2008). It reflects expert risk-bearing attitude. Let $B = (c, a, b)$ be a normal triangular fuzzy number. The index of rating attitude of an individual expert is defined as $Y = (a - c)/(b - c)$ (Chang and Chen, 1994). If $Y > 0.5$, it implies that the expert is a risk lover (Chang et al., 2008). If $Y < 0.5$, the expert is a risk averter (Chang et al., 2008). If $Y = 0.5$, the attitude of expert is neutral to the risk (Chang et al., 2008). Thus, the total index of rating attitude, R, with the evaluation data of individuals can be shown as

$$R = \left\{ \sum_{i=1}^{k} \sum_{j=1}^{n} (a_{tj} - c_{tj}) \middle/ (b_{tj} - c_{tj}) + \sum_{i=1}^{m} \sum_{t=1}^{k} \sum_{j=1}^{n} (o_{itj} - q_{itj}) \middle/ p_{itj} - q_{itj} \right\} \middle/$$
$$(k \cdot n + m \cdot k \cdot n). \qquad (20.48)$$

From Equations (20.42), (20.46) and (20.48), the ranking values $Ut(Fi)$ can be approximately expressed as

$$U_t(F_i) \cong R[(Z_i - x_1)/(x_2 - x_1 - Q_i + Z_i)]$$
$$+ (1 - R)[1 - (x_2 - Y_i)/(x_2 - x_1 + Q_i + Y_i)] \qquad (20.49)$$

TABLE 20.8 **Ranking values of the different alternatives**

Alternatives	Ranking values
Site 1	0.786689
Site 4	0.580556
Site 5	0.371734
Site 3	0.342253
Site 6	0.340457
Site 7	0.310792
Site 2	0.266668

Source: Chang et al. (2008).

The ranking values of the fuzzy appropriateness indices for alternatives are presented in Table 20.8. Site 1 exhibits the highest potential in this site selection process. As the SDSS may strengthen the generation and evaluation of alternatives by providing an insight of the problem among the varied objectives and granting essential support to the process of decision-making under uncertainty (Malczewki, 1999; Sharifi and van Herwijnen, 2003), it is concluded that "site 1" located near highway 77 closer to Cameron–Willacy boundary is the most suitable site for landfill based on an integrated GIS and FMCDM analysis in this SDSS.

20.3 FAIR FUND REDISTRIBUTION AND ENVIRONMENTAL JUSTICE WITH GIS-BASED FUZZY AHP METHOD

SWM facing a multifaceted crisis in many of the world's largest urban areas as economic development continuously draws populations to cities. This global change has led to ever increasing quantities of municipal solid waste (MSW) while space for landfill disposal decreases. Some of the municipal managers are looking for the development of municipal incinerators around the periphery of their cities as a first alternative as landfill space shrinks over time in many countries. This is especially true in those countries with relatively smaller land resources available, such as Japan and some of the Organization of Economic Cooperation and Development countries in Europe, such as Germany. Siting and construction of a municipal incinerator requires the acquisition of modern waste-to-energy (WTE) technology and good day-to-day operations in order to minimize possible environmental impacts based on an approved EIA or environmental impact statement (EIS). Burning MSW can generate energy and reduce the waste volume, which delivers benefits to society through resource conservation and energy recovery. But, the level of attention to environmental issues/problems during the project planning and design stage is low and environmental problems are often identified at later stage of the project life cycle, such as operational stage. Consequently, public reluctance with regard to accepting the incinerators as typical utilities often results in an intensive debate concerning how much quality of life is possibly lost for those residents living in the vicinity of an incinerator because of long-term exposure to environmental impacts during operation.

20.3.1   Fair Fund Distribution and Environmental Justice

Developers of incineration project armed with rosy financial forecasts based on the anticipated energy recovery profits and waste disposal fees can be found in all corners of the globe. Yet the inherent environmental impacts and hence welfare losses in local communities could lead to an operational burden for such day-to-day operations. Conflicts between local residents and the owners/operators of municipal incinerators are reported very often as headline news in some developing countries (Chang and Chang, 2000). These problems may include but are not limited to air pollution, traffic congestion, and noise impacts due to waste shipping and operation, wastewater treatment and disposal, ash disposal, etc.

"Fair Funds" that contain money recovered from a special avenue distribute the collected money to defrauded investors in business (Box 20.3). In this context, a fair fund was raised from energy recovery, material recycling incomes, and tipping fees of a municipal incinerator for compensation within the neighboring communities around a municipal incinerator. A disturbance occurring during the decision-making process for fair fund distribution could result from the fact that environmental impacts or risks caused by an incineration project might decay heterogeneously with the distance away from the facility; and marginal benefits and/or costs may vary spatially and add complexity in decision-making for compensation. To endorse a real "economically optimum" fee collection or fair fund distribution system, or the use of policy instruments in environmental management regime raises some of the basic ideas of allocation theory involving relocation of waste disposal fees as an equivalent step of the redistribution of a fair fund. In other words, the provision of waste treatment and disposal utilities for a specific region may encounter higher environmental impacts. A fair fund therefore, should be redirected to balance the external costs to communities in that region located in the vicinity of an incineration site. A flexible combination of these policy instruments can further be initiated and employed to set up integrated remedies for environmental externalities via a possible institutional arrangement. Although work on the use of policy instruments is underway to remedy the short- and long-term environmental externalities, little has been done to develop decision-making processes that tie together these issues and relationships into the context of environmental management at the societal level. More stringent regulatory requirements plus policy concerns are fueling the need for innovative decision analysis to cost-effectively remediate stressed communities confronting welfare losses due to regional pollution prevention and control actions. This brings up a new research need in the nexus of environmental management, environmental economics, and environmental policy.

BOX 20.3   THE DISTRIBUTION OF A FAIR FUND

A Fair Fund is a fund established by the US Securities and Exchange Commission (SEC) to distribute disgorgements (returns of wrongful profits) and penalties (fines) to defrauded investors. Fair Funds were established by the Sarbanes–Oxley Act of 2002. The Sarbanes–Oxley Act of 2002 (Pub.L. 107-204, 116 Stat.

745, enacted July 30, 2002), also known as the "Public Company Accounting Reform and Investor Protection Act" (in the Senate) and "Corporate and Auditing Accountability and Responsibility Act" (in the House) and more commonly called Sarbanes–Oxley, Sarbox or SOX, is a US federal law that set new or enhanced standards for all US public company boards, management, and public accounting firms (Jacobs, 2002). It is named after sponsors US Senator Paul Sarbanes and US Representative Michael G. Oxley.

The challenge is how to build upon the rules or policies which are the most appropriate or at least the most acceptable ones for stakeholders. Given that stakeholder participation cannot guarantee a smooth allocation, there is a need for developing a lucid procedure designed for a screening level assessment providing scientific clues to support the fair fund distribution. The proposed method in this case study aims to provide high level advice on science-based approaches that are self-supporting, environmentally responsible, and socially acceptable. It empowers the decision-makers to set up corresponding rules via a technocratic process.

20.3.2 The Strategies of Fair Fund Distribution

Current WTE practices are oftentimes not sustainable due to the associated environmental impacts from waste incineration in urbanized regions; these impacts have been a long-standing concern in local communities. Public reluctance to accept incinerators as typical utilities often results in an intensive debate concerning how much welfare is lost for those residents living in the vicinity of those incinerators. As the measures of welfare change with environmental quality constraints nearby these incinerators remain critical, new arguments related to how to allocate the fair fund among affected communities became a focal point in environmental management. Given the fact that most county fair fund rules allow a great deal of flexibility for redistribution, little is known about what type of methodology may be a good fit to determine the distribution of such a fair fund under uncertainty.

This section purports to demonstrate a system-based strategy that helps any fair fund distribution made with respect to residents' possible claim for fair damages due to the installation of a new incinerator. A case study using integrated GIS and a fuzzy analytic hierarchy process (FAHP) for finding out the most appropriate distribution strategy between two neighboring towns in Taipei County, Taiwan demonstrates the application potential. This case study in Taipei County was based on a methodology facilitating the essential decision-making process under uncertainty while fairness with respect to reciprocity and social exchange in a waste management service district was taken into account in a self-management process within those affected communities. Factors considered resembled a retrospective EIA after the plant has been built and included major environmental concerns in the context of an MCDM process under uncertainty. Participants determining the use of a fair fund entered a highly democratic procedure where all stakeholders involved eventually expressed a high level of satisfaction with the proposed method to facilitate the final decision-making process. Rules for fair fund redistribution developed herein may allow a great deal of flexibility for municipal managers to pursue long-term operation of incineration projects without irrational blockades by local communities. Such a decision-making process ensures that plans for the redistribution of a fair fund are carefully thought out and justified with a multifaceted nature that covers political, socioeconomic, technical, environmental, public health, and industrial aspects.

20.3.3 The Study Area

The Taipei County Government governs the largest administrative areas in Taiwan with three modern large-scale incinerators commissioned in the mid- and late-1990s. Before these three incinerators were in place, MSW had accumulated on the street without collection for months because landfill space had run out. The area was poised at the brink of social chaos in the mid- and late-1990s over the waste disposal issues. The incinerator of concern in this study is located in a valley that is close to the administrative boundary between Shu-Lin (Township A hereafter) and In-Kou (Township B hereafter) in Taipei County, Northern Taiwan. The service district of the Shu-Lin incinerator, operated by a private subcontractor since 1995, covers seven townships located in Southwest Taipei County. With an area of 4.5 hectares surrounded by small mountains, the Shu-Lin WTE plant is equipped with a modern mass burn water wall furnace that permits the routine processing of 1350 tonnes of waste per day using three treatment trains. The waste is fed from a storage pit into the furnace where combustion takes place on a Martin-type movable mechanical grate system. Within each treatment train, the flue gases generated in the furnace pass through a first furnace and are cooled down at the outlet of secondary furnace by a superheater installation, a boiler tube bank, an economizer, and a heat exchanger installed in the thermal cycle for the preheating of auxiliary combustion air. The flue gases are eventually led through a well-deigned air pollution control system, consisting of a conventional cyclone with dry sorbent injection followed by a fabric filter to absorb heavy metals, such as mercury, and toxic gases, such as dioxins/furans. A well-designed monofill including a leachate treatment plant was constructed permanently in parallel with the Shu-Lin WTE for the disposal of wastewater sludge and incineration ash.

20.3.4 The Integrative Approach for EIA and FAHP

The equity concerns the neighborhoods near an incineration site that bear a disproportionate share of environmental impacts from waste collection, traffic congestion, air pollution, and transportation/operation noise from incineration activities. The corresponding social welfare loss must be recovered via a redistribution of the disposal fees according to the reallocation of a particular fair fund. Therefore, in the mid-1990s, part of the disposal fees and energy recovery income from selling electricity at this WTE facility were raised as a fair fund for two towns, A and B. Since 1996, divergences in views concerning monetary compensation between neighboring communities temporarily suspended fair fund distribution. The fund, however, accumulated quickly, with over US$6.25 million in 1999 (US$ are United States dollars). The fund distribution was continuously obstructed by the arguments between these two towns (A and B) adjacent to the Shu-Lin incinerator. If a fair fund distribution fails, social turmoil against official actions for pollution prevention and control would inevitably impact the already fragile environment and disturb the social stability in that area. Even as the measures of welfare change with respect to environmental quality constraints were carried out carefully, the assessment procedures of fair fund distribution remain critical to society. New arguments related to assessment procedures shape the willingness-to-accept policy instruments—in this case, the fair fund.

Before the 1990s, no specific environmental management methods to support the important decisions in fair fund distribution for SWM had been developed and applied in Taiwan or elsewhere. After the 1990s, a worldwide change in environmental policy toward integrated pollution prevention, taking all environmental factors (air, water, land) into account, deepens the difficulties in environmental decision-making processes. While promoting good and universal norms of democracy and good governance in modern society, the civic transition in modern Taiwan from martial law to a more open environment in a post-martial-law era also had to be reflected in a democratization of the fair fund decision-making process. Strategies for solving these issues motivated the Taipei County Government to initialize a technocratic process— a holistic approach with the strong involvement of all region-wide stakeholders to deal with an assessment with a multifaceted nature that covers political, socioeconomic, technical, environmental, ecological, public health, and industrial aspects. It aims to harmonize discrepancies without causing unexpected chaos and anarchy due to environmental damage.

Environmental Impact Assessment   Integrated environmental vulnerability assessment in a structured AHP envelops objectives, criteria, and attribute layers. Since the choice of attributes and criteria is crucial to the results, prudence suggests the inclusion of the representative EIA metrics that can be feasibly assessed. It is essential to listing the environmental factors or impacts and associated attributes (Figure 20.12). Six major criteria including air pollution, noise impact, traffic congestion, solid waste impact (incineration ash and sludge), wastewater disposal, and socioeconomic impacts were taken into account. Two attributes associated with each criterion were listed for detailed elaboration. The research team drew up this list of criteria and attributes in consensus with the Taipei County Government (Chang et al., 2009). This assessment framework (Figure 20.12) also arrived at acceptance with the conflicting parties in local communities during an *ad hoc* committee meeting.

According to the structured hierarchy design as shown in Figure 20.12, the integrated environmental vulnerability assessment in the algorithm covers objectives, criteria, and attributes layers. A hierarchy descends from an objective layer, down to criteria layer, and finally to attribute layer from which the choice is to be made. The six criteria seen in Figure 20.12 will bridge between the objective and the attribute layers. By determining the priority weight of all criteria, the system structure and relative impact of components on the entire system are sketched. Similar applications can be found in the literature (Lai, 1995; Huizingh and Vrolijk, 1997). Specifically (Figure 20.12), solid waste impact was characterized based on both wastewater sludge

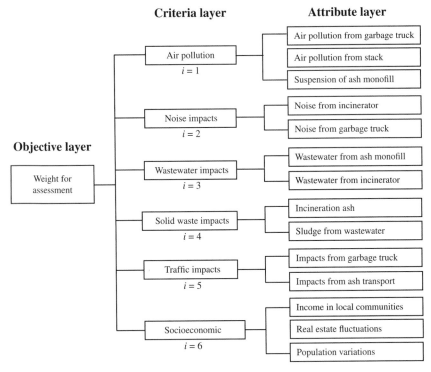

FIGURE 20.12   The structured AHP designed for this analysis. *Source*: Chang et al. (2009)

and incineration ash in a more qualitative than quantitative way. In fact, the wastewater treatment plant with the WTE facility was designed for biological treatment processes with a "zero discharge" design strategy as confirmed by the incineration plant. This implies all the treated wastewater effluents are to be recycled and reused as environmental resources to achieve the managerial goals of the plant. Nevertheless, the leachate treatment plant in the monofill must discharge the treated effluent into a local drainage system. Databases for routine shipping schedules and frequencies were investigated and aggregated for noise analysis. Impacts from traffic noise due to waste shipping were not evaluated within the criterion of "traffic impacts" but in a separate criterion of "noise impacts." Noise impact due to the traffic flow of truck fleets was estimated using a nonlinear function in terms of the distance from the noise sources (Chang et al., 1996). The background noise level was taken into account in order to identify the net noise impact due to waste incineration. The noise impact on neighboring subdivisions has to do with their geographical location. Noise control regulations and policy enable us to evaluate the gap between the existing and the allowable impacts through legislation. The attribute of "income to local communities" is defined as the income from job creation due to this waste incineration project. The attribute "population variations" is defined as population changes due to the implementation of this waste incineration project.

Since the incinerator is located in a "nonattainment area" (Box 20.4), the air pollution issue has long been regarded as a critical environmental impact on local communities starting in the late 1990s. Emissions from the stack (stationary source) and garbage trucks (mobile source) are identified on a yearly basis. Values for these quantities (i.e., ground level concentrations) are more often estimated through the use of atmospheric dispersion models in comparison to the ambient air quality standards (Yambert et al., 1991). The industrial source complex model (ISC) has been specifically developed to simulate air pollution from an industrial plant, taking into account accurately the effects from high stacks on the behavior of the pollutant plume. The ISC model may be applied in urban or rural environments with moderately complex terrain. When determining exposure to pollutants emitted from the incineration facility on a yearly basis, there has been a shift in the usage of the industrial source complex short-term dispersion model (ISCST3) since the 1990s, such that it requires minimal computing power and site specific information (USEPA, 1979). The model is based on a Gaussian plume formulation and contains only limited provisions for dealing with complicated atmospheric processes such as pollutant deposition and flows across complex terrain (i.e., terrain exceeding the height of the stack). In this case study, the high rise stack is actually higher than the mountains around the valley. While terrain elevation is the chief attribute of complex terrain that affects predictions made by ISCST3, such predictions are also influenced by a year-round meteorological pattern. The meteorological database in 1998 used in this analysis was collected by the plant operator. In general, except for July and August, an east-north-eastern wind predominates. With the aid of digital terrain modeling techniques, a continuous emission monitoring system, and a year-round meteorological database, ISCST3 was employed to estimate the air pollution impacts between these two towns with regard to eight pollutants, consisting of hydrogen chloride (HCl), nitrogen oxides (NO_X), sulfur dioxide (SO_2), carbon oxide (CO), particulate matter (PM_{10}), lead (Pb), mercury (Hg), and polychlorinated dioxins (PCDD)/dibenzofuran (PCDF). Selected receptor locations analyzed by ISCST3 in these two towns show the discrepancies in terms of air pollution impacts. Receptor distances for these model runs varied from 100 to 8000 m while elevations ranging from that of the stack base to 300 m were considered. Integrating the GIS grid-based data with ground level concentrations across the eight pollutants made the final assessment possible. A regular grid with 100 m^2 in each cell in our GIS program Arc/View® also helped organize all data sources related to each type of environmental impact.

BOX 20.4   ATTAINMENT VERSUS NONATTAINMENT AREA IN AIR POLLUTION CONTROL

Based upon levels of air pollutants, geographic areas are classified by USEPA (USEPA, 2013) as attainment or nonattainment areas.

- A geographic area that meets or has pollutant levels below the National Ambient Air Quality Seastrands (NAAQS) is called an **attainment area**.

> • An area with persistent air quality problems is designated a **nonattainment area**. This means that the area has violated federal health-based standards for outdoor air pollution.
>
> Each nonattainment area is declared for a specific pollutant. Nonattainment areas for different pollutants may overlap each other or share common boundaries.

Integrated EIA and FAHP Algorithm   The MCDM was introduced as a promising tool for decision analysis in the early 1970s. The key philosophical departure for MCDM lies in the representation of several conflicting criteria (Stewart, 1992). Since then, the number of theoretical contributions has continued to grow at a steady rate. A key extension for MCDM trying to bridge the gap between decision science and environmental management was made in this study by developing an integrated EIA and FAHP algorithm in support of such a decision analysis for fair fund redistribution. Such integration enables us to overcome the intricacy of environmental issues which may disrupt practical implementation. Regarding to the case of waste incineration, however, the impact assessment of cross-media aspects, considering transmedia problems from one environmental media to another, remains a difficult and challenging task. Risk assessment in terms of public health and ecological criteria relating to waste incineration is not still at a mature stage of development in the scientific community. The collection of accurate data for all criteria based on the EIA/EIS techniques poses further difficulties due to heterogeneous databases and measurements from different agencies at different times. In addition, the steps for impact assessment in an EIA approach may differ from case to case due to the variations of domain knowledge (Geldermann et al., 2000). For example, estimation of the potential human health risks associated with waste incineration is by all means a multidisciplinary task. The EIA may not be able to cover such human health risks based on the impact assessment underpinnings unless the assessment procedures engage with epidemiological data. This complexity necessitates AHP evaluation techniques.

Many of the generally applicable AHP evaluation techniques as related to environmental assessment are based on the underlying EIA/EIS theory (Weiss and Rao, 1987). The AHP, one multiattribute decision-making (MADM) method, has found widespread application in decision-making problems involving multiple criteria in systems with many levels designed for priority identification in complex systems. Earlier attempts in theoretical AHP development can be found in Belton and Gear (1983); Dyer (1990); Saaty (1986, 1990, 1994); Harker and Vargas (1987, 1990); Basak and Saaty (1993); Ramanathan and Ganesh (1995); Chang (1996); Lipovetsky (1996); Tung and Tang (1998); Ramanathan (2001); and Chang et al. (2008). Such kinds of evaluation renders a complex system into a structured hierarchy from the lowest level (sets of alternatives), through the intermediate levels (subcriteria and criteria), to the highest level (general objective). Using such a holistic AHP may help determine the priorities for alternatives independently for every criterion at each level. The weight (or priority) of each criterion can be defined by the same AHP procedure. Afterward, summing the priorities of every alternative with the weights of every criterion creates a composite priority for the highest level yielding the overall priorities of alternatives on each successively higher level as a linear combination of the subpriorities derived for the previous level. Such summing through the whole hierarchical structure produces a synthesized judgment for all alternatives found under the stated prerequisite goal (Saaty, 1980b).

Within the integrated EIA and AHP algorithm developed in this study, environmental assessments highlight the significant environmental ramifications of a waste incineration project. However, oftentimes there is considerable uncertainty surrounding the factors being evaluated (Tran et al., 2002; Noh and Lee, 2003). One of the critical problems of AHP is the means to deal with the uncertainty resulting from measurement errors, modeling bias, information availability, and imprecision or ambiguity embedded in the paired comparison procedure. Different ways of admitting approximate preferences have attracted much attention in the decision analysis literature. Since some comparisons do not hold with the two-valued logic (true/false), the concept of fuzzy sets is a way to systematically address unsharp figures gaining a better realization of the reality. Improved assessment of approximate reasoning and fuzzy logic can be considered as a bridge between the EIA and MCDM approaches. Theoretical work on using fuzzy set theory to modify the AHP has been done to extend methods in this regard so that decision-makers can express approximate preference statements through flexible judgments using fuzzy numbers (Zahedi, 1986; Webber et al., 1996). These fuzzy numbers describe human goals that matter in the case of decision processes, and therefore bridge the wide gap existing between theory and practice in decision analysis.

The FAHP analysis with slightly different forms was widely used to deal with a variety of decision-making problems involving multicriteria evaluation/selection of alternatives (Zahedi, 1986; Webber et al., 1996). The practical applications have even shown advantages in handling unquantifiable/qualitative criteria (Altrock and Krause, 1994; Teng and Tzeng, 1996; Baas and Kwakernaak, 1997; McIntyre and Parfitt, 1998; Chiou and Tzeng, 2001; Chen et al., 2008). Given the fact that a time constraint was a crucial factor in Taipei County Government at the time, the adoption of MCDM techniques in conjunction with fuzzy set theory seemed to be a more realistic choice to help enlighten the decision-making process via a retrospective EIA practice for these two townships within the time constraint. Hence, there are calculations for air pollution impact which may be factored into the FAHP that take population density into account for both townships, thus reflecting part of the public health concerns. Such considerations, to some extent, may help assess human health damage. Although these impact categories are still disputable, insights for the most environmentally important aspects of production and consequently for environmental improvement may still be anticipated. Finally, such a ranking process may be achieved by using FAHP techniques to deal with the several, possibly contradictory, preferences with respect to EIA outputs (Geldermann et al., 2000).

To achieve such an integrated EIA and AHP algorithm, the extent of environmental impacts largely relies on the spatial distribution of the effects either of proposed action or of the affected receptors (Antunes et al., 2001). The first efforts involved using databases available from all sources and environmental models suitable to create an

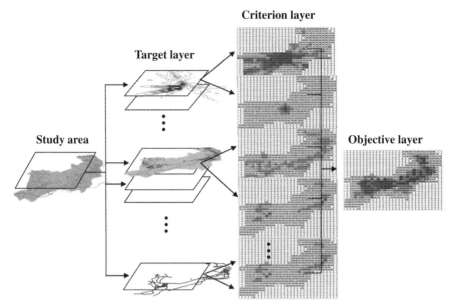

FIGURE 20.13   Process to build the spatial decision support system by raster-based GIS. *Source*: Chang et al. (2009)

array of environmental quality metrics, and building all aspects on a GIS platform. The use of a GIS platform renders a complex system into a simplified structured hierarchy that focuses on six major criteria in the criteria layer. With the aid of GIS, the integrated EIA and FAHP algorithm developed in this study may thus help analyze these multifaceted impacts in a raster-based GIS system (see Figure 20.13). The GIS software Arc/View® was used for proceeding spatial analysis data and summarizing environmental impacts for those EIA factors in the way to identify the relative importance via FAHP. Using this spatial information technology, each designated cell had six corresponding values with respect to the six major criteria considered, ending up with 317 cells for analysis (Figure 20.13). A synthesis of these multifaceted environmental impacts in the GIS system is carried out at the grid level. For example, a grid labeled for "traffic impacts" can be added to the grid labeled for "background traffic flow" to come up with the total traffic impact. However, the chosen scale for noise was the Decibel scale, dB. Decibel values cannot be added, since they are on a logarithmic scale. Hence, a nonlinear equation was applied for noise superimposition. The same was made for the air pollution impact assessment. The algorithm relating FAHP to EIA was presented in great detail and illustrated with a practical application in Chang et al. (2009).

Before letting the committee members assign weights in the FAHP process based on the results of the spatial model, there is a need to explore the use of fuzzy numbers and fuzzy arithmetic to model the imprecision in these domains. TFN and trapezoidal fuzzy numbers in particular are attractive to use in this fuzzy model. Each shape affects how well a fuzzy system of designated if–then rules may approximate a function. The TFN were adopted eventually because they have an intuitive appeal and are easily specified by experts in simulation. Further, the triangular membership functions adopted to represent environmental impacts in this analysis have three main properties, peak, minimum, and maximum, which describe the apex, left corner, and right corner of the triangle. The triangle is normalized in the vertical direction, so membership values will be between 0.0 and 1.0. The degree of membership can then be computed by the aggregate membership function as shown in Figure 20.14. Hence, the membership calculations in the first-stage fuzzy reasoning analysis in the algorithm may be performed to address the uncertainties embedded in EIA and paired comparison in FAHP collectively. The realization of a paired comparison, the derivation of prioritized weights, and evaluation of inconsistency happened in the lowest layer (sets of alternatives or attributes) which can then be translated sequentially throughout the algorithm.

More to the point, the questionnaire survey went directly to the experts and stakeholders. The social, political, and technical dispositions of the various interests involved in possible compensation were invited to attend several public hearings during the project time period. Members were drawn from the participating groups to form an *ad hoc* committee comprised of 26 people deemed capable of representing widely felt attitudes to fairness and trust. The total number of members drawn for both towns was equal. In addition, these official representatives and professionals were invited from different fields (e.g., air, water, solid waste, ecology, public health, economics) in order to avoid bias. Understanding the background of these members helped to avoid the inclusion of any local political influence as well. With the aid of a well-designed questionnaire tailored to estimate the dominance of one element over another pairwise element throughout the hierarchical structure, investigators selected in the committee would be able to pinpoint the relative weights stepwise to generate the final matrix of decision weights across all of the 91 pairings of the 14 criteria (see Chang et al., 2009). When this trial phase was over, we mastered the use of fuzzy arithmetic to derive a final fuzzy weight. The survey outcome supported the retrieval of fuzzy weights across six major criteria that stand out among the others as the most urgent concerns.

Comparative Impact Assessment and Decision Analysis for Fair Fund Distribution

After the derivation of fuzzy weights in the context of FAHP, the algorithm leads to final differentiation of the relative environmental impacts between these two towns. In fact, during the decision-making process human cognitive-evaluative structure of the final fund distribution was rather inexact. The impreciseness and uncertainties are also a part of impact assessment procedure (e.g., environmental modeling and data analysis). Thus, the membership functions of the fuzzy fraction of welfare loss due to environmental change or impact may be formulated in several different ways depending on the subjective choice (Figure 20.15 lists a few). They may vary from Z function, to an N function, to an S function, to a trapezoidal function, to a triangular function (e.g., B and C in Figure 20.15), and to a singleton (e.g., A in Figure 20.15). The use of triangular membership function types for the fuzzy

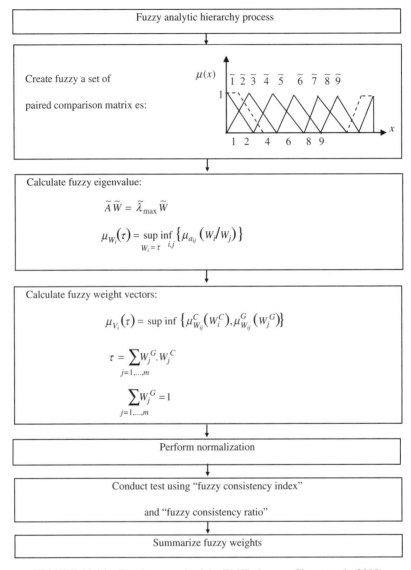

FIGURE 20.14   The framework of the FAHP. *Source:* Chang et al. (2009)

fractions of the corresponding environmental impacts in this case study was eventually applied for final decision-making. Hence, the membership calculations associated
with the final decision-making illustrate the second-stage fuzzy reasoning analysis
in the algorithm. In case C in Figure 20.15 can be selected, the final assessment may
be performed as a kind of fuzzy decomposition analysis based on the fuzzy weights
derived from FAHP (i.e. $\tilde{W}_i$ in Figure 20.16 in which i = town 1 or 2), in order
to complete the regional differentiation (i.e., $\tilde{F}_{i1}$ and $\tilde{F}_{i2}$ in Figure 20.16) for the

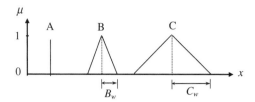

FIGURE 20.15 The degree of fuzziness in different membership functions. *Source*: Chang et al. (2009)

environmental changes (i.e., $\tilde{E}_{i1}$ and $\tilde{E}_{i2}$ in Figure 20.16) in these two towns due to this incineration project.

In real-world applications, more than one environmental impacts have to be factored into a holistic exposure assessment for a particular receptor (i.e., a township). Hence, Equation (20.50) associated with the fuzzy decomposition process described in Figure 20.16 expresses the integrated environmental impact for each town, derived with respect to the six criteria via the use of the integrated EIA and FAHP algorithm. Thus, the amount of a fair fund applicable to each town can be estimated according to Equation (20.51).

$$\tilde{L}_k = \sum_i \left[\tilde{W}_i \sum_j \left(\tilde{A}_{ij} \cdot \tilde{E}_{ijk} \right) \right] \quad \forall\, k \tag{20.50}$$

$$F_k = F_{\text{total}} \left(\frac{\tilde{L}_k}{\sum_k \tilde{L}_k} \right) \quad \forall k \tag{20.51}$$

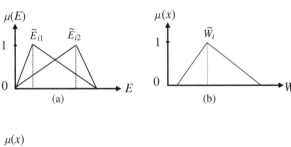

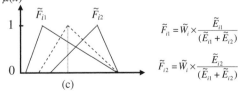

FIGURE 20.16 The illustration related to fuzzy decomposition. *Source*: Chang et al. (2009)

where $\tilde{W}_i$ is the fuzzy weight of an individual environmental impact (criteria) in the criteria layer of designated AHP; $\tilde{A}_{ij}$ is the fuzzy weight of an individual attribute j within each type of environment impact i; $\tilde{E}_{ijk}$ is the fuzzy fraction of welfare loss associated with attribute j pertaining to environmental impact category i in town k; $\tilde{L}_k$ is the total welfare loss (i.e., due to environmental change or impact), which is expressed as a fuzzy number, in town k; F_k is the amount of a fair fund apportioned to town k; and F_{total} is the total fair fund available. Both $\tilde{W}_i$ and $\tilde{A}_{ij}$ can be retrieved from the questionnaire survey; and $\tilde{E}_{ijk}$ may be estimated by using a GIS. Between Equations (20.50) and (20.51), $\tilde{L}_k$ carries the fuzziness embedded in the FAHP and deliver it to the final fair fund distribution.

Zimmermann (1987) emphasized the importance of selecting an appropriate operator for membership functions to be aggregated or to be decomposed. Any decomposed components must provide complete information (or at least substantially complete information) about the original set. From practical aspect, different operators should be selected for different phenomenon in order to build adequate models. In the past, many kinds of fuzzy operators such as Zimmermann's γ-family of operators, ordered weighted averaging aggregation operators, and polynomial composition were successfully adopted in fuzzy membership redistribution (Yang, 1997; Despic and Simonvic, 2000). In this study, the product operator and algebraic sum shown in Equations (20.52)–(20.57) were utilized to decompose the fuzzy membership functions involved.

$$\tilde{A}_\alpha = \left[a_L^\alpha, a_R^\alpha \right], \tag{20.52}$$

$$\tilde{B}_\alpha = \left[b_L^\alpha, b_R^\alpha \right], \tag{20.53}$$

$$\tilde{A}_\alpha(+)\tilde{B}_\alpha = \left[a_L^\alpha + b_L^\alpha, a_R^\alpha + b_R^\alpha \right], \tag{20.54}$$

$$\tilde{A}_\alpha(-)\tilde{B}_\alpha = \left[a_L^\alpha - b_R^\alpha, a_R^\alpha + b_L^\alpha \right], \tag{20.55}$$

$$\tilde{A}_\alpha(\times)\tilde{B}_\alpha = \left[a_L^\alpha b_L^\alpha, a_R^\alpha b_R^\alpha \right], \tag{20.56}$$

$$\tilde{A}_\alpha(/)\tilde{B}_\alpha = \left[a_L^\alpha / b_R^\alpha, a_R^\alpha / b_L^\alpha \right], \tag{20.57}$$

in which $\tilde{A}_\alpha$ and $\tilde{B}_\alpha$ stands for the fuzzy numbers of $\tilde{A}$ and $\tilde{B}$ at level α, and $(+)$, $(-)$, $(\times)$, and $(/)$ denote the addition, subtraction, multiplication, and division operators. Therefore, $\tilde{F}_{i1}$ and $\tilde{F}_{i2}$, as shown by C in Figure 20.16, can be estimated based on such operation according to fuzzy membership functions of $\tilde{E}_{i1}$ and $\tilde{E}_{i2}$, as shown by A in Figure 20.16, and $\tilde{W}_i$, as shown by B in Figure 20.16. The adoption of the suggested FAHP algorithm for decision support may eventually end up answering the scientific and managerial questions discussed previously.

Therefore, as experience with the application of EIA accumulates, ways in which FAHP methods for integrated impact assessment of all six criteria and associated attributes, incongruent with the workings of nature, are characterized collectively. The FAHP process must call upon the *ad hoc* committee for decision analysis of what relative importance might be influential in terms of the final decision weights derived from above. Thus, a desired solution via such a group decision-making process is that final fund distribution may minimize social welfare loss in totality, while being as equitable as possible in distributing environmental costs (i.e., external cost) among these two townships.

20.3.5   Decisions for Fair Fund Redistribution

Environmental Impact Assessment   The following descriptions add more detail on this front without considering human health risk. Noise impacts mainly result from the routine operation of waste incineration and shipping. Noise impacts from the operation of an incinerator were constantly monitored by the subcontractor responsible for the operation of the waste incinerator. Figure 20.17a shows the location of those monitoring sites as well as the off-site monitoring stations in the vicinity of the incineration plant. Figure 20.17b shows the noise control zone in the study area providing clues about the noise the neighboring communities experienced, thus helping FAHP implementation. This integrated analysis resulted in Figure 20.18 that describes the collective impacts of noise from all possible sources including waste incineration and shipping. Aside from the noise impact, a series of traffic congestion indices were developed to address the traffic impact given the inclusion of truck fleet flow for waste shipping, region wide; the criteria "traffic impacts" does not include transportation noise. Traffic impact analysis was accomplished by the determination of percentage of garbage trucks in the existing traffic stream based on the procedure for criteria mapping using the "addition operator" as indicated in Figure 20.14.

Since the ash monofill is located not very far away from the Shu-Lin incinerator, the attribute "impacts from ash transport" apparently only describes a long-term traffic congestion issue. Yet SWM for transporting both sludge and incineration ash creates a lot of psychological impacts in local communities due to the locally crowded road system. Although the bottom liner at the monofill may prevent the leachate from leaking, the monofill operation could result in a long-term threat to the environment once the liner is broken after decades of operation. Delivery of those waste streams, including the ash and sludge flows, in terms of mass flow rate was tracked down during this study. They were all recognized by the committee as significant decision elements and thus must be factored into the final decision analysis matrix. During this study, the recycle and reuse of wastewater was tracked down to support a "no impact" conclusion in the final assessment. This means that the criteria "wastewater impacts" is of little or no importance because it is recycled in the environment, providing clues for the impact assessment.

The long-term impacts with respect to each of the eight pollutants in the exposure assessment are summarized in Figure 20.19a. The findings show that Township B receives the majority of impact due to meteorological pattern and landscape. In order to aggregate the impact derived from various air pollution sources to form an ultimate impact assessment matrix, this analysis applied Equation (20.58). It suggests that all regulatory emission standards are equally important.

$$F_1 = \sum_m \left(\frac{1}{C_m} \cdot P_m \right), \tag{20.58}$$

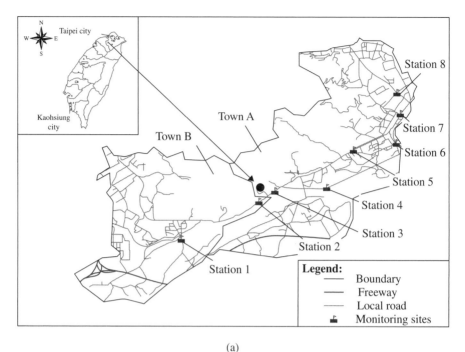

(a)

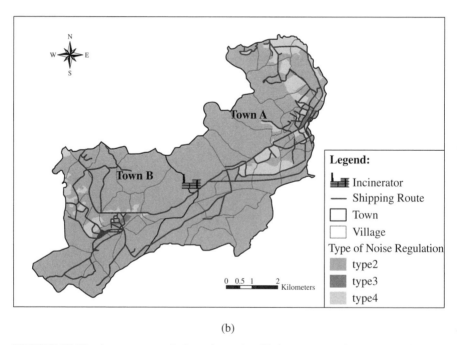

(b)

FIGURE 20.17   Assessment tools for noise and traffic impacts. (a) The locations of monitoring stations for noise and traffic impacts. (b) The distribution of noise control zones in the study area. *Source*: Chang et al. (2009)

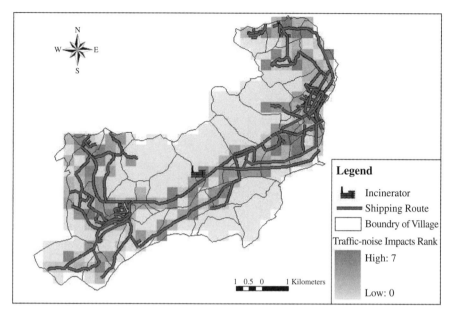

FIGURE 20.18 The estimated noise and traffic impacts in the study area. *Source*: Chang et al. (2009)

where P_m is the individual pollutant impact with respect to pollutant m, which is estimated by ISCST3 model; C_m is the regulatory emission standard for incinerators in Taiwan with respect to pollutant m. The more stringent the standard, the smaller is the C_m. Thus, $1/C_m$ was used to stand for the risk level to human beings with respect to air pollutant m. This aids in the generation of Figure 20.19a for decision analysis. Of course, F_1 summarizes the total air pollution impact with respect to the township 1 as an example. The population in local communities receives all levels of total air pollution impact imposed on them, but only a fraction of that was caused by the incinerator. Given the fact that this area fell into the category of a "nonattainment" area, the marginal impact above and beyond the regulations resulting from the incinerator is critical to be characterized quantitatively using the ISCST3 model in which SO_2, HCL, CO, PM_{10}, NOx, PCDD/PCDF, Pb, and Hg are included.

Socioeconomic Analysis Population density was taken into account in the subsection. As local communities continue to grow, local officials and community members are constantly challenged by the need to provide waste incineration facilities (Edwards, 2013). For example, a proposed waste management facility may increase employment in the community and create demand for more affordable housing. Both effects are easily quantifiable. Also of importance, however, are the perceptions of community members about whether the proposed waste incineration facility is consistent with a commitment to preserving the local character of the community. In

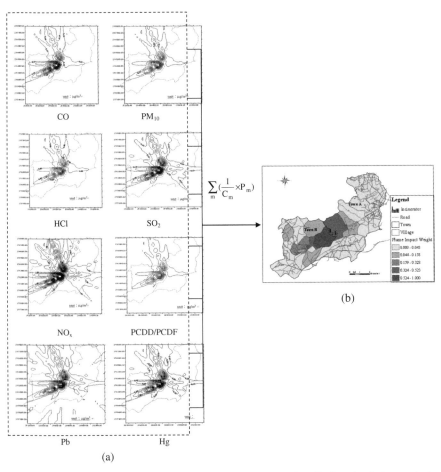

FIGURE 20.19 The environmental impact derived from the air pollution simulation analysis. (a) The isopleths of annual average concentration of air pollutants. (b) The environmental impact derived from air pollution. *Source*: Chang et al. (2009)

general, assessing community perceptions about development requires the use of methods capable of revealing often complex and unpredictable community values. But, they are hard to quantify. Hence, the socioeconomic impact assessment in this study only focused on the employment and housing changes in local communities. In addition to the population growth trend over the past decades, the changes in income and the price levels of real estate before and after the commission of this incinerator were also identified. Observations confirmed that increased population resulted in increased average income and real estate prices. The positive trend observed was partially due to the construction of the facility. Therefore, socioeconomic impact would not be a negative one, and it was eliminated eventually in the final decision analysis for fund distribution. Because the fund does not take into account any benefits that the plant generates, final decision was made only to compensate for the burdens due to waste incineration.

Decision Analysis   For a fair fund distribution absolute damages caused by the incinerator and inflicted on the inhabitants are relevant. However, relative differences in impacts play an important role in final decision-making. This is the essence of using FAHP which takes care of not only the relative differences in impacts but also the uncertainty embedded when comparing some quantitative and qualitative impacts simultaneously. The paired comparison judgment matrixes drawn from 26 selected committee members at different levels presented their insights and were used to determine the unique priority vectors of decision weights for those attributes and criteria at each level of hierarchy. In our case, for example, if the incinerator causes 5000 units of damage in Township A and 4900 units of damage in Township B, the fund might not be divided almost 50:50 given that the associated weight acquired in FAHP might not be 1:1. This is the influence of human judgment embedded in the decision-making process (i.e., FAHP) under uncertainty.

The column (a) in Tables 20.9 and 20.10 summarizes the aggregate weights associated with each selected group of decision-maker by AHP and FAHP, respectively. The attributes "noise from incinerator" and "wastewater from incinerator" mentioned in Figure 20.12 and included in Table 20.9 have zero weights. It is due to the fact that both of them are well managed by the subcontractor so that there is no weight that was assigned by the committee members. Final assessment via FAHP confirms that the impacts from air pollution due to waste incineration and potential leachate impact due to monofill operation (i.e., heavy metal contamination in groundwater aquifer) were perceived as the most important attributes. It is also worth noticing that the siting and building the monofill for disposal of incineration ash present obvious impacts

TABLE 20.9   The weights of various influential attributes by AHP

	Weights of attributes			
Influential attributes	Committee	Township A[a]	Township B[a]	Average
Suspension from monofill	0.0355	0.0563	0.1099	0.0672
Air pollution from stack	0.0683	0.1236	0.3008	0.1642
Air pollution from garbage truck	0.0793	0.1079	0.0637	0.0836
Noise from garbage truck	0.1386	0.0870	0.0797	0.1018
Noise from incineration plant	0.00	0.00	0.00	0.00
Sludge from wastewater	0.1498	0.1185	0.2321	0.1668
Incineration ash	0.1819	0.1381	0.0841	0.1347
Wastewater from incinerator	0.00	0.00	0.00	0.00
Wastewater from monofill	0.1523	0.1673	0.0738	0.1311
Traffic impacts from ash truck	0.0704	0.0511	0.0136	0.0450
Traffic impacts from garbage truck	0.1240	0.1504	0.0424	0.1056

Source: Chang et al. (2009).
[a]Weights were drawn from representative residents and officials in Townships A and B.

TABLE 20.10 The weights of various influential attributes by FAHP

Influential attributes	Weights of attributes (FAHP)			
	Committee	Town A[a]	Town B[a]	Average
Suspension from monofill	0.0010, 0.0345, 0.0619	0.0019, 0.0603, 0.0786	0.0000, 0.0998, 0.1093	0.0010, 0.0649, 0.0775
Air pollution from stack	0.0040, 0.0672, 0.1190	0.0498, 0.1203, 0.1846	0.0295, 0.2942, 0.3940	0.0278, 0.1606, 0.1942
Air pollution from garbage truck	0.0213, 0.0754, 0.0918	0.0123, 0.0920, 0.1206	0.0094, 0.0521, 0.0782	0.0143, 0.0732, 0.1010
Noise from garbage truck	0.0192, 0.1258, 0.1789	0.0091, 0.0831, 0.0842	0.0057, 0.0894, 0.1604	0.0114, 0.0994, 0.1469
Sludge from wastewater	0.0739, 0.1519, 0.1583	0.0229, 0.1324, 0.2201	0.0823, 0.2258, 0.3531	0.0597, 0.1700, 0.2240
Incineration ash	0.0572, 0.1732, 0.2883	0.0423, 0.1295, 0.1973	0.0421, 0.0914, 0.0921	0.0472, 0.1314, 0.2004
Wastewater from monofill	0.0239, 0.1614, 0.2459	0.0175, 0.1537, 0.2382	0.0181, 0.0825, 0.1379	0.0199, 0.1325, 0.2034
Truck impacts from ash truck	0.0390, 0.0985, 0.1646	0.0065, 0.0563, 0.0568	0.0154, 0.0314, 0.0514	0.0203, 0.0621, 0.0765
Truck impacts from garbage truck	0.0180, 0.1121, 0.1982	0.0407, 0.1724, 0.2849	0.0164, 0.0334, 0.0388	0.0250, 0.1060, 0.1522

Source: Chang et al. (2009).
[a]Weights were drawn from representative residents and officials in Townships A and B.

which are almost the same impacts as the discharge of wastewater from the leachate treatment plant to the local drainage system during the survey period. It is of relevance and of significance to know that due to the addition of environmental impacts, the siting and building of an ash monofill along with an incinerator is no longer an option anymore although it was popular before in Taiwan. Table 20.9 summarizes the associated weights of environmental impacts between these two towns. In an attempt to reach the final fund distribution policy, the EIA information listed in Table 20.10 must be prepared based on the average weights listed in Tables 20.9 and 20.10 to generate the initial version of the fund distribution policy between Townships A and B.

Through summing over those average weights in terms of all salient attributes identified in the EIA process, the AHP algorithm would suggest Townships A and B may have a 82.70% and 19.30% share of a fair fund, respectively. While the population density is taken into account, the shares then become 81.76% and 18.24%. The FAHP algorithm would suggest finally using 79.55% and 20.45%. Table 20.11 summarizes all the information from the decision analysis. This result is mainly due to the fact that Township B has a little bit higher population density, and suffers more from air pollution impact. Judging from the weights given, however, Township A suffers more from wastewater discharge from the leachate treatment plant and traffic congestion. Hence, the FAHP ended up with more funds allocated to Township A.

It is desirable that the final decision be reached in the political arena, even if the solution does not optimally address social welfare redistribution. What might be curious is the rationale in support of making an AHP decision (variables are discreet). In essence, why "fuzzify" the human judgment to an AHP process? Given that some of the judgments were based on a combination between quantitative and qualitative information in EIA, the degree of "good" or "bad" becomes vague in a linguistic sense. This is truly the essence of fuzzy decision-making. In most cases, such a decision may differ from a rational model due to limited knowledge, time, money, or motivation, on one hand, and a desire to reach a compromised decision as quickly as possible on the other hand. But several participants in the committee pointed out that the impacts from monofill might be overestimated in the FAHP process. With such analysis as described here, all participants in the decision-making arena in September 1999 eventually reached an agreement to support the decision that Township A receive 80% of the fund and Township B receives the rest of the fund. Such an agreement was formally implemented after the year 2000. Both towns have had few arguments in the past 10 or more years.

20.4 FINAL REMARKS

In the context of multiattribute decision-making, various sources of uncertainty have to be quantified by tailored mathematical algorithms to tackle the natural variability, incomplete knowledge, and even communication breakdown. In this chapter, an important fuzzy decision-making was carried out via a participatory process in which the stakeholders were called on based on their considerable professional

TABLE 20.11 The associated weights of environmental impacts between these two townships

Town Impacts	Crisp		Fuzzy numbers	
	Township A[a]	Township B[a]	Township A	Township B
Suspension from the monofill	1	0	0.718, 0.900, 1.057	0.034, 0.100 0.169
Air pollution from stack	0.235	0.765	0.086, 0.126, 0.348	0.654, 0.874, 0.912
Air pollution from garbage truck	0.926	0.074	0.677, 0.841, 1.380	0.113, 0.159, 0.192
Noise impacts from garbage truck	0.580	0.420	0.363, 0.511, 0.674	0.380, 0.489, 0.527
Sludge from wastewater	1	0	0.630, 0.900, 1.344	0.081, 0.100, 0.131
Incineration ash	1	0	0.436, 0.900, 1.276	0.076, 0.100, 0.152
Wastewater from monofill	1	0	0.569, 0.900, 1.193	0.061, 0.100, 0.178
Traffic impacts from ash truck	1	0	0.480, 0.900, 0.946	0.062, 0.100, 0.133
Traffic impact from garbage truck	0.926	0.074	0.696, 0.841, 1.494	0.099, 0.159, 0.253

Source: Chang et al. (2009).

Note: [a]Weights were drawn from representative residents and officials in Townships A and B.

background in relation to differing sustainability criteria. No matter if it was a fair fund issue or landfill siting issue, the responses collected from the involved stakeholders highlighted the difficulties when comparing an environmental impact against an economic criterion, since both are much different in nature, making final scoring challenging. When incorporating the EIA results into the fuzzy decision analysis, some of the stakeholders felt it to be difficult to comprehend the implications or identify the unintended consequences of the environmental impacts associated with relevant alternatives. In any circumstance, a possible way to minimize this gap could be the enhancement of risk communication for retrieval of the objective scores.

REFERENCES

Akash, B. A., Mamlook, R., and Mohsen, M. S. 1999. Multicriteria selection of electric power plants using analytical hierarchy process. *Electric Power Systems Research*, 52(1), 29–35.

Altrock, C. V. and Krause, B. 1994. Multicriteria decision making in German automotive industry using fuzzy logic. *Fuzzy Sets and Systems*, 63(3), 375–380.

Anderson, R. F. and Greenberg, M. R. 1982. Hazardous waste facility siting: a role for planners. *Journal of the American Planning Association*, 48(2), 204–218.

Antunes, P., Santos, R., and Jordão, L. 2001. The application of geographic information systems to determine environmental impact significance. *Environmental Impact Assessment Review*, 21(6), 511–535.

Baas, S. M. and Kwakernaak, H. 1997. Rating and ranking of multiple aspect alternative using fuzzy sets. *Automatica*, 13(1), 47–58.

Baban, S. M. J. and Flannagan, J. 1998. Developing and implementing GIS-assisted constraints criteria for planning landfill sites in the UK. *Planning Practice and Research*, 13(2), 139–151.

Basak, I. and Saaty, T. 1993. Group decision making using the analytical hierarchy process. *Mathematical and Computer Modeling*, 17(4–5), 101–109.

Belton, V. and Gear, T. 1983. On a short-coming of Saaty's method of analytic hierarchies. *Omega – International Journal of Management Science*, 11(3), 228–230.

Chan, F. T. S., Chan, M. H., and Tang, N. K. H. 2000. Evaluation methodologies for technology selection. *Journal of Materials Processing Technology*, 107(1–3), 330–337.

Chang, D. Y. 1996. Applications of the extent analysis method on fuzzy AHP. *European Journal of Operational Research*, 95(3), 649–655.

Chang, N. B. and Chang, Y. H. 2000. Assessing the redistribution strategy of a fair fund based on an integrated AHP and EIA approach for a municipal incinerator. In: Proceedings of 2nd International Conference on Solid Waste Management, Taipei, Taiwan, ROC, pp. 175–190.

Chang, P.-L. and Chen, Y. C. 1994. A fuzzy multi-criteria decision making method for technology transfer strategy selection in biotechnology. *Fuzzy Sets and Systems*, 63(2) 131–139.

Chang, N. B., Yang, Y. C., and Wang, S. F. 1996. Solid waste management system analysis with noise control and traffic congestion limitations. *Journal of Environmental Engineering ASCE*, 122(2), 122–131.

Chang, N. B., Parvathinathan, G., and Breeden, J. B. 2008. Combining GIS with fuzzy multiple criteria decision making for landfill siting in a fast growing urban region. *Journal of Environmental Management*, 87(1), 139–153.

Chang, N. B., Chang, Y. H., and Chen, H. W. 2009. Fair fund distribution for a municipal incinerator using a GIS-based fuzzy analytic hierarchy process. *Journal of Environmental Management*, 90(1), 441–454.

Charnpratheep, K., Zhou, Q., and Garner, B. 1997. Preliminary landfill site screening using fuzzy geographic information systems. *Waste Management & Research*, 15(2), 197–215.

Chen, S. M. 1996. Evaluating weapon systems using fuzzy arithmetic operations. *Fuzzy Sets and Systems*, 77(3), 265–276.

Chen, M. F., Tzeng, G. H., and Ding, C. G. 2008. Combining fuzzy AHP with MDS in identifying the preference similarity of alternatives. *Applied Soft Computing*, 8(1), 110–117.

Cheng, C. H. 1997. Evaluating naval tactical missile systems by fuzzy AHP based on the grade value of membership function. *European Journal of Operational Research*, 96(2), 343–350.

Chiou, H. K. and Tzeng, G. H. 2001. Fuzzy hierarchical evaluation with grey relation model of green engineering for industry. *International Journal Fuzzy System*, 3(3), 466–475.

del Valle, F. 2005. Both sides confident at landfill hearing. Valley Morning Star. Available at: http://www.valleystar.com/archives (accessed January 2005). No longer available afterwards.

Despic, O. and Simonvic, S. 2000. Aggregation operators for soft decision in water resources. *Fuzzy Sets and Systems*, 115(1), 11–33.

Dikshit, A. K., Padmavathi, T., and Das, R. K. 2000. Locating potential landfill sites using geographic information systems. *Journal of Environmental Systems*, 28(1), 43–54.

Dimitru, V. and Luban, F. 1982. Membership functions, some mathematical programming models and production scheduling. *Fuzzy Sets and Systems*, 8(1), 19–33.

Dimitru, V. and Luban, F. 1986. On some optimization problems under uncertainty. *Fuzzy Sets and Systems*, 18(3), 257–272.

Dombi, J. 1990. Membership function as an evaluation. *Fuzzy Sets and Systems*, 35(1), 1–21.

Dubois, D. and Prade, H. 1988. *Possibility Theory*, Plenum Press, New York and London.

Dyer, J. S. 1990. A clarification of remarks on the analytic hierarchy process. *Management Science*, 36(3), 274–275.

Edwards, M. 2013. Socio-economic impact analysis. Available at: http://www.lic.wisc.edu/shapingdane/facilitation/all_resources/impacts/analysis_socio.htm (accessed July 2013).

Fanti, M. P., Maione, B., Naso, D., and Turchiano, B. 1998. Genetic multi-criteria approach to flexible line scheduling. *International Journal of Approximate Reasoning*, 19(1–2), 5–21.

Geldermann, J., Spengler, T., and Rentz, O. 2000. Fuzzy outranking for environmental assessment: case study-iron and steel making industry. *Fuzzy Sets and Systems*, 115(1), 45–65.

Giles, R. 1988. The concept of grade of membership. *Fuzzy Sets and Systems*, 25(3), 297–323.

Hannan, E. L. 1981. Linear programming with multiple fuzzy goals. *Fuzzy Sets and Systems*, 6(3), 235–248.

Harker, P. T. and Vargas, L. G. 1987. The theory of ratio scale estimation: Saaty's analytical hierarchy process. *Management Science*, 33(11), 1383–1403.

Harker, P. T. and Vargas, L. G. 1990. Reply to "remarks on the analytical hierarchy process". *Management Science*, 36(3), 269–273.

Hauser, D. and Tadikamalla, P. 1996. The analytic hierarchy process in an uncertain environment: a simulation approach. *European Journal of Operational Research*, 91(1), 27–37.

Huizingh, E. K. R. E. and Vrolijk, H. C. J. 1997. Extending the applicability of the analytic hierarchy process. *European Journal of Operational Research*, 31(1), 29–39.

Jacobs, J. 2002. Public company accounting reform and investor protection act of 2002. Available at: https://www.govtrack.us/congress/bills/107/hr5070 (accessed July 2013).

Jensen, J. R. and Christensen, E. J. 1986. Solid and hazardous waste disposal site selection using digital geographic information system techniques. *Science of the Total Environment*, 56, 265–276.

Kao, J., Lin, H., and Chen, W. 1997. Network geographic information system for landfill siting. *Waste Management & Research*, 15(3), 239–253.

Kaufmann, A. and Gupta, M. M. 1988. *Fuzzy Mathematical Models in Engineering and Management Science*, Elsevier, Amsterdam, The Netherlands.

Keir, A. W., Doucett, J. A., and Oliveri, T. 1993. Landfill siting using GIS technology: the Case of the Peel Landfill Site Search. In: Proceedings of the Canadian Conference on GIS-1993, Ottawa, Canada, pp. 13–24.

Labib, A. W., O'Connor, R. F., and Williams, G. B. 1998. Effective maintenance system using the analytic hierarchy process. *Integrated Manufacturing Systems*, 9(2), 87–98.

Lai, S. K. 1995. A preference-based interpretation of AHP. *Omega – International Journal of Management Science*, 23(4), 453–462.

Leberling, H. 1981. On finding compromise solutions in multicriteria problems using the fuzzy min-operator. *Fuzzy Sets and Systems*, 6(2), 105–118.

Levary, R. R. and Ke, W. 1998. A simulation approach for handling uncertainty in the analytic hierarchy process. *European Journal of Operational Research*, 106(1), 116–122.

Lin, H. Y. and Kao, J. J. 1999. Enhanced spatial model for landfill siting analysis. *Journal of Environmental Engineering*, 125(9), 845–851.

Lipovetsky, S. 1996. The synthetic hierarchy method: an optimizing approach to obtaining priorities in the AHP. *European Journal of Operational Research*, 93(3), 550–564.

Lower Rio Grande Valley Development Council (LRGVDC). 2002. Regional solid waste management plan amendment 2002–2020.

Malcezwki, J. 1999. Spatial multicriteria decision analysis. In: *Spatial Multicriteria Decision Making and Analysis* (Ed. Thill, J. C.), Ashgate Publishing, Hampshire, pp. 377.

McIntyre, C. and Parfitt, M. K. 1998. Decision support system for residential land development site selection process. *Journal of Architecture Engineering*, 4(4), 125–131.

Nagar, B. B. and Mizra, U. K. 2002. Hydrogeological environmental assessment of sanitary landfill project at Jammu City, India. *Green Journal*, 17(8), 223–245.

Noh, J. and Lee, K. M. 2003. Application of multiattribute decision making methods for the determination of relative significance factor of impact categories. *Environmental Management*, 31(5), 633–641.

O'Leary, P. R., Canter, L., and Robinson, W. D. 1986. Land disposal. In: *The Solid Waste Handbook: A Practical Guide* (Ed. Robinson, W. D.), John Wiley & Sons, New York, pp. 271.

Paek, J. H., Lee, Y. W., and Napier, T. R. 1992. Selection of design/build proposal using fuzzy logic system. *Journal of Construction Engineering and Management*, 118(2), 303–317.

Ramanathan, R. 2001. A note on the use of the analytic hierarchy process for environmental impact assessment. *Journal of Environmental Management*, 63(1), 27–35.

Ramanathan, R. and Ganesh, L. S. 1995. Using AHP for resource allocation problems. *European Journal of Operational Research*, 80(2), 410–417.

Saaty, T. L. 1980a. *The Analytic Hierarchy Process*, McGraw-Hill, New York.

Saaty, T. L. 1980b. *The Analytic Hierarchy Process–Planning, Priority Setting, Resource Allocation*, McGraw-Hill, New York.

Saaty, T. L. 1986. Axiomatic foundation of the analytic hierarchy process. *Management Science*, 23(7), 841–855.

Saaty, T. L. 1990. An exposition of AHP in reply to the paper "Remarks on the Analytic Hierarchy Process". *Management Science*, 36(3), 259–268.

Saaty, T. L. 1994. *Fundamentals of Decision Making and Priority Theory with the Analytical Hierarchy Process*, RWS Publications, Pittsburgh, PA.

Sakawa, M. and Yumine, T. 1983. Interactive fuzzy decision-making for multi-objective linear fractional programming problems. *Large Scale Systems*, 5(2), 105–113.

Şener, B., Süzen, L., and Doyuran, V. 2006. Landfill site selection by using geographic information systems. *Environmental Geology*, 49(3), 376–388.

Sharifi, M. A. and van Herwijnen, M. 2003. *Spatial Decision Support System*, ITC Lecture Series, Enschede.

Siddiqui, M. Z., Everett, J. W., and Vieux, B. E. 1996. Landfill siting using geographical information systems: a demonstration. *Journal of Environmental Engineering*, 122(6), 515–523.

Smets, P. and Magrez, P. 1988. The measure of the degree of truth and of the grade of membership. *Fuzzy Sets and Systems*, 25(1), 67–72.

Stewart, T. J. 1992. A critical survey on the status of multiple criteria decision making theory and practice. *OMEGA – International Journal of Management Science*, 20(5–6), 569–586.

Svarovski, S. G. 1987. Usage of linguistic variable concept for human operator modelling. *Fuzzy Sets and Systems*, 22(1–2), 107–114.

Tagaris, E., Sotiropolou, R. E., Pilinis, C., and Halvadakis, C. P. 2003. A methodology to estimate odors around landfill sites: the use of methane as an odor index and its utility in landfill siting. *Journal of the Air and Waste Management Association*, 53(5), 629–634.

Tanaka, H., Uejima, S., and Asai, K. 1982. Linear regression analysis with fuzzy model. *IEEE Transactions on Systems, Man, and Cybernetics*, 12(6), 903–907.

Tang, M, T., Tzeng, G. H., and Wang, S. W. 1999. A hierarchy fuzzy MCDM method for studying electronic marketing strategies in the information service industry. *Journal of International Information Management*, 8(1), 1–22.

Teng, J. Y. and Tzeng, G. H. 1996. Fuzzy multicriteria ranking of urban transportation investment alternatives. *Transportation Planning Technology*, 20(1), 15–31.

Texas State Historical Association (TSHA). 2003. *The Handbook of Texas Online*, Rio Grande Valley, TX. Available at: https://www.tshaonline.org/handbook/online/articles/ryr01 (accessed September 2003).

Tran, L. T., Knight, C. G., O'Neill, R. V. L, Smith, E. R., Ritters, K. H., and Wickham, J. 2002. Fuzzy decision analysis for integrated environmental vulnerability assessment of the Mid-Atlantic region. *Environmental Management*, 29(6), 845–859.

Tung, S. L. and Tang, S. L. 1998. A comparison of the Saaty's AHP and modified AHP for right and left eigenvector inconsistency. *European Journal of Operational Research*, 106(1), 123–128.

United States Department of Agriculture (USDA). 1977. Soil survey of Cameron County Texas. Soil Conservation Service.

United States Department of Agriculture (USDA). 1982. Soil survey of Willacy County Texas. Soil Conservation Service.

United States Environmental Protection Agency (USEPA). 1979. *Industrial Source Complex (ISC) Dispersion Model User's Guide*, Vol. I, EPA-450/4-79-030.

United States Environmental Protection Agency (USEPA). 2013. Air pollution monitoring-attainment/nonattainment. Available at: http://www.epa.gov/airquality/montring.html# attainment (accessed January 2013).

Webber, S. A., Apostolou, B., and Hassell, J. M. 1996. The sensitivity of the analytic hierarchy process to alternative scale and cue presentations. *European Journal of Operational Research*, 96(2), 351–362.

Weiss, E. N. and Rao, V. R. 1987. AHP design issues for large-scale systems. *Decision Sciences*, 18(1), 43–61.

Yambert, M. W., Belcher, G. D., and Travis, C. C. 1991. Evaluation of flat versus complex terrain models in estimating pollutant transport and deposition in complex terrain. In: *Municipal Waste Incineration Risk Assessment* (Ed. Travis, C. C.), Plenum Press, New York, pp. 1–19.

Yang, R. R. 1997. On the inclusion of importance in OWA aggregation. In: *The Ordered Weighted Averaging Operators: Theory and Applications* (Eds. Yanger, R. R. and Kacprzyk, J.), Kluwer Academic Publishers, Boston, MA, pp. 41–59.

Zadeh, L. A. 1962. From Circuit Theory to Systems Theory. *Proceedings of Institute of Radio Engineering*, 50, 856–865.

Zadeh, L. A. 1965. Fuzzy sets. *Information and Control*, 8(3), 338–353.

Zadeh, L. A. 1971. Quantitative fuzzy semantics. *Information Sciences*, 3(2), 159–176.

Zadeh, L. A. 1972. A fuzzy set theoretic interpretation of linguistic hedges. *Journal of Cybernetics*, 2(3), 4–34.

Zahedi, F. 1986. The analytic hierarchy process – a survey of the method and its applications. *Interface*, 16(4), 96–108.

Zeiss, C. and Lefsrud, L. 1995. Analytical framework for facility waste siting. *Journal of Urban Planning and Development*, 121(4), 115–145.

Zimmermann, H. J. 1978. Fuzzy programming and linear programming with several objective functions. *Fuzzy Sets and Systems*, 1(1), 45–55.

Zimmermann, H. J. 1987. *Fuzzy Sets, Decision Making and Expert Systems*, Kluwer Academic Publishers, Boston, MA.

Zimmermann, H. J. and Zysno, P. 1985. Quantifying vagueness in decision models. *European Journal of Operational Research*, 22(2), 148–158.

CHAPTER 21

FUZZY MULTIATTRIBUTE DECISION-MAKING FOR SOLID WASTE MANAGEMENT WITH TECHNOLOGICAL COMPLICATIONS

Fuzzy set theory has been widely used to account for uncertainty in decision-making. Yet the membership function of a type 1 fuzzy set as defined in Chapter 20 has no explicit uncertainty associated with it except the direct assignment of membership values, something that seems to contradict the word "fuzzy." The type 2 fuzzy set that is also known as interval-valued fuzziness (IVF) is an extension of the traditional fuzzy set theory (i.e., type 1 fuzzy set). IVF that has advanced consideration of uncertainty based on interval values of a membership function is more flexible, general, and applicable than the type 1 fuzzy set in nature, regardless of the complexity involved in the determination of interval values. This chapter aims to provide readers with an extended insight into the fundamentals of fuzzy set theory and existing fuzzy multiattribute decision-making problems. A case study in Portugal was carried out to explore possible applications of IVF in dealing with screening recycling alternatives in Setúbal region area. In this exercise, the essence of IVF was emphasized in solid waste management (SWM) decision-making.

21.1 INTEGRATED FUZZY TOPSIS AND AHP METHOD FOR SCREENING SOLID WASTE RECYCLING ALTERNATIVES

Recent challenges for SWM in Europe are intimately tied to the fulfillment of the prescribed recycling and organic waste recovery targets required by European Directives. Challenges in characterizing and propagating uncertainty, as well as validating predictions, permeate the decision-making process. To address the societal

Sustainable Solid Waste Management: A Systems Engineering Approach, First Edition. Ni-Bin Chang and Ana Pires.
© 2015 The Institute of Electrical and Electronics Engineers, Inc. Published 2015 by John Wiley & Sons, Inc.

ramifications implicit in decision-making, this chapter integrates the analytic hierarchy process (AHP) and the technique for order performance by similarity to ideal solution (TOPSIS) to rank competitive alternatives and help decision makers in a Portuguese municipal solid waste (MSW) management system. To underscore the role of uncertainty in decision-making for alternative ranking, a new method, the fuzzy interval multiattribute decision-making (FIMADM) analysis, was developed by integrating fuzzy TOPSIS and AHP to aid in environmental policy analysis. In essence, AHP was used to determine the essential weighting factors over different criteria; then, screening and ranking those alternatives were carried out by TOPSIS under uncertainty expressed by an IVF method, namely type 2 fuzzy membership functions. Such an IVF–TOPSIS approach, driven by a set of AHP-based weighting factors associated with the selected criteria, has proven to be useful for final ranking via an iterative procedure. A practical implementation was assessed in a case study of Setúbal Peninsula, Portugal, for the selection of the best waste management practices under an uncertain environment. This program was geared toward the future fulfillment of European Directive targets.

21.1.1   System Planning with Uncertainty Concerns

In Portugal, it was vital to ensure the full compliance with the targets required by the European Directives for SWM, such as the Packaging and Packaging Waste Directive 2004/12/EC (European Parliament and Council, 2004) and Landfill Directive 1999/31/EC (Council, 1999). For example, Portugal was challenged in the early 2000s to comply with packaging recycling targets before 2011. For organic waste, the recycling targets established for 2009 and 2013 aimed to divert 50% and 65% of organic waste produced based on the 1995 generation basis, respectively, were delayed until 2013 and 2020. In addition to compliance with Landfill and Packaging Directives, a new challenge arose from the Waste Framework Directive 2008/98/EC (European Parliament and Council, 2008), making it imperative that the waste management systems (WMS) of Member States (MS) take into account general environmental protection principles with regard to precaution and sustainability, technical feasibility and economic viability, protection of resources as well as overall environmental, human health, social, and economic impacts. In other words, waste management practices are considered as a series of trade-offs among different stakeholders with different objectives, making it more difficult for decision makers to reach an amicable decision. These trade-offs involve technical, economic, environmental, and social criteria that are either quantitative or qualitative. Such challenges in the decision-making arena might best be addressed by a more scientifically credible approach to find out a sustainable solution.

Within this context, several sources of uncertainties can be addressed which can affect the compliance with Directive targets and subsequently the choice of the best waste management solution. The Directive targets are information that will promote innovations in the way as to how waste is handled under national law or regulations. However, as science and technology evolve over time, there is never

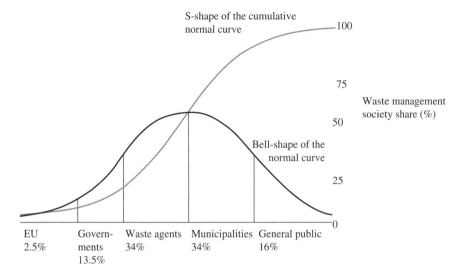

FIGURE 21.1 Information diffusion through waste management stakeholders. *Source*: From Pires et al. (2011a)

perfect knowledge ensuring the right choice. This fact makes implementation of waste management practices an educational process in the society. No matter what choice is made, government agencies have to convert the expected changes into direct impacts on the affected entities, mainly WMS. These direct impacts must also be translated into changes in the way waste is managed by a WMS. Entities besides WMS are also affected. These include producer responsibility organizations and extended producer responsibility schemes (like the Green Dot System for packaging waste in Portugal named *Sociedade Ponto Verde*) and relevant private sectors which will use products from WMS such as recyclables, compost, and electricity. Other changes can be induced by pay-as-you-throw (PAYT) (Box 21.1) systems, a successful economic instrument but not yet applied in Portugal in the 2000s. The information diffusion process related to this type of movement has obvious time lags (see Figure 21.1) as social education promotes PAYT while at the same time unintended consequences or complications of such a promotion may affect future electoral outcomes at both regional and local levels.

The process of information diffusion considered in waste management directives can be compared to the diffusion of innovation proposed by Rogers (1962). To allow innovation to get well-diffused, it needs to be adopted by the society. Rogers (1962) defined five categories of adopters: innovators (the first to adopt the innovation), early adopters (the second fastest who adopt an innovation), early majority (which are the individuals after a varying degree of time), late majority (which will adopt the innovation after the average of society have done it, presenting some level of skepticism), and laggards (the last one adopting the innovation). A parallelism can be done with waste management promoters, as shown in Figure 21.1. Using the same curve defined by Rogers (1962), it can be mentioned that the European Union (EU), specifically the Parliament and the Council, are the ones who elaborate and promulgate European legislation, which can be regarded as the innovators. Then, legislation has to be transposed to each MS in the EU by the governments (the early adopters). All the waste management agents which can be MSW management systems, recyclers, waste operators, and producer responsibility organizations are the next adopters to achieve the mandatory targets for recycling and organic waste diversion from landfills (the early majority). This group is the one responsible for the EU legislation implementation in the field, including planning strategies and measures to be taken. Municipalities are the next adopters (late majority) which will provide the conditions to comply with the new regulations. This group is away from the law elaboration process: for that reason they take a longer time to implement measures. Such longer time is also determined by the transposition time of the EU directives into national law, and the time period from law promulgation to law enforcement in each MS of the EU. Citizens or general public are the laggards in the information diffusion chain, where information reach them through environmental campaigns promoting recyclables sorting at home, or to promote home composting or other measures concerning waste recycling, reduction, and prevention. In consequence, distribution of information from the EU to general public in each MS is uneven. All those uncertainties at different scales compound the uncertainty quantification (UQ).

In general, the new measure such as PAYT is seen to cause uncertainty in the minds of potential adopters (Berlyne, 1962; Rogers, 1962; Nimmo, 1985). Challenges arise from the fact that the local government or general public receiving information about new waste regulation or initiatives like PAYT has to respond as quickly as possible in a short period of time to be able to comply with the waste management targets. How to model such information diffusion process through waste management stakeholders toward achieving this prescribed goal in the field might not be easy. This is due to that the evaluation metrics for mathematical constructs useful for describing such uncertainties in decision-making as a whole are lacking. This discrepancy provides a driving force for developing a new methodology introduced in this chapter to provide a holistic approach for suitable UQ for information diffusion in a WMS.

BOX 21.1   PAYT AND ECONOMIC INSTRUMENTS

Economic instruments enclose a variety of policy tools, including pollution taxes, marketable permits, deposit–refund systems, and performance bonds (UNEP, 2004). The PAYT is such a system to present some managerial scenarios in a nontechnical fashion, providing information on the practical use of specific economic instruments in varying SWM systems to achieve the prescribed targets in comparison with those alternatives without economic instruments. More information can be found in Chapter 3.

In such UQ, the spectrum of uncertainties may include but is not limited to (1) the projections of possible impact due to PAYT implementation, (2) the uncertainties arising from model parameters, (3) types of model, (4) inherent uncertainties in a modeling process, (5) the uncertainties due to lack of knowledge about a specific process or processes, and (6) uncertainties embedded in decision-making. All of them can affect the final outcomes of modeling analysis. This necessitates creating a new spectrum of UQ taking all of the uncertainties into account in one shot. It is now recognized as a critical element—integrated information diffusion function and fuzzy set theory—necessary for continued advancements in the handling of waste management with societal sustainability implications.

An information diffusion function can be defined as a fuzzy classifying function (Huang, 1997) in which fuzzy set theory can be applied to cope with social and technical complexity, to some extent simultaneously (Pires et al., 2011a). This thrust envelops a diversity of approaches to deal with uncertainty from different disciplines, reflecting differences in the underlying literatures. The general framework of fuzzy reasoning makes much of this uncertainty more manageable; fuzzy systems employing type 1 fuzzy sets represent uncertainty with numbers in the range [0, 1] (Pires et al., 2011a). When something is uncertain, like a measurement, it is difficult to determine its exact membership value, and type 2 fuzzy sets or IVF sets (Figure 21.2) make more sense than using traditional algebraic sets and type 1 fuzzy sets (Zadeh, 1975a, 1975b).

Due to the uncertain nature of information, indicators used in SWM analyses to address a unique topology of uncertainties have to cover unpredictability, structural uncertainty, and value/preference uncertainty in decision-making, such as random (due to natural variability) and epistemic (due to lack of or incomplete knowledge) uncertainties (Pires et al., 2011a). As a consequence, using type 2 fuzzy sets (Figure 21.2) to expand the credibility of uncertainty analyses might become a norm in the future (Karnik et al., 1999). According to Liang and Mendel (2000), applying type 2 fuzziness has been regarded as a way to increase the fuzziness of a relation and, according to Hisdal (1981), "increased fuzziness in a description means an increased ability to handle inexact information in a locally correct manner." Our disposition in this case study for handling decision analysis is to construct suitable IVF sets or type 2 fuzzy sets so as to characterize and quantify the unique topology of uncertainty encountered in SWM systems (Pires et al., 2011a).

As part of the companion studies found in Chapters 10 and 16, this chapter shows an integrated AHP and fuzzy TOPSIS to help decision makers in a Portuguese WMS-build priority setting (Section 10.4). To underscore the role of uncertainty in decision-making for alternative ranking, an FIMADM process was carried out to aid in environmental policy decisions. While AHP was used to determine the essential weighting factors, screening and ranking were carried out with a fuzzy TOPSIS under uncertainty expressed by using an IVF method. According to TOPSIS method, developed by Hwang and Yoon (1981), a chosen alternative should be the one nearest to the positive ideal solution and farthest to the negative ideal solution (Chen, 2000).

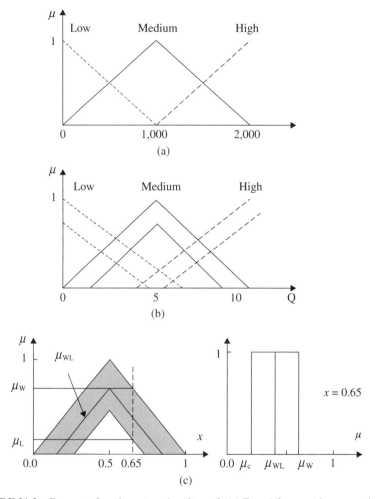

FIGURE 21.2 Fuzzy set functions: type 1 and type 2. (a) Type 1 fuzzy set has a membership grade that is a crisp number in [0, 1]. (b) Type 2 fuzzy sets are an extension of type 1 fuzzy sets in which uncertainty is represented by an additional dimension. (c) Special case: a type 2 membership function is an interval set where the secondary membership function is either zero or one.

The same concept can be applied to fuzzy sets, where the distance is calculated to the fuzzy positive ideal solution and the fuzzy negative ideal solution. Such a new decision-making process eventually led to the screening and ranking of 18 management alternatives in order to improve the sustainability of SWM in the Setúbal region, Portugal. Through the use of an FIMADM process under uncertainty, the chosen UQ levels help illustrate the sensitivity of various sources of uncertainty in decision-making (Pires et al., 2011a).

21.2 THE ALGORITHM OF FIMADM METHOD

The proposed method for the evaluation of MSW alternatives consists of two basic stages(Pires et al., 2011a): (1) AHP computations to discern criteria weights and (2) evaluation of alternatives with IVF TOPSIS, where the best results may be expressed as an interval rather than an exact ideal solution. IVF TOPSIS begins with an initial guess regarding the uncertainty range likely disturbing the determination of a specific solution closer to the ideal solution expressed by the interval that might possibly reflect the fluctuations. A schematic diagram of the proposed method can be seen in Figure 21.3. The process starts with an interval-value fuzzification of crisp data for attributes concerning each criterion, some of them obtained from life cycle inventory (LCI) and life cycle assessment (LCA), resulting in a membership function. The membership functions were defined using an interval scale. Then, the intervals, echoing the UQ associated with different sources, were iteratively tested. Starting with a defined interval value, the IVF matrix was analyzed by TOPSIS. The rest of the TOPSIS procedure was conducted until the final results were reached. Then, a new interval value is tested repeatedly until the final result cannot be changed, which indicates that uncertainty can no longer considerably influence those results. Iteration can be terminated when all types of uncertainty are fully taken into account.

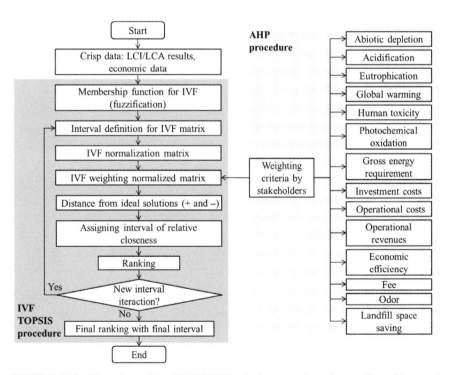

FIGURE 21.3 Flowchart of the IVF TOPSIS solution procedure. *Source*: From Pires et al. (2011a)

21.2.1 The AHP Method

In the first stage, criteria defined for the assessment of the alternatives have to be integrated in a decision hierarchy to support the AHP model formulation that is structured making the objective, criteria, and SWM alternatives be layered on the first, second, and third levels, respectively (Pires et al., 2011a). A weighting factor associated with each of the criteria can be derived by a hierarchical process through pairwise comparison matrices formed to determine the criteria weights. Computing the geometric mean of the values obtained by individual evaluation can lead to the identification of the final pairwise comparison matrix. The weights of the criteria can then be calculated based on this final comparison matrix (Pires et al., 2011a). The theoretical concepts behind AHP were discussed in detail in Chapter 8.

21.2.2 IVF TOPSIS Method

With the aid of the derived weighting factors through AHP, ranking of waste management alternatives can be determined by the IVF TOPSIS method in the second stage. The first action to be taken is to present TOPSIS, an MADM method developed by Hwang and Yoon (1981). In TOPSIS, the chosen alternative should have the shortest distance concerning the positive-ideal solution and be the farthest concerning the negative-ideal solution, having a geometric system concept as background. Whereas the negative-ideal solution is the composite of the worst performance values, the positive-ideal solution is formed as a composite of the best performance values exhibited in the decision matrix relative to any alternative in terms of each attribute of concern (Kahraman, 2008).

Consider an MCDM problem with n alternatives $A_1, A_2, \ldots, A_n$ and m criteria $C_1, C_2, \ldots, C_m$. Criteria are used to characterize and evaluate alternatives. A decision matrix $X(x_{ij})_{n \times m}$ is built to rank alternatives and their values. Let $W = (w_1, w_2, \ldots, w_m)$ be the weight vector. The TOPSIS procedure is described as follows (Hwang and Yoon, 1981; Chen, 2000; Ashtiani et al., 2009):

Step 1. Calculate the normalized decision matrix. The normalized value n_{ij} is calculated as

$$n_{ij} = \frac{x_{ij}}{\sqrt{\sum_{j=1}^{m} x_{ij}^2}}, i = 1, \ldots, n, j = 1, \ldots, m. \tag{21.1}$$

Step 2. Calculate the weighted normalized decision matrix. The weighted normalized value v_{ij} is calculated as

$$v_{ij} = w_i n_{ij}, i = 1, \ldots, n, j = 1, \ldots, m, \tag{21.2}$$

where w_i is the weight of the criterion, and $\sum_{i=1}^{n} w_i = 1$.

Step 3. Determine the positive ideal (A^+) and negative ideal (A^-) solutions.

$$A^+ = \{v_1^+, \ldots, v_n^+\} = \left\{ \left(\max_j v_{ij} | i \in I \right), \left(\min_j v_{ij} | i \in J \right) \right\}$$

$$A^- = \{v_1^-, \ldots, v_n^-\} = \left\{ \left(\min_j v_{ij} | i \in I \right), \left(\max_j v_{ij} | i \in J \right) \right\},$$

(21.3)

where I is the set of benefit criteria and J is the set of cost criteria.

Step 4. Calculate the separation measures, using the n-dimensional Euclidean distances. The distance of each alternative for the positive ideal solution (d_j^+) and for the negative ideal solution (d_j^-) are given as, respectively,

$$d_j^+ = \left\{ \sum_{i=1}^n \left(v_{ij} - v_i^+ \right)^2 \right\}^{1/2}, j = 1, \ldots m.$$

$$d_j^- = \left\{ \sum_{i=1}^n \left(v_{ij} - v_i^- \right)^2 \right\}^{1/2}$$

(21.4)

Step 5. Calculate the relative closeness to the ideal solution R_j.

$$R_j = \frac{d_j^-}{d_j^+ + d_j^-}, j = 1, \ldots, m,$$

(21.5)

since $d_j^- \geq 0$ and $d_j^+ \geq 0$, then $R_j \in [0, 1]$.

The TOPSIS procedure presented so far is applied to deal with crisp values only. Moreover, it is necessary to adapt it to fit IVF properly. According to Ashtiani et al. (2009), a fuzzy number is a convex fuzzy set, defined by a real number interval, each with a membership value between 0 and 1. Considering the uncertainty addressed in this case study and UQ development (presented in Section 21.1.1), using a type 1 fuzziness is not enough to bring all uncertainty into the analysis. To solve this issue, membership can be expressed as an interval of real numbers, like in type 2 fuzziness or IVF. In IVF, an upper and lower bound for membership are identified, while the spread of membership distribution is ignored under the assumption that membership values between upper and lower values are uniformly distributed or scattered with membership value of "1" on the $\mu(\mu(.))$ axis (Türksen, 2006). Thus, the upper and lower bounds of interval-valued type 2 (or IVF) fuzziness specify the range of uncertainty about the membership values. The triangular IVF number can be represented by $\tilde{x}$ (Ashtiani et al., 2009):

$$\tilde{x} = \begin{cases} (x_1, x_2, x_3) \\ (x_1', x_2, x_3') \end{cases},$$

(21.6)

which can be visualized in Figure 21.4. $\tilde{x}$ can also be represented as $\tilde{x} = [(x_1, x_1'), x_2; (x_3', x_3)]$.

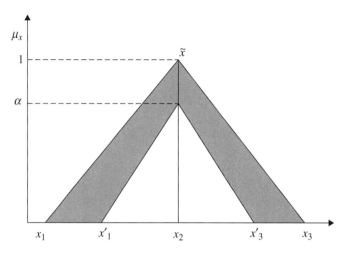

FIGURE 21.4   Illustration of an IVF membership function. *Source*: From Pires et al. (2011a)

The following IVF TOPSIS method, modified from Ashtiani et al. (2009), was applied in this case study. The method contains the following steps:

Step 1. Given $\tilde{x} = [(a_{ij}, a'_{ij}); b_{ij}; (c'_{ij}, c_{ij})]$, and $\tilde{x}$ is an IVF set, the normalized performance rating $\tilde{n}_{ij}$ (the normalized decision matrix) as an extension of Chen (2000) can be calculated as

$$\tilde{n}_{ij} = \left[\left(\frac{a_{ij}}{c_j^+}, \frac{a'_{ij}}{c_j^+} \right); \frac{b_{ij}}{c_j^+}; \left(\frac{c'_{ij}}{c_j^+}, \frac{c_{ij}}{c_j^+} \right) \right], i = 1, \ldots, n, j \in I \quad (21.7a)$$

for benefit criteria and,

$$\tilde{n}_{ij} = \left[\left(\frac{a_j^-}{c'_{ij}}, \frac{a_j^-}{c_{ij}} \right); \frac{a_j^-}{b_{ij}}; \left(\frac{a_j^-}{a_{ij}}, \frac{a_j^-}{a'_{ij}} \right) \right], i = 1, \ldots, n, j \in J, \quad (21.7b)$$

for cost criteria, where c_j^+ is the maximum element of the IVF set, a_j^- is the minimum element of the IVF set, I is the set of benefit criteria, and J is the set of cost criteria.

$$\begin{aligned} c_j^+ &= \max c_{ij}, j \in I \\ a_j^- &= \min a_{ij}, j \in J \end{aligned} \quad (21.7c)$$

Hence, the normalized matrix $\tilde{R} = [\tilde{n}_{ij}]_{n \times m}$ can be obtained.

Step 2. By considering the different importance of each criterion obtained from the AHP method, the weighted normalized fuzzy decision matrix $\tilde{v}_{ij}$ is constructed as

$$\tilde{V} = [\tilde{v}_{ij}]_{n \times m}, \text{ where } \tilde{v}_{IJ} = \tilde{n}_{ij} \times w_j, i = 1, \ldots, m, j = 1, \ldots, n$$

$$\tilde{v}_{ij} = w_j \left[\left(\tilde{n}_{1_{ij}}, \tilde{n}'_{1_{ij}} \right); \tilde{n}_{2_{ij}}; \left(\tilde{n}'_{3_{ij}}, \tilde{n}_{3_{ij}} \right) \right] = \left[\left(g_{ij}, g'_{ij} \right); h_{ij}; \left(l'_{ij}, l_{ij} \right) \right] \tag{21.8}$$

Step 3. Positive-ideal and negative-ideal solutions can be defined as

$$A^+ = [(1, 1); 1; (1, 1)], j \in I$$
$$A^- = [(0, 0); 0; (0, 0)], j \in J. \tag{21.9}$$

Step 4. Now it is time to calculate the distance to positive-ideal solution and negative-ideal solution. According to Chen (2000), the Euclidean distance $d(\tilde{m}, \tilde{n})$ between two triangular fuzzy numbers $\tilde{m} = (m_1, m_2, m_3)$ and $\tilde{n} = (n_1, n_2, n_3)$ as

$$d(\tilde{m}, \tilde{n}) = \sqrt{\frac{1}{3} \left[\left(m_1 - n_1 \right)^2 + \left(m_2 - n_2 \right)^2 + \left(m_3 - n_3 \right)^2 \right]}, \tag{21.10}$$

which can be applied to calculate both the positive ideal solutions and negative ideal solutions. Because these are IVF numbers (g, g'); $h(l', l)$, it is necessary to calculate the distance of the primary distant measure $d^-(\tilde{m}, \tilde{n})$ and secondary distant measure $d^+(\tilde{m}, \tilde{n})$, which can be described as

$$d^+(\tilde{m}, \tilde{n}) = \sqrt{\frac{1}{3} \sum_{i=1}^{3} \left[\left(n^+_{x_i} - m^+_{y_i} \right)^2 \right]}$$

$$d^-(\tilde{m}, \tilde{n}) = \sqrt{\frac{1}{3} \sum_{i=1}^{3} \left[\left(n^-_{x_i} - m^-_{y_i} \right)^2 \right]} \tag{21.11}$$

Thereby, distance of each alternative from the ideal alternative $[d^+_{i1}, d^+_{i2}]$ can be currently calculated, where

$$d^+_{i1} = \sum_{j=1}^{m} \sqrt{\frac{1}{3} [(g_{ij} - 1)^2 + (h_{ij} - 1)^2 + (l_{ij} - 1)^2]}$$

$$d^+_{i2} = \sum_{j=1}^{m} \sqrt{\frac{1}{3} \left[(g'_{ij} - 1)^2 + (h_{ij} - 1)^2 + (l'_{ij} - 1)^2 \right]}, \tag{21.12}$$

where d_{i1}^+ is the distance between upper limit of IVF to the positive ideal solution and d_{i2}^+ is the distance of the lower limit to the positive ideal solution. Similarly, the separation from the negative ideal solution is given by $[d_{i1}^-, d_{i2}^-]$, thus

$$d_{i1}^- = \sum_{j=1}^m \sqrt{\frac{1}{3}[(g_{ij} - 0)^2 + (h_{ij} - 0)^2 + (l_{ij} - 0)^2]}$$

$$(21.13)$$

$$d_{i2}^- = \sum_{j=1}^m \sqrt{\frac{1}{3}\left[(g_{ij}' - 0)^2 + (h_{ij} - 0)^2 + (l_{ij}' - 0)^2\right]},$$

where d_{i1}^- is the distance between upper limit of IVF to the negative ideal solution and d_{i2}^- is the distance of the lower limit to the negative ideal solution. A better visualization of the distances calculated is presented in Figure 21.5; distances calculated by Equations (21.12) and (21.13) are represented. Equations (21.12) to (21.13) are used to determine the distance from the ideal and negative ideal alternatives in interval values, losing less information (data values) than just converting immediately to crisp values (Ashtiani et al., 2009).

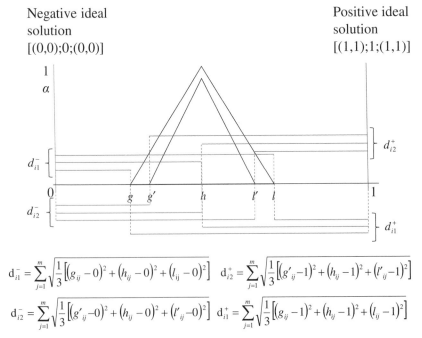

$$d_{i1}^- = \sum_{j=1}^m \sqrt{\frac{1}{3}\left[(g_{ij} - 0)^2 + (h_{ij} - 0)^2 + (l_{ij} - 0)^2\right]} \quad d_{i2}^+ = \sum_{j=1}^m \sqrt{\frac{1}{3}\left[(g_{ij}' - 1)^2 + (h_{ij} - 1)^2 + (l_{ij}' - 1)^2\right]}$$

$$d_{i2}^- = \sum_{j=1}^m \sqrt{\frac{1}{3}\left[(g_{ij}' - 0)^2 + (h_{ij} - 0)^2 + (l_{ij}' - 0)^2\right]} \quad d_{i1}^+ = \sum_{j=1}^m \sqrt{\frac{1}{3}\left[(g_{ij} - 1)^2 + (h_{ij} - 1)^2 + (l_{ij} - 1)^2\right]}$$

FIGURE 21.5   Distance of IVF numbers to positive ideal and negative ideal solutions schematic representation

Step 5. The relative closeness of the upper limit (RC1) and of the lower limit
(RC_2) can be calculated as follows:

$$RC_1 = \frac{d_{i1}^-}{d_{i1}^+ + d_{i1}^-}$$

$$RC_2 = \frac{d_{i2}^-}{d_{i2}^+ + d_{i2}^-}$$

(21.14)

The final value of RC_i^* is determined as

$$RC_i^* = \frac{RC_1 + RC_2}{2}.$$

(21.15)

21.3 THE SOLID WASTE MANAGEMENT SYSTEM

The Setúbal Peninsula is located in the district of Setúbal with an area of 1522 km^2
and has 714,589 inhabitants (Amarsul, 2009). The area is divided into nine municipalities. Amarsul, the company responsible for MSW since 1997, is owned by the
local municipalities on a regional basis (Pires et al., 2011a). The Amarsul MSW
management system is composed of nine recycling centers, two material recovery
facilities (MRF), two landfills, one transfer station, and one aerobic mechanical biological treatment (MBT) as described in Section 10.4. Currently, Amarsul promotes
the separation of paper/cardboard, glass and light packaging (plastics, metals, and
composite packaging) waste with curbside recycling systems (Pires et al., 2011a).
Each type of waste is collected separately in three specific containers at a site, and then
directly sent to the MRF for recycling, material recovery, and reuse. The remaining
fractions of waste from households are then collected through a door-to-door and/or
bin collection scheme, normally destined for final disposal at landfills (Pires et al.,
2011a). In the case of the Sesimbra municipality, the waste stream is first sent to a
transfer station, and then to final disposal at sanitary landfills. The residual waste collected from Setúbal municipality after waste separation and recycling is transported
to an aerobic MBT plant, where the "stabilized residue" is produced as fertilizer to
be applied as agriculture soil-amendment materials (Pires et al., 2011a).

The case study is devoted to the comparison of 18 alternatives plus current situation for managing MSW in the area of the Setúbal Peninsula in the Setúbal district, Portugal, considering environmental, economic, social, and technical criteria
for environmental sustainability. These alternatives, presented in Table 21.1, are
related to different combinations of technologies, like landfill with energy recovery
(ER), aerobic MBT, anaerobic digestion MBT, anaerobic digestion of biodegradable
municipal waste (BMW), and MRF. The current situation, named alternative "Base,"
is compared with remaining alternatives. As mentioned in Chapter 10, given the new
regulations that Amarsul must comply with, these 18 alternatives have been proposed
and elaborated with respect to the preselected waste management technologies. The
creation of Table 21.1 (the same as Table 10.12) is based on the total amount of waste

TABLE 21.1 **The distribution of waste streams associated with each alternative in the SWM system**

Fraction option (%)	Alternatives							
	$0/0^a/0^b$	$1/1^a$	$2/2^a/2^b$	$3/3^a$	$4/4^a/4^b$	$5/5^a/5^b$	$6/6^a$	Base
MRF	12.4	12.4	12.4	12.4	12.4	12.4	12.4	4.8
Anaerobic digestion BMW	5.4	0	0	13.3	0	7.5	28.7	0
Anaerobic digestion MBT	28.2	0	33.9	0	49.6	38.9	0	0
Aerobic MBT	13.2	49.7	15.8	32.6	0	0	0	13.8
Landfill with ER	40.8	37.9	37.9	41.7	38.0	41.2	58.9	81.4

Source: From Pires et al. (2011a).
[a]Alternatives considering RDF production plus incineration of high-calorific fraction.
[b]Alternatives not considering RDF production but considering incineration of high-calorific fraction from MBT.

produced in 2008, which was 421,726 metric tons (tonnes) (Pires et al., 2011a). Based on the average waste composition data region wide, the waste stream is composed of 31.69% putrescibles, 14.13% paper and cardboard, 11.35% plastics, 5.83% glass, 4.14% composites, 1.82% metals, 2.07% wood, 11.72% textiles, 15.33% fine particles, and 1.92% of other materials (Pires et al., 2011a). All the alternatives can assist with the actual need to reach the targets as prescribed in the European Directives. However, to reach the directive targets simultaneously requires a behavioral change in Portuguese society (Pires et al., 2011a).

Within this MSW system, there is a need to make some changes in order to comply with the Packaging and Packaging Waste Directive (European Parliament and Council, 2004) and Landfill Directive (Council, 1999). The National Plan for MSW (i.e., designated as PERSU II) stipulated the construction of several more MBT units (Pires et al., 2011a). An anaerobic digestion MBT unit, with a mechanical treatment to separate recyclables and high calorific material to produce refuse-derived fuel (RDF), was under the planning stage (Pires et al., 2011a). It is expected that this unit will work with two separate lines, one is related to the BMW and the other for the residual waste streams (Pires et al., 2011a). The RDF would be combusted in an incinerator to generate electricity. The existing aerobic MBT plant will be maintained as usual (Pires et al., 2011a). It is expected that both MRF plants with manual sorting will be replaced with two automatic sorting units. The alternatives were assessed by considering two scenarios: a baseline scenario, where external policies will be reinforced to reach the recycling and landfill targets required by European Directives, and PAYT scenario, where targets will be met solely through the application of an economic instrument.

21.3.1 Criteria and Decision Matrix

The sustainability criteria in AHP may include different evaluation criteria (Table 21.2). Seven stakeholders represented diversified areas of expertise as decision

TABLE 21.2   The sustainability criteria in AHP

Evaluation criteria	Description
Environmental criteria	
Abiotic depletion (AD)	Extraction of natural nonliving resources. It is the difference between resources consumed during waste life cycle and resource consumption avoided from materials and energy substitution, in kg Sb eq.
Acidification (Ac.)	Referent to acidifying pollutants emitted during waste life cycle. The calculation is the impacts from waste life cycle less the avoided impact from substituted materials and energy, in kg SO_2 eq.
Eutrophication (Eut.)	The consequence of high levels of macronutrients, such as nitrogen and phosphorous. It is the difference between the eutrophication substances potential impact during the waste life cycle and avoided impacts from substituted materials and energy, in kg PO_4^{3-} eq.
Global warming potential (GWP)	Represents the impact of greenhouse gas emissions on the radiative forcing of the atmosphere, inducing climate change. It is obtained from greenhouse gas (GHG) potential impact from waste life cycle less the GHG impact from substituted materials, kg CO_2 eq.
Human toxicity (HT)	The difference from impacts on human health of toxic substances emitted less the avoided impacts from substituted materials and energy life cycle, in kg p-DCB eq.
Photochemical oxidation (PO)	Represents the formation of reactive chemical compounds, such as ozone, by the action of sunlight on certain primary air pollutants. It is calculated from impact difference between waste life cycle and materials and energy-substituted life cycles, in kg C_2H_4 eq.
Gross energy requirement (GER)	Amount of commercial energy required directly and indirectly for the process of making a good or service. It is the difference between energy consumed and energy produced, in kJ.
Economic criteria	
Investment costs (IC)	The amount to be spent to implement the alternative (in infrastructure, equipment, vehicles, land).
Operational costs (OC)	The amount related to the amount to be spent during alternative operation, in material, electricity, maintenance, labor, and financial costs such as annuities.
Operational revenues (OR)	The amount related to the profits obtained from selling products (energy, recyclables, compost) or the avoidance of landfilling (RDF, recyclables).
Social criteria	
Economic efficiency (EE)	The ratio between the disposal fees collected from inhabitants and the net cost of MSW management system, as a percentage.
Fees	The amount paid by population to finance MSW management system.
Odor	The impact of odor substances emitted during waste life cycle.
Technical criteria	
Landfill space saving (LSS)	The ratio between waste not landfilled and total waste generated in a year, as a percentage.

Source: From Pires et al. (2011b).

makers, technicians, environmentalists, inhabitants, and experts invited to respond to inquiries for the retrieval of weighting factors through the AHP algorithm.

The criteria presented were selected considering the requirements of the new waste management philosophy proposed in the Waste Framework Directive 2008/98/EC (European Parliament and Council, 2008). This document justifies the application of technical, environmental, economic, and social aspects in MSW management. The technical aspect considered was LSS, since this is the major aspect that waste managers may control, while more costs will be needed to construct a new landfill. Environmental criteria are linked with LCA made for the alternatives elaborated in Chapter 10, referent to the work developed by Pires et al., (2011b). The use of LCA is justified by Waste Framework Directive 2008/98/EC (European Parliament and Council, 2008), in which a suggested waste management plan must conform with the waste hierarchy from waste prevention to waste recycling and reuse, to incineration and energy recovery, and to landfill sequentially. However, when applying LCA, MS in the EU must take measures to encourage the options which deliver the best overall environmental outcome. This may require specific waste streams departing from the hierarchy, and justifies life cycle thinking on the overall impacts of the production and management of such waste (European Parliament and Council, 2008). The LCA software used in this study was Umberto 5.5 to generate quantitative information. The environmental impact categories assessed were AD, Ac., Eut., GWP, HT, and PO. Another important environmental criterion is GER calculated for each alternative based on life cycle inventory data. Since Portugal does not produce fossil fuels, it is wise to look for waste management solutions keeping net energy demand as low as possible. All the data being used to perform the LCA may be seen in Pires et al. (2011b).

There are three criteria for addressing the economic aspects: IC, OC, and OR. Initial investment costs represent the amount needed to implement the WMS. Costs and benefits during the life cycle of MSW facilities also matter when choosing the best alternative. Several local entities provided information for calculating each category of costs and benefits, as summarized in Table 21.3.

Three social criteria were selected, including EE, fees, and odor. Odor information was obtained from LCA (Pires et al., 2011b). Since odor impact can be considered a public health issue, it was therefore classified as social criterion. Fees are the price paid by population to ensure the service of MSW disposal. The not-in-my-backyard (NIMBY) syndrome is a specific social impact that links odor to new facility siting. Fees are dependent on costs and revenues during a specific time frame. The importance of this criterion was justified by the fact that fees related to waste disposal are not popular in Portugal; therefore, the municipalities and Amarsul would not favor this option. Without the polluter pays principle, however, MSW facilities must be financed by municipalities from other sources. This justifies the use of economic sufficiency criteria. Economic sufficiency corresponds to the ratio between the amount paid by municipalities to Amarsul to manage the waste streams and the total cost required. Overall, some criteria are self-explanatory, but others may require further elaboration to avoid ambiguity and ensure sound understanding among the respondents in AHP analysis.

TABLE 21.3   The entities related to benefit cost analysis in relation to economic criteria

Types of data	Sources of data
Infrastructures and equipment	
Collection and transport of MSW and recyclables	Local data from collection companies; Hogg (2001), Empresa Geral do Fomento (EGF) (personal communication, 2009) and Piedade and Aguiar (2010)
MRF unit	Amarsul (2009), EGF (personal communication, 2009), InCI (2010), and Piedade and Aguiar (2010)
Aerobic MBT unit	Tsilemou and Panagiotakopoulos (2004, 2005), Amarsul (2009), and EGF (personal communication, 2009)
Anaerobic MBT unit with/without BMW line unit	Tsilemou and Panagiotakopoulos (2004, 2005), Amarsul (2009), and EGF (personal communication, 2009)
Landfill	Tsilemou and Panagiotakopoulos (2004, 2005), Amarsul (2009), and EGF (personal communication, 2009)
Products	
Recyclables	SPV (2010)
Compost	Amarsul (2009), and EGF (personal communication, 2009)
Electricity	MEI (2007)

Source: From Pires et al. (2011a).

All the criteria and associated values retrieved for each alternative are presented in Tables 21.4 and 21.5, which together represent the decision matrix. Those alternatives with capital P as the leading letter in the alternative column represent the alternative considering the PAYT option. The economic and social criteria for base scenario are overestimated since in the last few years, and biogas collection at landfills for producing electricity has not yet been lucrative because of insufficient biogas production. In the context of the environmental criteria, the situation was similar; hence, the base scenario is also overestimated for the best possible environmental performance.

21.3.2   First Stage: The AHP Method

The weights of the criteria to be used in the evaluation process were calculated using the AHP method. In this phase, the selected stakeholders were given the task of forming individual pairwise comparison matrices. The geometric means of these values were retrieved to form a pairwise comparison matrix with a consensus across all stakeholders (Table 21.6). The HT and GWP are the two relatively important criteria in the selection of waste management alternatives for the Amarsul system, as given in Table 21.7. The AHP allows calculation of a decision maker's inconsistency,

TABLE 21.4 Evaluation outcome of environmental criteria

				Environmental criteria			
Alternatives	AD (kg Sb eq)	Ac. (kg SO$_2$ eq)	Eut. (kg PO$_4^{3-}$ eq)	GWP (kg CO$_2$ eq)	HT (kg p-DCB eq)	PO (kg C$_2$ H$_2$ eq)	GER (kJ)
A0	−2.2E+05	−1.9E+05	4.9E+03	3.0E+08	3.8E+06	7.7E+04	−1.3E+12
A0*	−5.4E+05	−3.9E+05	−1.0E+04	1.6E+08	3.1E+06	5.4E+04	−2.6E+12
A0′	−5.6E+05	−4.0E+05	−1.1E+04	1.4E+08	3.1E+06	4.6E+04	−2.8E+12
A1	−1.7E+05	−2.0E+05	−4.5E+03	2.5E+08	6.2E+06	6.2E+04	−1.4E+12
A1*	−1.9E+05	−2.1E+05	−5.5E+03	2.4E+08	6.2E+06	6.0E+04	−1.7E+12
A2	−1.9E+05	−2.1E+05	−6.1E+02	3.2E+08	4.8E+06	8.2E+04	−1.5E+12
A2*	−5.7E+05	−4.5E+05	−1.8E+04	1.5E+08	4.0E+06	4.6E+04	−2.7E+12
A2′	−5.9E+05	−4.6E+05	−1.9E+04	1.4E+08	4.0E+06	4.4E+04	−2.8E+12
A3	−2.1E+05	−1.3E+05	1.4E+04	2.3E+08	3.5E+06	6.0E+04	−1.7E+12
A3*	−2.3E+05	−1.6E+05	1.3E+04	2.2E+08	3.5E+06	5.8E+04	−1.8E+12
A4	−2.4E+05	−2.4E+05	−7.4E+02	3.1E+08	4.0E+06	8.0E+04	−1.8E+12
A4*	−7.5E+05	−5.5E+05	−2.4E+04	1.0E+08	3.0E+06	4.0E+04	−3.7E+12
A4′	−7.8E+05	−5.7E+05	−2.6E+04	8.7E+07	2.9E+06	3.6E+04	−3.8E+12
A5	−2.4E+05	−1.9E+05	8.3E+03	3.3E+08	3.0E+06	8.7E+04	−1.7E+12
A5*	−6.8E+05	−4.6E+05	−1.2E+04	1.2E+08	2.0E+06	4.4E+04	−3.4E+12
A5′	−7.0E+05	−4.7E+05	−1.3E+04	1.1E+08	2.0E+06	4.2E+04	−3.5E+12
A6	−2.5E+05	−5.7E+04	3.4E+04	2.3E+08	−2.7E+06	6.1E+04	−1.7E+12
A6*	−2.7E+05	−7.2E+04	3.3E+04	2.1E+08	−2.8E+06	5.9E+04	−1.8E+12

P.A0	−2.2E+05	−1.9E+05	4.9E+03	3.0E+08	3.8E+06	7.7E+04	−1.3E+12
P.A0*	−5.4E+05	−3.9E+05	−1.0E+04	1.6E+08	3.1E+06	5.4E+04	−2.6E+12
P.A0′	−5.6E+05	−4.0E+05	−1.1E+04	1.4E+08	3.1E+06	4.6E+04	−2.8E+12
P.A1	−1.7E+05	−2.0E+05	−4.5E+03	2.5E+08	6.2E+06	6.2E+04	−1.4E+12
P.A1*	−1.9E+05	−2.1E+05	−5.5E+03	2.4E+08	6.2E+06	6.0E+04	−1.7E+12
P.A2	−1.9E+05	−2.1E+05	−6.1E+02	3.2E+08	4.8E+06	8.2E+04	−1.5E+12
P.A2*	−5.7E+05	−4.5E+05	−1.8E+04	1.5E+08	4.0E+06	4.6E+04	−2.7E+12
P.A2′	−5.9E+05	−4.6E+05	−1.9E+04	1.4E+08	4.0E+06	4.4E+04	−2.8E+12
P.A3	−2.1E+05	−1.3E+05	1.4E+04	2.3E+08	3.5E+06	6.0E+04	−1.7E+12
P.A3*	−2.3E+05	−1.6E+05	1.3E+04	2.2E+08	3.5E+06	5.8E+04	−1.8E+12
P.A4	−2.4E+05	−2.4E+05	−7.4E+02	3.1E+08	4.0E+06	8.0E+04	−1.8E+12
P.A4*	−7.5E+05	−5.5E+05	−2.4E+04	1.0E+08	3.0E+06	4.0E+04	−3.7E+12
P.A4′	−7.8E+05	−5.7E+05	−2.6E+04	8.7E+07	2.9E+06	3.6E+04	−3.8E+12
P.A5	−2.4E+05	−1.9E+05	8.3E+03	3.3E+08	3.0E+06	8.7E+04	−1.7E+12
P.A5*	−6.8E+05	−4.6E+05	−1.2E+04	1.2E+08	2.0E+06	4.4E+04	−3.4E+12
P.A5′	−7.0E+05	−4.7E+05	−1.3E+04	1.1E+08	2.0E+06	4.2E+04	−3.5E+12
P.A6	−2.5E+05	−5.7E+04	3.4E+04	2.3E+08	−2.7E+06	6.1E+04	−1.7E+12
P.A6*	−2.7E+05	−7.2E+04	3.3E+04	2.1E+08	−2.8E+06	5.9E+04	−1.8E+12
Base	−1.4E+05	−3.7E+04	−2.7E+03	9.5E+08	5.8E+06	2.5E+05	−1.3E+12

Source: From Pires et al. (2011a).

TABLE 21.5   Evaluation outcome of technical, economic, and social criteria

Criteria alternatives	Economic			Social		Technical	
	IC (10^6 US\$)	OC (US\$·y^{-1})	OR (US\$·y^{-1})	Fee (US\$· $tonne^{-1}$)	EE (%)	Odor (m^3)	LSS (%)
A0	1.72E+02	5.30E+07	2.12E+07	55	91	1.5E+13	30
A0*	1.59E+02	5.17E+07	2.25E+07	58	87	1.2E+13	43
A0′	1.59E+02	5.17E+07	2.25E+07	58	86	1.2E+13	44
A1	1.33E+02	4.51E+07	1.99E+07	67	70	1.4E+13	42
A1*	1.33E+02	4.51E+07	1.99E+07	67	70	1.3E+13	43
A2	1.72E+02	5.30E+07	2.25E+07	56	85	1.6E+13	23
A2*	1.59E+02	5.17E+07	2.25E+07	57	84	1.2E+13	43
A2′	1.59E+02	5.17E+07	2.25E+07	58	82	1.2E+13	44
A3	1.46E+02	4.51E+07	2.25E+07	73	77	1.3E+13	43
A3*	1.46E+02	4.51E+07	2.25E+07	73	77	1.2E+13	44
A4	1.72E+02	5.17E+07	2.39E+07	61	77	1.6E+13	21
A4*	1.59E+02	5.04E+07	2.25E+07	63	76	1.2E+13	45
A4′	1.46E+02	5.04E+07	2.25E+07	64	74	1.1E+13	46
A5	1.72E+02	5.04E+07	2.39E+07	63	82	1.6E+13	25
A5*	1.59E+02	5.04E+07	2.39E+07	65	80	1.2E+13	44
A5′	1.59E+02	4.90E+07	2.39E+07	66	78	1.2E+13	45
A6	1.72E+02	4.37E+07	2.39E+07	87	81	1.2E+13	38
A6*	1.59E+02	4.37E+07	2.39E+07	87	81	1.1E+13	38
P.A0	1.59E+02	5.30E+07	2.12E+07	100	91	1.5E+13	30
P.A0*	1.46E+02	5.17E+07	2.25E+07	100	86	1.2E+13	43
P.A0′	1.46E+02	5.17E+07	2.25E+07	100	86	1.2E+13	44
P.A1	1.72E+02	4.51E+07	1.99E+07	100	70	1.4E+13	42
P.A1*	1.59E+02	4.51E+07	1.99E+07	100	70	1.3E+13	43
P.A2	1.72E+02	5.30E+07	2.25E+07	100	85	1.6E+13	23
P.A2*	1.33E+02	5.17E+07	2.25E+07	100	82	1.2E+13	43
P.A2′	1.33E+02	5.17E+07	2.25E+07	100	82	1.2E+13	44
P.A3	1.72E+02	4.51E+07	2.25E+07	100	77	1.3E+13	43
P.A3*	1.59E+02	4.51E+07	2.25E+07	100	76	1.2E+13	44
P.A4	1.59E+02	5.04E+07	2.39E+07	100	76	1.6E+13	21
P.A4*	1.46E+02	5.04E+07	2.25E+07	100	74	1.2E+13	45
P.A4′	1.46E+02	4.90E+07	2.25E+07	100	74	1.1E+13	46
P.A5	1.72E+02	5.04E+07	2.39E+07	100	82	1.6E+13	25
P.A5*	1.59E+02	5.04E+07	2.39E+07	100	78	1.2E+13	44
P.A5′	1.59E+02	4.90E+07	2.39E+07	100	78	1.2E+13	45
P.A6	1.72E+02	4.37E+07	2.39E+07	100	81	1.2E+13	38
P.A6*	1.59E+02	4.37E+07	2.39E+07	100	81	1.1E+13	38
Base	1.59E+02	3.05E+07	1.59E+07	107	41	3.8E+13	13

Source: From Pires et al. (2011a).

TABLE 21.6 The pairwise comparison matrix in AHP analysis

	AD	Ac.	Eut.	GWP	HT	PO	GER	IC	OC	OR	EE	Fee	Odor	LSS
AD	1.00	0.96	2.94	2.26	0.54	0.27	0.43	1.24	1.78	1.18	0.71	0.64	0.57	0.63
Ac.	1.04	1.00	2.35	2.14	0.47	0.21	0.37	0.77	0.79	0.63	0.42	0.40	0.33	0.58
Eut.	0.34	0.43	1.00	1.55	0.27	0.22	0.40	0.63	0.74	0.78	0.42	0.39	0.54	0.54
GWP	0.44	0.47	0.64	1.00	0.28	0.19	0.32	0.47	0.48	0.43	0.42	0.38	0.37	0.50
HT	1.85	2.14	3.65	3.55	1.00	0.64	1.76	1.38	2.52	2.14	2.00	2.24	1.83	1.62
PO	3.77	4.74	3.73	5.38	1.56	1.00	3.09	4.34	4.68	4.68	4.88	4.27	5.10	3.29
GER	2.33	2.70	3.36	3.14	0.57	0.32	1.00	1.70	1.83	1.85	1.76	1.92	1.73	2.14
IC	0.81	1.30	1.70	2.12	0.72	0.23	0.59	1.00	1.41	1.47	1.41	0.89	0.71	0.79
OC	0.56	1.26	1.26	2.08	0.40	0.21	0.55	0.71	1.00	0.42	0.33	0.28	0.35	0.77
OR	0.85	1.59	1.28	2.33	0.47	0.21	0.54	0.68	2.38	1.00	0.64	0.33	0.34	0.71
EE	1.40	2.40	2.38	2.38	0.50	0.20	0.57	0.71	3.03	1.57	1.00	0.55	0.40	0.52
Fee	1.57	2.52	2.54	2.64	0.45	0.23	0.52	1.12	3.53	2.99	1.82	1.00	1.51	0.73
Odor	1.74	3.03	1.85	2.69	0.55	0.19	0.58	1.41	2.82	2.90	2.49	0.66	1.00	1.16
LSS	1.59	1.71	1.85	2.00	0.62	0.30	0.47	1.26	1.41	1.40	1.92	1.37	0.86	1.00

Source: From Pires et al. (2011a).

TABLE 21.7   Final outcome of AHP analysis

Criteria	Weights (w)	Criteria	Weights (w)	λ_{max}, CI, RI	CR
AD	0.040	IC	0.036	$\lambda_{max} = 14.63$	0.03
Ac.	0.031	OC	0.046	CI = 0.05	
Eut.	0.025	OR	0.058	RI = 1.57	
GWP	0.111	EE	0.079		
HT	0.220	Fee	0.080		
PO	0.098	Odor	0.067		
GER	0.056	LSS	0.053		

Source: From Pires et al. (2011a).

the consistency index (CI). This parameter is used to determine whether decisions violate the transitivity rule, and by how much (Bello-Dambatta et al., 2009). CI is defined by CI = $(\lambda_{max} - n)/(n - 1)$, where λ_{max} as above, n is the dimension. Based on the CI, it is possible to calculate consistency ratio, CR = CI/RI, where RI is the random index, being, at this case, for matrix order 14, RI is 1.57 (Lin and Yang, 1996). The weights obtained during AHP are consistent, since the CR of the pairwise comparison matrix is 0.03, valid because it is smaller than <0.1, in accordance with Saaty (1980).

21.3.3 Second Stage: The IVF TOPSIS Method

To develop the second stage, the applied methodology was started by defining membership functions to support the conversion of crisp values presented in the decision matrices (Tables 21.4 and 21.5) into triangular fuzzy numbers. To further convert triangular fuzzy functions into IVF with upper and lower membership functions, interval values were defined. The IVF TOPSIS cannot be fully initialized until all relevant decision matrices may contain essential IVF numbers. Given the IVF matrices, the normalization process may be carried out by multiplying IVF numbers by the weights listed in Table 21.7. Then, the distances to positive and negative ideal solutions can be calculated and, in the end, alternatives can be ranked. These numerical efforts can be sequentially described below.

Definition of Membership Functions and Conversion of Crisp into Triangular Numbers   In order to apply IVF TOPSIS, it is necessary to convert crisp values from Tables 21.4 and 21.5 (the decision matrices) into fuzzy membership functions and then from fuzzy membership into IVF membership functions. This is so-called a fuzzification process. The membership function shown in Figure 21.4 is presented for each criterion, with five linguistic variables including very good (VG), good (G), medium (M), poor (P), and very poor (VP). The selected types of membership functions are triangular in this study, since that has been commonly used for representing fuzzy numbers (Ding and Liang, 2005). A triangular fuzzy number $\tilde{a}$ can

TABLE 21.8 **Membership function of AD**

AD	Very good	Good	Medium	Poor	Very poor
−8.0E+5	0.95	0			
−6.3E+5	0	0.95	0		
−4.6E+5		0	0.95	0	
−3.0E+5			0	0.95	0
−1.3E+5				0	0.95

be defined as a triplet (a_1, a_2, a_3), and such representation of membership functions $\mu_{\tilde{a}}(x)$ can be realized by Figure 21.4.

$$\mu_{\tilde{a}}(x) = \begin{cases} 0, x < a_1 \\ \dfrac{x - a_1}{a_2 - a_1}, a_1 \leq x \leq a_2 \\ \dfrac{x - a_3}{a_2 - a_3}, a_2 \leq xa_3 \\ 0, x > a_3 \end{cases} \tag{21.16}$$

It is noticeable that a membership value does not reach 1, since the triangular IVF number would be situated between [0, 1]. To illustrate the procedure, the alternative A0 case for the criterion of AD is demonstrated. The conversion from crisp into linguistic variables is presented in Table 21.8.

The crisp value of A0 for addressing AD is −21,900 kg Sb eq. Looking at the first graphic shown in Figure 21.6, the triplet would be (−297,500, −130,000, −130,000) corresponding to linguist variable "Very Poor." The crisp value of A0* is −542,000 kg Sb eq., corresponding to linguist variable "Good", with the triplet of (−800,000, −632,500, −465,000).

Definition of Upper and Lower Membership Functions and Conversion from Triangular Numbers into IVF
A decision matrix was needed to construct the membership functions for each criterion, within [0, 1] for the upper bound and [0, 0.9] for the lower bound. A relative degree of uncertainty, such as 5% in this study, was proposed to address the independent impact associated with different types of uncertainty in an iterative process (UQ levels) for subsequent testing based on the interval between linguistic classes. To gain a better understanding, the case presented is related to the logical setting for AD and alternatives A0 and A0* for the purpose of demonstration. The upper and lower limits of IVF membership values are presented in Table 21.9.

In the case of A0, the triangular membership function may be defined by three numbers of (−297,500, −130,000, −130,000) which can be modified to be (−305,875, −289,125, −130,000, −130,000, −130,000). By the same logic, in the case of A0*, the triangular membership function may be defined by three numbers of (−800,000,

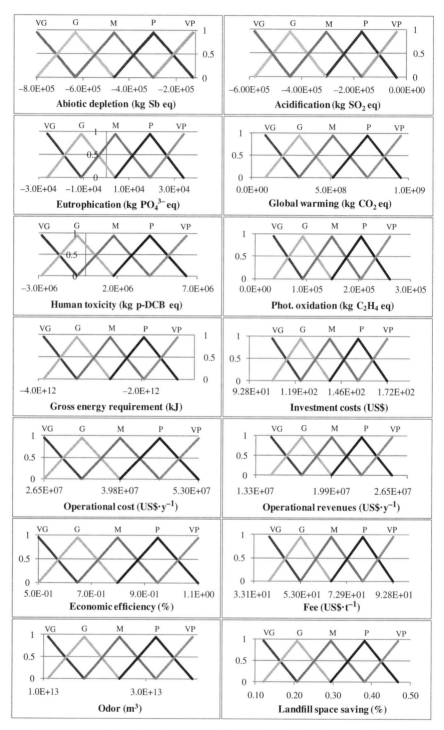

FIGURE 21.6 Membership function for IVF. *Source*: From Pires et al. (2011a)

TABLE 21.9 Upper and lower limits of IVF membership for AD

Upper limit	VG upper	G upper	M upper	P upper	VP upper
−8.0E+5	1				
−6.2E+5	0				
−8.1E+5		0			
−6.3E+5		1			
−4.6E+5		0			
−6.4E+5			0		
−4.6E+5			1		
−2.9E+5			0		
−4.7E+5				0	
−3.0E+5				1	
−1.2E+5				0	
−3.0E+5					0
−1.3E+5					1

Lower limit	VG lower	G lower	M lower	P lower	VP lower
−8.0E+5	0.9				
−6.4E+5	0				
−7.9E+5		0			
−6.3E+5		0.9			
−4.7E+5		0			
−6.2E+5			0		
−4.6E+5			0.9		
−3.0E+5			0		
−4.6E+5				0	
−3.0E+5				0.9	
−1.4E+5				0	
−2.9E+5					0
−1.3E+5					0.9

−632,500, −465,000) which can be modified to be (−808,375, −791,625, −632,500, −473,375, −456,625). Both cases are represented in Figure 21.7.

IVF TOPSIS: Normalized Matrix of IVF The weights obtained in AHP enable us to proceed with the normalization of IVF TOPSIS procedure. The normalized IVF matrix may be used in subsequent steps of the IVF TOPSIS procedure for distance-based assessment. Within the IVF TOPSIS procedure, if the criterion is cost-related and the value is negative, then the benefit formula must be used and vice versa. The case of AD exactly exhibits such outcome. In this case, the IVF for AD associated with option A0 is ((0.38, 0.36); 0.16; (0.16, 0.16)). By the same principle, the IVF for environmental criterion of AD associated with option A0* is ((1, 0.98); 0.78; (0.59, 0.56)). We are able to present the summary of the results reached for A0 and A0* so far, in the context of IVF TOPSIS (see Table 21.10). We may follow the procedure to create the IVF associated with each criterion for all the rest of the alternatives.

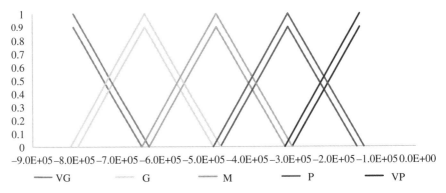

FIGURE 21.7   Graphic representation of the IVF membership functions for AD

TABLE 21.10   IVF membership for AD and normalized values for alternatives A0 and A0*

Alternatives	Triangular fuzzy number	IVF	Normalization
A0	(−297,500, −130,000, −130,000)	((−305,875, −289,125); −130,000; (−130,000, −130,000))	((0.38, 0.36); 0.16; (0.16, 0.16))
A0*	(−800,000, −632,500, −465,000)	((−808,375, −791,625); −632,500; (−473,375, −456,625))	((1, 0.98); 0.78; (0.59, 0.56))

Source: From Pires et al. (2011a).

Distance from Ideal Solutions   By applying Equations (21.8)–(21.15), the results for distance-based ideal solutions and relative closeness were calculated and are presented in Table 21.11 for demonstration purposes.

21.3.4   Overall Assessment

In an iterative procedure, the calculations with respect to varying degrees of uncertainty may be obtained to explore the results via UQ. The iterative procedure can be stopped once we are sure that all types of uncertainty are addressed. The degrees of uncertainty being tested in this study were based on the interval between linguistic classes, such as 5%, 50%, 75%, and 100% (Pires et al., 2011a). When the iteration can be made possible by simply reducing the interval gradually after the initial selection

TABLE 21.11   A demonstration of the TOPSIS outcome

Alternatives	d^+	d^-	RC_1	RC_2	RC_j^*
A0	13.69	13.69	0.05	0.05	0.048
A0*	13.67	13.68	0.05	0.05	0.048

Source: From Pires et al. (2011a).

TABLE 21.12   Final ranking of the AHP-based IVF TOPSIS analysis

			Uncertainty tested and rankings				
5%	Rank	50%	Rank	75%	Rank	100%	Rank
0.0619	P.A5*	0.0505	P.A5*	0.0479	P.A5*	0.0492	P.A1
0.0618	P.A5′	0.0504	P.A5′	0.0478	P.A5′	0.0484	P.A1*
0.0611	P.A2	0.0494	A5*	0.0469	A5*	0.0482	A1
0.0608	A5*	0.0493	A5′	0.0467	A5′	0.0474	A1*
0.0606	A5′	0.0468	P.A2	0.0467	P.A1	0.0466	P.A5*
0.0598	P.A4	0.0460	P.A1	0.0459	P.A1*	0.0465	P.A5′
0.0593	P.A1	0.0456	P.A4	0.0456	A1	0.0456	A5*
0.0585	P.A1*	0.0452	P.A1*	0.0448	A1*	0.0454	A5′
0.0583	A2	0.0449	A1	0.0439	P.A2	0.0452	P.A2
0.0582	A1	0.0442	A2	0.0426	P.A4	0.0440	P.A4
0.0574	A4	0.0440	A1*	0.0415	P.A4*	0.0429	A2
0.0573	A1*	0.0433	A4	0.0414	A2	0.0426	P.A3
0.0525	P.A4*	0.0432	P.A4*	0.0413	P.A4′	0.0422	P.A3*
0.0523	P.A4′	0.0430	P.A4′	0.0404	A4	0.0421	P.A0
0.0511	P.A2*	0.0419	P.A2*	0.0403	P.A2*	0.0419	A4
0.0511	P.A2′	0.0419	P.A2′	0.0403	P.A2′	0.0410	P.A6′
0.0507	P.A0*	0.0417	P.A0*	0.0401	P.A0*	0.0410	P.A4*
0.0502	P.A0′	0.0411	P.A0′	0.0396	P.A0′	0.0409	A3
0.0501	P.A0	0.0405	A4*	0.0393	P.A3	0.0408	P.A4′
0.0496	A4*	0.0405	A4′	0.0389	A4*	0.0406	A6*
0.0496	A4′	0.0404	Base	0.0389	A4′	0.0405	A3*
0.0494	Base	0.0396	A2*	0.0388	P.A3*	0.0405	A0
0.0489	P.A3	0.0393	A0*	0.0382	Base	0.0403	P.A6
0.0487	A2*	0.0388	A0′	0.0380	A2*	0.0399	A6
0.0484	P.A3*	0.0388	A2′	0.0378	A0*	0.0399	P.A2*
0.0483	A0*	0.0376	P.A3	0.0375	A3	0.0399	P.A2′
0.0482	A0	0.0372	P.A3*	0.0375	P.A6*	0.0397	P.A0*
0.0477	A2′	0.0362	P.A0	0.0374	P.A0	0.0392	P.A0′
0.0477	A0′	0.0358	P.A6*	0.0373	A0′	0.0386	A4*
0.0469	A3	0.0357	A3	0.0373	A2′	0.0386	A4′
0.0468	P.A6*	0.0356	P.A5	0.0371	A3*	0.0377	A2*
0.0465	A3*	0.0353	A3*	0.0371	A6*	0.0375	A0*
0.0464	P.A5	0.0353	A6*	0.0368	P.A6	0.0372	Base
0.0463	A6*	0.0349	P.A6	0.0363	A6	0.0370	A0′
0.0459	P.A6	0.0345	A6	0.0356	A0	0.0370	A2′
0.0454	A6	0.0344	A0	0.0342	P.A5	0.0369	P.A5
0.0440	A5	0.0333	A5	0.0319	A5	0.0347	A5

Source: From Pires et al. (2011a).

across all fuzzy variables, it is possible to observe that only when the interval is one time bigger than the linguistic classes is exactly when there is a change in ranking (Table 21.12) (Pires et al., 2011a). The best solution for the Amarsul system would be implementation of anaerobic digestion MBT and an anaerobic digestion plant for biodegradable municipal waste followed by RDF production, managed by a PAYT

TABLE 21.13 Final ranking of the IVF TOPSIS analysis—without weighted criteria

Uncertainty tested and rankings			
5%, weighted criteria	Rank	5%, without weighted criteria	Rank
0.0619	P.A5*	0.0578	P.A5*
0.0618	P.A5'	0.0575	P.A5'
0.0611	P.A2	0.0567	A5*
0.0608	A5*	0.0564	A5'
0.0606	A5'	0.0531	P.A4*
0.0598	P.A4	0.0528	P.A4'
0.0593	P.A1	0.0517	P.A1
0.0585	P.A1*	0.0508	P.A2*
0.0583	A2	0.0508	P.A2'
0.0582	A1	0.0506	A1
0.0574	A4	0.0506	P.A1*
0.0573	A1*	0.0505	P.A2
0.0525	P.A4*	0.0497	A4*
0.0523	P.A4'	0.0497	A4'
0.0511	P.A2*	0.0495	P.A0*
0.0511	P.A2'	0.0495	A1*
0.0507	P.A0*	0.0493	A2*
0.0502	P.A0'	0.0485	P.A0'
0.0501	P.A0	0.0484	P.A4
0.0496	A4*	0.0479	A2
0.0496	A4'	0.0477	A2'
0.0494	Base	0.0473	A0*
0.0489	P.A3	0.0471	P.A0
0.0487	A2*	0.0463	A0'
0.0484	P.A3*	0.0461	A4
0.0483	A0*	0.0459	P.A6*
0.0482	A0	0.0455	A6*
0.0477	A2'	0.0453	A0
0.0477	A0'	0.0452	P.A5
0.0469	A3	0.0449	P.A6
0.0468	P.A6*	0.0448	P.A3
0.0465	A3*	0.0444	A6
0.0464	P.A5	0.0440	P.A3*
0.0463	A6*	0.0432	Base
0.0459	P.A6	0.0430	A3
0.0454	A6	0.0429	A5
0.0440	A5	0.0422	A3*

Source: From Pires et al. (2011a).

program. As a consequence, A5 is the best option aided by a PAYT program (i.e., designated as P.A5* in this chapter) (Pires et al., 2011a).

If criteria weights are deemed equally important without the involvement of AHP and 5% uncertainty is assumed with the same degree of information diffusion among stakeholders in this practice, the decision analysis would turn out to be different. The options obtained in this situation are presented in Table 21.13. The same best alternative is reached. The change of weights would signify the higher importance

TABLE 21.14   Final ranking of the IVF TOPSIS analysis—uncertainty tested without weighted criteria

		Uncertainty tested and rankings without weights					
5%	Rank	50%	Rank	75%	Rank	100%	Rank
0.0619	P.A5*	0.5711	P.A5*	0.5413	P.A5*	0.5167	P.A5*
0.0618	P.A5′	0.5682	P.A5′	0.5385	P.A5′	0.5140	P.A5′
0.0611	P.A2	0.5587	A5*	0.5292	A5*	0.5052	A5*
0.0608	A5*	0.5558	A5′	0.5265	A5′	0.5025	A5′
0.0606	A5′	0.5536	P.A4*	0.5225	P.A4*	0.4978	P.A4*
0.0598	P.A4	0.5506	P.A4′	0.5196	P.A4′	0.4951	P.A4′
0.0593	P.A1	0.5246	P.A2*	0.4957	P.A2*	0.4736	P.A2*
0.0585	P.A1*	0.5246	P.A2′	0.4957	P.A2′	0.4736	P.A2′
0.0583	A2	0.5118	A2*	0.4827	A2*	0.4612	A4*
0.0582	A1	0.5109	A4*	0.4826	A4*	0.4612	A4′
0.0574	A4	0.5109	A4′	0.4826	A4′	0.4605	A2*
0.0573	A1*	0.5029	P.A0*	0.4767	P.A0*	0.4571	P.A0*
0.0525	P.A4*	0.4922	P.A0′	0.4665	P.A0′	0.4477	P.A0′
0.0523	P.A4′	0.4901	A2′	0.4630	A2′	0.4429	A2′
0.0511	P.A2*	0.4775	A0*	0.4550	P.A1	0.4424	P.A1
0.0511	P.A2′	0.4748	P.A1	0.4522	A0*	0.4339	A0*
0.0507	P.A0*	0.4668	A0′	0.4440	A1	0.4324	A1
0.0502	P.A0′	0.4629	A1	0.4435	P.A1*	0.4320	P.A1*
0.0501	P.A0	0.4623	P.A1*	0.4421	A0′	0.4245	A0′
0.0496	A4*	0.4605	P.A2	0.4324	A1*	0.4220	A1*
0.0496	A4′	0.4504	A1*	0.4301	P.A2	0.4160	P.A2
0.0494	Base	0.4435	P.A6*	0.4249	P.A6*	0.4149	P.A6*
0.0489	P.A3	0.4407	P.A4	0.4200	A6*	0.4104	A6*
0.0487	A2*	0.4381	A6*	0.4141	P.A3	0.4054	P.A3
0.0484	P.A3*	0.4335	A2	0.4139	P.A6	0.4051	P.A6
0.0483	A0*	0.4334	Base	0.4127	Base	0.4007	A6
0.0482	A0	0.4313	P.A6	0.4112	P.A4	0.3988	P.A3*
0.0477	A2′	0.4312	P.A3	0.4089	A6	0.3986	P.A4
0.0477	A0′	0.4307	P.A5	0.4068	P.A3*	0.3973	Base
0.0469	A3	0.4259	A6	0.4045	A2	0.3929	A2
0.0468	P.A6*	0.4233	P.A3*	0.4024	P.A5	0.3927	P.A0
0.0465	A3*	0.4220	P.A0	0.3990	P.A0	0.3918	P.A5
0.0464	P.A5	0.4167	A4	0.3957	A3	0.3890	A3
0.0463	A6*	0.4111	A3	0.3883	A3*	0.3823	A3*
0.0459	P.A6	0.4057	A5	0.3882	A4	0.3775	A4
0.0454	A6	0.4031	A3*	0.3812	A0	0.3771	A0
0.0440	A5	0.4025	A0	0.3786	A5	0.3701	A5

of economic considerations although this change still cannot completely alter the final option.

Given the equal weight assumption, varying degrees of uncertainty may still be assumed regarding the differing degrees of information diffusion among stakeholders to signify the sensitivity of fuzzy classes. Changes can be reported based on the most sensitive retardation of information diffusion so that the final option may be altered. With this consideration, A5 is still the best option, as presented in Table 21.14, and is based on the implementation of anaerobic and aerobic MBT units, including a PAYT program.

21.4 FINAL REMARKS

The selection of waste management strategies to improve sustainability in the Amarsul system is a challenging issue particularly when trying to reach the targets at the national level as set by European Directives. There are many alternatives that can be geared toward reaching such goals, but how the policy information can be propagated from government to all stakeholders of the general public and how the stakeholders respond to this urgency remains uncertain. If new measures like PAYT are considered in relation to 18 alternatives in the decision-making process to promote the odds of success, a scientific methodology (i.e., UQ) to assess waste management alternatives must be available. In the context of FIMADM, various sources of uncertainty have to be quantified using tailored mathematical algorithms to tackle natural variability, incomplete knowledge, and even communication breakdown. Through the use of interval-valued triangular fuzzy numbers to express linguistic classes embedded in the decision-making process, the expanded MADM model described in this study provides us with an objective screening and ranking procedure with respect to environmental, economic, technical, and social criteria partially supported by a stand-alone LCA. Both AHP and TOPSIS are seamlessly integrated and applied to retrieve criteria weights for alternative selection. Whereas IVF TOPSIS is employed to determine the priorities of the alternatives, the weights derived from AHP reveal the impacts in a societal context. This expanded MADM model was shown to be adequate in this case study for decision-making, since many sources of uncertainty can be collectively characterized by the IVF scheme. The final success of this thrust in the Amarsul system is linked to the proper handling of recycling programs, PAYT, and the choice of the best solution.

REFERENCES

Amarsul. 2009. Annual report 2009 [Portuguese: Relatório e contas 2009]. Amarsul. Available at: http://www.amarsul.pt/relatorios e contas.php (accessed July 2009).

Ashtiani, B., Haghighirad, F., Makui, A., and Ali Montazer, G. 2009. Extension of fuzzy TOPSIS method based on interval-valued fuzzy sets. *Applied Soft Computing*, 9(2), 457–461.

Bello-Dambatta, A., Farmani, R., Javadi, A. A., and Evans, B. M. 2009. The analytical hierarchy process for contaminated land management. *Advanced Engineering Informatics*, 23(4), 433–441.

Berlyne, D. E. 1962. Uncertainty and epistemic curiosity. *British Journal of Psychology*, 53(1), 27–34.

Chen, C. T. 2000. Extensions of the TOPSIS for group decision-making under fuzzy environment. *Fuzzy Sets and Systems*, 114(1), 1–9.

Council. 1999. Council Directive 1999/31/EC of April 1999 on the landfill of waste. *Official Journal of European Communities*, L182, 1–19.

Ding, J. F. and Liang, G. S. 2005. Using fuzzy MCDM to select partners of strategic alliances for liner shipping. *Information Sciences*, 173(1–3), 197–225.

European Parliament and Council. 2004. Directive 2004/12/EC of the European Parliament and of the Council of 11 February 2004 amending Directive 94/62/EC on packaging and packaging waste. *Official Journal of European Communities*, L047, 26–31.

European Parliament and Council. 2008. Directive 2008/98/EC of the European Parliament and of the Council of 19 November 2008 on waste and repealing certain Directives. *Official Journal of European Communities*, L312, 3–30.

Hisdal, E. 1981. The IF THEN ELSE statement and interval-valued fuzzy sets of higher type. *International Journal of Man-Machine Studies*, 15(4), 385–455.

Hogg, D. 2001. Costs of municipal waste management in the EU. Final report to Directorate General Environment. European Commission. Available at: http://europa .eu.int/comm/environment/waste/studies/eucostwastemanagement.html (accessed October 2010).

Huang, C. 1997. Principle of information diffusion. *Fuzzy Sets and Systems*, 91(1), 69–90.

Hwang, C. L. and Yoon, K. S. 1981. *Multiple Attribute Decision Making: Method and Application*, Springer-Verlag, New York.

Instituto da Construção e do Imobiliário, (InCI). (2010). Database: online public contracts [Portuguese: Base: contratos públicos online]. InCI. Available at: http://www.base.gov.pt/ Paginas/Default.aspx (accessed July 2010).

Kahraman, C. 2008. Multi-criteria decision making methods and fuzzy sets. In: *Fuzzy Multi-Criteria Decision Making: Theory and Applications with Recent Developments* (Ed. Kahraman, C.), Springer, New York, pp. 1–18.

Karnik, N. N., Mendel, J. M., and Liang, Q. 1999. Type-2 fuzzy logic systems. *IEEE Transactions on Fuzzy Systems*, 7(6), 643–658.

Liang, Q. and Mendel, J. M. 2000. Interval type-2 fuzzy logic systems: theory and design. *IEEE Transactions on Fuzzy Systems*, 8(5), 535–550.

Lin, Z.-C. and Yang, C.-B. 1996. Evaluation of machine selection by the AHP method. *Journal of Materials Processing Technology*, 57(3–4), 253–258.

Ministério da Economia e da Inovação (MEI). 2007. Law Decret 225/2007, May 31 [Portuguese: Decreto-Lei n. 225/2007, de 31 de Maio]. *Diário da República 1-Série,* 105, 3630–3638.

Nimmo, D. 1985. Information and political behavior. In: *Information and Behavior* (Ed. Ruben, B. D.), Transaction Publishers, New Brunswick, NJ, pp. 343–368.

Piedade, M. and Aguiar, P. 2010. *Urban Waste Management Options* [Portuguese: Opcões de Gestão de Resíduos Urbanos], ERSAR – Guias técnicos 15, Lisbon.

Pires, A., Chang, N. B., and Martinho, G. 2011a. An AHP-based fuzzy interval TOPSIS assessment for sustainable expansion of the solid waste management system in Setúbal Peninsula, Portugal. *Resources Conservation and Recycling*, 56(1), 7–21.

Pires, A., Chang, N. B., and Martinho, G. 2011b. Reliability-based life cycle assessment for future solid waste management alternatives in Portugal. *The International Journal of Life Cycle Assessment*, 16(4), 316–337.

Rogers, E. M. 1962. *Diffusion of Innovations*, Free Press, New York.

Saaty, T. L. 1980. *The Analytical Hierarchy Process*, McGraw-Hill, New York.

Sociedade Ponto Verde (SPV). 2010. Green Dot System—Financial counterparts [Portuguese: Sociedade PontoVerde – Contrapartidas financeiras]. SPV. Available at: http://www.pontoverde.pt/; 2010 (accessed July 2010).

Tsilemou, K. and Panagiotakopoulos, D. 2004. Estimating costs for solid waste treatment facilities. In: Proceedings of ISWA World Environmental Congress and Exhibition, Rome, Italy.

Tsilemou, K. and Panagiotakopoulos, D. 2005. A statistical methodology for generating cost functions for solid waste treatment facilities. In: Proceedings of the 5th International Exhibition and Conference in Environmental Technology (HELECO'05), Athens, Greece.

Türksen, I. B. 2006. *An Ontological and Epistemological Perspective of Fuzzy Set Theory*, Elsevier, Netherlands.

United Nations Environment Programme (UNEP). 2004. *The Use of Economic Instruments in Environmental Policy: Opportunities and Challenges*, United Nations Publications, Geneva, Switzerland.

Zadeh, L. A. 1975a. The concept of a linguistic variable and its application to approximate reasoning, Part I. *Information Sciences*, 8(3), 199–249.

Zadeh, L. A. 1975b. The concept of a linguistic variable and its application to approximate reasoning, Part II. *Information Sciences*, 8(4), 301–357.

CHAPTER 22

FUZZY MULTIOBJECTIVE DECISION-MAKING FOR SOLID WASTE MANAGEMENT

In a multiobjective programming model, the principal ingredients include a set of alternatives, a set of objective functions that describe the gain or loss associated with each alternative resulting from the choice of that alternative, and a set of constraints on the choice between different alternatives. Both the multiple engineering goals and managerial targets in the constraint set can be tied together by using fuzzy descriptions. It is the aim of this chapter to provide readers with a holistic insight into fuzzy multiobjective decision-making problems when dealing with a specific type of solid waste management (SWM) issue. A case study of balancing recycling and incineration goals via a fuzzy goal programming model for SWM illuminates the essence of two problems: (1) to what extent are the recycling and incineration compatible? and (2) what are the subsequent economic impacts on the private or the public sectors in various types of management scenarios? In any case, the technique of fuzzy multiobjective decision-making provides an avenue to elucidate the complexity of challenging issues.

22.1 FUZZY LINEAR PROGRAMMING

22.1.1 Fuzzy Decision and Operators

Fuzzy decision-making frequently requires assessing the interactions between different membership functions reflecting vague features in a knowledge domain of interest. To make such interactions logical and meaningful, set-theoretic operators (aggregators) are defined for various types of decision analysis. Therefore, the

Sustainable Solid Waste Management: A Systems Engineering Approach, First Edition. Ni-Bin Chang and Ana Pires.
© 2015 The Institute of Electrical and Electronics Engineers, Inc. Published 2015 by John Wiley & Sons, Inc.

context-dependent semantic interpretation inherent to fuzzy decision-making will lead to search for different options when using different set-theoretic operators. A system analyst helping decision-making has to select a final set-theoretic operator, deemed the most germane and appropriate for a real-world problem. The search for an appropriate operator itself independently could become a challenge when linked to societal implications. Oftentimes, neither the concept of membership value itself nor the set-theoretic operator finally selected can uniquely present semantic interpretation, since such a process is context dependent. When these managerial scenarios are tied with mathematical programming models, such as the fuzzy linear programming or fuzzy multiobjective programming models, it is likely that deepened insight is present within the optimal solution under uncertainty.

Solving these fuzzy linear programming or fuzzy multiobjective programming models to obtain an optimal solution is by no means an easy task. Zadeh (1965) proposed max and min operators to define the intersection and union of fuzzy sets, assuming that objective(s) and constraints can be represented by fuzzy sets when they exhibit imprecise information. When a decision-making process can be delineated in a fuzzy environment to mimic the interactions among the stakeholders involved, the symmetry between goals and constraints is the most important feature (Bellman and Zadeh, 1970). With such a symmetry property, a fuzzy decision can be defined as the fuzzy set of alternatives resulting from the intersection of the objectives and constraints in a unique mathematical construct.

We have a fuzzy objective O and a fuzzy constraint C in a decision space. Given the symmetry assumption, a decision can be made in the situation when both the objective and constraint can be satisfied. The key factor is that the fuzzy sets O and C are connected by an operator "and." This operation corresponds to the "intersection" of both fuzzy sets implying that their combined effect on the choice of alternatives can be represented by $O \cap C$. That is,

$$\mu_{O \cap C}(x) = \mu_O(x) \wedge \mu_C(x) = \min\{\mu_O(x), \mu_C(x)\} \quad \forall x \in X.$$

If the decision is to maximize the minimum membership value, then

$$\mu_D(x') = \max \mu_D(x) \text{ for } x \in X.$$

Figure 22.1 describes such an operation (max–min operator).

On the other hand, if fuzzy sets O and C are connected by an operator "or," this operation corresponds to the "union" of both fuzzy sets implying that their combined effect on the choice of alternatives can be represented by $O \cup C$. That is:

$$\mu_{O \cup C}(x) = \mu_O(x) \vee \mu_C(x) = \mu_O(x) + \mu_C(x) - \mu_O(x)\mu_C(x)$$
$$= \max\{\mu_O(x), \mu_C(x)\} \quad \forall x \in X.$$

If the decision is to minimize the maximum membership value, then

$$\mu_D(x') = \min \mu_D(x) \text{ for } x \in X.$$

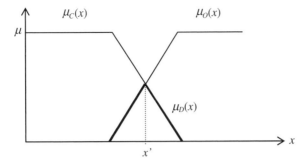

FIGURE 22.1   Fuzzy decision-making with respect to an operator "and"

Figure 22.2 describes such an operation (min–max operator).

From a logical point of view, interpretation of the intersection as "logical and" and the union as "logical or" could meet the requirement for axiomatic justification. The former is consistent with a decision that maximizes the minimal possible return and the latter is consistent with a decision that minimizes the maximal possible penalty (or regret) (Chen and Hwang, 1991). Both exhibit the common but compelling merits of computational tractability and simplicity in association with complex decision-making processes. Chen and Hwang (1991) further summarized all the compensatory operators for different applications as follows.

1. **Compensatory min operators**
    - **Algebraic product:** $\mu_D(x) = \mu_O(x)\mu_C(x)$.
    - **Bound product:** $\mu_D(x) = \max[0, \mu_O(x) + \mu_C(x) - 1]$.
    - **Hamacher's min operator (Γ-operator):** for $\Gamma \in [0, 1]$,

$$\mu_D(x) = \frac{\mu_O(x)\mu_C(x)}{\Gamma + (1 + \Gamma)[\mu_O(x) + \mu_C(x) - \mu_O(x)\mu_C(x)]}.$$

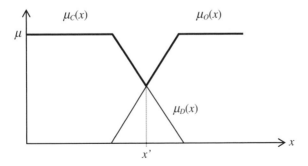

FIGURE 22.2   The relationship between O, C, and D with respect to an operator "or"

- **Yager's min operator**: for $q \geq 0$,

$$\mu_D(x) = 1 - \min\{1, [(1 - \mu_O(x))^q + (1 - \mu_C(x))^q]^{1/q}\}.$$

- **Dubois and Prade's min operator**: for $r \in [0, 1]$,

$$\mu_D(x) = \frac{\mu_O(x)\mu_C(x)}{\max[\mu_O(x), \mu_C(x), r]}.$$

- **Werner's "fuzzy and" operator**: for $r \in [0, 1]$,

$$\mu_D(x) = r \min[\mu_O(x), \mu_C(x)] + (1 - r)[\mu_O(x) + \mu_C(x)]/2.$$

2. **Compensatory max operators**
   - **Algebraic sum**: $\mu_D(x) = \mu_O(x) + \mu_C(x) - \mu_O(x)\mu_C(x)$.
   - **Bound sum**: $\mu_D(x) = \min[1, \mu_O(x) + \mu_C(x)]$.
   - **Hamacher's max operator (Γ-operator)**: for $\Gamma \in [0, 1]$,

$$\mu_D(x) = \frac{(1 - \Gamma)\mu_O(x)\mu_C(x) + \Gamma[\mu_O(x) + \mu_C(x)]}{\Gamma + \mu_O(x)\mu_C(x)}.$$

- **Yager's max operator**: for $q \geq 1$,

$$\mu_D(x) = \min\{1, [\mu_O(x)^q + \mu_C(x)^q]^{1/q}\}.$$

- **Dubois and Prade's max operator**: for $r \in [0, 1]$,

$$\mu_D(x) = \frac{\mu_O(x) + \mu_C(x) - \mu_O(x)\mu_C(x) - \min[1 - r, \mu_O(x), \mu_C(x)]}{\max[r, 1 - \mu_O(x), 1 - \mu_C(x), r]}.$$

- **Werner's "fuzzy or" operator**: for $r \in [0, 1]$,

$$\mu_D(x) = r \max[\mu_O(x), \mu_C(x)] + (1 - r)[\mu_O(x) + \mu_C(x)]/2.$$

22.1.2 The Formulation of Fuzzy Linear Programming

Two types of fuzzy linear programming problems are included in this section in which the right-hand side values of the constraint are taken into account as fuzzy resources. In most cases, environmental resources are considered as fuzzy resources in decision-making. For simplicity, a linear membership function and max–min operator is employed in the following discussion. So the following linear programming problem can be formulated in terms of fuzzy sets.

$$\begin{array}{ll} \text{Max} & \tilde{z} = cx \\ \text{s.t.} & (Ax)_i \leq \tilde{b}_i, \quad i = 1, 2, \ldots, m \\ & x \geq 0. \end{array}$$

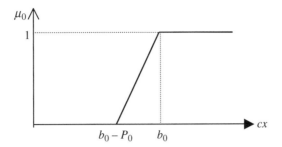

FIGURE 22.3   The membership of the fuzzy objective and its definition

The Zimmermann approach is a symmetric approach (Zimmermann, 1978; Zimmermann and Zysno, 1985). It assumes that both objective and constraints are considered fuzzy simultaneously. The target value b_0 and the corresponding tolerance P_0 of the fuzzy objective are given initially, so are the fuzzy resources in the constraint. The right-hand side b_i and its corresponding tolerance P_i are given beforehand. Thus, a fuzzy linear programming model can be reformulated as

$$\text{find } x$$
$$\text{such that} \quad cx \geq \tilde{b}_0$$
$$(Ax)_i \leq \tilde{b}_i, \quad \forall i$$
$$x \geq 0.$$

The fuzzy objective function and the fuzzy constraints are defined by their respective membership functions. Let us assume the membership function of fuzzy objective function μ_0 is a nondecreasing continuous linear function (see Figure 22.3) and the membership function of fuzzy constraint μc is a nonincreasing continuous linear function (see Figure 22.4).

$$\mu_0(x) = \begin{cases} 1 & \text{if} & x \geq b_0 \\ 1 - \dfrac{b_0 - cx}{P_0} & \text{if} & b_0 - P \leq cx \leq b_0. \\ 0 & \text{if} & cx \leq b_0 - P_0 \end{cases}$$

$$\mu_i(x) = \begin{cases} 1 & \text{if} & (Ax)_i < b_i \\ 1 - \dfrac{(Ax)_i - b_i}{P_i} & \text{if} & b_i \leq (Ax)_i \leq b_i + P_i. \\ 0 & \text{if} & (Ax)_i > b_i - P_i \end{cases}$$

The optimal solution can be found by using max–min operator,

$$\max \mu_D = \max\{\min[\mu_0(x), \mu_1(x), \dots, \mu_m(x)]\},$$

where μ_D is the membership function of the decision space D (Figure 22.5).

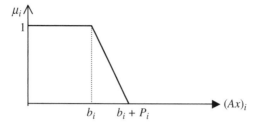

FIGURE 22.4   The membership of the fuzzy constraint and its definition

Therefore, if $\alpha = \mu_D(x)$, the model can be expressed as follows:

$$\max \alpha$$

$$\text{s.t.} \quad \mu_0(x) = 1 - \frac{b_0 - cx}{P_0} \geq \alpha$$

$$\mu_i(x) = 1 - \frac{(Ax)_i - b_i}{P_i} \geq \alpha \quad \forall i$$

$$\alpha \in [0, 1]$$

$$x \geq 0$$

or

$$\max \alpha$$

$$\text{s.t.} \quad cx \geq b_0 - (1 - \alpha)P_0$$

$$(Ax)_i \leq b_i + \alpha P_i \quad \forall i$$

$$x \geq 0$$

$$\alpha \in [0, 1].$$

22.2 FUZZY MULTIOBJECTIVE PROGRAMMING—FUZZY GLOBAL CRITERION METHOD

Based on the understanding in Section 22.1, fuzzy global criterion can be regarded as an extension of the global criterion approach (i.e., similar to the concept of TOPSIS

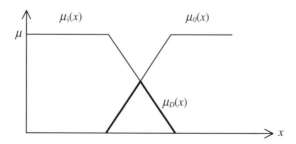

FIGURE 22.5   The max–min operator and decision-making

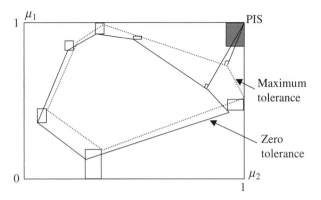

FIGURE 22.6 A set of compromised solutions with respect to fuzzy multiobjective programming

described in Chapter 21). In this case, only a positive ideal solution (PIS) will form a fuzzy set. Conflicting objectives will be resolved via the derivation of a set of compromise solutions on the basis of the fuzzy PIS only. Let us consider the following model:

$$\begin{aligned} \max \quad & f_1(x), \dots \dots, f_k(x) \\ \text{s.t.} \quad & g_i(x) \le \tilde{b}_i \quad \forall i \\ & x \ge 0, \end{aligned}$$

where $\tilde{b}_i$ are fuzzy resources in a system available associated with tolerance p_i in a nonincreasing membership function. In this case, the PIS could be a crosshatched area instead of a single point in a decision domain as indicated in Figure 22.6 (Lai and Hwang, 1994). Leung (1983) proposed the following distance-based equation:

$$d_a = \left\{ \sum_k [1 - \mu_k(f_k(x))]^a \right\}^{1/4} \quad \text{for } p \ge 1.$$

where $\mu_k(f_k(x))$ is the membership function of the distance of $f_k(x)$ from the fuzzy PIS. The model can be solved to search for the compromise solution by the following steps (Leung, 1983).

Step 1: Find PIS $f^{1*} = (f_1^{1*}, f_2^{1*}, \dots \dots, f_k^{1*})$ where f^{1*} is the optimal solution of the following problem where no tolerance is considered:

$$\begin{aligned} \max \quad & f_k(x) \\ \text{s.t.} \quad & g_i(x) \le b_i \quad \forall i \\ & x \ge 0. \end{aligned}$$

Step 2: Obtain a compromise solution corresponding to f^{1*} by solving the following problem:

$$\min \quad d_a^1 = \left\{ \sum_k [f_k^{1*} - f_k(x)/f_k^{1*}]^a \right\}^{1/a}$$

$$\text{s.t.} \quad x \in X \quad \forall_i.$$
$$X = \{x | g_i(x) \le b_i, \quad \forall_i, \quad x \ge 0\},$$

where $f_k^{1*} - f_k(x) = 1 - \mu_k(f_k(x))$ if $\mu_k(f_k(x)) = f_k(x)/f_k^{1*}$, and a can be any natural number. For practical reason, the value of "a" could be 1, 2, or ∞ in which $a = 1$ implies there is strong compensatory effect and $a = \infty$ implies there is weak compensatory effect. For simplicity we only consider two cases—$a = 1$ and $a = \infty$—that may turn this problem into a linear programming model.

When $a = 1$, the model is

$$\min \quad d_1^1 = \sum_k f_k(x)$$
$$\text{s.t.} \quad x \in X.$$

When $a = \infty$, the model is

$$\min \quad d_1^\infty$$
$$\text{s.t.} \quad [f_k^{1*} - f_k(x)] / f_k^{1*}] \le d_1^\infty \quad \forall k$$
$$\text{s.t.} \quad x \in X.$$

Both compromise solutions shown above would serve as bounds of the compromise solution set for $1 \le a \le \infty$.

Step 3: Similar to Step 1, find the PIS by using the constraint with fuzzy tolerance P_i. Find PIS $f^{0*} = (f_1^{0*}, f_2^{0*}, \dots, f_k^{0*})$, where f_k^{0*} is the solution of the following problem:

$$\min \quad f_k(x)$$
$$\text{s.t.} \quad g_i(x) \le b_i + P_i \quad \forall i$$
$$x \ge 0.$$

Step 4: Similar to Step 2, find the compromise solution by solving the following problem:

$$\min \quad d_a^0 = \left\{ \sum_k [f_k^{0*} - f_k(x)/f_k^{0*}]^a \right\}^{1/a} \quad d_a^0 = \left\{ \sum_k [(f_k^{0*} - f_k(x)) / f_k^{0*}]^a \right\}^{1/a}$$

$$\text{s.t.} \quad x \in X'.$$
$$X' = \{x | g_i(x) \le b_i + P_i, \quad \forall_i, \quad x \ge 0\}.$$

Again, we may solve the cases with the value of "a" equal to 1 and ∞ to find the bounds. Therefore the PIS may become fuzzy, which can be identified by the following range:

$$\left\{ \left[f_1(x), \dots, f_k(x) | f_k^{1*} \leq f_k(x) \leq f_k^{0*} \right], \quad k = 1, \dots, k \right\}.$$

For each single objective we have $f_k(x) \in [f_k^{1*}, f_k^{0*}]$, for all k under the impact of fuzzy constraints.

Step 5: Derive the most appropriate PIS by solving the following k single-objective fuzzy linear programming model:

$$f_k(x) \geq f_k^{1*}; f_k^{0*}$$
$$g_i(x) \leq b_i; b_i + P_i \quad \forall i.$$

For simplicity, let us assume all membership functions of fuzzy objectives and fuzzy constraints are linear, such as

$$\mu_k(x) = \begin{cases} 1 & \text{if} & f_k^{0*} < f_k(x) \\ 1 - \left[f_k^{0*} - f_k(x) \right] / \left(f_k^{0*} - f_k^{1*} \right) & \text{if} & f_k^{1*} \leq f_k(x) \leq f_k^{0*} \\ 0 & \text{if} & f_k(x) < f_k^{1*} \end{cases}$$

for all k and

$$\mu_k(x) = \begin{cases} 1 & \text{if} & g_i(x) < b_i \\ 1 - [g_i(x) - b_i]/P_i & \text{if} & b_i \leq g_i(x) \leq b_i + P_i \\ 0 & \text{if} & g_i(x) < b_i + P_i \end{cases}$$

for all i.

By using the Bellman and Zadeh (1970) max–min operator, the model becomes

$$\max_{x \in X} \min_i \{ \mu_k(f_k(x)) \mu_i(x) \}$$

or

$$\max \alpha$$
$$\text{s.t.} \quad 1 - \left[f_k^{0*} - f_k(x) \right] / \left(f_k^{0*} - f_k^{1*} \right) \geq \alpha$$
$$1 - [g_i(x) - b_i]/P_i \geq \alpha$$
$$\alpha \in [0, 1]$$
$$x \geq 0$$

whose solution is assumed to be $f_k^* = f_k(x^*)$ and α_k^*. After solving the above model, the most appropriate PIS can be $f^* = (f_1^*, \dots \dots, f_k^*)$ with $\alpha^* = \max_k \alpha_k^*$.

Step 6: Obtain a compromise solution from the original multiobjective programming model when the most appropriate PIS is f^* and α^*. With α^*, the new constraint will be

$$g_i(x) \leq b_i + (1 - \alpha^*)P_i.$$

Thus, by using the global criterion approach, the compromise solution can be obtained by solving the following problem:

$$
\min \quad d_a = \left\{ \sum_k [f_k^* - f_k(x)/f_k^*]^a \right\}^{1/a} \quad d_a^0 = \left\{ \sum_k [(f_k^* - f_k(x))/f_k^*]^a \right\}^{1/a}
$$

$$
\text{s.t.} \quad x \in X
$$

$$
X = \{x | g_i(x) \leq b_i + (1 - \alpha^*)P_i, \quad \forall_i, \quad x \geq 0\},
$$

where "a" can be any nature number. Again, we may only choose $a = 1$ or ∞ to identify the interval for decision-making.

Leung (1983) pointed out that the merit of using this method as described is that it allows decision makers to be more flexible when generating alternatives in decision-making when the fuzziness of constraints is versatile.

22.3 FUZZY GOAL PROGRAMMING

As discussed in Section 8.2.1, the goal programming (GP) model is one type of multiobjective programming model. According to the priority of the goals, GP can be classified as non-preemptive (weighted) or preemptive (lexicographic). However, there exists a hierarchy of priority levels for the goals in the preemptive structure. To apply preemptive GP, a decision maker must rank his or her goals from most important to least important. Solution techniques for both types of deterministic GP focus on the minimization of the deviations from each goal, subjective to the goal constraints and other functional constraints.

The integrated use of GP and fuzzy set theory has already been widely reported in the literature. For example, an additive model, using the relevant decision function for solving the fuzzy goal programming (FGP) model, was formulated by Tiwari et al. (1987). The theory of linear programming with multiple fuzzy goals was discussed by Hannan (1981), in which linear piecewise membership functions were applied. Tiwari et al. (1986) introduced priority structure in FGP. A fuzzy approach to aspiration levels, or the values of the objective function that would satisfy a decision maker, was proposed by Rao et al. (1988). Moreover, a solution algorithm for solving fuzzy linear programming with piecewise linear membership functions was considered by Inuiguchi et al. (1990). Lai and Hwang (1992, 1994) integrated several fuzzy linear and multiobjective programming techniques, and applied the methods to real-world problems. The approach chosen in this case study to be described in the next section is similar to the method used by Zimmermann (1978) in the formulation of the FGP problem. The following discussions introduce ways to represent fuzzy information and a solution method for fuzzy mathematical programming.

Since a solution for fuzzy mathematical programming must satisfy the fuzzy objective and constraints, a decision in a fuzzy environment is thus defined as the intersection of those membership functions corresponding to the fuzzy objective and constraints (Zimmermann, 1978; Zimmermann and Zysno, 1985; Lai and Hwang, 1992, 1994). Therefore, according to a preemptive GP model or a non-preemptive model as defined in Section 8.2.1, if $(\mu_{G1}, \mu_{G2}, \ldots, \mu_{Gm})$ and $(\mu_{C1}, \mu_{C2}, \ldots, \mu_{Cm})$ are denoted as membership functions for the fuzzy goals $\{G_1, G_2, \ldots, G_m\}$ and fuzzy constraints $(C_1, C_2, \ldots, C_p)$, respectively, in a decision space X, all the membership functions of G_m and C_p may then be combined to form a decision D, which is a fuzzy set resulting from the intersection of all related G_m and C as follows:

$$D = G_1 \cap G_2 \cap \ldots \cap G_m \cap C_1 \cap C_2 \cap \ldots \cap C_p. \tag{22.1}$$

Since the decision D is defined as a fuzzy set, the optimal decision is any alternative $x \in X$ that can maximize the minimum attainable aspiration levels in decision-making, represented by those corresponding membership functions, in the decision set $\mu_D(x)$. Thus, the max–min convolution requires maximization of the minimum membership values of those elements, as shown in Zimmermann and Zysno (1985):

$$\max_x \mu_D = \max_x [\min(\mu_{G_1}, \mu_{G_2}, \cdots \mu_{G_m}, \mu_{C_1}, \mu_{C_2}, \ldots, \mu_{C_p})]. \tag{22.2}$$

Both fuzzy objective functions and/or constraints may thus be integrated to form a fuzzy mathematical programming model. Such an operation is actually an analogy to the nonfuzzy environment as the selection of activities satisfies the objective and constraints simultaneously. However, only the fuzziness of the goals involved in SWM decision-making is considered in this analysis.

In general, the nonincreasing and nondecreasing linear membership functions are frequently used for fuzzy descriptions of minimization and maximization, respectively, as shown in Figure 22.7. It is assumed that the membership values are linearly increasing over the tolerance interval δ_i for those goals and constraints (denoted

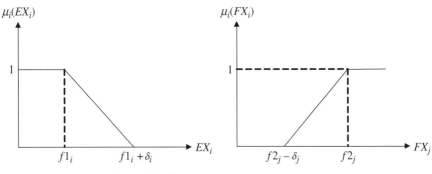

(a) definition of a non-increasing linear
membership function

(b) definition of a non-decreasing linear
membership function

FIGURE 22.7   Expressions of fuzzy linear membership functions

as EX_i here) with "the higher, the better" implication, and linearly decreasing over the tolerance interval δ_j for those goals and constraints (denoted as FX_j here) with "the lower, the better" implication, respectively. Hence, the most sensitive parts of the decision maker's aspiration or preference levels are the values of tolerance intervals δ_i and δ_j corresponding to these fuzzy goals. More complicated nonlinear membership functions can be found in the literature.

In the solution procedure, an intermediate control variable α (i.e., the level of satisfaction in fuzzy programming), a decision variable corresponding to the minimum membership value associated with a specific goal achieved in the maximization process, is typically employed as a unique term in the objective function, subject to a set of fuzzy goal constraints and original functional constraints. Such a mathematical programming framework will be applied to the following system analysis problem for SWM:

$$
\mu_i(EX_i) = \begin{cases} 1 & \text{if} & EX_i \leq f1_i \\ 1 - \dfrac{(EX_i - f1_i)}{\delta_i} & \text{if} & f1_i < EX_i < f1_i + \delta_i \\ 0 & \text{if} & EX_i \geq f1_i + \delta_i \end{cases}
$$

$$
\mu_j(FX_j) = \begin{cases} 1 & \text{if} & FX_i \leq f2_j \\ 1 - \dfrac{(f2_j - FX_j)}{\delta_j} & \text{if} & f2_j - \delta_j < FX_j < f2_j \\ 0 & \text{if} & FX_j \geq f2_j + \delta_j \end{cases}.
$$

22.4 CASE STUDY

22.4.1 Background

The reasons for choosing the FGP approach rather than other deterministic or stochastic programming techniques to address a multiobjective SWM issue can be summarized as follows (Chang and Wang, 1996a, 1996b): (1) the conventional deterministic GP model cannot reflect the strength of a decision maker's preferences associated with planning goals; (2) the imprecise information found in SWM may not be fully identified by conventional probability; hence, stochastic programming techniques are not sufficient for such a system analysis; (3) scaling various incommensurable descriptions of different goals before solving the model is a difficult problem in these traditional solution procedures in deterministic GP (Romero, 1991); but this problem does not exist in the FGP solution procedure; (4) the FGP formulation is simpler than nonfuzzy GP and so is the solution method; no deviational variables or probability distributions have to be defined and thus the total number of decision variables is reduced; and 5) it is not necessary to check the Pareto optimal condition (i.e., the condition of economic efficiency in the trade-off process) in FGP because the unique solution procedure in the FGP algorithm relaxes the crisp trade-off process in the context of optimization analysis.

From a practical sense, in an SWM system with an emphasis on both recycling and incineration, multiple conflicting objectives signify complexity because of the inherent trade-offs between recycling materials with high heating value and the need for energy recovery through complete combustion. These two goals in connection with recycling and incineration are fuzzy and both are tied to social recognition. First, in response to the public health and safety concerns, good combustion practice (GCP) regulations require combustion temperature control to reduce potential emissions of trace organic compounds such as dioxins and furans during incineration (USEPA, 1989; Environment Canada, 1991). Second, waste reduction and recycling motivate the recovery of recyclables such as paper and plastics, both of which have high heating value. Although the removal of noncombustible recyclables, such as metal and glass, may improve an incinerator's performance, recycling paper and plastics results in a lower average heating value from the waste stream destined for incineration, a higher possibility of toxic substance emissions due to incomplete combustion, and lower energy recovery potential. In this context, the traditional focus of economic-oriented allocation of waste streams can be integrated with social relevancy and environmental considerations for planning waste shipping, household recycling, and regional material recovery facilities simultaneously in an evolving SWM system (Chang and Wang, 1997; Davila and Chang, 2005).

One important strategy for resolving the conflicting goals is to optimally allocate various types of recycling programs across different service or administrative districts with respect to the combustion temperature requirements during incineration (Chang and Wang, 1996a, 1996b). When the tasks of waste collection and incineration are performed by different management, trade-offs among social, economic, and environmental factors, including waste recycling, waste generation, hauling distance, varying heating value and composition of waste streams, and the price of recyclables, can exist within different decision-making groups (Chang and Wang, 1996a, 1996b). At least two critical environmental science and engineering questions are worth exploring (Chang and Wang, 1996a, 1996b). First, to what extent are the recycling and incineration compatible? Second, what are the subsequent economic impacts on the private or the public sectors in various types of management scenarios?

However, uncertainty plays an important role in most SWM problems. Fuzziness is one type of random characteristic, linguistic in nature, and generally cannot be described by traditional probability distributions (Chang and Wang, 1996a, 1996b). Such an imprecision refers to the absence of sharp boundaries in information and frequently exists in the decision-making process. The corresponding fuzzy mathematical programming, where the parameters or goals are modeled as fuzzy sets, is viewed as an alternative to stochastic programming. Koo et al. (1991) accomplished location planning for a regional hazardous waste treatment center in Korea using a fuzzy multiobjective programming technique. Methods for combining environmental goals, such as air pollution, leachate, noise, and traffic impact controls, into a location/allocation model for SWM planning were established by Chang et al. (1996a), 1996b) in Taiwan. Yet, there have been few studies in the literature involving the use of fuzzy multiobjective programming models for tackling the compatibility issues for recycling and incineration programs in SWM (MDEP, 1989; Chang and Wang, 1996a, 1996b).

The following case study illustrates the use of an FGP model to facilitate evaluation of the compatibility issue for waste recycling and incineration programs based on several SWM planning scenarios. It starts with a case study demonstration of a deterministic GP model but an inherent scaling issue hampers further application. This case study is followed by the formulation of an improved FGP model that is deemed a mathematical transformation of the prior deterministic GP model. Net benefits, recycling targets, and anticipated combustion temperature during incineration are considered simultaneously as three fuzzy goals. In particular, several new management disciplines affecting the privatization of SWM are also explored with the proposed model in this systems analysis. Subsequent feasibility studies and therefore policy analyses for selected waste collection, shipping, treatment, and disposal alternatives are achievable. It is envisioned that fuzzy optimal outputs may generate more realistic and flexible policies for solving complex real-world SWM problems, especially considering the compatibility issue between material recycling and energy recovery in a metropolitan region.

22.4.2 Formulation of a Fuzzy Goal Programming Model

Deterministic GP Model Formulation In a typical SWM system, an incineration facility may be designated to handle solid waste streams for several service areas or administrative districts. Daily operations might be run by either a private business or public agency. While optimization analysis may suggest a better solid waste flow control pattern, this does not imply that solid waste flow control will be followed up in such a system. Considering the fuzziness in a system, such a flow control scenario could change over time in response to the modification of overarching management plans, market competition, and so on, although flow control requires regional consensus unilaterally. To elevate systemic performance, market-driven strategies and environmental constraints may be closely linked with the proposed optimal flow control patterns in the context of sustainable development over time (Chang et al., 1996a, 1996b; Chang and Wang, 1996a).

Waste generation rate, physical composition, and heating value of the waste stream usually differ from one administrative district to another. From an integrated management perspective, an adequate recycling level in such a system is constrained by many engineering, economic, social, and environmental factors. The most significant factors consist of the price of recyclables in the secondary material market, expected heating value of the waste stream, hauling distance, required combustion temperature during incineration, essential tipping fees/subsidy, and the selling price of electricity to a power grid. Recycling paper and plastics, however, may result in a lower average heating value of the waste stream destined for incineration, a higher possibility of toxic substance emissions due to incomplete combustion, and a lower energy recovery potential. As a consequence, the adverse effects from recycling paper and plastics on the combustion temperature during incineration must be properly addressed to achieve the desired level of sustainable development.

In such an optimization framework, three goals corresponding to economic, environmental, and managerial considerations are of paramount importance to the present analysis: (1) to establish economic requirements, so that the total benefits are greater than, or at least equal to, total costs for private agencies; by the same token, total costs should be as close as possible to total income if the system is managed by public agencies; (2) to accomplish the environmental goals, so that the estimated combustion temperature in an incinerator is close to or above the required level proposed by the engineering design criteria (i.e., GCP) to avoid hazardous air emissions; and (3) to pursue the goal of waste reduction and material conservation, so that paper and plastics are recovered at a rate as close as possible to the prescribed recycling targets. Hence, the proposed mathematical formulation must accommodate the needs of either private or public agencies that choose recycling or treatment/ disposal options.

Formulating a deterministic GP model with this philosophy in mind may help us gain some insights necessary for the evaluation of the compatibility issue for household recycling and incineration programs. In other words, the spatiotemporal patterns of household recycling programs may be factored into a systems analysis such that the overall shipping strategies, after recycling, are amenable for municipal solid waste treatment and disposal. Thus, the promotion of a regional household recycling program can be tuned to the characteristics of an SWM system responding to rapid urbanization with changing waste composition and quantity (Chang and Wang, 1997).

In the model formulation, only the recycling goal, bounded by the recyclable ratio and the resident participation rate, is formulated as a lower, one-sided goal (see Section 8.3.2). The priorities for the three goals considered in the model formulation are assumed to be roughly equivalent for decision makers in the initial run. Nonlinear characteristics may occur from the interactions between recycling efforts and the impacts on the after recycling heating value. Therefore, the GP model can be stated as (Chang and Wang, 1997):

$$\min \quad z = P_1 d_1^- + P_2(w_{22}^- d_2^- + w_{22}^+ d_2^+) + P_2(w_{23}^- d_3^- + w_{23}^+ d_3^+) + \sum_{i=1}^{N'} P_3(a_i^- + b_i^-),$$

$$(22.3)$$

where P_1, P_2, and P_3 are the priority order of these three goals (unitless); d_1^- is the deviational variable of combustion temperature (°C); d_2^- and d_2^+ are deviational variables of system benefits obtained from waste collection, in US\$; d_3^- and d_3^+ are deviational variables of system benefits obtained from waste incineration (US\$); a_i^- and b_i^- are deviational variables of paper and plastics recycling in district i (%); N' is total number of administrative districts considered in the service area of the incinerator (unitless); w_{22}^- and w_{22}^+ are managerial weights of the collection efficiency corresponding to the goal of cost/benefit (unitless); and w_{23}^- and w_{23}^+ are managerial weights of the combustion efficiency and emission control corresponding to the goal of cost/benefit (unitless).

Subject to:

1. **Goal constraints**: The required combustion temperature was defined by (22.4); total net incomes were defined by (22.5) and (22.6); and recycling targets of paper and plastics were defined by (22.7) and (22.8).

$$T_f - d_1^+ + d_1^- = MINT. \tag{22.4}$$

$$TB1 - d_2^+ + d_2^- = TC1. \tag{22.5}$$

$$TB2 - d_3^- + d_3^+ = TC2. \tag{22.6}$$

$$PA_i + a_i^- = MAXPA_i \quad \forall i \in J. \tag{22.7}$$

$$PL_i + b_i^- = MAXPL_i \quad \forall i \in J. \tag{22.8}$$

2. **Capacity constraints**: Capacity limitation constraints of an incinerator facility and of collection equipment were defined by (22.9) and (22.10), respectively.

$$G' = \sum_{i=1}^{N'} G_i' \leq CAP1 \quad \forall i \in J \tag{22.9}$$

$$G_i' \leq CAP2_i \quad \forall i \in J \tag{22.10}$$

3. **Nonnegativity constraints**: Defined by (22.11) and (22.12), respectively, all decision variables were set nonnegative.

$$T_f, TB1, TB2, TC1, TC2, d_1^+ d_1^-, d_2^+ d_2^-, d_3^- d_3^+ \geq 0 \tag{22.11}$$

$$PA_i, PL_i, G_i', a_i^-, b_i^- \geq 0 \quad \forall i \tag{22.12}$$

4. **Complementary constraints**: Defined by (22.13) and (22.14), respectively, deviational variables were set equal to either zero or nonzero dichotomously.

$$d_1^+ \cdot d_1^- = 0; d_2^+ \cdot d_2^- = 0; d_3^+ \cdot d_3^- = 0 \tag{22.13}$$

$$a_i^+ \cdot a_i^- = 0; b_i^+ \cdot b_i^- = 0, \tag{22.14}$$

where *MINT* is required minimum combustion temperature in engineering design criteria for emission control of trace organic compounds during incineration (°C); T_f is the estimated combustion temperature based on waste stream after recycling (°C); *TB*1 and *TB*2 = system benefit (i.e., incomes) obtained from operating waste collection and incineration business, respectively (US$); *TC*1 and *TC*2 are the system costs incurred by shipping waste and operating incinerator, respectively (US$); PA_i and PL_i are the target levels of recycling paper and plastics, respectively, in district i (%, on wet basis); J is the set of administrative districts (unitless). G_i' is the solid waste generation rate after recycling in district i (wet basis, tonnes·per day); G' is the total solid waste generation rate after recycling (on wet basis, tonnes·per day); $CAP1$ is the design capacity of an incinerator (tonne·per day); $CAP2_i$ is the hauling capacity of waste collection in district i (tonne·per day); and d_i^+ are deviational variables in the GP model.

Several sub-models are required to support the effective operation of the above optimization model. They are described as in Equations (22.15)–(22.28).

- The following equation of combustion temperature proposed by Tillman et al. (1989):

$$T_f(°C) = (5/9)(0.108HHV + 3467K - 4.554M + 0.59(T_a - 77) - 319), \tag{22.15}$$

where M is the moisture content of waste stream after recycling (%); HHV is the high heating value of waste stream after recycling (kcal·kg^{-1}); K is the proposed equivalence ratio used in estimation of waste combustion temperature (unitless) (i.e., the equivalence ratio, is the ratio of the actual fuel–air ratio to the theoretical fuel–air ratio); and T_a is the preheated temperature of auxiliary air in combustion (°F) (assume the ambient temperature is 25°C or 77°F).

- The following equation of heating value in relation to waste recycling was established by an empirical formula (Chang and Wang, 1997):

$$HHV(\text{kcal} \cdot \text{kg}^{-1}) = 1587 + 7.63 + 13.66\,R', \tag{22.16}$$

where P' and R' are gross paper and plastics content in solid waste after recycling, respectively (on dry basis, %).

- The following equation of electricity production ratio was established by an empirical formula:

$$E(\text{kWh} \cdot \text{tonne}^{-1}) = 0.2HHV(\text{kcal} \cdot \text{kg}^{-1}), \tag{22.17}$$

where E is the conversion factor of power generation (kWh·tonne^{-1}).

- The following equations of waste production rate after recycling, corresponding to the dry or wet basis, were defined by mass balance principles:

$$G = \sum_{i=1}^{N'} [T_i - T_i \cdot R'_i \cdot PL_i - T_i \cdot P'_i \cdot PA_i] \quad \text{(dry basis)} \tag{22.18}$$

$$G' = \sum_{i=1}^{N'} [S_i - S_i \cdot R_i \cdot PL_i - S_i \cdot P_i \cdot PA_i] \quad \text{(wet basis)}, \tag{22.19}$$

where P'_i and R'_i are the paper and plastics content in solid waste stream before recycling in district i, respectively (on dry basis,%); G is the solid waste generation rate after recycling (on dry basis, tonnes·per day); G' is the solid waste generation rate after recycling (on wet basis, tonnes per·day); T_i is the solid waste generation rate before recycling in district i (dry basis, tonnes·per day); and S_i the solid waste generation rate before recycling in district i (wet basis, tonnes·per day).

- The equation of moisture content in the waste after recycling was defined by the mass balance principle:

$$M = \sum_{i=1}^{N'}[S_i \cdot W_i - (S_i \cdot R_i \cdot PL_i - S_i \cdot P_i \cdot PA_i)f]/G', \qquad (22.20)$$

where f is the average water content of paper and plastics (%); and W_i is the original water content in the solid waste stream in district i (on wet basis, %).

- The equations of waste composition of paper and plastics, after recycling, corresponding to the dry and wet basis, respectively:

$$P' = 100 \times \left[\sum_{i=1}^{N'} T_i \cdot P'_i(1 - PA_i)\right]/G \text{ (paper, dry basis)} \qquad (22.21)$$

$$R' = 100 \times \left[\sum_{i=1}^{N'} T_i \cdot R'_i(1 - PL_i)\right]/G \text{ (paper, dry basis)} \qquad (22.22)$$

$$P = 100 \times \left[\sum_{i=1}^{N'} S_i \cdot P_i(1 - PA_i)\right]/G' \text{ (paper, wet basis)} \qquad (22.23)$$

$$R = 100 \times \left[\sum_{i=1}^{N'} S_i \cdot R_i(1 - PL_i)\right]/G', \text{(paper, wet basis)} \qquad (22.24)$$

where P and R are the gross paper and plastics content in solid waste after recycling, respectively (on wet basis, %).

- The equations of benefit and cost expressions were established as below:

$$TB1 = \left[PPA \cdot \sum_{i=1}^{N'} S_i \cdot PA_i + PPL \cdot \sum_{i=1}^{N'} S_i \cdot PL_i\right]$$

$$+ \left[TIP \cdot \sum_{i=1}^{V'} S_i\right] \text{(for shipping waste)} \qquad (22.25)$$

$$TB2 = \left[SU \sum_{i=1}^{V'} G'_i \right] + \left[PE \cdot E \cdot \sum_{i=1}^{N'} G'_i \right] \text{ (for hauling waste to incinerator)}$$

(22.26)

$$TC1 = \left[\sum_{i=1}^{N'} G'_i \cdot CT_i \right] + \left[SU \sum_{i=1}^{N'} G'_i \right] \text{ (for shipping waste)} \quad (22.27)$$

$$TC2 = \left[\sum_{i=1}^{N'} G'_i \cdot CO \right] \text{ (for hauling waste to incinerator),} \quad (22.28)$$

where *PPL* and *PPA* are the prices of paper and plastics in the secondary material market, respectively (US\$·tonne^{-1}); *SU* is the government subsidy for the treatment of waste (US\$·tonne^{-1}); and *TIP* is the tipping fees charged to the citizens (US\$·tonne^{-1}); CT_i is the unit transportation cost from district i to incinerator (US\$·tonne^{-1}); *CO is the* unit operation cost of incineration (US\$·tonne^{-1}); and *PE is the* price of electricity (US\$·kWh^{-1}).

FGP Model Formulation According to Chang and Wang (1996b), the scaling issue with regard to non-commensurate units of those goals from Equations (22.4) to (22.8) could disturb the integrity of a traditional GP solution procedure. The nonlinear GP model with multiple fuzzy goals for managing solid waste is thus formulated as follows from (22.29) to (22.56). In such an FGP model, the objective function is formulated for the maximization of the minimum membership values of each fuzzy goal simultaneously. The information needed in these fuzzy goal constraints consists of the combustion temperature achieved during incineration, as shown in (22.30)–(22.39), net income for operating waste management system, as shown in (22.42)–(22.47), and predicted recycling levels of paper and plastics, as shown in (22.50) and (22.51). On the other hand, the functional constraint set, in (22.53) and (22.54), expresses the capacity limitations of the designated incinerator and shipping equipment, possible recycling upper bounds, and nonnegativity requirements. Hence, in the present analysis, the optimization criterion used is a qualitative parameter (decision variable)—satisfaction (aspiration) level α—to quantify various uncertain measures in (22.29) and (22.52), such as net income for waste management, combustion temperature after recycling, and recycling rate for plastics and paper.

$$\text{max} \quad \alpha \quad (22.29)$$

Subject to:

1. **A fuzzy goal constraint for combustion temperature control**. This constraint provides a normalized measure of the amount the actual combustion

temperature falls short of the required combustion temperature due to the effect of recycling.

$$\mu(T) = 1 - \frac{MAXT - T}{\delta_1} \geq \alpha \tag{22.30}$$

$$T = \left(\frac{5}{9}\right)[(a_0 + a_1 \cdot HHV + a_2 \cdot K + a_3 \cdot M + a_4 \cdot TP - 32] \tag{22.31}$$

$$HHV = b_0 + b_1 \cdot P_1 + b_2 \cdot R_1 \tag{22.32}$$

$$P1 = 100 \sum_{i=1}^{N} \frac{[T_i \cdot P1_i(1 - PA_i)]}{G2} \tag{22.33}$$

$$R1 = 100 \sum_{i=1}^{N} \frac{[T_i \cdot R1_i(1 - PL_i)]}{G2} \tag{22.34}$$

$$M = 100 \sum_{i=1}^{N} \frac{[S_i W_i - (S_i \cdot R2_i \cdot PL_i \cdot f_R + S_i \cdot P2_i \cdot PA_i \cdot f_P)]}{G1} \tag{22.35}$$

$$G1 = \sum_{i=1}^{N} G1_i = \sum_{i=1}^{N} (S_i - S_i \cdot R2_i \cdot PL_i - S_i \cdot P2_i \cdot PA_i) \tag{22.36}$$

$$G2 = \sum_{i=1}^{N} G2_i = \sum_{i=1}^{N} (T_i - T_i \cdot R1_i \cdot PL_i - T_i \cdot P1_i \cdot PA_i) \tag{22.37}$$

$$P2 = 100 \sum_{i=1}^{N} \frac{[S_i \cdot P2_i(1 - PA_i)]}{G1} \tag{22.38}$$

$$R2 = 100 \sum_{i=1}^{N} \frac{[S_i \cdot R2_i(1 - PL_i)]}{G1}, \tag{22.39}$$

where $\mu(T)$ is the membership function corresponding to estimated combustion temperature; δ_i is the tolerance interval, a subjectively chosen constant for possible deviations of the combustion control levels during incineration (°C); *MAXT* is the required combustion temperature in engineering design criteria for emission control of trace organic compounds during incineration (°C); *T* is the estimated combustion temperature based on waste stream after recycling (°C); *M* is the moisture content of waste stream after recycling (%); *HHV* is the high heat value of waste stream after recycling (kcal·kg^{-1}); *K* is the proposed equivalence ratio used in estimation of waste combustion temperature (unitless); *TP* is the proposed preheated temperature of auxiliary air used in estimation of waste combustion temperature (°C or °F); a_0, a_1, a_2, a_3, a_4 are regression coefficients for estimation of combustion temperature during incineration (with units °C, °C/kcal/kg, °C, °C/%, and °C/°F, respectively); *N* is the total number of administrative districts considered in the service area of an incinerator (unitless); *P1* and *R1* are the gross paper and plastics contents, respectively, in the waste stream after recycling (%, on dry basis); b_0, b_1, and b_2 are the regression coefficients for estimation of high heating value from the waste stream after recycling (with unit $kcal \cdot kg^{-1}$, $kcal \cdot \% \cdot kg^{-1}$, and $kcal \cdot \% \cdot kg^{-1}$, respectively); $G2$ is the solid waste generation rate after recycling (tonne·per day on a dry basis); $G2_i$ is the solid waste generation rate after recycling in district i (tonne·per day, on a dry basis); $G1_i$ is the solid waste generation rate after recycling in district i (tonne·per day, on a wet basis); $G1$ is the solid waste generation rate after recycling (tonne·per day, on a wet basis); T_i is the solid waste generation rate before recycling in district i (tonne·per day, on a dry basis); $P1_i$ and $R1_i$ are the paper and plastics contents in a waste stream before recycling in district i (%, on a dry basis); PA_i and PL_i are the target levels of recycling paper and plastics, respectively, in district i (%, on a wet basis); $P2_i$ and $R2_i$ are the paper and plastics contents in a waste stream before recycling in district i (%, on a wet basis); W_i is the water content in a waste stream in district i (%, on a wet basis); S_i is the solid waste generation rate before recycling in district i (tonne·per day, on a wet basis); $P2$ and $R2$ are the gross paper and plastics contents, respectively, in a waste stream after recycling (%, on a wet basis); and f_p and f_R are the average water content of recycled paper and plastics (%, on a wet basis). In Equations (22.22)–(22.31), the constants of 5/9 and 32 are used for conversion between celsius and fahrenheit.

To evaluate how well combustion temperature is controlled, a mathematical illustration of combustion temperature must be defined. In Equation (22.31), a semiempirical formula (Tillman et al., 1989), based on a regression analysis of combustion temperature in terms of heating value and the moisture content of a solid waste stream after recycling, equivalence ratio in combustion, and the temperature of preheated air during incineration, may be applied. This equation is valid even if the higher rates of recycling occur. The inclusion of heating value for the solid waste stream in the foregoing equation is essential because the effects of paper and plastics recycling on combustion temperature can be directly accounted for. Information on the composition of solid wastes is of importance in the evaluation of the impacts by recycling. Hence, Equation (22.32), a regression description of high heating value in terms of paper and plastics contents in the waste stream, is required. It is also known that moisture adds weight to the solid waste without having any net heat value, and the evaporation of water will reduce the heat released from the solid waste. The descriptive function of high heating value of the solid waste stream is usually defined based on the dry basis. However, the physical composition or chemical analysis of solid waste is generally conducted based on the wet basis. It is the reason that some equations need to be differentiated based on the physical or chemical property from a dry to wet basis in the process of modeling analysis. Hence, the equations depicting gross paper and plastics contents of the solid waste stream as well as the waste generation rate have to be defined separately according to wet or dry basis in the model formulations. Furthermore, the paper, plastic, and water content of the waste stream are altered by recycling. Such a phenomenon is specifically illustrated by Equations (22.33)–(22.39).

Thus, the achieved membership value of combustion temperature is expressed in Equation (22.30) with respect to these supporting expressions compiled together in Equations (22.31)–(22.39).

2. **A fuzzy goal constraint for net income of private and public agencies in the system**. This constraint illustrates the possible achievement of the economic goal for handling a specific type of SWM system.

$$\mu(NB1) = 1 - \frac{MAXB_1 - NB1}{\delta_2} \geq \alpha \tag{22.40}$$

$$\mu(NB2) = 1 - \frac{MAXB_2 - NB2}{\delta_3} \geq \alpha \tag{22.41}$$

$$TB1 = \left(PPA \sum_{i=1}^{N} S_i \cdot PA_i + PPL \sum_{i=1}^{N} S_i \cdot PL_i \right) + \left(TIP \cdot \sum_{i=1}^{N} S_i \right) \tag{22.42}$$

$$TB2 = \left(SU \cdot \sum_{i=1}^{N} G1_i \right) + \left(PE \cdot E \cdot \sum_{i=1}^{N} G1_i \right) \tag{22.43}$$

$$TC1 = \sum_{i=1}^{N} (G1_i \cdot CT_i) + \left(SU \cdot \sum_{i=1}^{N} G1_i \right) \tag{22.44}$$

$$TC2 = \sum_{I=1}^{N} (G1_i \cdot AC) \tag{22.45}$$

$$NB1 = TB1 - TC1 \tag{22.46}$$

$$NB2 = TB2 - TC2, \tag{22.47}$$

where $NB1$ and $NB2$ are the net benefits obtained from operating waste collection and incineration business, respectively (US$); $MAXB_1$ and $MAXB_2$ are the highest aspiration levels of the tolerance intervals in the net benefit membership functions corresponding to operation of waste collection and incineration businesses, respectively (US$); δ_2 and δ_3 are the tolerance intervals, subjectively chosen constants for possible deviations of cost/benefit aspiration levels corresponding to waste collection and incineration (US$); TIP is the tipping fees charged to residents (US$·tonne^{-1}); PE is the price of electricity (US$·kWh^{-1}); E is the conversion factor between amount of waste stream and power generation (kWh·tonne^{-1}); CT_i is the average transportation cost from district i to a designated incinerator (US$·tonne^{-1}); AC is the average cost of constructing and operating an incinerator (US$·tonne^{-1}); SU is the government subsidies for treatment of waste (US$·tonne^{-1}, on wet basis); and PPL and PPA are the prices of paper and plastics in secondary material market, respectively (US$·tonne^{-1}).

In the foregoing formulation, system benefits associated with private or public agencies responsible for the collection and incineration of solid waste are characterized by the income from recycling, the tipping fees charged to the residents, the possible subsidies from the municipality, and the income from selling electricity, as defined by Equations (22.42) and (22.43). On the other hand, system costs corresponding to the private or public agency in charge of the collection and incineration of solid waste primarily include the transportation costs for hauling waste and the construction and operating costs for handling the incineration facilities, as defined in Equations (22.44) and (22.45). The net benefits, which are decision variables, for managing solid waste collection and incineration can then be obtained from (22.46) and (22.47). Therefore, based on a set of cost/benefit terms, the membership values of the decision maker's aspiration levels for economically managing the proposed patterns in an SWM system can be fully expressed by Equations (22.40) and (22.41).

3. **The goal constraints for the levels of paper and plastics recycling:**

$$\mu(PAi) = 1 - \frac{MAXPAi - PAi}{\delta PAi} \geq \alpha \quad \forall i \tag{22.48}$$

$$\mu(PLi) = 1 - \frac{MAXPLi - PLi}{\delta PLi} \geq \alpha \quad \forall i \tag{22.49}$$

where δ_{PAi} and δ_{PLi} are the tolerance intervals, subjectively chosen constants for the recycling levels of paper and plastics (%, on wet basis). *MAXPAi* and *MAXPLi* are the upper bounds of recyclables of paper and plastics in the waste stream in district i. The degree of recycling paper and plastics is constrained by their physical upper bounds, which could be directly related to the solid waste composition, resident participation rate, and recycling efficiency. Once they are explicitly decided, the membership values corresponding to the levels of paper and plastics recycling can then be defined for decision analysis, as listed in Equations (22.48) and (22.49).

4. **Functional constraints of the upper bounds for paper and plastics recycling:**

$$PA_i \leq MAXPA_i \quad \forall i. \tag{22.50}$$

$$PL_i \leq MAXPL_i \quad \forall i. \tag{22.51}$$

5. **A functional constraint for membership value:**

$$0 \leq \alpha \leq 1. \tag{22.52}$$

6. **A functional constraint for capacity limitation of a designated incinerator:**

$$G1 = \sum_{i=1}^{N} G1_i \leq CAP1. \tag{22.53}$$

7. **A functional constraint of capacity limitations of shipping equipment in each district:**

$$G1_i = CAP2_i \quad \forall i. \tag{22.54}$$

8. **Nonnegativity constraints (all decision variables are nonnegative)**:

$$T, NB1, NB2, TB1, TB2, TC1, TC2, G1, G2 \geq 0. \qquad (22.55)$$

$$PAi, PLi, G1i, G2i \geq 0 \quad \forall i. \qquad (22.56)$$

22.4.3 Modeling Structures

The analytical settings of this FGP model allow the work of waste collection and treatment to be handled by different management agencies so that comparison between public and private ownership is achievable. The modeling structure is designed to flexibly express as many types of managerial scenarios as possible according to various types of combination associated with cost and benefit terms to be considered in private and/or public partnerships. When the entire SWM system is managed by a single agency, either private or public, the cost and benefit terms defined for different partners as defined in those goal constraints, in Equations (22.42)–(22.45), may be combined together and the total number of constraints is reduced accordingly. The possibility of privatization for SWM can be assessed with respect to differing scenarios as those cost and benefit terms may be redefined as needed at any time. Any future managerial scenarios related to differing public or private ownerships/partnerships may be flexibly formulated to reflect specific requirements.

On one hand, if an incineration facility is operated by a private sector, the local municipality may subsidize such an operation, and residents' tipping fees may support part or all of the subsidies in the system. Although tipping fees and government subsidies are transfers, they still can be viewed as external sources of benefits to the private operators. On the other hand, if both collection and incineration tasks are operated by a single public agency, consideration of such subsidies should be excluded in the model. In this situation, the tipping fees charged to the residents directly can be defined as a decision variable in the model formulation to reflect the budget limitation. In cases where the incinerator is privately operated and the waste collection task is performed by a public agency, both tipping fees and subsidies can be defined as decision variables in the model formulation. If the entirely system is run by a private sector, the level of subsidy may become a decision variable. Overall, whether the tipping fees and/or subsidies should be defined as decision variables depends on the management settings, operation strategies, and public policies selected in advance by decision makers. Therefore, the related operation strategies and public policies of the privatization for SWM can be well evaluated by this FGP model. It appears that different trade-offs among different goals through the prescribed membership functions exist. Such an application is significant for not only the related accounting stance but also the managerial disciplines involved in this unique system analysis.

Modeling such a complex system requires us to (Chang and Wang, 1996b): (1) propose management alternatives in terms of the concerns of privatization, (2) define membership functions for model parameters, (3) solve the model, (4) perform sensitivity analysis, (5) conduct policy analysis, and (6) carry out reoptimization after model modification if necessary.

22.4.4 Data Analysis

Tainan City, located in the southern part of Taiwan, was divided into seven administrative districts in the late 1990s (Figure 22.8) (i.e., the administrative districts were reorganized later on after 2000). The task of waste shipping in each district was handled by a public agency—the Bureau of Environmental Protection of the City Government. The city government had built a high-end waste incineration plant, which started commercial operations in 1999. The plant is equipped with modern facilities and a steam–electricity cogeneration system for the purpose of recycling thermal resources, which can accordingly solve pollution impacts. However, various waste recycling programs had been in effect for several years, and the Environmental Protection Administration in Taiwan is highly likely to keep subsidizing the local government to promote such household recycling activities. While household recycling enhancement is anticipated, an increasing public concern over environmental quality due to the potential emissions of trace organic compounds from incineration has brought many similar programs under intense scrutiny in Taiwan. Many studies had been conducted for the separate planning of waste recycling and energy recovery. To achieve the sustainable management of solid waste in this city, it is of significance to

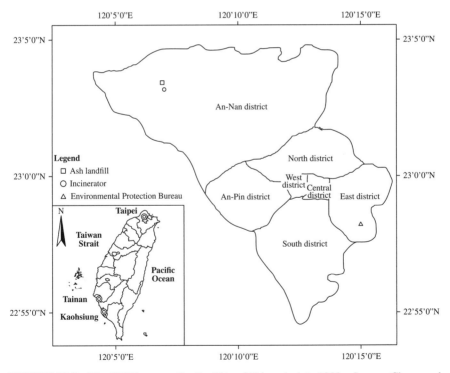

FIGURE 22.8   The SWM system for the City of Tainan in late 1990s. *Source*: Chang and Wang (1996b) (with permission from ASCE)

explore integrated optimal planning for waste recycling and energy recovery simultaneously. With this FGP modeling analysis, it is envisioned that current and future planning scenarios can be fully coordinated, expressed, and evaluated smoothly.

As mentioned in Section 22.4.2, three system goals were considered in this case study: (1) total income, that is, direct benefits, should be greater than, or at least equal to, total costs for private agencies; but total costs should be as close as possible to total income if the system is managed by public agencies; (2) estimated combustion temperature should be close to 982°C, as suggested in the GCP criteria (Clark, 1988; Schindler, 1989), although higher than that level is also acceptable; and (3) paper and plastics should be recovered as closely as possible to the prescribed recycling targets (50% of the plastics and 70% of the paper in the waste stream) as estimated according to the waste composition, resident participation rate, and collection efficiency in this area. However, part of the planning scenarios could result in a combustion temperature that is lower than the goal of 982°C due to a level of recycling greater than the allowable level. The final choice of a management alternative has something to do with the public perception of risk, which is considered through the corresponding membership values in fuzzy decision-making.

Site-specific information about the generation rate (kg·per day) and physical composition of waste (%) in the City of Tainan was collected and is shown in Tables 22.1 and 22.2 (Chang and Wang, 1996b). It appears that the waste composition associated with each district is quite different, particularly in terms of paper and plastic fractions. The Sino Environmental Service Corporation since 1999 has been operating the incineration plant. The plant can process about 200,000 metric tons (tonnes) of municipal and industrial waste per year. The solid waste streams are assumed to be well mixed at the incineration site. The different delivery schedules for varying waste streams do not affect this assumption. In addition, many cost and benefit parameters, according to the managerial structure, have to be handled properly for this modeling analysis. The information for average shipping costs is summarized in Table 22.3. As reported by the responsible government agency, the average operating cost for

TABLE 22.1 Waste generation rate in Tainan City (1992 averages)

Location (1)	East (2)	South (3)	West (4)	North (5)	An-Pin (6)	An-Nan (7)	Central (8)
Waste generation	136,125[a]	121,327[a]	32,322[a]	104,516[a]	19,698[a]	118,171[a]	45,365[a]
–	66,565[b]	68,428[b]	20,230[b]	46,301[b]	10,026[b]	58,376[b]	23,050[b]
Plastics content	21,535[a]	24,654[a]	8,300[a]	8,920[a]	2,949[a]	28,726[a]	15,186[a]
–	14,644[b]	17,765[b]	5,644[b]	6,056[b]	2,005[b]	19,556[b]	10,326[b]
Paper content	33,968[a]	21,535[a]	7,438[a]	22,742[a]	4,216[a]	18,715[a]	11,288[a]
–	23,098[b]	14,644[b]	5,058[b]	15,465[b]	2,867[b]	12,726[b]	7,676[b]

Source: From Chang and Wang (1996b) (with permission from ASCE).
[a]Wet basis
[b]Dry basis.

TABLE 22.2   Waste composition in Tainan City (1992 averages)

Location (1)	East (2)	South (3)	West (4)	North (5)	An-Pin (6)	An-Nan (7)	Central (8)
(a) Physical composition[a]							
Paper	34.7	21.4	25.0	33.4	28.6	21.8	33.3
Textiles	7.2	18.3	15.1	35.0	3.7	5.6	5.4
Wood	1.4	2.5	3.3	3.3	1.4	8.2	2.3
Food waste	21.0	11.5	15.8	3.3	16.4	14.0	4.1
Plastics	22.0	24.5	27.9	13.1	20.0	33.5	44.8
Leather	1.0	0.0	0.0	0.0	0.0	0.0	0.0
Others	0.8	1.2	2.4	1.4	2.0	3.9	1.8
[Subtotal of combustibles]	88.1	79.4	89.5	89.5	72.1	87.0	91.7
Metal	5.5	11.0	4.8	4.8	5.8	5.2	7.9
Glass	4.8	4.3	2.3	2.3	10.0	3.7	0.5
Ceramics	0.0	0.0	2.1	2.1	4.8	2.0	0.0
Sand/stone	1.6	5.3	1.3	1.3	7.3	2.2	0.0
(b) Chemical composition[b]							
Water	51.1	43.6	52.2	55.7	49.1	50.6	49.2
Ash content	17.9	25.7	19.6	15.24	21.3	17.8	17.1
Combustibles	31.0	30.7	28.2	29.06	29.6	31.6	33.7

Source: From Chang and Wang (1996b) (with permission from ASCE).
[a]Wet basis.
[b]Dry basis.

handling an incinerator was a little bit over US$33.33 per tonne (i.e., based on a currency ratio between NT$ and US$ that was about 30 in the late 1990s). However, the facility capital cost was temporarily excluded in this case study because the construction budget was fully financed by the Taiwan Provincial Government. Local residents had to pay a tipping fee of approximately US$28 per tonne for waste collection and treatment. Part of the additional expense for SWM was financed by the City Government in the 1990s. Based on a market investigation of secondary materials at that time, US$40 and US$66.67 per tonne were selected as the average prices for recycled paper and plastics, respectively. The selling price of electricity was about US$0.05 per kWh.

TABLE 22.3   Average shipping cost from each district to the incinerator

Location (1)	East (2)	South (3)	West (4)	North (5)	An-Pin (6)	An-Nan (7)	Central (8)
Cost (US$·tonne^{-1})	50	60	40	27	43	17	40

Source: From Chang and Wang (1996b) (with permission from ASCE).

22.4.5 Decision Analysis

The planning scenarios proposed in this analysis are summarized in Table 22.4 (Chang and Wang, 1996b). The base case, and cases 1 and 2 illustrate the situation in a partially privatized system where both private and public sectors take part in the work of SWM (Chang and Wang, 1996b). Cases 3 and 4 are specifically arranged for an examination of similar scenarios in which both emphasize public ownership of collection and treatment (Chang and Wang, 1996b). Furthermore, cases 5 and 6 are designed to evaluate a possible future scenario when the waste collection and incineration are both operated by private agencies (Chang and Wang, 1996b). In these cases where the tipping fees and subsidies are defined as separate decision variables, the financing issues become crucial in system planning. The minimum levels of tipping fees and subsidies necessary to keep the system in effective operation can be predicted by this FGP model (Chang and Wang, 1996b). The FGP model formulation may provide decision support with strategic intelligence making the potential for privatization strategies advisable (Chang and Wang, 1996b).

As stated previously in the descriptions of three system goals, the values of 982°C for waste incineration, 70% for the paper content, and 50% for the plastics content were selected as the highest aspiration levels for the tolerance intervals in the corresponding membership functions. However, observations were obtained in several initial runs to determine the reasonable tolerance intervals of the net system benefits for the corresponding membership functions. This analysis also assumes that linear membership functions are reasonable for such applications. Debt is allowed in the cases where the task of waste collection is performed by a government agency and the tipping fees are limited. Hence, the initial settings of these membership functions are established as follows:

$$\mu(T) = \begin{cases} 1 & \text{if} \quad T > 982 \\ 1 - \dfrac{(982 - T)}{20} & \text{if} \quad 962 \le T \le 982 \\ 0 & \text{if} \quad T < 0 \end{cases} \tag{22.57}$$

$$\mu(PL_i) = \begin{cases} 1 & \text{if} \quad 0.5 < PL_i \le 1 \\ 1 - \dfrac{(0.5 - PL_i)}{0.5} & \text{if} \quad 0 \le PL_i \le 0.5 \end{cases} \tag{22.58}$$

$$\mu(PA_i) = \begin{cases} 1 & \text{if} \quad 0.7 < PA_i \le 1 \\ 1 - \dfrac{(0.7 - PA_i)}{0.7} & \text{if} \quad 0 \le PA_i \le 0.7 \end{cases} \tag{22.59}$$

$$\mu(NB1) = \begin{cases} 1 & \text{if} \quad NB1 > 0 \\ 1 - \dfrac{(20,000 + NB1)}{20,000} & \text{if} \quad -20,000 \le NB1 \le 0 \\ 0 & \text{if} \quad NB1 < -20,000 \end{cases} \tag{22.60}$$

$$\mu(NB2) = \begin{cases} 1 & \text{if} \quad NB2 > 16,667 \\ 1 - \dfrac{(16,667 - NB2)}{16,667} & \text{if} \quad 0 \le NB2 \le 16,667 \\ 0 & \text{if} \quad NB2 < 0 \end{cases} \tag{22.61}$$

TABLE 22.4 Summary of planning scenarios in this modeling analysis

Condition	Base Case	Case 1	Case 2	Case 3	Case 4	Case 5	Case 6
Collection is operated by public sector	X	X	X	X	X		
Collection is operated by private sector						X	X
Incinerator is operated by public sector	X	X	X	X	X		
Incinerator is operated by private sector						X	X
System is operated by single company						X	X
Fixed subsidy for incinerator (50 US$·tonne⁻¹)	X		X				
Subsidy is defined as a decision variable (<67 US$·tonne⁻¹)		X		X			X
Fixed tipping fee for residents (28 US$·tonne⁻¹)	X		X				
Subsidy is defined as a decision variable (<50 US$·tonne⁻¹)		X	X		X		
HAL[a] of net benefit for collection (US$·tonne⁻¹)	0	0	0				
LAL[b] of net benefit for collection (US$·per day)	20,000	20,000	20,000				
HAL of net benefit for incinerator (US$·per day)	16,667	16,667	16,667				
LAL of net benefit for incinerator (US$ per·day)	0	0	0				
HAL of net system benefit (US$·per day)				0	0	16,667	16,667
LAL of net system benefit (US$·per day)				−33,333	−33,333	0	0

Source: From Chang and Wang (1996b) (with permission from ASCE).
[a]HAL, highest aspiration level; [b]LAL, lowest aspiration level.

In the solution procedure, the GINO software package was used to generate optimal solutions, but the GINO software package does not guarantee a global optimal solution automatically since it uses the "reduced gradient algorithm (LINDO Systems Inc., 2000)." In order to obtain a global optimal solution, a special algorithm for linearization of nonlinear constraints was employed. With the aid of GINO, a suite of optimal solutions for these tested cases is summarized in Tables 22.5 and 22.6. In the base case, only the plastics in the An-Nan district (i.e., the closest district to the incinerator) cannot be fully recycled. This is primarily due to the combined effects of the higher heating value of plastics as compared to paper, the relatively short hauling distance, and the combustion temperature requirement in the incineration facility. It also appears that system benefit is nonpositive because of insufficient tipping fees, as they were limited to 28 US\$·tonne^{-1} as shown in Table 22.5. In cases 1, 3, 4, and 6, the highest aspiration level achieved is 0.6865 to come up with the minimum total recycling ratio (22.2%), implying that the system is dominated by the combustion temperature requirement. In the situation of case 5 where the system is handled by a private sector only, the recycling target becomes the major driving force in the trade-off process such that a higher recycling level and greater positive net benefits are anticipated. If the subsidy was formulated as a decision variable, as demonstrated in case 6, much higher net benefits could result.

Overall, recycling targets and combustion temperatures are highly likely to be achieved with respect to different types of ownership in varying planning scenarios. The cost/benefit metrics on the other hand, illuminate the possible economic trade-offs in decision-making. In this context, the attainable recycling targets range from 22% to 30% under different planning scenarios while the achieved combustion temperature varies from 964 to 976°C due to the possible impacts of paper and plastics recycling. The suggested recycling levels of paper and plastics provided by the FGP model cannot reach the target values in many districts in most of the cases. This is primarily due to the trade-offs between the combustion temperature requirement, the economic value of recyclables in the secondary material market, and the potential savings from hauling to treatment in the optimization process. The net system benefit varies tremendously from US\$914 in the base case to US\$11,442 in case 6. Nevertheless, this analytical framework still drove us to maximize profits in all cases as much as possible.

Once the related cost terms are modified, the evaluation of various planning scenarios associated with privatization for SWM can further be established. For example, some cities in the United States have structured collection service bidding and contracting to encourage performance and price competition between collectors (public and private sector) who service different zones within the city. These factors can be included in this FGP model. In addition, many private or public agencies are in charge of both construction and operation of the incineration facility, such as Waste Management, Inc., and Ogden Martin, Inc., in the United States, and may have their own investment benefit/cost profiles. Such a profile can be easily factored into a proposed FGP model formulation based on the principles of engineering economics.

TABLE 22.5 Optimal solutions of cases based on various planning scenarios for base case and cases 1 and 2

Parameter	Base Case		Case 1		Case 2	
Aspiration level (α)	0.1276		0.6865		0.5952	
Combustion temperature (°C)	964 (1,768°F)		976 (1,788°F)		974 (1,785°F)	
Respective recycling rate (%)	PL_i	PA_i	PL_i	PA_i	PL_i	PA_i
East district	50	70	34	48	30	70
South district	50	70	34	48	30	70
West district	50	70	34	48	30	70
North district	50	70	34	48	30	55
An-Pin district	50	70	34	48	30	70
An-Nan district	27	70	34	48	30	42
Central district	50	70	34	48	30	70
Gross recycling rate (%)	30.7%		22.2%		25.1%	
Gross recycling rate of paper (%)	70.0%		48.1%		62.8%	
Gross recycling rate of plastics (%)	44.1%		34.3%		30.0%	
System income for collection (US$·per day)	23,052		34,206		34,577	
System costs for collection (US$·per day)	40,501		40,258		42,672	
Net system income for collection (US$·per day)	−17,449		−6,052		−8,095	
System income for incineration (US$·per day)	31,700		30,813		33,692	
System cost for incineration (US$·per day)	15,165		16,402		15,983	
Net income for incineration (US$·per day)	16,535		14,411		17,711	
Net system benefit (US$·per day)	−914		8,359		962	
Tipping fee (US$·tonne⁻¹)	28[a]		50[a]		50[a]	
Subsidy (US$·tonne⁻¹)	50[a]		42[a]		50[a]	

Source: From Chang and Wang (1996b) (with permission from ASCE).

[a]Predetermined value (upper bound).

TABLE 22.6 Optimal solutions of cases based on various planning scenarios for the rest of the cases

Parameter	Case 3		Case 4		Case 5		Case 6	
Aspiration level	0.6865		0.6865		0.3082		0.6865	
Combustion temperature (°C)	976 (1,788°F)		976 (1,788°F)		967 (1,774°F)		975 (1,788°F)	
Respective recycling rate (%)	PL_i	PA_i	PL_i	PA_i	PL_i	PA_i	PL_i	PA_i
East district	34	48	34	48	50	70	34	48
South district	34	48	34	48	50	70	34	48
West district	34	48	34	48	50	70	34	48
North district	34	48	34	48	50	70	34	48
An-Pin district	34	48	34	48	50	70	34	48
An-Nan district	34	48	34	48	18	22	34	48
Central district	34	48	34	48	50	70	34	48
Gross recycling rate (%)	22.20		22.20		28.00		22.20	
Gross recycling rate of paper (%)	48.10		48.10		62.40		48.10	
Gross recycling rate of plastics (%)	34.30		34.30		41.70		34.30	
System benefit (US$·per day)	31,223		39,550		38,639		47,232	
System cost (US$·per day)	35,789		35,789		33,503		35,789	
Net system benefit (US$·tonne^{-1})	−4,566		3,760		5,136		11,442	
Subsidy (US$·tonne^{-1})	–		–		1,500[a]		1,979[a]	
Tipping fee (US$·tonne^{-1})	28[a]		42[a]		–		–	

Source: From Chang and Wang (1996b) (with permission from ASCE).
[a]Predetermined value (upper bound).

22.4.6 Sensitivity Analysis

Sensitivity analysis was performed with the following settings to explore (Chang and Wang, 1996b): (1) the impacts of different tolerance intervals in the membership functions, (2) the impacts of different tipping fees charged, and (3) the impacts of the subsidies, such as described in case 5. We may list the tested scenarios by the varying tolerance intervals (i.e., upper bound) of membership functions in association with net system benefits in case 5 to examine the corresponding changes of the optimal solutions (Table 22.7) (Chang and Wang, 1996b). In cases 5A to 5E, all extensions of case 5, the results indicate that the greater the increase in the upper level of the tolerance interval in the membership function corresponding to the net system benefit, the higher the total recycling levels and the lower the aspiration level α achieved in the optimal solutions. It is insightful to visualize the dynamic interaction among social, economic, and environmental goals in a three-dimensional representation (Figure 22.9). The interactive factors included are the total recycling level, the achieved combustion temperature, and the system cost/benefit attained. It can be seen that a higher total recycling level would compromise combustion efficiency to some extent and furthermore, such an interaction appears approximately linear (Figure 22.9) (Chang and Wang, 1996b).

To understand the impact of varying tipping fees and subsidies in case 5, an additional sensitivity analysis was performed that might be useful for future policy making. It is observed that the total recycling level would reach a steady state after a reasonable increase in subsidies (Figure 22.10) (Chang and Wang, 1996b). This implies that the combustion temperature requirement becomes influential once the economic goal is achieved. Such information may provide valuable management guidelines for both government agencies and private enterprises responsible for SWM (Chang and Wang, 1996b).

22.5 FINAL REMARKS

The proposed FGP model detailed in this chapter is an effective tool for generating a set of near optimal solutions for a real-world SWM recycling in which interactions between incineration and recycling programs conflict with each other. Although the inherent complexity when comparing the social and environmental impacts to economic benefits is difficult to understand, the proposed FGP model can successfully integrate these factors and elucidate possible profiles in a systematic way. The optimal solutions and subsequent sensitivity analyses capture clearly the inherent complexity and demonstrate the interactions between the goals of combustion temperature and recycling, thereby providing a set of operational guidelines with respect to sustainable management strategies for SWM in an urban setting. The results in the case study suggest that a nonlinear FGP approach may fully illuminate the optimal extent to which recycling and combustion are compatible, while at the same time, help in an investigation of the associated economic impacts on the private/public sectors and ratepayers under various scenarios by adjusting the proper settings to carry out different managerial scenarios.

TABLE 22.7 Optimal solutions by varying the tolerance intervals associated with the membership function of net system benefit

Reference basis	Case 5		Case 5		Case 5		Case 5		Case 6	
Test number	5 A		5B		5 C		5D		5E	
Net system benefit corresponding to highest aspiration level	100,000		300,000		400,000		700,000		900,000	
Net system benefit corresponding to lowest aspiration level	0		0		0		0		0	
Aspiration level	0.6865		0.6865		0.3082		0.6865		0.6865	
Combustion temperature (°C)	976 (1,788 °F)		976 (1,788 °F)		967 (1,774 °F)		976 (1,788 °F)		976 (1,788 °F)	
Respective recycling level (%)	PL_i	PA_i	PL_i	PA_i	PL_i	PA_i	PL_i	PA_i	PL_i	PA_i
East district	34	48	50	70	50	70	50	70	50	70
South district	34	48	50	70	50	70	50	70	50	70
West district	34	48	50	70	50	70	50	70	50	70
North district	34	48	24	33	19	63	50	70	50	70
An-Pin district	34	48	50	70	50	70	50	70	50	70
An-Nan district	34	48	24	33	19	26	33	16	41	12
Central district	34	48	29	70	50	70	50	70	50	70
Total recycling level (%)	22.2		25.7		27.3		28.8		29.2	
Total recycling level of paper (%)	48.1		57.3		61.9		61.5		61.0	
Total recycling level of plastics (%)	34.3		38.1		39.3		45.5		47.7	
System benefit (US$·per day)	39,375		38,930		38,723		38,604		38,578	
System cost (US$·per day)	35,790		34,195		33,710		33,349		33,261	
Net system benefit (US$·tonne⁻¹)	3,585		4,735		5,013		5,256		5,317	

Source: From Chang and Wang (1996b) (with permission from ASCE).

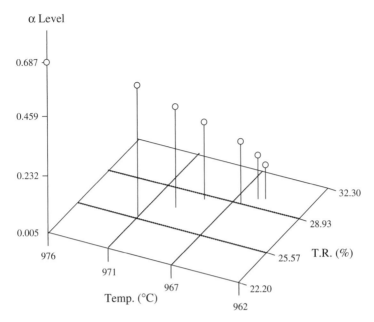

FIGURE 22.9 Dynamic interactions between changes in combustion temperature, total recycling level, and achieved aspiration level. *Source*: Chang and Wang (1996b) (with permission from ASCE). Note: Temp, combustion temperature; α Level, level of satisfaction; T.R, total recycling level

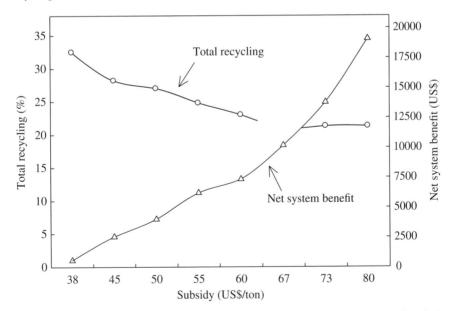

FIGURE 22.10 A comparison between total recycling level, subsidy, and system benefit in Case 5. *Source*: Chang and Wang (1996b) (with permission from ASCE)

REFERENCES

Bellman, R. E. and Zadeh, L. A. 1970. Decision-making in a fuzzy environment. *Management Science*, 17(4), 141–164.

Chang, N. B. and Wang, S. F. 1996a. Comparative risk analysis of solid waste management alternatives in a metropolitan region. *Environmental Management*, 20(1), 65–80.

Chang, N. B. and Wang, S. F. 1996b. Managerial fuzzy optimal planning for solid waste management systems. *Journal of Environmental Engineering, ASCE*, 122(7), 649–658.

Chang, N. B. and Wang, S. F. 1997. Integrated analysis of recycling and incineration programs by goal programming techniques. *Waste Management & Research*, 15(2), 121–136.

Chang, N. B., Shoemaker, C. A., and Schuler, R. E. 1996a. Solid waste management system analysis with air pollution control and leachate impact limitations. *Waste Management & Research*, 14(5), 463–481.

Chang, N. B., Yong, Y. C., and Wang, S. F. 1996b. Solid waste management system analysis with noise control and traffic congestion limitations. *Journal of Environmental Engineering ASCE*, 122(2), 122–131.

Chen, S. J. and Hwang, C. L. 1991. *Fuzzy Multiple Attribute Decision Making: Methods and Application*, Springer-Verlag, Heidelberg, Germany.

Clark, M. J. 1988. *Improving Environmental Performance of MSW Incinerators*, Industrial Gas Cleaning Institute Forum, New York.

Davila, E. and Chang, N.-B. 2005. Sustainable pattern analysis of publicly-owned material recovery facility under uncertainty. *Journal of Environmental Management*, 75(4), 337–352.

Environment Canada. 1991. *The National Incinerator Testing and Evaluation Program (NITEP)*, Environment Canada, Ontario, Canada.

Hannan, E. L. 1981. Linear programming with multiple fuzzy goals. *Fuzzy Sets and Systems*, 6(3), 235–248.

Inuiguchi, M., Ichihashi, H., and Kume, Y. 1990. A solution algorithm for fuzzy linear programming with piecewise linear membership functions. *Fuzzy Sets and Systems*, 34(1), 15–31.

Koo, J.-K., Shin, H.-S., and Yoo, H.-C. 1991. Multi-objective siting planning for a regional hazardous waste treatment center. *Waste Management & Research*, 9(3), 205–218.

Lai, Y. J. and Hwang, C. L. 1992. *Fuzzy Mathematical Programming: Methods and Applications*, Springer-Verlag, Heidelberg, Germany.

Lai, Y. J. and Hwang, C. L. 1994. *Fuzzy Multiple Objective Decision Making: Methods and Applications*, Springer-Verlag, Heidelberg, Germany.

Leung, Y. 1983. A concept of a fuzzy ideal for multi-criteria conflict resolution. In: *Advance in Fuzzy Sets, Possibility Theory and Applications* (Ed. Wang, P. P.), Plenum, New York, pp. 387–402.

LINDO Systems Inc. 2000. Available at: http://www.lindo.com/ (accessed November 2000).

Massachusetts Department of Environmental Protection (MDEP). 1989. *Toward a System of Integrated Solid Waste Management. Commonwealth's Master Plan for Solid Waste Management*, Boston, MA.

Rao, J. R., Tiwari, R. N., and Mohanty, B. K. 1988. A preference structure on aspiration levels in a goal programming problem—a fuzzy approach. *Fuzzy Sets and Systems*, 25(2), 175–182.

Romero, C. 1991. *Handbook of Critical Issues in Goal Programming*, Pergamon Press, New York.

Schindler, P. J. 1989. *Municipal Waste Combustion Assessment Combustion Control at New Facilities*, EPA-60018-89-057, Washington, DC.

Tillman, D. A., Rossi, A. J., and Vick, K. M. 1989. *Incineration of Municipal and Hazardous Solid Wastes*, Academic Press, New York.

Tiwari, R. N., Dhannar, S., and Rao, J. R. 1986. Priority structure in fuzzy goal programming. *Fuzzy Sets and Systems*, 19(3), 251–259.

Tiwari, R. N., Dhannar, S., and Rao, J. R. 1987. Fuzzy goal programming-an additive model. *Fuzzy Sets and Systems*, 1, 45–55.

United States Environmental Protection Agency (USEPA). 1989. *Environmental News: EPA Proposes Air Emissions Standards for Municipal Waste Incinerators*, R233, 1–4, Washington, DC.

Zadeh, L. A. 1965. Fuzzy sets. *Information and Control*, 8(3), 338–353.

Zimmermann, H. J. 1978. Fuzzy programming and linear programming with several objective functions. *Fuzzy Sets and Systems*, 1(1), 45–55.

Zimmermann, H. J. and Zysno, P. 1985. Quantifying vagueness in decision models. *European Journal of Operational Research*, 22(2), 148–158.

CHAPTER 23

GREY SYSTEMS THEORY FOR SOLID WASTE MANAGEMENT

Given the intensive data requirements for stochastic or probabilistic studies, one need is to simplify the uncertainty description in situations where only a very few samples exist. Using interval numbers to uniquely address this problem will allow such uncertainty propagation to happen throughout the optimization process. This chapter introduces the idea of grey systems and discusses the traditional rationale and philosophy of interval linear programming (ILP) or grey linear programming (GLP) models with uncertainties expressed by interval numbers, considered in terms of both objectives and constraints. The stability issues related to GLP or ILP are delineated with numerical examples to demonstrate current barriers in this field. A final summary of hybrid approaches for uncertainty quantification (UQ) helps elevate the level of UQ for future applications in solid waste management (SWM), with respect to various sources of uncertainty.

23.1 GREY SYSTEMS THEORY

Grey systems theory was developed by Deng in 1984 (Deng, 1984a, 1984b), in which all systems are divided into three categories: white, grey, and black parts. While the white part yields completely certain and clear messages in a system, the black part has totally unknown characteristics. The messages released by the grey part are in-between. Understanding social or natural phenomena actually rests upon a dynamic process of knowledge production evolving from the black stage to the

Sustainable Solid Waste Management: A Systems Engineering Approach, First Edition. Ni-Bin Chang and Ana Pires.
© 2015 The Institute of Electrical and Electronics Engineers, Inc. Published 2015 by John Wiley & Sons, Inc.

white stage. In reality, many phenomena encountered in environmental management are in the grey stage due to an insufficient amount of information or intricate social implications. Therefore, a grey number or interval number may be used instead of a random variable to describe the uncertain parameters in many social, engineering, and natural systems where only a very few samples exist. In real-world applications, the grey or interval numbers can be viewed as an alternative description of system parameters in light of the concept of a "confidence interval" in probability theory. Such an approach dramatically simplifies the expression of system uncertainties whenever the probability density functions cannot be fully identified with sufficient samples.

With limited observations, management strategies for traditional deterministic programming models are not robust enough to be applied for solving real-world problems. The grey systems theory developed by Deng in 1984 (Deng, 1984a, 1984b, 1986) was therefore proposed as a supplemental tool in uncertainty analysis. A grey or interval number $a^\pm$ is simply an open interval with upper and lower limits as expressed by $[a^-, a^+]$, in which a^- is the lower bound and a^+ is the upper bound. Moore (1979) described interval analysis and interval programming techniques with no well-confirmed solution procedure. The solution procedure of grey mathematical programming, in which all or part of the input parameters are represented by interval numbers, was further developed by Huang et al. (1992) and was applied to several SWM systems in the early 1990s (Huang et al., 1992, 1993, 1995a; Huang and Moore, 1993). Huang et al. (1992, 1993, 1994) define the following definitions of grey numbers.

Definition 1: Let x denote a closed and bounded set of real numbers. A grey number $x^\pm$ is defined as an interval with known upper and lower bounds but unknown distribution information for x:

$$x^\pm = [x^-, x^+] = \{t \in x^\pm | x^- \leq t \leq x^+\},$$

where x^- and x^+ are the lower and upper bounds of $x^\pm$, respectively.

Definition 2: Let $* \in \{+, -, \times, \div\}$ be a binary operation on grey numbers. For grey numbers $x^\pm$ and $y^\pm$, we have

$$x^\pm * y^\pm = [\min\{x * y\}, \max\{x * y\}], x^- \leq x \leq x^+, y^- \leq y \leq y^+.$$

Then we have

$$x^\pm + y^\pm = [x^- + y^-, x^+ + y^+],$$
$$x^\pm - y^\pm = [x^- - y^+, x^+ - y^-],$$
$$x^\pm \times y^\pm = [\min\{x \times y\}, \max\{x \times y\}],$$
$$x^\pm \div y^\pm = [\min\{x \div y\}, \max\{x \div y\}].$$

Definition 3: For a grey number $x^\pm$, we have

$$x^\pm \geq 0 \text{ if } x^+ \geq 0 \text{ and } x^- \geq 0,$$
$$x^\pm \leq 0 \text{ if } x^+ \leq 0 \text{ and } x^- \leq 0.$$

Definition 4: For $x^\pm = [x^-, x^+]$ and $y^\pm = [y^-, y^+]$, their order relations are as follows:

$$x^\pm \leq y^\pm \text{ if } x^- \leq y^- \text{ and } x^+ \leq y^+,$$
$$x^\pm < y^\pm \text{ if } x^\pm < y^\pm \text{ and } x^\pm \neq y^\pm.$$

Definition 5: For a grey number $x^\pm$, $\text{sign}(x^\pm)$ is defined as follows:

$$\text{sign}(x^\pm) = 1 \text{ if } x^\pm \geq 0$$
$$= -1 \text{ if } x^\pm < 0.$$

Definition 6: For a grey number $x^\pm$, its grey absolute value $|x^\pm|$ is defined as follows:

$$|x^\pm| = x^\pm \text{ if } x^\pm \geq 0$$
$$= x^\pm \text{ if } x^\pm < 0.$$

Thus,

$$|x^-| = x^- \text{ if } x^\pm \geq 0$$
$$= x^- \text{ if } x^\pm < 0$$

and

$$|x^\pm| = x^\pm \text{ if } x^\pm \geq 0$$
$$= -x^\pm \text{ if } x^\pm < 0.$$

Definition 7: For a grey number $x^\pm = [x^-, x^+]$, its whitened mid-value x^M can be defined as follows:

$$x^M = [x^- + x^+]/2.$$

Definition 8: The degree of greyness of $x^\pm$ is defined as

$$\deg(x^\pm) = [x^+ - x^-]/x^M.$$

23.2 GREY LINEAR PROGRAMMING

Linear programming is a classical optimization tool to derive an optimal solution under the complete information assumption. This assumption means that all the coefficients and right-hand sides in the linear programming model should be perfectly known before a decision can be made. However, most real-world problems may violate this assumption for different types of reasons. A decision may be made by a group of people who may have different recognitions of a problem resulting in vagueness for the parameters in the problem. Some parameters in a proposed problem can be random variants that may or may not follow some underlying distributions. Or those parameters in a proposed problem are extremely difficult, or unable to be obtained such that decision makers are forced to make a decision based on the incomplete information.

23.2.1 Formulation of a GLP Model

A GLP or ILP model proposed by Huang et al. (1992, 1993, 1994) can be given in the following standard format:

$$\text{Max} f^{\pm} = C^{T\pm} X^{\pm} \tag{23.1}$$

Subject to

$$A^{\pm} X^{\pm} \leq B^{\pm} \tag{23.2}$$

$$x_j^{\pm} \geq 0, x_j^{\pm} \in X^{\pm}, \forall j = 1, \ldots, n \tag{23.3}$$

where

$$C^{T\pm} = \left[c_1^{\pm}, c_2^{\pm}, \ldots, c_n^{\pm} \right],$$
$$X^{T\pm} = \left[x_1^{\pm}, x_2^{\pm}, \ldots, x_n^{\pm} \right],$$
$$B^{T\pm} = \left[b_1^{\pm}, b_2^{\pm}, \ldots, b_m^{\pm} \right],$$
$$A^{\pm} = \left\{ a_{ij}^{\pm} \right\}, \quad \forall i = 1, \ldots, m, j = 1, \ldots, n.$$

For the grey numbers $c_j^{\pm}, a_{ij}^{\pm}$, and $b_i^{\pm}$, we have

$$c_j^{\pm} = \left[c_j^{-}, c_j^{+} \right], \quad \forall j \tag{23.4}$$

$$a_{ij}^{\pm} = \left[a_{ij}^{-}, a_{ij}^{+} \right], \quad \forall ij \tag{23.5}$$

$$b_i^{\pm} = \left[b_i^{-}, b_i^{+} \right], \quad \forall j \tag{23.6}$$

Since some grey parameters exist as an objective function and constraints, the optimal solution of model equations (23.1)–(23.3) will be

$$f^{*\pm} = [f^{*-}, f^{*+}], \tag{23.7}$$

$$X^{*+} = \left[x_1^{*\pm}, x_2^{*\pm}, \ldots, x_n^{*\pm} \right], \tag{23.8}$$

$$x_j^{*\pm} = \left[x_j^{*-}, x_j^{*+} \right], \quad \forall j. \tag{23.9}$$

23.2.2 Solution Procedure of a GLP Model

Model equations from (23.1) to (23.3) can be converted from a grey problem (uncertain) to a white problem (certain) in the following way:

$$\text{Max} f_m^\pm = C_m^{T\pm} X_m^\pm \tag{23.10}$$

$$\text{Subject to } A_m^\pm X_m^\pm \leq B_m^\pm \tag{23.11}$$

$$x_{jm}^\pm \geq 0, x_{jm}^\pm \in x_m^\pm, j = 1, \ldots, n \tag{23.12}$$

$$C_m^{T\pm} = \left[c_{1m}^\pm, c_{2m}^\pm, \ldots, c_{nm}^\pm \right],$$

$$X_m^{T\pm} = \left[x_{1m}^\pm, x_{2m}^\pm, \ldots, x_{nm}^\pm \right],$$

$$B_m^{T\pm} = \left[b_{1m}^\pm, b_{2m}^\pm, \ldots, b_{nm}^\pm \right],$$

$$A_m^\pm = \left\{ a_{ijm}^\pm \right\}, \quad \forall i = 1, \ldots, m, j = 1, \ldots, n.$$

where c_{jm}, a_{ijm}, and b_{1m} are the whitening values of $c_{jm}^\pm, a_{ijm}^\pm$, and $b_{1m}^\pm$, respectively. Therefore, a set of whitening solution $f_m^{*\pm}$ and $x_m^{*\pm}$, which are included in the optimal grey solutions $f_m^{*\pm}$ and $x_m^{*\pm}$, can be derived by solving the model defined in Equations (23.10)–(23.12).

For n grey coefficients $c_j^\pm (j = 1, 2, \ldots, n)$ in the objective function, if k_1 of them are positive, and k_2 coefficients are negative, $c_j^\pm \leq 0$ $(j = 1, 2, \ldots, k_2)$, where $k_1 + k_2 = n$ (the model does not include the situation where the two bounds of $c_j^\pm$ have different signs). Thus, we can develop the following expressions for the upper and lower bounds of $f^\pm$:

$$f^+ = c_1^+ x_1^+ + c_2^+ x_2^+ + \cdots + c_{k_1}^+ x_{k_1}^+ + c_{k_1+1}^+ x_{k_1+1}^- + \cdots + c_n^+ x_n^- \tag{23.13}$$

$$f^- = c_1^- x_1^- + c_2^- x_2^- + \cdots + c_{k_1}^- x_{k_1}^- + c_{k_1+1}^- x_{k_1+1}^+ + \cdots + c_n^- x_n^+ \tag{23.14}$$

Based on Equation (23.13), relevant constraints can be given as

$$a_{i1}^- x_1^+ + a_{i2}^- x_2^+ + \ldots + a_{ik_1}^- x_{k_1}^+ + a_{ik_1+1}^+ x_{k_1+1}^- + \cdots + a_{in}^+ x_n^- \leq b_i^+. \tag{23.15}$$

Similarly, based on Equation (23.14), relevant constraints are

$$a_{i1}^+ x_1^- + a_{i2}^+ x_2^- + \cdots + a_{ik_1}^+ x_{k_1}^- + a_{ik_1+1}^- x_{k_1+1}^+ + \cdots + a_{in}^- x_n^+ \leq b_i^- \tag{23.16}$$

For whitening solutions $x_{jm}^{*\pm}$, we have $x_{jm}^{*\pm} \in x^{*\pm}$. Therefore,

$$x_j^+ \geq x_{jm}^{*\pm}, \quad j = 1, 2, \ldots, k_1 \tag{23.17}$$

$$x_j^- \leq x_{jm}^{*\pm}, \quad j = k_1 + 1, k_1 + 2, \ldots, n \tag{23.18}$$

Thus, the model defined by Equations (23.1)–(23.3) can be divided into two submodels:

$$\text{Max } f^+ \tag{23.19}$$

$$\text{Subject to (23.13), (23.15), (23.17), and (23.18)} \tag{23.20}$$

$$\text{Max } f^- \tag{23.21}$$

$$\text{Subject to (23.14), (23.16), (23.17), and (23.18)} \tag{23.22}$$

The model defined by Equations (23.19)–(23.20) and (23.21)–(23.22) are linear programming models with a single objection function. Therefore, $f^{*+}, x_j^{*+}(j = 1, 2, \ldots, k_1)$ and $x_j^{*-}(j = k_1 + 1, k_1 + 2, \ldots, n)$ can be solved by model equations (23.21)–(23.22), and $f^{*-}, x_j^{*+}(j = k_1 + 1, k_1 + 2, \ldots, n)$ and $x_j^{*-}(j = 1, 2, \ldots, k_1)$ can be solved by model equations (23.23)–(23.24). Thus, the solutions of the GLP model equations from (23.1) to (23.3) are

$$f^{*\pm} = \left[f^{*-}, f^{*+}\right], \tag{23.23}$$

$$x_j^{*\pm} = \left[x_j^{*-}, x_j^{*+}\right], \quad \forall j \tag{23.24}$$

where $f^{*\pm}$ and $x_j^{*\pm}$ are all grey numbers.

Solutions of the GLP model following the "two-step method" from Equations (23.19)–(23.22) include decision variables ($x_j^{*\pm}$, $\forall j$) and the relevant objective value ($f^{*\pm}$). The solutions of decision variable are expressed as $x_j^{*\pm} = [x_j^{*-}, x_j^{*+}]$, $\forall j$, which means that the maximum possible value of $x_j^{*\pm}$ is x_j^{*+} (upper limit), and the minimum is x_j^{*-} (lower limit). The solutions can be directly applied to decision-making, with the values being adjusted within the intervals in the final decision scheme.

The solution of the objective function is expressed as $f^{*\pm} = [f^{*-}, f^{*+}]$, which means that the maximum objective value is f^{*+} (upper limit), and the minimum is f^{*-} (lower limit). The upper and lower limits of the objective function value correspond to different distributions of decision variables that are important for assessing decision efficiencies. The adjustment of decision variables within their intervals will lead to the variation of objective function value within its corresponding interval.

23.2.3 Applications for Solid Waste Management

In an optimization analysis, uncertainties embedded in a linear programming model may exist in model coefficients and stipulations (right-hand side constraints). These uncertainties can propagate through the optimization analysis and generate uncertainties in the final optimal solutions. Previous probabilistic or stochastic methods dealing with uncertainty were too complicated to be applied to many real-world problems, while fuzzy sets were unable to reflect completely the uncertainties of the input and output information. In this section, the application of a GLP model demonstrates that this method allows uncertainties in the model inputs to be communicated into the optimization process, thereby connecting the initial uncertainty associated with input information to the inherent uncertainties in the final optimal solutions. A GLP problem with equivalent deterministic forms (i.e., Equations (23.19)–(23.22)) can be solved easily by running a simplex program several times. The modeling approach in the following example is applied to a hypothetical problem of waste flow allocation planning within a municipal SWM system. The results indicate that final optimal solutions can be generated for both the lower and upper limits of the objective function and final optimal solutions.

Example 23.1   In a regional SWM system, two municipalities share a landfill and a waste-to-energy (WTE) facility to serve municipal solid waste treatment/disposal needs, as summarized in Figure 23.1. Three time periods with a time interval of 1 year are considered. Over 3 years of planning horizon, the landfill has an existing capacity of 584,000–657,000 metric tons (tonnes) while WTE facility has a design capacity of 200–300 tonnes per day. The WTE facility generates ash residues of approximately 10% (on a mass basis) of incoming waste stream. There is a monofill cell within the current landfill for ash disposal. The revenue from the WTE facility is approximately US$70–75 per tonne combusted. Table 23.1 summarizes the waste generation rate, shipping costs, and the operating costs of two SWM facilities to be used in the three planning time periods. It is noticeable that the waste generation rates and cost of waste transportation vary spatially and temporally. With grey uncertainty, the problem is to generate the waste flows in the study region with respect to a number of economic and

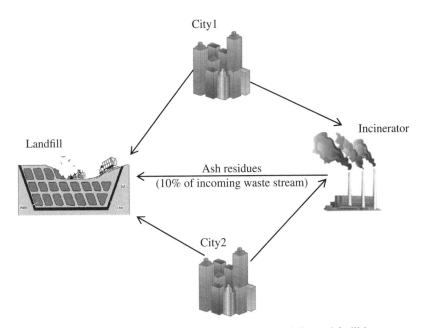

FIGURE 23.1   Municipalities, waste streams, and disposal facilities

TABLE 23.1   Waste generation rates, shipping costs, and operating costs

		Planning time period		
		1	2	3
Waste generation rate (tonne·d^{-1})	City 1	[100,200]	[180,250]	[220,350]
	City 2	[80,100]	[120,150]	[220, 220]
Cost of transportation to landfill	City 1	[13.4,16.7]	[15.2,18.3]	[17.7,20.4]
(US$·tonne^{-1})	City 2	[10.3,11.7]	[10.6,12.8]	[14.4,16.3]
Cost of transportation to incinerator	City 1	[10.5,12.7]	[12.3,15.4]	[13.7,16.9]
(US$·tonne^{-1})	City 2	[11.5,14]	[13,15.8]	[14.6,16.3]
Operational costs (US$·tonne^{-1})	Incinerator	[45,64]	[52,71]	[65,85]
	Landfill	[20,35]	[28,43]	[36,57]
Cost of transportation to ash monofill (US$·tonne^{-1})		10	14	16
Inflation rate (%)		6	6	6
Market interest rate (%)		8	8	8

treatment/disposal capacity availability constraints in order to minimize the overall system costs or maximize the overall system benefits.

Solution:
The GLP model can be formulated as follows:

1. **Objective function:**
   Max (total income from electricity sales) − (total transportation and operating costs + costs of transportation to ash monofill + operating costs of incoming ash stream)

$$\text{Max}(f)^{\pm} = \sum_{i=1}^{n} \sum_{t=1}^{T} 365 \cdot (x_{i2t})^{\pm} \cdot (\text{RE}_t)^{\pm} \cdot (1+r)^{-(t-1)}$$

$$- \sum_{i=1}^{n} \sum_{j=1}^{m} \sum_{t=1}^{T} 365[(x_{ijt})^{\pm} \cdot ((\text{TR}_{ijt})^{\pm} + (\text{OP}_{jt})^{\pm})](1+r)^{-(t-1)}$$

$$- \sum_{i=1}^{n} \sum_{t=1}^{T} 365[(x_{i2t})^{\pm} \cdot \text{FR} \cdot \text{TC}_t](1+r)^{-(t-1)}$$

$$- \sum_{i=1}^{n} \sum_{t=1}^{T} 365[(\text{OP}_{1t})^{\pm} \cdot (x_{i2t})^{\pm} \cdot \text{FR}](1+r)^{t-1}$$

where

FR: Residue flow from incinerator to landfill facility (percent of incoming waste stream into incinerator).

TC_t: Transportation cost for ash residue flow from incinerator to landfill facility during time period t (US\$ per tonne).

OP_{it}: Operating cost for facility i in time period t (US\$ per tonne).

RE_t: Revenues from incinerator facility during time period t (US\$ per tonne).

WEC: Operating capacity of WTE facility (tonne per day).

TR_{ijt}: Transportation cost for shipping raw waste from facility i to municipality j during time period t (US\$ per tonne), where $i = 1$ for the landfill facility, and $i = 2$ for the WTE facility.

x_{ijt}: Waste flow from facility i to municipality j during time period t (US\$ per tonne).

r: Real interest rate calculated by the following mathematical relationship between the inflation, real, and market interest rates $r = (\text{MIR} - f)/1 + f$, where f is the inflation rate, and MIR is the market interest rate.

Subject to

1. Capacity limitation constraint for landfill

$$365 \left[\sum_{i=1}^{n} \sum_{t=1}^{T} x_{i1t}^{\pm} + \sum_{i=1}^{n} \sum_{t=1}^{T} x_{i2t}^{\pm} \cdot \text{FR} \right] \leq \text{LC}^{\pm}$$

2. Capacity limitation constraint for incinerator

$$\sum_{i=!}^{n} x_{i2t}^{\pm} \leq \text{WEC}^{\pm} \quad \forall t$$

3. Mass balance constraints

$$\sum_{i=!}^{m} x_{ijt}^{\pm} = \text{WG}_{jt}^{\pm} \quad \forall i, t$$

4. Non-negativity constraints

$$x_{ijt}^{\pm} \geq 0 \quad \forall i, j, t$$

where

WG_{jt}: Waste generation rate of municipality j during time period t (US\$ per tonne), in which $j = 1, 2$, and $t = 1, 2$, and 3.

LC: Landfill capacity (tonne per day)

Two submodels of GLP are formulated, separately, as follows:

Submodel I
Since the problem is to maximize the objective function, the submodel related to the upper bound of $(f)^\pm$, which has been expressed as $(f)^\pm$, may be solved first.

$$\text{Max}(f)^+ = \sum_{i=1}^{n} \sum_{t=1}^{T} 365 \cdot (x_{i2t})^+ \cdot (\text{RE}_t)^+ \cdot (1+r)^{-(t-1)}$$

$$- \sum_{i=1}^{n} \sum_{j=1}^{m} \sum_{t=1}^{T} 365[(x_{ijt})^- \cdot ((\text{TR}_{ijt})^- + (\text{OP}_{jt})^-)](1+r)^{-(t-1)}$$

$$- \sum_{i=1}^{n} \sum_{t=1}^{T} 365[(x_{i2t})^- \cdot \text{FR} \cdot \text{TC}_t](1+r)^{-(t-1)}$$

$$- \sum_{i=1}^{n} \sum_{t=1}^{T} 365[(\text{OP}_{1t})^- \cdot (x_{i2t})^- \cdot \text{FR}](1+r)^{t-1}$$

Subject to:

$$365 \left[\sum_{i=1}^{n} \sum_{t=1}^{T} (x_{i1t})^- + \sum_{i=1}^{n} \sum_{t=1}^{T} (x_{i2t})^+ \cdot \text{FR} \right] \le \text{LC}^+$$

$$\sum_{i=!}^{n} (x_{i2t})^+ \le (\text{WEC})^+ \quad \forall t$$

$$(x_{i2t})^+ + (x_{i1t})^- = (\text{WG}_{it})^+ \quad \forall i, t$$

$$(x_{ijt})^\pm \ge 0 \quad \forall i, j, t$$

Submodel II
Then, the submodel related to the lower bound of $(f)^\pm$, which has been expressed as $(f)^-$, may be solved.

$$\text{Max}(f)^- = \sum_{i=1}^{n} \sum_{t=1}^{T} 365 \cdot (x_{i2t})^- \cdot (\text{RE}_t)^- \cdot (1+r)^{-(t-1)}$$

$$- \sum_{i=1}^{n} \sum_{j=1}^{m} \sum_{t=1}^{T} 365 \cdot [(x_{ijt})^+ \cdot ((\text{TR}_{ijt})^+ + (\text{OP}_{jt})^+)] \cdot (1+r)^{-(t-1)}$$

$$- \sum_{i=1}^{n} \sum_{t=1}^{T} 365 \cdot [(x_{i2t})^+ \cdot \text{FR} \cdot \text{TC}_t] \cdot (1+r)^{-(t-1)}$$

$$- \sum_{i=1}^{n} \sum_{t=1}^{T} 365 \cdot [(\text{OP}_{1t})^+ \cdot (x_{i2t})^+ \cdot \text{FR}] \cdot (1+r)^{t-1}$$

Subject to:

$$365\left[\sum_{i=1}^{n}\sum_{t=1}^{T}(x_{i1t})^{+}+\sum_{i=1}^{n}\sum_{t=1}^{T}(x_{i2t})^{-}\cdot FR\right]\le LC^{-}$$

$$\sum_{i=!}^{n}(x_{i2t})^{-}\le(WEC)^{-}\quad\forall t$$

$$(x_{i1t})^{+}+(x_{i2t})^{-}=(WG_{it})^{-}\quad\forall i,t$$

$$(x_{ijt})^{\pm}\ge0\quad\forall i,j,t$$

The obtained decision variables of the first submodel should be used as the constraints of the second submodel. Therefore, the following three additional constraints can be considered to make sure that the uncertainty propagation can be in a legitimate order.

$$(x_{i2t})^{-}\le(x_{i2t_{opt}})^{+}\quad\forall it$$
$$(x_{ijt})^{+}\ge(x_{ijt_{opt}})^{-}\quad\forall ijt$$
$$(x_{ijt})^{+}\ge(x_{ijt})^{-}\quad\forall ijt$$

The optimal shipping strategies is shown in Figure 23.2, and the minimum cost is [9,594,758, 14,345,720] US\$. In addition, the maximum and minimum waste flows to the landfill and incinerator facilities are indicated with the upper and lower bound

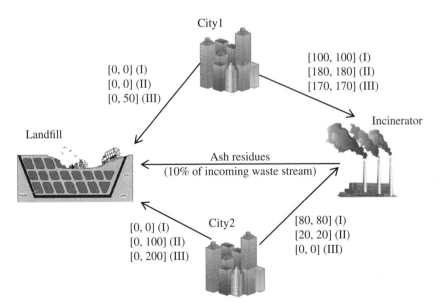

FIGURE 23.2 The optimal solution of the GLP model for SWM

values in this figure. It should be mentioned that the parameter is deterministic in case the upper and lower bound values are equal. The shipping strategy reveals that two municipalities mainly count on one incinerator facility simultaneously while sending the residual amount of raw waste to landfill.

More important issues, such as siting the SWM facilities in a network system using grey integer programming under uncertainty, may follow the same two-step method (Huang et al., 1995a). An inexact multistage integer programming approach involving the use of a substantial number of integer variables was developed to explore more complicated siting issues spatially and temporally (Li and Huang, 2009). However, those multistage planning issues can be formulated and solved using grey dynamic programming model as well (Huang et al., 1994).

23.3   THE STABILITY ISSUES OF GREY PROGRAMMING MODELS

Classical sensitivity analysis is a tool for postoptimality analysis that provides ranges for coefficients in the objective function where the right-hand side in which the changes occur are allowed to keep the optimal, if only one right-hand side varies at a time. For changes in more than one coefficient in the objective or right-hand side at a time, the 100% rule (Bradley et al., 1977) provides a sufficient condition to keep the optimal. These postoptimality analysis tools are derived from the simplex method and hence cannot be used to analyze the uncertainties of coefficients in the constraints, because the inverse of an uncertain matrix is NP-hard (the acronym of nondeterministic polynomial-time hard) (see Box 15.1) (Coxson, 1999). Furthermore, these postoptimality analysis methods do not suggest solutions other than the optimal, based on incomplete information. These reasons motivated the development of the linear programming model under uncertainty.

Continuous effort has been made by previous researchers to address the uncertainties in single or multiple objective linear programming models. For example, there are studies on only the uncertainties in a single-objective function (Rommelfanger et al., 1989; Inuiguchi et al., 1990) or in constraints (Mráz, 1998; Kuchta, 2008). Uncertainties can be embedded in either single-objective or multiobjective programming models (Huang et al., 1992; Urli and Nadeau, 1992; Shaocheng, 1994). Uncertain parameters can be stochastic and based on underlying probability distributions, as discussed in Chapter 19, fuzzy numbers based on underlying membership functions, as discussed in Chapters 20–22, or interval numbers that only specify the lower and upper bounds as described in Sections 23.1 and 23.2. Stochastic programming (Kall and Wallace, 1994; Birge and Louveaus, 1997; Ruszczyński and Shapiro, 2003), fuzzy programming (Zimmerman, 1978; Inuiguchi, et al., 1990; Inuiguchi and Ramik, 2000), interval programming (Chinneck and Ramadan, 2000; Oliveira and Antunes, 2007) as well as combinations of these methods (Liu and Iwamura, 1998; Huang et al., 2001; Nie et al., 2007) were developed to address those uncertainties.

Due to limited space, this section only focuses on GLP with the following general-ized form to delineate the stability issues in GLP, as described in Equation (23.25).

$$\text{Min } Z_p = \sum_{j=1}^{n} c_{pj}^{\pm} x_j \tag{23.25}$$

$$\text{Subject to } \sum_{j=1}^{n} a_{ij}^{\pm} x_j \leq b_i^{\pm}, \, x_j \geq 0$$

$$i = 1, 2, \ldots m, \, j = 1, 2, \ldots n, \, p = 1, 2, \ldots P$$

The parameters $c_{pj}^{\pm}$, $a_{ij}^{\pm}$, and $b_i^{\pm}$ are interval numbers with their lower and upper bounds known. So that $c_{pj}^{\pm} \in \{c_{pj} | c_{pj}^{-} \leq c_{pj} \leq c_{pj}^{+}\}$, $a_{ij}^{\pm} \in \{a_{ij} | a_{ij}^{-} \leq a_{ij} \leq a_{ij}^{+}\}$, and $b_i^{\pm} \in \{b_i | b_i^{-} \leq b_i \leq b_i^{+}\}$.

From the first perspective, a two-step method (Huang et al., 1992; Huang, 1994) and a similar method (Shaocheng, 1994) were proposed for a single-objective GLP to obtain a possibly optimal solution set. Both methods suggest transforming the original GLP into two subproblems. One will have the most favorable version of the objective function and the maximum value range inequality and the other will have the least favorable version of the objective function and the minimum value range inequality (i.e., Equations (23.19)–(23.22)). The maximum and minimum value range inequalities are the largest and smallest possible feasible regions determined by the nondeterministic constraints (Chinneck and Ramadan, 2000). The derived solutions from these two methods are interval solutions with the expectation that they include all possibly optimal solutions. A possibly optimal solution to a single-objective GLP problem is an optimal solution to at least one deterministic linear programming problem with the uncertain parameters selected within their admissible ranges of variance. The solution can be obtained fast and thus it is popularly referenced and applied to many real-world examples (Maqsood et al., 2005; Cheng et al., 2009; Cao et al., 2010). However, the rationality of the solution to the original GLP is highly doubtable. Thus, we refer back to the original numeric example used in Huang et al. (1992).

$$\text{Max } f = [50, 60]x_1 - [70, 90]x_2 \tag{23.26}$$

Subject to

$$[4, 6]x_1 + x_2 \leq 150$$
$$6x_1 + [5, 7]x_2 \leq 280$$
$$x_1 + [3, 4]x_2 \leq 90$$
$$[1, 2]x_1 - 10x_2 \leq -1$$
$$x_1, x_2 \geq 0$$

The interval solution that is derived from the two-step method is $x_1 = [24.18, 36.56]$, $x_2 = [3.76, 4.94]$, and $f = [764.71, 1930.73]$. When the interval solution is

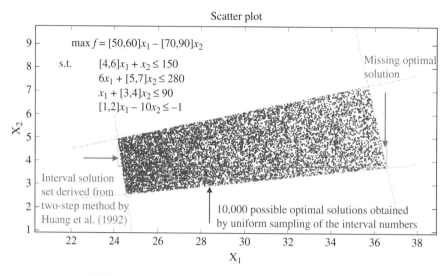

FIGURE 23.3   An illustrative example of the two-step method

checked as in Problem (23.26), the expectation was that the final optimal interval solution will cover all possibly optimal solutions associated with any sub-GLP in Problem (23.26). This expectation was not satisfied. To illustrate this first stability issue, Figure 23.3 plots 10,000 possible optimal solutions for Problem (23.26), each of which is solved from an equivalent deterministic optimization problem associated uniformly with those grey or interval coefficients. However, more than half of the possible optimal solutions in this example are out of the interval solution set derived from the two-step method as the final possibly optimal interval solution. Therefore, not all the elements contained in the derived optimal interval solution set are true optimal solutions. Huang and Cao (2011) later recognized this second issue and proposed a three-step method, which adds an extra step to the two-step method to shrink the possibly optimal interval solution set to the q $(0 < q < 1)$ level so that all elements in the derived optimal interval solution are possibly optimal. However, this approach makes the first stability issue even more severe since more final optimal solutions are excluded from the optimal interval solution set.

Shown in this example is that the true possibly optimal solution set in the solution space is hardly a rectangle-like shape so that it cannot be assumed to have an interval solution pattern as $x_{opt} = [x_{opt}^-, x_{opt}^+]$. More likely, as illustrated in Figure 23.4, the possibly optimal solutions can be dispersed in a solution space based on the random number simulation approach. As this case shows, there is almost no way for an interval number to cover such two detached areas. Thus, an interval solution set for an ILP will inevitably cause either the issue of not including all the possibly optimal solutions or the issue of having not all the elements as possibly optimal interval solutions or even that both stability issues appear at the same time. Regardless of this discrepancy, the optimal solution associated with a GLP model such as the one shown

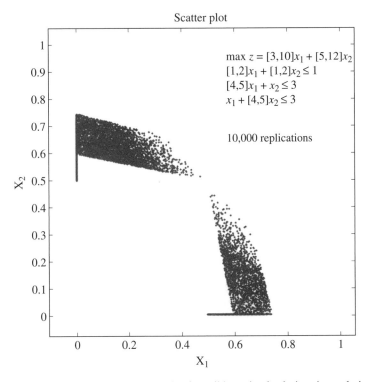

Scatter plot

$\max z = [3,10]x_1 + [5,12]x_2$
$[1,2]x_1 + [1,2]x_2 \le 1$
$[4,5]x_1 + x_2 \le 3$
$x_1 + [4,5]x_2 \le 3$

10,000 replications

FIGURE 23.4   Another illustrative example of possibly optimal solutions in a solution space

in Figure 23.3 might still give people a general sense about where the optimal solution could be located, which reduces a feasible region if a deeper search is undertaken.

23.4 THE HYBRID APPROACH FOR VARIOUS CASES OF UNCERTAINTY QUANTIFICATION

Even with the presence of the stability issues, the minimum and maximum possibly objective values derived from the two-step method are still useful to a certain degree (Chinneck and Ramadan, 2000). At least it gives decision makers some insights into a probable way for making a better final option where fuzzy set theory may come to play to address epistemological gaps. This need opens a new avenue for developing some hybrid approaches to combine fuzzy set theory and grey systems theory to address two types of uncertainty in systems planning for SWM (Huang et al., 1995b; Chang et al., 1997; Sun et al., 2010a). On the one hand, when decision-making issues involve two-stage decision analyses or games within the GLP, the inclusion of probability analysis, chance-constrained formulation, or even its inherent stochastic nature could contribute to a deepened delineation of uncertainty in SWM (Sun et al., 2010b). On the other hand, fuzzy two-stage programming may address both sources of fuzziness and randomness for planning SWM under uncertainty (Li and Huang, 2007). These cross-cutting scenarios add to the demand for systems scientists to explore an advanced hybrid approach for addressing three types of uncertainty simultaneously. From interval numbers to fuzzy sets and to stochastic uncertainties, an interval-valued fuzzy-stochastic programming approach may be developed to account for an all-inclusive case study for SWM (Tan et al., 2010; Wang et al., 2012).

For multiobjective linear programming, the optimal solution for all the objective functions usually does not exist. Instead, efforts were made to find a Pareto optimal solution (or efficient solution) set. A solution is efficient if there is no other feasible solution available to improve at least one of the objective functions without compromising the others (Zimmermann, 1978). When there are only uncertainties in the objective functions, Bitran (1980) proposed the concept of a necessarily and possibly efficient solution set and provided a test method to determine whether a feasible solution belongs to those sets. A necessarily efficient solution is a feasible solution that is efficient in any deterministic multiobjective linear programming model with the uncertain parameters selected within their admissible range of variance. A possibly efficient solution is a feasible solution that is efficient for at least one deterministic multiobjective linear programming model with the uncertain parameters selected within their admissible range of variance. The test method was later extended for uncertain parameters as fuzzy numbers (Inuiguchi and Sakawa, 1996). In addition, available research publications have shown necessarily efficient and possibly efficient solution sets in regard to Pareto frontiers (Steuer, 1986; Ida, 2005). A grey compromise programming approach in relation to stability concerns was discussed by Chang et al. (1999). However, all these approaches are limited to the cases with a certain feasible region after proper transformation.

23.5 FINAL REMARKS

The uncertainties of parameters in a system can be described as intervals, fuzzy numbers, or random variables. The focus of fuzzy set theory is placed upon nonstatistical characteristics in nature that refer to the absence of sharp boundaries in the available information (Zedah, 1965). A subjective description of continuous membership function is usually used for the description of such type information. In an interval analysis, with limited samples during investigation, a parameter can be defined as a closed interval with upper and lower limits while the vagueness of its intrinsic characteristics remains. With an iterative experimental process, the range of an interval number could differ over time and a probabilistic distribution could be identified. In a decision analysis, however, the corresponding samples may not always be sufficient to characterize randomness for identifying the probability distribution. Although interval numbers and fuzzy numbers do not look alike, some have argued that interval numbers are simply rectangular fuzzy numbers; or interval sets are actually fuzzy sets of type 2 (Kaufmann and Gupta, 1991). Yet, fuzzy information can also be viewed as one type of grey message when the associated membership interval can be approximately identified for decision-making. Since everything is uncertain in real-world applications, interval numbers are a different representation of uncertainty as compared to fuzzy sets. All of these expressions associated with different types of uncertainty may be integrated into a single framework if needed. As the stability issues will not go away anyhow in the short term, how to communicate such grey information systematically in concert with fuzziness and randomness throughout an optimization process becomes an interesting research topic in systems science and engineering community in the future.

REFERENCES

Birge, J. R. and Louveaus, F. 1997. *Introduction to Stochastic Programming*, Springer, Berlin, Germany.

Bitran, G. R. 1980. Linear multiple objective problems with interval coefficients. *Management Science*, 26(7), 694–706.

Bradley, S. P., Hax, A. C., and Magnanti, T. L. 1977. *Applied Mathematical Programming*, Addison-Wesley, Reading, MA.

Cao, M. F., Huang, G. H., and Lin, Q. G. 2010. Integer programming with random-boundary intervals for planning municipal power systems. *Applied Energy*, 87(8), 2506–2516.

Chang, N. B., Chen, Y. L., and Wang, S. F. 1997. A fuzzy interval multiobjective mixed integer programming approach for the optimal planning of metropolitan solid waste management system. *Fuzzy Sets and Systems*, 89(1), 35–60.

Chang, N. B., Yeh, S. C., and Wu, G. C. 1999. Stability analysis of grey compromise programming and its applications. *International Journal of Systems Science*, 30(6), 571–589.

Cheng, G. H., Huang, G. H., Li, Y. P., Cao, M. F., and Fan, Y. R. 2009. Planning of municipal solid waste management systems under dual uncertainties: a hybrid interval stochastic programming approach. *Stochastic Environmental Research and Risk Assessment*, 23(6), 707–720.

Chinneck, J. W. and Ramadan, K. 2000. Linear programming with interval coefficients. *Journal of Operation Research Society*, 51(2), 209–220.

Coxson, G. E. 1999. Computing exact bounds on elements of an inverse interval matrix is NP-hard. *Reliable Computing*, 5(2), 137–142.

Deng, J. 1984a. *The Theory and Methods of Socio-economic Grey Systems* (in Chinese), Science Press, Beijing, China.

Deng, J. 1984b. *Contributions to Grey Systems and Agriculture* (in Chinese), Science and Technology Press, Taiyuan, Shanxi, China.

Deng, J. 1986. *Grey Prediction and Decision* (in Chinese), Huazhong Institute of Technology Press, Wuhan, China.

Huang, G. H. 1994. Grey mathematical programming and its application to municipal solid waste management planning. PhD Dissertation, Department of Civil Engineering, McMaster University, Ontario, Canada.

Huang, G. H. and Cao, M. F. 2011. Analysis of solution methods for interval linear programming. *Journal of Environmental Informatics*, 17(2), 54–64.

Huang, G. H. and Moore, R. D. 1993. Grey linear programming, its solving approach, and its application. *International Journal of Systems Science*, 24(1), 159–172.

Huang, G. H., Baetz, B. W., and Patry, G. G. 1992. A grey linear programming approach for municipal solid waste management planning under uncertainty. *Civil Engineering Systems*, 9(4), 319–335.

Huang, G. H., Baetz, B. W., and Patry, G. G. 1993. A grey fuzzy linear programming approach for municipal solid waste management planning under uncertainty. *Civil Engineering Systems*, 10(2), 123–146.

Huang, G. H., Baetz, B. W., and Patry, G. G. 1994. Grey dynamic programming for solid waste management planning under uncertainty. *Journal of Urban Planning and Development*, 120(3), 132–156.

Huang, G. H., Baetz, B. W., and Patry, G. G. 1995a. Grey integer programming: an application to waste management planning under uncertainty. *European Journal of Operational Research*, 83(3), 594–620.

Huang, G. H., Baetz, B. W., and Patry, G. G. 1995b. Grey fuzzy integer programming: an application to regional waste management planning under uncertainty. *Socio-Economic Planning Sciences*, 29(1), 17–38.

Huang, G. H., Sae-Lim, N., Liu, L., and Chen, Z. 2001. An interval-parameter fuzzy-stochastic programming approach for municipal solid waste management and planning. *Environmental Modeling and Assessment*, 6(4), 271–283.

Ida, M. 2005. Efficient solution generation for multiple objective linear programming based on extreme ray generation method. *European Journal of Operation Research*, 160(1), 242–251.

Inuiguchi, M. and Ramik, J. 2000. Possibilistic linear programming: a brief review of fuzzy mathematical programming and a comparison with stochastic programming in portfolio selection problem. *Fuzzy Sets and Systems*, 111(1), 3–28.

Inuiguchi, M. and Sakawa, M. 1996. Possible and necessary efficiency in possibilistic multiobjective linear programming problems and possible efficiency test. *Fuzzy Sets and Systems*, 78(2), 231–241.

Inuiguchi, M., Ichihashi, H., and Tanaka, H. 1990. Fuzzy programming: a survey of recent developments. In: *Stochastic Versus Fuzzy Approaches to Multiobjective Programming under Uncertainty* (Eds. Slowinski, R. and Teghem, J.), Kluwer Academic Publishers, Dordrecht, pp. 45–68.

Kall, P. and Wallace, S. W. 1994. *Stochastic Programming*, John Wiley & Sons, Chichester.

Kaufmann, K. and Gupta, M. 1991. *Introduction to Fuzzy Arithmetic*, International Thomson Computer Press, Boston, MA.

Kuchta, D. 2008. A modification of a solution concept of the linear programming problem with interval coefficients in the constraints. *Central European Journal of Operation Research*, 16(3), 307–316.

Li, Y. P. and Huang, G. H. 2007. Fuzzy two-stage quadratic programming for planning solid waste management under uncertainty. *International Journal of Systems Science*, 38(3), 219–233.

Li, Y. P. and Huang, G. H. 2009. Dynamic analysis for solid waste management systems: an inexact multistage integer programming approach. *Journal of the Air & Waste Management Association*, 59(3), 279–292.

Liu, B. and Iwamura, K. 1998. Chance constrained programming with fuzzy parameters. *Fuzzy Sets and Systems*, 94(2), 227–237.

Maqsood, I., Huang, G. H., and Yeomans, J. S. 2005. An interval-parameter fuzzy two-stage stochastic program for water resources management under uncertainty. *European Journal of Operation Research*, 167(1), 208–255.

Moore, R. E. 1979. *Method and Applications of Interval Analysis*, SIAM, Philadelphia, PA.

Mráz, F. 1998. Calculating the exact bounds of optimal values in LP with interval coefficients. *Annals of Operations Research*, 81, 51–62.

Nie, X. H., Huang, G. H., Li, Y. P., and Liu, L. 2007. IFRP: a hybrid interval-parameter fuzzy robust programming approach for waste management planning under uncertainty. *Journal of Environmental Management*, 84(1), 1–11.

Oliveira, C. and Antunes, C. H. 2007. Multiple objective linear programming models with interval coefficients—an illustrated overview. *European Journal of Operation Research*, 181(3), 1434–1463.

Rommelfanger, H., Hanuscheck, R., and Wolf, J. 1989. Linear programming with fuzzy objectives. *Fuzzy Sets and Systems*, 29(10), 31–48.

Ruszczyński, A. and Shapiro, A. 2003. Stochastic programming models. In: *Handbooks in Operations Research and Management Science, 10: Stochastic Programming* (Eds. Ruszczyński, A. and Shapiro, A.), Elsevier, Amsterdam, pp. 1–64.

Shaocheng, T. 1994. Interval number and fuzzy number linear programming. *Fuzzy Sets and Systems*, 66(3), 301–306.

Steuer, R. E. 1986. *Multiple Criteria Optimization: Theory, Computation and Application*, John Wiley & Sons, New York.

Sun, Y., Li, Y. P., and Huang, G. H. 2010a. Development of a fuzzy-queue-based interval linear programming model for municipal solid waste management. *Environmental Engineering Science*, 27(6), 451–468.

Sun, Y., Huang, G. H., and Li, Y. P. 2010b. ICQSWM: an inexact chance-constrained quadratic solid waste management model. *Resources, Conservation and Recycling*, 54(10), 641–657.

Tan, Q., Huang, G. H., and Cai, Y. P. 2010. A superiority-inferiority-based inexact fuzzy-stochastic programming approach for solid waste management under uncertainty. *Environmental Modeling & Assessment*, 15(5), 381–396.

Urli, B. and Nadeau, R. 1992. An interactive method to multiobjective linear programming problems with interval coefficients. *Information Systems and Operational Research*, 30(2), 127–137.

Wang, S., Huang, G. H., and Yang, B. T. 2012. An interval-valued fuzzy-stochastic programming approach and its application to municipal solid waste management. *Environmental Modelling & Software*, 29(1), 24–36.

Zedan, L. A. 1965. Fuzzy sets. *Information and Control*, 8, 338–353.

Zimmermann, H.-J. 1978. Fuzzy programming and linear programming with several objective functions. *Fuzzy Sets and Systems*, 1(1), 45–55.

CHAPTER 24

SYSTEMS ANALYSIS FOR THE FUTURE OF SOLID WASTE MANAGEMENT: CHALLENGES AND PERSPECTIVES

In many metropolitan regions, solid waste management (SWM) systems involve complex and multifaceted trade-offs among a plethora of technological alternatives, economic instruments, and regulatory frameworks. These challenges have resulted in various environmental, economic, social, and regulatory impacts on waste management practices. Such influences not only complicate sustainable urban development, but also reshape the regional planning and policy analysis. Systems analysis, a discipline that harmonizes these integrated solid waste management (ISWM) strategies, uniquely provides interdisciplinary support for SWM decision-making. Since the 1970s, systems engineering models and system assessment tools, both of which enrich the analytical framework for ISWM, were specifically generated to handle particular types of SWM problems. Over the past two decades, advancements in environmental informatics technologies have also aided short-term and long-term capacity for planning, design, and operations of SWM, all requiring the consideration of various sustainability criteria from time to time. This chapter provides a holistic review of the development of systems engineering models and system assessment tools in relation to advances in systems engineering technologies pertaining to SWM. Following a retrospective review of main trends, the challenges for and perspectives of systems analysis in SWM are discussed in sequence. The streamlined goal of this chapter is to deepen insights into the role of environmental informatics for ISWM and thereby improve decision-making reliability. Final remarks in the end will not only suggest a clear evolutionary pathway for this scientific domain, but also lay down some foundations to guide future directions in ISWM research and applications.

Sustainable Solid Waste Management: A Systems Engineering Approach, First Edition. Ni-Bin Chang and Ana Pires.
© 2015 The Institute of Electrical and Electronics Engineers, Inc. Published 2015 by John Wiley & Sons, Inc.

24.1 THE EVOLUTION OF SYSTEMS ANALYSIS FOR SOLID WASTE MANAGEMENT

Modeling SWM systems is highly multidisciplinary and by no means an easy task. The mechanisms to smooth out barriers preventing appropriate systems synthesis and integration of these models and tools to aid decision-making under uncertain conditions remain challenging. Thus, a thorough review of relevant systems engineering models and system assessment tools associated with informatics technologies as a whole is a necessary task. A retrospective review of the relevant work was conducted chronologically in this section. A holistic summary of key historical trends as sequentially outlined will clarify some of the facts that encompass the pros and cons of waste management practices. Illuminating the possible overlapping boundaries of relevant models and tools aided by different informatics technologies will promote better ISWM strategies in relation to sustainable urban development (Chang et al., 2011; Pires et al., 2011; Lu et al., 2013).

24.1.1 Systems Analysis for Solid Waste Management in the 1970s and Before

The first-generation systems engineering models using linear programming (LP) with a single-objective optimization scheme (i.e., cost minimization) were developed around the end of the 1960s (Chang et al., 2011). These LP models tend to characterize waste flow patterns as simply flows from transfer stations to landfill sites, aiming to minimize the total or partial costs involved in an SWM system (Anderson and Nigam, 1967; Anderson, 1968). Later on, fixed-charge problems concerning the siting of new facilities were decomposed into fixed and variable costs. The former, fixed costs are defined as site acquisition and preparation costs to basic facility capital, since costs are incurred regardless of the level of activity at a site (Chang et al., 2011). The latter, variable cost, involves the parts of facility capital and those operating costs that can be defined as a function that is linearly dependent on facility capacity. Such a model formulation enables the use of integer programming models for site screening and selection across several candidate sites. Marks and Liebman (1970) considered the selection problem for new transfer stations in relation to transportation cost, including both fixed and variable costs. Rossman (1971) extended the work of Marks and Liebman (1970, 1971) by adding incinerators to the set of potential facilities in the context of cost minimization. In addition, Esmaili (1972) developed a transfer station locational model to choose the combinatorial options for processing or disposal facilities, or both, from among a number of alternative facilities that would minimize the overall cost of shipping, processing, and disposal of SWM operations over an extended period of time (Chang et al., 2011). Starting from such theoretical foundations, Greenberg et al. (1976a, 1976b) applied LP techniques to plan a real-world waste management system in the United States with respect to the cost minimization principles and technical constraints. Later, Clark (1973) discussed some regional planning models for SWM, formulated as fixed-charge problems via a mixed-integer programming (MIP) model. Helms and Clark (1974) presented an LP model to aid in selecting alternatives in regard to incinerators and landfills considering fixed and variable costs to be minimized together. Chang et al. (2011) also highlighted the use of mixed-integer linear programming (MILP) techniques for solving real-world SWM issues, related to single network planning (Anderson, 1968; Fuertes et al., 1974; Helms and Clark, 1974; Kuhner and Harrington, 1975; Clayton, 1976; Jenkins, 1979), and dynamic, multi-period investment for SWM regionalization (Marks et al., 1970; Marks and Liebman, 1971). Walker et al. (1974) and Walker (1976) developed the simplex with forcing trials algorithm for helping decide the number, type, size, and location of the disposal sites in a region simultaneously, which was adopted by the United States Environmental Protection Agency (USEPA) (Chang et al., 2011). On its own, however, the USEPA (1977) developed another model, the Waste Resources Allocation Program (WRAP). The WRAP model contains static and dynamic MILP modules. In parallel with these two advances, other types of optimization models (OMs), such as the dynamic programming (DP) approach, for SWM planning were also developed and applied as by Rao (1975). From the regional to local scale, Truitt et al. (1969) and Liebman et al. (1975) developed OM to solve waste vehicle routing, since a few researchers had concerned themselves with local scale analysis of vehicle routing at this time.

In concert with these optimization efforts in system analysis, some independent simulation models (SM) and forecasting models (FM) have been carried out for waste generation and flow pattern prediction. The first FMs were developed for SWM in the early 1970s by Niessen and Alsobrook (1972) and Grossman et al. (1974), in which the extended per-capita coefficients were fixed over time while change was projected by including population, income level, and dwelling unit size effects via a linear regression model. Clark and Gillean (1974) proposed modeling solid waste generation with a management information system (MIS). They applied this idea to solve vehicle routing issues using data from the USEPA. "This was the first attempt to apply computational tools to SWM planning. The USEPA conducted the first attempt to carry out planning and management at the system level using a prototype MIS. Even at this early stage, interdependencies between the various components of SWM were recognized" (Chang et al., 2011). One SM applied to SWM planning was found in the early 1970s (Bodner et al., 1970), in which a practical simulation was developed to determine the optimal routes for refuse collection vehicles. This program yielded exact routes suitable for use by municipalities. Crew size, vehicle capacity, and pickup times could be varied to permit efficient labor and equipment use. "The program computes overtime, incentive time, vehicular capacity utilization, mileage traveled, weight hauled, and productive time" (Chang et al., 2011). Based upon these findings, a strategic evaluation of waste management that summarized these practices was suggested by Wilson (1977).

In parallel with the progress of applied systems engineering models for SWM, the earliest explorations of environmental informatics technologies to improve the efficiency and effectiveness of SWM appeared during the 1970s. The earliest systematic environmental informatics tool, referred to as an environmental information system (EIS), was developed and commissioned in 1977. This was a response to the recommendation for global exchange of environmental information and experiences via electronic data interchange, after an United Nations Conference on the Human

Environment (UNHEC) occurring in 1972 (United Nations Environment Programme, 2004). Subsequently, EIS were applied at different scales of use, driven by various international or national policies (Haklay, 1999).

24.1.2 Systems Analysis for Solid Waste Management in the 1980s

The 1980s were a decade in which several programs with a wide range of configurations for SWM became available for experimentation. "The need to make models more realistic by using a hierarchical approach became an emphasis during the 1980s" (Chang et al., 2011). This hierarchical approach was intended to add complexity to models at the system level. According to Chang et al. (2011), the spectrum of these OMs developed in the 1980s includes a standard operational procedure (Gottinger, 1986, 1988), multiobjective evaluation for disposal planning (Perlack and Willis, 1985), MILP approaches including more types of constraints for SWM planning (Jenkins, 1980, 1982; Hasit and Warner, 1981), pure MIP models (Kirka and Erkip, 1988), and DP models (Baetz et al., 1989). Single objective for vehicle routing and scheduling OM models were still influential at the operational levels (Chiplunkar et al., 1981; Brodie and Waters, 1988).

These more sophisticated models were developed to some extent as a result of higher computer accessibility, while the complexity of SWM issues prompted interest in the use of computational tools, especially electronic spreadsheets (MacDonald, 1996a). "In fact, computational accessibility allowed the development of specific tools either free or proprietary" (Chang et al., 2011). Specific proposed tools were the Resource Recovery Planning (RRPLAN) described by Chapman and Yakowitz (1984), a model that uses LP techniques to size and site facilities, and a cost accounting system to incorporate economies of scale and estimate the effects of decisions" (Chang et al., 2011). Rushbrook (1987) and Rushbrook and Pugh (1987) described the Harbinger waste management planning system using an optimizing model. Practically, it showed potential for SWM planning applications for Hong Kong. The WRAP model was applied in the United States by Hasit and Warner (1981). The Route Optimization and Management Allocation (ROMA) model developed by Beture, an engineering and design company, for the City of Paris to optimize the collection routes is also identified as a milestone in this field (Burelle and Monterrat,1985; Light, 1990). The MIMES/WASTE that stands for "Model for description and optimization of Integrated Material flows and Energy Systems" was developed by Sundberg (1989, 1993) and Sundberg et al. (1994), which utilized a nonlinear programming model for energy recovery assessment in response to the increased complexity of SWM in a case study of Goteborg, Sweden. At this time, most SWM models were still not widely used, because they were difficult for the non-specialists to understand and required a large initial investment of time and/or investment and often were proprietary systems (Anex et al., 1996).

Surveys at that time revealed a wide variation in household waste generation multipliers. Forecasting models for waste generation prediction exhibited progress at a less intensive though promising pace. Khan and Burney (1989) presented a regression model for forecasting solid-waste composition with respect to the consideration of recycling and resource recovery. This model utilized data from 28 international cities for predictions that illustrated a strong correlation with actual observed data. Other types of forecasting models were developed by Rufford (1984) and Rhyner and Green (1988). Estimates of residential, industrial, and commercial solid waste quantities were computed for some regional SWM systems.

SWM systems planning is not solely governed by technical issues. Environmental concerns about system components, such as leachate from landfills and emissions from incineration plants, resulted in more regulations and compliance obligations (Chang et al., 2011). Highly specialized and ever more expensive pollution control technologies came into favor. Public wariness about environmental impacts forced SWM models to evolve not only in terms of technical advancement and economic incentives, but also in relation to environmental quality constraints in the context of optimization (Chang, 1989). Along these lines, cost–benefit analysis (CBA) from the environmental economics regime came into play with regards to recycling for the first time. Environmental and economic assessment became integrated cohesively with each other by the end of 1980s (Glenn, 1988). Besides, expert systems (ES) and artificial intelligence (AI) technologies were also deployed to help assess possible contamination of aquifers from dumpsites and landfills, given the interplay of SWM and environmental concerns. Salient examples include the DEMOTOX model developed by Ludvigsen and Dupont (1988). The application of ES for SWM was elucidated in greater detail regarding its cost-effectiveness in relation to the choice of waste treatment and disposal alternatives (Thomas et al., 1990).

"From the 1980s, some branches of EIS were employed for SWM, mostly concerning radioactive waste" (Lu et al., 2013). Database systems (DBS) were leveraged for managerial understanding of process relationships, inventory management, and prediction (Aquilina et al., 1982; Notz et al., 1984). Decision support systems (DSS) were developed for SWM to improve the level of decision-making, such as evaluation of waste transport alternatives (Bowen et al., 1989), and optimization for waste treatment plant operation (Westrom et al., 1989). Geographic information systems (GIS) began to be used for site selection or ranking by means of the salient analysis functions found in a GIS (Jensen and Christensen, 1986; Stewart, 1988; Scott et al., 1989). "Criteria for GIS analysis were derived from regulations and domain expertise for applications" (Lu et al., 2013). GIS was also used as a framework for managing and displaying information about waste sites and processes (Wong and TenBroek, 1989). Additionally, ES were developed to provide expertise advice for waste regulation and treatment process evaluation according to rules (Anandalingam, 1987; Barrow, 1988; Heydinger and Jennings, 1988; Jennings and Heydinger, 1989), and to aid in site selection for SWM (Rouhani and Kangari, 1987). ES shell programs, such as Insight^{TM} 2+ and DIKE^{TM} 2.0, were, by this time, already employed for developing ES (Rouhani and Kangari, 1987; Jennings and Heydinger, 1989).

24.1.3 Systems Analysis for Solid Waste Management in the 1990s

In the 1990s, the improvement of OM started with the inclusion of green infrastructures such as recycling centers, to reflect sustainability goals (Englehardt and

Lund, 1990), source separation and curbside recycling programs, and material recovery facilities (MRFs) (Morris, 1991). Optimal scheduling for landfill operation with the recycling effect was also evaluated by Jacobs and Everett (1992). Efforts were also directed toward evaluating and scheduling a given set of recycling measures to help achieve least-cost landfilling with extended lifetime (Lund, 1990). Optimization analyses for locating recycling facilities also became a big concern (Chang et al., 2011). Hsieh and Ho (1993) and Lund et al. (1994) discussed the optimization of solid waste disposal and recycling systems by using LP techniques for economic optimization. Huhtala (1997) emphasized the use of an OM to assess recycling rate at the most economically effective option. Daskalopoulos et al. (1998) included net cost and environmental impacts into a model to assess an SWM system. Integrated modeling systems may include studying locational theory for siting recycling centers (Highfill et al., 1994), transfer stations (Rahman and Kuby, 1995; Chang and Lin, 1997a), MRF (Lund et al., 1994), optimal allocation of trucks for SWM (Bhat, 1996), waste collection (Kulcar, 1996), and vehicle routing systems (Ong et al., 1990). With the aid of GIS, Chang et al. (1997c) and Chang and Lin (1997a) conducted local scale optimization for collection vehicle routing and scheduling and siting transfer stations, respectively. Chang and Lin (1997b) further applied GIS to siting transfer stations as an integral part of a bigger regional assessment model for screening and sequencing dynamic operations among a set of waste management facilities. The possibilities for combining environmental impacts, such as air pollution, leachate, noise control, and traffic congestion, as a set of Environmental Impact Assessment (EIA) constraints in a series of economics-oriented locational models were explored using the MILP models (Chang et al., 1993a, 1996, 1997a, 1997b; Chang and Wang, 1994, 1996a, 1996b). Yet, the ISWM and waste management hierarchy (WMH) (i.e., the WMH refers to the sequence of source reduction and reuse, recycling, combustion/incineration, and landfilling in a priority order, also known as waste hierarchy principle) may result in dilemmas in policy decision-making and applications involving multiple objectives and criteria.

At the same time, multicriteria decision-making (MCDM) emerged as an approach to support decision makers faced with making numerous and conflicting evaluations; deriving ways to come to a compromised solution in a transparent process (Chang et al., 2011). Caruso et al. (1993) developed a location-allocation MCDM model for SWM that reflects environmental issues like resource and environmental impacts as well as costs. Courcelle et al. (1998) formulated an MCDM model to assess economic and environmental performance of municipal multimaterial waste collection and sorting programs as applied to nine such programs in European municipalities. Fawcett et al. (1993) and Alidi (1998) applied a goal-programming (GP) model with predetermined target values to aid in an ISWM, using an analytic hierarchy process (AHP) for determining the weights and priorities for a given set of goals. Other types of MCDM models were later developed by Hokkanen and Salminen (1994, 1997), Karagiannidis and Moussiopoulos (1997), and Chung and Poon (1996). The inclusion of multiple objectives in decision-making within a dedicated ISWM analysis involves various trade-off problems among conflicting objectives (Haastrup et al., 1998). Such complexity is tied to costs, environmental aspects like discharge coefficients, impact factors, and planning objectives, and may affect the simulation and optimization processes; thus the generated solutions in modeling stages (Huang et al., 2002; Fiorucci et al., 2003; Costi et al., 2004). In particular, some waste management practices applied a compromise programming technique to harmonize potential conflicts when siting landfills, incinerators, and transfer stations in a growing metropolitan region (Chang et al., 2011). A GP model, a simplified form of a multiobjective programming model, was also applied to assess the compatibility issues between recycling and incineration, considering economic efficiency and environmental protection goals as achieved during numerous tradeoffs (Chang and Wang, 1997b). The nonlinearity embedded in the modeling process was specifically handled by using a nonlinear goal programming model for urban SWM (Sudhir et al., 1996; Chang and Chang, 1998). The use of modeling-to-generate-alternatives approach developed by Chang and Li (1997) aimed at generating SWM alternatives with specific cost constraints. Specifically, Rubenstein-Montano and Zandi (1999) applied a genetic algorithm for an SWM policy planning.

System integration requires concatenating external functions of FM and/or SM step by step with OM providing dynamic information about waste generation and shipping over time (Chang et al., 2011). For example, Chang et al. (1993b) presented a time series FM (geometric lag model) of solid waste generation to meet such goals. When reviewed by Beigl et al. (2008), during the 1990s, about 20 forecasting models overall were being developed. Lawver et al. (1990) evaluated integrated SWM systems with an SM related to discrete event simulation. Similar work was conducted by Anex et al. (1996) producing the Garbage In, Garbage Out (GIGO) model to support large-scale optimization analysis. Tanskanen and Melanen (1999) developed an SM Tool for Analyzing Separation Actions and Recovery (TASAR) to study the recovery levels reached by different separation strategies in Finland. Baetz (1990) developed an integrated simulation and OM for determining the optimal capacity expansion patterns associated with waste-to-energy and landfill facilities over time. Salient examples of advanced system synthesis, such as the ORganic WAste REsearch (ORWARE) model, include SM developed to support system assessment tools, such as life cycle assessment (LCA) and strategic environmental assessment (SEA) (Dalemo et al., 1997; Björklund et al., 1999, 2000). Further, Powell et al. (1996 1999) synthesized LCA and MCDM models to examine environmental impacts from alternative waste management scenarios for the city of Bristol, United Kingdom. Weitz et al. (1999) assembled an economic assessment tool covering life cycle management of municipal solid waste (MSW).

In parallel with advancements in models and tools, it was recognized that uncertainty plays an important role in decision-making. "In response to this challenge, systems engineering models for SWM also evolved at that time from deterministic to probabilistic approaches, from certain to uncertain concerns, and from affirmative to risk-based attitudes compounding the analytical framework at different levels, from data to models, to management" (Chang et al., 2011). There are three types of theories for uncertainty analysis, including probability theory, grey system theory, and fuzzy set theory, helpful for addressing various types of uncertainty. For example, uncertainty is relevant to the randomness governing solid waste generation and the estimation errors in some parameter values (Chang et al., 1997a). Such uncertainties can also be related to technical maintenance and are generally difficult to quantify as exact assessment data (Seo et al., 2003). Moreover, the waste-generation rate in a community varies both temporally and spatially (Huang and Chang, 2003). Furthermore, the vagueness of planning objectives as well as external constraints create even more uncertainty on decision-making (Chang et al., 1997a).

To fully address uncertainties in decision-making, fuzzy set theory and interval (grey or inexact) programming techniques received wide attention in the 1990s. A series of extended optimization analyses dealing with a hypothetical SWM problem include the use of grey LP, grey fuzzy LP, grey fuzzy dynamic programming, and grey integer programming (GIP) approaches for identifying the optimal location and capacity of waste treatment facilities in Canada (Huang et al., 1992, 1993a, 1994, 1995). Chang and Wang (1996c, 1997a) applied fuzzy goal programming in dealing with several specific issues for the ISWM in Taiwan.

As informatics technologies boomed in the 1990s, both DSS and ES received wide and ongoing attention and were promoted for ISWM with extensive studies combining various information science theories and informatics technologies. Charnpratheep and Garner (1997) combined fuzzy set theory and AHP into a raster-based GIS for a preliminary screening of landfill sites in Thailand. Other use cases utilized decision support knowledge for waste processing and economic assessment associated with SWM (Barlishen and Baetz, 1996; Chang and Wang, 1996a, 1996b, 1996c, 1996d; Bhargava and Tettelbach, 1997; Haastrup et al., 1998). The USEPA also developed several ES to enhance computer-aided design of a leachate collection system, final cover and vegetative cover plans for landfills. A review article documented more applications of SWM during this decade (Basri and Stentiford, 1995). Knowledge-based models developed by Boyle (1995) were used to compare components and parameters of the inputs and wastes from different industries and determined the potential for reuse, recycling, or disposal and respective treatment. Haastrup et al. (1998) concentrated on costs, air, water and soil pollution, road congestion, and technological reliability, but did not consider noise, employment, health impacts, and recycling goals in their model.

Evolving methodologies in systems analysis were extended from systems engineering models. These enveloped system assessment tools, leading to more versatile model synthesis practices. In this regard, individual life cycle impact assessment models compared recycling routes for particular objects (Song et al., 1999), to assess packaging alternatives (Tillman et al., 1991), and to screen out waste treatment options such as landfill versus recycling and associated systems (Kirkpatrick, 1993; Craighill and Powell, 1996; Rieradevall et al., 1997; Powell et al., 1998). White et al. (1995) presented a life cycle inventory (LCI) model specific to an SWM system, where the product system was held constant and evaluation was done based on the performance of alternatives for solid waste disposal, being the first LCI dedicated to waste management. Interest in LCA increased in the late 1990s against a background of comprehensive environmental legislation including Integrated Pollution Prevention Control and Best Available Techniques Not Entailing Excessive Cost (also known as BATNEEC), the growth of the green consumer market, and in response to pressure from voluntary green groups (Chang et al., 2011). Corporate interest was aroused by the introduction of the British Standard BS 7750: Environmental Management Systems, and the European Commission (EC) Eco-Management and Audit Regulation in 1993, criteria for the EC eco-labeling scheme were based on the results of partial life cycle studies (Craighill and Powell, 1996).

At this stage, the assessment of policies (including WMH) and directives through CBA (Hanley and Slark, 1994; Litvan, 1994; Touche Ross Management, 1994; Brisson, 1997; AEA Technology, 1998; Bruvoll, 1998) became popular, including both tangible and intangible cost and benefit terms in association with the necessary environmental and ecological assessment (Chang et al., 2011). This was linked to systems analysis locating pollutant sources and selecting wastes to be allocated so to determine possible treatment options through material flow analysis (MFA) methods (Brunner and Rechberger, 2003). Significantly, SEA practices during 1990s were all developed by European countries (Chang et al., 2011). The purpose of such practices is assessment of national and regional waste management plans (Salhofer et al., 2007). The EIA promulgated by many developed countries in the 1990s was the gold standard for specific installations. For example, through Directive 85/337/EEC (Council, 1985) in Europe, EIA reports became required for new waste management facilities (Barker and Wood, 1999). Countries like Germany and Ireland, therefore, had to produce EIA for waste disposal (Chang et al., 2011). In Portugal, EIA was required for all incineration plants (Coutinho et al., 1998). Risk assessment (RA) with the aid of the exposure assessment in the context of an EIA was developed in the 1990s for MSW composting as well as incineration in order to examine various issues related to toxic substance emissions (Calabrese and Kenyon, 1991; Travis, 1991), the comparative risks when handling the SWM planning (Chang and Wang, 1996a), and the safety of employees and environment (CWMI, 1999).

During the 1990s, the term environmental informatics was presented in a formalized way as more communication, sensing, and informatics technologies became available over the Internet (Page, 1992; Schütt and Hofestädt, 1992; Avouris and Page, 1995). "The infusion of informatics into SWM was enhanced by a wide range of applications, from strategic planning to operation control" (Lu et al., 2013). More potential facets of SWM were considered and included in the environmental informatics applications, such as environmental impact assessment (Lo Porto et al., 1997), potential risk analysis (Brainard et al., 1996), and public health assessment (Fay and Mumtaz, 1996). These all involved waste generation, characteristics, hauling, recycling, treatment, and consequent environmental impacts to be managed by a DBS (Interrante et al., 1991; Stacey et al., 1995; Adams et al., 1996; Fay and Mumtaz, 1996). Automatic data acquisition was accelerated with various kinds of techniques, such as remote sensing (RS) was used for building geospatial data structures (Bresnahan, 1998; Fischer and Hermsmeyer, 1999) and identifying waste dumping sites or repositories (Irvine et al., 1997; Bresnahan, 1998); global positioning systems (GPS) were used for recording sampling locations (Kaletsky et al., 1996) and monitoring landfill deformation (MacDoran et al., 1992); sensing devices were used for rapid detection (Dieckman et al., 2000) and identification (van den Broek et al., 1997), and even for ambient monitoring (Holland, 1997). In addition, barcoding was employed for source separation during the collection of solid waste (Ogawa, 1998).

Both systems engineering methods and system assessment tools can be flexibly woven with several information technologies, such as GIS, to enhance the application potential. As were oftentimes applied for spatial analysis, GIS plays a role in spatial DSS (SDSS) when selecting appropriate sites for landfilling, incineration, composting, and waste transfer. Systems engineering models, like MIP models, were combined with GIS for the screening of candidate sites determined by GIS (Chang and Chang, 1996; Kao, 1996). With the aid of SDSS, fuzzy evaluation method was introduced to deal with the uncertainty of landfill site selection (Charnpratheep and Garner, 1997), while a heuristic algorithm was used to reduce time consumption in computation (Muttiah et al., 1996). Extended GIS practices were applied for the optimization of routing and scheduling schemes for waste collection (Chang et al., 1997a) and risk analysis of hazardous waste transport (Brainard et al., 1996). Frameworks were produced for coupling EIA models (Fraisse et al., 1996; Lo Porto et al., 1997), creating ES (Davies and Lein, 1991; Lin and Kao, 1998) for utilizing the capacity of visualization and interaction through GIS operations.

Later on, DSS were also developed and successfully implemented in many areas for SWM in the 1990s (Chang and Wang, 1996d). At that time, DSS were principally used for strategic planning, for siting new plants (Fujita and Tamura, 1999), or for semi-structured decision-making in SWM, for example, landfill cover design (Paige et al., 1996). Several researchers described the structure of DSS suitable for SWM (Reitsma and Sullivan, 1992; Chang and Wang, 1996d; Haastrup et al., 1998), and the exclusionary and preference criteria used in DSS were emphasized for SWM (Manoliadis et al., 2001). Potential environmental impacts of SWM, simulated by many existing models, received attention in early applications of DSS for SWM (Paige et al., 1996; Fujita and Tamura, 1999). Moreover, group DSS for SWM programs was brought into this regime to support decision-making regarding possible environmental impacts (Shirland and Kraushaar, 1991). As compared to DSS, ES applied for SWM emphasizes expertise-based determination, predominantly for strategy selection (Wei and Weber, 1996) and automatic operations (Barnett, 1992). For instance, ES for strategy selection was developed for screening and selecting treatment alternatives (Chen, 1994) and site remediation (Hushon and Read, 1991; Staudinger et al., 1997; Akladios et al., 1998; Basri, 1998). ES for automatic operation were developed for odor detection processes (Kordon et al., 1996), vitrification of radioactive waste (Arakali, 1991), automatic control of anaerobic digesters (Pullammanappallil, 1998) and incinerators (Yang and Okrent, 1991), radioactive waste classification (Williamson, 1990; Hodges et al., 1999), and fuel rod consolidation process (Kim et al., 1993). ES were called knowledge-based expert system, interchangeably (Barnett, 1992; Fonseca et al., 1997; Akladios et al., 1998). All of these applications of ES, however, depend on prior knowledge acquisition. In addition to questionnaires, phone interviews were employed as a good means for knowledge acquisition (Coursey et al., 1993). Prolog language (Chen, 1994; Wei and Weber, 1996) and Gensym G2 software environment (Kordon et al., 1996) were introduced in ES to ease the design of ES. Furthermore, knowledge-based DSS (KB-DSS) was proposed for integrating the advantages of both DSS and ES. Smith et al. (1997) devised a KB-DSS for SWM. GIS may be combined with DSS and ES to evaluate candidate locations for landfills and determine site suitability (Davies and Lein, 1991). Boyle and Baetz (1998a, 1998b) developed and tested a KB-DSS to determine the recycling potential of waste streams. However, easy access web integration was seldom involved in applications of the 1990s, except in a few cases (Bhargava and Tettelbach, 1997; Akladios et al., 1998).

24.1.4 Systems Analysis for Solid Waste Management in the 2000s

The challenges of the 1990s encouraged systems engineering models and system assessment tools for SWM to become more realistic and multifaceted based on different configurations and purposes. Yet, modelers must keep in mind their purpose—helping to choose the best technologies and/or management alternatives to make SWM systems more sustainable. New perspectives are gradually entering into the domain like constraints violation for siting waste management facilities and waste flow allocation (Huang et al., 2002), and waste generation step (den Boer et al., 2007). The social dimensions are increasingly important aspects in connection to environmental and economic assessments (Kijak and Moy, 2004; Contreras et al., 2008), with social interactions modeled in terms of game theory under uncertainty (Davila et al., 2005), and societal responses through minimax regret criteria (Chang and Davila, 2006, 2007). Financial management is also a consideration since government bonds are associated with economic development and population growth (Davila and Chang, 2005; Dyson and Chang, 2005) for, siting and sizing material recovery facilities (Chang et al., 2005). There must be closer linkages between water characteristics and management strategies and policy (Chang and Davila, 2007, 2008), and between policy concerns with economic incentives (Su et al., 2007; Chang, 2008). Landfill siting procedures now include more stakeholders (Chang et al., 2008), integrated simulaton and forecasting models (Beigl et al., 2008), and stakeholder-driven decision-making processes aided by spatial decision support technologies (Chang et al., 2009). Recent efforts compare different technologies and options to improve existing assessment methodologies, like LCA, by constructing more friendly tools (Environment Agency of England and Wales, 2000; EPIC and CSR, 2000; Diaz and Warith, 2006; Kirkeby et al., 2006), and emphasizing energy recovery goals (Chang and Chang, 2001).

The 2000s were a decade with fast developments in environmental informatics for SWM. This dynamic growth is evident in the growing number of publications, applications, and diffusion of web-based technologies dedicated to environmental informatics. The total number of publications increased three times over 1990s and preceding years. The spectrum of applications at that time encompassed DBS, GIS, DSS, and ES simultaneously (1) facilitating public participation (Carver et al., 2000; Higgs, 2006; Huang and Sheng, 2006; Liu et al., 2006), (2) emphasizing ecological benefits with sustainability implications (Manoliadis et al., 2001; Shmelev and Powell, 2006; Chen et al., 2007; Page and Wohlgemuth, 2010), and (3) coupling different technologies (Arebey et al., 2009; Atkins et al., 2009; Luo et al., 2009; Tsai et al., 2011). Integrative terms such as SDSS and web-based SDSS became more widely diffused. The rising popularity of internet technologies motivated a plethora of web-based applications including web-based DBS, web GIS, web-based DSS, and web-based ES. These, in turn, triggered more opportunities for public participation in SWM (Carver et al., 2000; Rinner, 2003; Dantzler et al., 2008).

Under any circumstances, broadening public participation in decision-making processes is helpful to alleviate the not in my backyard (NIMBY) syndrome. Collaboration is beneficial as it contributes to the quality of decision-making. One of the salient use cases included the use of web-based SDSS to increase public participation in online surveys from the inception to the final phases of decision-making for SWM site selection (Liu et al., 2006). More and more heterogeneous applications have made simple text-based or fact-based databases obsolete; they can no longer meet societal requirements. Integrated databases, such as chemical reaction databases, composed of radioactive waste thermodynamics and sorption (Yui et al., 2001; Wieland et al., 2003), bottom ash leaching characteristics (Jeong et al., 2005), bio-methane potential (Riscoa and Dubourguier, 2010), and problem-oriented nuclear waste databases (Korovin et al., 2007), were developed and implemented.

The concept of eco-efficiency and sustainability in SWM decision-making and planning became a focus in the 2000s (Shrivastava et al., 2005; Löfgren et al., 2006; John, 2010). Environmental, ecological, and social impacts simultaneously received greater consideration alongside the more traditional economic factors and engineering feasibility concerns (Lu et al., 2013). SMs for EIA and post environmental assessment were seamlessly integrated into GIS or DSS creating synergistic functionalities (Lu et al., 2013). For example, the ecological benefits from rehabilitation of coal mine waste areas were evaluated (Chen et al., 2007). Stability and landfill risks were analyzed with three-dimensional analysis tools found in a GIS (Stormont and Farfan, 2005; Caiti et al., 2006). Public health and infection rates around waste disposal sites received higher attention and triggered statistical analyses in a GIS environment (Altavista et al., 2004; WHO, 2007; Vinceti et al., 2009). A fair fund (e.g., compensatory funds) for residents living in the proximity of a municipal incinerator was redistributed through an MCDM process with the aid of a GIS (Chiueh et al., 2008; Chang et al., 2009). Environmental racism and injustice in the distribution of hazardous waste facilities were also assessed through geo-statistical and social analyses (Mohai and Saha, 2007; Sicotte, 2008). The scope of site selection factors expanded from treatment/disposal sites to transfer stations and waste collection points in a hierarchical framework. Of these two kinds of problems in site selection, the latter (i.e., waste collection) was oftentimes solved through cost-effective optimization analyses at the microscale (Kao and Lin, 2002; Tralhão et al., 2010), while the former (i.e., treatment and disposal) was confirmed by engineering feasibility studies (Basnet et al., 2001) and environmentally benign assessment (Manoliadis et al., 2001; Lee, 2003; Şener et al., 2011) at the macroscale. Thus, both single-objective OM (Fiorucci et al., 2003) and MCDM models (Huang, 2006; John, 2010) were implemented at a regional scale to tune waste allocation strategies. Waste prevention strategies maximizing recovery and reducing harmful impacts due to the disposal of solid waste were explored to expose their sustainability implications (Guzman et al.,

2010). Vehicle routing strategies at a local scale were investigated for operational control, including two thematic areas: (1) safety routing based on the minimization of risk (Lazar et al., 2001; Huang, 2006), and (2) economic routing based on the shortest distance or the lowest cost (Kim et al., 2006; Apaydin and Gonullu, 2007). Waste collection optimization between the source and waste containers for a source separation system can be viewed as an extension of these contemporary advances (Zamorano et al., 2009).

Sustainability concerns in complex SWM systems oftentimes compound SWM problems which must be tackled by a combination of modern technologies. MCDM have been widely applied to different types of SWM problems involving multiple objectives and criteria, such as a selection between alternative management strategies (Higgs, 2006), more rigorous routing (Chen et al., 2008), and more challenging site selection (Gómez-Delgado and Tarantola, 2006; Sumathi et al., 2008). Criteria are extracted from a wide range of available information not only quantitative but also qualitative (Ahmadi et al., 2010). Nevertheless, most of them are tied to waste composition. The diversity of protocols for testing waste materials (Dahlén and Lagerkvist, 2008), however, prevents a comparison of waste characteristic data from different countries, except for a few exploratory studies using statistical inference analysis (Chang and Davila, 2008).

To respond to more complex decision-making processes, a few analytical tools, such as AHP and fuzzy logic, were combined with GIS or DSS to help elucidate linguistic variables and generate criterion weights though various extended methods under uncertainty. These include but are not limited to fuzzy AHP and fuzzy comprehensive evaluation (Kunsch and Fortemps, 2000; Zeng and Trauth, 2005; Karadimas et al., 2006; Lotfi et al., 2007; Chang et al., 2008; Chen and Li, 2008; Karadimas and Loumos, 2008; Alves et al., 2009; Chang et al., 2009; Ahmadi et al., 2010). Expert domain knowledge contributes to the study of uncertainty in MCDM. Sensitivity analysis though Monte Carlo simulations (Chang et al., 2008) or scenario analysis (Zeng and Trauth, 2005) was carried out to quantify uncertainty.

During this decade, greater computational power and more efficient software packages permitted more complicated optimization schemes and simulation runs (Lu et al., 2013). With these sophistications, understanding uncertainty with modern computing power and software, such as GIS analyses, could greatly expand our knowledge base. For instance, combining GIS vector data for estimating current waste generation and composition to predict future waste generation by building spatial relationships among land-use classes derived by and RS images, solid waste characteristics, and socioeconomic indices may be tied together to aid in more effective routing planning (Karadimas and Loumos, 2008; Lara-Valencia et al., 2009; Katpatal and Rao, 2011). Network analyst, a function extension of ArcGIS® or Arc/Info® software, helped solve routing problems for waste shipping and allocation for treatment and disposal (Karadimas et al., 2008; Chalkias and Lasaridi, 2009a, 2009b; Hřebíček and Soukopová, 2010). Several AI algorithms, including a genetic algorithm and ant colony algorithm, were developed for routing optimization and vehicle scheduling to reduce the time consumption of waste collection and shipping (Chang and Wei, 2000; Cortés et al., 2000; Karadimas et al., 2007a, 2007b; Karadimas et al., 2008;

Fan et al., 2010). Routing scheme can be dependent on siting philosophy of treatment facilities as well. Practices such as the use of an artificial neural network for training a prototype ES using partial hazardous scores were explored for site selection of sanitary landfill, which could in turn affect routing scheme (Chau, 2006).

The proliferation of automatic data acquisition technologies, such as radio frequency identification (RFID) and sensor technologies (Tsai et al., 2011), brought new vitality into SWM. In combination with the GIS, RS, and GPS implementations developed in the last two decades, a wide range of SWM applications such as those for waste sorting, measurement, exhaust gas detection, ambient monitoring, and fine control of linear displacement were developed (van Kessel et al., 2002; Khijwania et al., 2003; Stegemiller et al., 2003; Cox, 2008; Fuchs et al., 2008). Specifically, RFID was applied for waste flow tracking, source reduction, management of pay-as-you-throw systems, and even for law enforcement (Chowdhury and Chowdhury, 2007; Wyatt, 2008; Atkins et al., 2009; Gillispie, 2010; Zhang et al., 2010). GIS, RFID, digital image recognition, web-based applications, and global systems for mobile communications (GSM) networks were utilized to create a highly cohesive EIS for waste monitoring and managerial control (Arebey et al., 2009, 2011; Luo et al., 2009). At this stage, challenges arose from the proper handling of massive data streams in EIS, requiring an effective mechanism for data storage, management, and search (Voigt and Welzl, 2001; Liao, 2011).

24.2 TREND ANALYSIS

24.2.1 Trend Analysis for Solid Waste Management in the 1970s and Before

The historical narrative presented above clearly indicates that most available models at this stage were developed interactively over time, in response to external pressures. The work done during the 1970s for the planning and management of SWM systems dealt with applying and refining various optimization techniques and heuristic algorithms for more realistic representations of SWM practices (MacDonald, 1996b). Planners and decision makers used systems engineering models for achieving the basic short-term and long-term SWM planning with respect to cost minimization principles and technical constraints. The types of real-world SWM issues investigated varied from dynamic, multi-period investment for SWM regionalization to local scale analysis of vehicle routing. In the mid-1970s, when forecasting models initially appeared, some researchers considered endogenous variables, such as the effects of population, income level, and the dwelling unit size, when characterizing waste generation.

Berger et al. (1999) pointed out the shortcomings in the models developed during 1970s. These included models considering, in most cases, only one time period, recyclables rarely were taken into account, and only one processing option was available for each type, with only a single generating source. These limitations effectively made these models unsuitable for large-scale long-term planning, according to Sudhir et al.

(1996). Another issue is that the models developed in the 1970s did not promote the WMH, even though the waste hierarchy concept actually originated in the European Union (EU) 2nd Environmental Framework Program of 1977. However, WMH emerged as one of the foci for waste management research in the 1980s.

The earliest EIS (Infoterra since 1977) marked the formative stage in the area of environmental informatics. It appeared a few years later than the sporadic computer-based applications for environmental information management that began nearly four decades ago (Clark and Gillean, 1974; Hilty et al., 2006). Then, over a 15-year time span things evolved from the earliest prototype of EIS to the emergence of a formal terminology for "environmental informatics" in 1992 (Page, 1992). Overall, environmental informatics applied for SWM faced an even slower evolution from the application perspective relative to systems engineering tools and models.

24.2.2 Trend Analysis for Solid Waste Management in the 1980s

What actually motivated the boom in such models during the 1980s? First, it is necessary to understand the context to justify it. Increased waste generation, difficulties in labor-management relations, rising costs, and uncertainties in technological evolution were problems that drove model development, as pointed out by Clark and Gilleann (1974). According to Gottinger (1988), attention was directed at the efficiency and effectiveness of waste management operations and the economy of scale effects in large waste treatment facilities. The concept of ISWM and WMH emerged in 1980s as two strands of research that together simultaneously promoted thinking in a systems-oriented way (Chang et al., 2011). Furthermore, suitable and user friendly mathematical OM emerged gradually as more powerful computational resources and comparative modeling skills provided an extra impetus for the construction of such models (Chang et al., 2011). Cost–benefit concerns appeared in OM with respect to more types of constraints contemporaneously, such as environmental and recycling constraints. FM for the waste generation prediction in the 1980s moved on with a less intensive yet promising pace to include waste composition (Chang et al., 2011). Further, the inclusion of MIS and ES gradually allowed the implementation of systems assessment framework using AI technology.

Uncertainty began playing an important role in decision-making in the 1980s. Until the early 1980s, probability was the only kind of uncertainty handled by mathematics. The idea of probability, as symbolized by the concept of randomness, is based on the "chance" or "opportunity" that exists in a real-world event. However, fuzziness takes in another aspect of uncertainty. In reality, fuzziness is the ambiguity that can be found in the linguistic description of a concept or a feeling. For instance, the uncertainty in expressions like "the air around the incinerator is dirty" or "the smell in the landfill is bad" is called fuzziness. The degree of fuzziness to be recognized in such questions is "how dirty is dirty?" or "how bad is bad?" Therefore, randomness and fuzziness differ in nature. In mathematics, the probability density function and membership function are used to illustrate the concepts of "probability" and "fuzziness," respectively. Usually a large number of samples are generally required to identify the probability density function for the occurrence of a phenomenon, whereas a subjective description of the fuzzy membership function is usually applied in the determination of fuzziness.

On the other hand, grey systems theory emphasizes a different way to address uncertainty in which all systems are divided into three categories, white, grey, and black (Deng, 1984a, 1984b). The white part shows a completely certain and clear message in a system, the black part has totally unknown characteristics. A message released in the grey part is in-between, which can be expressed by an interval or a grey number. A grey or interval number is simply an open interval with upper and lower limits (Moore, 1979). In reality, any phenomenon encountered in environmental management is most likely in the grey stage due to the insufficient amount of available information. Environmental management rests upon a dynamic process with uncertainty evolving from the black stage to the white stage. Therefore, a grey number or interval number may be used instead of random variable to describe the uncertain parameters in many social, engineering, and natural systems where only a very few samples exist. In real-world applications, the grey or interval numbers can be viewed as an alternative description of system parameters in light of the concept of "confidence interval." Such an approach dramatically simplifies the expression of system uncertainties whenever the probability density functions cannot be fully identified with sufficient samples and a membership function is unclear in formation. Moore (1979) further described interval analysis and interval programming techniques. Later on, the development of GIP actually integrated the concepts in grey systems theory with MILP (Deng, 1986) that at that time provided a brand new avenue to account for uncertainty in SWM.

In the late 1970s and early 1980s, various kinds of information systems and their prototypes were born in succession, but their applications for SWM lagged behind. The idea of relational DBS was proposed in 1970 (Codd, 1970), the term of DSS coined in 1971 (Scott-Morton, 1971), the well-known ES MYCIN (an acronym for Microbial Infection Therapy System) was introduced in 1974 (Yu et al., 1979) and the first GIS to utilize the personal computer became operational in 1982 (University of Northern British Columbia, 2011). However, the first applications of these new technologies for SWM were not reported until the 1980s. In fact, Lu et al. (2013) mentioned that a DBS was applied to SWM in 1982, DSS was applied in 1989, GIS was applied in 1988, and ES was applied to SWM problems in 1987. At the very beginning, these applications were mostly designed in the United States. Later on, many projects appeared in a handful of developed countries (e.g., Italy, Spain, and Greece) and developing countries (e.g., China, India, and Turkey).

24.2.3 Trend Analysis for Solid Waste Management in the 1990s

In the 1990s, decision makers came to rely on both ISWM and WMH with respect to different occasions and regions. ISWM were oftentimes used to rank a few treatment options in order of preference with regard to scientific or technical evidence while WMH highlighted a community-based approach trying to bring all stakeholders into the problem solving process (Chang et al., 2011). Technology evolved to meet new challenges at beginning of the 1990s. Innovations covered a diverse range of technical issues for composting plants (odor control), landfills (leachate control and gas recovery), and incineration facilities (toxic gaseous emissions), forcing decision makers to consider a variety of waste management alternatives at an enhanced level of sophistication (Chang et al., 2011). Environmental emissions control at waste treatment and disposal facilities also became mandatory requirements. Decision-making for siting new waste incineration and disposal facilities became a contentious question for authorities, with big societal impacts like the syndromes NIMBY, build absolutely nothing anywhere near anyone (BANANA), and locally unacceptable land use (LULU). After experiencing resources scarcity, population growth and environmental deterioration, the emergence of the sustainable development concept in the Brundtland report of 1987 (WCED, 1987) triggered more concerns about SWM issues. During this period, the utility of increased capacity for modeling complexity and improved simulation skills were confirmed for optimization, forecasting, and control in SWM (Chang et al., 2011). Consciousness and consensus were to some degree reached with regard to the integration of waste management options based on technical, economic, and environmental factors. This in turn suggested the potential value of systems analysis for SWM.

In the 1990s, the criteria of interest for dynamic optimization enveloped the simultaneous interactions among the effects of waste generation, source reduction, and curbside recycling, collection and transfer, processing and transformations, site selection, waste disposal, tipping fee evaluation, and environmental impacts like air pollution and leachates. A single integrated modeling system (IMS) normally requires step-by-step integration and concatenation of several external functions associated with FM or SM in concert with OM at a higher level. Together, these form a powerful functional structure with interactive or hierarchical relationships, leading to more sophisticated practices.

Obviously, the work done in the 1990s demonstrates that ISWM strategies can reach sustainability goals, given the complex conditions must strike a balance among options for incineration, composting, and recycling. It was shown that this balance must maximize the social welfare and minimize the public health impacts simultaneously, adjusting to increased waste generation and limited land and resources availability. Over this decade, many ISWM strategies were developed by integrating various types of systems engineering models; as indispensable tools to possibly rank alternatives and reach a sustainable management goal for waste minimization, cleaner production, and resource conservation and recovery. This thrust of ISWM drove a wealth of unique system synthesis and system integration in the context of IMS, gradually becoming the norm in the field in the late 1990s.

This evolution generated many uncertainties related to the complexities of the decision-making process, triggering a reformulation of OM in the 1990s (Huang et al., 2002). At the same time, systems engineering models formulated to illustrate the source of uncertainties via the use of fuzzy set theory, grey systems theory, and probability theory experienced explosive growth in relation to SWM planning. While ISWM became a common acronym that was almost mandatory in many industrialized countries at the government level, this new concept in management of SWM systems deepened the insights needed for conceptual development and resulted in a profound impact on the methodological foundations in systems analysis (Chang et al., 2011). This observation is confirmed by the increasing array of waste management strategies that considered uncertainty with high levels of complexity and subjectivity. This development actually opened the way for a renewal of systems engineering model evolution. Armed with these theoretical advances, GIS and DSS could now be integrated with each other to support decision-making challenges with varying levels of system synthesis and system integration on a long-term basis (Chang et al., 2011).

Observation also confirms that improvements were extended from systems engineering models to system assessment tools, including LCA or LCI, EIA, SEA, MFA, socioeconomic assessment (SoEA), and RA to formalize uncertainty quantification. Such endeavors were intended to supply decision makers with the necessary information about how environmental impacts related to SWM systems can be better understood and managed as a whole under uncertainty. The strengths at this stage rested upon the capability to integrate a variety of SM, FM with OM to fulfill multifaceted assessment needs, whereas system assessment tools may provide background information to narrow down options via appropriate system synthesis. Uncertainty embedded in various system assessment tools at the local level can support uncertainty quantification at the system level. Yet, insufficient system integration and/or system synthesis for more representative ISWM strategies at different levels still limits an all-inclusive exploration for SWM due to computation complexity.

In the 1990s, large capital investments in municipal infrastructure planning became increasingly required for environmental services but planning decisions are laden with uncertainty. Thus, uncertainty analysis using the interval values for solid waste facility expansion and waste flow allocation within an MSW management system became one of the foci of research in the 1990s (Huang et al., 1992, 1993a, 1993b, 1994, 1995). Waste management on the basis of uncertainty analysis improved systematic reliability using the concepts of fuzzy sets and grey intervals (Chang et al., 1997a). Grey mathematical programming techniques, in which all or part of the input parameters are represented as interval numbers, was applied to tackle SWM systems in the 1990s (Huang and Moore, 1993; Huang et al., 1992, 1993a, 1994, 1995). Grey integer variables were used to help screen decision alternatives for solid waste facility selection/expansion. For system planners, this added value to grey interval concepts and techniques for SWM (Huang et al., 1993b). Risk may be compounded when using forecasting trends, working with a collection of small datasets, or planning around event-based uncertainty, a key concern on the minds of decision makers in pursuit of a compromised solution.

The evolutionary pathways of fundamental EIS technologies, as described in Section 24.1.2, greatly influenced trends in environmental informatics and its applications for SWM. The development of EIS to address public concerns with forward-looking solutions was promoted in two international government negotiations, UNHEC in 1972 and United Nations Conference on Environment and Development in 1992 (Haklay, 1999). Data and knowledge, and supporting applications, came from statistics, surveying, and regulations. Authorities elaborated most of the large-scale databases, not only because they had the capacity, but also because they must act to balance competing societal interests. Many applications reflect the obligations and needs of authorities; site selection must conform to the public interest, and route optimization saves on public costs. Before the early 1990s, efforts focused mostly on industrial SWM, including radioactive and hazardous solid waste streams. Later on in the decade, the fast growth of MSW related to population growth and economic development challenged SWM since improperly managed MSW posed a risk to human health and the environment (USEPA, 2002). Therefore, MSW was the target for research during the late 1990s when the USEPA established an ISWM concept and a WMH.

From a retrospective point of view, the early 1990s were the incipient stage of environmental informatics formally as applied to SWM. During that period, environmental informatics was introduced into SWM, with applications aimed to improve management efficiency. At the local level, a comparison of historical data was restricted to chart analysis, such as the USEPA's annual report on MSW (USEPA, 2011). As Haklay (1999) pointed out, the EIS were helpful for finding, analyzing, monitoring, and learning about environmental problems rather than solving any environmental problems in reality. To promote the highly interdisciplinary nature of environmental informatics for SWM, curricula or courses on environmental informatics became offered in some universities under EC-funded projects and EU–Canada cooperation programs (Tsankova and Damianova, 1995; Swayne and Denzer, 2000; Gerbilsky et al., 2001; Denzer, 2003) in late 1990s and early 2000s. Due to the limited nature of the associated information technologies and, evolving computing power, the modeling ability, concepts of informatics, and data assurance meant that complicated problems were reduced to simplified and compromised solutions. Research efforts in that period, however, provided the foundation for follow-up developments.

24.2.4 Trend Analysis for Solid Waste Management in the 2000s

From a retrospective point of view, the need to comply with regulatory aspects was already imposed during the 1990s, though it still received wide attention in systems engineering models and system assessment even into the early 2000s. During this time period, regulations focused not only on environmental emissions but also targeted recycling, recovery and incineration, waste diversion from landfill targets, as well as market-based instruments for SWM. New system assessment tools, such as scenario development (SD), sustainability assessments (SA), MFA, SoEA, and SEA, were evolving quickly for more versatile waste management plans. In parallel, the heightened sophistication of SWM strategies can be seen in the development of DSS (Fiorucci et al., 2003; Costi et al., 2004), LCA (Bovea and Powell, 2006), OM (Ljunggren, 2000; Chang and Davila, 2006), and the inclusion of market-based instruments with models and tools (Nilsson et al., 2005). To account for ever increasing scope of considerations in ISWM systems, uncertainty analysis turned out to be a critical factor for improving the reliability of decision-making.

To account for uncertainty in decision-making, the minimax regret (MMR) criterion is useful when alternate outcomes are possible. The emergence of MMR helped frame the loss of opportunity concept in decision-making (Luce and Raiffa, 1957; Taylor, 1996; López-Cuñat, 2000). Applications of MMR in multidisciplinary systems planning and optimal scheduling are numerous. These include operations research and management science considering multifacility location problems under complete or partial uncertainty (Love et al., 1988; Drezner and Guyse, 1999; Averbakh, 2000). To extend the potential of MMR in decision analysis under uncertainty, there are also examples using interval variables and, in turn, interval objectives and constraints. This can be seen as an analogy to grey programming (Inuiguchi and Sakawa, 1995). The MMR criterion has also been applied for various pollution abatement planning purposes coping with stochastic uncertainty (Loulou and Kanudia, 1999; Chevé and Congar, 2002). In 2005, the MMR criterion was first used for SWM system planning with the inclusion of uncertain scenarios (Chang and Davila, 2006). However, uncertainty that can be addressed by the state of nature for possible events in SWM has not yet been included into optimization analyses within an MMR structure.

In the 2000s and after, systems analysis for SWM problems became more sophisticated. Meanwhile, the integration of models and informatics technologies became more seamless. Resulting from the improvement of cyberinfrastructure, the 21st century has also brought more powerful computers and more distributed data storage capacity to support systems analysis for SWM (Chang et al., 2011). Given that the Internet-based information technologies are more distributed than ever, the increasing number of applications includes web-based GIS along with electronic data exchange throughput (Chang et al., 2001). Automatic data acquisition techniques further promote management efficiency. These advances extend and empower environmental informatics applications for SWM to pursue better management effectiveness, with higher computing power, more perfect modeling and deeper public involvement. At the top of these advances, the technology of "internet of things (IoTs)" is becoming quite popular since IBM (an acronym of International Business Machines Company) proposed the strategy of "Smarter Planet" in late 2000s. IoTs empower the Internet extending into people's everyday lives through a wireless network uniquely identifiable objects (Welbourne et al., 2009). It is predictable that the IoTs will be very helpful for the entire SWM process and support better sustainable decision-making (Lu et al., 2013).

Although IoTs is a new concept, it is derived from internet technologies (Lu et al., 2013). The whole framework, according to IBM's proposition, is divided into three layers, being explained in Lu et al. (2013): "The bottom layer is a perception layer that collects data about waste amounts, source producers, vehicle positions and states, facility states, environmental impacts, images and surveillance videos. Barcode-based manifests may be used to record waste flow and exchange data among different operation companies. Data collection devices are abstracted as data acquisition nodes, with their own intercommunications to be accessed by the internet directly or via sink nodes (Figure 24.1). The immediate layer is a network layer that couples the internet, cloud storage and semantic web. Data acquisition nodes access the network layer in a way similar to the existing internet. Heterogeneous data can be exchanged in response to demands. Users can access the network layer to get what they want, regardless of the storage site. The semantic web makes the network layer become a network

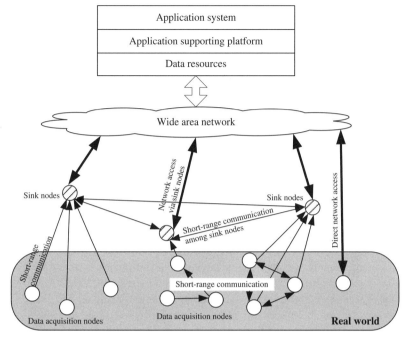

FIGURE 24.1   The schematic diagram illustrating data acquisition and communication. *Source:* From Lu et al. (2013)

of data and knowledge, enabling machines to learn to understand the human mind including it's' data, information and knowledge. The top layer is an application layer that is adaptive to heterogeneous data environments. The application layer is actually a super integrated EIS. This kind of EIS combines the advantages of all kinds of EIS, is able to solve both well-structured and poorly-structured SWM problems and maintains itself independently. Moreover, the application layer provides other layers with instructions for diagnosis and feedback control."

24.3 TECHNICAL BARRIERS AND SOCIOECONOMIC CHALLENGES

ISWM oftentimes demands high-end operational control, managerial control, and even strategic planning. Therefore, the complex nature of ISWM problems is a great challenge to producing systematic, cost-effective, risk-informed, and forward-looking solutions. A few technical barriers and socioeconomic challenges are listed below for future endeavors:

- **Complexities involved in large-scale systems synthesis and system integration**: There are specific and unique limitations constraining the smoothing of the interfaces among these models and tools. While these decision-making techniques may improve ISWM to some extent, ISWM is time-consuming and data-intensive with respect to varying boundaries set forth in different models and tools with their attendant technical, environmental, economic, and social complexities. Ideal solution procedures, however, may yield a balance between simplification of the analysis and soundness when capturing the essential features, but might also result in additional complications for SWM systems analysis.

- **Advancements in individual sensor network**: Self-weighing containers with RFID at the source point is the precondition for weight-based billing. There are two methods for source weighing at present. One is in-situ weighing by an onboard scale installed within the loader of vehicle. The embedded sensor detects pressure changes in hydraulic circuit for measurement. When a waste container is loaded for tipping, its weight is measured and recorded. The precision of weighing is about 1%. If an RFID tag is embedded in the waste container, the source producer is also recorded. The other is *ex-situ* weighing through RFID tag embedded in a waste container and an *ex-situ* platform scale. A vehicle delivers dispersive waste containers with RFID tags to a transfer station, where weighing can be done first and then waste is tipped. Both methods, however, cannot yield the precise amount of waste generated from individual households or apartments in densely populated cities or the industrial waste generated from individual factory. Pressure sensors based on stainless steel sheet or micro-electromechanical systems can be embedded into waste containers to form self-weighing containers, which also have embedded RFID tags for identification and communication. The sensors and tags are integrated so that the measured values can be temporarily stored in the RFID tags and transmitted through sensor networks. With the aid of systems engineering models and system assessment tools, advanced vehicle routing models with proper configurations via an integrated GIS, GPS, and RFID could improve operational control to the maximum extent.

- **Innovation for next-generation efficient and ubiquitous sensor networks**: Widely spread waste generation points require numerous sensor nodes for constructing efficient and ubiquitous sensor networks for data collection. This presents difficulties such as the energy supply for sensor nodes, signal attenuation when crossing buildings, signal interference, routing algorithm issues, and self-adaptation problems encountered after node failure or new node addition. It is highly likely that an array of similar sensors will be used to improve the reliability of measurement, and an array of unlike sensors will have self-diagnostic capabilities. Self-adaptive and scaleable wireless sensor networks are vital for control and efficient operation of future SWM systems.

- **Advancements in environmental informatics**: At this stage, challenges arose from massive data streams (big data), which need effective mechanisms for data storage, management, and search. To meet the goals for sophisticated decision-making required in ISWM, informatics technologies in concert with grid and parallel computational capability are becoming more important for SWM, due to the increasing needs for large-scale complex data storage, communication, analysis, and applications. High performance computing environments and multiple terabyte-to-petabyte scale data storage systems are required for real-time processing for data-intensive SWM applications. Virtualization-based allocation of computing resources is required to improve calculation performance. Effective models for mining massive data, matched to the new computing environments, will influence decision-making support research directions.

- **Robust algorithms**: As an example in optimal vehicle routing an increasing number of stops and more realistic simulations are causing an exponential growth in computational complexity. It is necessary, therefore, to combine advantages of mathematical programming and AI to develop robust new algorithms with fast convergence rate and optimal results. Benchmark datasets for typical SWM applications must be constructed so as to permit more generalized conversations about algorithms.

- **Temporal-spatial evolution of regional ecosystems**. Carbon life cycle inventories for different functional areas must be constructed for assessing the carrying capacity of regional ecosystems. Data mining historical statistics is helpful for building the background field of a regional ecosystem. On that basis, the carbon or ecological footprint impacts and the sustainability of SWM strategies can be assessed or simulated.

- **Synergistic effects of public involvement**: Policy makers, experts, and the public are stakeholders with different decision weights in SWM systems. Abundant interfaces for public involvement must fuse all stakeholders, all computing platforms, and all EIS into a seamless decision-making environment whenever and wherever. For example, an interactive questionnaire completed through the Internet would influence the selection of location of a new composting or anaerobic sludge digestion plant. The emerging technologies, including automatic knowledge acquisition, machine learning, auto-reasoning, the semantic web, and web-based knowledge acquisition, can be employed to construct a more effective ISWM in the future.

- **Rapid development of municipal utility parks (MUPs)**: An MUP is defined as a park that combines several utility components for an urban region, such as SWM, drinking water treatment, wastewater treatment, and stormwater treatment. These utility components work well together based on the concept of industrial symbiosis. It is a highly complex task to construct appropriate SMs to capture the true essence of an MUP. Moreover, the behavioral changes of consumers are poorly understood as a means of reducing environmental impacts. Human factors must frame fundamental research on the evolution of social processes. Investigating the formation of MUP has to be connected to the study of important social phenomena such as firms, markets, governments, and social movements.

- **Industrial ecology, pollution prevention, and green products**: Large-scale systems analysis is required for a proper evaluation regarding how innovation of green products could impact SWM from local to regional, and to global scales. Achieving this goal requires a delicate LCI that might not be available or complete.

- **Uncertainty analysis faces stringent limitations**. Probability theory requires vast amounts of available data for characterization of probability density functions. In many occasions, such large amounts of data are simply not available. Interval analysis solves this problem found when dealing with data-intensive probability theory, but suffers from a stability issue. Fuzzy set theory can provide a flexible way to address the uncertainty but faces challenges as to how to retrieve and construct much needed membership functions. None of these methods are perfect in a theoretical sense.

- **Urban mining, recycling, and circular economy dealing with waste as economic resources**. The knowledge discovery of the urban system planning is not yet accomplished, which limits the mining of secondary resources to be used as an integral part of the circular economy. Systems analysis is needed to assess where secondary raw materials are available in various anthropogenic sources, like the case of landfills in cities where recycled materials can be recovered and used further to replace virgin materials. Urban mining is not focused on landfill mining solely, and it could be broadened to other possible sources resulting from the urban metabolism stocks like buildings, electrical and electronic equipment, vehicles, just to name a few. Industrial symbiosis is vital for the development of a circular economy, whereby secondary materials (which are included in green products) are recycled and reused for promoting the closed-loop processes, resulting in a zero waste society. Systemic and cost-effective solutions need to be explored with the aid of information and communications technologies as well as MFA.

24.4 FUTURE PERSPECTIVES

24.4.1 Systems Engineering Models and System Assessment Tools for SWM

The development of models and tools for SWM systems over the last few decades was fully reviewed in Chapters 8 and 13 and this chapter. Fourteen main categories of models and tools were clearly classified and discussed in Chapter 8, including CBA, FM, SM, OM, IMS, MIS/DSS/ES, SD, MFA, LCA, RA, EIA, SEA, SoEA, and SA. It is noticeable that the systems engineering models developed in the early stages are SM, OM, FM, and CBA. These were followed by IMS, where market-based instruments and regulatory requirements can be included in the decision analysis. The emphasis on the concept of sustainability with system assessment tools associated with MIS/DSS/ES, collectively, or separately, supports more sustainable development for SWM.

According to Chang et al. (2011), future systems analysis requires interdisciplinary and policy-relevant research related to SWM systems, with an emphasis on enhancing the sustainability of SWM systems challenged by rapid changes of societal environment and/or extreme events associated with global climate change. IMS will be applied for different types of issues at different scales in combination with different system assessment tools, more often than not. Gaps in the knowledge base as to how to create better system integration and system synthesis between systems engineering models and system assessment tools must be overcome. When dealing with such complexities of MUP, there is a need to create functioning agent-based computer SMs to allow us to investigate some classic problems in business–environment relations as well as the technical intricacy embedded in an MUP design.

Regarding the impact of industrial ecology, pollution prevention, green products, and urban mining on SWM, high level simulation and forecasting analyses are required before optimization and control analyses can be performed. Salient examples may include but are not limited to green supply chain relationships, pollution-control incentives within organizations and across different organizations, and incentive incompatibilities between organizations and society as a whole. The introduction of environmentally benign products to the marketplace under varying information propagation conditions, and understanding the pre-conditions and post-conditions for radical innovations of a new product, will reduce adverse environmental impacts in SWM. The same goal can be achieved via collecting secondary raw materials from urban mining. With such research needs, a sound system integration and system synthesis with a holistic approach will become prevalent over the next decade. All of these efforts will certainly promote a sustainable decision-making, where risk is considered (risk-informed decision-making), with a progressive-looking perspective, where cost-effectiveness and environmentally benign solutions will be preferable.

24.4.2 Environmental Informatics for SWM

Five EIS, including DBS, GIS, DSS, ES, and integrated environmental information systems (IEIS) are in the systematic form of environmental informatics, were discussed in Chapter 13. In the 1990s, various applications of EIS set up the basic milestones for SWM environmental informatics. Since the 2000s, more techniques for automatic data acquisition have been enriching environmental informatics. Informatics can address a series of problems that ISWM are facing, including the lack of source separation and reduction, poor waste collection efficiency and transport, and monitoring faultiness. In the future, the development of environmental informatics for SWM systems needs greater depth and scope. On the one hand, the entire waste management cycle from source separation and reduction to final disposal should be cohesively considered; on the other hand, high-level integrated, intelligent, and sustainable systems must be developed for practical purposes in the context of industrial ecology and sustainable engineering. It is asserted without question that the 2010s is the decade of IoTs, as reflected by search frequency from Google Trend. IoTs can be understood as a smart internet possessing ubiquitous tentacles that detect and connect all people and/or things by technologies including RFID, sensors, and so on. IoTs reflect the developmental tendencies of informatics in the next few years, for example, the development of hysteric modeling and control of a solid waste collection system via an integrated GIS, GPS, and RFID operations. In any circumstance, IoTs are different from supervisory control and data acquisition because it is smart enough to improve both waste management efficiency and effectiveness.

Hannan et al. (2010) developed a waste collection, monitoring, and management system built and operated based on RFID, GPS, GSM, and cameras. However, this system cannot realize source weighing and the optimal vehicle scheduling. Tsai et al. (2010) demonstrated a nuclear material tracking and monitoring system based on RFID, GPS, general packed radio service (GPRS), and GIS but encountered a shortfall. Nevertheless, a brand new system was directed to optimize waste logistics by both constraint-based DSS and RFID-based management (Nielsen et al., 2010), which can be considered as rudimentary IoTs application for SWM. More power databases that can mix complex database queries, transformations, geospatial, and real-time analytics (NoSQL Archive, 2011) may be developed and applied for SWM to support the systematic needs of integrated RFID, GPS, GPRS, GIS, and systems engineering models and tools. More advanced SWM system along these lines can be anticipated.

24.4.3 High Level System Synthesis and Integration for SWM

Modern data acquisition and communication systems combined with the flexible philosophy for system integration and system synthesis of models and tools can aid advances for managing input of data and models, storage, human–machine interaction, computing processes, and information output and representation. The achievement of synthesis and integration goals also demands greater effort and understanding of the human mind's data, information, and knowledge. For example, some innovations are possible by using data mining and GIS spatial analysis to discover new knowledge or patterns from existing non-spatial and spatial data, and by employing inference engines to learn new knowledge from existing rules for improving ISWM systems. Moreover, it is now feasible to consult domain experts and accumulating wisdom from professionals through web-based knowledge discovery systems via various models and tools, informing the public to accommodate closer partnerships, and making the public a stakeholder. The public and the experts who are not the domain experts may learn from the system and give their opinions via a shared vision modeling framework. These opinions will, however, influence decision-making with different weights. Design weights may be assigned and adjusted dynamically according to the human–machine interactive learning and progressive improvements.

24.5 FINAL REMARKS

This chapter reviewed developments in systems engineering models and system assessment tools as well as environmental informatics technologies applicable for ISWM over the past four decades as a conclusive chapter of the book. The developmental history of systems analysis for SWM can be regarded as the evolutionary history of the collision and fusion of multiple disciplines, including computer science, electronics, communication science, software engineering, AI, geographical science, regional planning, systems engineering, urban science, decision science, and environmental engineering. In the incipient stage, research and applications were tentative, so concepts were mixed and the defined tasks were relatively thin. As systems analysis migrated to and was implemented in SWM, development of integrated applications with respect to three independent regimes, including systems engineering models, system assessment tools, and environmental informatics, was accelerated. The applications of systems analysis to deal with several typical SWM problems, including regional planning, site selection, and arrangement of vehicle routing, are relatively mature. The demand for sustainability was reflected in recent applications, which take economic, environmental, social impacts and public involvement into account thoughtfully. The systemic structures of ISWM involving environmental informatics technologies for data acquisition, communication, storage, deep processing, and utilization patterns can be further elaborated as advancements in systems engineering models and system assessment tools move at an unprecedented pace.

However, low-level system integration and system synthesis hamper further development of source separation and waste minimization to some extent. The EIS and related techniques have their limitations and drawbacks. In the future, it will be indispensable to keep pace with latest developments in environmental informatics and seek more high-end synergistic innovations in systems engineering models and system assessment tools. Sensing techniques will collect large-scale and accurate data, while advanced identification techniques will be applied to promote waste separation and source reduction. Gaps between society perceptions and decision-making will be minimized as IoTs technology and cloud sourcing develop to maturity. More sophisticated systems will combine the various advantages of different EIS and new models/tools will be developed. All of these efforts will certainly push forward an intelligent ISWM.

REFERENCES

Adams, S. K., Even, J. C., and Blewett, T. J. 1996. MARTIN: a midwest regional database for recycling materials from post-consumer solid waste. *Resources, Conservation and Recycling*, 17(3), 153–167.

AEA Technology. 1998. Computer-based models in integrated environmental assessment. Technical Report, No. 14. European Environmental Agency.

Ahmadi, M., Zade, P. S., Pouryani, S. B. M., and Gilak, S. 2010. Site selection of waste storage locations with fuzzy logic and analytic hierarchy process in a GIS framework, case study: Si-Sakht town in Dena city, Kohgilouye O Boyer Ahmad province, Iran. In: Proceedings of the 10th International Multidisciplinary Scientific Geoconference. SGEM, vol. 1, Curran Associates, Inc., pp. 1143–1150.

Akladios, M., Gopalakrishnan, B., Bird, A., Carr, M., Garcia, R., McMullin, D., Myers, W. R., Vennetti, V., Zayas, J., Becker, P. E., and McCullom, D. 1998. Development of an expert system to help design for worker safety. In: *Intelligent Systems in Design and Manufacturing (Proceedings of SPIE)* (Eds. Gopalakrishnan, B. and Murugesan, S.), SPIE Press, pp. 240–250.

Alidi, A. 1998. A goal programming model for an integrated solid waste management system. *Arabian Journal of Science and Engineering*, 23(1B), 3–16.

Altavista, P., Belli, S., Bianchi, F., Binazzi, A., Comba, P., Del Giudice, R., Fazzo, L., Felli, A., Mastrantonio, M., Menegozzo, M., Musmeci, L., Pizzuti, R., Savarese, A., Trinca, S., and Uccelli, R. 2004. Cause-specific mortality in an area of Campania with numerous waste disposal sites. *Epidemiologia e Prevenzione*, 28(6), 311–321.

Alves, M. C., Lima, B. S., Evsukoff, A. G., and Vieira, I. N. 2009. Developing a fuzzy decision support system to determine the location of a landfill site. *Waste Management & Research*, 27(7), 641–651.

Anandalingam, G. 1987. REGWASTE: an expert system for regulating hazardous wastes. In: Proceedings of the 1987 IEEE International Conference on Systems, Man and Cybernetics (Cat. No.87CH2503-1), Institute of Electrical and Electronics Engineers, New York.

Anderson, L. 1968. A mathematical model for the optimization of a waste management system. SERL Report. Sanitary Engineering Research Laboratory, University of California, Berkeley, CA.

Anderson, L. E., and Nigam, A. K. 1967. A mathematical model for the optimization of a waste management system, ORC 67-25, Operations Research Center, University of California, Berkeley, CA.

Anex, R., Lawver, R., Lund, J., and Tchobanoglous, G. 1996. GIGO: spreadsheet-based simulation for MSW systems. *Journal of Environmental Engineering, ASCE*, 122(4), 259–262.

Apaydin, O. and Gonullu, M. T. 2007. Route optimization for solid waste collection: Trabzon (Turkey) case study. *Global Nest Journal*, 9(1), 6–11.

Aquilina, C. A., Everette, S. E., Jouse, C. A., and Smiley, L. J. 1982. Low-level waste management data base system. *Transactions of the American Nuclear Society*, 41, 86.

Arakali, V. S. 1991. VITEX: an expert system to evaluate vitrification of nuclear waste. *Transactions of the American Nuclear Society*, 63, 76.

Arebey, M., Hannan, M. A., Basri, H., and Abdullah, H. 2009. Solid waste monitoring and management using RFID, GIS and GSM. In: Proceedings of the IEEE Student Conference on Research and Development (SCOReD), IEEE – Institute of Electrical and Electronic Engineers, pp. 37–40.

Arebey, M., Hannan, M. A., Basri, H., Begum, R. A., and Abdullah, H. 2011. Integrated technologies for solid waste bin monitoring system. *Environmental Monitoring and Assessment*, 177(1), 399–408.

Atkins, A. S., Zhang, L., Yu, H., and Miao, W. 2009. Application of intelligent systems using knowledge hub and RFID technology in healthcare waste management in the UK and China. In: *Proceedings of the International Conference on E-Business (ICE-B 2009)* (Eds Filipe, J., Marca, D. A., Shishkov, B., and van Sinderen, M.), INSTICC Press, pp. 44–49.

Averbakh, I. 2000. Minimax regret solutions for minimax optimization problems with uncertainty. *Operations Research Letters*, 27(2), 57–65.

Avouris, N. M. and Page, B. 1995. *Environmental Informatics: Methodology and Applications of Environmental Information Processing*, Kluwer Academic Publishers, New York, NY.

Baetz, B. 1990. Optimization/simulation modeling for waste management capacity planning. *Journal of Urban Planning and Development, ASCE*, 116(2), 59–79.

Baetz, B., Pas, E., and Neebe, A. 1989. Trash management: sizing and timing decisions for incineration and landfill facilities. *Interfaces*, 19(6), 52–61.

Barker, A. and Wood, C. 1999. An evaluation of EIA system performance in eight EU countries. *Environmental Impact Assessment Review*, 19(4), 387–404.

Barlishen, K. and Baetz, B. 1996. Development of a decision support system for municipal solid waste management systems planning. *Waste Management & Research*, 14(1), 71–86.

Barnett, M. W. 1992. Knowledge-based expert system applications in waste treatment operation and control. *ISA Transactions*, 31(1), 53–60.

Barrow, P. A. 1988. An expert system applied to the regulation of hazardous waste. In: Simulators V: Proceedings of the SCS Simulators Conference.

Basnet, B. B., Apan, A. A., and Raine, S. R. 2001. Selecting suitable sites for animal waste application using a raster GIS. *Environmental Management*, 28(4), 519–531.

Basri, H. 1998. An expert system for planning landfill restoration. *Water Science and Technology*, 37(8), 211–217.

Basri, H. B. and Stentiford, E. I. 1995. Expert systems in solid waste management. *Waste Management & Research*, 13(1), 67–89.

Beigl, P., Lebersorger, S., and Salhofer, S. 2008. Modelling municipal solid waste generation: a review. *Waste Management*, 28(1), 200–214.

Berger, C., Savard, G., and Wizere, A. 1999. EUGENE: an optimization model for integrated regional solid waste management planning. *International Journal of Environment and Pollution*, 12(2/3), 280–307.

Bhargava, H. and Tettelbach, C. 1997. A web-based decision support system for waste disposal and recycling. *Computers, Environment and Urban Systems*, 21(1), 47–65.

Bhat, V. 1996. A model for the optimal allocation of trucks for solid waste management. *Waste Management & Research*, 14(1), 87–96.

Björklund, A., Dalemo, M., and Sonesson, U. 1999. Evaluating a municipal waste management plan using ORWARE. *Journal of Cleaner Production*, 7(4), 271–280.

Björklund, A., Bjuggren, C., Dalemo, M., and Sonesson, U. 2000. Planning biodegradable waste management in Stockholm. *Journal of Industrial Ecology*, 3(4), 43–58.

Bodner, R., Cassell, A., and Andros, P. 1970. Optimal routing of refuse collection vehicles. *Journal of the Sanitary Engineering Division*, 96(SA4), 893–903.

Bovea, M. and Powell, J. 2006. Alternative scenarios to meet the demands of sustainable waste management. *Journal of Environmental Management*, 79(2), 115–132.

Bowen, W. M., Weeks, K. D., Batra, D., and Hill, T. R. 1989. A deep knowledge planning decision support system for aiding nuclear waste transportation decisions. *Computers Environment and Urban Systems*, 13(1), 15–27.

Boyle, C. 1995. *Integrated Waste Management: A Knowledge-Based Decision Support System Prototype for Developed and Developing Countries*. ETD Collection for McMaster University.

Boyle, C. A. and Baetz, B. W. 1998a. A prototype knowledge-based decision support system for industrial waste management: Part I. The decision support system. *Waste Management*, 18(2), 87–97.

Boyle, C. A. and Baetz, B. W. 1998b. A prototype knowledge-based decision support system for industrial waste management: Part II. Application to a Trinidadian industrial estate case study. *Waste Management*, 17(7), 411–428.

Brainard, J., Lovett, A., and Parfitt, J. 1996. Assessing hazardous waste transport risks using a GIS. *International Journal of Geographical Information Systems*, 10(7), 831–849.

Bresnahan, P. J. 1998. Identification of potential hazardous waste units using aerial radiological measurements. *Photogrammetric Engineering and Remote Sensing*, 64(10), 995–1001.

Brisson, I. 1997. Assessing the waste hierarchy—a social cost-benefit analysis of municipal solid waste management in European Union. AKF—Danish Institute of Governmental Research. Available at: http://www.akf.dk/udgivelser_en/container/2005/udgivelse_157./ (accessed November 2008).

Brodie, G. R. and Waters, C. D. J. 1988. Integer linear programming formulation for vehicle routing problems. *European Journal of Operational Research*, 34(3), 403–404.

Brunner, P. H. and Rechberger, H. 2003. *Practical Handbook of Material Flow Analysis*, CRC Press, Boca Raton, FL.

Bruvoll, A., 1998. *The Costs of Alternative Policies for Paper and Plastic Waste*, Statistical Central Office (Statistisk sentralbyrå), Oslo.

Burelle, J. and Monterrat, C. 1985. Mechanization of the collection of household refuse and data processing follow-up in the City of Paris. *Waste Management & Research*, 3(2), 119–126.

Caiti, A., Minciardi, R., Palmese, M., and Robba, M. 2006. GIS tools application for risk assessment of toxic waste buried in seafloor sediments. *Chemistry and Ecology*, 22(Suppl 1), S145–S161.

Calabrese, E. J. and Kenyon, E. M. 1991. *Air Toxics and Risk Assessment*. Lewis Publishers, Chelsea, MI.

Caruso, A., Colorni, A., and Paruccini, M. 1993. The regional urban solid waste management system: a modelling approach. *European Journal of Operational Research*, 70(1), 16–30.

Carver, S., Evans, A., Kingston, R., and Turton, I. 2000. Accessing geographical information systems over the World Wide Web: improving public participation in environmental decision-making. *Information Infrastructure and Policy*, 6(3), 157–170.

Chalkias, C. and Lasaridi, K. 2009a. A GIS based model for the optimisation of municipal solid waste collection: the case study of Nikea, Athens, Greece. *WSEAS Transactions on Environment and Development*, 10(5), 640–650.

Chalkias, C. and Lasaridi, K. 2009b. Optimizing municipal solid waste collection using GIS. In: Proceedings of the Energy, Environment, Ecosystems, Development and Landscape Architecture Conference (Eds Mastorakis, N., Helmis, C., Papageorgiou, C. D., Bulucea, C. A., and Panagopoulos, T.), Athens, Greece, pp. 45–50.

Chang, N. B. 1989. Solid waste management system planning with environmental quality constraints. MSc Thesis, Cornell University, Ithaca, NY.

Chang, N. B. 2008. Economic and policy instrument analyses in support of the scrap tires recycling program in Taiwan. *Journal of Environmental Management*, 86(3), 435–450.

Chang, Y. H. and Chang, N. B. 1996. Fuzzy optimal operation of solid waste management systems. In: Proceeding of the 7th ISWA International Congress, vol. 2 (Eds Salama, K., Taplin, D. M. R., Rama Rao, P., and Ravi-Chandar, K.), International Solid Waste Association, pp. 295–296.

Chang, Y. H. and Chang, N. B. 1998. Optimization analysis for the development of short-term solid waste management strategies using presorting process prior to incinerators. *Resources, Conservation and Recycling*, 24(1), 7–32.

Chang, N. B. and Chang, Y. H. 2001. Optimal shipping strategy of solid waste streams with respect to throughput and energy recovery goals of incineration facilities. *Civil Engineering and Environmental Systems*, 18(3), 193–214.

Chang, N. B. and Davila, E., 2006. Siting and routing assessment for solid waste management under uncertainty using grey minimax regret criteria. *Environmental Management*, 38(4), 654–672.

Chang, N. B. and Davila, E., 2007. Minimax regret optimization analysis for a regional solid waste management system. *Waste Management*, 27(8), 820–832.

Chang, N. B. and Davila, E., 2008. Municipal solid waste characterizations and management strategies for the Lower Rio Grande Valley, Texas. *Waste Management*, 28(5), 776–794.

Chang, S. Y., and Li, Z., 1997. Use of a computer model to generate solid waste disposal alternatives. *Journal of Solid Waste Technology and Management*, 24(1), 9–18.

Chang, N. B., and Lin, Y., 1997a. Optimal siting of transfer station locations in a metropolitan solid waste management system. *Journal Environmental Science and Health*, A32(8), 2379–2401.

Chang, N. B. and Lin, Y., 1997b. Economic evaluation of a regionalization program for solid waste management in a metropolitan region. *Journal of Environmental Management*, 51(3), 241–274.

Chang, N. B. and Wang, S. F. 1994. A locational model for the site selection of solid waste management facilities with traffic congestion constraint. *Civil Engineering Systems*, 11(4), 287–306.

Chang, N. B. and Wang, S. F. 1996a. Comparative risk analysis of solid waste management alternatives in a metropolitan region. *Environmental Management*, 20(1), 65–80.

Chang, N. B. and Wang, S. F. 1996b. Solid waste management system analysis by multiobjective mixed integer programming model. *Journal of Environmental Management*, 48(1), 17–43.

Chang, N. B. and Wang, S. F. 1996c. Managerial fuzzy optimal planning for solid waste management systems. *Journal of Environmental Engineering, ASCE*, 122(7), 649–658.

Chang, N.-B. and Wang, S. F. 1996d. The development of an environmental decision support system for municipal solid waste management. *Computers, Environment and Urban Systems*, 20(3), 201–212.

Chang, N. B. and Wang, S. F. 1997a. A fuzzy goal programming approach for the optimal planning of metropolitan solid waste management systems. *European Journal of Operational Research*, 99(2), 303–321.

Chang, N. B. and Wang, S. F. 1997b. Integrated analysis of recycling and incineration programs by goal programming techniques. *Waste Management & Research*, 15(2), 121–136.

Chang, N. B. and Wei, Y. L. 2000. Siting recycling drop-off stations in urban area by genetic algorithm-based fuzzy multiobjective nonlinear integer programming modeling. *Fuzzy Sets and Systems*, 114(1), 133–149.

Chang, N. B., Schuler, R. E., and Shoemaker, C. A. 1993a. Environmental and economic optimization of an integrated solid waste management system. *Journal of Resource Management and Technology*, 21(2), 87–100.

Chang, N. B., Pan, Y., and Huang, S. 1993b. Time series forecasting of solid waste generation. *Journal of Resource Management and Technology*, 21(1), 1–10.

Chang, N. B., Yang, Y., and Wang, S. F. 1996. Solid-waste management system analysis with noise control and traffic congestion limitations. *Journal of Environmental Engineering, ASCE*, 122(2), 122–131.

Chang, N. B., Chen, Y. L., and Wang, S. F. 1997a. A fuzzy interval multiobjective mixed integer programming approach for the optimal planning of solid waste management systems. *Fuzzy Sets and Systems*, 89(1), 35–60.

Chang, N. B., Chang, Y. H., and Chen, Y. 1997b. Cost-effective and equitable workload operation in solid-waste management systems. *Journal of Environmental Engineering, ASCE*, 123(2), 178–190.

Chang, N. B., Lu, H., and Wei, Y., 1997c. GIS technology for vehicle routing and scheduling in solid waste collection systems. *Journal of Environmental Engineering, ASCE*, 123(9), 901–910.

Chang, Y. C., Chang, N. B., and Ma, G. D. 2001. Internet web-based information system for scrap vehicle disposal in Taiwan. *Environmental Modeling and Assessment*, 6(4), 237–248.

Chang, N. B., Davila, E., Dyson, B., and Brown, R. 2005. Optimal site selection and capacity planning of a municipal solid waste material recovery facility in an urban setting. *Waste Management*, 25(8), 833–846.

Chang, N. B., Parvathinathan, G., and Breeden, J. 2008. Combining GIS with fuzzy multicriteria decision-making for landfill siting in a fast-growing urban region. *Journal of Environmental Management*, 87(1), 139–153.

Chang, N. B., Chang, Y. H., and Chen, H. W. 2009. Fair fund distribution for a municipal incinerator using GIS-based fuzzy analytic hierarchy process. *Journal of Environmental Management*, 90(1), 441–454.

Chang, N. B., Pires, A., and Martinho, G. 2011. Empowering systems analysis for solid waste management: challenges, trends and perspectives. *Critical Reviews in Environmental Science and Technology*, 41(16), 1449–1530.

Chapman, R. and Yakowitz, H. 1984. Evaluating the risks of solid waste management programs: a suggested approach. *Resources and Conservation*, 11(2), 77–94.

Charnpratheep, K. S. and Garner, B. 1997. Preliminary landfill site screening using fuzzy geographical information systems. *Waste Management & Research*, 15(2), 197–215.

Chau, K. W. 2006. An expert system on site selection of sanitary landfill. *International Journal of Environment and Pollution*, 28(3–4), 402–411.

Chen, T. H. 1994. An expert system for pig waste management in Taiwan. In: Computers in Agriculture, Proceedings of the 5th International Conference, American Society of Agricultural & Biological Engineers, Orlando, FL, pp. 757–761.

Chen, Y. Y. and Li, D. L. 2008. A web-GIS based decision support system for revegetation in coal mine waste land. In: Proceedings of the 7th WSEAS International Conference on Applied Computer and Applied Computational Science, Hangzou, China, pp. 579–584.

Chen, Y. Y., Jiang, Y. J., and Li, D. L. 2007. A decision support system for evaluation of the ecological benefits of rehabilitation of coal mine waste areas. *New Zealand Journal of Agricultural Research*, 50(5), 1205–1211.

Chen, Y. W., Wang, C. H., and Lin, S. J. 2008. A multi-objective geographic information system for route selection of nuclear waste transport. *Omega-International Journal of Management Science*, 36(3), 363–372.

Chevé, M. and Congar, R. 2002. Managing environmental risks under scientific uncertainty and controversy. In: *Proceedings of International Conference on Risk and Uncertainty in Environmental and Resource Economics* (Eds van Ierland, E. C., Weikard, H.-P., and Wesseler, J.), Wageningen University, Wageningen, The Netherlands, pp. 122–132.

Chiplunkar, A. V., Mehndiratta, S. L., and Khanna, P. 1981. Optimization of refuse collection systems. *Journal of Environmental Engineering Division, Division, ASCE*, 107(EE6), 1203–1211.

Chiueh, P. T., Lo, S. L., and Chang, C. L. 2008. A GIS-based system for allocating municipal solid waste incinerator compensatory fund. *Waste Management*, 28(12), 2690–2701.

Chowdhury, B. and Chowdhury, M. U. 2007. RFID-based real-time smart waste management system. In: Proceedings of the Australasian Telecommunication Networks and Applications Conference, IEEE – Institute of Electrical and Electronic Engineers, pp. 64–69.

Chung, S. and Poon, C. 1996. Evaluating waste management alternatives by the multiple criteria approach. *Resources, Conservation and Recycling*, 17(3), 189–210.

Clark, R. 1973. Solid waste: management and models. In: *Models for Environmental Pollution Control* (Ed. Deininger, R.), Ann Arbor Science Publishers Inc., Ann Arbor, MI, pp. 269–305.

Clark, R. M. and Gillean, J. I. 1974. Systems analysis and solid waste planning. *Journal of the Environmental Engineering Division*, 100(1), 7–24.

Clayton, K. 1976. A planning model for regional solid waste management systems. Unpublished Ph.D. Dissertation, Purdue University, Lafeyatte, IN.

Codd, E. F. 1970. A relational model of data for large shared data banks. *Communications of the ACM*, 13(6), 377–387.

Contreras, F., Hanaki, K., Aramaki, T., and Connors, S. 2008. Application of analytical hierarchy process to analyze stakeholders preferences for municipal solid waste management plans, Boston, USA. *Resources, Conservation and Recycling*, 52(7), 979–991.

Cornell Waste Management Institute (CWMI). 1999. Risk assessment methodology in municipal risk solid waste composting. CWMI.

Cortés, U., Sànchez-Marrè, M., Ceccaroni, L., R-Roda, I., and Poch, M. 2000. Artificial intelligence and environmental decision support systems. *Applied Intelligence*, 13(1), 77–91.

Costi, P., Minciardi, R., Robba, M., Rovatti, M., and Sacile, R. 2004. An environmentally sustainable decision model for urban solid waste management. *Waste Management*, 24(3), 277–295.

Council. 1985. Council Directive of 27 June 1985 on the assessment of the effects of certain public and private projects on the environment 85/337/EEC. *Official Journal*, L175, 40–48.

Courcelle, C., Kestmont, M., and Tyteca, D. 1998. Assessing the economic and environmental performance of municipal solid waste collection and sorting programmes. *Waste Management & Research*, 16(3), 253–263.

Coursey, D., Bretschneider, S., and Blair, J. 1993. IWSAS: expert system phone survey assistance for collecting data on hazardous waste generation. *Interfaces*, 23(3), 79–90.

Coutinho, M., Conceição, M., Borrego, C., and Nunes, M. 1998. Atmospheric impact assessment and monitoring of dioxin emissions of municipal solid waste incinerators in Portugal. *Chemosphere*, 37(9–12), 2119–2126.

Cox, B. 2008. Sensor helps speed waste collection. *Hydraulics and Pneumatics*, 61(6), 20.

Craighill, A. and Powell, J. 1996. Lifecycle assessment and economic evaluation of recycling: a case study. *Resources, Conservation and Recycling*, 17(2), 75–96.

Dahlén, L. and Lagerkvist, A. 2008. Methods for household waste composition studies. *Waste Management*, 28(7), 1100–1112.

Dalemo, M., Sonesson, U., Bjorklund, A., Mingarini, K., Frostell, B., Nybrant, T., Jonsson, H., Sundqvist, J. O., and Thyselius, L. 1997. ORWARE—a simulation model for organic waste handling systems. Part 1: model description. *Resources, Conservation and Recycling*, 21(1), 17–37.

Dantzler, D. W., Gering, L. R., Straka, T. J., and Yarrow, G. K. 2008. Creating a destination for tourism, recreation, and education on an active solid waste landfill site. *Natural Areas Journal*, 28(4), 410–413.

Daskalopoulos, E., Badr, O., and Probert, S. 1998. An integrated approach to municipal solid waste management. *Resources, Conservation and Recycling*, 24(1), 33–50.

Davies, R. E. and Lein, J. K. 1991. Applying an expert system methodology for solid waste landfill site selection. In: Proceedings of the Annual Conference of the Urban and Regional Information Systems Association, Association of American Geographers, San Francisco, CA, pp. 40–53.

Davila, E. and Chang, N. B. 2005. Sustainable pattern analysis of publicly-owned material recovery facility under uncertainty. *Journal of Environmental Management*, 75(4), 337–352.

Davila, E., Chang, N. B., and Diwakaluni, S. 2005. Dynamic landfill space consumption assessment in the Lower Rio Grande Valley, South Texas by GIP-based game theory. *Journal of Environmental Management*, 75(4), 353–366.

den Boer, J., den Boer, E., and Jager, J. 2007. LCA-IWM: a decision support tool for sustainability assessment of waste management systems. *Waste Management*, 27(8), 1032–1045.

Deng, J. 1984a. *The Theory and Methods of Socio-economic Grey Systems* (in Chinese), Science Press, Beijing, China.

Deng, J. 1984b. *Contributions to Grey Systems and Agriculture* (in Chinese), Science and Technology Press, Taiyuan, Shanxi, China.

Deng, J. 1986. *Grey Prediction and Decision*, Huazhong Institute of Technology Press, Wuhan, China.

Denzer, R. 2003. A computing program for scientists and engineers – What is the core of computing? In: *Informatics Curricula and Teaching Methods* (Eds. Cassel, L. and Reis, R. A.), Kluwer Academic Publishers, Norwell, MA, pp. 69–75.

Diaz, R. and Warith, M. 2006. Life-cycle assessment of municipal solid wastes: development of the WASTED model. *Waste Management*, 26(8), 886–901.

Dieckman, S. L., Jendrzejczyk, J. A., and Raptis, A. C. 2000. *Development of a Magnetic Resonance Sensor for On-line Monitoring of 99-Tc and 23-Na in Tank Waste Cleanup Processes: Final Report and Implementation Plan*, Argonne National Laboratory, Argonne, WI.

Drezner, Z. and Guyse, J. 1999. Application of decision analysis techniques to the Weber facility location problem. *European Journal of Operational Research*, 116(1), 69–79.

Dyson, B. and Chang, N. B. 2005. Forecasting municipal solid waste generation in a fast-growing urban region with system dynamics modeling. *Waste Management*, 25(7), 669–679.

Englehardt, J., and Lund, J. 1990. Economic analysis of recycling for small municipal waste collectors. *Journal of Resource Management and Technology*, 18(2), 84–96.

Environment Agency of England and Wales. 2000. Waste-integrated systems assessment for recovery and disposal (WISARD). PricewaterhouseCoopers.

Environment and Plastics Industry Council and Corporations Supporting Recycling (EPIC and CSR). 2000. Integrated solid waste management tools. IWM-model. Available at: http://www.iwm-model.uwaterloo.ca/iswm_booklet.pdf (accessed January 2008).

Esmaili, H. 1972. Facility selection and haul optimisation model. *Journal of the Sanitary Engineering Division*, 98(6), 1005–1021.

Fan, X., Zhu, M., Zhang, X., He, Q., and Rovetta, A. 2010. Solid waste collection optimization considering energy utilization for large city area. In: *International Conference on Logistics Systems and Intelligent Management (ICLSIM 2010)*, Vol. 1–3, IEEE – Institute of Electrical and Electronic Engineers, Harbin, China, pp. 1905–1909.

Fawcett, T., Holland, M., Holmes, J., and Powell, J. 1993. Evaluation of multicriteria analysis as an aid to decision making in waste management. Report AEA-EE-0426 ETSU. Harwell.

Fay, R. M. and Mumtaz, M. M. 1996. Development of a priority list of chemical mixtures occurring at 1188 hazardous waste sites, using the hazdat database. *Food and Chemical Toxicology*, 34(11–12), 1163–1165.

Fiorucci, P., Minciardi, R., Robba, M., and Sacile, R. 2003. Solid waste management in urban areas. Development and application of a decision support system. *Resources, Conservation and Recycling*, 37(4), 301–328.

Fischer, C. and Hermsmeyer, D. 1999. Digital photogrammetry and GIS in tailings and mine waste management. In: Proceedings of the 6th International Conference Tailings and Mine Waste. Taylor & Francis, Fort Collins, Colorado, USA, pp. 77–88.

Fonseca, D. J., Seals, R. K., Knapp, G. M., and Metcalf, J. B. 1997. Expert system for industrial residuals application assessment. *Journal of Computing in Civil Engineering*, 11(3), 201–205.

Fraisse, C. W., Campbell, K. L., Jones, J. W., and Boggess, W. G. 1996. GIDM: a GIS-based model for dairy waste management analysis. In: Proceedings of the AWRA Annual Symposium on GIS and Water Resources, American Water Resources Association, Fort Lauderdale, FL, pp. 155–164.

Fuchs, A., Zangl, H., Holler, G., and Brasseur, G. 2008. Design and analysis of a capacitive moisture sensor for municipal solid waste. *Measurement Science & Technology*, 19(2), 1–9.

Fuertes, L., Hudson, J., and Mark, D. 1974. Solid waste management: equity trade-off-models. *Journal of Urban Planning and Development, ASCE*, 100(2), 155–171.

Fujita, S. and Tamura, H. 1999. A decision support system for siting a refuse incineration plant. In: Proceedings of the Large Scale Systems Conference: Theory and Applications 1998 (Lss'98), Pergamon, Rion, Greece, vol. 1, Kidlington, Oxford [England]; Tarrytown, N.Y.: Pergamon, pp. 587–592.

Gerbilsky, L., Staroseletska, O., and Tissen, L. 2001. Integrating environmental informatics and environmental education for sustainable development. In: Proceedings of the 15th International Symposium Informatics for Environmental Protection (Eds Hilty, L. M. and Gilgen, P. W., EMPA), Metropolis-Verlag, Marburg, Zurich, Switzerland, pp. 957–963.

Gillispie, M. 2010. High-tech carts will tell on Cleveland residents who don't recycle and they face $100 fine. Available at: http://blog.cleveland.com/metro/2010/08/city_of_cleveland_to_use_high-.html (accessed October 2013).

Glenn, J. 1988. Encouraging yard water utilization. *Biocycle*, 29(7), 49–52.

Gómez-Delgado, M. and Tarantola, S. 2006. Global sensitivity analysis, GIS and multi-criteria evaluation for a sustainable planning of a hazardous waste disposal site in Spain. *International Journal of Geographical Information Science*, 20(4), 449–466.

Gottinger, H. 1986. A computational model for solid waste management with applications. *Applied Mathematical Modelling*, 10(5), 330–338.

Gottinger, H. 1988. A computational model for solid waste management with application. *European Journal of Operational Research*, 35(3), 350–364.

Greenberg, M., Bottge, M., Caruana, J., Horowitz, D., Krugman, B., Masucci, N., Milewski, A., Nebenzahl, L., O'Neill, T., Skypeck, J., and Valente, N. 1976a. *Solid Waste Planning In Metropolitan Regions*, Center for Urban Policy Research, New Brunswick, NJ.

Greenberg, M., Caruana, J., and Krugman, B. 1976b. Solid-waste management: a test of alternative strategies using optimization techniques. *Environment and Planning*, A8(5), 587–597.

Grossman, D., Hudson, J., and Marks, D. 1974. Waste generation methods for solid waste collection. *Journal of Environmental Engineering, ASCE*, 100(6), 1219–1230.

Guzman, J. B., Paningbatan, E. P., and Alcantara, A. J. 2010. A geographic information systems-based decision support system for solid waste recovery and utilization in Tuguegarao City, Cagayan, Philippines. *Journal of Environmental Science and Management*, 13(1), 52–66.

Haastrup, P., Maniezzo, V., Mattarelli, M., Rinaldi, F., Mendes, I., and Paruccini, M. 1998. A decision support system for urban waste management. *European Journal of Operational Research*, 109(2), 330–341.

Haklay, M. 1999. *From Environmental Information Systems to Environmental Informatics: Evolution and meaning*, Centre for Advanced Spatial Analysis, University College London, London, UK.

Hanley, N., and Slark, R. 1994. Cost-Benefit Analysis of paper recycling: a case study and some general principles. *Journal of Environmental Planning and Management*, 37(2), 189–197.

Hannan, M. A., Arebey, M., Basri, H., and Begum, R. A. 2010. Intelligent solid waste bin monitoring and management system. *Australian Journal of Basic and Applied Sciences*, 4(10), 5314–5319.

Hasit, Y. and Warner, D. B. 1981. Regional solid waste planning with WRAP. *Journal of Environmental Engineering, ASCE*, 107(3), 511–526.

Helms, B. and Clark, R. 1974. Locational models for solid waste management. *Journal of Urban Planning and Development, ASCE*, 97(1), 1–13.

Heydinger, A. G. and Jennings, A. A. 1988. Stability routines for expert system permit application review of hazardous waste surface impoundment dikes. *Environmental Software*, 3(4), 162–170.

Higgs, G. 2006. Integrating multi-criteria techniques with geographical information systems in waste facility location to enhance public participation. *Waste Management & Research*, 24(2), 105–117.

Highfill, J., McAsey, M., and Weinstein, R. 1994. Optimality of recycling and the location of a recycling center. *Journal of Regional Science*, 34(4), 583–597.

Hilty, L., Page, B., and Hrebicek, J. 2006. Environmental informatics. *Environmental Modelling & Software*, 21(11), 1517–1518.

Hodges, J., Bridges, S., Sparrow, C., Wooley, B., Tang, B., and Jun, C. 1999. The development of an expert system for the characterization of containers of contaminated waste. *Expert Systems with Applications*, 17(3), 167–181.

Hokkanen, J. and Salminen, P. 1994. The choice of a solid waste management system by using the ELECTRE III decision aid method. In: *Applying Multiple Criteria Aid for Decision to Environmental Management* (Ed. Paruccini, M.). Springer.

Hokkanen, J. and Salminen, P. 1997. Choosing a solid waste management system using multicriteria decision analysis. *European Journal of Operational Research*, 98(1), 19–36.

Holland, J. M. 1997. An intelligent semiautonomous waste inspection robot-ARIES. *Transactions of the American Nuclear Society*, 77, 406.

Hřebíček, J. and Soukopová, J. 2010. Modelling integrated waste management system of the Czech Republic. In: Proceedings of the 14th WSEAS International Conference on Systems Latest Trends on Systems (Part of the 14th WSEAS CSCC Multiconference), Corfu, Greece, pp. 510–515.

Hsieh, H. and Ho, K. 1993. Optimization of solid waste disposal system by linear programming technique. *Journal of Resource Management and Technology*, 21(4), 194–201.

Huang, B. 2006. GIS-based route planning for hazardous material transportation. *Journal of Environmental Informatics*, 8(1), 49–57.

Huang, G. H. and Chang, N. B. 2003. Perspectives of environmental informatics and systems analysis. *Journal of Environmental Informatics*, 1(1), 1–6.

Huang, G. H. and Moore, R. D. 1993. Grey linear programming, its solving approach, and its application. *International Journal of Systems Science*, 24(1), 159–172.

Huang, L. X. and Sheng, G. 2006. Web-services-based spatial decision support system to facilitate nuclear waste siting. In: Geoinformatics 2006 – Geospatial Information Technology (Proceedings of the SPIE) (Eds. Wu, H., and Zhu, Q.), 64, 2115.

Huang, G. H., Baetz, B. W., and Patry, G. G. 1992. A grey linear programming approach for municipal solid waste management planning under uncertainty. *Civil Engineering Systems*, 9(4), 319–335.

Huang, G. H., Baetz, B. W., and Patry, G. G. 1993a. A grey fuzzy linear programming approach for municipal solid waste management planning under uncertainty. *Civil Engineering Systems*, 10(2), 123–146.

Huang, G. H., Baetz, B. W., and Patry, G. G. 1993b. Grey integer programming: an application to waste management planning under uncertainty. *Journal of Operational Research*, 83(3), 594–620.

Huang, G. H., Baetz, B. W., and Patry, G. G. 1994. Grey fuzzy dynamic programming: application to municipal solid waste management planning problems. *Civil Engineering Systems*, 11(1), 43–73.

Huang, G. H., Baetz, B. W., and Patry, G. G. 1995. Grey integer programming: an application to management planning under uncertainty. *European Journal of Operational Research*, 83(3), 594–620.

Huang, Y. F., Baetz, B. W., Huang, G. H., and Liu, L. 2002. Violation analysis for solid waste management systems: an interval fuzzy programming approach. *Journal of Environmental Management*, 65(4), 431–446.

Huhtala, A. 1997. A post-consumer waste management model for determining optimal levels of recycling and landfilling. *Environmental and Resource Economics*, 10(3), 301–314.

Hushon, J. M. and Read, M. W. 1991. Defense priority model: experience of developing an environmental expert system for remedial site ranking. In: Proceedings of International Symposium on Artificial Intelligence, Cancun, Mexico, pp. 252–258.

Interrante, C. G., Messina, A., and Fraker, A. C. 1991. A review process and a database for waste-package documents. In: Proceedings of the Scientific Basis for Nuclear Waste Management XIV Symposium (Ed. Abrajano, T.), Materials Research Society (April 1991), pp. 917–922.

Inuiguchi, M. and Sakawa, M. 1995. Minimax regret solution to linear programming problems with an interval objective function. *European Journal of Operational Research*, 86(3), 526–536.

Irvine, J. M., Evers, T. K., Smyre, J. L., Huff, D., King, A. L., Stahl, G., and Odenweller, J. 1997. The detection and mapping of buried waste. *International Journal of Remote Sensing*, 18(7), 1583–1595.

Jacobs, T. and Everett, J. 1992. Optimal scheduling of consecutive landfill operations with recycling. *Journal of Environmental Engineering, ASCE*, 118(3), 420–429.

Jenkins, A. 1979. Optimal location of facilities for recycling municipal solid waste in Southern Ontario. Unpublished Ph.D. Dissertation, University of Toronto, Toronto, ON, Canada.

Jenkins, L. 1980. The Ontario waste management systems model. Technical Report, Ontario Ministry of Environment.

Jenkins, L. 1982. Developing a solid waste management model for Toronto. *INFOR*, 20(3), 237–247.

Jennings, A. A. and Heydinger, A. G. 1989. An expert system for regulatory review of hazardous waste surface impoundment dikes. *Microcomputers in Civil Engineering*, 4(1), 29–38.

Jensen, J. R. and Christensen, E. J. 1986. Solid and hazardous waste disposal site selection using digital geographic information system techniques. *Science of the Total Environment*, 56, 265–276.

Jeong, S. M., Osako, M., and Kim, Y. J. 2005. Utilizing a database to interpret leaching characteristics of lead from bottom ashes of municipal solid waste incinerators. *Waste Management*, 25(7), 694–701.

John, S. 2010. Sustainability-based decision-support system for solid waste management. *International Journal of Environment and Waste Management*, 6(1–2), 41–50.

Kaletsky, K., Earle, J. R., and Schneider, T. A. 1996. Integrating GIS and GPS in environmental remediation oversight. In: Proceedings of Eco-Informa '96. Global Networks for Environmental Information, vol. 10, Environmental Research Institute of Michigan, pp. 173–178.

Kao, J. J. 1996. A raster-based C program for siting a landfill with optimal compactness. *Computers & Geosciences*, 22(8), 837–847.

Kao, J. J. and Lin, T. I. 2002. Shortest service location model for planning waste pickup locations. *Journal of the Air & Waste Management Association*, 52(5), 585–592.

Karadimas, N. V. and Loumos, V. G. 2008. GIS-based modelling for the estimation of municipal solid waste generation and collection. *Waste Management & Research*, 26(4), 337–346.

Karadimas, N. V., Loumos, V., and Orsoni, A. 2006. Municipal solid waste generation modelling based on fuzzy logic. In: Proceedings of the 20th European Conference on Modelling and Simulation ECMS 2006 (Eds Borutzky, W. and Orsoni, A.), Bonn, Germany, pp. 309–314.

Karadimas, N. V., Kolokathi, M., Defteraiou, G., and Loumos, V. 2007a. Ant colony system VS ArcGIS network analyst: the case of municipal solid waste collection. In: Proceedings of the 5th WSEAS International Conference on Environment, Ecosystems and Development, World Science and Engineering Academy and Society, pp. 133–139.

Karadimas, N. V., Papatzelou, K., and Loumos, V. G. 2007b. Optimal solid waste collection routes identified by the ant colony system algorithm. *Waste Management & Research*, 25(2), 139–147.

Karadimas, N. V., Doukas, N., Kolokathi, M., and Defteraiou, G. 2008. Routing optimization heuristics algorithms for urban solid waste transportation management. *WSEAS Transactions on Computers*, 7(12), 2022–2031.

Karagiannidis, A. and Moussiopoulos, N. 1997. Application of ELECTRE III for the integrated management of municipal solid wastes in the Greater Athens Area. *European Journal of Operational Research*, 97(3), 439–449.

Katpatal, Y. B. and Rao, B. 2011. Urban spatial decision support system for municipal solid waste management of Nagpur urban area using high-resolution satellite data and geographic information system. *Journal of Urban Planning and Development, ASCE*, 137(1), 65–76.

Khan, M. and Burney, F. 1989. Forecasting solid waste composition—an important consideration in resource recovery and recycling. *Resources, Conservation and Recycling*, 3(1), 1–17.

Khijwania, S. K., Kumar, A., Yueh, F. Y., and Singh, J. P. 2003. Raman sensor to monitor the nitrate and nitrite in the nuclear waste tank. In: Proceedings of the SPIE Conference on Chemical and Biological Standoff Detection. Providence, NY, 5268, pp. 47–52

Kijak, R. and Moy, D. 2004. A decision support framework for sustainable waste management. *Journal of Industrial Ecology*, 8(3), 33–50.

Kim, H. D., Kim, K. J., and Yoon, W. K. 1993. Development of a prototype expert system for intelligent operation aids in rod consolidation process nuclear materials. *Journal of the Korean Nuclear Society*, 25(1), 1–7.

Kim, B. I., Kim, S., and Sahoo, S. 2006. Waste collection vehicle routing problem with time windows. *Computers & Operations Research*, 33(12), 3624–3642.

Kirka, O. and Erkip, N. 1988. Selecting transfer station locations for large solid waste systems. *European Journal of Operational Research*, 35(3), 339–349.

Kirkeby, J., Birgisdottir, H., Hansen, T., Christensen, T., Bhander, G., and Hauschild, M. 2006. Environmental assessment of solid waste systems and technologies: EASEWASTE. *Waste Management & Research*, 24(1), 3–15.

Kirkpatrick, N. 1993. Selecting a waste management option using a LCA approach. *Packaging Technology and Science*, 6(3), 159–172.

Kordon, A. K., Dhurjati, P. S., and Bockrath, B. J. 1996. On-line expert system for odor complaints in a refinery. *Computers & Chemical Engineering*, 20(Supplement 2), S1449–S1454.

Korovin, Y. A., Artisyuk, V. V., Ignatyuk, A. V., Pilnov, G. B., Stankovsky, A. Y., Titarenko, Y. E., and Yavshits, S. G. 2007. Transmutation of radioactive nuclear waste — present status and requirement for the problem-oriented nuclear data base. *Pramana*, 68(2), 181–191.

Kuhner, J. and Harrington, J. J. 1975. Mathematical models for developing regional solid waste management policies. *Engineering Optimization*, 1(4), 237–256.

Kulcar, T. 1996. Optimizing solid waste collection in Brussels. *European Journal of Operational Research*, 5(1), 71–77.

Kunsch, P. L. and Fortemps, P. H. 2000. Practical implementation of a fuzzy decision support system for the economic calculus in radioactive waste management. In: *Intelligent Techniques and Soft Computing in Nuclear Science and Engineering (Proceedings of the 4th International FLINS Conference)* (Eds Ruan, D., Abderrahim, H. A., D'hondt, P., and Kerre, E. E.), World Scientific Publishing, pp. 414–420.

Lara-Valencia, F., Harlow, S. D., Lemos, M. C., and Denman, C. A. 2009. Equity dimensions of hazardous waste generation in rapidly industrialising cities along the United States-Mexico border. *Journal of Environmental Planning and Management*, 52(2), 195–216.

Lawver, R., Lund, J., and Tchobanoglous, G. 1990. GIGO – A solid waste management model for municipalities. In: *Proceedings of the Sixth International Conference on Solid Waste Management and Secondary Materials* (Ed. Shiech, W. K.), University of Pennsylvania, Philadelphia, PA, pp. 8.

Lazar, R. E., Dumitrescu, M., and Stefanescu, I. 2001. Risk assessment of hazardous waste transport-perspectives of GIS application. In: Proceedings of International Conference Nuclear Energy in Central Europe (CD-ROM), Portorož, Slovenia, pp. 737–744.

Lee, S. 2003. Evaluation of waste disposal site using the DRASTIC system in Southern Korea. *Environmental Geology*, 44(6), 654–664.

Liao, L. 2011. Project proposal: framework and demonstration of IoT for integrated municipal solid waste management (SQ2011SF09C02839). China's Ministry of Education (unpublished).

Liebman, J., Male, J., and Wathne, M. 1975. Minimum cost in residential refuse vehicle routes. *Journal of Environmental Engineering*, 101(3), 399–411.

Light, G. L. 1990. *Microcomputer Software in Municipal Solid Waste Management: A Review of Programs and Issues for Developing Countries*. UNDP–World Bank Water and Sanitation Program.

Lin, H. Y. and Kao, J. J. 1998. A vector-based spatial model for landfill siting. *Journal of Hazardous Materials*, 58(1–3), 3–14.

Litvan, D. 1994. *Analysis of the Environmental Benefits of the Directive Proposal on the Emissions of Incineration Plants*, European Commission, Brussels, Belgium.

Liu, Z. R., Sheng, G., and Wang, L. 2006. Involving the public in spatial decision-making using Internet GIS. In: *Geoinformatics 2006—Geospatial Information Technology (Proceedings of the SPIE)*, vol. 6421 (Eds Wu, H. and Zhu, Q.), International Society of Optical Engineering.

Ljunggren, M. 2000. Modelling national solid waste management. *Waste Management & Research*, 18(6), 525–537.

Löfgren, A., Miliander, S., Truvé, J., and Lindborg, T. 2006. Carbon budgets for catchments across a managed landscape mosaic in southeast Sweden: contributing to the safety assessment of a nuclear waste repository. *Ambio*, 35(8), 459–468.

López-Cuñat, J. M. 2000. Adverse selection under ignorance. *Economic Theory*, 16(2), 379–399.

Lo Porto, A., Garnier, M., Marini, R., and Leone, A., 1997. The use of GIS in assessing the land disposal capacity for animal waste and preventing groundwater pollution. In: Freshwater Contamination Proceedings of a symposium held during the Fifth IAHS Scientific Assembly. IAHS Press, Rabat, Morocco, 243, pp. 375–383.

Lotfi, S., Habibi, K., and Koohsari, M. J. 2007. Integrating GIS and fuzzy logic for urban solid waste management (a case study of Sanandaj city, Iran). *Pakistan Journal of Biological Sciences*, 10(22), 4000–4007.

Loulou, R. and Kanudia, A. 1999. Minimax regret strategies for greenhouse gas abatement: methodology and application. *Operations Research Letters*, 25(5), 219–230.

Love, R., Morris, J. G., and Wesolowsky, G. O. 1988. *Facilities Location: Models and Methods*, North-Holland Publishing Company, New York-Amsterdam-London.

Lu, G. W., Chang, N. B. and Liao, L. 2013. Environmental informatics for solid and hazardous waste management: advances, challenges, and perspectives. *Critical Reviews in Environmental Science and Technology*, 43(15), 1557–1656.

Luce, R. D. and Raiffa, H. 1957. *Games and Decisions: Introduction and Critical Survey*, John Wiley & Sons, Inc., New York, NY.

Ludvigsen, P. and Dupont, R. 1988. Formal evaluation of the expert system DEMOTOX. *Journal of Computing in Civil Engineering*, 2(4), 398–412.

Lund, J. 1990. Least-cost scheduling of solid waste recycling. *Journal of Environmental Engineering, ASCE*, 116(1), 182–197.

Lund, J., Tchobanoglous, G., Anex, R., and Lawver, R. 1994. Linear programming for analysis of material recovery facilities. *Journal of Environmental Engineering, ASCE*, 120(5), 1082–1094.

Luo, S., Zhong, N., Cao, X., Zhao, M., and Chen, Y. 2009. Medical waste management system based on RFID and GPRS. *Journal of Chongqing Institute of Technology*, 23(12), 106–110.

MacDonald, M. 1996a. Solid waste management models: a state of the art review. *Journal of Solid Waste Technology and Management*, 23(2), 73–83.

MacDonald, M. 1996b. A multi-attribute spatial decision support system for solid waste planning. *Computers, Environment and Urban Systems*, 20(1), 1–17.

MacDoran, P. F., Feuerstein, R. J. and Schreiner, W. S. 1992. GPS spread spectrum signal transmission over fiber optic links. *Proceedings of the IEEE Transactions on Geoscience and Remote Sensing*, 30(5), 1073–1076.

Manoliadis, O., Baronos, A., Tsolas, I., and Sawides, S. 2001. A multicriteria decision support system for landfill site selection. *Journal of Environmental Protection and Ecology*, 2(2), 273–278.

Marks, D., and Liebman, J. 1970. *Mathematical Analysis of Solid Waste Collections*. USPHS, Bureau of Solid Waste Management.

Marks, D., and Liebman, J. 1971. Location models: solid waste collection example. *Journal of the Urban Planning and Development Division, ASCE*, 97(1), 15–30.

Marks, D. H., ReVelle, C. S., and Liebman, J. C. 1970. Mathematical models of location: a review. *Journal of the Urban Planning and Development Division, ASCE*, 96(1), 81–93.

Mohai, P. and Saha, R. 2007. Racial inequality in the distribution of hazardous waste: a national-level reassessment. *Social Problems*, 54(3), 343–370.

Moore, R. E. 1979. *Method and Applications of Interval Analysis*, SIAM, Philadelphia, PA.

Morris, J. 1991. Source separation vs centralised processing: an avoided cost optimisation model provides some intriguing answers. *Journal of Resource Management and Technology*, 19(3), 133–140.

Muttiah, R. S., Engel, B. A., and Jones, D. D. 1996. Waste disposal site selection using GIS-based simulated annealing. *Computers & Geosciences*, 22(9), 1013–1017.

Nielsen, I., Ming, L., and Nielsen, P. 2010. Optimizing supply chain waste management through the use of RFID technology. In: Proceedings of the IEEE International Conference on RFID-Technology and Applications (RFID-TA), Institute of Electrical and Electronics Engineers (IEEE), pp. 296–301.

Niessen, W. R., and Alsobrook, A. F. 1972. Municipal and industrial refuse: composition and rates. In: Proceedings of the 1972 ASME Incin. Conference, ASME, New York.

Nilsson, M., Bjorklund, A., Finnveden, G., and Johansson, J. 2005. Testing a SEA methodology for the energy sector: a waste incineration tax proposal. *Environmental Impact Assessment Review*, 25(1), 1–32.

NoSQL Archive. 2011. List of NoSQl databases. Available at: http://nosql-database.org/ (accessed May 2014).

Notz, K. J., Forsberg, C. W., and Mastal, E. F. 1984. *Spent Fuel and Radioactive Waste: An Integrated Data Base of Inventories, Projections, and Characteristics*, Transactions of the American Nuclear Society, Oak Ridge National Lab, Tucson, AZ.

Ogawa, M., 1998. Waste treatment management information system using SCMS. Patent (JP10095505-A; JP3772463-B2). Nippon Tokushu Kogyo Kk, Fuji Denki Techno Eng Kk.

Ong, H., Goh, T., and Lim, C. 1990. A computerised vehicle routing system for refuse collection. *Advances in Engineering Software*, 12(2), 54–58.

Page, B. 1992. Environmental protection as a challenge to applied informatics: a workshop introduction. In: Proceedings of the IFIP 12th World Computer Congress on Education and Society – Information Processing '92, vol. 2 (Ed. Aiken, R. M.), International Federation for Information Processing (IFIP), Madrid, Spain, pp. 595–604.

Page, B. and Wohlgemuth, V. 2010. Advances in environmental informatics: integration of discrete event simulation methodology with ecological material flow analysis for modelling eco-efficient systems. *Procedia Environmental Sciences*, 2, 696–705.

Paige, G. B., Stone, J. J., Lane, L. J., Yakowitz, D. S., and Hakonson, T. E. 1996. Evaluation of a prototype decision support system for selecting trench cap designs. *Journal of Environmental Quality*, 25(1), 127–135.

Perlack, R. and Willis, C. 1985. Multiobjective decision-making in waste disposal planning. *Journal of Environmental Engineering, ASCE*, 111(3), 373–385.

Pires, A., Martinho, G., and Chang, N. B. 2011. Solid waste management: in European countries: a review of systems analysis techniques. *Journal of Environmental Management*, 92(4), 1033–1050.

Powell, J., Craighill, A., Parfitt, J., and Turner, R. 1996. A lifecycle assessment and economic valuation of recycling. *Journal of Environmental Planning and Management*, 39(1), 97–112.

Powell, J., Steele, A., Sherwood, N., and Robson, T. 1998. Using life cycle inventory analysis in the development of waste management strategy for Gloucestershire, UK. *Environmental and Waste Management*, 1(4), 97–112.

Powell, J., Sherwood, N., Dempsey, M., and Steele, A. 1999. Life Cycle Inventory Analysis of Alternative Waste Management options for Bristol City Council: Summary Report. University of Glocestershire, Environmental Management Research Group.

Pullammanappallil, P. C. 1998. Expert system for control of anaerobic digesters. *Biotechnology and Bioengineering*, 58(1), 13–22.

Rahman, M. and Kuby, M. 1995. A multiobjective model for locating solid waste transfer facilities using an empirical opposition function. *INFOR*, 33(1), 34–49.

Rao, D. 1975. A dynamic model for optimal planning of regional solid waste management. Ph.D. Thesis, Clarkson College of Technology, Potsdam, NY.

Reitsma, R. F. and Sullivan, J. F. 1992. Application of decision support systems (DSS) to the management of radioactive wastes. In: Proceedings of the Third International Conference High Level Radioactive Waste Management, Las Vegas, NV, vol. 1–2, pp. 469–479.

Rhyner, C. and Green, B. 1988. The predictive accuracy of published solid waste generation factors. *Waste Management & Research*, 6(1), 329–338.

Rieradevall, J., Domènech, X., and Fullana, P. 1997. Application of life cycle assessment to landfill. *The International Journal of Life Cycle Assessment*, 2(3), 141–144.

Rinner, C. 2003. Web-based spatial decision support: status and research directions. *Journal of Geographic Information and Decision Analysis*, 7(1), 14–31.

Riscoa, M. A. L. D. and Dubourguier, H. C. 2010. A web-based database on methanogenic potential of crops and wastes. *Environmental Modelling & Software*, 25(8), 970–971.

Rossman, L. 1971. *A General Model for Solid Waste Management Facility Selection*, Department of Civil Engineering, University of Illinois.

Rouhani, S. and Kangari, R. 1987. Landfill site selection, a microcomputer expert system. *Microcomputers in Civil Engineering*, 2(1), 47–54.

Rubenstein-Montano, B. and Zandi, I. 1999. Application of a genetic algorithm to policy planning: the case of solid waste. *Environment and Planning B: Planning and Design*, 26(6), 893–907.

Rufford, N. 1984. The analysis and prediction of the quantity and composition of household refuse. Unpublished Ph.D. Dissertation, University of Aston, UK.

Rushbrook, P. 1987. The benefits of forward planning and the role of computer assistance. In: Proceedings of the HARBINGER Symposium, Llandrindod Wells, UK.

Rushbrook, P. and Pugh, M. 1987. Waste management planning: an illustrated description of 'HARBINGER' the Harwell Waste Management model. *Wastes Management*, 77, 348–361.

Salhofer, S., Wassermann, G., and Binner, E. 2007. Strategic environmental assessment as an approach to assess waste management systems. Experiences from an Austrian case study. *Environmental Modelling & Software*, 22(5), 610–618.

Schütt, D. and Hofestädt, R., 1992. Bioinformatics and environmental informatics-new aspects and tasks for computer science [German: Bioinformatik und Umweltinformatik – neue Aspekte und Aufgaben der Informatik]. *Informatik Forschung ud Entwicklung*, 7, 4.

Scott, M., Thompson, S. N., Anderson, W. A., and Williams, J. S. 1989. Status of Maine's low-level radioactive waste program. In: Proceedings of the Waste Management Symposium: Waste Processing, Transportation, Storage and Disposal, Technical Programs and Public Education, 2 (Eds Post, R. G., Wacks, M. E., and McComb, D.), Arizona Board of Regents, Tucson, Arizona, pp. 39–42.

Scott-Morton, M. S. 1971. *Management Decision Systems: Computer-Based Support for Decision Making*, Harvard University Press, Cambridge, MA.

Şener, Ş. E., Sener, E., and Karagüzel, R. 2011. Solid waste disposal site selection with GIS and AHP methodology: a case study in Senirkent-Uluborlu (Isparta) Basin, Turkey. *Environmental Monitoring and Assessment*, 173(1–4), 533–554.

Seo, S., Aramaki, T., Hwang, Y., and Hanaki, K. 2003. Evaluation of solid waste management system using fuzzy composition. *Journal of Environmental Engineering, ASCE*, 129(6), 520–531.

Shirland, L. E. and Kraushaar, J. M. 1991. A group decision support system for large group processes. In: Proceedings of the Technology Management: The New International Language, Institute of Electrical and Electronics Engineers (IEEE), Portland, OR, pp. 589.

Shmelev, S. E. and Powell, J. R. 2006. Ecological-economic modelling for strategic regional waste management systems. *Ecological Economics*, 59(1), 115–130.

Shrivastava, P., Zhang, H. C., Li, J., and Whitely, A. 2005. Evaluating obsolete electronic products for disassembly, material recovery and environmental impact through a decision support system. In: Proceedings of the IEEE International Symposium on Electronics & the Environment, Institute of Electrical and Electronics Engineers (IEEE), New Orleans, LA, pp. 221–225.

Sicotte, D. 2008. Dealing in toxins on the wrong side of the tracks: lessons from a hazardous waste controversy in Phoenix. *Social Science Quarterly*, 89(5), 1136–1152.

Smith, E. G., Lindwall, C. W., Green, M., and Pavlik, C. K. 1997. PARMS: a decision support system for planting and residue management. *Computers and Electronics in Agriculture*, 16(3), 219–229.

Song, H.-S., Moon, K.-S., and Hyun, J. 1999. A life-cycle assessment study on the various recycle routes of pet bottles. *Korean Journal of Chemical Engineering*, 16(2), 202–207.

Stacey, W. M., Hertel, N. E., and Hoffman, E. A. 1995. Radioactive waste produced by demonstration and commercial fusion reactors extrapolated from ITER and advanced databases. *Fusion Engineering and Design*, 29(3), 198–206.

Staudinger, J., Oralkan, G. A., Levitt, R. E., and Roberts, P. V. 1997. The Haztimator knowledge-based (expert) system: providing design and time/cost estimates for hazardous waste remediation. *Environmental Progress*, 16(2), 82–87.

Stegemiller, M. L., Heineman, W. R., Seliskar, C. J., Ridgway, T. H., Bryan, S. A., Hubler, T., and Sell, R. L. 2003. Spectroelectrochemical sensing based on multimode selectivity simultaneously achievable in a single device. 11. Design and evaluation of a small portable sensor for the determination of ferrocyanide in Hanford waste samples. *Environmental Science & Technology*, 37(1), 123–130.

Stewart, J. C. 1988. The application of low-level waste siting criteria to geographic information systems. *Transactions of the American Nuclear Society*, 56, 52.

Stormont, J. C. and Farfan, E. 2005. Stability evaluation of a mine waste pile. *Environmental & Engineering Geoscience*, 11(1), 43–52.

Su, J. P., Chiueh, P. T., Hung, M. L., and Ma, H. W. 2007. Analyzing policy impact potential for municipal solid waste management decision-making: a case study of Taiwan. *Resources, Conservation and Recycling*, 51(2), 418–434.

Sudhir, V., Muraleedharan, V., and Srinivasan, G. 1996. Integrated solid waste management in Urban India: a critical operational research framework. *Socio-Economic Planning Sciences*, 30(3), 163–181.

Sumathi, V. R., Natesan, U., and Sarkar, C. 2008. GIS-based approach for optimized siting of municipal solid waste landfill. *Waste Management*, 28(11), 2146–2160.

Sundberg, J. 1989. MIMES—a model for integrating the material flow with an energy system. In: Proceedings of the KT Symposium on Non-Waste Technology, Technical Research Centre of Finland, Espoo, Finland.

Sundberg, J. 1993. A system approach to municipal solid waste management: results from a case study of Goteborg – Part 1. In: Proceedings of the International Conference on Integrated Energy and Environmental Management, Air & Waste Management Association, New Orleans, LA.

Sundberg, J., Gipperth, P., and Wene, C. 1994. A systems approach to municipal solid waste management: a pilot study of Goteborg. *Waste Management & Research*, 12(1), 73–91.

Swayne, D. A. and Denzer, R. 2000. Teaching EIS development – The EU Canada curriculum on environmental Informatics In: *Environmental Software Systems—Environmental Information and Decision Support*, vol. 39 (Eds Denzer, R., Swayne, D. A., Purvis, M., and Schimak, G.), Kluwer Academic Publishers, pp. 152–156.

Tanskanen, J. H. and Melanen, M. 1999. Modelling separation strategies of municipal solid waste in Finland. *Waste Management & Research*, 17(2), 80–92.

Taylor, B. W. 1996. *Introduction to Management Science*, Prentice-Hall, Englewood Cliffs, NJ.

Thomas, B., Tamblyn, D., and Baetz, B. 1990. Expert systems in municipal solid waste management planning. *Journal of Urban Planning and Development, ASCE*, 116(3), 150–155.

Tillman, A. M., Baumann, H., Eriksson, E., and Rydberg, T. 1991. Lifecycle analysis of selected packaging materials. Report commissioned by Swedish National Commission on Packaging.

Touche Ross Management, 1994. *Cost-Benefit Analysis of the Proposed Council Directive on the Landfill of Waste*, European Commission, Brussels.

Tralhão, L., Coutinho-Rodrigues, J., and Alcada-Almeida, L. 2010. A multiobjective modeling approach to locate multi-compartment containers for urban-sorted waste. *Waste Management*, 30(12), 2418–2429.

Travis, C. C. (Ed.), 1991. *Municipal Waste Incineration Risk Assessment*. Plenum Press, New York.

Truitt, M., Liebman, J., and Kruse, C. 1969. Simulation model of urban refuse collection. *Journal of the Sanitary Engineering Division*, 95, 289–298.

Tsai, H. C., Chen, K., Liu, Y. Y., and Shuler, J. M. 2010. Demonstration (DEMO) of radiofrequency identification (RFID) system for tracking and monitoring of nuclear materials. *Packaging, Transport, Storage & Security of Radioactive Material*, 21(2), 91–102.

Tsai, H. C., Liu, Y. Y., and Shuler, J. 2011. RFID technology for environmental remediation and radioactive waste management, In: Proceedings of the 13th International Conference on Environmental Remediation and Radioactive Waste Management, ASME, 1, Tsukuba, Japan, pp. 511–518.

Tsankova, R. and Damianova, T. 1995. Computer based environmental education as a mutual challenge. In: Proceedings of the Sixth IFIP World Conference on Computers in Education, Birmingham, pp. 487–794.

United Nations Environment Programme. 2004. UNEP-Infoterra: the global environmental information exchange network. Available at: http://www.unep.org/infoterra./ (accessed October 2013).

United States Environmental Protection Agency (USEPA). 1977. *WRAP: A Model for Solid Waste Management Planning User's Guide*. USEPA.

United States Environmental Protection Agency (USEPA). 2002. *Solid Waste Management: A Local Challenge with Global Impacts*. USEPA.

United States Environmental Protection Agency (USEPA). 2011. *The Municipal Waste Management in the United States: Facts and Figures*. Available at http://www.epa.gov/osw/nonhaz/municipal/msw99.htm (accessed November 2013).

University of Northern British Columbia, 2011. *GEOG 300—Introduction to Geographic Information System: History of GIS*. Available at: http://gis.unbc.ca/courses/geog/lectures/ (accessed October. 2013).

van den Broek, W. H. A. M., Wienke, D., Melssen, W. J., Feldhoff, R., Huth-Fehre, T., Kantimm, T., and Buydens, L. M. C. 1997. Application of a spectroscopic infrared focal plane array sensor for on-line identification of plastic waste. *Applied Spectroscopy*, 51(6), 856–865.

van Kessel, L. B. M., Leskens, M., and Brem, G. 2002. On-line calorific value sensor and validation of dynamic models applied to municipal solid waste combustion. *Process Safety and Environmental Protection*, 80(B5), 245–255.

Vinceti, M., Malagoli, C., Fabbi, S., Teggi, S., Rodolfi, R., Garavelli, L., Astolfi, G., and Rivieri, F. 2009. Risk of congenital anomalies around a municipal solid waste incinerator: a GIS-based case-control study. *International Journal of Health Geographics*, 8, 8.

Voigt, K. and Welzl, G. 2001. Evaluation of search engines concerning environmental terms. In: Proceedings of the 15th International Symposium Informatics for Environmental Protection, Zürich, pp. 685–691.

Walker, W. 1976. A heuristic adjacent extreme point algorithm for the fixed charge problem. *Management Science*, 22(5), 587–596.

Walker, W., Aquilina, M., and Schur, D. 1974. Development and use of a fixed charge programming model for regional solid waste planning. In: Proceedings of the 46th Joint Meeting of the operational Research Society of America and the Institute of Management Sciences, Puerto Rico.

Wei, M. S. and Weber, F. 1996. An expert system for waste management. *Journal of Environmental Management*, 46(4), 345–358.

Weitz, K., Barlaz, M., Ranji, R., Brill, D., Thorneloe, S., and Ham, R. 1999. Life cycle management of municipal solid waste. *International Journal of Life Cycle Assessment*, 4(4), 195–201.

Welbourne, E., Battle, L., Cole, G., Gould, K., Rector, K., Raymer, S., Balazinska, M., and Borriello, G. 2009. Building the Internet of things using RFID: the RFID ecosystem experience. *IEEE Internet Computing*, 13(3), 48–55.

Westrom, G., Vance, J. N., and Gelhaus, F. E. 1989. Radwaste decision support system (functional specification). In: Proceedings of the Waste Management '89 Symposium Processing, Transportation, Storage and Disposal, Technical Programs and Public Education, vol. 2, pp. 507–511.

White, P., Franke, M., and Hindle, P. 1995. *Integrated Solid Waste Management: A Life-Cycle Inventory*, Blackie Academic & Professional, Glasgow, UK.

Wieland, E., Bradbury, M. H. and van Loon, L. 2003. Development of a sorption data base for the cementitious near-field of a repository for radioactive waste. *Czechoslovak Journal of Physics*, 53(1), A629–A638.

Williamson, D. A. 1990. An expert system for greater-than-class-C waste classification. *Transactions of the American Nuclear Society*, 61, 79.

Wilson, D. C. 1977. Strategy evaluation in planning of waste management to land—a critical review of the literature. *Applied Mathematical Modelling*, 1(4), 205–217.

Wong, A. K. and TenBroek, M. 1989. PC-based GIS framework for the management and analysis of hazardous waste site remediation projects. In: Annual Conference of the Urban and Regional Information Systems Association, vol. 5, pp. 358–366.

World Commission on Environment and Development (WCED). 1987. In: *Our Common Future* (Ed. Brundtland, G. H.), Oxford University Press, Oxford, UK.

World Health Organization (WHO). 2007. *Population Health and Waste Management: Scientific Data and Policy Option*, World Health Organization Regional Office for Europe, Rome, Italy.

Wyatt, J. 2008. *Maximizing Waste Management Efficiency Through the Use of RFID*, Texas Instrument.

Yang, X. and Okrent, D. 1991. A diagnostic expert system for helping the operation of hazardous waste incinerators. *Journal of Hazardous Materials*, 26(1), 27–46.

Yu, V. L., Fagan, L. M., Wraith, S. M., Clancey, W. J., Scott, A. C., Hannigan, J., Blum, R. L., Buchanan, B. G., and Cohen, S. N. 1979. Antimicrobial selection by a computer. *JAMA: The Journal of the American Medical Association*, 242(12), 1279–1282.

Yui, M., Shibutani, T., Shibata, M., Rai, D., and Ochs, M. 2001. A plutonium geochemical database for performance analysis of high-level radioactive waste repositories. *Radioactivity in the Environment*, 1, 159–174.

Zamorano, M., Molero, E., Grindlay, A., Rodríguez, M. L., Hurtado, A., and Calvo, F. J. 2009. A planning scenario for the application of geographical information systems in municipal waste collection: a case of Churriana de la Vega (Granada, Spain). *Resources Conservation and Recycling*, 54(2), 123–133.

Zeng, Y. and Trauth, K. M. 2005. Internet-based fuzzy multicriteria decision support system for planning integrated solid waste management. *Journal of Environmental Informatics*, 6(1), 1–15.

Zhang, Y., Huang, Q., Qu, T., and Jiang, P. 2010. Implementation of real-time shop floor manufacturing using RFID technologies. *International Journal of Manufacturing Research*, 5(1), 74–86.

INDEX

Abiotic depletion, 374, 569, 573, 774, 780–781, 783
Apartments curbside, 577–579
AC, *see* Apartments curbside
Adaptive management strategies, 15, 193, 195–196, 198, 200, 202, 204, 206, 208–210, 212, 418
Aerobic MBT, 63, 172, 364, 366, 368, 771–772
Aerobic processes, 60, 62
AHP, *see* Analytical hierarchy process
AHP method, 276, 769, 775
Allocation problem, 333–336
Aluminum, 22, 66, 83, 85, 201, 448
Anaerobic digestion, 33, 60–63, 76–77, 106–107, 172, 364, 368, 373–374, 771, 783
Analytical hierarchy process, 271, 274, 276, 556, 562, 564, 715, 738–739, 749, 760, 766, 772, 788, 854
Analytical network processes, 271, 276–277, 565
ANN, *see* Artificial neural network

ANP, *see* Analytical network processes
Application layer, 646, 869
Approaches
    damage-oriented, 346, 393, 395
    problem-oriented, 346, 395
    semantic, 706–707
    true-value, 707, 709
Artificial neural network, 637, 640–641, 862
Attributes, 216–217, 249–250, 252, 258, 271–273, 276, 523, 560, 562, 735–736, 741, 744–745, 749, 765–766

Batteries, 27, 85, 150–151, 153, 330
Behavior, 10, 130–132, 156, 160, 217, 223, 376, 479, 490, 737
    recycling, 47, 116, 131–132
Best available techniques, 856
Biodegradable municipal waste, 38, 42, 59, 62, 77–79, 121, 197, 364–365, 367–368, 374, 561, 619, 771–772, 785

Sustainable Solid Waste Management: A Systems Engineering Approach, First Edition. Ni-Bin Chang and Ana Pires.
© 2015 The Institute of Electrical and Electronics Engineers, Inc. Published 2015 by John Wiley & Sons, Inc.

IEEE PRESS SERIES ON
SYSTEMS SCIENCE AND ENGINEERING

Editor:
MengChu Zhou, *New Jersey Institute of Technology and Tongji University*

Co-Editors:
Han-Xiong Li, *City University of Hong-Kong*
Margot Weijnen, *Delft University of Technology*

The focus of this series is to introduce the advances in theory and applications of systems science and engineering to industrial practitioners, researchers, and students. This series seeks to foster system-of-systems multidisciplinary theory and tools to satisfy the needs of the industrial and academic areas to model, analyze, design, optimize and operate increasingly complex man-made systems ranging from control systems, computer systems, discrete event systems, information systems, networked systems, production systems, robotic systems, service systems, and transportation systems to Internet, sensor networks, smart grid, social network, sustainable infrastructure, and systems biology.

Reinforcement and Systemic Machine Learning for Decision Making
Parag Kulkarni

Remote Sensing and Actuation Using Unmanned Vehicles
Haiyang Chao and YangQuan Chen

Hybrid Control and Motion Planning of Dynamical Legged Locomotion
Nasser Sadati, Guy A. Dumont, Kaveh Akbari Hamed, and William A. Gruver

Modern Machine Learning: Techniques and Their Applications in Cartoon Animation Research
Jun Yu and Dachen Tao

Design of Business and Scientific Workflows: A Web Service-Oriented Approach
MengChu Zhou and Wei Tan

Operator-based Nonlinear Control Systems: Design and Applications
Mingcong Deng

System Design and Control Integration for Advanced Manufacturing
Han-Xiong Li and XinJiang Lu

Sustainable Solid Waste Management: A Systems Engineering Approach
Ni-Bin Chang and Ana Pires